D1070836

SOLAR ENGINEERING OF THERMAL PROCESSES

St. Olaf College

OCT 18 1985

Science Library

Solar Engineering
of Thermal Processes

JOHN A. DUFFIE
Professor of Chemical Engineering

WILLIAM A. BECKMAN
Professor of Mechanical Engineering

Solar Energy Laboratory
University of Wisconsin-Madison

A WILEY-INTERSCIENCE PUBLICATION

JOHN WILEY & SONS
New York • Chichester • Brisbane • Toronto • Singapore

TJ
810
.D8
1980
Science
Library

Copyright © 1980 by John Wiley & Sons, Inc.

All rights reserved. Published simultaneously in Canada.

Reproduction or translation of any part of this work
beyond that permitted by Sections 107 or 108 of the
1976 United States Copyright Act without the permission
of the copyright owner is unlawful. Requests for
permission or further information should be addressed to
the Permissions Department, John Wiley & Sons, Inc.

Library of Congress Cataloging in Publication Data:

Duffie, John A
  Solar engineering of thermal processes.

  "A Wiley-Interscience publication."
  Includes index.
  1. Solar energy. 2. Heat—Transmission.
I. Beckman, William A., joint author. II. Title.
TJ810.D8 1980   621.47   80-13297
ISBN 0-471-05066-0

Printed in the United States of America

10 9 8 7 6 5 4

**To
Pat and Sylvia**

Preface

When we started to revise our earlier book, *Solar Energy Thermal Processes*, it quickly became evident that the years since 1974 had brought many significant developments in our knowledge of solar processes. What started out to be a second edition of the 1974 book quickly grew into a new work, with new analysis and design tools, new insights into solar process operation, new industrial developments, and new ideas on how solar energy can be used. The result is a new book, substantially broader in scope and more detailed than the earlier one. Perhaps less than 20 percent of this book is taken directly from *Solar Energy Thermal Processes*, although many diagrams have been reused and the general outline of the work is similar. Our aim in preparing this volume has been to provide both a reference book and a text. Throughout it we have endeavored to present quantitative methods for estimating solar process performance.

In the first two chapters we treat solar radiation, radiation data, and the processing of the data to get it in forms needed for calculation of process performance. The next set of three chapters is a review of some heat transfer principles that are particularly useful and a treatment of the radiation properties of opaque and transparent materials. Chapters 6 through 9 go into detail on collectors and storage, as without an understanding of these essential components in a solar process system it is not possible to understand how systems operate. Chapters 10 and 11 are on system concepts and economics. They serve as an introduction to the balance of the book, which is concerned with applications and design methods.

Some of the topics we cover are very well established and well understood. Others are clearly matters of research, and the methods we have presented can be expected to be outdated and replaced by better methods. An example of this situation is found in Chapter 2; the methods for estimating the fractions of total radiation which are beam and diffuse are topics of current research, and procedures better than those we suggest will probably become available. In these situations we have included in the text extensive literature citations so the interested reader can easily go to the references for further background.

Collectors are at the heart of solar processes, and for those who are starting a study of solar energy without any previous background in the subject, we suggest reading Sections 6.1 and 6.2 for a general description of these unique heat transfer devices. The first half of the book is aimed entirely at development of the ability to calculate how collectors work, and a reading of the description will make clearer the reasons for the treatment of the first set of chapters.

Our emphasis is on solar applications to buildings, as they are the applications developing most rapidly and are the basis of a small but growing industry. The same ideas that are the basis of applications to buildings also underlie applications to industrial process heat, thermal conversion to electrical energy generation, and evaporative processes, which are all discussed briefly. Chapter 15 is a discussion of passive heating, and uses many of the same concepts and calculation methods for estimating solar gains that are developed and used in active heating systems. The principles are the same; the first half of the book develops these principles, and the second half is concerned with their application to active, passive, and nonbuilding processes.

New methods of simulation of transient processes have been developed in recent years, in our laboratory and in others. These are powerful tools in the development of understanding of solar processes and in their design, and in the chapters on applications the results of simulation studies are used to illustrate the sensitivity of long-term performance to design variables. Simulations are the basis of the design procedures described in Chapters 14 and 18. Experimental measurements of system performance are still scarce, but in several cases we have made comparisons of predicted and measured performance.

Since the future of solar applications depends on the costs of solar energy systems, we have included a discussion of life cycle economic analysis, and concluded it with a way of combining the many economic parameters in a life cycle savings analysis into just two numbers which can readily be used in system optimization studies. We find the method to be highly useful, but we make no claims for the worth of any of the numbers used in illustrating the method, and each user must pick his own economic parameters.

In order to make the book useful, we have wherever possible given useful relationships in equation, graphical and tabular form. We have used the recommended standard nomenclature of the *Journal of Solar Energy* (**21**, 69, 1978), except for a few cases where additional symbols have been needed for clarity. For example, G is used for irradiance (a rate, W/m^2), H is used for irradiation for a day (an integrated quantity, MJ/m^2), and I is used for irradiation for an hour (MJ/m^2), which can also be thought of as an average rate for an hour. A listing of nomenclature appears in Appendix B, and includes page references to discussions of the meaning of symbols where there might be confusion. SI units are used throughout, and Appendix C provides useful conversion tables.

Numerous sources have been used in writing this book. The *Journal of Solar Energy*, a publication of the International Solar Energy Society, is very useful, and contains a variety of papers on radiation data, collectors of various types, heating and cooling processes, and other topics. Publications of ASME and ASHRAE have provided additional sources. In addition to these journals, there exists a very large and growing body of literature in the form of reports to and by government agencies which are not reviewed in the usual sense but which contain useful information not readily available elsewhere. These materials are not as readily available as journals, but they are referenced where we have not found the material in journals. We also call the reader's attention to *Gelio-*

tekhnika (Applied Solar Energy), a journal published by the Academy of Sciences of the UZSSR which is available in English and the *Revue Internationale d'Heliotechnique*, published by COMPLES in Marseille.

Many have contributed to the growing body of solar energy literature on which we have drawn. Here we note only a few of the most important of them. The work of H. C. Hottel and his colleagues at MIT, that of A. Whillier at MIT and McGill University, and that of B. Y. H. Liu and R. C. Jordan at Minnesota continues to be of basic importance. In space heating, the publications of G. O. G. Löf, S. Karaki and their colleagues at Colorado State University provide much of the quantitative information we have on that application.

Individuals who have helped us with the preparation of this book are many. Our graduate students and staff at the Solar Energy Laboratory have provided us with ideas, useful information and reviews of parts of the manuscript. Their constructive comments have been invaluable, and references to their work are included in the appropriate chapters. The help of students in our course on Solar Energy Technology is also acknowledged; the number of errors in the manuscript is substantially lower as a result of their good-natured criticisms.

Critical reviews are imperative, and we are indebted to S. A. Klein for his reading of the manuscript. He has been a source of ideas, a sounding board for a wide range of concepts, the author of many publications on which we have drawn, and a constructive critic of the best kind.

High on any list of acknowledgements for support of this work must be the College of Engineering and the Graduate School of the University of Wisconsin–Madison. The College has provided us with support while the manuscript was in preparation, and the Graduate School made it possible for each of us to spend a half year at the Division of Mechanical Engineering of the Commonwealth Scientific and Industrial Research Organization, Australia, where we made good use of their library and developed some of the concepts of this book. Our Laboratory at Wisconsin has been supported by the National Science Foundation, the Energy Research and Development Administration, and now the Department of Energy, and the research of the Laboratory has provided ideas for the book.

It is again appropriate to acknowledge the inspiration of the late Farrington Daniels. He kept interest in solar energy alive in the 1960s and so helped to prepare for the new activity in the field during the 1970s.

Generous permissions have been provided by many publishers and authors for the use of their tables, drawings and other materials in this book. The inclusion of these materials makes the book more complete and useful, and their cooperation is deeply appreciated.

A book such as this takes more than authors and critics to bring it into being. Typing and drafting help are essential and we are pleased to note the help of Shirley Quamme and her co-workers in preparing the manuscript. We have been through several drafts of the book which have been typed by our student helpers at the laboratory; it has often been difficult work, and their persistence, skill and good humor have been tremendous.

Not the least, we thank our patient families for their forbearance during the lengthy process of putting this book together.

JOHN A. DUFFIE
WILLIAM A. BECKMAN

Madison, Wisconsin
June 1980

Contents

Appendices   653

SOLAR ENGINEERING OF THERMAL PROCESSES

CHAPTER 1

Solar Radiation

The sun's structure and characteristics determine the nature of the energy it radiates into space. This chapter notes the characteristics of this energy outside of the earth's atmosphere and the effects of the atmosphere in attenuating the radiation. Then the characteristics of the resulting energy resource available at the earth's surface are outlined, that is, its intensity, spectral distribution, and its directional characteristics. We are concerned primarily with radiation in a wavelength range of 0.3 to 3.0 μm, the portion of the spectrum that includes most of the energy radiated by the sun.

In general, it is not practical to start from knowledge of extraterrestrial radiation and predict the intensity and spectral distribution to be expected on the ground. Adequate meteorological data for such calculations are seldom available, and recourse usually is made to measurements. However, an understanding of the nature of extraterrestrial radiation, atmospheric attenuation, and the effects of orienting a receiving surface is important in understanding and using solar radiation data.

1.1 THE SUN

The sun is a sphere of intensely hot gaseous matter with a diameter of 1.39×10^9 m and is, on the average, 1.5×10^{11} m from the earth. As seen from the earth, the sun rotates on its axis about once every four weeks. However, it does not rotate as a solid body; the equator takes about 27 days and the polar regions take about 30 days for each rotation.

The sun has an effective blackbody temperature of 5762 K.* The temperature in the central interior regions is variously estimated at 8×10^6 to 40×10^6 K and the density at about 100 times that of water. The sun is, in effect, a continuous fusion reactor with its constituent gases as the "containing vessel" retained by gravitational forces. Several fusion reactions have been suggested to supply the

---

* This effective blackbody temperature of 5762 K is the temperature of a blackbody radiating the same amount of energy as does the sun. Other effective temperatures can be defined, for example, that corresponding to the blackbody temperature giving the same wavelength of maximum radiation as solar radiation (about 6300 K).

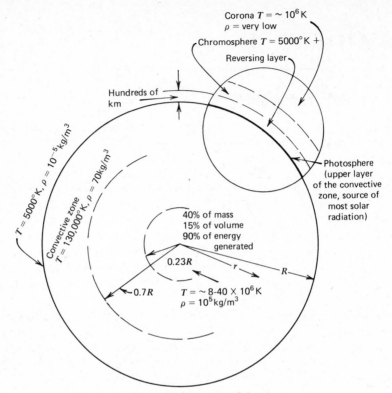

Figure 1.1.1   The structure of the sun.

energy radiated by the sun; the one considered the most important is a process in which hydrogen (i.e., four protons) combines to form helium (i.e., one helium nucleus); the mass of the helium nucleus is less than that of the four protons, mass having been lost in the reaction and converted to energy.

   This energy is produced in the interior of the solar sphere, at temperatures of many millions of degrees. It must be transferred out to the surface and then be radiated into space. A succession of radiative and convective processes must occur, with successive emission, absorption, and reradiation; the radiation in the sun's core must be in the x-ray and gamma-ray parts of the spectrum with the wavelengths of the radiation increasing as the temperature drops at larger radial distances.

   A schematic of the structure of the sun is shown in Figure 1.1.1. It is estimated that 90% of the energy is generated in the region of 0 to $0.23R$ (where R is the radius of the sun), which contains 40% of the mass of the sun. At a distance $0.7R$ from the center, the temperature has dropped to about 130,000 K and the density has dropped to 70 kg/m³; here convection processes begin to become important and the zone from 0.7 to $1.0R$ is known as the *convective zone*. Within this zone, the temperature drops to about 5000 K and the density to about 10^{-5} kg/m³.

The sun's surface appears to be composed of granules (irregular convection cells), with dimensions of cells from 1000 to 3000 km and with cell lifetime of a few minutes. Other features of the solar surface are small dark areas called pores, which are of the same order of magnitude as the convective cells, and larger dark areas called sunspots, which vary in size. The outer layer of the convective zone is called the *photosphere*. The edge of the photosphere is sharply defined, even though it is of low density (about 10^{-4} that of air at sea level). It is essentially opaque, as the gases of which it is composed are strongly ionized and able to absorb and emit a continuous spectrum of radiation. The photosphere is the source of most solar radiation.

Outside of the photosphere is a more or less transparent solar atmosphere, which is observable during total solar eclipse or by instruments that occult the solar disk. Above the photosphere is a layer of cooler gases several hundred kilometers deep called the *reversing layer*. Outside of that is a layer referred to as the *chromosphere*, with a depth of about 10,000 km. This is a gaseous layer with temperatures somewhat higher than that of the photosphere and with lower density. Still further out is the *corona*, of very low density and of very high (10^6 K) temperature. For further information on the sun's structure see Thomas (1958) or Robinson (1966).

This simplified picture of the sun, its physical structure, and its temperature and density gradients, will serve as a basis for appreciating that the sun does not, in fact, function as a blackbody radiator at a fixed temperature. Rather, the emitted solar radiation is the composite result of the several layers that emit and absorb radiation of various wavelengths. The resulting extraterrestrial solar radiation and its spectral distribution have now been measured by various methods in several experiments; the results are noted in the following two sections.

1.2 THE SOLAR CONSTANT

Figure 1.2.1 shows schematically the geometry of the sun-earth relationships. The eccentricity of the earth's orbit is such that the distance between the sun and the earth varies by 1.7%. At a distance of one astronomical unit, 1.495×10^{11} m, the mean earth-sun distance, the sun subtends an angle of 32'. The radiation emitted by the sun and its spatial relationship to the earth result in a nearly fixed intensity of solar radiation outside of the earth's atmosphere. The *solar constant*, G_{sc}, is the energy from the sun, per unit time, received on a unit area of surface perpendicular to the direction of propagation of the radiation, at the earth's mean distance from the sun, outside of the atmosphere.

Until recently, estimates of the solar constant had to be made from ground-based measurements of solar radiation after it had been transmitted through the atmosphere, and thus in part absorbed and scattered by components of the atmosphere. Extrapolations from the terrestrial measurements, which were

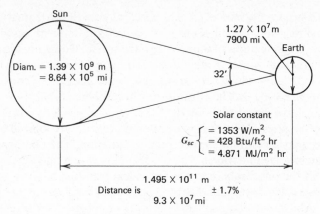

Figure 1.2.1 Sun-earth relationships.

made from high mountains, were based on estimates of atmospheric transmission in various portions of the solar spectrum. Pioneering studies were done by C. G. Abbot and his colleagues at the Smithsonian Institution. These studies and later measurements from rockets were summarized by Johnson (1954); Abbot's value of the solar constant of 1322 W/m^2 was revised upward by Johnson to 1395 W/m^2.

More recently, the availability of very-high-altitude aircraft, balloons, and spacecraft has permitted direct measurements of solar radiation outside most or all of the earth's atmosphere. These measurements were made with a variety of instruments in nine separate experimental programs. They resulted in a value of the solar constant, G_{sc}, of 1353 W/m^2 (1.940 cal/cm² min, 428 Btu/ft² hr, or 4.871 MJ/m^2 hr). The estimated error was ± 1.5 percent. For discussions of these experiments, see Thekaekara (1976) or Thekaekara and Drummond (1971). This standard value was accepted by NASA [see NASA (1971)] and by the American Society for Testing Materials.

The data on which the 1353 W/m^2 value was based, have been reexamined by Frohlich (1977), and reduced to a new pyrheliometric scale* based on comparisons of the instruments with absolute radiometers. Data from Nimbus and Mariner satellites have also been included in the analysis, and as of 1978, Frohlich recommends a new value of the solar constant of $G_{sc} = 1373$ W/m^2, with a probable error of 1 to 2 percent. This is 1.5 percent higher than the earlier value, and 1.2 percent higher than the best available determination of the solar constant by integration of spectral measurements. Thus there remains some uncertainty about the value of G_{sc}, but the uncertainty is of the order of 1 percent. (As will be seen in Chapter 2, uncertainties in most terrestrial solar radiation measurements are an order of magnitude larger than that.) In this book we use the value of 1353 W/m^2.

---

* Pyrheliometric scales are discussed in Section 2.2.

1.3 SPECTRAL DISTRIBUTION OF EXTRATERRESTRIAL RADIATION

In addition to the total energy in the solar spectrum (i.e., the solar constant) it is useful to know the spectral distribution of the extraterrestrial radiation, that is, the radiation that would be received in the absence of the atmosphere. A standard spectral irradiance curve has been compiled, based on high altitude and space measurements. This NASA/ASTM standard is shown in Figure 1.3.1. The averaged energy $G_{sc, \lambda}$ over small bandwidths centered at wavelength λ and the integrated fraction of the energy, $f_{0-\lambda}$, at wavelengths less than λ for the standard curve are indicated in Table 1.3.1. This is a condensed table; more detailed tables are available [e.g., see Thekaekara (1976)].

Example 1.3.1

Calculate the fraction of the extraterrestrial solar radiation and the amount of that radiation in the ultraviolet ($\lambda < 0.38 \ \mu$m), the visible ($0.38 \ \mu$m $< \lambda < 0.78 \ \mu$m), and the infrared ($\lambda > 0.78 \ \mu$m) portions of the spectrum.

Solution

From Table 1.3.1, the fractions $f_{0-\lambda}$ corresponding to wavelengths of 0.38 and 0.78 μm are 0.0700 and 0.5429 (interpolated). Thus, the fraction in the ultraviolet is 0.0700, the fraction in the visible range is $(0.5429 - 0.0700) = 0.4729$, and the fraction in the infrared is $(1.0 - 0.5429) = 0.4571$. Applying these fractions to a solar constant of 1353 W/m² and tabulating the results, we have:

Wavelength range (μm)	0–0.38	0.38–0.78	0.78–∞
Fraction in range	0.0700	0.4729	0.4571
Energy in range (W/m²)	95	640	618 ∎

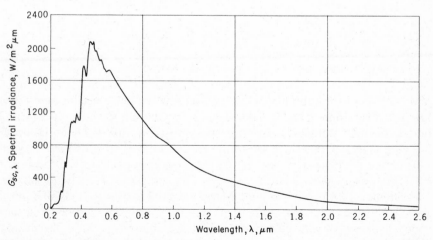

Figure 1.3.1 The NASA/ASTM standard spectral irradiance at the mean sun-earth distance and a solar constant of 1353 W/m².

Table 1.3.1   Extraterrestrial Solar Irradiance (Solar Constant = 1353 W/m²)

λ	$G_{sc,\lambda}{}^{a}$	$f_{0-\lambda}{}^{b}$	λ	$G_{sc,\lambda}{}^{a}$	$f_{0-\lambda}{}^{b}$	λ	$G_{sc,\lambda}{}^{a}$	$f_{0-\lambda}{}^{b}$
0.24	63.0	0.0014	0.47	2033	0.1817	1.0	748	0.6949
0.25	70.9	0.0019	0.48	2074	0.1968	1.2	485	0.7840
0.26	130	0.0027	0.49	1950	0.2115	1.4	337	0.8433
0.27	232	0.0041	0.50	1942	0.2260	1.6	245	0.8861
0.28	222	0.0056	0.51	1882	0.2401	1.8	159	0.9159
0.29	482	0.0081	0.52	1833	0.2538	2.0	103	0.9349
0.30	514	0.0121	0.53	1842	0.2674	2.2	79	0.9483
0.31	689	0.0166	0.54	1783	0.2808	2.4	62	0.9586
0.32	830	0.0222	0.55	1725	0.2938	2.6	48	0.9667
0.33	1059	0.0293	0.56	1695	0.3065	2.8	39	0.9731
0.34	1074	0.0372	0.57	1712	0.3191	3.0	31	0.9783
0.35	1093	0.0452	0.58	1715	0.3318	3.2	22.6	0.9822
0.36	1068	0.0532	0.59	1700	0.3444	3.4	16.6	0.9850
0.37	1181	0.0615	0.60	1666	0.3568	3.6	13.5	0.9872
0.38	1120	0.0700	0.62	1602	0.3810	3.8	11.1	0.9891
0.39	1098	0.0782	0.64	1544	0.4042	4.0	9.5	0.9906
0.40	1429	0.0873	0.66	1486	0.4266	4.5	5.9	0.9934
0.41	1751	0.0992	0.68	1427	0.4481	5.0	3.8	0.9951
0.42	1747	0.1122	0.70	1369	0.4688	6.0	1.8	0.9972
0.43	1639	0.1247	0.72	1314	0.4886	7.0	1.0	0.9982
0.44	1810	0.1373	0.75	1235	0.5169	8.0	0.59	0.9988
0.45	2006	0.1514	0.80	1109	0.5602	10.0	0.24	0.9994
0.46	2066	0.1665	0.90	891	0.6337	50.0	3.9×10^{-4}	1.0000

[a] $G_{sc,\lambda}$ is the solar spectral irradiance in W/m² μm averaged over a small bandwidth centered at λ.

[b] $f_{0-\lambda}$ is the fraction of the solar constant associated with wavelengths shorter than λ. From Thekaekara (1974).

1.4   VARIATION OF EXTRATERRESTRIAL RADIATION

Two sources of variation in extraterrestrial radiation must be considered. The first is the variation in the radiation emitted by the sun. There are conflicting reports in the literature on periodic variations of intrinsic solar radiation. It has been suggested that there are small variations (less than ± 1.5 percent) with different periodicities and variation related to sunspot activities. Others consider the measurements to be inconclusive or not indicative of regular variability. Measurements from Nimbus and Mariner satellites over periods of several months showed variations within limits of ± 0.2 percent over a time when sunspot activity was very low [Frohlich (1977)]. See Coulson (1975) or Thekaekara (1976) for further discussion of this topic. For engineering purposes, in view of the uncertainties and variability of atmospheric transmission, and until reliable

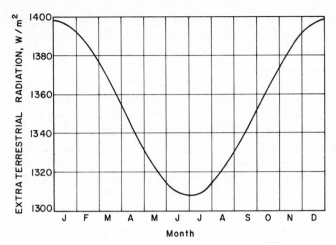

Figure 1.4.1 Variation of extraterrestrial solar radiation with time of year.

measurements indicate otherwise, the energy emitted by the sun can be considered to be fixed.

Variation of the earth-sun distance, however, does lead to variation of extraterrestrial radiation flux in the range of $\pm 3\%$. The dependence of extraterrestrial radiation on time of year is indicated by Equation 1.4.1 and Figure 1.4.1.

$$G_{on} = G_{sc}\left(1 + 0.033 \cos \frac{360n}{365}\right) \qquad (1.4.1)$$

where G_{on} is the extraterrestrial radiation, measured on the plane normal to the radiation on the nth day of the year.

1.5 DEFINITIONS; SOLAR TIME

Several definitions will be useful in understanding the balance of this chapter.

Zenith Angle, θ_z The angle subtended by a vertical line to the zenith (i.e., the point directly overhead) and the line of sight to the sun.

Air Mass, m The ratio of the optical thickness of the atmosphere through which beam radiation passes to the optical thickness if the sun were at the zenith. Thus at sea level, $m = 1$ when the sun is at the zenith, and $m = 2$ for a zenith angle θ_z, of 60°. For zenith angles from 0° to 70° at sea level,

$$m = (\cos \theta_z)^{-1}$$

For higher zenith angles, the effect of the earth's curvature becomes significant and must be taken into account. For a more complete discussion of air mass, see Robinson (1966) or Kondratyev (1969).

Beam Radiation   The solar radiation received from the sun without having been scattered by the atmosphere. (Beam radiation is often referred to as direct solar radiation; to avoid confusion between subscripts for direct and diffuse, we use the term beam radiation.)

Diffuse Radiation   The solar radiation received from the sun after its direction has been changed by scattering by the atmosphere. (Diffuse radiation is referred to in some meteorological literature as sky radiation or solar sky radiation; the definition used here will distinguish the diffuse solar radiation from radiation emitted by the atmosphere.)

Total Solar Radiation   The sum of the beam and the diffuse radiation on a surface.* (The most common measurements of solar radiation are total radiation on a horizontal surface, often referred to as *global radiation*).

Additional radiation terminology used in this book includes the following terms:

Irradiance, W/m^2   The rate at which radiant energy is incident on a surface, per unit area of surface. The symbol G is used, with appropriate subscripts, for beam or diffuse radiation.

Irradiation or Radiant Exposure, J/m^2   The incident energy per unit area on a surface, found by integration of irradiance over a specified time, usually an hour or a day. (*Insolation* is a term applying specifically to solar energy irradiation.) The symbol H is used for insolation for a day (or other period if specified). The symbol I is used for insolation for an hour. H and I can be beam, diffuse, or total and can be on surfaces of any orientation.†

Radiosity or Radiant Exitance, W/m^2   The rate at which radiant energy leaves a surface, per unit area, by combined emission, reflection, and transmission.

Emissive Power or Radiant Self Exitance, W/m^2   The rate at which radiant energy leaves a surface per unit area, by emission only. The symbol E is used, with appropriate subscripts.

Any of these terms, except insolation, can apply to any specified wavelength range (such as the solar energy spectrum) or to monochromatic radiation. Insolation refers only to irradiation in the solar energy spectrum.

---

* Total solar radiation is sometimes used to indicate quantities integrated over all wavelengths of the solar spectrum.

† Subscripts on G, H, and I are as follows: o refers to radiation above the earth's atmosphere, referred to as extraterrestrial radiation; b and d refer to beam and diffuse radiation; T and n refer to radiation on tilted or normal planes. If neither T nor n appear, the radiation is on a horizontal plane.

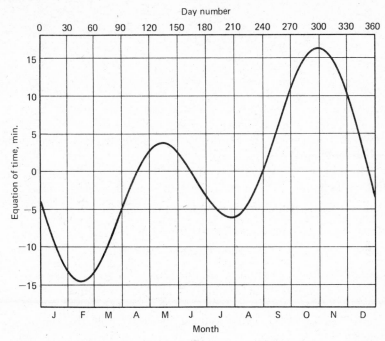

Figure 1.5.1 The equation of time, E, in minutes, as a function of time of year.

Solar Time Time based on the apparent angular motion of the sun across the sky, with *solar noon* the time the sun crosses the meridian of the observer.

Solar time is the time specified in all of the sun angle relationships; it does not coincide with local clock time. It is necessary to convert standard time to solar time by applying two corrections. First, there is a constant correction for the difference in longitude between the observer's meridian location and the meridian on which the local standard time is based*; the sun takes four minutes to transverse 1° of longitude. The second correction is from the *equation of time*, which takes into account the perturbations in the earth's rate of rotation, which affect the time the sun crosses the observer's meridian.† Solar time is related to standard time by

$$\text{solar time} = \text{standard time} + 4\,(L_{st} - L_{loc}) + E \qquad (1.5.1)$$

where E is the equation of time from Figure 1.5.1 or Equation 1.5.2 [from Whillier (1979)] in minutes, L_{st} is the standard meridian for the local time zone, and L_{loc} is the longitude of the location in question in degrees west.

$$E = 9.87 \sin 2B - 7.53 \cos B - 1.5 \sin B \qquad (1.5.2)$$

* Standard meridians for continental U. S. time zones are Eastern, 75°W; Central, 90°W; Mountain, 105°W; and Pacific, 120°W.

† There may also be an additional 1 hr correction for daylight saving time.

where

$$B = \frac{360(n - 81)}{364}$$

n = day of the year, $1 \leq n \leq 365$

Example 1.5.1

At Madison, WI, what is the solar time corresponding to 10:30 A.M. central standard time on February 2?

Solution

In Madison, where the longitude is 89.4°, Equation 1.5.1 gives

$$\text{solar time} = \text{standard time} + 4(90 - 89.4) + E$$
$$= \text{standard time} + 2.48 + E$$

On February 2, E is -13.5 min, so the correction to standard time is -11 min. Thus 10:30 A.M. central standard time is 10:19 A.M. solar time.   ■

In this book, all times are assumed to be solar times unless indication is given otherwise.

1.6 DIRECTION OF BEAM RADIATION

The geometric relationships between a plane of any particular orientation relative to the earth at any time (whether that plane is fixed or moving relative to the earth) and the incoming beam solar radiation, that is, the position of the sun relative to that plane, can be described in terms of several angles (Benford and Bock (1939)). These angles and the relationships between them are as follows:

ϕ Latitude, that is, the angular location north or south of the equator, north positive. $-90° \leq \phi \leq 90°$.

δ Declination, that is, the angular position of the sun at solar noon with respect to the plane of the equator, north positive. $-23.45° \leq \delta \leq 23.45°$.

β Slope, that is, the angle between the plane surface in question and the horizontal. $0 \leq \beta \leq 180°$ ($\beta > 90°$ means that the surface has a downward facing component).

γ Surface azimuth angle, that is, the deviation of the projection on a horizontal plane of the normal to the surface from the local meridian, with zero due south, east negative, west positive.* $-180° \leq \gamma \leq 180°$.

---

* The sign convention used here for γ and ω is the reverse of that in *Solar Energy Thermal Processes*. This convention is consistent with Hottel and Woertz (1942) and other authors in the solar energy literature. Either is correct; it is necessary to be consistent.

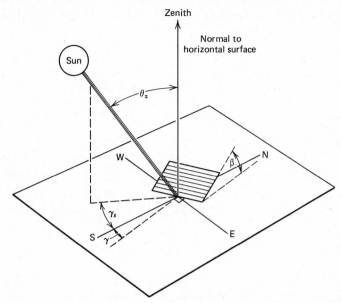

Figure 1.6.1 Zenith angle, slope, surface azimuth angle and solar azimuth angle for a tilted surface.

ω Hour angle, that is, the angular displacement of the sun east or west of the local meridian due to rotation of the earth on its axis at 15° per hour, morning negative, afternoon positive.

θ Angle of incidence, that is, the angle between the beam radiation on a surface and the normal to that surface.

Zenith angle, slope, and surface azimuth angle are shown in Figure 1.6.1. The declination, δ, can be found from the equation of Cooper (1969):

$$\delta = 23.45 \sin\left(360 \frac{284 + n}{365}\right) \tag{1.6.1}$$

where n is the day of the year; n can be conveniently obtained with the help of Table 1.6.1 (or from Figure 1.6.3).

Thus at 2:30 P.M. solar time on October 15 at Madison (latitude 43°N) for a surface tilted 45° from the horizontal and facing 20° west of south, $\phi = 43°$, $n = 288$, $\delta = -9.60$ (from Equation 1.6.1), $\beta = 45°$, $\gamma = 20°$, and $\omega = 37.5°$.

The equation relating the angle of incidence of beam radiation, θ, and the other angles is:

$$\begin{aligned}
\cos \theta = {} & \sin \delta \sin \phi \cos \beta - \sin \delta \cos \phi \sin \beta \cos \gamma \\
& + \cos \delta \cos \phi \cos \beta \cos \omega \\
& + \cos \delta \sin \phi \sin \beta \cos \gamma \cos \omega \\
& + \cos \delta \sin \beta \sin \gamma \sin \omega
\end{aligned} \tag{1.6.2}$$

Table 1.6.1 Recommended Average Day^a for Each Month [from Klein (1976)] and Values of n by Months

Month	n for ith Day of Month^b	For the Average Day of the Month		
		Date	n, Day of Year^b	δ, Declination
January	i	17	17	−20.9
February	$31 + i$	16	47	−13.0
March	$59 + i$	16	75	−2.4
April	$90 + i$	15	105	9.4
May	$120 + i$	15	135	18.8
June	$151 + i$	11	162	23.1
July	$181 + i$	17	198	21.2
August	$212 + i$	16	228	13.5
September	$243 + i$	15	258	2.2
October	$273 + i$	15	288	−9.6
November	$304 + i$	14	318	−18.9
December	$334 + i$	10	344	−23.0

^{a} The average day is that day which has the extraterrestrial radiation closest to the average for the month. See Section 1.8.
^{b} These do not account for leap year; values of n from March onward for leap years can be corrected by adding 1. Declination values will also shift slightly.

Example 1.6.1

Calculate the angle of incidence of beam radiation on a surface located at Madison, WI at 10:30 (solar time) on February 13, if the surface is tilted 45° from the horizontal and is pointed 15° west of south.

Solution

Under these conditions, the declination is −14°, the hour angle is −22.5°, and the surface azimuth angle is 15°. Using the slope of 45° and Madison's latitude of 43°N, Equation 1.6.2 is

$$\cos \theta = \sin(-14)\sin 43 \cos 45$$
$$- \sin(-14)\cos 43 \sin 45 \cos 15$$
$$+ \cos(-14)\cos 43 \cos 45 \cos(-22.5)$$
$$+ \cos(-14)\sin 43 \sin 45 \cos 15 \cos(-22.5)$$
$$+ \cos(-14)\sin 45 \sin 15 \sin(-22.5)$$
$$\cos \theta = -0.117 + 0.121 + 0.464 + 0.418 - 0.068$$
$$\cos \theta = 0.817$$
$$\theta = 35°$$

Additional angles are also defined. The *solar azimuth angle* γ_s, is the angular displacement from south of the projection of the beam radiation on the horizontal plane, as shown in Figure 1.6.1. In architectural and illumination practice, other angles are defined, such as the profile angle and the solar altitude angle $(90 - \theta_z)$. Care must be exercised in the use of any source of information on these angles so that definitions and sign conventions are understood and followed.

There are several commonly occurring cases for which Equation 1.6.2 is simplified. For fixed surfaces sloped toward the south or north, that is, with a surface azimuth angle, γ, of $0°$ or $180°$ (a very common situation for fixed flat-plate collectors), the last term drops out. For vertical surfaces, $\beta = 90°$ and the equation becomes

$$\cos \theta = -\sin \delta \cos \phi \cos \gamma + \cos \delta \sin \phi \cos \gamma \cos \omega$$
$$+ \cos \delta \sin \gamma \sin \omega \qquad (1.6.3)$$

For horizontal surfaces, $\beta = 0°$, and the angle of incidence is the zenith angle of the sun, θ_z. Equation 1.6.2 becomes

$$\cos \theta_z = \cos \delta \cos \phi \cos \omega + \sin \delta \sin \phi \qquad (1.6.4)$$

Example 1.6.2

Calculate the zenith angle of the sun at Madison at 9:30 on February 13.

Solution

For this date, declination is $-14°$. From Equation 1.6.4

$$\cos \theta_z = \cos(-14)\cos 43 \cos(-37.5) + \sin(-14)\sin 43$$
$$\cos \theta_z = 0.398$$
$$\theta_z = 66° \qquad \blacksquare$$

Useful relationships for the angle of incidence on surfaces sloped to the north or south can be derived from the fact that surfaces with slope β to the north or south have the same angular relationship to beam radiation as a horizontal surface at an artificial latitude of $(\phi - \beta)$. The relationship is shown in Figure 1.6.2, for the northern hemisphere. Modifying Equation 1.6.4,

$$\cos \theta = \cos(\phi - \beta)\cos \delta \cos \omega + \sin(\phi - \beta)\sin \delta \qquad (1.6.5)$$

For the southern hemisphere the equation is modified by replacing $(\phi - \beta)$ by $(\phi + \beta)$, consistent with the sign conventions on ϕ and δ

$$\cos \theta = \cos(\phi + \beta)\cos \delta \cos \omega + \sin(\phi + \beta)\sin \delta \qquad (1.6.6)$$

Equation 1.6.4 can be solved for the *sunset hour angle*, ω_s, when $\theta_z = 90°$

$$\cos \omega_s = -\frac{\sin \phi \sin \delta}{\cos \phi \cos \delta}$$

$$\cos \omega_s = -\tan \phi \tan \delta \qquad (1.6.7)$$

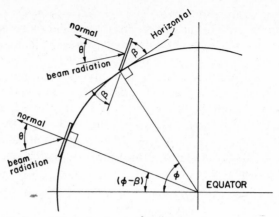

Figure 1.6.2 Section of Earth showing β, θ, ϕ and ($\phi - \beta$) for a south-facing surface.

It also follows that the number of daylight hours is given by

$$N = \tfrac{2}{15} \cos^{-1}(-\tan \phi \tan \delta) \tag{1.6.8}$$

A convenient nomogram for determining day length has been devised by Whillier (1965), and is shown in Figure 1.6.3. Information on latitude and declination leads directly to times of sunrise and sunset and day length, for either hemisphere.

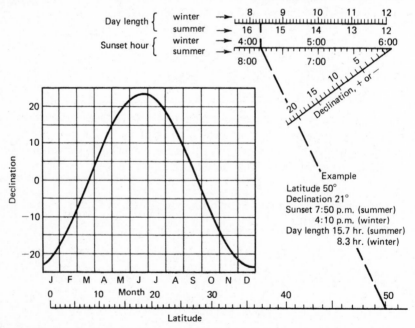

Figure 1.6.3 Nomogram to determine time of sunset and day length. Adapted from Whillier, Solar Energy **9**, 164 (1965).

Solar azimuth and altitude angles are tabulated as functions of latitude, declination, and hour angle by the U. S. Hydrographic Office (1940). Information on the position of the sun in the sky is also available with less precision but easier access in various types of charts. Examples of these are the Sun Angle Calculator (1951) and diagrams in a paper by Hand (1948). (Care is necessary in interpreting information from sources such as these, since definitions of angles and sign conventions may vary from those used here.) Brooks and Miller (1963) also present a discussion of these geometrical relationships.

Equation 1.6.2 is a generally applicable relationship for the angle of incidence of beam radiation on a surface of any orientation, and reduces to simpler forms for special cases such as horizontal surfaces (Equation 1.6.4) and vertical surfaces (Equation 1.6.3). By far the most common case is a surface tilted toward the equator (Equations 1.6.5 and 1.6.6). There are other special cases of interest, for example, the angle of incidence on planes which are moved in prescribed ways. Some collectors are moved to track the beam radiation to varying degrees.* In this section we show the forms of Equation 1.6.2 applicable to some of the more common modes of tracking.

Tracking systems can be classified by the mode of their motion. Motion can be about a single axis (which can be oriented east-west, north-south, or parallel to the earth's axis), or it can be about two axes. The following equations are derivable from Equation 1.6.2, and apply to planes moved as indicated [see Eibling et al. (1953)].

For a plane rotated about a horizontal east-west axis with a single daily adjustment so that its surface-normal coincides with the solar beam at noon each day,

$$\cos \theta = \sin^2 \delta + \cos^2 \delta \cos \omega \tag{1.6.9}$$

For a plane rotated about a horizontal east-west axis with continuous adjustment to minimize the angle of incidence,

$$\cos \theta = (1 - \cos^2 \delta \sin^2 \omega)^{1/2} \tag{1.6.10}$$

For a plane rotated about a horizontal north-south axis with continuous adjustment to minimize the angle of incidence

$$\cos \theta = [(\sin \phi \sin \delta + \cos \phi \cos \delta \cos \omega)^2 + \cos^2 \delta \sin^2 \omega]^{1/2} \tag{1.6.11}$$

For a plane rotated about a north-south axis parallel to the earth's axis, with continuous adjustment

$$\cos \theta = \cos \delta \tag{1.6.12}$$

A two-axis tracking surface continuously oriented to face the sun will at all times have

$$\cos \theta = 1 \tag{1.6.13}$$

---

* Tracking solar collectors are almost always of the concentrating type, which are discussed in Chapter 8.

1.7 RATIO OF BEAM RADIATION ON TILTED SURFACE TO THAT ON HORIZONTAL SURFACE

For purposes of solar process design and performance calculations, it is often necessary to calculate the hourly radiation on a tilted surface of a collector from measurements or estimates of solar radiation on a horizontal surface. The most commonly available data are total radiation for hours or days on the horizontal surface, whereas the need is for information on the plane of a collector, either total or beam and diffuse.

The geometric factor, R_b, the ratio of beam radiation on the tilted surface to that on a horizontal surface at any time, can be calculated exactly by appropriate use of Equation 1.6.2. Figure 1.7.1 indicates the angle of incidence of beam radiation on the horizontal and tilted surfaces. The ratio G_{bT}/G_b* is given by

$$R_b = \frac{G_{bT}}{G_b} = \frac{G_{bn} \cos \theta}{G_{bn} \cos \theta_z} = \frac{\cos \theta}{\cos \theta_z} \tag{1.7.1}$$

and $\cos \theta$ and $\cos \theta_z$ are both determined from Equation 1.6.2 (or from the equations derived from equation 1.6.2).

Example 1.7.1

What is the ratio of beam radiation for the surface and time specified in Example 1.6.1 to that on a horizontal surface?

Solution

Example 1.6.1 shows the calculation for $\cos \theta$. For the horizontal surface, from Equation 1.6.4:

$$\cos \theta_z = \sin(-14)\sin 43 + \cos(-14)\cos 43 \cos(-22.5)$$
$$= 0.491$$

And from Equation 1.7.1

$$R_b = \frac{\cos \theta}{\cos \theta_z} = \frac{0.817}{0.491} = 1.67 \qquad\blacksquare$$

The optimum azimuth angle for flat plate collectors is usually 0° in the northern hemisphere (or 180° in the southern hemisphere). Thus it is a common situation that $\gamma = 0°$ (or 180°). In this case, Equations 1.6.4 and 1.6.5 can be used to determine $\cos \theta_z$ and $\cos \theta$, respectively, leading in the northern hemisphere for $\gamma = 0°$ to

$$R_b = \frac{\cos(\phi - \beta)\cos \delta \cos \omega + \sin(\phi - \beta)\sin \delta}{\cos \phi \cos \delta \cos \omega + \sin \phi \sin \delta} \tag{1.7.2}$$

* The symbol G is used in this book to denote rates, while the symbol I is used for energy quantities integrated over an hour. The original development of R_b was by Hottel and Woertz (1942) for hourly periods; for an hour (using angles at the midpoint of the hour), $R_b = I_{bT}/I_b$.

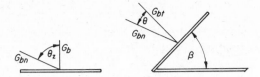

Figure 1.7.1 Beam radiation on horizontal and tilted surfaces.

In the southern hemisphere, $\gamma = 180°$ and the equation is

$$R_b = \frac{\cos(\phi + \beta)\cos \delta \cos \omega + \sin(\phi + \beta)\sin \delta}{\cos \phi \cos \delta \cos \omega + \sin \phi \sin \delta} \qquad (1.7.3)$$

Hottel and Woertz (1942) pointed out that this equation provides a convenient method for calculating R_b for the most common cases. They also showed a graphical method for solving these equations. This graphical method has been revised by Whillier (1975), and an adaptation of Whillier's curves are given here. Figure 1.7.2a–e are plots of both $\cos \theta_z$ as a function of ϕ, and $\cos \theta$ as a function of $(\phi - \beta)$, for various dates (i.e., declinations). By plotting the curves for sets of dates having (nearly) the same absolute value of declination, the curves "reflect back" on each other at latitude $= 0°$. Thus each set of curves, in effect, covers the latitude range of $-60°$ to $60°$.

As will be seen in later chapters, solar process performance calculations are very often done on an hourly basis. The $\cos \theta_z$ plots are shown for the midpoints of hours before and after solar noon, and the values of R_b found from them are applied to those hours. (This procedure is satisfactory for most hours of the day, but in hours that include sunrise and sunset, unrepresentative values of R_b may be obtained. This is usually of little consequence, as solar collection in those hours is most often zero or a negligible part of the total daily collector output. However, care must be taken that unrealistic products of R_b and radiation are not used.)

To find $\cos \theta_z$, enter the chart for the appropriate time with the date and latitude of the location in question. Cos θ is found for the same date and latitude by entering with abscissa corresponding to $(\phi - \beta)$. R_b is then found from Equation 1.7.1. The dates on the sets of curves are shown in two sets, one for north (positive) latitudes and the other for south (negative) latitudes.

Two situations arise, for positive values or for negative values of $(\phi - \beta)$. For positive values, the charts are used directly.

If $(\phi - \beta)$ is negative (which frequently occurs when collectors are sloped for optimum performance in winter, or with vertical collectors), the procedure is modified. Determine $\cos \theta_z$ as before. Determine $\cos \theta$ from the appropriate absolute value of $(\phi - \beta)$, using the curve for the other hemisphere, that is, with the sign on the declination reversed.

Example 1.7.2

Calculate R_b for a surface at latitude 40°N, at a tilt 30° toward the south, for the hour 9 to 10, solar time, on February 16.

Solution

Use Figure 1.7.2c, for the hour ± 2.5 hours from noon, as representative of the hour from 9 to 10. To find $\cos \theta_z$, enter at a latitude of 40° for the north latitude date of February 16. $\cos \theta_z = 0.45$. To find $\cos \theta$, enter at a latitude of $\phi - \beta = 10°$ for the same date. $\cos \theta = 0.73$.

Then

$$R_b = \frac{\cos \theta}{\cos \theta_z} = \frac{0.73}{0.45} = 1.62.$$

The ratio can also be calculated using Equation 1.7.2. The declination on February 16 is $-13°$.

$$R_b = \frac{\cos 10 \cos(-13)\cos(-37.5) + \sin 10 \sin(-13)}{\cos 40 \cos(-13)\cos(-37.5) + \sin 10 \sin(-13)} = 1.61 \qquad \blacksquare$$

Example 1.7.3

Calculate R_b for a latitude 40°N at a tilt of 50° toward the south, for the hour 9 to 10 solar time, on February 16.

Solution

Cos θ_z is found as in the previous example, and is 0.45. To find $\cos \theta$, enter at an abscissa of 10°, using the curve for February 16 for south latitudes. The value of $\cos \theta$ from the curve is 0.80. Thus $R_b = 0.80/0.45 = 1.78$. Equation 1.7.2 can also be used.

$$R_b = \frac{\cos(-10)\cos(-13)\cos(-37.5) + \sin(-10)\sin(-13)}{\cos 40 \cos(-13)\cos(-37.5) + \sin 40 \sin(-13)} = 1.80 \qquad \blacksquare$$

It is possible using Equation 1.7.2 or Figures 1.7.2, to build up plots showing the effects of collector tilt on R_b for various times of the year and day. Figure 1.7.3 shows such a plot for a latitude of 40° and a slope of 50°. It illustrates that very large gains in incident beam radiation are to be had by tilting a receiving surface toward the equator.

Equation 1.7.1 can also be applied to other than fixed flat plate collectors. Equations 1.6.9 to 1.6.13 give $\cos \theta$ for surfaces moved in prescribed ways in which concentrating collectors may move to track the sun. If the beam radiation on a horizontal surface is known or can be estimated, the appropriate one of these equations can be used in the numerator of Equation 1.7.1 for $\cos \theta$. For example, for a plane rotated continuously about a horizontal east-west

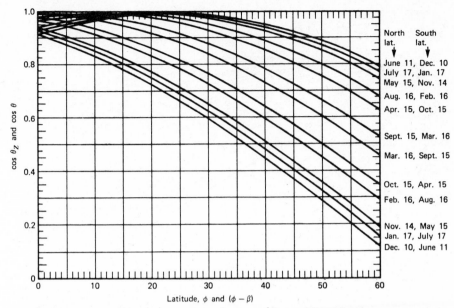

Figure 1.7.2a Cos θ vs $(\phi - \beta)$ and cos θ_z vs ϕ, for hours 11 to 12 and 12 to 1, for surfaces tilted toward the equator. The columns on the right show dates for the curves for north and south latitudes. Figures 1.7.2a to e are adapted from Whillier (1975).

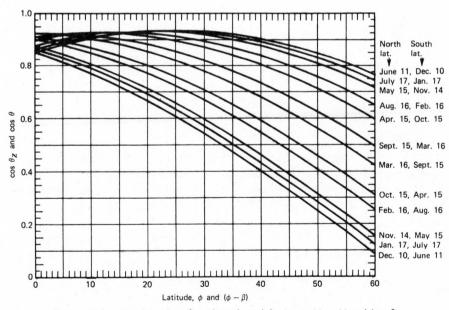

Figure 1.7.2b Cos θ vs $(\phi - \beta)$ and cos θ_z vs ϕ for hours 10 to 11 and 1 to 2.

19

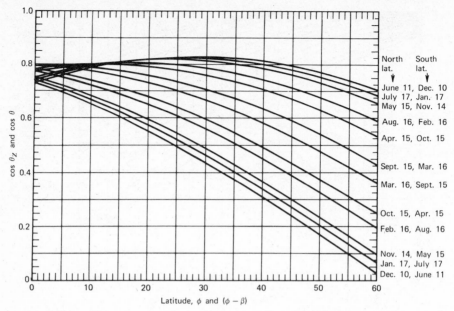

Figure 1.7.2c Cos θ vs ($\phi - \beta$) and cos θ_z vs ϕ for hours 9 to 10 and 2 to 3.

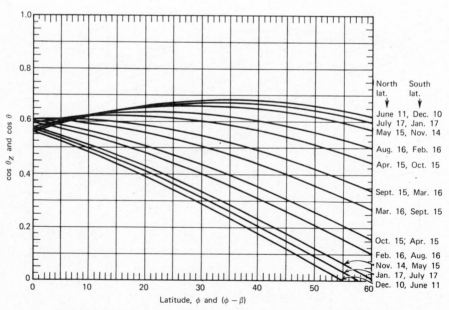

Figure 1.7.2d Cos θ vs ($\phi - \beta$) and cos θ_z vs ϕ, for hours 8 to 9 and 3 to 4.

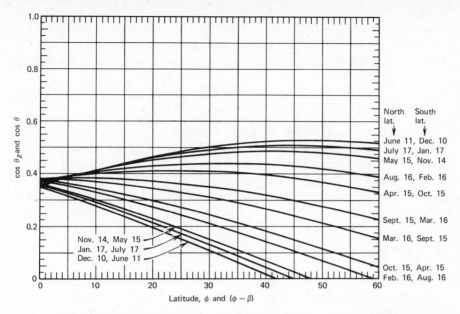

Figure 1.7.2e Cos θ vs $(\phi - \beta)$ and cos θ_z vs ϕ, for hours 7 to 8 and 4 to 5.

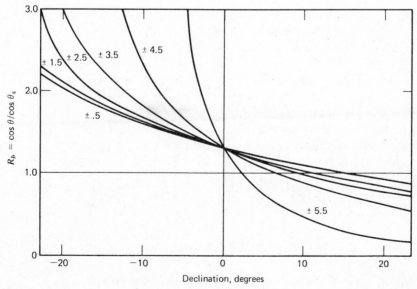

Figure 1.7.3 Ratio R_b, cos θ/cos θ_z, for a surface with slope 50° to south at latitude 40°, for various hours from solar noon.

axis to maximize the beam radiation on the plane, from Equation 1.6.10, the ratio of beam radiation on the plane to that on a horizontal surface at any time is:

$$R_b = \frac{(1 - \cos^2 \delta \sin^2 \omega)^{1/2}}{\sin \phi \sin \delta + \cos \phi \cos \delta \cos \omega} \tag{1.7.4}$$

Some of the data available are beam radiation on a surface normal to the radiation, as measured by a pyrheliometer.* In this case the denominator of Equation 1.7.1 becomes unity and $R_b' = \cos \theta$ from Equations 1.6.9 to 1.6.13. (In this case, we use R_b' to mean the ratio of beam radiation on the plane to that on a surface normal to the radiation.)

1.8 EXTRATERRESTRIAL RADIATION ON HORIZONTAL SURFACE

Several types of radiation calculations are most conveniently done using normalized radiation levels, that is, the ratio of radiation level to the theoretically possible radiation that would be available if there were no atmosphere. For these calculations, which are discussed in Chapter 2, we need a method of calculating the extraterrestrial radiation.

At any point in time, the solar radiation outside the atmosphere incident on a horizontal plane is

$$G_o = G_{sc}\left[1 + 0.033 \cos\left(\frac{360n}{365}\right)\right]\cos \theta_z \tag{1.8.1}$$

where G_{sc} is the solar constant, and n is the day of the year. Equation 1.6.4 gives $\cos \theta_z$. Combining with Equation 1.8.1, G_o for a horizontal surface at any time between sunrise and sunset is given by

$$G_o = G_{sc}\left[1 + 0.033 \cos\left(\frac{360n}{365}\right)\right](\sin \phi \sin \delta + \cos \phi \cos \delta \cos \omega) \tag{1.8.2}$$

It is often necessary for calculations of daily solar radiation to have the integrated daily extraterrestrial radiation on a horizontal surface, H_o. This is obtained by integrating Equation 1.8.2 over the period from sunrise to sunset. If G_{sc} is in watts per square meter, H_o in Joules per square meter is

$$H_o = \frac{24 \times 3600 G_{sc}}{\pi}\left[1 + 0.033 \cos\left(\frac{360n}{365}\right)\right]$$

$$\times \left[\cos \phi \cos \delta \sin \omega_s + \frac{2\pi \omega_s}{360} \sin \phi \sin \delta\right] \tag{1.8.3}$$

where ω_s = sunset hour angle, in degrees, from Equation 1.6.7.

* Instruments and the radiation measured by them are discussed in Sections 2.2 and 2.3.

The monthly mean* daily extraterrestrial radiation, $\bar{H}_o$ is a useful quantity. It can be calculated with Equation 1.8.3 using n and δ for the mean day of the month from Table 1.6.1.† H_o is plotted as a function of latitude, for the northern and southern hemisphere, in Figure 1.8.1. The curves are for dates that give the mean radiation for the month and thus show $\bar{H}_o$. Values of H_o for any day can be estimated by interpolation. Values of $\bar{H}_o$ are also tabulated for latitudes $-60°$ to $60°$ in Table 1.8.1.

Example 1.8.1

What is the day's solar radiation on a horizontal surface in the absence of the atmosphere, H_0, at latitude 43°N, on April 15?

Solution

For these circumstances, $n = 105$ (from Table 1.6.1), $\delta = 9.4°$ (from Equation 1.6.1), and $\phi = 43°$. From Equation 1.6.7

$$\cos \omega_s = -\tan 43 \tan 9.4 = -0.154$$
$$\omega_s = 98.9°$$

Thus, from Equation 1.8.3

$$H_o = \frac{24 \times 3600}{\pi} \times 1353 \left[1 + 0.033 \cos\left(\frac{360 \times 105}{365}\right) \right]$$

$$\times \left[\cos 43 \cos 9.4 \sin 98.9 + \frac{2\pi \times 98.9}{360} \sin 43 \sin 9.4 \right]$$

$$= 33.4 \text{ MJ/m}^2$$

From Figure 1.8.1a, for the curve for April, we read $H_o = 33.0$ MJ/m² and from Table 1.8.1 we obtain $H_o = 33.5$ MJ/m² by interpolation. ∎

It is also of interest to calculate the extraterrestrial radiation on a horizontal surface for an hour period. Integrating Equation 1.8.2 for a period defined by hour angles ω_1 and ω_2 which define an hour (where ω_2 is the larger):

$$I_o = \frac{12 \times 3600}{\pi} G_{sc} \left[1 + 0.033 \cos\frac{360n}{365} \right]$$

$$\times \left[\cos \phi \cos \delta(\sin \omega_2 - \sin \omega_1) + \frac{2\pi(\omega_2 - \omega_1)}{360} \sin \phi \sin \delta \right] \quad (1.8.4)$$

(The limits ω_1 and ω_2 may define a time other than an hour.)

---

* An overbar is used throughout this book to indicate a monthly average of the quantity.

† The mean days were determined as the day having H_o closest to $\bar{H}_o$ for the month.

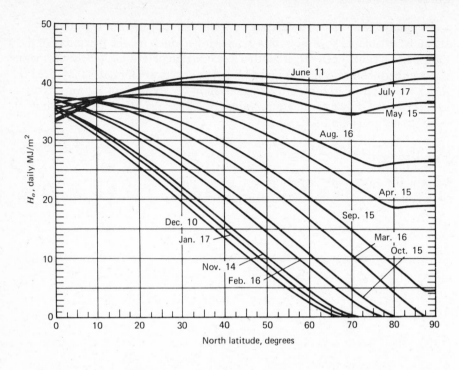

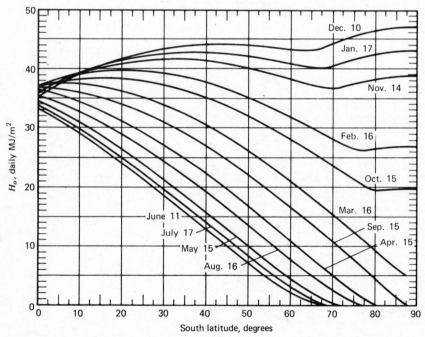

Figure 1.8.1 Extraterrestrial daily radiation on a horizontal surface, for $G_{sc} = 1353$ W/m². The curves are for the mean days of the month from Table 1.6.1.

Table 1.8.1 Monthly Average Daily Extraterrestrial Radiation, $\bar{H}_o$, MJ/m², for G_{sc} = 1353 W/m²

Average Daily Extraterrestrial Radiation

Latitude	Jan.	Feb.	Mar.	Apr.	May	June	July	Aug.	Sep.	Oct.	Nov.	Dec.
60	3.5	8.2	16.7	27.3	36.3	40.6	38.4	30.6	20.3	10.7	4.5	2.3
55	6.1	11.2	19.6	29.3	37.2	40.8	39.0	32.2	22.9	13.6	7.2	4.8
50	9.1	14.2	22.3	31.2	38.1	41.1	39.6	33.7	25.3	16.6	10.2	7.6
45	12.1	17.2	24.8	32.9	38.8	41.3	40.0	35.0	27.5	19.4	13.2	10.5
40	15.1	20.1	27.2	34.3	39.3	41.3	40.2	36.1	29.5	22.1	16.2	13.6
35	18.1	22.8	29.3	35.5	39.5	41.1	40.2	36.9	31.3	24.7	19.1	16.7
30	21.1	25.5	31.2	36.4	39.6	40.7	40.0	37.5	32.9	27.1	22.0	19.7
25	23.9	27.9	32.9	37.1	39.4	40.0	39.6	37.8	34.2	29.3	24.8	22.6
20	26.7	30.2	34.4	37.5	38.9	39.1	38.9	37.8	35.3	31.3	27.4	25.5
15	29.3	32.3	35.5	37.6	38.1	38.0	37.9	37.6	36.1	33.1	29.8	28.2
10	31.7	34.1	36.4	37.5	37.1	36.6	36.7	37.1	36.6	34.6	32.1	30.8
5	33.9	35.7	37.1	37.1	35.9	35.0	35.3	36.3	36.8	35.9	34.1	33.1
0	35.9	37.0	37.4	36.4	34.4	33.2	33.6	35.3	36.8	36.9	36.0	35.3
−5	37.6	38.1	37.5	35.4	32.7	31.1	31.7	34.1	36.5	37.7	37.5	37.3
−10	39.1	38.9	37.3	34.2	30.7	28.9	29.6	32.6	35.9	38.1	38.9	39.0
−15	40.4	39.4	36.8	32.7	28.6	26.5	27.4	30.8	35.0	38.3	39.9	40.4
−20	41.4	39.6	36.0	31.0	26.3	23.9	24.9	28.8	33.9	38.2	40.7	41.7
−25	42.1	39.6	35.0	29.0	23.8	21.3	22.3	26.7	32.5	37.8	41.3	42.6
−30	42.5	39.3	33.7	26.9	21.2	18.5	19.7	24.3	30.9	37.2	41.5	43.3
−35	42.7	38.7	32.1	24.5	18.4	15.7	16.9	21.8	29.0	36.3	41.5	43.8
−40	42.7	37.8	30.3	22.0	15.6	12.8	14.0	19.2	27.0	35.1	41.3	44.0
−45	42.4	36.7	28.3	19.4	12.8	9.9	11.2	16.5	24.7	33.7	40.8	44.0
−50	41.9	35.3	26.1	16.6	9.9	7.1	8.3	13.6	22.2	32.0	40.1	43.8
−55	41.3	33.8	23.6	13.7	7.1	4.5	5.6	10.8	19.6	30.2	39.2	43.5
−60	40.6	32.1	21.0	10.8	4.4	2.1	3.1	7.9	16.8	28.1	38.3	43.2

Example 1.8.2

What is the solar radiation on a horizontal surface in the absence of the atmosphere at latitude 43°N on April 15, between the hours of 10:00 and 11:00?

Solution

The declination is 9.4° (from the previous example). Using Equation 1.8.4, with $\omega_1 = -30°$ and $\omega_2 = -15°$

$$
I_o = \frac{12 \times 3600}{\pi} (1353)\left[1 + 0.033 \cos \frac{360 \times 105}{365} \right]
$$

$$
\times \left\{ \frac{2\pi(-15 - (-30))}{360} \sin 43 \sin 9.4 + \cos 43 \cos 9.4 \right.
$$

$$
\left. \times \left[\sin(-15) - \sin(-30) \right] \right\} = 3.75 \text{ MJ/m}^2 \qquad \blacksquare
$$

The hourly extraterrestrial radiation can also be approximated by using Equation 1.8.2 at ω for the midpoint of the hour. For the circumstances of Example 1.8.2, the hour's radiation so estimated is 3.76 MJ/m². Differences between the hourly radiation calculated by these two methods will be slightly larger at times near sunrise and sunset, but are still small. For larger time spans, the differences become larger. For example, for the same circumstances as in Example 1.8.2 but for the 2 hour span from 7:00 to 9:00, the use of Equation 1.8.4 gives 4.53 MJ/m², and Equation 1.8.2 for 8:00 gives 4.56 MJ/m².

1.9 SUMMARY

In this chapter, we have outlined the basic characteristics of the sun and the radiation it emits, noting that the solar constant, the mean radiation flux density outside of the earth's atmosphere, is 1353 W/m² (or slightly higher, based on new measurements), with most of the radiation in a wavelength range of 0.3 to 3 μm. This radiation has directional characteristics that are defined by a set of angles that determine the angle of incidence of the radiation on a surface. We have included in this chapter those topics that are based on extraterrestrial radiation and the geometry of the earth and sun. This is background information for Chapter 2, which is concerned with effects of the atmosphere, radiation measurements, and data manipulation.

REFERENCES

Benford, F. and J. E. Bock, *Transactions of the American Illumination Engineering Society*, **34**, 200 (1939). "A Time Analysis of Sunshine."

Brooks, F. A. and W. Miller, paper in *Introduction to the Utilization of Solar Energy*, A. M. Zarem and D. D. Erway, Eds., New York, McGraw-Hill (1963). "Availability of Solar Energy."

Cooper, P. I., *Solar Energy*, **12**, 3 (1969). "The Absorption of Solar Radiation in Solar Stills."

Coulson, K. L., *Solar and Terrestrial Radiation*, Academic Press, New York (1975).

Eibling, J. A., R. E. Thomas, and B. A. Landry, Report to the Office of Saline Water, U.S. Department of the Interior (1953). "An Investigation of Multiple-Effect Evaporation of Saline Waters by Steam from Solar Radiation."

Frohlich, C., paper in *The Solar Output and its Variation*, O. R. White (ed.) Colorado Associated University Press, Boulder (1977). "Contemporary Measures of the Solar Constant."

Hand, I. F., *Heating and Ventilating*, **45**, 86 (October 1948). "Charts to Obtain Solar Altitudes and Azimuths."

Hottel, H. C. and B. B. Woertz, *Transactions of the American Society of Mechanical Engineers*, **64**, 91 (1942). "Performance of Flat-Plate Solar Heat Collectors."

Johnson, F. S., *Journal of Meteorology*, **11**, 431 (1954). "The Solar Constant."

Klein, S. A., *Solar Energy*, **19**, 325 (1977). "Calculation of Monthly Average Insolation on Tilted Surfaces."

Kondratyev, K. Y., *Radiation in the Atmosphere*, Academic Press, New York and London (1969).

NASA SP-8005, National Aeronautics and Space Administration, May (1971). "Solar Electromagnetic Radiation."

Robinson, N. (ed.), *Solar Radiation*, Elsevier, Amsterdam (1966).

Sun Angle Calculator, Libby-Owens-Ford Glass Company (1951).

Thekaekara, M. P., *Solar Energy*, **18**, 309 (1976). "Solar Radiation Measurement: Techniques and Instrumentation."

Thekaekara, M. P., *Supplement to the Proceedings of the 20th Annual Meeting of the Institute for Environmental Science*, 21 (1974). "Data on Incident Solar Energy."

Thekaekara, M. P. and A. J. Drummond, *National Physical Science*, **229**, 6 (1971). "Standard Values for the Solar Constant and Its Spectral Components."

Thomas, R. N., *Transactions of the Conference on Use of Solar Energy*, **1**, 1, University of Arizona Press (1958). "The Features of the Solar Spectrum as Imposed by the Physics of the Sun."

U. S. Hydrographic Office Publications No. 214 (1940). "Tables of Computed Altitude and Azimuth."

Whillier, A., Notes on Solar Energy prepared at McGill University (1965).

Whillier, A., Personal communication (1975 and 1979).

Whillier, A., *Solar Energy*, **9**, 164 (1965). "Solar Radiation Graphs."

CHAPTER 2

Available Solar Radiation

In this chapter we describe instruments for solar radiation measurements, the solar radiation data that are available, and the calculation of needed information from the available data. It is generally not practical to base predictions or calculations of solar radiation on attenuation of the extraterrestrial radiation by the atmosphere, as adequate meteorological information is seldom available. Instead, to predict solar performance, we use past measurements of solar radiation at the location in question or from a nearby similar location.

Solar radiation data are used in several forms and for a variety of purposes. The most detailed information we have is beam and diffuse solar radiation on a horizontal surface, by hours, which is useful in simulations of solar processes. Daily data are more often available and hourly radiation can be estimated from daily data. Monthly total solar radiation on a horizontal surface can be used in some process design methods. However, as process performance is generally not linear with solar radiation, the use of averages may lead to serious errors if nonlinearities are not taken into account. It is also possible to reduce radiation data to more manageable forms by statistical methods.

2.1 DEFINITIONS

Figure 2.1.1 shows the primary radiation fluxes on a surface at or near the ground that are important in connection with solar thermal processes. It is convenient to consider radiation in two wavelength ranges*:

1 **Solar, or short-wave radiation.** Radiation originating from the sun, in the wavelength range of 0.3 to 3.0 μm. In the terminology used throughout this book, solar radiation includes both beam and diffuse components unless otherwise specified.
2 **Long-wave radiation.** Radiation originating from sources at temperatures near ordinary ambient temperatures and thus substantially all at

---

* We will see in Chapters 3, 4, and 6 that the wavelength ranges of incoming solar radiation and emitted radiation from flat-plate solar collectors overlap to a negligible extent, and for many purposes the distinction noted here is very useful. For collectors operating at high enough temperatures that there is significant overlap, more precise distinctions are needed.

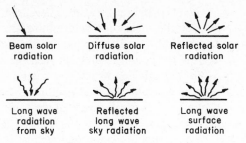

Figure 2.1.1 The radiant energy fluxes of importance in solar thermal processes. Short-wave solar radiation is shown by →. Long-wave radiation is shown by ↝.

wavelengths greater than 3 μm. Long-wave radiation is emitted by the atmosphere, by a collector, or by any other body at ordinary temperatures. (This radiation, if originating from the ground, is referred to in some literature as "terrestrial" radiation.)

Instruments for measuring solar radiation are of two basic types. The accepted terms for these are as follows:

1 **Pyrheliometer.** An instrument using a collimated detector for measuring solar radiation from the sun and from a small portion of the sky around the sun (i.e., beam radiation) at normal incidence.
2 **Pyranometer.** An instrument for measuring total hemispherical solar (beam + diffuse) radiation, usually on a horizontal surface. If shaded from the beam radiation by a shade ring or disc, a pyranometer measures diffuse radiation.
3 In addition, the terms **solarimeter** and **actinometer** are encountered; solarimeter can generally be interpreted to mean the same as pyranometer, whereas actinometer usually refers to a pyrheliometric instrument.

In the following sections we discuss briefly the two basic radiation instruments and the pyrheliometric scales that are used in solar radiometry. More detailed discussions of instruments, their use and the associated terminology are found in WMO (1969), Robinson (1966), Kondratyev (1969), Coulson (1975), Thekaekara (1976), and Yellott (1977).

2.2 PYRHELIOMETERS AND PYRHELIOMETRIC SCALES

The standard and secondary standard solar radiation instruments are all pyrheliometers. The water flow pyrheliometer, designed by Abbot in 1905, was an early standard instrument. This instrument uses a cylindrical black body cavity to absorb radiation, which is admitted through a collimating tube. Water flows around and over the absorbing cavity, and measurement of its temperature and flow rate provide the means for determining the absorbed energy. The design was modified by Abbot in 1932 to include the use of two thermally identical chambers, dividing the cooling water between them, and heating one chamber electrically while the other is heated by solar radiation; when the instrument is adjusted so as to make the heat produced in the two chambers identical, the electrical power input is a measure of the solar energy absorbed.

Standard pyrheliometers are not easy to use. As a result, secondary standard instruments have been devised that are calibrated against the standard instruments, and which in turn serve to calibrate field instruments. Robinson (1966) and Coulson (1975) provide detailed discussion and bibliography on this topic. Two of these secondary standard instruments, both of which have a pyrheliometric scale associated with them, are of importance.

The Abbot silver disc pyrheliometer, first built by Abbot in 1902 and modified in 1909 and 1927, uses a silver disc 38 mm in diameter and 7 mm thick as the radiation receiver. The side exposed to radiation is blackened, and the bulb of a precision mercury thermometer is inserted in a hole in the side of the disc and is in good thermal contact with the disc. The silver disc is suspended on wires at the end of a collimating tube, which in later models has dimensions such that 0.0013 of the hemisphere is "seen" by the detector, that is, any point on the detector sees an aperture angle of 5.7°. The disc is mounted in a copper cylinder, which in turn is in a cylindrical wood box that insulates the copper and the disc from the surroundings. A shutter alternately admits radiation and shades the detector, at regular intervals; the corresponding changes in disc temperature are measured and provide the means to calculate the absorbed radiation. A section drawing of the pyrheliometer is shown in Figure 2.2.1.

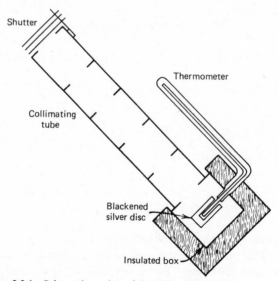

Figure 2.2.1 Schematic section of the Abbot Silver Disc Pyrheliometer.

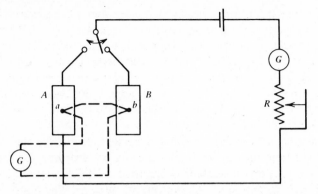

Figure 2.2.2 Circuit diagram for the Angstrom Compensation Pyrheliometer. *A* and *B* are blackened manganin strips to which are attached thermocouples *a* and *b*. *R* is adjusted so the irradiated strip is at the same temperature as the electrically heated strip.

The other secondary standard of particular importance is the Angstrom compensation pyrheliometer, first constructed by K. Angstrom in 1893, and modified in several developments since then. In this instrument, two identical blackened manganin strips are arranged so that either one can be exposed to radiation at the base of collimating tubes by moving a reversible shutter. Each strip can be electrically heated and each is fitted with a thermocouple, as shown in the circuit in Figure 2.2.2.

With one strip shaded and one strip exposed to radiation, a current is passed through the shaded strip to heat it to the same temperature as the exposed strip. When there is no difference in temperature, the electrical energy to the shaded strip must equal the solar radiation absorbed by the exposed strip.

Solar radiation is then found by equating the electrical energy to the product of incident solar radiation, strip area, and absorptance. After a determination is made, the position of the shutter is reversed, interchanging the electrical and radiation heating, and a second determination is made. Alternating the shade and the functions of the two strips compensates for minor differences in the strips, such as edge effects and lack of uniformity of electrical heating.

The Angstrom instrument serves, in principle, as an absolute or primary standard. However, there are difficulties in applying correction factors in its use, and in practice there are several particular primary standard Angstrom instruments to which those in use as secondary standards are compared.

The Abbot and Angstrom instruments are used as secondary standards, for calibration of other instruments, and there is a *pyrheliometric scale* associated with each of them. The first scale, based on measurements with the Angstrom instrument, was established in 1905 (the Angstrom scale of 1905 or AS05). The second, based on the Abbot silver disc pyrheliometer (which was in turn calibrated with a standard water flow pyrheliometer) was established in 1913 (the Smithsonian Scale of 1913 or SS13).

Reviews of the accuracy of these instruments and intercomparisons of them led to the conclusion that measurements made on SS13 were 3.5 percent higher than those on AS05, that SS13 was 2 percent too high, and AS05 was 1.5 percent too low. As a result, in 1956, a compromise was reached, reflecting these differences, and the *International Pyrheliometric Scale 1956* (IPS 56) was adopted. Measurements made before 1956 on the scale AS05 were increased by 1.5 percent, and those of SS13 were decreased by 2 percent to correct them to IPS56.

IPS56 has been the basis for most measurements reported since its inception, including data from the International Geophysical Year program. In 1970 and 1975, International Pyrheliometer Comparisons (IPC III and IV) were made under World Meteorological Organization auspices; the result was a new pyrheliometric scale, the Solar Constant Reference Scale (SCRS), which is very nearly 2 percent higher than the IPS56 scale. (SS13 is very close to SCRS.)

The Abbot and Angstrom instruments described above are secondary standard instruments. Operational or field instruments are calibrated against these standards and are the source of most of the data on which solar process engineering designs must be based. We include here brief descriptions of two of these, the Eppley Normal Incidence and the Linke-Feussner pyrheliometers. The Eppley Normal Incidence pyrheliometer is the instrument in most common use in the United States for measuring beam solar radiation. A cross-section diagram of a recent model is shown in Figure 2.2.3. The instrument mounted on a tracking mechanism is shown in Figure 2.2.4.

The detector is at the end of the collimating tube, which contains several diaphragms and which is blackened on the inside. The detector is a multijunction thermopile coated with Parsons optical black. Temperature compensation to minimize sensitivity to variations in ambient temperature is

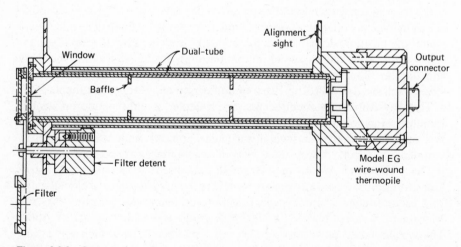

Figure 2.2.3 Cross-section of the Eppley Normal Incidence Pyrheliometer. Courtesy Eppley Laboratories.

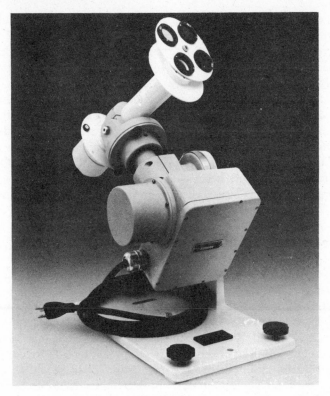

Figure 2.2.4 An Eppley Normal Incidence Pyrheliometer mounted on an altazimuth tracking mount. Courtesy Eppley Laboratories.

provided. The aperture angle of the instrument is 5.7°, so the detector receives radiation from the sun and from an area of the circumsolar sky two orders of magnitude larger than that of the sun.

The Linke-Feussner pyrheliometer uses a constantan-manganin thermopile with hot junctions heated by radiation and cold junctions in good thermal contact with the case. In this instrument the assembly of copper diaphragms and case has very large thermal capacity, orders of magnitude more than the hot junctions, and on exposure to solar radiation the hot junctions rise quickly to temperatures above the cold junction, the difference in the temperatures providing a measure of the radiation. Other pyrheliometers were designed by Moll-Gorczynski, Yanishevskiy, and Michelson.

In the previous discussions, it has been pointed out that the geometry of pyrheliometers, that is, the dimensions of the collimating systems, are such that the detectors are exposed to radiation from the sun and from a portion of the sky around the sun. Since the detectors do not distinguish between forward-scattered radiation, which comes from the circumsolar sky, and beam radiation, the instruments are, in effect, defining beam radiation. An experimental study by Jeys and Vant-Hull (1976), which utilized several lengths of collimating tubes so that the aperture angles were reduced in step from 5.72 to 2.02°, indicated that for cloudless conditions this reduction in aperture angle resulted in insignificant changes in the measurements of beam radiation. On a day of thin uniform cloud cover, however, with a solar altitude angle of less than 32°, as much as 11 percent of the measured intensity was received from the circumsolar sky between aperture angles of 5.72 and 2.02°. It is difficult to generalize from the few data available, but it appears that thin clouds or haze can affect the angular distribution of radiation within the field of view of standard pyrheliometers. The World Meteorological Organization (WMO) recommends that calibration of pyrheliometers only be undertaken on days in which atmospheric clarity meets or exceeds a minimum value.

2.3 PYRANOMETERS

Instruments for measuring total (beam + diffuse) radiation are referred to as pyranometers, and it is from these instruments that most of the available data on solar radiation are obtained. The detectors for these instruments must have a response independent of wavelength of radiation over the solar energy spectrum. In addition, they should have a response independent of the angle of incidence of the solar radiation. The detectors of most pyranometers are covered with hemispherical glass covers to protect them from wind and other extraneous effects; the covers must be very uniform in thickness so as not to cause uneven distribution of radiation on the detectors. These factors are discussed in more detail by Coulson (1975).

Commonly used pyranometers in the United States are the Eppley and Spectrolab instruments, in Europe the Moll-Gorczynski, in the USSR the Yanishevskiy, and in Australia the Trickett-Norris (Groiss) pyranometer.

The most common instrument in the United States has been the Eppley 180° pyranometer, shown in Figure 2.3.1. It uses a detector consisting of two concentric silver rings; the outer ring is coated with magnesium oxide, which has a high reflectance for radiation in the solar energy spectrum, and the inner ring is coated with Parson's black, which has a very high absorptance for solar radiation.

The temperature difference between these rings is a measure of absorbed solar radiation; it is detected by 50 (or 10) junction thermopiles. A central disc inside the black ring is coated with magnesium oxide, but is not part of the detector. The circular symmetry of the detector minimizes the effects of surface azimuth angle on instrument response. The detector assembly is placed in a nearly spherical glass bulb, which contains dry air, with the upper surface of the receiver placed at the center of curvature of the bulb. The glass used is soda-lime glass, which has a transmittance greater than 0.90 over most of the solar radiation spectrum, so the instrument response is nearly independent of wavelength except at the extremes of the spectrum.

Figure 2.3.1 The Eppley 180° Pyranometer. Courtesy Eppley Laboratories.

The response of this Eppley is dependent on ambient temperature, with sensitivity decreasing by 0.05 to 0.15%/C [Coulson (1975)]; much published data taken with these instruments was not corrected. It is possible to add temperature compensation to the external circuit and remove this source of error. It is estimated that carefully used Eppleys of this type could produce data with less than 5% errors, but that errors of twice this could be expected from poorly maintained instruments. The theory of this instrument has been carefully studied by MacDonald (1951).

The Eppley 180° pyranometer is no longer manufactured and has been replaced by other instruments. One is the Eppley "black and white pyranometer" utilizing Parsons-black- and barium-sulfate-coated hot and cold thermopile junctions. This instrument has better angular (cosine) response, an optically ground glass envelope, and temperature compensation to maintain calibration within ±1.5% over a temperature range of −20 to +40°C. It is shown in Figure 2.3.2.

The second new Eppley instrument is the Eppley Precision Spectral pyranometer. It utilizes a thermopile detector, two concentric hemispherical optically ground covers, and temperature compensation that results in temperature dependence of 0.5% from −20 to +40°C. It is shown in Figure 2.3.3.

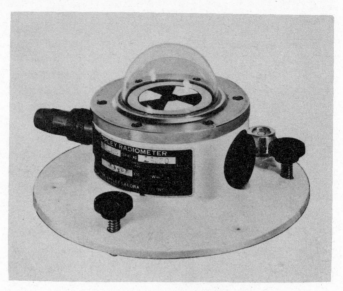

Figure 2.3.2 The Eppley Black and White Pyranometer. Courtesy Eppley Laboratories.

Figure 2.3.3 The Eppley Precision Spectral Pyranometer. Courtesy Eppley Laboratories.

Figure 2.3.4 The WEATHERTronics model 3015 Pyranometer. Courtesy WEATHERTronics Co.

The 1978 Weather Service standard pyranometer was the Spectrosun model SR-75 instrument; it is now manufactured by WEATHERTronics (Model 3015). The sensor is a thermopile. The pyranometer has a high degree of linearity, low deviations from cosine response, low sensitivity to azimuth angle, fast response, and low temperature sensitivity. The detector is covered by double hemispherical windows. The instrument is shown in Figure 2.3.4.

The Moll-Gorczynski pyranometer uses a Moll thermopile to measure the temperature difference of the black detector surface and the housing of the instrument. The thermopile assembly is covered with two concentric glass hemispherical domes to protect it from weather and is rectangular in configuration with the thermocouples aligned in a row (which results in some sensitivity to the azimuth angle of the radiation).

Pyranometers are usually calibrated against standard pyrheliometers. A standard method has been set forth in the Annals of the International Geophysical Year (1958), which requires that readings be taken at times of clear skies, with the pyranometer shaded and unshaded at the same time as readings are taken with the pyrheliometer. Shading is recommended to be accomplished by means of a disc held 1 m from the pyranometer, with the disc just large enough to shade the glass envelope. The calibration constant is then the ratio of the difference in the output of the shaded and unshaded pyranometer to the output of the pyrheliometer, multiplied by the calibration constant of the pyrheliometer and $\cos \theta_z$, the angle of incidence of beam radiation on the horizontal pyranometer. Care and precision are required in these calibrations.

It is also possible, as described by Norris (1973), to calibrate pyranometers against a secondary standard pyranometer such as the Eppley Precision pyranometer. This secondary standard pyranometer is thought to be good to ± 1 percent when calibrated against a standard pyrheliometer. Direct comparison of the precision Eppley and field instruments can be made to determine the calibration constant of the field instruments.

A pyranometer (or pyrheliometer) produces a voltage from the thermopile detectors that is a function of the incident radiation. It is necessary to use a potentiometer to detect and record this output. Radiation data usually must be integrated over some period of time, such as an hour or a day. Integration can be done by means of planimetry or by electronic integrators. It has been estimated that with careful use and reasonable frequent pyranometer calibration radiation measurements should be good within ± 5 percent; integration errors would increase this number. Much of the available radiation data prior to 1975 is probably not this good, largely because of infrequent calibration and in some instances because of inadequate integration procedures.

Another class of pyranometers, originally designed by Robitzsch, utilizes detectors that are bimetallic elements heated by solar radiation; mechanical motion of the element is transferred by a linkage to an indicator or recorder pen. These instruments have the advantage of being entirely spring-driven and thus requiring no electrical energy. Variations of the basic design are manufactured by several European firms (Fuess, Casella, and SIAP). They are widely used in isolated stations, and are a major source of the solar radiation data that are available for locations outside of Europe, Australia, Japan, and North America. Data from them are generally not as accurate as that from thermopile-type pyranometers.

Pyranometers have also been based on photovoltaic (solar cell) detectors, (e.g., the *Yellott solarimeter*). Silicon cells are the most common for solar energy measurements, although cadmium sulfide and selenium cells have been used, for example, for measurements of visible light in photography. Silicon solar cells have the property that their light current (approximately equal to the short-circuit current at normal radiation levels) is a linear function of the incident solar radiation. They have the disadvantage that their spectral response is not linear, so instrument calibration is a function of the spectral distribution of the incident radiation. Curves of the spectral response of silicon solar cells and distribution of solar radiation are shown in Figure 2.3.5. Also, unless the cell is provided with a plastic diffuser, the calibration varies substantially with the angle of incidence of the radiation [see Coulson (1975)].

The above discussion dealt entirely with measurements of total radiation on a horizontal surface. Two additional kinds of measurements are made with pyranometers: measurement of diffuse radiation on horizontal surfaces and measurement of solar radiation on inclined surfaces.

Measurements of diffuse radiation can be made with pyranometers by shading the instrument from beam radiation. This is usually done by means of a shading ring, as shown in Figure 2.3.6. The ring is used to allow continuous recording

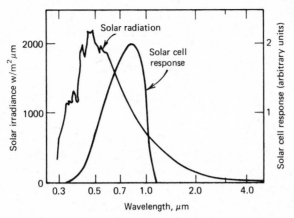

Figure 2.3.5 Spectral distribution of extraterrestrial solar radiation and spectral response of a silicon solar cell. From Coulson, *Solar and Terrestrial Radiation*, Academic Press, New York (1975).

Figure 2.3.6 Pryanometer with shading ring to eliminate beam radiation. Courtesy Eppley Laboratories.

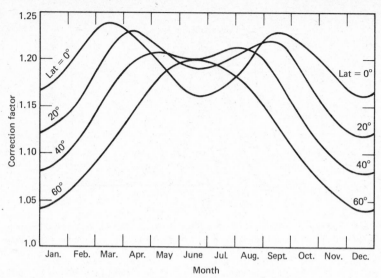

Figure 2.3.7 Typical shade ring correction factors to account for shading of the detector from diffuse radiation. Adapted from Coulson, **Solar and Terrestrial Radiation**, Academic Press, New York (1975).

of diffuse radiation without the necessity of continuous positioning of smaller shading devices; adjustments need to be made for changing declination only and can be made every few days. The ring shades the pyranometer from part of the diffuse radiation, and a correction for this shading must be estimated and applied to the observed diffuse radiation [see Drummond (1956, 1964), IGY (1958) or Coulson (1975)]. The corrections are based on assumptions of the distribution of diffuse radiation over the sky, and typically are factors of 1.05 to 1.2. An example of shade ring correction factors, to illustrate their trends and magnitudes, is shown in Figure 2.3.7.

Measurements of solar radiation on inclined planes is important in determining the input to solar collectors. There is evidence that the calibration of pyranometers changes if the instrument is inclined to the horizontal. The reason for this appears to be changes in the convection patterns inside the glass dome, which changes the manner in which heat is transferred from the hot junctions of the thermopiles to the cover and other parts of the instrument. The Eppley 180° pyranometer has been variously reported to show a decrease in sensitivity on inversion from 5.5 percent to no decrease. Norris (1974) measured the response of four pyranometers, when subject to radiation from an incandescent lamp source, at various inclinations. The results are shown in Figure 2.3.8. The two curves for the Kipp pyranometer represent measurements with the axis of its Moll thermopile (which is rectangular in configuration) aligned on the inclined or horizontal plane.

It is evident from these data and other published results that the calibration of pyranometers is, to some degree, dependent on inclination, and that experi-

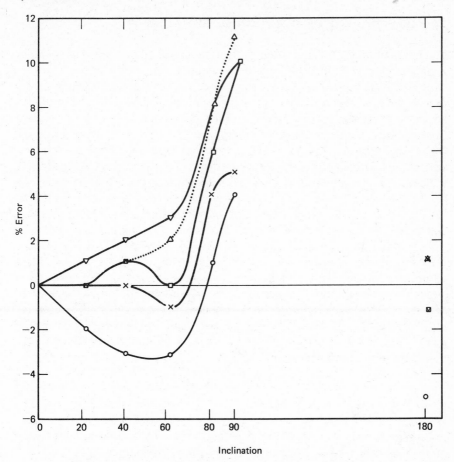

Figure 2.3.8 Effects of inclination of pyranometers on calibration. Instruments are: O, Eppley 180°; $\triangle$, Precision Eppley; X, Trickett–Norris (CSIRO); $\square$ Kipp, thermopile axis inclined; ∇, Kipp, thermopile axis horizontal. Adapted from Norris, Solar Energy, **16**, 53 (1974).

mental information is needed on a particular pyranometer in any orientation to adequately interpret information from it.

One other type of pyranometer is worth noting. The *Bellani spherical distillation pyranometer* uses a spherical container of alcohol connected to a calibrated condenser receiver tube. The quantity of alcohol condensed is a measure of integrated solar energy on the spherical receiver. Data on the total energy received by a body, as represented by the sphere, are of interest in some biological processes.

Measurements of irradiance in spectral bands can be made by use of band-pass filters. The Eppley Precision Spectral pyrheliometer can be fitted with hemispherical domes of filter glass for this purpose.

2.4 MEASUREMENT OF DURATION OF SUNSHINE

The hours of bright sunshine, that is, the time in which the solar disc is visible, is of some use in estimating long-term averages of solar radiation.* Two instruments have been or are widely used. The *Campbell-Stokes sunshine recorder* uses a solid glass sphere of approximately 10 cm diameter as a lens that produces an image of the sun on the opposite surface of the sphere. A strip of standard treated paper is mounted around the appropriate part of the sphere, and the solar image burns a mark on the paper whenever the beam radiation is above a critical level. The lengths of the burned portions of the paper provide an index of the duration of "bright sunshine." These measurements are uncertain on several counts: the interpretation of what constitutes a burned portion is uncertain, the instrument does not respond to low levels of radiation early and late in the day, and the condition of the paper may be dependent on humidity.

A photoelectric sunshine recorder, the *Foster Sunshine switch* [Foster and Foskett (1953)], is now in use by the United States Weather Service. It incorporates two photovoltaic cells, one of which is shaded from beam radiation and one exposed to it. In the absence of beam radiation, the two detectors indicate (nearly) the same radiation level. When beam radiation is incident on the unshaded cell, the output of that cell is higher than that of the shaded cell. The duration of a critical radiation difference detected by the two cells is a measure of the duration of bright sunshine.

2.5 SOLAR RADIATION DATA

Solar radiation data are available in several forms. The following information about radiation data is important in its understanding and use:

1 Whether they are instantaneous measurements (irradiance) or values integrated over some period of time (irradiation) (usually hour or day).
2 The time or time period of the measurements.
3 Whether the measurements are of beam, diffuse or total radiation, and the instruments used.
4 The receiving surface orientation (usually horizontal, sometimes inclined at a fixed slope, or normal to the beam radiation).
5 If averaged, the period over which they are averaged (e.g., monthly averages of daily radiation).

Most radiation data available are for horizontal surfaces, include both direct and diffuse radiation, and were measured with thermopile pyranometers (or, in some cases, Robitzsch-type instruments). Most of these instruments provide radiation records as a function of time, and do not themselves provide a means

---

* The relationship between sunshine hours and solar radiation will be discussed in Section 2.6.

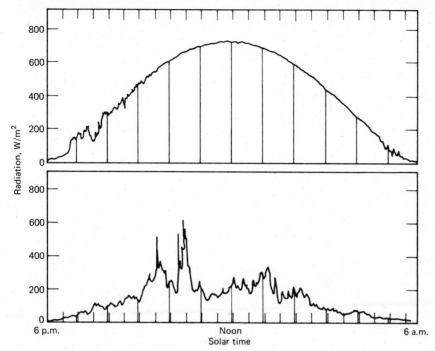

Figure 2.5.1 Total (beam and diffuse) solar radiation on a horizontal surface vs time for clear and largely cloudy days, latitude 43°, for days near the equinox.

of integrating the records. The data are usually recorded in a form similar to that shown on Figure 2.5.1 by recording potentiometers, and are integrated graphically or electronically. Errors in integration thus add to errors in pyranometer response.

Examples of radiation data are shown in Tables 2.5.1 and 2.5.2. Table 2.5.1 indicates daily total radiation on horizontal surface, averaged by months, for several locations. This is the most common form in which solar radiation data are available. The traditional units have been cal/cm^2; these are shown in the more useful MJ/m^2. Extensive data of this kind, measured with pyranometers, are available from weather services [e.g., Cinquimani et al. (1978)] and the literature [e.g., from Löf et al. (1966a, b)]. Appendix G includes monthly average data from many North American stations. The accuracy of these data is generally less than desirable, as standards of calibration and care in use of instruments and integration has not always been adequate.* The averages are probably no better than ±10 percent in many cases, and for some stations

---

* The SOLMET (1978) program of the United States Weather Service has addressed this problem by careful study of the history of individual instruments and their calibrations, and subsequent "rehabilitation" of the data to correct for identifiable errors. The United States data in Appendix G has all been processed in this way.

Table 2.5.1 Monthly Average Daily Solar Radiation on Horizontal Surface,[a] MJ/m²

Station	Latitude	Elevation (m)	Radiation, MJ/m²												
			Annual	Jan.	Feb.	Mar.	Apr.	May	June	July	Aug.	Sept.	Oct.	Nov.	Dec.
Albuquerque	35N	1620	20.7	11.5	15.2	20.1	25.3	28.8	30.4	28.2	26.0	22.4	17.6	12.9	10.5
Bismarck	47N	502	14.2	5.3	8.8	13.3	16.6	21.0	23.4	24.8	21.3	15.4	10.3	5.8	4.2
Boston	42N	194	12.5	5.4	8.1	11.5	15.1	18.4	20.6	10.9	16.9	14.3	10.1	5.7	4.6
Charleston	33N	18	15.3	8.5	11.3	15.2	19.7	21.1	20.9	20.4	18.0	15.8	13.5	10.6	8.2
Columbia	38N	248	15.1	6.9	9.9	13.4	17.3	21.3	23.7	24.0	21.3	16.5	12.5	8.0	5.9
El Paso	32N	1194	21.6	12.8	16.8	21.7	26.8	29.5	30.4	27.8	25.9	22.6	18.6	14.1	11.7
Fresno	36N	110	19.4	7.4	11.5	17.8	23.7	28.2	31.0	30.5	27.5	22.5	16.2	10.1	6.5
Madison	43N	271	13.5	5.9	9.1	12.9	15.9	19.8	22.1	22.0	19.4	14.8	10.3	5.7	4.4
Miami	26N	2	16.7	12.0	14.9	18.2	21.1	20.9	19.4	20.0	18.5	16.5	14.8	12.7	11.6
North Omaha	41N	404	15.0	7.2	10.1	13.9	17.7	21.3	24.1	23.9	21.1	15.6	11.9	7.3	5.8
Santa Maria	35N	72	18.2	18.2	13.0	18.0	21.8	24.3	26.7	26.6	23.9	19.6	15.4	11.1	9.1
Seattle	47N	122	11.9	3.0	5.6	9.6	14.7	19.4	20.5	25.5	18.3	13.0	7.4	3.8	2.4
Pretoria	26S	1418	19.9	25.6	21.8	20.5	17.2	15.1	14.2	15.1	18.0	21.0	22.2	23.9	24.3
Canberra	34S	17	17.7	25.9	22.9	18.6	14.4	11.1	8.6	9.6	12.7	17.9	21.2	24.9	26.7
Tokyo	36N	s.l.	10.9	8.0	9.7	11.5	13.1	14.4	12.7	14.1	14.2	10.6	8.5	7.8	7.1
Stockholm	59N	s.l.	10.1	1.2	3.3	8.4	12.9	19.6	21.7	21.0	16.4	10.2	4.7	1.3	0.8

[a] From Klein et al. (1978), Löf et al. (1966a), and Cinquimani et al. (1978).

Table 2.5.2 Hourly Radiation, Air Temperature, and Wind Speed Data for a January Week, Boulder, Colorado[a]

Day	Hour	I	°C	V	Day	Hour	I	°C	V
8	1	0	−1.7	3.1	8	13	1105	2.8	8.0
8	2	0	−3.3	3.1	8	14	1252	3.8	9.8
8	3	0	−2.8	3.1	8	15	641	3.3	9.8
8	4	0	−2.2	3.1	8	16	167	2.2	7.2
8	5	0	−2.8	4.0	8	17	46	0.6	7.6
8	6	0	−2.8	3.6	8	18	0	−0.6	7.2
8	7	0	−2.2	3.6	8	19	0	−1.1	8.0
8	8	17	−2.2	4.0	8	20	0	−1.7	5.8
8	9	134	−1.1	1.8	8	21	0	−1.7	5.8
8	10	331	1.1	3.6	8	22	0	−2.2	7.2
8	11	636	2.2	1.3	8	23	0	−2.2	6.3
8	12	758	2.8	2.2	8	24	0	−2.2	5.8
9	1	0	−2.8	7.2	9	13	1185	−2.2	2.2
9	2	0	−3.3	7.2	9	14	1009	−1.3	1.7
9	3	0	−3.3	6.3	9	15	796	−0.6	1.3
9	4	0	−3.3	5.8	9	16	389	−0.6	1.3
9	5	0	−3.9	4.0	9	17	134	−2.2	4.0
9	6	0	−3.9	4.5	9	18	0	−2.8	4.0
9	7	0	−3.9	1.8	9	19	0	−3.3	4.5
9	8	4	−3.9	2.2	9	20	0	−5.6	5.8
9	9	71	−3.9	2.2	9	21	0	−6.7	5.4
9	10	155	−3.3	4.0	9	22	0	−7.8	5.8
9	11	343	−2.8	4.0	9	23	0	−8.3	4.5
9	12	402	−2.2	4.0	9	24	0	−8.3	6.3
10	1	0	−9.4	5.8	10	13	1872	2.2	7.6
10	2	0	−10.0	6.3	10	14	1733	4.4	6.7
10	3	0	−8.9	5.8	10	15	1352	6.1	6.3
10	4	0	−10.6	6.3	10	16	775	6.7	4.0
10	5	0	−8.3	4.9	10	17	205	6.1	2.2
10	6	0	−8.3	7.2	10	18	4	3.3	4.5
10	7	0	−10.0	5.8	10	19	0	0.6	4.0
10	8	33	−8.9	5.8	10	20	0	0.6	3.1
10	9	419	−7.2	6.7	10	21	0	0.0	2.7
10	10	1047	−5.0	9.4	10	22	0	0.6	2.2
10	11	1570	−2.2	8.5	10	23	0	1.7	3.6
10	12	1805	−1.1	8.0	10	24	0	0.6	2.7
11	1	0	−1.7	8.9	11	13	138	−5.0	6.7
11	2	0	−2.2	4.9	11	14	96	−3.9	6.7
11	3	0	−2.2	4.5	11	15	84	−4.4	7.6
11	4	0	−2.8	5.8	11	16	42	−3.9	6.3
11	5	0	−4.4	5.4	11	17	4	−5.0	6.3
11	6	0	−5.0	4.5	11	18	0	−5.6	4.5
11	7	0	−5.6	3.6	11	19	0	−6.7	4.5

(*continued*)

Table 2.5.2 (Cont.)

Day	Hour	I	°C	V	Day	Hour	I	°C	V
11	8	4	−6.1	5.8	11	20	0	−7.8	3.1
11	9	42	−5.6	5.4	11	21	0	−9.4	2.7
11	10	92	−5.6	5.4	11	22	0	−8.9	3.6
11	11	138	−5.6	9.4	11	23	0	−9.4	4.0
11	12	163	−5.6	8.0	11	24	0	−11.1	3.1
12	1	0	−11.7	4.0	12	13	389	−2.2	5.8
12	2	0	−12.8	3.1	12	14	477	−0.6	4.0
12	3	0	−15.6	7.2	12	15	532	2.8	2.2
12	4	0	−16.7	6.7	12	16	461	−0.6	2.2
12	5	0	−16.7	6.3	12	17	33	−1.7	3.1
12	6	0	−16.1	6.3	12	18	0	−4.4	1.3
12	7	0	−17.2	3.6	12	19	0	−7.8	2.7
12	8	17	−17.8	2.7	12	20	0	−7.8	4.0
12	9	71	−13.3	8.0	12	21	0	−8.9	4.9
12	10	180	−11.1	8.9	12	22	0	−10.6	4.9
12	11	247	−7.8	8.5	12	23	0	−12.8	4.9
12	12	331	−5.6	7.6	12	24	0	−11.7	5.4
13	1	0	−10.6	4.0	13	13	1926	5.6	5.4
13	2	0	−10.6	5.4	13	14	1750	7.2	4.5
13	3	0	−10.0	4.5	13	15	1340	8.3	4.9
13	4	0	−11.1	3.1	13	16	703	8.9	4.5
13	5	0	−10.6	3.6	13	17	59	6.7	5.4
13	6	0	−9.4	3.1	13	18	0	4.4	3.6
13	7	0	−7.2	3.6	13	19	0	1.1	3.6
13	8	17	−10.6	4.0	13	20	0	0.0	3.1
13	9	314	−8.3	5.8	13	21	0	−2.2	6.7
13	10	724	−1.7	6.7	13	22	0	2.8	7.2
13	11	1809	1.7	5.4	13	23	0	1.7	8.0
13	12	2299	3.3	6.3	13	24	0	1.7	5.8
14	1	0	−0.6	7.2	14	13	1968	6.7	1.8
14	2	0	−1.1	7.6	14	14	1733	6.7	2.7
14	3	0	−0.6	6.3	14	15	1331	7.2	3.1
14	4	0	−3.9	2.7	14	16	837	6.7	3.1
14	5	0	−1.7	4.9	14	17	96	7.2	2.7
14	6	0	−2.8	5.8	14	18	4	3.3	2.7
14	7	0	−2.8	4.0	14	19	0	0.0	3.6
14	8	38	−5.0	3.1	14	20	0	3.9	5.4
14	9	452	−5.0	4.9	14	21	0	−3.9	3.6
14	10	1110	−1.7	4.5	14	22	0	−3.9	5.8
14	11	1608	2.8	3.1	14	23	0	−6.1	5.4
14	12	1884	3.8	3.6	14	24	0	−6.7	6.3

[a] Solar Radiation, I, is kJ/m^2 for the hour ending at the indicated time; wind speed, V, is in m/s. See Example 10.9.1.

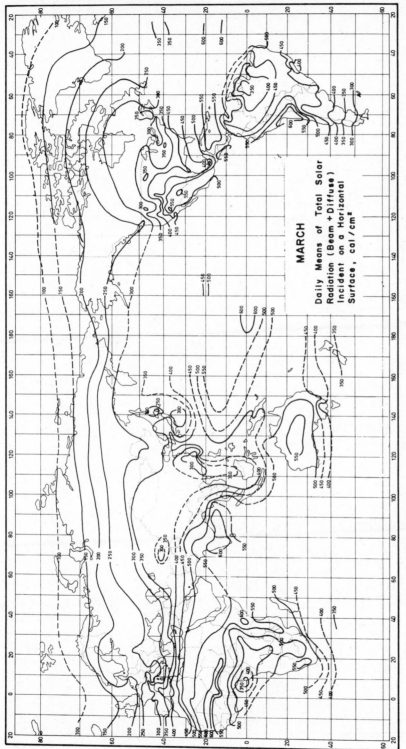

Figure 2.5.2 Average daily radiation for March. Figures 2.5.2 to 2.5.5 are adapted from deJong (1973) and Löf et al. (1966*a*).

47

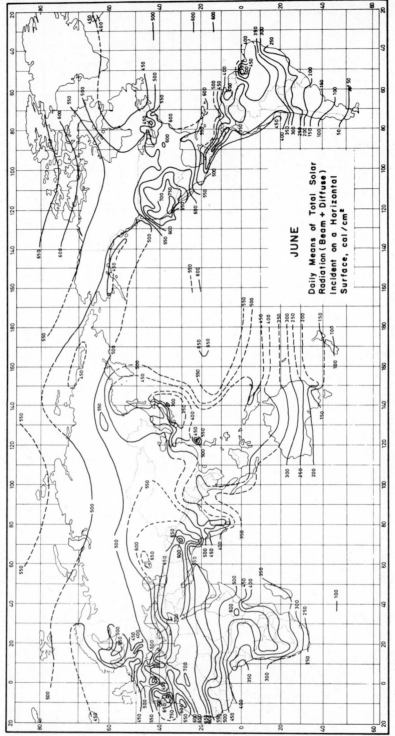

Figure 2.5.3 Average daily radiation for June.

48

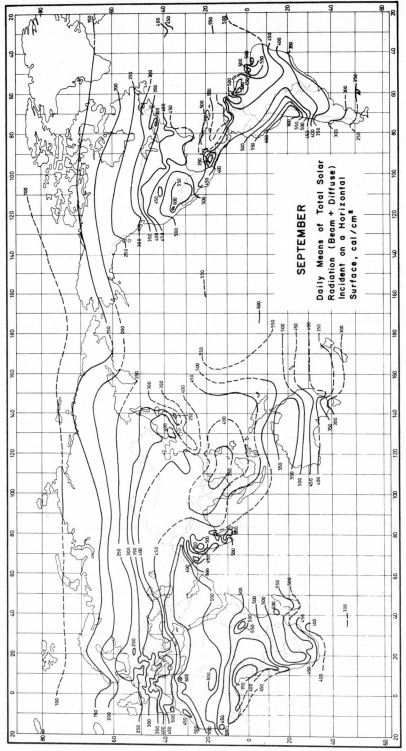

SEPTEMBER

Daily Means of Total Solar
Radiation (Beam + Diffuse)
Incident on a Horizontal
Surface, cal/cm²

Figure 2.5.4 Average daily radiation for September.

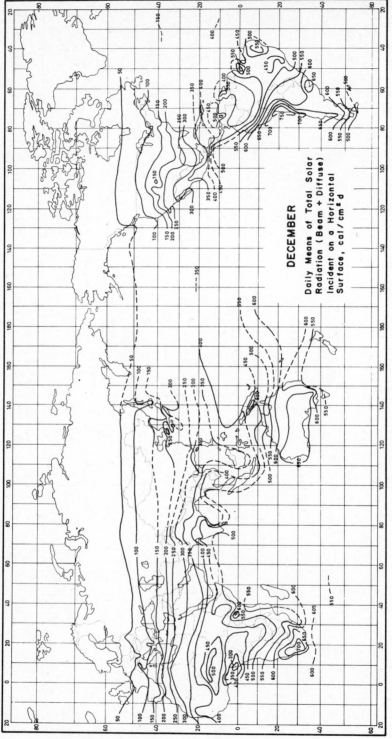

DECEMBER

Daily Means of Total Solar Radiation (Beam + Diffuse) Incident on a Horizontal Surface, cal / cm² d

Figure 2.5.5 Average daily radiation for December.

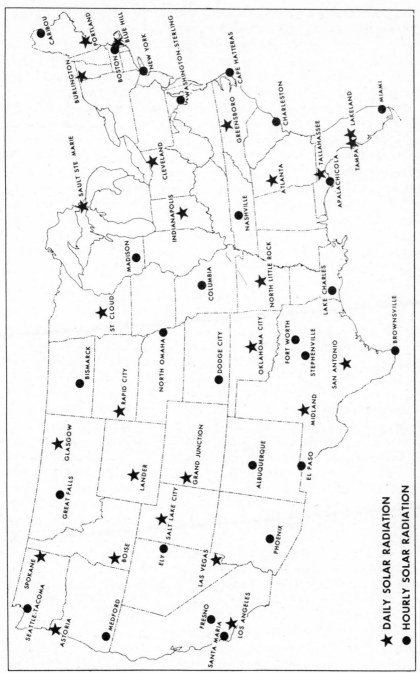

Figure 2.5.6 United States stations for which rehabilitated solar radiation data are available.

★ DAILY SOLAR RADIATION

● HOURLY SOLAR RADIATION

51

a better estimate may be ± 20 percent. Substantial inconsistencies will be found in data for the same location from different sources.

Average solar radiation data are also available from maps that indicate general trends and averages. For example, world maps are shown in Figures 2.5.2 to 2.5.5 [Löf et al. (1966a, b)].* These are useful in predicting areas of potential solar energy applications. In some geographical areas where climate does not change abruptly with distance (i.e., away from major influences such as mountains or large industrial cities), maps can be used as a source of average radiation if data are not available. However, large-scale maps must be used with care, because they do not show local physical or climatological conditions that may greatly affect local solar energy availability.

For calculating the dynamic behavior of solar energy equipment and processes, and for simulations of long-term process operation, more detailed data of solar radiation (and related meteorological information) are needed. An example of this type of data, that is, hourly integrated radiation, ambient temperature, and wind speed, is shown in Table 2.5.2, for a January week in Boulder, Colorado. Additional useful information may also be included, such as wet bulb temperature and wind direction.

There has been, in the United States, a network of stations recording solar radiation on a horizontal surface and reporting it as daily values. A few of these stations also have hourly records. In recent years the U. S. National Oceanic and Atmospheric Administration (NOAA) has undertaken a program to upgrade the number and quality of the radiation measuring stations, to rehabilitate past data to account for sensor deterioration, calibration errors, and changes in pyrheliometric scales, and to make these data available (with related meteorological data) on magnetic tapes. The result is that in 1978 corrected data tapes are available of hourly meteorological information (including solar radiation on a horizontal surface based on SCRS) for 26 stations over a period of 23 years. The 26 stations are shown in Figure 2.5.6. These tapes are referred to as the SOLMET tapes, and are described in detail in the SOLMET User's Manual (1978). The monthly average data for United States stations shown in Table 2.5.1 are derived from the SOLMET data. In addition to the stations for which hourly data are available, daily data are available for an additional group of stations as shown on Figure 2.5.6.

The SOLMET data program of NOAA includes the estimation of radiation from related meteorological data (a topic to be discussed in Section 2.8) for an additional 214 stations. These estimates are on data tapes in the same format as the rehabilitated measured data.

2.6 ATTENUATION OF SOLAR RADIATION BY THE ATMOSPHERE

Solar radiation at normal incidence received at the surface of the earth is subject to variations due to change in the extraterrestrial radiation as noted

* Figures 2.5.2 to 2.5.5 are reproduced from deJong (1973), who redrew maps originally published by Löf et al. deJong has compiled maps and radiation data from many sources. These maps show the traditional units, cal/cm².

in Chapter 1 and to two additional and more significant phenomena, (1) atmospheric scattering by air molecules, water vapor, and dust, and (2) atmospheric absorption by O_3, H_2O, and CO_2.

Scattering, which results in attenuation of the beam radiation by air molecules, water vapor, and dust, has been the subject of a number of studies, and approximate methods have been developed to estimate the magnitude of the effect. This question and the pertinent literature up to 1940 are treated by Moon (1940). A more recent discussion, including the effects of clouds, is given by Fritz (1958). Thekaekara (1974) reviews effects of scattering and includes an extensive bibliography.

Air molecules are very small compared to the wavelengths of radiation significant in the solar energy spectrum. Scattering of this radiation by molecules occurs in accordance with the theory of Rayleigh, which indicates that the scattering coefficient would vary approximately as λ^{-4}, where λ is the wavelength of the radiation. Dust scattering from particles that are much larger than air molecules and that vary in size and concentration from location to location, according to height, and from time to time, is more difficult to assess. Moon developed a scattering coefficient varying approximately as $\lambda^{-0.75}$. Water vapor scattering depends on the amount of precipitable water (the amount of water vapor in the air column above the observer), and a scattering coefficient can also be developed for water vapor that varies as λ^{-2}. The total effect of scattering on the beam radiation is the product of three exponential terms, each a function of wavelength and of the amount of molecules, dust, and precipitable moisture through which the radiation is transmitted. Although the theory exists, it has limited practical value in that data are seldom available on which to base calculation of beam radiation intensity.

Absorption of radiation in the atmosphere in the solar energy spectrum is due largely to ozone in the ultraviolet and water vapor in bands in the infrared. There is almost complete absorption of short-wave radiation by ozone in the upper atmosphere at wavelengths below 0.29 μm. Ozone absorption decreases as λ increases above 0.29 μm, until at 0.35 μm there is no absorption. There is also a weak ozone absorption band near $\lambda = 0.6$ μm.

Water vapor absorbs strongly in bands in the infrared part of the solar spectrum, with strong absorption bands centered at 1.0 μm, 1.4 μm, and 1.8 μm. Beyond 2.3 μm, the transmission of the atmosphere is very low due to absorption by H_2O and CO_2, the energy in the extraterrestrial solar energy spectrum is less than 5% of the total solar spectrum, and energy received at the ground at $\lambda > 2.3$ μm is very small.

Calculations of scattering and absorption were summarized by Moon (1940), based on the best data then available on the spectral distribution of solar radiation outside the atmosphere and on transmission factors for typical atmospheres, to generate a series of "proposed standard curves" for beam solar radiation as a function of wavelength at air masses of 0 to 5. He also tabulated the spectral irradiance for a standard atmosphere for various wavelengths for air mass 2, and proposed that his spectral incidence be used as a standard spectral distribution curve for beam solar radiation for locations near sea level.

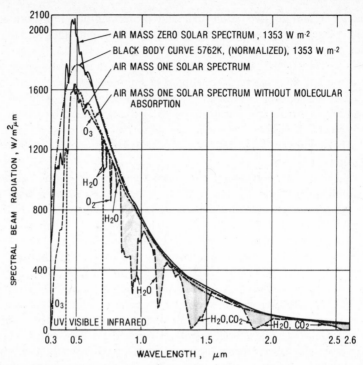

Figure 2.6.1 Spectral distribution of solar radiation for air mass zero and one, and 5762°K blackbody. From Thekaekara (1974).

More recent studies of Johnson and of Thekaekara and Drummond have indicated that Moon's intensities in the short-wave end of the spectrum for $m = 0$ are too low. Moon used a solar constant of 1322 W/m^2 (in contrast to the value of 1353 W/m^2 used in this book). Thekaekara (1974) presents new spectral distribution curves for beam radiation, based on the extraterrestrial distribution shown in Figure 1.3.1, for very clear and relatively clear atmospheres, for air masses of 1, 4, 7, and 10. Figure 2.6.1 from Thekaekara, shows the extraterrestrial distribution, its comparison with a normalized blackbody distribution curve for a source temperature of 5762°K, and for the very clear atmosphere the air mass two solar spectrum without and with molecular absorption. The latter shows absorption bands due to O_3, H_2O, and a small band due to O_2.

Figure 2.6.2 shows computed spectral beam radiation curves for extraterrestrial radiation and for four air masses, for an atmosphere containing 20 mm of precipitable water vapor, 3.4 mm of ozone, and very clear air.

For purposes of calculating properties of materials (absorptance, reflectance, and transmittance), which depend on the spectral distribution of solar radiation, it is convenient to have the distribution of terrestrial radiation in tabular form. Wiebelt and Henderson (1979) have prepared such tables for several air masses (zenith angles) and atmospheric conditions, based on Thekaekara's spectral

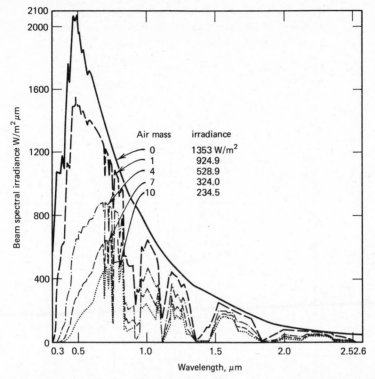

Figure 2.6.2 Spectral distribution of beam radiation for various air masses, assuming a very clear atmosphere with 20 mm precipitable water vapor and 3.4 mm ozone. From Thekaekara (1974).

distribution curves and a solar constant of 1353 W/m^2. Table 2.6.1 shows the terrestrial spectrum divided into twenty equal increments of energy, with a mean wavelength for each increment that divides the increment into two equal parts. This table is for a relatively clear atmosphere and air mass two. It is recommended that it be used as a typical distribution of terrestrial beam radiation.

The spectral distribution of total radiation depends also on the spectral distribution of the diffuse radiation. Some measurements are available in the ultraviolet and visible portions of the spectrum [see Robinson (1966) and Kondratyev (1969)], which have led to the conclusion that in the wavelength range 0.35 to 0.80 μm the distribution of the diffuse radiation is similar to that of the total beam radiation.* Figure 2.6.3 shows relative data on spectral distribution of total and diffuse radiation for a clear sky. The diffuse component has a distribution similar to the total, but shifted slightly toward the short-wave

---

* Scattering theory predicts that shorter wavelengths are scattered most, and hence diffuse radiation tends to be at shorter wavelengths. Thus, clear skies are blue.

Table 2.6.1 Spectral Distribution of Terrestrial Beam
Radiation at Air Mass 2 and 23 km Visibility, in Twenty
Equal Increments of Energy[a]

Energy Band Number	Wavelength Range μm	Midpoint Wavelength, μm
1	0.300–0.434	0.402
2	0.434–0.479	0.458
3	0.479–0.517	0.498
4	0.517–0.557	0.537
5	0.557–0.595	0.576
6	0.595–0.633	0.614
7	0.633–0.670	0.652
8	0.670–0.710	0.690
9	0.710–0.752	0.730
10	0.752–0.799	0.775
11	0.799–0.845	0.820
12	0.845–0.894	0.869
13	0.894–0.975	0.923
14	0.975–1.035	1.003
15	1.035–1.101	1.064
16	1.101–1.212	1.170
17	1.212–1.310	1.258
18	1.310–1.603	1.532
19	1.603–2.049	1.689
20	2.049–5.000	2.292

[a] From Wiebelt and Henderson (1979).

end of the spectrum; this is consistent with scattering theory which indicates more scatter at shorter wavelengths. Fritz (1958) suggests that the spectrum of an overcast sky is similar to that for a clear sky. Thus the distributions shown in Figure 2.6.2 can be considered as first approximations to that of total solar radiation.

In summary, the normal solar radiation incident on the earth's atmosphere has a spectral distribution indicated by Figure 1.3.1. The x-rays and other very short-wave radiation of the solar spectrum are absorbed high in the ionosphere by nitrogen, oxygen, and other atmospheric components. Most of the ultraviolet is absorbed by ozone. At wavelengths longer than 2.5 μm, a combination of low extraterrestrial radiation and strong absorption by CO_2 and H_2O means that very little energy reaches the ground. Thus, from the viewpoint of terrestrial applications of solar energy, only radiation of wavelengths between 0.29 and 2.5 μm need be considered.

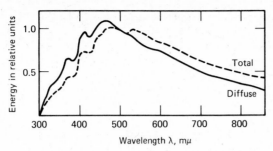

Figure 2.6.3 Relative energy distribution of total and diffuse radiation, for a clear sky. From Kondratyev (1965).

2.7 ESTIMATION OF AVERAGE SOLAR RADIATION

Radiation data are the best source of information for estimating average incident radiation. Lacking these or data from nearby locations of similar climate, it is possible to use empirical relationships to estimate radiation from hours of sunshine or cloudiness. Data on average hours of sunshine, or average percent of possible sunshine hours, are widely available from many hundreds of stations in many countries, and are usually based on data taken with Campbell–Stokes instruments. Examples are shown in Table 2.7.1. Cloud cover data (i.e., cloudiness) are also widely available, but are based on visual estimates and are probably less useful than hours of sunshine data.

The original Angstrom-type regression equation related monthly average daily radiation to clear day radiation at the location in question, and average fraction of possible sunshine hours:

$$\frac{\bar{H}}{\bar{H}_c} = a' + b' \frac{\bar{n}}{\bar{N}} \tag{2.7.1}$$

where $\bar{H}$ = the monthly average daily radiation on a horizontal surface
$\bar{H}_c$ = the average clear sky daily radiation for the location and month in question
a', b' = empirical constants
$\bar{n}$ = monthly average daily hours of bright sunshine
$\bar{N}$ = monthly average of the maximum possible daily hours of bright sunshine (i.e., the day length of the average day of the month)

A basic difficulty with Equation 2.7.1 lies in the ambiguity of the terms $\bar{n}/\bar{N}$ and $\bar{H}_c$. The former is an instrumental problem (records from sunshine recorders are open to interpretation), whereas the latter stems from uncertainty in the definition of a clear day.* Page (1964) and others have modified the

---

* Section 2.8 describes a recent method for calculating clear sky radiation. The regression constants a' and b' would have to be determined for the location in question using the same $\bar{H}_c$ as is to be used in the estimation of $\bar{H}$.

Table 2.7.1   Examples of Monthly Averages of Hours Per Day of Sunshine

Station	Latitude	Alt., m	Annual	Jan.	Feb.	Mar.	Apr.	May	June	July	Aug.	Sept.	Oct.	Nov.	Dec.
Hong Kong	22°N	s.l.	5.3	4.7	3.5	3.1	3.8	5.0	5.3	6.7	6.4	6.6	6.8	6.4	5.6
Paris	48°N	50	5.1	2.1	2.8	4.9	7.4	7.1	7.6	8.0	6.8	5.6	4.5	2.3	1.6
Bombay	19°N	s.l.	7.4	9.0	9.3	9.0	9.1	9.3	5.0	3.1	2.5	5.4	7.7	9.7	9.6
Sokoto (Nigeria)	13°N	107	8.8	9.9	9.6	8.8	8.9	8.4	9.5	7.0	6.0	7.9	9.6	10.0	9.8
Perth (Australia)	32°S	20	7.8	10.4	9.8	8.8	7.5	5.7	4.8	5.4	6.0	7.2	8.1	9.6	10.4
Madison	43°N	270	7.3	4.5	5.7	6.9	7.5	9.1	10.1	9.8	10.0	8.6	7.2	4.2	3.9

method to base it on extraterrestrial radiation on a horizontal surface, rather than on clear day radiation:

$$\frac{\overline{H}}{\overline{H}_0} = \left(a + b\,\frac{\bar{n}}{\overline{N}}\right) \qquad (2.7.2)$$

where $\overline{H}_0$ is the radiation outside of the atmosphere for the same location, averaged over the time period in question, and a and b are modified constants depending on location. The ratio $\overline{H}/\overline{H}_0$ is termed the clearness index, $\overline{K}_T$, and will be used frequently in later sections and chapters.

$\overline{H}_0$ can be calculated from Equation 1.8.3 using day numbers from Table 1.6.1 for the mean days of the month. It can also be obtained from Table 1.8.1, or from Figure 1.8.1. $\overline{N}$ can be calculated from Equation 1.6.8, or it can be obtained from Figure 1.6.3, for the mean day of the month as indicated in Table 1.6.1. Löf et al. (1966a) developed sets of constants a and b for various climate types and locations, based on radiation data then available. These are given in Table 2.7.2.

Table 2.7.2   Climatic Constants for Use in Equation 2.7.2

Location	Climate[a]	Veg.[b]	Sunshine Hours in Percentage of Possible Range	Sunshine Hours in Percentage of Possible Avg.	a	b
Albuquerque, N.M.	BS-BW	E	68–85	78	0.41	0.37
Atlanta, Ga.	Cf	M	45–71	59	0.38	0.26
Blue Hill, Mass.	Df	D	42–60	52	0.22	0.50
Brownsville, Tex.	BS	GDsp	47–80	62	0.35	0.31
Buenos Aires, Arg.	Cf	G	47–68	59	0.26	0.50
Charleston, S.C.	Cf	E	60–75	67	0.48	0.09
Darien, Manchuria	Dw	D	55–81	67	0.36	0.23
El Paso, Tex.	BW	Dsi	78–88	84	0.54	0.20
Ely, Nevada	Bw	Bzi	61–89	77	0.54	0.18
Hamburg, Germany	Cf	D	11–49	36	0.22	0.57
Honolulu, Hawaii	Af	G	57–77	65	0.14	0.73
Madison, Wisconsin	Df	M	40–72	58	0.30	0.34
Malange, Angola	Aw-BS	GD	41–84	58	0.34	0.34
Miami, Fla.	Aw	E-GD	56–71	65	0.42	0.22
Nice, France	Cs	SE	49–76	61	0.17	0.63
Poona, India (Monsoon)	Am	S	25–49	37	0.30	0.51
(Dry)			65–89	81	0.41	0.34
Kisangani, Zaire	Af	B	34–56	48	0.28	0.39
Tamanrasset, Algeria	BW	Dsp	76–88	83	0.30	0.43

(Footnotes for Table 2.7.2 on next page)

The following example is based on Madison data (although the procedure is not recommended for a station where there are data) and includes comparisons of the estimated radiation with SOLMET data and estimates for Madison based on the Blue Hill constants (those which would have been used in the absence of constants for Madison).

Example 2.7.1

Estimate the monthly averages of total solar radiation on a horizontal surface for Madison, Wisconsin, latitude 43°, based on the average duration of sunshine hours data of Table 2.7.1.

---

[a] Climatic classification based on Trewartha's climate map (1954, 1961), where climate types are:

Af Tropical forest climate, constantly moist; rainfall all through the year.
Am Tropical forest climate, monsoon rain; short dry season, but total rainfall sufficient to support rain forest.
Aw Tropical forest climate, dry season in winter.
BS Steppe or semiarid climate.
BW Desert or arid climate.
Cf Mesothermal forest climate; constantly moist; rainfall all through the year.
Cs Mesothermal forest climate; dry season in winter.
Df Microthermal snow forest climate; constantly moist; rainfall all through the year.
Dw Microthermal snow forest climate; dry season in winter.

[b] Vegetation classification based on Küchler's map, where vegetation types are:

B Broadleaf evergreen trees.
Bzi Broadleaf evergreen, shrubform, minimum height 3 feet, growth singly or in groups or patches.
D Broadleaf deciduous trees.
Dsi Broadleaf deciduous, shrubform, minimum height 3 feet, plants sufficiently far apart that they frequently do not touch
Dsp Broadleaf deciduous, shrubform, minimum height 3 feet, growth singly or in groups or patches.
E Needleleaf evergreen trees.
G Grass and other herbaceous plants.
GD Grass and other herbaceous plants; broadleaf deciduous trees.
GDsp Grass and other herbaceous plants; broadleaf deciduous, shrubforms, minimum height 3 feet, growth singly or in groups or patches.
M Mixed: broadleaf deciduous and needleleaf evergreen trees.
S Semideciduous: broadleaf evergreen and broadleaf deciduous trees.
SE Semideciduous: broadleaf evergreen and broadleaf deciduous trees; needleleaf evergreen trees.

Note: These constants are based on radiation data available before 1966, and do not reflect improvements in data processing and interpretation made since then. The results of estimations for United States stations will be at variance with SOLMET data. It is recommended that these correlations be used *only* when there are no radiation data available.

Solution

The estimates are based on Equation 2.7.2 using constants $a = 0.30$ and $b = 0.34$ from Table 2.7.2. Values of $\bar{H}_0$ are obtained from Figure 1.8.1 and day lengths $\bar{N}$ from Equation 1.6.8, each for the mean days of the month. The desired estimates are obtained in the following table, which shows daily $\bar{H}$ in MJ/m^2. (For comparison, data for Madison from Table 2.5.1 are shown, and in the last column estimates of Madison radiation made using constants a and b for Blue Hill.)

Month	$\bar{H}_0$, MJ/m^2	$\bar{N}$, hr	$\bar{n}/\bar{N}$	$\bar{H}$, est. MJ/m^2	$\bar{H}$, measured MJ/m^{2} [a]	$\bar{H}$, est. [b]
Jan.	13.2	9.2	0.489	6.2	5.9	7.2
Feb.	18.6	10.3	0.553	9.1	9.1	10.7
Mar.	25.8	11.7	0.590	12.9	12.9	15.4
Apr.	33.4	13.2	0.568	16.5	15.9	19.5
May	39.0	14.5	0.628	20.0	19.8	23.9
June	41.4	15.2	0.665	21.8	22.1	26.2
July	40.1	14.0	0.658	21.0	22.0	25.2
Aug.	35.6	13.8	0.725	19.5	19.4	23.6
Sept.	28.5	12.3	0.699	15.3	14.8	18.5
Oct.	20.7	19.8	0.667	10.9	10.3	13.1
Nov.	14.5	9.5	0.442	6.5	5.7	7.6
Dec.	11.8	8.8	0.443	5.3	4.4	6.2

[a] From SOLMET data
[b] Using constants for Blue Hill

The agreement between measured and calculated radiation is reasonably good, even though the constants a and b for Madison were derived from a different data base from the measured data. If we did not have constants for Madison and had to choose a climate close to that of Madison, Blue Hill would be a reasonable choice. The estimated averages using the Blue Hill constants are shown in the last column. The trends are shown, but the agreement is not as good. This is the more typical situation in the use of Equation 2.7.2. ∎

Data are also available on mean monthly cloud cover, $\bar{C}$, expressed as tenths of the sky obscured by clouds. Empirical relationships have been derived to relate monthly average daily radiation, $\bar{H}$, to monthly average cloud cover, $\bar{C}$. These are of a form such as:

$$\frac{\bar{H}}{\bar{H}_0} = a'' - b''\bar{C} \qquad (2.7.3)$$

Norris (1968) reviews several attempts to develop such a correlation. Bennett (1965) compared correlations of $\bar{H}/\bar{H}_0$ with $\bar{C}$, with $\bar{n}/\bar{N}$, and with a combination of the two variables and found the best correlation to be with $\bar{n}/\bar{N}$, that is, Equation 2.7.2. Cloud cover data are estimated visually, and there is not necessarily a direct relationship between the presence of partial cloud cover and solar radiation at any particular time. Thus there may not be as good a statistical relationship between $\bar{H}/\bar{H}_0$ and $\bar{C}$ as there is between $\bar{H}/\bar{H}_0$ and $\bar{n}/\bar{N}$. Many surveys of solar radiation data [e.g., Bennett (1965) and Löf et al. (1966)] have been based on correlations of radiation with sunshine hour data. However, Paltridge and Proctor (1976) have used cloud cover data to modify clear sky data for Australia and derived therefrom monthly averages of $\bar{H}_0$ which are in good agreement with measured average data.

2.8 ESTIMATION OF CLEAR SKY RADIATION

The effects of the atmosphere in scattering and absorbing radiation are variable with time, as atmospheric conditions and air mass change. It is useful to define a standard "clear" sky, and calculate the hourly and daily radiation which would be received on a horizontal surface under these standard conditions.

Hottel (1976) has presented a convenient method for estimating the beam radiation transmitted through clear atmospheres which takes into account zenith angle and altitude for a standard atmosphere and for four climate types. The atmospheric transmittance for beam radiation, τ_b, is G_{bn}/G_o and is given in the form

$$\tau_b = a_o + a_1 e^{-k/\cos\theta_z} \tag{2.8.1}$$

The constants a_o, a_1, and k for the standard atmosphere with 23 km visibility are found from a_o^*, a_1^*, and k^*, which are given for altitudes less than 2.5 km by

$$a_o^* = 0.4237 - 0.00821\,(6 - A)^2 \tag{2.8.2}$$

$$a_1^* = 0.5055 + 0.00595\,(6.5 - A)^2 \tag{2.8.3}$$

$$k^* = 0.2711 + 0.01858\,(2.5 - A)^2 \tag{2.8.4}$$

where A is the altitude of the observer in kilometers. Plots of these coefficients are shown in Figure 2.8.1. (Hottel also gives equations for a_o^*, a_1^*, and k^* for a standard atmosphere with 5 km visibility.)

Table 2.8.1 Correction Factors for Climate Types

Climate Type	r_o	r_1	r_k
Tropical	0.95	0.98	1.02
Mid-Latitude Summer	0.97	0.99	1.02
Subarctic Summer	0.99	0.99	1.01
Mid-Latitude Winter	1.03	1.01	1.00

[a] From Hottel (1976).

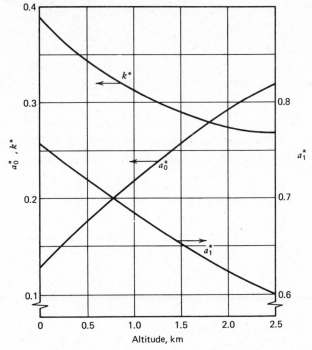

Figure 2.8.1 Constants a_0^*, a_1^*, and k^* for the 23 km visibility standard atmosphere. Adapted from Hottel (1976).

Correction factors are applied to a_0^*, a_1^*, and k^* to allow for changes in climate types. The correction factors $r_o \equiv a_o/a_o^*$, $r_1 \equiv a_1/a_1^*$ and $r_k \equiv k/k^*$ are given in Table 2.8.1. Thus, the transmittance of this standard atmosphere for beam radiation can be determined for any zenith angle and any altitude up to 2.5 km. The clear sky beam normal radiation is then

$$G_{cnb} = G_{on}\tau_b \qquad (2.8.5)$$

where G_{on} is obtained from Equation 1.4.1. The clear sky horizontal beam radiation is

$$G_{cb} = G_{on}\tau_b \cos\theta_z \qquad (2.8.6)$$

For periods of an hour, the clear sky horizontal beam radiation is

$$I_{cb} = I_{on}\tau_b \cos\theta_z \qquad (2.8.7)$$

Example 2.8.1

Calculate the transmittance for beam radiation of the standard clear atmosphere at Madison (altitude 270 m) on August 22 at 11:30 a.m. solar time. Estimate the intensity of beam radiation at that time and its component on a horizontal surface.

Solution

On August 22, n is 234, the declination is $11.4°$, and from Equation 1.6.4 the cosine of the zenith angle is 0.846.

The next step is to find the coefficients for Equation 2.8.1. First, the values for the standard atmosphere are obtained from Equations 2.8.2–2.8.4 for an altitude of 0.27 km:

$$a_o^* = 0.4237 - 0.00821 (6 - 0.27)^2 = 0.154$$
$$a_1^* = 0.5055 + 0.00595 (6.5 - 0.27)^2 = 0.736$$
$$k^* = 0.2711 + 0.01858 (2.5 - 0.27)^2 = 0.363$$

The climate-type correction factors are obtained from Table 2.8.1 for mid-latitude summer. Equation 2.8.1 becomes

$$\tau_b = 0.154(0.97) + 0.736(0.99)e^{-0.363(1.02)/0.846}$$
$$= 0.62$$

The extraterrestrial radiation is given by Equation 1.4.1. For the solar constant of 1353 W/m² G_o is 1325 W/m². The beam radiation is then

$$G_{cbn} = 1325 \times 0.62 = 822 \text{ W/m}^2$$

The component on a horizontal plane is

$$822 \times 0.846 = 695 \text{ W/m}^2 \qquad \blacksquare$$

It is also necessary to estimate the clear sky diffuse radiation on a horizontal surface to get the total radiation. Liu and Jordan (1960) developed an empirical relationship between the transmission coefficient for beam and diffuse radiation for clear days:

$$\tau_d = 0.2710 - 0.2939\tau_b \qquad (2.8.7)$$

where τ_d is G_d/G_o (or I_d/I_o) the ratio of diffuse radiation to the extraterrestrial radiation on a horizontal plane. The equation is based on data for three stations. The data used by Liu and Jordan predated that used by Hottel and may not be entirely consistent with it; until better information becomes available, it is suggested that Equation 2.8.7 be used to estimate diffuse clear sky radiation, which can then be added to the beam radiation predicted by Hottel's method to obtain a clear day total. (For purposes of correlating radiation data, it is necessary to have a well-defined standard (clear) day. This definition of a standard clear sky radiation is used in later sections.)

These calculations can be repeated for each hour of the day, based on the midpoints of the hours, to obtain the standard clear day's radiation, H_c.

Example 2.8.2

Estimate the standard clear day radiation on a horizontal surface, for Madison, on August 22.

Solution

For each hour, based on the midpoints of the hour, the transmittances of the atmosphere for beam and diffuse radiation are estimated. The calculation of τ_b is illustrated for the hour 11–12 (i.e., at 11:30) in Example 2.8.1, and the beam radiation for a horizontal surface for the hour is 2.50 MJ/m^2.

The calculation of τ_d is based on Equation 2.8.7

$$\tau_d = 0.2710 - 0.2939(0.62) = 0.089$$

The diffuse irradiance on the horizontal plane is obtained from G_o, which on August 22 is 1325 W/m^2, and cos θ_z for 11:30 is 0.846.

$$G_{cd} = 1325 \times 0.089 \times 0.846 = 100 \text{ W/m}^2$$

Then the diffuse radiation for the hour is 0.36 MJ/m^2. The total radiation on a horizontal plane for the hour is $2.50 + 0.36 = 2.86$ MJ/m^2. These calculations are repeated for each hour of the day. The result is shown in the tabulation.

Hours	τ_b	I_{cb} MJ/m^2 normal	I_{cb} MJ/m^2 horizontal	τ_d	I_{cd} MJ/m^2	I_c MJ/m^2
11–12, 12–1	0.620	2.96	2.50	0.089	0.36	2.86
10–11, 1–2	0.608	2.90	2.31	0.092	0.35	2.66
9–10, 2–3	0.580	2.77	1.95	0.101	0.34	2.29
8–9, 3–4	0.531	2.53	1.44	0.115	0.31	1.75
7–8, 4–5	0.445	2.12	0.87	0.140	0.27	1.14
6–7, 5–6	0.290	1.38	0.31	0.186	0.20	0.51
5–6, 6–7	0.150	0.72	0.03	0.227	0.04	0.07

The beam for the day, H_{cb}, is twice the sum of column 4, giving 18.8 MJ/m^2. The day's total radiation, H_c, is twice the sum of column 7, or 22.6 MJ/m^2. ∎

A simpler method for estimating clear sky radiation by hours is to use data for the ASHRAE standard atmosphere. Farber and Morrison (1977) provide tables of beam normal radiation and total radiation on a horizontal surface as a function of zenith angle. These are plotted on Figure 2.8.2. For a given day, hour-by-hour estimates of I can be made, based on midpoints of the hours.

Example 2.8.3

For August 22 for Madison, estimate the hour by hour insolation on a horizontal surface for the ASHRAE clear sky.

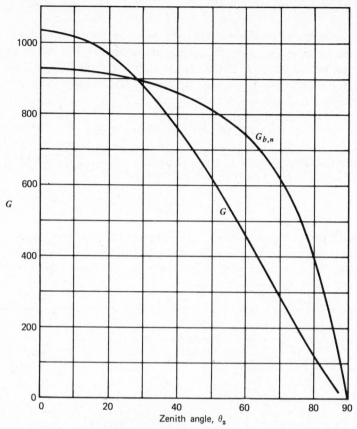

Figure 2.8.2 Total horizontal radiation and beam normal radiation for the ASHRAE standard atmosphere. Data are from Farber and Morrison (1977).

Solution

The hour-by-hour estimates are obtained from Figure 2.8.2 from the curve for horizontal radiation, based on the mid-points of the hours as indicated in the

Hour	θ_z, degrees	I_c, MJ/m^2
11–12, 12–1	32.3	3.1
10–11, 1–2	37.1	2.9
9–10, 2–3	45.3	2.5
8–9, 3–4	55.2	1.95
7–8, 4–5	65.8	1.3
6–7, 5–6	76.8	0.6
5–6, 6–7	87.6	0.05

table. θ is calculated by Equation 1.6.4. The sum of the hourly estimates is 24.8 MJ/m^2. ∎

This method estimates the "clear sky" day's radiation as 10 percent greater than the Hottel and Liu and Jordan "standard" day method of Example 2.8.3. The difference lies in the definition of a standard (clear) day. While the ASHRAE data are easier to use, the Hottel and Liu and Jordan method provides a means of taking into account climate type and altitude.

2.9 DISTRIBUTION OF CLEAR AND CLOUDY DAYS AND HOURS

The frequency of occurrence of periods of various radiation levels, for example, of good and bad days, is of interest in two contexts. First, information on the frequency distribution is the link between two kinds of correlations, that of the daily fraction of diffuse with daily radiation, and that of the monthly average fraction of diffuse with monthly average radiation. Second, in a later chapter* we develop the concept of utilizability, which depends on these frequency distributions.

$\bar{K}_T$, the monthly average clearness index,† is the ratio of monthly average radiation on a horizontal surface to the monthly average daily extraterrestrial radiation. In equation form,

$$\bar{K}_T = \frac{\bar{H}}{\bar{H}_0} \tag{2.9.1}$$

We can also define a daily clearness index, K_T, as the ratio of a particular day's radiation to the extraterrestrial radiation for that day. In equation form

$$K_T = \frac{H}{H_0} \tag{2.9.2}$$

An hourly clearness index k_T, can also be defined

$$k_T = \frac{I}{I_0} \tag{2.9.3}$$

The data $\bar{H}$, H, and I are from measurements of total solar radiation on a horizontal surface, that is, the commonly available pyranometer measurements. $\bar{H}_0$, H_0, and I_0 can be calculated by the methods of Section 1.8.

If for locations with a particular value of $\bar{K}_T$, the frequency of occurrence of days with various values of K_T is plotted, the resulting distribution would

* See Chapter 18.

† These ratios were originally referred to by Liu and Jordan (1960) as cloudiness indexes. As their values approach unity with increasing atmospheric clearness, they are also referred to as clearness indexes, the terminology used here.

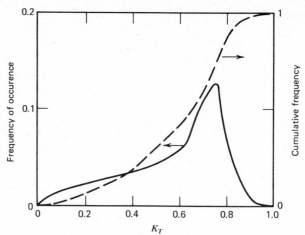

Figure 2.9.1 An example of the frequency of occurrence of days with various clearness indexes and the cumulative frequency of occurrence of those days.

appear like Figure 2.9.1. For intermediate $\bar{K}_T$ values, days with very low K_T or very high K_T occur relatively infrequently, and most of the days have K_T values intermediate between the extremes. The shape of this curve depends on the average clearness index; if it is high the distribution must be skewed toward high K_T values, and if it is low the curve must be skewed toward low K_T values. For low or high values of K_T the distribution can be bimodal.

The data used to construct Figure 2.9.1 can also be plotted as cumulative distribution, that is, as the fraction of the days that are less clear than K_T versus K_T. The result is shown as the dashed line on Figure 2.9.1. In practice, following the precedent of Whillier (1956), the plots are usually shown as K_T versus f, the fraction of days for which the day's radiation is less than K_T.

Liu and Jordan found that the cumulative distribution curves are very nearly identical for locations having the same values of $\bar{K}_T$, even though the locations varied widely in latitude and elevation. On the basis of this information, they developed a set of generalized distribution curves of K_T versus f which are functions of $\bar{K}_T$, the monthly clearness index. These are shown in Figure 2.9.2. The coordinates of the curves are given in Table 2.9.1. Thus if a location has a $\bar{K}_T$ of 0.6, 19% of the days will have K_T equal to or less than 0.40.*

Similar distribution functions have been developed for hourly radiation. Whillier (1953) observed that when the hourly and daily curves for a location are plotted, the curves are very similar. Thus the distribution curves of daily occurrences of K_T can also be applied to hourly clearness indexes. The ordinate on Figure 2.9.2 can thus be replaced by k_T and the curves will approximate the cumulative distribution of hourly clearness. Thus for a climate with $\bar{K}_T = 0.4$, 0.493 of the hours will have k_T equal to or less than 0.40.

---

* Recent research indicates that there may be some seasonal dependence of these distributions.

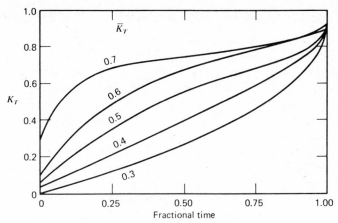

Figure 2.9.2 Generalized distribution of days with various values of K_T, as a function of $\bar{K}_T$.

Table 2.9.1   Coordinates of the Generalized Monthly K_T Cumulative Distribution Curves[a, b]

K_T	Value of f for $\bar{K}_T =$				
	0.3	0.4	0.5	0.6	0.7
0.04	0.073	0.015	0.001	0.000	0.000
0.08	0.162	0.070	0.023	0.008	0.000
0.12	0.245	0.129	0.045	0.021	0.007
0.16	0.299	0.190	0.082	0.039	0.007
0.20	0.395	0.249	0.121	0.053	0.007
0.24	0.496	0.298	0.160	0.076	0.007
0.28	0.513	0.346	0.194	0.101	0.013
0.32	0.579	0.379	0.234	0.126	0.013
0.36	0.628	0.438	0.277	0.152	0.027
0.40	0.687	0.493	0.323	0.191	0.034
0.44	0.748	0.545	0.358	0.235	0.047
0.48	0.793	0.601	0.400	0.269	0.054
0.52	0.824	0.654	0.460	0.310	0.081
0.56	0.861	0.719	0.509	0.360	0.128
0.60	0.904	0.760	0.614	0.410	0.161
0.64	0.936	0.827	0.703	0.467	0.228
0.68	0.953	0.888	0.792	0.538	0.295
0.72	0.967	0.931	0.873	0.648	0.517
0.76	0.979	0.967	0.945	0.758	0.678
0.80	0.986	0.981	0.980	0.884	0.859
0.84	0.993	0.997	0.993	0.945	0.940
0.88	0.995	0.999	1.000	0.985	0.980
0.92	0.998	0.999		0.996	1.000
0.96	0.998	1.000		0.999	
1.00	1.000			1.000	

[a] From Liu and Jordan (1960).

2.10   BEAM AND DIFFUSE COMPONENTS OF HOURLY RADIATION

In this and the following two sections we review methods for estimation of the fractions of total horizontal radiation that are beam and diffuse. In these sections, which deal in succession with hourly, daily, and monthly splits, there remain important questions of the best method for doing the estimations. A broader data base and improved understanding of the data will undoubtedly lead to improved methods. In each section we review the methods that have been published and then suggest one that can be used. It will be evident that the suggested methods are not entirely consistent with one another. This is an area of further research.

The split of total solar radiation on a horizontal surface into its beam and diffuse components is of interest in two contexts. First, methods for calculating total radiation on surfaces of other orientation from data on a horizontal surface require separate treatments of beam and diffuse radiation (see Section 2.15). Second, estimates of the long-time performance of most concentrating collectors must be based on estimates of availability of beam radiation. The present methods for estimating the distribution are based on studies of available measured data; they are possibly adequate for the first purpose, but less than adequate for the second.

Two approaches have been used to estimate I_d/I, the fraction of the hourly radiation on a horizontal plane which is diffuse. Orgill and Hollands (1977) have used data from Canadian stations to correlate I_d/I with the hourly clearness index, k_T, the ratio of the total radiation to the extraterrestrial radiation for the

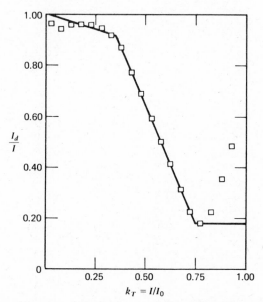

Figure 2.10.1 I_d/I as a function of hourly clearness index. Adapted from Orgill and Hollands (1977).

hour. The correlation is shown in Figure 2.10.1. The equations for the correlation are

$$\frac{I_d}{I} = \begin{cases} 1.0 - 0.249k_T & \text{for } k_T < 0.35 \\ 1.557 - 1.84k_T & \text{for } 0.35 < k_T < 0.75 \\ 0.177 & \text{for } k_T > 0.75 \end{cases} \qquad (2.10.1)$$

In a related approach described in the SOLMET manual (1978), a similar relationship is used but values of I_d/I are modified by a restricted random number that adds a statistical variation to the correlation.

The second approach has been to correlate I_d/I with I/I_c, the ratio of an hour's radiation to the standard clear sky radiation for that hour. This method was developed by Bugler (1977) based on data for Melbourne. In using I_c as the base for a modified clearness index, it implicitly takes into account air mass. One of Bugler's data sets (in this case for zenith angles of 50 to 70°) and his recommended correlation are shown on Figure 2.10.2; data for other zenith angle ranges are similar.

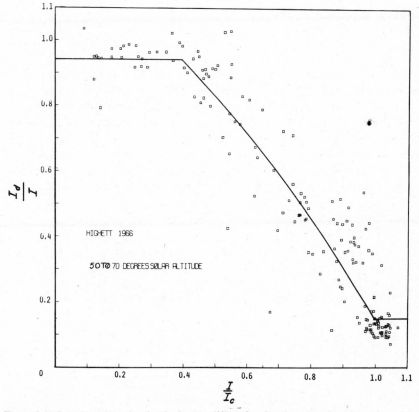

Figure 2.10.2 I_d/I vs I/I_c, that is, the hourly diffuse fraction vs ratio of total to clear sky total radiation. Plots of other ranges of solar zenith angles are similar. From Bugler (1977).

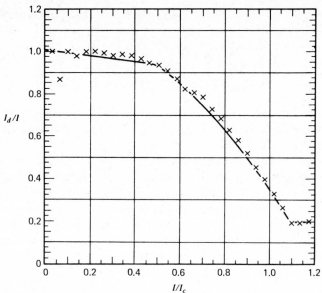

Figure 2.10.3 Correlation of I_d/I with I/I_c, where clear sky radiation is defined as in Section 2.8. From Stauter and Klein (1979).

Stauter and Klein (1979) used pyrheliometer and pyranometer data for five U. S. cities and the definition of beam and diffuse atmospheric transmission given by Equations 2.8.1 and 2.8.7 to develop a correlation similar to that of Bugler. The correlation is shown on Figure 2.10.3 with the data points for one city. Each point represents the average of many observations. The spread of the data is less than that when plotted like Figure 2.10.2, but it is still large. This arises from the fact that an hour with an intermediate value of I/I_c can be one of constant thin cloud with a high percentage of diffuse, or one of intermittent opaque clouds with a high percentage of beam.

An equation representing this correlation is

$$\frac{I_d}{I} = \begin{cases} 1.00 - 0.1\dfrac{I}{I_c} & \text{for } 0 \leq \dfrac{I}{I_c} < 0.48 \\[3mm] 1.11 + 0.0396\left(\dfrac{I}{I_c}\right) - 0.789\left(\dfrac{I}{I_c}\right)^2 & \text{for } 0.48 \leq \dfrac{I}{I_c} < 1.10 \\[3mm] 0.20 & \text{for } 1.10 \leq \dfrac{I}{I_c} \end{cases} \qquad (2.10.2)$$

Example 2.10.1

For the hour 11–12, for an August 22 in Madison, the radiation on a horizontal surface was 1.75 MJ/m². Estimate the beam and diffuse radiation on a horizontal surface for this hour.

Solution

From Example 2.8.2 the clear sky radiation is 2.86 MJ/m^2 for this hour. I/I_c is then 1.75/2.86 = 0.61 and from Figure 2.10.3 (or Equation 2.10.2) I_d/I = 0.84. Therefore, the estimated diffuse radiation is 0.84 × 1.75 = 1.47 MJ/m^2 and the beam radiation is 1.75 − 1.47 = 0.28 MJ/m^2.   ∎

2.11 BEAM AND DIFFUSE COMPONENTS OF DAILY RADIATION

Studies of available daily radiation data have shown that the average fraction which is diffuse, H_d/H, is a function of K_T. The original correlation of Liu and Jordan (1960) is shown in Figure 2.11.1; the data were for Blue Hill, Massachusetts. Also shown on the graph are plots of data for Israel from Stanhill (1966), for New Delhi from Choudhury (1963), for Canadian stations from Ruth and Chant (1976) and Tuller (1976), for Highett (Melbourne), Australia, from Bannister (1969) and from Collares-Pereira and Rabl (1979) for four United States stations. There is obvious disagreement, with differences probably due

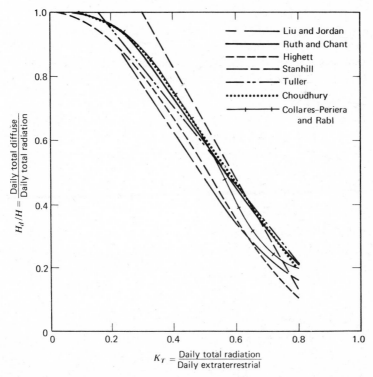

Figure 2.11.1 Correlations of daily diffuse fraction with daily clearness index. Adapted from Klein and Duffie (1978).

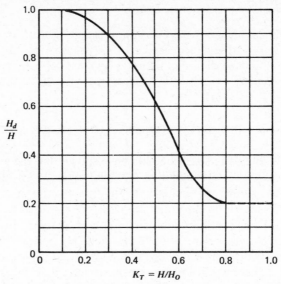

Figure 2.11.2 Suggested correlation of daily diffuse fraction with K_T. Adapted from Collares-Pereira and Rabl (1979).

in part to instrumental difficulties such as shading ring corrections, and possibly in part due to air mass and/or seasonal effects. Pending further work, the recommended correlation is that of Collares-Pereira and Rabl; this is shown on Figure 2.11.2. An equation representing this correlation is:

$$\frac{H_d}{H} \begin{cases} = 0.99 & \text{for } K_T \leq 0.17 \\ \begin{aligned} &= 1.188 - 2.272\, K_T + 9.473\, K_T^2 \\ &\quad - 21.865\, K_T^3 + 14.648\, K_T^4 \end{aligned} & \text{for } 0.17 < K_T < 0.75 \\ = -0.54\, K_T + 0.632 & \text{for } 0.75 < K_T < 0.80 \\ = 0.2 & \text{for } K_T \geq 0.80 \end{cases} \qquad (2.11.1)$$

Example 2.11.1

The day's total radiation on a horizontal surface for St. Louis, Missouri (latitude 38.6°N) on Sept. 3 is 23.0 MJ/m². Estimate the fraction and amount which is diffuse.

Solution

For September 3, the declination is 7°. From Equation 1.6.7, the sunset hour angle ω_s is 95.6°. From Equation 1.8.3, the day's extraterrestrial radiation is 32.9 MJ/m². Then

$$\frac{H}{H_o} = 23.0/32.9 = 0.70$$

From Figure 2.11.2 or Equation 2.11.1, H_d/H is 0.26, so an estimated 26% of the day's radiation is diffuse. The diffuse energy is $0.26 \times 23.0 = 5.9$ MJ/m^2.

∎

2.12 BEAM AND DIFFUSE COMPONENTS OF MONTHLY AVERAGE RADIATION

Charts similar to Figures 2.11.1 and 2.11.2 have been derived to show the distribution of monthly average daily radiation into its beam and diffuse components. In this case, the monthly fraction that is diffuse, $\bar{H}_d/\bar{H}$, is plotted as a function of monthly average clearness index, $\bar{K}_T(=\bar{H}/\bar{H}_0)$. The data for these plots can be obtained from daily data in either of two ways. First, monthly data can be plotted by summing the daily data of diffuse and total radiation. Second, as shown by Liu and Jordan, a generalized daily H_d/H versus K_T curve can be used with a knowledge of the distribution of good and bad days (the cumulative distribution curves of Figure 2.9.2) to develop the monthly average relationships.

Figure 2.12.1 shows several correlations of $\bar{H}_d/\bar{H}$ versus $\bar{K}_T$. The curve of Page, (1964) and Collares-Pereira and Rabl (1979) are based on summations of daily total and diffuse radiation. The original curve of Liu and Jordan (modified to correct for a small error in $\bar{H}_d/\bar{H}$ at low $\bar{K}_T$) and those labeled Bannister, Stanhill, Choudhury and Ruth and Chant (which are essentially identical), and Tuller are based on daily correlations by the various authors (as in Figure 2.11.1) and the distribution of days with various K_T as shown in Figure 2.9.2.

The Collares-Pereira and Rabl curve on Figure 2.12.1 is for their all-year correlation. They found a seasonal dependence of the relationship, that they expressed in terms of the sunset hour angle of the mean day of the month, which is a function of latitude and declination as given by Equation 1.6.7. The dependence of $\bar{H}_d/\bar{H}$ on $\bar{K}_T$ is shown for sunset hour angles of 80°, 90°, and 100° by the three curves of Figure 2.12.2. An equation for $\bar{H}_d/\bar{H}$ (with ω_s in degrees) is:

$$\frac{\bar{H}_d}{\bar{H}} = 0.775 + 0.00653(\omega_s - 90)$$

$$- [0.505 + 0.00455(\omega_s - 90)]\cos[115\bar{K}_T - 103] \qquad (2.12.1)$$

There is considerable disagreement among the various correlations of Figure 2.12.1. Instrumental problems may contribute to the differences, and atmospheric variables (air mass, season, or other) may have to be taken into account. It is suggested that the Collares-Pereira and Rabl correlation of Figure 2.12.2 and Equation 2.12.1 be used until a better data base or better methods of correlation indicate otherwise. (Note, however, that the seasonal dependence shown in Figure 2.12.2 is not shown in the corresponding curve for individual days, Figure 2.11.2.)

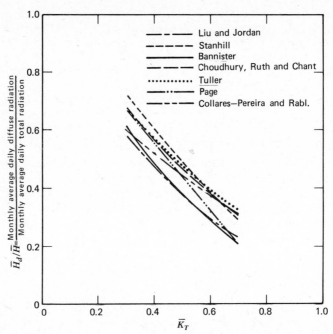

Figure 2.12.1 Correlations of average diffuse fractions with average clearness index. Adapted from Klein and Duffie (1978).

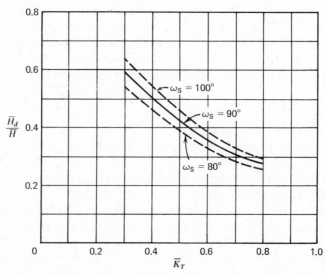

Figure 2.12.2 Suggested correlation of $\bar{H}_d/\bar{H}$ vs $\bar{K}_T$ and ω_s. Adapted from Collares–Pereira and Rabl (1979).

Example 2.12.1

Estimate the fraction of the average June radiation on a horizontal surface which is diffuse, in Madison, Wisconsin.

Solution

From Table 2.5.1, the June average daily radiation, $\bar{H}$, for Madison is 22.1 MJ/m^2. From Equation 1.8.3, for June 11, (the mean day of the month, $n = 162$, from Table 1.6.1), when the declination is 23.1°, H_o is 41.3 MJ/m^2. From Equation 1.6.7, $\omega_s = 113.4°$. Thus,

$$\bar{K}_T = \frac{22.1}{41.3} = 0.54$$

From Figure 2.12.2, at $\bar{K}_T = 0.54$, and $\omega_s = 113.4°$ the average fraction of solar energy which is diffuse is estimated as 0.48. Using Equation 2.12.1, where $\omega_s = 113°$, $\bar{H}_d/\bar{H} = 0.47$. ∎

2.13 ESTIMATION OF HOURLY RADIATION FROM DAILY DATA

When hour-by-hour (or other short time base) performance calculations for a system are to be done, it may be necessary to start with daily data and estimate hourly values from daily numbers. As with the estimation of diffuse from total radiation, this is not an exact process. For example, daily total radiation values in the middle range between clear day and completely cloudy day values can arise from various circumstances, such as intermittent heavy clouds, continuous light clouds, or heavy cloud cover for part of the day. There is no way to determine these circumstances from the daily totals. However, the methods presented here work best for clear days, and those are the days that produce most of the output of solar processes (particularly those processes that operate at temperatures significantly above ambient). And, they tend to produce conservative estimates of long-time process performance.

Statistical studies of the time distribution of total radiation on horizontal surfaces through the day, using monthly average data from a number of stations, have led to generalized charts of r_t, the ratio of hourly total to daily total radiation, as a function of day length and the hour in question. Figure 2.13.1 shows such a chart, adapted from Liu and Jordan (1977) and based on Whillier (1956, 1965) and Hottel and Whillier (1958). The hours are designated by the time for the midpoint of the hour, and days are assumed to be symmetrical about solar noon. A curve for the hour centered at noon is also shown. Day length can be calculated from Equation 1.6.8, or it can be estimated from Figure 1.6.3. Thus from a knowledge of day length (a function of latitude ϕ and declination δ) and daily total radiation, the hourly total radiation for symmetrical days can be estimated.

A study of New Zealand data by Benseman and Cook (1969) indicates that the curves of Figure 2.13.1 represent the New Zealand data in a satisfactory way. The figure is based on long-term averages and is intended for use in determining averages of hourly radiation. Whillier (1956) recommends that it be used for individual days only if they are clear days. Benseman suggests that it is adequate for individual days, with best results for clear days and increasingly uncertain results as daily total radiation decreases. For lack of a better alternative, it is applied to individual days.

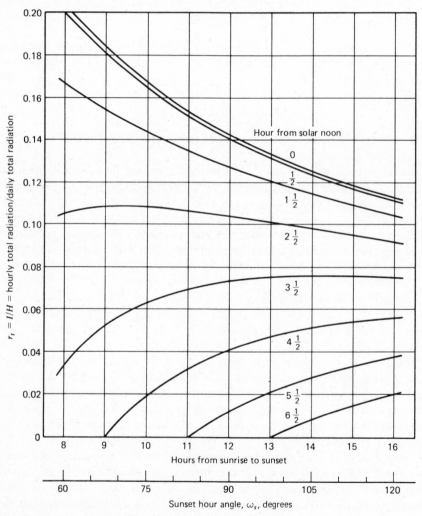

Figure 2.13.1 Relationship between hourly and daily total radiation on a horizontal surface as a function of day length. Adapted from Liu and Jordan (1960).

The curves of Figure 2.13.1 are represented by the following equation from Collares-Pereira and Rabl (1979).

$$r_t = \frac{\pi}{24}(a + b \cos \omega)\frac{\cos \omega - \cos \omega_s}{\sin \omega_s - (2\pi\omega_s/360)\cos \omega_s} \tag{2.13.1}$$

The coefficients a and b are given by

$$a = 0.409 + 0.5016 \sin(\omega_s - 60)$$
$$b = 0.6609 - 0.4767 \sin(\omega_s - 60) \tag{2.13.2}$$

In these equations ω is the hour angle in degrees for the time in question (e.g., the midpoint of the hour for which the calculation is made), and ω_s is the sunset hour angle.

Example 2.13.1

What is the fraction of the average January daily radiation that is received at Melbourne, Australia, in the hour between 8:00 and 9:00?

Solution

For Melbourne, $\phi = -38°$. From Table 1.6.1 the mean day for January is the 17th. From Equation 1.6.1 the declination is $-20.9°$. From Equation 1.6.8 the day length is 14.3 hours. From Figure 2.13.1, using the solid curve for 3.5 hours from solar noon, at a day length of 14.3 hours, approximately 7.8 percent of the day's radiation will be in that hour. Or, Equations 2.13.1 and 2.13.2 can be used, with $\omega_s = 107°$ and $\omega = -52.5°$; the result is $r_t = 0.076$. ∎

Example 2.13.2

The total radiation for Madison on an August 23 was 31.4 MJ/m². Estimate the radiation received between 1 and 2 p.m.

Solution

For that date, $\delta = 11°$, ϕ for Madison is 43°N. From Figure 1.6.3, sunset is at 6:45 p.m. and day length is 13.4 hr. Then from Figure 2.13.1, at day length of 13.5 hr and mean of 1.5 hr from solar noon, the ratio of hourly total to daily total is 0.118. The estimated radiation in the hour from 1 to 2 p.m. is then 3.7 MJ/m². (The measured value for that hour was 3.47 MJ/m².) ∎

Figure 2.13.2 shows a related set of curves for r_d, the ratio of hourly diffuse to daily diffuse radiation, as a function of time and day length. It is based on monthly average data for two stations. In conjunction with Figure 2.11.2, it can be used to estimate hourly averages of diffuse radiation if the average daily total radiation is known.

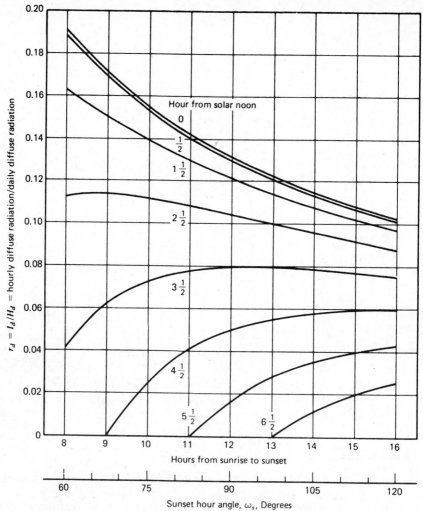

Figure 2.13.2 Relationship between hourly diffuse and daily diffuse radiation on a horizontal surface as a function of day length. Adapted from Liu and Jordan (1960).

The curves of Figure 2.13.2 can be represented by the following equation from Liu and Jordan (1960):

$$r_d = \frac{\pi}{24}\left[\frac{\cos\omega - \cos\omega_s}{\sin\omega_s - (2\pi\omega_s/360)\cos\omega_s}\right] \qquad (2.13.3)$$

Example 2.13.3

From Table 2.5.1 the average daily June total radiation on a horizontal plane in Madison is 22.1 MJ/m². Estimate the average diffuse, the average beam, and the average total radiation for the hours 10 to 11 and 1 to 2.

Solution

The mean daily extraterrestrial radiation $\bar{H}_o$ for June for Madison is 41.3 MJ/m^2 (from Table 1.8.1 or Equation 1.8.3 with $n = 162$), $\omega_s = 113°$ and the day length is 15.1 hr (from Equation 1.6.8). Then (as in Example 2.12.1), $\bar{K}_T = 0.54$. From Equation 2.12.2, $\bar{H}_d/\bar{H} = 0.47$, and the average daily diffuse radiation is $0.47 \times 22.1 = 10.4 \ MJ/m^2$. Entering Figure 2.13.2 for an average day length of 15.1 hr and for 1.5 hr from solar noon, we find $r_d = 0.102$. (Alternatively, Equation 2.13.3 can be used with $\omega = 22.5°$ and $\omega_s = 113°$ to obtain $r_d = 0.102$. Thus the average diffuse for those hours is $0.102 \times 10.4 = 1.06 \ MJ/m^2$.

From Figure 2.13.1 (or from Equations 2.13.1 and 2.13.2) from the curve for 1.5 hr from solar noon, for an average day length of 15.1 hr, $r_t = 0.108$, and average hourly total $= 0.108 \times 22.1 = 2.38 \ MJ/m^2$. The average beam radiation is the difference between the total and diffuse, or $2.38 - 1.06 = 1.32 \ MJ/m^2$. ∎

Other analytical functions have been used to represent a day's radiation. For example, Close (1967) and Cooper (1973) use a sine function for the day's radiation and superimpose on it an additional cosine function to model intermittent clouds. Sheridan et al. (1967) used a truncated sine function (of absorbed radiation) in an early analog study of solar processes. Hirschmann (1974) outlines the use of cosine functions to analytically express the variation of incident radiation and the output of solar processes. Brinkworth (1978) shows a general, empirically derived function for radiation variation through the day.

2.14 DIRECTION OF DIFFUSE RADIATION

We now turn to the general problem of calculation of radiation on tilted surfaces when only the total radiation on a horizontal surface is known. For this we need the direction from which both the beam and diffuse components reach the surface in question. Section 1.6 dealt with the geometrical problem of the direction of beam radiation. The directions from which diffuse solar radiation is received, that is, its distribution over the sky dome, is a strong function of sky conditions of cloudiness and atmosphere clarity, which are highly variable. A few data are available, for example, from Kondratyev (1969) for the particular circumstances of clear sky and sky with a solid cover of transparent strato-cumulus clouds. These are shown in Figures 2.14.1 and 2.14.2, respectively. These examples show a concentration of diffuse solar radiation around the sun and also near the horizon.

Coulson (1975) shows computed profiles of diffuse radiation across the principal plane (the plane including the sun, the zenith, and the observer) for clear skies that are very similar to that shown in Figure 2.14.1. Figure 2.14.3 adapted from Coulson, shows measurements, at a wavelength of 0.365 μm, of distribution in the principal plane, for Los Angeles, for clear and smog conditions in September. The high turbidity of the smog case leads to more

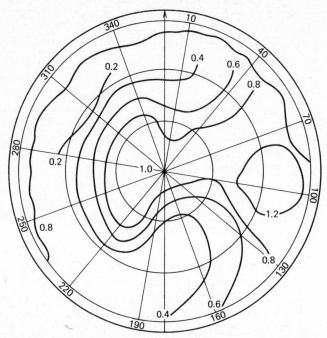

Figure 2.14.1 Relative distribution of diffuse radiation over the sky dome for a cloudless sky, for a solar altitude angle of 39°. Adapted from Kondratyev (1965).

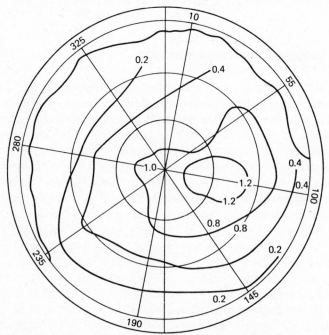

Figure 2.14.2 Relative distribution of diffuse solar radiation over the sky dome, for solid strato-cumulus clouds, for a solar altitude of 48°. Adapted from Kondratyev (1965).

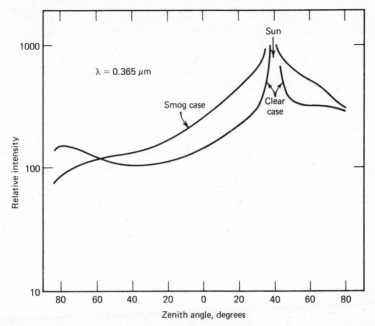

Figure 2.14.3 Relative intensity of solar radiation at $\lambda = 0.365$ μm as a function of zenith angle*
in the principal plane for Los Angeles, for a relatively clear sky and for smog. Adapted from
Coulson, *Solar and Terrestrial Radiation*, Academic Press, New York (1975).

scattering and higher diffuse intensities except near the horizon. The clear
sky curve indicates increased intensity of diffuse radiation in the vicinity of
the sun and (to a lesser extent) at the horizon. This is also noted by Temps and
Coulson (1977). Apart from the circumsolar and horizon brightening of the
sky, the clear sky curve lends some support to a commonly made assumption
that diffuse radiation for practical purposes is nearly isotropic.

The angular distribution and intensity of diffuse radiation are to some degree
functions of the albedo (reflectance) of the ground. A high surface albedo (such
as that of fresh snow, approximately 0.7) results in reflection of solar radiation
back to the sky; some of this in turn may be reflected back to the ground to be
received as additional diffuse radiation.

2.15 TOTAL RADIATION ON FIXED SLOPED SURFACES

Flat-plate solar collectors absorb both beam and diffuse components of solar
radiation. To use horizontal total radiation data to estimate radiation on the

* Zenith angle, as used here, is the angle from any point in the sky to the zenith in contrast to the
zenith angle of the sun, which is the angle from the sun to the zenith.

tilted plane of a collector of fixed orientation, it is necessary to know R, the ratio of total radiation on a tilted surface to that on the horizontal surface.*

R is, by definition:

$$R \equiv \frac{\text{total radiation on a tilted surface}}{\text{total radiation on a horizontal surface}}$$

$$= \frac{I_T}{I} \tag{2.15.1}$$

It can be expressed in terms of the contributions of the beam and diffuse radiation:

$$R_b = \frac{\text{beam radiation on a tilted surface}}{\text{beam radiation on a horizontal surface}} = \frac{I_{bT}}{I_b} \tag{2.15.2}$$

and

$$R_d = \frac{\text{diffuse radiation on a tilted surface}}{\text{diffuse radiation on a horizontal surface}} = \frac{I_{dT}}{I_d} \tag{2.15.3}$$

Then

$$R = \frac{I_b}{I} R_b + \frac{I_d}{I} R_d \tag{2.15.4}$$

The angular correction for the beam component is straightforward and can be calculated by methods of Section 1.7. The correction for the diffuse component depends on the distribution of diffuse radiation over the sky, which generally is not well known (see Section 2.14); this distribution depends on the type, extent, and location of clouds and also on the amounts and spacial distribution of other atmospheric components that scatter solar radiation. Also, some solar radiation may be reflected from the ground to the surface. Two distributions are often assumed as a basis for angular correction for the diffuse radiation.

First, it can be assumed that most of the diffuse radiation comes from an apparent origin near the sun, from the circumsolar sky, that is, the scattering of solar radiation is mostly forward scattering, with the result that most of the radiation comes from the direction of the sun. This approximation will be best on very clear days. The angular correction factor to be applied to the diffuse component is then the same as that for the beam component. In other words, the horizontal radiation is treated as though it is all beam radiation, and $R = R_b$.

Example 2.15.1

Estimate the ratio of total radiation on a surface sloped 60° toward the equator, at a latitude of 40°N, at 9:30 a.m. on February 20, assuming the diffuse radiation is concentrated from the area of the sky near the sun.

---

* We usually write these in terms of irradiation over an hour, I. The ratios might also be written for any point in time in terms of irradiance, for example, $R = G_T/G$.

Solution

Since $R = R_b$ under this assumption, Equation 1.7.2 can be used. On February 20, $\delta = -11.6°$. Then

$$R_b = \frac{\cos(40-60)\cos(-11.6)\cos(-37.5) + \sin(40-60)\sin(-11.6)}{\cos 40 \cos(-11.6)\cos(-37.5) + \sin 40 \sin(-11.6)}$$

$$R = R_b = 1.71 \qquad \blacksquare$$

Second, it can be assumed (as suggested by Hottel and Woertz (1942)) that the diffuse component is isotropic, i.e., uniformly distributed over the sky. This may be a reasonable approximation when, for example, there is a uniform cloud cover or when the atmosphere is hazy. If it is further assumed that the properties of the ground or other surfaces that reflect solar radiation to a tilted surface reflect the radiation in such a way as to be a source of diffuse solar radiation equivalent to the sky, then the surface will receive the same diffuse radiation no matter what its orientation. Under this assumption, R_d is always unity, and the irradiation on a tilted surface for an hour is given by

$$I_T = I_b R_b + I_d \qquad (2.15.5)$$

The effective ratio of solar energy on the tilted surface to that on the horizontal surface is

$$R = \frac{I_T}{I} = \frac{I_b}{I} R_b + \frac{I_d}{I} \qquad (2.15.6)$$

Example 2.15.2

For the conditions of Example 2.15.1, if 0.75 of the total solar radiation is beam radiation and 0.25 is diffuse (i.e., skies are quite clear), estimate R, assuming the diffuse component is uniformly incident on the surface from sky and ground.

Solution

From Equation 2.15.6, using the R_b calculated from Example 2.15.1

$$R = 0.75(1.71) + 0.25 = 1.53 \qquad \blacksquare$$

An improvement on this model has been derived by Liu and Jordan (1963) by considering the radiation on the tilted surface to be made up of three components: beam radiation, diffuse solar radiation, and solar radiation diffusely reflected from the ground. A surface tilted at slope β from the horizontal has a view factor to the sky given by $(1 + \cos \beta)/2$. If the diffuse solar radiation is isotropic,* this is also R_d. The surface has a view factor to the ground of

---

* Kuchler (1979) propose a combination of the Liu and Jordan and the Temps and Coulson (1977) models to account for sky brightness near the sun and near the horizon.

$(1 - \cos \beta)/2$, and if those surroundings have reflectance of ρ for the total solar radiation, the reflected radiation from the surroundings on the surface from the solar radiation is $(I_b + I_d)\rho(1 - \cos \beta)/2$. The total solar radiation on the tilted surface for an hour* is then the sum of three terms:

$$I_T = I_b R_b + I_d\left(\frac{1 + \cos \beta}{2}\right) + (I_b + I_d)\rho\left(\frac{1 - \cos \beta}{2}\right) \qquad (2.15.7)$$

and by definition of R

$$R = \frac{I_b}{I}R_b + \frac{I_d}{I}\left(\frac{1 + \cos \beta}{2}\right) + \left(\frac{1 - \cos \beta}{2}\right)\rho \qquad (2.15.8)$$

Liu and Jordan suggest values of diffuse ground reflectance of 0.2 when there is no snow and 0.7 when there is a fresh snow cover. The last two terms of Equation 2.15.7 are sometimes considered together as the diffuse radiation incident on the surface.

Example 2.15.3

What is R for the conditions of Example 2.15.2, if reflectance of the ground is taken into account? Estimate it for bare ground and for fresh snow.

Solution

Again using the beam radiation correction factor of Example 2.15.1 and Equation 2.15.8 for the condition of no snow, where $\rho = 0.2$:

$$R = 0.75(1.71) + 0.25\frac{1 + \cos 60}{2} + \frac{(1 - \cos 60)0.2}{2}$$

$$= 1.28 + 0.19 + 0.05 = 1.52$$

For the condition of snow, where $\rho = 0.7$

$$R = 0.75(1.71) + 0.25\frac{1 + \cos 60}{2} + \frac{(1 - \cos 60)0.7}{2}$$

$$= 1.28 + 0.19 + 0.18 = 1.65 \qquad \blacksquare$$

The use of $R = R_b$ generally leads to higher estimates of I_T. The use of Equation 2.15.7 or 2.15.8 leads to more conservative estimates of I_T, allows an approximation of effects of ground cover, and provides a basis for estimating the gains to be had from diffuse reflectors placed horizontally in front of receiving surfaces. Although none of these approximations is entirely satisfactory, the use of Equations 2.15.7 and 2.15.8 is recommended.

* Equations 2.15.5 to 2.15.8 could be written in terms of G if irradiance on a tilted surface is desired.

We are now in a position to estimate the hourly beam and diffuse components of solar radiation on fixed sloped surfaces such as windows or collectors. Two starting points may be encountered. First, the day's total radiation may be known. In this case, the day's diffuse radiation can be estimated from Figure 2.11.2 (or Equation 2.11.1), the day's total and diffuse can be distributed into hours by use of Figure 2.13.1 (or Equations 2.13.1 and 2.13.2) and Figure 2.13.2 (or Equation 2.13.3), and then Equation 2.15.7 can be used to determine I_T.

The second case is the more common, and starts with hourly radiation on a horizontal surface. This is the kind of radiation data illustrated in Table 2.5.2, and is widely available and used in solar process calculations. The procedure now suggested with this data base is to use the Stauter and Klein method of splitting the hour's totals into beam and diffuse (Section 2.10), and then use Equation 2.15.7 to estimate the radiation on the tilted surface.

Example 2.15.4

A flat-plate collector is to be installed at a location at latitude 40° and altitude 1650 m, with a slope of 60° and a surface azimuth angle of 0°. On January 13, what will be the hourly beam, diffuse, and ground-reflected components of radiation on this collector if the hourly horizontal total radiation data are as shown in the second column of the table, and the ground reflectance is 0.7?

Solution

A set of steps are required for each of the hours of the day. The calculations are shown for only one hour, for 9 to 10. The same calculations are done for other hours and the results are tabulated. For this day, $n = 13$ and $\delta = -21.6°$. The zenith angle of the sun for this hour is calculated from Equation 1.6.4 for $\omega = 37.5°$.

$$\cos \theta_z = \sin 40 \sin(-21.6) + \cos 40 \cos(-21.6)\cos(-37.5)$$
$$\cos \theta_z = 0.328, \text{ and } \theta_z = 70.8°$$

The constants a_0, a_1, and k in Equation 2.8.1 are calculated from Equations 2.8.2–2.8.4 with correction factors for midlatitude winter from Table 2.8.1.

$$a_0 = 1.03a_0^* = 1.03[0.4237 - 0.00821(6 - 1.65)^2] = 0.276$$
$$a_1 = 1.01a_1^* = 1.01[0.5055 + 0.00595(6.5 - 1.65)^2] = 0.652$$
$$k = k^* = 0.2711 + 0.01858(2.5 - 1.65)^2 = 0.258$$

The clear sky hour's beam radiation on a horizontal surface is calculated with a combination of Equations 1.8.1 and 2.8.1.

$$I_{cb} = 1353 \times 3600\left(1 + 0.033 \cos \frac{360 \times 13}{365}\right)\cos 70.8(0.276$$
$$+ 0.652e^{-0.258/\cos 70.8}) = 1.65 \times 10^6 \times 0.574 = 0.949 \text{ MJ/m}^2$$

The clear sky hour's diffuse radiation is found from Equations 2.8.7 and 1.8.1

$$I_{cd} = 1.65 \times 10^6(0.271 - 0.2939 \times 0.574) = 1.65 \times 10^6 \times 0.102$$
$$= 0.169 \text{ MJ/m}^2$$

The hour's clear sky total radiation on a horizontal surface is obtained by the sum of I_{cb} and I_{cd}

$$I_c = 0.949 + 0.169 = 1.118 \text{ MJ/m}^2$$

Now I/I_c and I_d/I can be determined. I/I_c is $0.724/1.118 = 0.648$, and from Equation 2.10.2

$$I_d/I = 1.11 + 0.0396(0.649) - 0.789(0.649)^2 = 0.80$$

The diffuse radiation, I_d, for the 9 to 10 hour is thus $0.724 \times 0.80 = 0.58 \text{ MJ/m}^2$, and the beam, I_b is the balance, or 0.14 MJ/m^2. The radiation on a horizontal surface can now be corrected to the 60° sloped surface. R_b is obtained from Equation 1.7.2, and is 2.49 for this hour. Thus the beam radiation on the surface is

$$I_{Tb} = 2.49 \times 0.14 = 0.35 \text{ MJ/m}^2$$

The diffuse radiation (combined sky and ground-reflected) is estimated from the second and third terms of Equation 2.15.7

$$I_{Td} = 0.58\left(\frac{1 + \cos 60}{2}\right) + 0.724 \times 0.7 \frac{(1 - \cos 60)}{2} = 0.56 \text{ MJ/m}^2.$$

The total radiation on the tilted surface is then $0.35 + 0.56 = 0.91 \text{ MJ/m}^2$.

This procedure is repeated for each of the hours of the day. The results are summarized (with several intermediate steps) in Table 2.15.1. (The process is not as tedious as it appears. Once the coefficients in Hottel's Equation 2.9.1 are determined, the clear sky radiation is a function only of ω, the hour angle. A programmable calculator reduces the problem to an almost trivial one.) ∎

In this example we have carried three significant figures up to the point of estimation of I_d/I, and then reduced to two figures. It is difficult to put limits of accuracy on the final result, but an argument can be made that the estimates are no better than 0.1 MJ/m^2, and they may not be that good. Two kinds of uncertainties arise. First, the original data base may only be good to $\pm 10\%$. Second, there are uncertainties in the correlations for obtaining beam and diffuse components of the radiation, and in the equations for calculating R. However, it is very often necessary to make these kinds of estimates and the best possible methods should be used for doing so. Fortunately, the method works best when skies are clearest, which is when solar processes collect most energy. The results of calculations of the type in this example are used as the basis for calculations of absorbed solar radiation and collector performance in later chapters.*

---

* Calculation of absorbed radiation on a collector is discussed in Chapter 5, collector performance in Chapter 6, and window and collector-storage wall performance in Chapter 15.

Table 2.15.1 Hour-by-hour Radiation Calculations for Example 2.15.4. Energy Quantities are MJ/m² for the hour Indicated.

Hour	I	θ_z	τ_b	I_c	I/I_c	I_d/I	I_d	I_b	θ_T	I_{TB}	I_{Td}	I_T
7–8	0.02	87.9	0.277	0.086	0.233	0.98	0.02	0	62.6	—	0.02	0.02
8–9	0.314	78.6	0.453	0.587	0.535	0.91	0.29	0.03	48.9	0.10	0.27	0.37
9–10	0.724	70.8	0.574	1.118	0.648	0.80	0.58	0.14	35.0	0.35	0.56	0.91
10–11	1.809	65.1	0.629	1.514	1.195	0.20	0.36	1.45	21.1	3.20	0.59	3.79
11–12	1.790	62.0	0.652	1.726	1.037	0.30	0.54	1.25	7.2	2.64	0.72	3.36
12–1	1.926	62.0	0.652	1.726	1.116	0.20	0.39	1.54	7.2	3.25	0.63	3.88
1–2	1.750	65.1	0.629	1.514	1.156	0.20	0.35	1.40	21.1	3.09	0.57	3.66
2–3	1.010	70.8	0.574	1.118	0.903	0.50	0.51	0.50	35.0	1.25	0.56	1.81
3–4	0.520	78.6	0.453	0.587	0.886	0.53	0.28	0.24	48.9	0.80	0.30	1.10
4–5	0.06	87.9	0.277	0.086	0.698	0.75	0.05	0	62.6	—	0.05	0.05

Other approaches have been developed. For example, Heywood (1966) has made extensive simultaneous measurements in the London, England, area of total radiation on horizontal and inclined planes under cloudless sky conditions. He developed an empirical method, based on those measurements, for estimating an angular correction factor for total radiation.

2.16   AVERAGE RADIATION ON FIXED SLOPED SURFACES

In the previous section, we considered the ratio of total radiation on tilted surfaces to radiation on horizontal surfaces at any time or over a short span of time such as an hour. For use in solar heating system design procedures,* we also need $\bar{R}$, the ratio of the monthly average daily radiation on the tilted surface to that on a horizontal surface. The procedure for calculating $\bar{R}$ is like that for R, that is, by adding the contributions of the beam radiation, the diffuse radiation and the reflected solar radiation from the ground. The method presented here is outlined by Liu and Jordan (1962), and extended by Klein (1977). If the diffuse and reflected radiation are each assumed to be isotropic, then, in a manner analogous to Equation 2.15.8, the monthly mean ratio $\bar{R}$ can be expressed as

$$\bar{R} = \frac{\bar{H}_T}{\bar{H}} = \left(1 - \frac{\bar{H}_d}{\bar{H}}\right)\bar{R}_b + \frac{\bar{H}_d}{\bar{H}}\left(\frac{1 + \cos\beta}{2}\right) + \rho\left(\frac{1 - \cos\beta}{2}\right) \quad (2.16.1)$$

and†

$$\bar{H}_T = \bar{H}\left(1 - \frac{\bar{H}_d}{\bar{H}}\right)\bar{R}_b + \bar{H}_d\left(\frac{1 + \cos\beta}{2}\right) + \bar{H}\rho\left(\frac{1 - \cos\beta}{2}\right) \quad (2.16.2)$$

$\bar{H}_d/\bar{H}$ is the ratio of monthly average daily diffuse radiation to monthly average daily total radiation on a horizontal surface, a function of $\bar{K}_T$ as shown in Figure 2.12.2. $\bar{R}_b$ is the ratio of the average daily beam radiation on the tilted surface to that on a horizontal surface, for the month, $\bar{H}_{bT}/\bar{H}_b$. It is a function of transmittance of the atmosphere, but Liu and Jordan suggest that it can be estimated by assuming that it has the value which would be obtained if there were no atmosphere.

For surfaces sloped toward the equator in the northern hemisphere, that is, for surfaces with $\gamma = 0°$,

$$\bar{R}_b = \frac{\cos(\phi - \beta)\cos\delta\sin\omega_s' + (\pi/180)\omega_s'\sin(\phi - \beta)\sin\delta}{\cos\phi\cos\delta\sin\omega_s + (\pi/180)\omega_s\sin\phi\sin\delta} \quad (2.16.3a)$$

---

* See Chapters 14 and 18.

† An equation for $\bar{H}_T$ for vertical surfaces partially shaded by overhangs is given in Section 15.4.

where ω'_s is the sunset hour angle for the tilted surface for the mean day of the month, which is given by

$$\omega'_s = \min\left[\begin{array}{l} \cos^{-1}(-\tan\phi\tan\delta), \\ \cos^{-1}(-\tan(\phi-\beta)\tan\delta) \end{array}\right] \qquad (2.16.3b)$$

Where "min" means the smaller of the two items in the bracket.

For surfaces in the southern hemisphere sloped toward the equator, with $\gamma = 180°$, the equations are

$$\bar{R}_b = \frac{\cos(\phi+\beta)\cos\delta\sin\omega'_s + (\pi/180)\omega'_s\sin(\phi+\beta)\sin\delta}{\cos\phi\cos\delta\sin\omega_s + (\pi/180)\omega_s\sin\phi\sin\delta} \qquad (2.16.4a)$$

and

$$\omega'_s = \min\left[\begin{array}{l} \cos^{-1}(-\tan\phi\tan\delta) \\ \cos^{-1}(-\tan(\phi+\beta)\tan\delta) \end{array}\right] \qquad (2.16.4b)$$

The numerator of Equation 2.16.3a or 2.16.4a is the extraterrestrial radiation on the tilted surface, and the denominator is that on the horizontal surface. Each of these is obtained by integration of Equation 1.6.2 over the appropriate time period, from true sunrise to sunset for the horizontal surface and from apparent sunrise to sunset on the tilted surface. For convenience, plots of $\bar{R}_b$ as a function of latitude for various slopes for $\gamma = 0°$ are shown in Figure 2.16.1.

The following example illustrates the kind of calculations that will be used in estimating monthly radiation on collectors as part of heating system design procedures.

Example 2.16.1

A collector is to be installed in Madison, latitude 43°, to be sloped 60° to the south. Average daily radiation data are shown in Table 2.5.1. If the ground reflectance is 0.2 for all months except December and March ($\rho = 0.4$), and January and February ($\rho = 0.7$), what will be the monthly average radiation incident on the collector?

Solution

The calculation is detailed below for January, and the results for the year are indicated in a table. The basic equation to be used is Equation 2.16.2. The steps are to obtain $\bar{H}_d/\bar{H}$ and then $\bar{R}_b$ for the $\bar{R}$ calculation. The ratio $\bar{H}_d/\bar{H}$ is a function of $\bar{K}_T$ and can best be obtained from Equation 2.12.2. For the mean January day, the 17th, from Table 1.6.1, $n = 17$, $\delta = -20.9°$, and the sunset hour angle calculated from Equation 1.6.7 is 69.1°. H_o for $n = 17$ is obtained by Equation 1.8.3 (or Figure 1.8.1 or Table 1.8.1), with $\omega_s = 69.1$, and is 13.22 MJ/m². Then $\bar{K}_T = 5.9/13.22 = 0.45$.

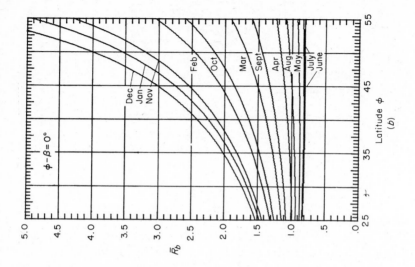

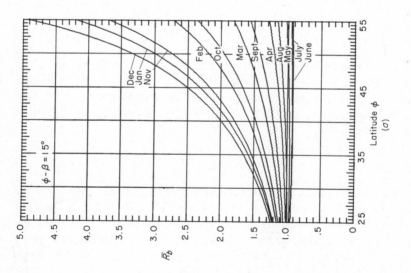

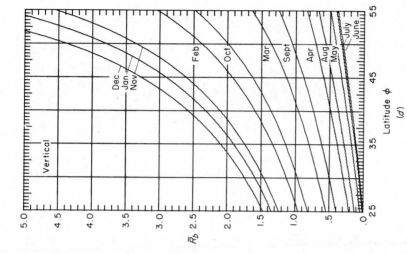

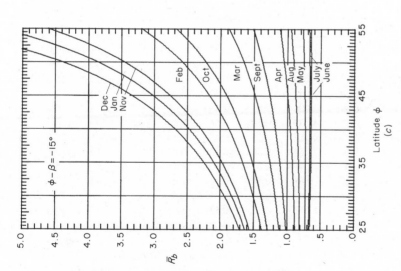

Figure 2.16.1 Estimated $\bar{R}_b$ for surfaces facing the equator as a function of latitude for various $(\phi - \beta)$, by months. (*a*) $(\phi - \beta) = 15°$; (*b*) $(\phi - \beta) = 0°$; (*c*) $(\phi - \beta) = -15°$; (*d*) $\beta = 90°$. For the southern hemisphere, interchange months as shown on Figure 1.7.2. From Beckman et al. (1977).

94

Equation 2.12.1 then is used to calculate $\bar{H}_d/\bar{H}$ from $\bar{K}_T$ and ω_s, and gives $\bar{H}_d/\bar{H} = 0.38$. The calculation of $\bar{R}_b$ requires the sunset hour angle on the sloped collector, from Equation 2.16.3b:

$$\cos^{-1}(-\tan(43 - 60)\tan(-20.8)) = 96.7°$$

ω_s was calculated as 69.1°, and is the lesser of the two, so $\omega'_s = 69.1°$. Then

$$\bar{R}_b = \frac{\cos(-17)\cos(-20.9)\sin 69.1 + (\pi \times 69.1/180)\sin(-17)\sin(-20.9)}{\cos 43 \cos(-20.9)\sin 69.1 + (\pi \times 69.1/180)\sin 43 \sin(-20.9)}$$

$$= 2.79$$

The equation for $\bar{R}$, Equation 2.16.1, can now be solved:

$$\bar{R} = (1 - 0.38)2.79 + 0.38\left(\frac{1 + \cos 60}{2}\right) + 0.7\left(\frac{1 - \cos 60}{2}\right)$$

$$= 2.19$$

Then $\bar{H}_T = \bar{H}\bar{R} = 5.9 \times 2.19 = 12.9 \text{ MJ/m}^2$. The results for the 12 months are shown in the table. Energy quantities are in megajoules per square meter. The effects of sloping the receiving plane 60° to the south on the average radiation (and thus on the total radiation through the winter season) are large indeed. The monthly results are:

Month	$\bar{H}$	$\bar{H}_0$	$\bar{K}_T$	$\bar{H}_d/\bar{H}$	$\bar{R}_b$	ρ	$\bar{R}$	$\bar{H}_T$
Jan.	5.9	13.22	0.45	0.38	2.79	0.7	2.19	12.9
Feb.	9.1	18.61	0.49	0.39	2.05	0.7	1.72	15.6
Mar.	12.9	25.75	0.50	0.42	1.42	0.4	1.24	16.0
Apr.	15.9	33.42	0.48	0.47	0.96	0.2	0.91	14.5
May	19.8	39.00	0.51	0.48	0.71	0.2	0.78	15.4
June	22.1	41.33	0.53	0.48	0.62	0.2	0.73	16.2
July	22.0	40.12	0.55	0.45	0.66	0.2	0.75	16.5
Aug.	19.4	33.54	0.55	0.43	0.84	0.2	0.85	16.5
Sept.	14.8	28.49	0.52	0.42	1.21	0.2	1.07	15.8
Oct.	10.3	20.68	0.50	0.39	1.81	0.2	1.45	14.9
Nov.	5.7	14.47	0.39	0.43	2.56	0.2	1.83	10.4
Dec.	4.4	11.78	0.37	0.43	3.06	0.4	2.17	9.5

The $\bar{H}_T$ values are shown to a tenth of a megajoule per square meter. The last place is uncertain, due to the combined uncertainties in the data and the correlations for $\bar{H}_d/\bar{H}$ and $\bar{R}$. It is difficult to put limits of accuracy on them; they may be no better than $\pm 10\%$. ∎

An equation for $\bar{R}_b$ analogous to Equation 2.16.3 has been developed by Klein (1977) for a surface of any orientation.*

$$\bar{R}_b = \frac{\begin{cases} (\cos \beta \sin \delta \sin \phi)(\omega_{ss} - \omega_{sr})(\pi/180) \\ - (\sin \delta \cos \phi \sin \beta \cos \gamma)(\omega_{ss} - \omega_{sr})(\pi/180) \\ + (\cos \phi \cos \delta \cos \beta)(\sin \omega_{ss} - \sin \omega_{sr}) \\ + (\cos \delta \cos \gamma \sin \phi \sin \beta)(\sin \omega_{ss} - \sin \omega_{sr}) \\ - (\cos \delta \sin \beta \sin \gamma)(\cos \omega_{ss} - \cos \omega_{sr}) \end{cases}}{2(\cos \phi \cos \delta \sin \omega_s + (\pi/180)\omega_s \sin \phi \sin \delta)} \qquad (2.16.5)$$

where γ is the surface azimuth angle, and ω_{sr} and ω_{ss} are sunrise and sunset hour angles on the tilted surface given by

$$\omega_{sr} = -\min\left\{\omega_s, \arccos\left[\frac{(AB + \sqrt{A^2 - B^2 + 1})}{(A^2 + 1)}\right]\right\}$$

$$\omega_{ss} = \min\left\{\omega_s, \arccos\left[\frac{(AB - \sqrt{A^2 - B^2 + 1})}{(A^2 + 1)}\right]\right\} \qquad \text{if } \gamma > 0 \quad (2.16.6)$$

$$\omega_{sr} = -\min\left\{\omega_s, \arccos\left[\frac{(AB - \sqrt{A^2 - B^2 + 1})}{(A^2 + 1)}\right]\right\}$$

$$\omega_{ss} = \min\left\{\omega_s, \arccos\left[\frac{(AB + \sqrt{A^2 - B^2 + 1})}{(A^2 + 1)}\right]\right\} \qquad \text{if } \gamma < 0 \quad (2.16.7)$$

$$A = \frac{\cos \phi}{[\sin \gamma \tan \beta]} + \frac{\sin \phi}{\tan \gamma} \qquad (2.16.8)$$

$$B = \tan \delta\left\{\frac{\cos \phi}{\tan \gamma} - \frac{\sin \phi}{[\sin \gamma \tan \beta]}\right\} \qquad (2.16.9)$$

Calculations with these equations are not difficult, but they can be tedious. They have been done for sets of latitude, surface azimuth angles, and differences between latitude and tilt; the results are tabulated in Appendix D. An example of the results of these calculations for a specific location is shown in Table 2.16.1.

Example 2.16.2

What is $\bar{H}_T$ for the surface of Example 2.16.1 for January if its surface azimuth angle is 30° (i.e., it faces 30° west of south)?

Solution

The solution is the same as that for the previous problem, except that $\bar{R}_b$ must be estimated from Equations 2.16.5 to 2.16.9, or from tables in Appendix D.

---

* This equation is not valid for surfaces that receive beam radiation more than once during a day; that is, for surfaces on which the sun sets and then rises between normal sunrise and sunset.

Table 2.16.1 Tabulation[a] of Estimated Monthly Average Daily Insolation on Surfaces of Various Orientation, for Madison, Wisconsin, Latitude 43.1°, Based on the Correlation of Page (1964), and $\rho_g = 0.2$.

	Jan.	Feb.	Mar.	Apr.	May	June	July	Aug.	Sept.	Oct.	Nov.	Dec.
Horiz. rad.	5.85	9.12	12.89	15.87	19.78	22.11	21.95	19.38	14.75	10.34	5.72	4.41
$\bar{K}_T$	0.44	0.49	0.50	0.48	0.51	0.53	0.55	0.55	0.52	0.50	0.40	0.38
Average temp.	−7.0	−6.0	0.0	7.0	13.0	19.0	21.0	20.0	15.0	10.0	1.0	−5.0
Degree-days	830	696	599	328	165	40	8	22	96	263	505	742

Average Daily Radiation on Tilted Surfaces (MJ/m^2)

Slope	Azimuth	Jan.	Feb.	Mar.	Apr.	May	June	July	Aug.	Sept.	Oct.	Nov.	Dec.
20	0	8.3	11.9	15.0	16.7	19.7	21.5	21.6	20.1	16.6	13.0	7.6	6.2
30	0	9.2	12.8	15.6	16.7	19.1	20.6	20.8	19.8	17.0	13.9	8.3	6.9
40	0	10.0	13.5	15.8	16.3	18.2	19.3	19.6	19.1	17.0	14.4	8.8	7.4
50	0	10.5	13.9	15.7	15.5	16.9	17.8	18.1	18.1	16.6	14.7	9.2	7.7
60	0	10.7	13.9	15.2	14.5	15.4	15.9	16.3	16.7	15.8	14.5	9.3	7.9
70	0	10.7	13.6	14.4	13.2	13.6	13.9	14.3	14.9	14.7	14.1	9.2	7.9
80	0	10.4	13.0	13.3	11.7	11.6	11.7	12.1	12.9	13.4	13.3	8.8	7.7
90	0	9.8	12.0	11.9	9.9	9.6	9.4	9.8	10.8	11.7	12.2	8.3	7.3
20	15	8.2	11.8	14.9	16.7	19.7	21.5	21.6	20.1	16.5	12.9	7.5	6.1
30	15	9.1	12.7	15.5	16.7	19.2	20.6	20.9	19.8	16.9	13.7	8.2	6.8
40	15	9.8	13.3	15.7	16.3	18.3	19.4	19.7	19.2	16.9	14.2	8.7	7.3
50	15	10.3	13.6	15.6	15.6	17.1	17.9	18.3	18.2	16.5	14.4	9.0	7.6
60	15	10.5	13.6	15.1	14.6	15.6	16.1	16.5	16.8	15.8	14.3	9.1	7.7
70	15	10.4	13.3	14.3	13.4	13.8	14.1	14.5	15.2	14.8	13.8	9.0	7.7
80	15	10.1	12.7	13.2	11.9	11.9	12.0	12.4	13.3	13.4	13.0	8.6	7.5
90	15	9.6	11.7	11.8	10.2	9.9	9.8	10.2	11.2	11.8	11.9	8.1	7.1

20	30	7.9	11.5	14.7	16.7	19.8	21.6	21.7	20.0	16.4	12.6	7.3	5.9
30	30	8.7	12.3	15.2	16.6	19.3	20.8	21.0	19.9	16.7	13.3	7.9	6.5
40	30	9.3	12.8	15.4	16.3	18.5	19.7	20.0	19.3	16.7	13.7	8.3	6.9
50	30	9.7	13.0	15.2	15.7	17.4	18.4	18.7	18.4	16.3	13.9	8.5	7.2
60	30	9.8	12.9	14.8	14.7	16.0	16.7	17.1	17.2	15.7	13.7	8.6	7.3
70	30	9.7	12.6	14.0	13.6	14.4	14.9	15.3	15.7	14.7	13.2	8.4	7.2
80	30	9.4	11.9	12.9	12.2	12.7	12.9	13.3	13.9	13.4	12.4	8.0	7.0
90	30	8.8	11.0	11.6	10.7	10.9	10.9	11.3	12.1	11.9	11.3	7.5	6.6
20	45	7.5	11.0	14.4	16.5	19.8	21.7	21.7	19.9	16.1	12.2	7.0	5.6
30	45	8.1	11.7	14.8	16.5	19.4	21.0	21.2	19.8	16.3	12.7	7.5	6.1
40	45	8.6	12.0	14.9	16.2	18.7	20.1	20.3	19.3	16.3	13.1	7.8	6.4
50	45	8.8	12.2	14.7	15.6	17.7	18.9	19.2	18.5	16.0	13.1	7.9	6.5
60	45	8.9	12.0	14.2	14.8	16.5	17.5	17.8	17.4	15.3	12.9	7.9	6.6
70	45	8.7	11.6	13.5	13.7	15.1	15.8	16.2	16.1	14.4	12.4	7.6	6.4
80	45	8.3	11.0	12.5	12.5	13.5	14.0	14.4	14.5	13.3	11.6	7.3	6.2
90	45	7.8	10.1	11.3	11.1	11.8	12.2	12.5	12.8	11.9	10.6	6.8	5.8

[a] From Klein et al. (1978).

From the tables, $\bar{R}_b$ is estimated for $(\phi - \beta) = -17°$, $\gamma = 30°$ and $\phi = 43°$, for January, as 2.51. This is then substituted in Equation 2.16.2,

$$\bar{R} = (1 - 0.38)2.51 + 0.38\left(\frac{1 + \cos 60}{2}\right) + 0.7\left(\frac{1 - \cos 60}{2}\right) = 2.02$$

The average radiation on the sloped plane is then $\bar{H}_T = \bar{H}\bar{R} = 5.9 \times 2.02 = 11.9$ MJ/m^2.   ∎

The uncertainties in estimating radiation on surfaces sloped to the east or west of south are greater than those for south-facing surfaces, as greater contributions to the daily totals occur early and late in the days when the air mass is greater, when atmospheric transmission is less certain, and when instrumental errors in measurements made on a horizontal plane may be larger than at times when the sun is nearer the zenith.

2.17 EFFECTS OF RECEIVING SURFACE ORIENTATION

The methods outlined in the previous section for estimating average radiation on surfaces of various orientations can be used to show the effects of slope and azimuth angle on total energy received on a monthly, seasonal, or annual basis.

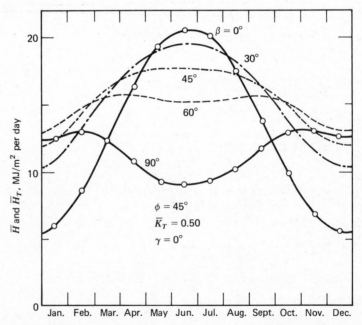

Figure 2.17.1 Variation in estimated average daily radiation on surfaces of various slopes as a function of time of year, for a latitude of 45°, $\bar{K}_T$ of 0.50, surface azimuth angle of 0° and a ground reflectance of 0.20.

(Optimization of collector orientation for any solar process that meets season-ally varying energy demands, such as space heating, must ultimately be done taking into account the time dependence of these demands. This will be discussed in later chapters on specific applications.)

To illustrate the effects of the receiving surface slope on monthly average daily radiation, we have used the methods of Section 2.16 to estimate $\bar{H}_T$ for surfaces of several slopes, for values of $\phi = 45°$, $\gamma = 0°$ and ground reflectance 0.2. $\bar{H}_T$ is a function of $\bar{H}_d/\bar{H}$, which in turn is a function of the average clearness index $\bar{K}_T$; the illustration is for $\bar{K}_T = 0.50$, constant through the year, a value typical of many temperate climates. Figure 2.17.1 shows the variations of $\bar{H}_T$ (and $\bar{H}$) through the year and shows the marked differences in energy received by surfaces of various slopes in summer and winter.

Figure 2.17.2 shows the total annual energy received as a function of slope and indicates a maximum at approximately $\beta = \phi$. The maximum is a broad one, and the changes in total annual energy are less than 5% for slopes of 20° more or less than the optimum. Figure 2.17.2 also shows total "winter" energy, taken as the total energy for the months of December, January, February,

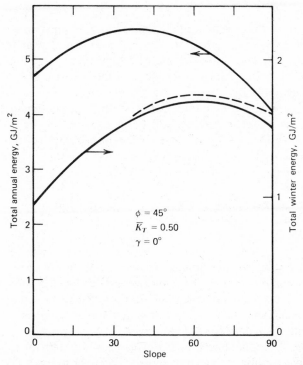

Figure 2.17.2 Variation of total annual energy and total winter (December to March) energy as a function of surface slope, for a latitude of 45°, $\bar{K}_T$ of 0.50 and a surface azimuth angle of 0°. Ground reflectance is 0.20 except for the dashed curve where it is taken as 0.60 for January and February.

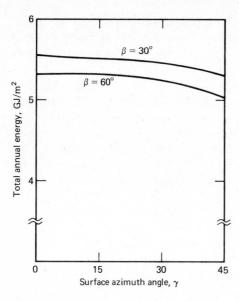

Figure 2.17.3 Variation of total annual energy with surface azimuth angle, for slopes of 30 and 60°, for latitude of 45°, for $\overline{K}_T$ of 0.50 and ground reflectance of 0.20.

and March, which would represent the time of the year when most space heating loads would occur under circumstances assumed in this example. The slope corresponding to the maximum estimated total winter energy is approximately 60°, or $\phi + 15°$. A 15° change in the slope of the collector from the optimum means a reduction of approximately 5 percent in the incident radiation. The dashed portion of the winter total curve is estimated assuming that there is substantial snow cover in January and February, resulting in a mean ground reflectance of 0.6 for those 2 months. Under this assumption, the total winter energy is less sensitive to slope than with $\rho = 0.2$. The vertical surface receives 8 percent less energy than does the 60° surface if $\rho = 0.6$, and 11 percent less if $\rho = 0.2$.

Calculation of estimates of total annual energy for $\phi = 45°$, $\overline{K}_T = 0.50$, and $\rho = 0.20$ for surfaces of slopes 30° and 60°, are shown as a function of surface azimuth angle in Figure 2.17.3. Note the expanded scale. The reduction in estimated annual energy is small for these examples, and the generalization can be made that facing collectors 20° to 30° east or west of south should make little difference in the annual energy received. (Not shown by annual radiation figures is the effect of azimuth angle γ on the diurnal distribution of radiation on the surface. Each shift of γ of 15° will shift the daily maximum of available energy by roughly an hour toward morning if γ is negative and toward afternoon if γ is positive. This could affect the performance of a system for which there are regular diurnal variations in energy demands on the process.)

Similar conclusions have been reached by others, for example, Morse and Czarnecki (1958), who estimated the relative total annual beam radiation on surfaces of variable slope and azimuth angle.

From studies of this kind, general "rules of thumb" can be stated. For maximum annual energy availability, a surface slope equal to latitude is best. For maximum summer availability, slope should be approximately 10 to 15° less than the latitude. For maximum winter energy availability, slope should be approximately 10 to 15° more than the latitude. The slopes are not critical; deviations of 15° result in reduction of the order of 5%. The expected presence of a reflective ground cover such as snow leads to higher slopes for maximizing wintertime energy availability. The best surface azimuth angles for maximizing incident radiation are 0° in the northern hemisphere or 180° in the southern hemisphere, that is, the surfaces should face the equator. Deviations of 20 or 30° have small effect on total annual energy availability.

2.18 RADIATION AUGMENTATION

It is possible to increase the radiation incident on an absorber by use of planar reflectors. In Equation 2.15.7 and 2.15.8, ground-reflected radiation was taken into account in the third term, with the ground assumed to be a diffuse reflector infinite in extent. With ground reflectances normally of the order of 0.2 and low slopes, the contributions of ground-reflected radiation are small. However, with ground reflectances of 0.6–0.7 typical of snow, and high slopes, the contribution of reflected radiation on surfaces may be substantial. In this section we show a more general case of the effects of diffuse reflectors of finite size, and comment on the use of specular reflectors.

Consider the geometry sketched in Figure 2.18.1. We are concerned with two intersecting planes, the receiving surface, C (i.e., a solar collector or passive absorber), and a diffuse reflector, R. The angle between the planes is ψ. The angle ψ is $180 - \beta$ if the reflector is horizontal, but the analysis is not restricted to a horizontal reflector.

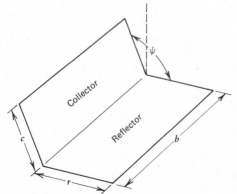

Figure 2.18.1 Geometric relationship of an energy receiving surface such as a collector and a reflecting surface.

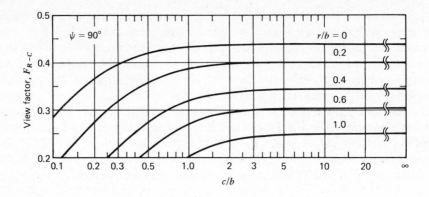

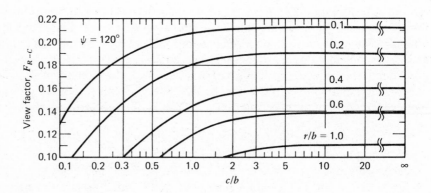

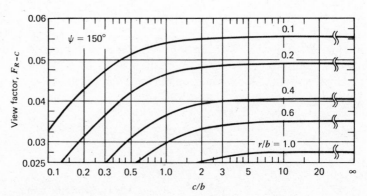

Figure 2.18.2 View factor, F_{R-C}, as a function of the relative dimensions of the reflecting surface and the collecting surface. Adapted from Hamilton and Morgan (1952).

102

The radiation incident on surface C at any time will be the sum of the beam component, the diffuse component from the portion of the sky "seen" by the surface, the diffusely reflected radiation from surface R, and possibly some ground-reflected radiation from ground outside of R. The radiation reflected to surface C from surface R is the product of the total radiation on R, its reflectance, and the view factor F_{R-C}. F_{R-C} is shown in Figure 2.18.2 as a function of the ratios c/b and r/b, for $\psi = 90°$, $120°$, and $150°$. If a vertical collector-wall is 3 m high and 6 m long, has a horizontal reflecting surface in front of it extending out 2.4 m, c/b is $3/6 = 0.5$, r/b is $2.4/6 = 0.4$ and from Figure 2.18.2a, the view factor is 0.28.

If the surfaces C and R are very long in extent (as might be the case with long arrays of collectors for large-scale solar applications), Hottel's "crossed-string" method gives the view factor as

$$F_{R-C} = \frac{c + r - s}{2r} \qquad (2.18.1)$$

where s is the distance from the upper edge of the collector to the outer edge of the reflector, measured in a plane perpendicular to planes C and R. This can be determined from

$$s = [c^2 + r^2 - 2cr \cos \psi]^{1/2} \qquad (2.18.2)$$

For a collector array noted above but very long in extent, $s = (3^2 + 2.4^2)^{0.5} = 3.84$ m, and $F_{R-C} = (3 + 2.4 - 3.84)/4.8 = 0.33$.

It is necessary to know the incident radiation on the plane of the reflector. The beam component is again calculated by use of $R_{b,R}$ but for the orientation of the reflector surface. The diffuse component must be estimated from the view factor F_{R-S}. For the general case, that is, for any orientation of surface R,

$$F_{R-S} + F_{R-C} + F_{R-G} = 1 \qquad (2.18.3)$$

Where the view factors are from surface R to sky, to surface C, and to ground. F_{R-C} is determined as noted above. F_{R-G} will be zero for a horizontal reflector and will be small for collectors that are long in extent. Thus as a first approximation for many practical cases (where there are no other obstructions) $F_{R-S} = (1 - F_{R-C})$.

There remains the questions of the angle of incidence of radiation reflected from surface R on surface C. As an approximation, an average angle of incidence can be taken as that of the radiation from the midpoint of surface R to the midpoint of surface C, as shown in Figure 2.18.3.

The angle of incidence, θ_r, is given by

$$\cos \theta_r = \frac{c \sin \psi}{s} \qquad (2.18.4)$$

The total radiation reflected from surface R with area A_R to surface C with area A_c if R has a diffuse reflectance of ρ is

$$A_c G_R = (G_b R_{b,R} + (1 - F_{R-C})G_d)\rho A_R F_{R-C} \qquad (2.18.5)$$

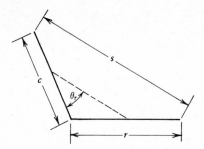

Figure 2.18.3 Section of reflector and collector surfaces, showing an approximate angle of incidence of reflected radiation, θ_r.

Example 2.18.1

A south-facing vertical surface is 4.5 m high and 12 m long. It has in front of it a horizontal diffuse reflector of the same length which extends out 4 m. The reflectance is 0.85. At solar noon, the total irradiance on a horizontal surface is 800 W/m² of which 200 is diffuse. The zenith angle of the sun is 50°. Estimate the total radiation on the vertical surface and the angle of incidence of that part of the total that is reflected from the diffuse reflector.

Solution

First, estimate F_{R-C} from Figure 2.18.2a. At $c/b = 4.5/12 = 0.38$ and $r/b = 4/12 = 0.33$, $F_{R-C} = 0.28$. The total radiation on the reflector will be the beam component, 600 W/m², plus the diffuse component, which is $G_d F_{R-S}$ or $G_d(1 - F_{R-C})$. The diffuse radiation on the reflector is thus $200(1 - 0.28) = 144$ W/m². The radiation reflected from the reflector that is incident on the vertical surface is estimated by Equation 2.18.5

$$G_R = (600 + 144)\frac{0.85 \times 48 \times 0.28}{54} = 160 \text{ W/m}^2$$

The beam component on the vertical surface is obtained by R_b, which in this case is cos 40/cos 50 = 1.19. Then $G_{bT} = 600 \times 1.19 = 715$ W/m². The diffuse component from the sky on the vertical surface is estimated as

$$G_{dT} = 200\frac{(1 + \cos 90)}{2} = 100 \text{ W/m}^2$$

The total radiation on the vertical surface (neglecting reflected radiation from ground areas beyond the reflector) is the sum of the three terms

$$G_T = 160 + 715 + 100 = 975 \text{ W/m}^2$$

An average angle of incidence of the reflected radiation on the vertical surface is estimated with Equation 2.18.4.

$$s = (4.0^2 + 4.5^2)^{0.5} = 6.02 \text{ m.,}$$

and

$$\cos \theta_r = 4.5 \sin 90/6.02 = 0.75$$
$$\theta_r = 42° \qquad\blacksquare$$

The contributions of diffuse reflectors may be significant, although they will not result in large increments in incident radiation. In the example above the contribution is approximately 160 W/m². If the horizontal surface in front of the vertical plane were ground with $\rho = 0.2$, the contribution from ground-reflected radiation would have been $0.2 \times 800(1 - \cos 90)/2$ or 80 W/m².

It has been pointed out by McDaniels et al. (1975), Grassie and Sheridan (1977), and others that a specular reflector can have more effect in augmenting radiation on a collector than a diffuse reflector. Hollands (1971) presents a method of analysis of some reflector-collector geometries, and Bannerot and Howell (1979) show effects of reflectors on average radiation on surfaces. The effects of reflectors that are partly specular and partly diffuse are treated by Grimmer et al. (1978). The practical problem is to maintain high specular reflectance, particularly on surfaces that are facing upward, are difficult to protect against weathering, and will accumulate snow in cold climates and become diffuse reflectors. Thus we do not cover the case of reflectors with significant specular component of reflectance, but suggest the above reference for further information on the problem.

2.19   BEAM RADIATION ON MOVING SURFACES

Sections 2.15–2.18 have dealt with estimation of total radiation on surfaces of fixed orientation, such as flat plate collectors or windows. It is also of interest to estimate the radiation on surfaces that move in various prescribed ways. Most concentrating collectors utilize beam radiation only and move to "track" the sun. This section is concerned with the calculation of beam radiation on these planes, which move about one or two axes of rotation. The tracking motions of interest are described in Section 1.6, and for each the relationships between the angle of incidence and latitude, declination, and hour angle are given.

At any time the beam radiation on a surface is given by

$$G_b = G_{on}\tau_b \cos \theta \qquad (2.19.1)$$

where $\cos \theta$ is given by Equations 1.6.9 to 1.6.13. The problem in the use of this equation is in the determination of the transmittance of the atmosphere for beam radiation, τ_b. For clear atmospheres that are like the standard atmospheres noted in Section 2.8, Equations 2.8.1 to 2.8.4 can be used to estimate τ_b. In general, however, it will be necessary to start with past radiation measurements and estimate what the beam radiation on the surface in question would have been for that set of measured data.

If the radiation data base is hourly total radiation on a horizontal surface, the procedures of Example 2.15.4 can be followed to estimate each hour's diffuse and beam radiation on a horizontal surface. From I_b, the beam irradiation on any surface can be estimated from the equations of Section 1.6, for example, by Equation 1.6.2.

Example 2.19.1

A cylindrical concentrating collector is to be oriented so that it rotates about a horizontal east-west axis in such a way as to constantly minimize the angle of incidence and thus maximize the incident beam radiation. For the location and day of Example 2.15.4, estimate the hour by hour beam radiation on the aperture of this collector.

Solution

As in Example 2.15.4, the calculations are shown for an hour, in this case 10 to 11, and a tabulation of the day's results is provided.

R_b is given by the ratio of Equations 1.6.10 and 1.6.4.

$$R_b = \frac{[1 - \cos^2(-21.6)\sin^2(-22.5)]^{1/2}}{\cos 40 \cos(-21.6)\cos 22.5 + \sin 40 \sin(-21.6)}$$

$$= 2.22$$

Then I_{ba}, the beam radiation on the moving plane (which will be the aperture of the collector), is

$$I_{ba} = R_b I_b = 2.22 \times 1.45 = 3.22$$

The table indicates the time, zenith angle and I_b from Example 2.15.4, and the R_b and I_{ba} for the plane moving as specified. The hourly energy quantities are in MJ/m^2. The hours 7 to 8 and 4 to 5 are omitted, as the calculation of I_b for those hours is highly uncertain.

Hour	θ_z	I_b	R_b	I_{ba}
8–9	78.6	0.03	3.43	0.10
9–10	70.8	0.14	2.56	0.36
10–11	65.1	1.45	2.22	3.22
11–12	62.0	1.25	2.11	2.64
12–1	62.0	1.54	2.11	3.25
1–2	65.1	1.40	2.22	3.11
2–3	70.8	0.50	2.51	1.26
3–4	78.6	0.24	3.43	0.82

The uncertainties in these estimations of beam radiation are greater than those associated with estimations of total radiation. The use of pyrheliometric data is preferred if available.

2.20 SUMMARY

In this chapter we have described how solar radiation is measured, and the standard instruments used to establish the calibrations of working pyrheliometers and pyranometers. Solar radiation data are available in several forms, with the most widely available being pyranometer measurements of total (beam plus diffuse) radiation on horizontal surfaces; these data are available on an hourly basis from a few stations, and on a daily basis for many additional stations.

Solar radiation information is needed in several different forms, depending on the kinds of calculations that are to be done. These calculations fall into two major categories. First (and most detailed), we may wish to calculate on an hour-by-hour basis the long time performance of a solar process system; for this we want hourly information of solar radiation and other meteorological measurements. Second, monthly average solar radiation is useful in estimating long-term performance of some kinds of solar processes.

Here we have presented methods (and commented on their limitations) for the estimation of solar radiation information in the desired format from the data that are available. This includes estimation of beam and diffuse radiation from total radiation, time distribution of radiation in a day, and radiation on surfaces other than horizontal.

In following chapters we treat the heat transfer and materials properties topics that are basic to an understanding of solar collectors. Chapters 6 through 8 will present analyses of the thermal performance of flat plate and focusing collectors, and in those chapters we will use the solar radiation data processing methods described in this chapter.

REFERENCES

Bannerot, R. B., and J. R. Howell, *Solar Energy*, **22**, 229 (1979). "Predicted Daily and Yearly Average Radiation Performance of Optimal Trapezoidal Groove Solar Energy Collectors."

Bannister, J. W., Solar Radiation Records, Division of Mechanical Engineering, Commonwealth Scientific and Industrial Research Organization, Australia (1966–1969).

Beckman, W. A., S. A. Klein, and J. A. Duffie, *Solar Heating Design by the f-Chart Method*, Wiley-Interscience, New York (1977).

Bennett, I., *Solar Energy*, **9**, 145 (1965). "Monthly Maps of Mean Daily Insolation for the United States."

Benseman, R. F. and F. W. Cook, *New Zealand Journal of Sciences*, **12**, 696 (1969). "Solar Radiation in New Zealand—The Standard Year and Radiation on Inclined Slopes."

Brinkworth, B. J., *Solar Energy*, **21**, 171 (1978). "Asymptotic Behaviour as a Guide to the Long Term Performance of Solar Water Heating Systems."

Bugler, J. W., *Solar Energy*, **19**, 477 (1977). "The Determination of Hourly Insolation on an Inclined Plane using a Diffuse Irrandiance Model Based on Hourly Measured Global Horizontal Insolation."

Choudhury, N. K. D., *Solar Energy*, **7**, 44 (1963). "Solar Radiation at New Delhi."

Cinquemani, V., J. R. Owenby, and R. G. Baldwin, NOAA report to Department of Energy (Nov. 1978). "Input Data for Solar Systems."

Close, D. J., *Solar Energy*, **11**, 112 (1967). "A Design Approach for Solar Processes."

Collares-Pereira, M. and A. Rabl, *Solar Energy*, **22**, 155 (1979). "The Average Distribution of Solar Radiation—Correlations Between Diffuse and Hemispherical and Between Daily and Hourly Insolation Values."

Cooper, P. I., *Solar Energy*, **14**, 451 (1973). "Digital Simulation of Experimental Solar Still Data."

Coulson, K. L., *Solar and Terrestrial Radiation*, Academic Press, New York (1975).

deJong, B., Monograph published by Delft University Press, Netherlands (1973). "Net Radiation Received by a Horizontal Surface at the Earth."

Drummond, A. J., *Archiv fur Meteorologie Geophysik Bioklimatologie*, Series B, **7**, 413 (1956). "On the Measurement of Sky Radiation."

Drummond, A. J., *Journal Applied Meteorology*, **3**, 810 (1964). "Comments on Sky Radiation Measurement and Corrections."

Farber, E. A. and C. A. Morrison, paper in *Applications of Solar Energy for Heating and Cooling of Buildings*, R. C. Jordan and B. Y. H. Liu (ed.), ASHRAE GRP-170, New York (1977). "Clear-Day-Design Values."

Foster, N. B., and L. W. Foskett, *Bulletin of the American Meteorological Society*, **34**, 212 (1953). "A Photoelectric Sunshine Recorder."

Fritz, S., *Transactions of the Conference on Use of Solar Energy*, **1**, 17, University of Arizona Press, Tucson (1958). "Transmission of Solar Energy Through the Earth's Clear and Cloudy Atmosphere."

Grassie, S. L., and N. L. Sheridan, *Solar Energy*, **19**, 663 (1977). "The Use of Planar Reflectors for Increasing the Energy Yield of Flat-Plate Collectors."

Grimmer, D. P., K. G. Zinn, K. C. Herr, and B. E. Wood, Report LA-7041 of Los Alamos Scientific Lab (1978). "Augmented Solar Energy Collection Using Various Planar Reflective Surfaces: Theoretical Calculations and Experimental Results."

Hamilton, D. C. and W. R. Morgan, National Advisory Committee for Aeronautics Technical Note 2836 (1952). "Radiant Interchange Configuration Factors."

Heywood, H., *Solar Energy*, **10**, 51 (1966). "The Computation of Solar Radiation Intensities."

Hirschmann, J. R., *Solar Energy*, **16**, 117 (1974). "The Cosine Function as a Mathematical Expression for the Processes of Solar Energy."

Hollands, K. G. T., *Solar Energy*, **13**, 149 (1971). "A Concentrator for Thin-Film Solar Cells."

Hottel, H. C., *Solar Energy*, **18**, 129 (1976). "A Simple Model for Estimating the Transmittance of Direct Solar Radiation Through Clear Atmospheres."

Hottel, H. C. and A. Whillier, *Transactions of the Conference on Use of Solar Energy*, **2**, 74, University of Arizona Press (1958). "Evaluation of Flat Plate Solar Collector Performance."

Hottel, H. C. and B. B. Woertz, *Transactions of the American Society of Mechanical Engineers*, **64**, 91 (1942). "Performance of Flat-Plate Solar-Heat Collectors."

IGY Instruction Manual, London, Pergamon Press, Vol. V, Pt. VI, pp. 426–429 (1958). "Radiation Instruments and Measurements."

Jeys, T. H. and L. L. Vant-Hull, *Solar Energy*, **18**, 343 (1976). "The Contribution of the Solar Aureole to the Measurements of Pyrheliometers."

Klein, S. A., *Solar Energy*, **19**, 325 (1977). "Calculation of Monthly Average Insolation on Tilted Surfaces."

Klein, S. A. and J. A. Duffie, *Proceedings of 1978 Annual Meeting American Section, International Solar Energy Society*, Denver **2.2**, 672 (1978). "Estimation of Monthly Average Diffuse Radiation."

Klein, S. A., W. A. Beckman, and J. A. Duffie, Report #44-2 of the Engineering Experiment Station, University of Wisconsin-Madison (1978). "Monthly Average Solar Radiation on Inclined Surfaces for 261 North American Cities."

Kondratyev, K. Y., *Actinometry* (translated from Russian) NASA TT F-9712 (1965); also *Radiation in the Atmosphere*, Academic Press, New York (1969).

Kuchler, T. M., *Solar Energy*, **23**, 111 (1979). "Evaluating Models to Predict Insolation on Tilted Surfaces."

Liu, B. Y. H. and R. C. Jordan, paper in *Applications of Solar Energy for Heating and Cooling of Buildings*, ASHRAE, New York (1977). "Availability of Solar Energy for Flat-Plate Solar Heat Collectors."

Liu, B. Y. H. and R. C. Jordan, *ASHRAE Journal*, **3** (10), 53 (1962). "Daily Insolation on Surfaces Tilted Toward the Equator."

Liu, B. Y. H. and R. C. Jordan, *Solar Energy*, **4** (3), (1960). "The Interrelationship and Characteristic Distribution of Direct, Diffuse and Total Solar Radiation."

Liu, B. Y. H. and R. C. Jordan, *Solar Energy*, **7**, 53 (1963). "The Long-Term Average Performance of Flat-Plate Solar Energy Collectors."

Löf, G. O. G., J. A. Duffie, and C. O. Smith, Engineering Experiment Station Report 21, University of Wisconsin, Madison (July 1966a). "World Distribution of Solar Radiation."

Löf, G. O. G., J. A. Duffie, and C. O. Smith, *Solar Energy*, **10**, 27 (1966b). "World Distribution of Solar Energy."

McDaniels, D. K., D. H. Lowndes, H. Mathew, J. Reynolds, and R. Gray, *Solar Energy*, **17**, 277 (1975). "Enhanced Solar Energy Collection Using Reflector-Solar Thermal Collector Combinations."

MacDonald, T. H., *Monthly Weather Review*, **79** (8) (1951). "Some Characteristics of the Eppley Pyrheliometer."

Moon, P., *Journal of the Franklin Institute*, **230**, 583 (1940). "Proposed Standard Solar Radiation Curves for Engineering Use."

Morse, R. N. and J. T. Czarnecki, Report E. E. 6 of Engineering Section (now Mechanical Engineering Division), Commonwealth Scientific and Industrial Research Organization, Melbourne, Australia (1958). "Flat Plate Solar Absorbers: The Effect on Incident Radiation of Inclination and Orientation."

Norris, D. J., *Solar Energy*, **14**, 99 (1973). "Calibration of Pyranometers."

Norris, D. J., *Solar Energy*, **16**, 53 (1974). "Calibration of Pyranometers in Inclined and Inverted Positions."

Norris, D. J., *Solar Energy*, **12**, 107 (1968). "Correlation of Solar Radiation with Clouds."

Orgill, J. F., and K. G. T. Hollands, *Solar Energy*, **19**, 357 (1977). "Correlation Equation for Hourly Diffuse Radiation on a Horizontal Surface."

Page, J. K., *Proceedings of the UN Conference on New Sources of Energy*, **4**, 378 (1964). "The Estimation of Monthly Mean Values of Daily Total Short-Wave Radiation on Vertical and Inclined Surfaces from Sunshine Records for Latitudes 40°N–40°S."

Paltridge, G. W. and D. Proctor, *Solar Energy*, **18**, 235 (1976). "Monthly Mean Solar Radiation Statistics for Australia."

Robinson, N. (ed.), *Solar Radiation*, Elsevier, Amsterdam (1966).

Ruth, D. W., and R. E. Chant, *Solar Energy*, **18**, 153 (1976). "The Relationship of Diffuse Radiation to Total Radiation in Canada."

Sheridan, N. R., K. J. Bullock, and J. A. Duffie, *Solar Energy*, **11**, 69 (1967). "Study of Solar Processes by Analog Computer."

SOLMET Manual, US National Climatic Center, Ashville, N. C., Vols. 1 and 2 (1978).

Stanhill, G., *Solar Energy*, **10** (2), 96 (1966). "Diffuse Sky and Cloud Radiation in Israel."

Stauter, R. and S. A. Klein, Personal communication (1979).

Temps, R. C. and K. L. Coulson, *Solar Energy*, **19**, 179 (1977). "Solar Radiation Incident upon Slopes of Different Orientations."

Thekaekara, M. P., *Supplement to the Proceedings of the 20th Annual Meeting of the Institute for Environmental Science*, 21 (1974). "Data on Incident Solar Energy."

Thekaekara, M. P., *Solar Energy*, **18**, 309 (1976). "Solar Radiation Measurement: Techniques and Instrumentation."

Trewartha, G. T., *An Introduction to Climate*, 3rd ed., McGraw-Hill, New York (1954).

Trewartha, G. T., *The Earth's Problem Climates*, University of Wisconsin Press, Madison (1961).

Tuller, S. E., *Solar Energy*, **18**, 259 (1976). "The Relationship between Diffuse, Total and Extraterrestrial Solar Radiation."

Whillier, A., *Arch. Met. Geoph. Biokl.* Series B, **7**, 197 (1956). "The Determination of Hourly Values of Total Radiation from Daily Summations."

Whillier, A., Ph.D. Thesis, MIT, Cambridge, Mass. (1953). "Solar Energy Collection and It's Utilization for House Heating."

Whillier, A., *Solar Energy*, **9**, 164 (1965). "Solar Radiation Graphs."

Wiebelt, J. A. and J. B. Henderson, *Transactions American Society of Mechanical Engineers, Journal of Heat Transfer*, **101**, 101 (1979). "Selected Ordinates for Total Solar Radiation Property Evaluation from Spectral Data."

WMO, *Guide to Meteorological Instruments and Observing Practices*, 3rd Edn., WMO and T.P.3, WHMO, Geneva (1969).

Yellott, J. I., paper in *Applications of Solar Energy for Heating and Cooling of Buildings*, ASHRAE, New York (1977). "Solar Radiation Measurement."

CHAPTER 3

Selected Topics in Heat Transfer

This chapter is intended to review those aspects of heat transfer that are important in the design and/or analysis of solar collectors and systems. It begins with a review of radiation heat transfer, which is often given cursory treatment in standard heat transfer courses. The next sections cover convection heat transfer between parallel plates. The final sections review some convection correlations for internal flow and wind-induced flow.

The role of convection and conduction heat transfer in the performance of solar systems is obvious. Radiation heat transfer plays a role in bringing energy to the earth, but not so obvious is the significant role radiation heat transfer plays in the operation of solar collectors. In the usual engineering practice radiation heat transfer is often negligible. In a solar collector the energy flux is often two orders of magnitude smaller than in conventional heat transfer equipment, and thermal radiation is a significant mode of heat transfer.

3.1 THE ELECTROMAGNETIC SPECTRUM

Thermal radiation is electromagnetic energy that is propagated through space at the speed of light. For most solar energy applications, only thermal radiation is important. Thermal radiation is emitted by bodies by virtue of their temperature; the atoms, molecules, or electrons are raised to excited states, return spontaneously to lower energy states, and in doing so, emit energy in the form of electromagnetic radiation. Because the emission results from changes in electronic, rotational, and vibrational states of atoms and molecules, the emitted radiation is usually distributed over a range of wavelengths.

The spectrum of electromagnetic radiation is divided into wavelength bands. These bands and the wavelengths representing their approximate limits, are shown in Figure 3.1.1. The wavelength limits associated with the various names and the mechanism producing the radiation are not sharply defined. There is no basic distinction between these ranges of radiation other than the wavelength λ; they all travel with the speed of light C and have a frequency v such that

$$C = \frac{C_0}{n} = \lambda v \tag{3.1.1}$$

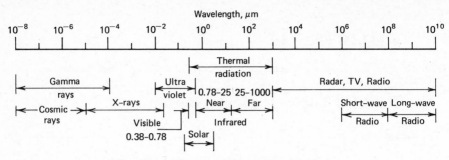

Figure 3.1.1   Spectrum of electromagnetic radiation.

where C_0 is the speed of light in a vacuum and n is the index of refraction.

The wavelengths of importance in solar energy and its applications are in the ultraviolet to near-infrared range, that is, from 0.3 to approximately 25 μm. This includes the visible spectrum, light being a particular portion of the electromagnetic spectrum to which the human eye responds. Solar radiation outside the atmosphere has most of its energy in the range of 0.3 to 3 μm, whereas solar energy received at the ground is substantially in the range of 0.29 to 2.5 μm, as noted in Chapters 1 and 2.

3.2   PHOTON RADIATION

For some purposes in solar energy applications, the classical electromagnetic wave view of radiation does not explain the observed phenomena. In this connection, it is necessary to consider the energy of a particle or photon, which can be thought of as an "energy unit" with zero mass and zero charge. The energy of the photon is given by

$$E = h\nu \qquad\qquad (3.2.1)$$

where h is Planck's constant (6.6256×10^{-34} Js). It follows that as the frequency ν increases (i.e., as the wavelength λ decreases) the photon energy increases. This fact is particularly significant where a minimum photon energy is needed to cause a required change (for example, the creation of a hole-electron pair in a photovoltaic device). There is thus an upper limit of wavelength of radiation that can cause the change.

3.3   THE BLACKBODY, A PERFECT ABSORBER AND EMITTER OF RADIATION

By definition, a blackbody is a perfect absorber of radiation. No matter what wavelengths or directions describe the radiation incident on a blackbody, all incident radiation will be absorbed. A blackbody is an ideal concept since all real opaque substances will reflect some radiation.

Even though a true blackbody does not exist in nature, some materials approach a blackbody. For example, a thick layer of carbon black can absorb approximately 99 % of all incident thermal radiation. This absence of reflected radiation is the reason for the name given to a blackbody. The eye would perceive a blackbody as being black. However, the eye is not a good indicator of the ability of a material to absorb radiation, since the eye is only sensitive to a small portion of the wavelength range of thermal radiation. White paints are good reflectors of visible radiation but most are good absorbers of infrared radiation.

A blackbody is also a perfect emitter of thermal radiation. In fact, the definition of a blackbody could have been put in terms of a body that emits the maximum possible radiation. A simple thought experiment can be used to show that if a body is a perfect emitter of radiation, then it must also be a perfect absorber of radiation. Suppose a small blackbody and a small non-black body are placed in a large evacuated enclosure made from a blackbody material. If the enclosure is isolated from the surroundings, then the blackbody, the real body, and the enclosure will in time come to the same equilibrium temperature. Now, the blackbody must, by definition, absorb all the radiation incident on it, and to maintain a constant temperature the blackbody must also emit an equal amount of energy. The non-black body in the enclosure must absorb less radiation than the blackbody and will consequently emit less radiation than the blackbody. Thus a blackbody both absorbs and emits the maximum amount of radiation.

3.4 PLANCK'S LAW AND WIEN'S DISPLACEMENT LAW

Radiation in the region of the electromagnetic spectrum from approximately 0.2 to approximately 1000 μm is called thermal radiation and is emitted by all substances by virtue of their temperature. The wavelength distribution of radiation emitted by a blackbody is given by Planck's law* (Richtmyer and Kennard, 1947):

$$E_{\lambda b} = \frac{2\pi h C_0^2}{\lambda^5 (e^{hC_0/\lambda kT} - 1)} \qquad (3.4.1)$$

where h is Planck's constant and k is Boltzmann's constant. The groups $2\pi h C_0^2$ and hC_0/k are often called Planck's first and second radiation constants and given the symbols C_1 and C_2, respectively.† Recommended values are, $C_1 = 3.7405 \times 10^{-16}$ W m^2 and $C_2 = 0.0143879$ m K.

---

* The symbol $E_{\lambda b}$ represents energy per unit area per unit time per unit wavelength interval at wavelength λ. The subscript b represents blackbody.

†Sometimes the definition of C_1 does not include the factor 2π.

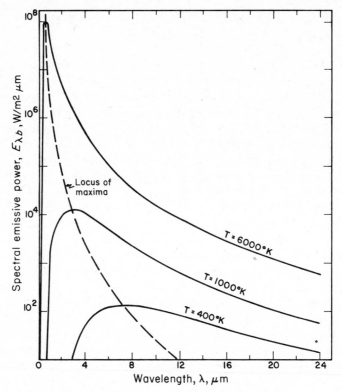

Figure 3.4.1   Spectral distribution of blackbody radiation.

It is also of interest to know the wavelength corresponding to the maximum intensity of blackbody radiation. By differentiating Planck's distribution and equating to zero, the wavelength corresponding to the maximum of the distribution can be derived. This leads to Wien's displacement law, which can be written

$$\lambda_{max} T = 2897.8 \; \mu m \; K \tag{3.4.2}$$

Planck's law and Wien's displacement law are illustrated in Figure 3.4.1, which shows spectral radiation distribution for blackbody radiation from sources at 6000, 1000, and 400 K. The shape of the distribution and the displacement of the wavelength of maximum intensity is clearly shown. Note that 6000 K represents an approximation of the surface temperature of the sun, and that the distribution shown for that temperature is an approximation of the distribution of solar radiation outside the earth's atmosphere. The other two temperatures are representative of those encountered in low- and high-temperature solar-heated surfaces.

The same information shown in Figure 3.4.1 has been replotted on a normalized linear scale in Figure 3.4.2. The ordinate on this figure, which ranges from

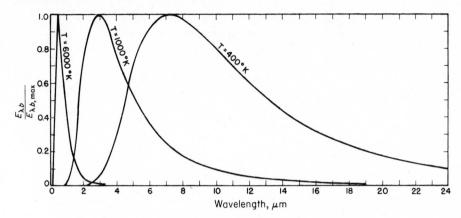

Figure 3.4.2 Normalized spectral distribution of blackbody radiation.

zero to one, is the ratio of the spectral emissive power to the maximum value at the same temperature. This clearly shows the wavelength division between a 6000 K source and lower temperature sources at 1000 and 400 K.

3.5 STEFAN-BOLTZMANN EQUATION

Planck's law gives the spectral distribution of radiation from a blackbody, but in engineering calculations the total energy is often of more interest. By integrating Planck's law over all wavelengths, the total energy emitted by a blackbody is found to be

$$E_b = \int_0^\infty E_{\lambda b}\, d\lambda = \sigma T^4 \tag{3.5.1}$$

where σ is the Stefan-Boltzmann constant and is equal to 5.6697×10^{-8} W/m^2 K^4.

3.6 RADIATION TABLES

Starting with Planck's law (Equation 3.4.1) of the spectral distribution of blackbody radiation, Dunkle (1954) has presented a method for simplifying blackbody calculations. Planck's law can be written as

$$E_{\lambda b} = \frac{C_1}{\lambda^5 (e^{C_2/\lambda T} - 1)} \tag{3.6.1}$$

Table 3.6.1a Fraction of Blackbody Radiant Energy Between Zero and λT for even increments of λT

λT, μm K	$f_{0-\lambda T}$	λT, μm K	$f_{0-\lambda T}$
1000	0.0003	6200	0.7541
1100	0.0009	6300	0.7618
1200	0.0021	6400	0.7692
1300	0.0043	6500	0.7763
1400	0.0077	6600	0.7831
1500	0.0128	6700	0.7897
1600	0.0197	6800	0.7961
1700	0.0285	6900	0.8022
1800	0.0393	7000	0.8080
1900	0.0521	7100	0.8137
2000	0.0667	7200	0.8191
2100	0.0830	7300	0.8244
2200	0.1009	7400	0.8295
2300	0.1200	7500	0.8343
2400	0.1402	7600	0.8390
2500	0.1613	7700	0.8436
2500	0.1831	7800	0.8479
2700	0.2053	7900	0.8521
2800	0.2279	8000	0.8562
2900	0.2506	8100	0.8601
3000	0.2732	8200	0.8639
3100	0.2958	8300	0.8676
3200	0.3181	8400	0.8711
3300	0.3401	8500	0.8745
3400	0.3617	8600	0.8778
3500	0.3829	8700	0.8810
3600	0.4036	8800	0.8841
3700	0.4238	8900	0.8871
3800	0.4434	9000	0.8899
3900	0.4624	9100	0.8927
4000	0.4829	9200	0.8954
4100	0.4987	9300	0.8980
4200	0.5160	9400	0.9005
4300	0.5327	9500	0.9030
4400	0.5488	9600	0.9054
4500	0.5643	9700	0.9076
4600	0.5793	9800	0.9099
4700	0.5937	9900	0.9120
4800	0.6075	10000	0.9141
4900	0.6209	11000	0.9318
5000	0.6337	12000	0.9450
5100	0.6461	13000	0.9550
5200	0.6579	14000	0.9628

Table 3.6.1a (cont.)

λT, μm K	$f_{0-\lambda T}$	λT, μm K	$f_{0-\lambda T}$
5300	0.6693	15000	0.9689
5400	0.6803	16000	0.9737
5500	0.6909	17000	0.9776
5600	0.7010	18000	0.9807
5700	0.7107	19000	0.9833
5800	0.7201	20000	0.9855
5900	0.7291	30000	0.9952
6000	0.7378	40000	0.9978
6100	0.7461	50000	0.9988

Table 3.6.1b   Fraction of Blackbody Radiation Energy Between Zero and λT for Even Fractional Increments

$f_{0-\lambda T}$	λT μm K	λT at energy midpoint
0.05	1880	1660
0.10	2200	2050
0.15	2450	2320
0.20	2680	2560
0.25	2900	2790
0.30	3120	3010
0.35	3350	3230
0.40	3580	3460
0.45	3830	3710
0.50	4110	3970
0.55	4410	4250
0.60	4740	4570
0.65	5130	4930
0.70	5590	5350
0.75	6150	5850
0.80	6860	6480
0.85	7850	7310
0.90	9380	8510
0.95	12500	10600
1.00	∞	16300

Equation 3.6.1 can be integrated to give the radiation between any wavelength limits. The total emitted from zero to any wavelength λ is given by

$$E_{0-\lambda,\,b} = \int_0^{\lambda} E_{\lambda b}\, d\lambda \tag{3.6.2}$$

Substituting Equation 3.6.1 into 3.6.2 and noting that by dividing by σT^4, the integral can be made to be only a function of λT, we have

$$f_{0-\lambda T} = \frac{E_{0-\lambda T}}{\sigma T^4} = \int_0^{\lambda T} \frac{C_1 d(\lambda T)}{\sigma (\lambda T)^5 (e^{C_2/\lambda T} - 1)} \tag{3.6.3}$$

The value of this integral is the fraction of the blackbody energy between zero and λT. Sargent (1972) has calculated values for convenient intervals and the results are given in Table 3.6.1. (Note that, when the upper limit of integration of Equation 3.6.3 is ∞, the value of the integral is unity.)

For use in a digital computer, the following polynomial approximations to Equation 3.6.3 have been given by Pivovonsky and Nagel (1961). For γ greater than or equal to 2

$$f_{0-\lambda T} = \frac{E_{0-\lambda T,\,b}}{\sigma T^4} = \frac{15}{\pi^4} \sum_{m=1,\,2,\,\ldots} \frac{e^{-m\gamma}}{m^4} \{[(m\gamma + 3)m\gamma + 6]m\gamma + 6\} \tag{3.6.4}$$

For γ less than 2

$$f_{0-\lambda T} = \frac{E_{0-\lambda T,\,b}}{\sigma T^4} = 1 - \frac{15}{\pi^4}\gamma^3\left(\frac{1}{3} - \frac{\gamma}{8} + \frac{\gamma^2}{60} - \frac{\gamma^4}{5040} + \frac{\gamma^6}{272{,}160} - \frac{\gamma^8}{13{,}305{,}600}\right)$$

$$\tag{3.6.5}$$

where $\gamma = C_2/\lambda T$.

Example 3.6.1

Assume that the sun is a blackbody at 5762 K. (A) What is the wavelength at which the maximum monochromatic emissive power occurs? (B) What is the energy from this 5762 K source that is in the visible part of the electromagnetic spectrum (0.38 to 0.78 μm)?

Solution

(A) The value of λT at which the maximum monochromatic emissive power occurs is 2897.8 μm K. Therefore, the desired wavelength is 2897.8/5762 or 0.503 μm. (B) From Table 3.6.1 the fraction of energy between zero and $\lambda T = 0.78 \times 5762 = 4494$ μm K is 56% and the fraction of the energy between zero and $\lambda T = 0.38 \times 5762 = 2190$ μm K is 10%. Therefore, the fraction of the energy in the visible is 56% minus 10% or 46%. *Note*: These numbers are close to the values obtained from the actual distribution of energy from the sun as calculated in Example 1.3.1.   ∎

3.7 RADIATION INTENSITY AND FLUX

Thus far, we have considered the radiation leaving a black surface in all directions; however, it is often necessary to describe the directional characteristics of a general radiation field in space. The radiation intensity is used for this purpose and is defined as the energy passing through an imaginary plane per unit area per unit time and per unit solid angle whose central direction is perpendicular to the imaginary plane. Thus, in Figure 3.7.1, if ΔE represents the energy per unit time passing through ΔA and remaining within $\Delta \omega$, then the intensity is*

$$I = \lim_{\substack{\Delta A \\ \Delta \omega} \to 0} \frac{\Delta E}{\Delta A \Delta \omega} \tag{3.7.1}$$

The intensity I has both a magnitude and a direction and can be considered as a vector quantity. It should be pointed out that for a given imaginary plane in space, we can consider two intensity vectors that are in opposite directions. These two vectors are often distinguished by the symbol I^+ and I^-.

The radiation flux is closely related to the intensity and is defined as the energy passing through an imaginary plane per unit area, per unit time, and in all directions on one side of the imaginary plane. Note that the difference between intensity and flux is that the differential area for intensity is perpendicular to the direction of propagation, whereas the differential area for flux lies in a plane that forms the base of a hemisphere through which the radiation is passing.

The intensity can be used to determine the flux through any plane. Consider an elemental area ΔA on an imaginary plane covered by a hemisphere of radius r as shown in Figure 3.7.2. The energy per unit time passing through an area $\Delta A'$ on the surface of the hemisphere from the area ΔA is equal to

$$\Delta Q = I \Delta A \cos \theta \frac{\Delta A'}{r^2} \tag{3.7.2}$$

Where $\Delta A'/r^2$ is the solid angle between ΔA and $\Delta A'$ and $\Delta A \cos \theta$ is the area perpendicular to the intensity vector. The energy flux per unit solid angle in the θ, ϕ direction can then be defined as

$$\Delta q = \lim_{\Delta A \to 0} \frac{\Delta Q}{\Delta A} = I \cos \theta \frac{\Delta A'}{r^2} \tag{3.7.3}$$

The radiation flux is then found by integrating over the hemisphere. The sphere incremental area can be expressed in terms of the angles θ and ϕ so that

$$q = \int_0^{2\pi} \int_0^{\pi/2} I \cos \theta \sin \theta \, d\theta \, d\phi \tag{3.7.4}$$

* The symbol I is used for intensity when presenting basic radiation heat transfer ideas and for solar radiation integrated over an hour period when presenting solar radiation ideas. The two will seldom be used together.

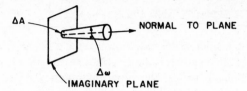

Figure 3.7.1 Schematic of radiation intensity.

It is convenient to define $\mu = \cos\theta$. Then we can write

$$q = \int_0^{2\pi} \int_0^1 I\mu \, d\mu \, d\phi \qquad (3.7.5)$$

Two important points concerning the radiation flux must be remembered. First, the radiation flux is, in general, a function of the orientation of the chosen imaginary plane. Second, the radiation flux will have two values corresponding to each of the two possible directions of the normal to the imaginary plane. When it is necessary to emphasize which of the two possible values of the radiation flux is being considered, the superscript $+$ or $-$ can be used along with a definition of the positive and negative direction.

Thus far, we have defined radiation flux and intensity at a general location in space. When it is desired to find the heat transfer between surfaces in a vacuum, or at least in radiative nonparticipating media, the most useful values of radiative flux and intensity occur at the surfaces. For the special case of a surface that has I independent of direction, the integration of Equation 3.7.5 yields

$$q = \pi I \qquad (3.7.6)$$

Surfaces that have the intensity equal to a constant are called either *Lambertonian* or diffuse surfaces. A blackbody emits in a diffuse manner and therefore the blackbody emissive power is related to the blackbody intensity by

$$E_b = \pi I_b \qquad (3.7.7)$$

The foregoing equations were written for total radiation, but apply equally well to monochromatic radiation. For example, Equation 3.7.7 could be written in terms of a particular wavelength, λ

$$E_{\lambda b} = \pi I_{\lambda b} \qquad (3.7.8)$$

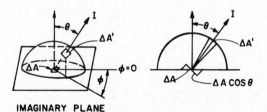

IMAGINARY PLANE

Figure 3.7.2 Schematic of radiation flux.

3.8   INFRARED RADIATION HEAT TRANSFER BETWEEN GRAY SURFACES

The general case of infrared radiation heat transfer between many gray surfaces having different temperatures is treated in a number of textbooks [e.g., Hottel and Sarofim (1967), Siegel and Howell (1972)]. The various methods all make the same basic assumptions which, for each surface, can be summarized as follows

1   The surface is gray. (All radiation properties are independent of wavelength.)
2   The surface is diffuse or specular-diffuse (see Section 4.3).
3   The surface temperature is uniform.
4   The incident energy over the surface is uniform.

Beckman (1971) also utilized these basic assumptions and defined a total exchange factor between pairs of surfaces of an N surface enclosure such that the net heat flux to a typical surface, i, is*

$$Q_i = \sum_{j=1}^{N} \varepsilon_i \varepsilon_j A_i \hat{F}_{ij}(T_j^4 - T_i^4) \tag{3.8.1}$$

The factor $\hat{F}_{ij}$ is the total exchange factor between surfaces i and j and is found from the matrix equation

$$[\hat{F}_{ij}] = [\delta_{ij} - \rho_j E_{ij}]^{-1}[E_{ij}] \tag{3.8.2}$$

where E_{ij}, the specular exchange factor, accounts for radiation going from surface i to surface j directly and by all possible specular (mirrorlike) reflections. Methods for calculating E_{ij} are given in advanced radiation texts. When the surfaces of the enclosure do not specularly reflect radiation, the specular exchange factors of Equation 3.8.2 reduce to the usual view factor (configuration factor).

The majority of heat-transfer problems in solar energy applications involve radiation between two surfaces. The solution of Equations 3.8.1 and 3.8.2 for diffuse surfaces with $N = 2$ is

$$Q_1 = -Q_2 = \frac{\sigma(T_2^4 - T_1^4)}{\dfrac{1 - \varepsilon_1}{\varepsilon_1 A_1} + \dfrac{1}{A_1 F_{12}} + \dfrac{1 - \varepsilon_2}{\varepsilon_2 A_2}} \tag{3.8.3}$$

Two special cases of Equation 3.8.3 are of particular interest. For radiation between two infinite parallel plates (i.e., as in flat-plate collectors) the area A_1

---

* The emittance, ε, is defined by Equation 4.1.8.

and A_2 are equal and the view factor F_{12} is unity. Under these conditions, Equation 3.8.3 becomes

$$\frac{Q}{A} = \frac{\sigma(T_2^4 - T_1^4)}{\dfrac{1}{\varepsilon_1} + \dfrac{1}{\varepsilon_2} - 1} \tag{3.8.4}$$

The second special case is for a small convex object (surface 1) surrounded by a large enclosure (surface 2). Under these conditions, the area ratio A_1/A_2 approaches zero, the view factor F_{12} is unity, and Equation 3.8.3 becomes

$$Q_1 = \varepsilon_1 A_1 \sigma(T_2^4 - T_1^4) \tag{3.8.5}$$

This result is independent of the surface properties of the large enclosure since virtually none of the radiation leaving the small object is reflected back from the large enclosure. In other words, the large enclosure absorbs all radiation from the small object and acts like a blackbody. Equation 3.8.5 also applies in the case of a flat plate radiating to the sky, as is found with a collector cover radiating to the surroundings.

3.9 SKY RADIATION

To predict the performance of solar collectors, it will be necessary to evaluate the radiation exchange between a surface and the sky. The sky can be considered as a blackbody at some equivalent sky temperature so that the actual net radiation between a horizontal flat plate and the sky is given by Equation 3.8.5. The net radiation to a surface with emittance ε and temperature T is thus found from

$$Q = \varepsilon A \sigma(T_{sky}^4 - T^4) \tag{3.9.1}$$

The equivalent blackbody sky temperature of Equation 3.9.1 accounts for the facts that the atmosphere is not at a uniform temperature and that the atmosphere radiates only in certain wavelength bands. The atmosphere is essentially transparent in the wavelength region from 8 to 14 μm, whereas outside the "window" the atmosphere has radiating bands covering much of the infrared spectrum. Several relations have been proposed to relate T_{sky}, for clear skies, to other measured meterological variables. Swinbank (1963) relates sky temperature to the local air temperature in the simple relationship

$$T_{sky} = 0.0552 T_a^{1.5} \tag{3.9.2}$$

where T_{sky} and T_a are both in degrees Kelvin. Brunt (1932) and Bliss (1961) relate the effective sky temperature to water vapor content of the air and/or air temperature.

The results of Bliss can be expressed in terms of the local air temperature and dew point temperature, T_{dp}, as

$$T_{sky} = T_a \left[0.8 + \frac{T_{dp} - 273}{250} \right]^{1/4} \tag{3.9.3}$$

Equations 3.9.2 and 3.9.3 give essentially the same results when the relative humidity is approximately 25%. From Equation 3.9.3 the range of the difference between the sky temperature and air temperature is from 10 C in a hot moist climate to 30 C in a cold dry climate.

It is not at all clear which relationship should be used in the absence of actual measurements. The influences of clouds and of the ground if the collector is tilted and can see the ground have not been included in either expression. It is certain that both the clouds and the ground will tend to increase the effective sky temperature over that for a clear sky. In any event, it is fortunate that it does not make much difference which expression is used in evaluating collector performance (see Section 6.4).

3.10   RADIATION HEAT TRANSFER COEFFICIENT

To retain the simplicity of linear equations it is convenient to define a radiation heat transfer coefficient. The heat transfer by radiation between two arbitrary surfaces is found from Equation 3.8.3. If we define a heat transfer coefficient so that the radiation between the two surfaces is given by

$$Q = A_1 h_r (T_2 - T_1) \tag{3.10.1}$$

then it is clear that

$$h_r = \frac{\sigma (T_2^2 + T_1^2)(T_2 + T_1)}{\dfrac{1 - \varepsilon_1}{\varepsilon_1} + \dfrac{1}{F_{12}} + \dfrac{(1 - \varepsilon_2)A_1}{\varepsilon_2 A_2}} \tag{3.10.2}$$

It is important to remember that if the areas A_1 and A_2 are not equal, then the numerical value of h_r depends on whether it is to be used with A_1 or with A_2.

The numerator of Equation 3.10.2 can be expressed as $4\sigma \overline{T}^3$.

$$4\overline{T}^3 = (T_1^2 + T_2^2)(T_1 + T_2) \tag{3.10.3}$$

When T_1 and T_2 are close together, it is not difficult to estimate $\overline{T}$ without actually knowing both T_1 and T_2. Once $\overline{T}$ is estimated, the equations of radiation heat transfer are reduced to linear equations that can be easily solved along with the linear equations of conduction and convection. If more accuracy is needed, a second or third iteration may be required. Most of the radiation calculation in this book use the linearized radiation coefficient.

3.11 NATURAL CONVECTION BETWEEN PARALLEL FLAT PLATES

The rate of heat transfer between two plates inclined at some angle to the
horizon is of obvious importance in the performance of flat-plate collectors.
Free convection heat transfer data are usually correlated in terms of two or
three dimensionless parameters, the Nusselt number, Nu, the Raleigh number,
Ra, and the Prandtl number, Pr. Some authors correlate data in terms of the
Grashof number which is the ratio of the Raleigh number to the Prandtl
number.

The Nusselt, Raleigh, and Prandtl numbers are given by*

$$\mathrm{Nu} = \frac{hL}{k} \tag{3.11.1}$$

$$\mathrm{Ra} = \frac{g\beta'\Delta T L^3}{\nu\alpha} \tag{3.11.2}$$

$$\mathrm{Pr} = \frac{\nu}{\alpha} \tag{3.11.3}$$

where

h = heat transfer coefficient
L = plate spacing
k = thermal conductivity
g = gravitational constant
β' = volumetric coefficient of expansion (for an ideal gas, $\beta' = 1/T$)
ΔT = temperature difference between plates
ν = kinematic viscosity
α = thermal diffusivity

Note that for parallel plates the Nusselt number is the ratio of a pure conduction resistance to a convection resistance [i.e., $\mathrm{Nu} = (L/k)/(1/h)$] so that a
Nusselt number of unity represents pure conduction.

Tabor (1958) examined the published results of a number of investigations
and concluded that the most reliable data for use in solar collector calculations
as of 1958 was contained in Report 32 published by the U.S. Home Finance
Agency (1954).

In a more recent experimental study using air, Hollands et al. (1976) give the

---

* Fluid properties in the convection relationships of this chapter should be evaluated at the mean
temperature.

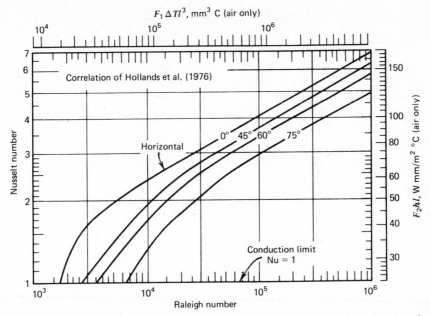

Figure 3.11.1 Nusselt Number as a function of Raleigh Number for free convection heat transfer between parallel flat plates at various slopes.

relationship between the Nusselt number and Raleigh number for tilt angles from 0 to 75° as

$$\text{Nu} = 1 + 1.44\left[1 - \frac{1708}{\text{Ra}\cos\beta}\right]^+ \left(1 - \frac{(\sin 1.8\beta)^{1.6}1708}{\text{Ra}\cos\beta}\right)$$

$$+ \left[\left(\frac{\text{Ra}\cos\beta}{5830}\right)^{1/3} - 1\right]^+ \tag{3.11.4}$$

where the meaning of the $+$ exponent is that only positive values of the terms in the square brackets are to be used (i.e., use zero if the term is negative).

For horizontal surfaces, the results presented by Tabor compare favorably with the correlation of Equation 3.11.4, whereas for vertical surfaces the data from Tabor approximate the 75° tilt data of Hollands et al. Actual collector performance will always differ from analysis but a consistent set of data is necessary to predict the trends to be expected from design changes. Since a common purpose of this type of data is to evaluate collector design changes, the correlation of Hollands et al. is considered to be the most reliable.

Equation 3.11.4 is plotted on Figure 3.11.1. In addition to the Nusselt number, this figure has a second scale on the ordinate giving the value of the heat transfer coefficient times the plate spacing for a mean temperature of 10 C. The scale of this ordinate is not dimensionless but is W mm/m² C. For temperatures other

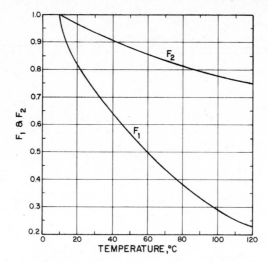

Figure 3.11.2   Air property corrections F_1 and F_2 for use with Figure 3.11.1. From Tabor (1958).

than 10 C, a factor F_2 has been plotted as a function of temperature in Figure 3.11.2. This factor is the ratio of the thermal conductivity of air at 10 C to that at any other temperature. Thus, to find hl^* at any temperature other than 10 C, it is only necessary to divide $F_2 hl$ as read from the chart by F_2 at the appropriate temperature.

   The abscissa also has an extra scale, $F_1 \Delta T l^3$. To find $\Delta T l^3$ at temperatures other than 10 C, it is only necessary to divide $F_1 \Delta T l^3$ by F_1, where F_1 is the ratio of $1/Tv\alpha$ at the desired temperature to $1/Tv\alpha$ at 10 C; F_1 is also plotted on Figure 3.11.2.

Example 3.11.1

Find the heat transfer coefficient between two parallel plates separated by 25 mm with a 45° tilt. The lower plate is at 70 C and the upper plate is at 50 C.

Solution

The mean temperature is 60 C so that air properties at 60 C are $k = 0.029$ W/m K, $T = 333$ K, $v = 1.88 \times 10^{-5}$ m²/s. $\alpha = 2.69 \times 10^{-5}$ m²/s. (Property data can be found in appendix E.) The Raleigh number is then

$$\mathrm{Ra} = \frac{9.81 \times 20 \times (0.025)^3}{333 \times 1.88 \times 10^{-5} \times 2.69 \times 10^{-5}} = 1.82 \times 10^4$$

---

* The lower case letter l is used as a reminder that the units are millimeters instead of meters.

From Equation 3.11.4 or Figure 3.11.1 the Nusselt number is 2.4. The heat transfer coefficient is found from

$$h = \frac{\text{Nu } k}{L} = \frac{2.4 \times 0.029}{0.025} = 2.7 \text{ W/m}^2 \text{ K}$$

As an alternative, the dimensional scales of Figure 3.11.1 can be used along with the property corrections from Figure 3.11.2. At 60 C $F_1 = 0.49$ and $F_2 = 0.86$. Therefore, $F_1 \Delta T l^3 = 0.49 \times 20 \times 25^3 = 1.53 \times 10^5 \text{ K/mm}^3$. From the 45° curve on Figure 3.11.1, $F_2 hl = 59$. Finally, $h = 59/(0.86 \times 25) = 2.7$ W/m² K ∎

The correlation given by Equation 3.11.1 does not cover the range from 75 to 90°. Raithby et al. (1977) have examined vertical surface convection data from a wide range of experimental investigations. They propose a correlation that includes the influence of aspect ratio, that is, the ratio of plate height to spacing. Their correlation is plotted on Figure 3.11.3 for aspect ratios of 5, 60, and ∞. For comparison, other correlations that do not show an aspect ratio effect are also plotted on this figure and correspond approximately to the Raithby et al. correlation with an aspect ratio of between 10 and 20.

It is not clear which correlation should be used. Most of the experiments utilize a guarded hot plate technique that measures the heat transfer only at the center of the test region. Consequently, the end effects are largely excluded. However, Randall et al. (1977) used an interferometric technique that allowed

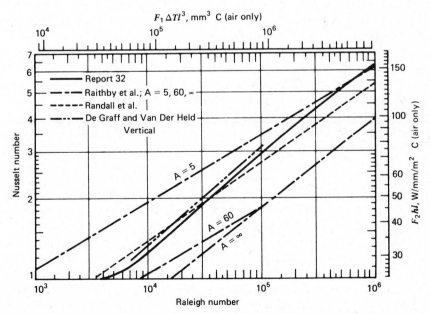

Figure 3.11.3 Nusselt number as a function of Raleigh number for free convection heat transfer between vertical flat plates.

determinations of local heat transfer coefficients from which averages were
determined. They could not find an aspect ratio effect although they covered a
range of aspect ratios from 9 to 36. The Raithby et al. (1977) correlation also
includes an angular correction for angles from 70 to 110°, which shows a slight
increase in Nusselt number as the tilt angle increases from 70° to vertical. Other
experimentors have shown a decrease in Nusselt number over this range of tilt
angles consistant with the trends of Figure 3.11.1 [see Randall et al. (1977)].

The angular dependence of convection heat transfer between 75 and 90° is not
important since it is reasonable to assume that solar collectors will not be tilted
at angles larger than 75° except for vertical collectors, which are common in high
latitudes, and in windows, and collector-storage walls. For vertical surfaces the
four correlations shown on Figure 3.11.3 (with $A \simeq 15$ for the Raithby et al.
result) agree within approximately 15% with the 75° correlation of Hollands
et al. on Figure 3.11.1. Vertical solar collectors will have an aspect ratio on the
order of 60, but at this aspect ratio the Raithby et al. result falls well below other
correlations. Consequently, the 75° correlation of Figure 3.11.1 will give reasonable or conservative predictions for vertical surfaces.

3.12 CONVECTION SUPPRESSION

One of the objectives in designing solar collectors is to reduce the heat loss
through the covers. This has led to studies of convection suppression by Hollands
(1965), Edwards (1969), Buchberg et al. (1976), Arnold et al. (1977, 1978),
Meyer et al. (1978), and others. In these studies the space between two plates,
with one plate heated, is filled with a transparent or specularly reflecting honeycomb to suppress the onset of fluid motion. Without fluid motion the heat
transfer between the plates is by conduction and radiation. Care must be
exercised since improper design can lead to increased rather than decreased
convection losses. This was first shown experimentally by Charters and Peterson
(1972) and later verified by others.

For slats, as shown in Figure 3.12.1, the results of Meyer et al. (1978) can be
expressed as the maximum of two numbers as:

$$Nu = \max[(1.1 C_1 C_2\, Ra_L^{0.28}), 1] \qquad (3.12.1)$$

Where C_1 and C_2 are given in Figure 3.12.2 and the subscript L indicates
that the plate spacing, L, is the characteristic length. Note that the coefficient
C_1 has a maximum near an aspect ratio of 2.

To assess the magnitude of the convection suppression with slats, it is possible
to compare Equation 3.12.1 with the correlation of Randall et al. (1977) obtained
from data taken on the same equipment. Although the Randall correlation uses
an exponent of 0.29 on the Raleigh number, the correlation can be slightly
modified to have an exponent of 0.28. The ratio of the two correlations is then

$$\frac{Nu_{slats}}{Nu_{no\ slats}} = \frac{\max[(1.1 C_1 C_2\, Ra^{0.28}), 1]}{\max[0.13\, Ra^{0.28}[\cos(\beta - 45)]^{0.58}, 1]} \qquad (3.12.2)$$

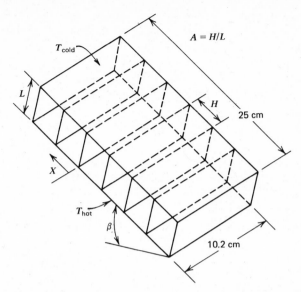

Figure 3.12.1 Slats for suppression of convection. From Meyer et al. (1978).

Note that as long as fluid motion is not suppressed (i.e., Nu greater than one) then the ratio of the two Nusselt numbers is independent of Raleigh number.

At a collector angle of 45°, the addition of slats will reduce convection as long as the aspect ratio is less than approximately 0.5 (i.e., $C_1 = 0.12$, $C_2 = 1.0$). At an aspect ratio of 0.25, the slats reduce convection by one-third. At a Raleigh number of 5800 and a tilt of 45°, fluid movement is just beginning with an aspect ratio of 0.25 and the Nusselt number is 1.0. From Randall's correlation without slats and with a Raleigh number of 5800, the Nusselt number is 1.47; a nearly 50% reduction in convection heat transfer.

Arnold et al. (1977, 1978) experimentally investigated cores with aspect ratios between 0.125 and 0.25 but with the addition of another partition that produced rectangular honeycombs having horizontal aspect ratios (width to plate spacing) ranging from 0.25 to 6.25. The results of these experiments, using silicone oil as the working fluid to suppress thermal radiation, can be correlated within $\pm 15\%$ with the following equation

$$\text{Nu} = 1 + 1.15\left[1 - \frac{\text{Ra}_1}{\text{Ra}\cos\beta}\right]^{+} + 1.25\left[1 - \frac{\text{Ra}_2}{\text{Ra}\cos\beta}\right]^{+} \quad (3.12.3)$$

for

$$0 \le \beta \le 60$$
$$\text{Ra}\cos\beta \le \text{Ra}_3$$
$$4 \le \frac{L}{H} \le 8$$
$$1 \le \frac{W}{H} \le 24$$

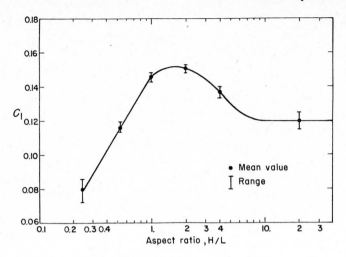

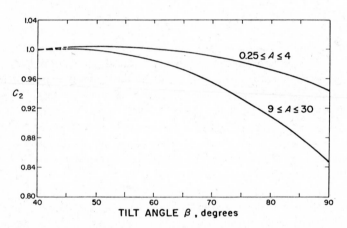

Figure 3.12.2   Coefficients C_1 and C_2 for use in Equation 3.12.1. From Meyer et al. (1978).

where

$$\mathrm{Ra}_k = \frac{(a_k + b_k)^3}{a_k}$$

$$a_k = a_0 + \frac{b_k}{2}$$

$$b_k = (k\pi + 0.85)^2$$

$$a_0 = \left[\frac{7}{1 + L/7D} + 18\left(\frac{H}{W}\right)^2\right]\left(\frac{L}{H}\right)^2$$

For vertical orientation ($\beta = 90$), the results can be correlated by

$$\text{Nu} = 1 + \frac{10^{-4}(H/L)^{4.65}}{1 + 5(H/W)^4}\,\text{Ra}^{1.3} \tag{3.12.4}$$

for the same L/H and W/H limits as given for Equation 3.12.3.

These equations show little effect on heat transfer of horizontal aspect ratios beyond unity. Consequently, the results of Meyer et al. for slats should be directly comparable. At an angle of 45° and a Raleigh number of 4×10^4 both experiments give a Nusselt number of approximately 1.7, but the slope of the data on a Nusselt-Raleigh plot from Arnold et al. is approximately 0.48 and the slope from Meyer et al. is 0.28. Since the Raleigh number range of the two experiments was not large, these two very different correlations give similar Nusselt numbers but extrapolation beyond the range of test data (i.e., Ra > 10^5) could lead to large differences.

This discrepancy points out a problem in estimating the effect of collector design options based on heat transfer data from two different experiments. It is the nature of heat transfer work that sometimes significant differences are observed in carefully controlled experiments using different equipment or techniques. Consequently, in evaluating an option, such as the addition of honeycombs, heat transfer data with and without honeycombs measured in the same laboratory will probably be the most reliable.

The addition of a honeycomb in a solar collector will modify the collectors radiative characteristics. The honeycomb will certainly decrease the solar radiation reaching the absorbing plate of the collector. Hollands et al. (1978) have analyzed the solar transmittance* of a square celled honeycomb and compared the results with measurements at normal incidence. For the particular polycarbonate plastic configuration tested, the honeycomb transmittance at normal incidence was 0.98. Its transmittance decreased in a nearly linear manner with incidence angle to approximately 0.90 at an angle of 70°.

The infrared radiation characteristics will also be affected in a manner largely dependent upon the honeycomb material. If the honeycomb is constructed of either an infrared transparent material or an infrared specularly reflective material then the infrared radiative characteristics of the collector will not be significantly changed. If the honeycomb material is constructed of a material that is opaque in the infrared then the radiative characteristics of the collector will approach that of a blackbody. As will be shown in the next two chapters, this is undesirable.

3.13 VEE-CORRUGATED ENCLOSURES

Vee-corrugated absorber plates with the corrugations running horizontally have been proposed for solar collectors to improve the radiative characteristics

---

* Transmittance of cover systems is discussed in Chapter 5.

Table 3.13.1   Constants for Use in Equation 3.13.1

β	A'	C	n
0	0.75	0.060	0.41
	1	0.060	0.41
	2	0.043	0.41
45	0.75	0.075	0.36
	1	0.082	0.36
	2	0.037	0.41
60	0.75	0.162	0.30
	1	0.141	0.30
	2	0.027	0.42

of the absorber plate (see Section 4.6). Also, this configuration approximates the shape of some concentrating collectors (see Chapter 8). One problem with this configuration is that the improved radiative properties are, at least in part, offset by increased convection losses. Elsherbiny et al. (1977) state that free convection losses from a vee-corrugated surface to a single plane above is as much as 50% greater than for two plane surfaces at the same mean spacing.

Randall (1978) investigated vee-corrugated surfaces and correlated the data in terms of the Nusselt and Raleigh numbers in the form

$$\mathrm{Nu} = \max[(C\,\mathrm{Ra}^n),\, 1] \qquad (3.13.1)$$

where the values of C and n are given in Table 3.13.1 as functions of the tilt angle and the vee aspect ratio, A' (ratio of mean plate spacing to vee height).

3.14   HEAT TRANSFER RELATIONS FOR INTERNAL FLOW

Heat transfer coefficients for common geometries are given in many heat transfer textbooks [e.g., McAdams (1954), Kays (1966), and Kreith (1973)]. For fully developed turbulent flow inside tubes (i.e., $\mathrm{Re} = \rho V D_h/\mu > 2200$) Webb (1971) suggests that the Petukhov and Popov equation be used

$$\mathrm{Nu} = \frac{(f/8)\mathrm{Re}\,\mathrm{Pr}}{1.07 + 12.7\sqrt{f/8}(\mathrm{Pr}^{2/3} - 1)} \qquad (3.14.1)$$

where the friction factor, f, for smooth pipes is given by

$$f = [0.79 \ln \mathrm{Re} - 1.64]^{-2} \qquad (3.14.2)$$

For noncircular tubes the hydraulic diameter can be used for the characteristic length in the above equation. The hydraulic diameter is defined as

$$D_h = \frac{4(\text{flow area})}{\text{wetted perimeter}} \qquad (3.14.3)$$

For short tubes with L/D greater than 1.0 and a sharp-edged entry, McAdams recommends that the Nusselt number be calculated from

$$\text{Nu}_{\text{short}} = \text{Nu}_{\text{long}}\left[1 + \left(\frac{D}{L}\right)^{0.7}\right] \qquad (3.14.4)$$

For laminar flow in tubes the thermal boundary condition is important. With fully developed hydrodynamic and thermal profiles, the Nusselt number is 3.7 for constant wall temperature and 4.4 for constant heat flux. In a solar collector the thermal condition is closely represented by a constant resistance between the flowing fluid and the constant temperature environment.* If this resistance is large, the thermal boundary condition approaches constant heat flux and if this resistance is small, the thermal boundary condition approaches constant temperature. Consequently, the theoretical performance of a solar collector should lie between the results for constant heat flux and constant temperature. Since a constant wall temperature assumption yields somewhat lower heat transfer coefficients, this is the recommended assumption for conservative design.

For short tubes the developing thermal and hydrodynamic boundary layers will result in a significant increase in the heat transfer coefficient near the entrance. Heaton et al. (1964) present local Nusselt numbers for the case of constant heat rate. Their data is well represented by an equation of the form

$$\text{Nu} = \text{Nu}_\infty + \frac{a(\text{Re Pr } D_h/L)^m}{1 + b(\text{Re Pr } D_h/L)^n} \qquad (3.14.5)$$

where the constants a, b, m, and n are given in Table 3.14.1.

Goldberg (1958), as reported by Rohsenow and Choi (1961), presents average Nusselt numbers for the case of constant wall temperature. The results for

Table 3.14.1 Constants for Equation 3.14.5 for Circular Tubes With Constant Heat Rate. Local Nusselt Numbers

Prandtl Number	a	b	m	n
0.7	0.00398	0.0114	1.66	1.12
10	0.00236	0.00857	1.66	1.13
∞	0.00172	0.00281	1.66	1.29
	$\text{Nu}_\infty = 4.4$			

* This will become apparent in Chapter 6.

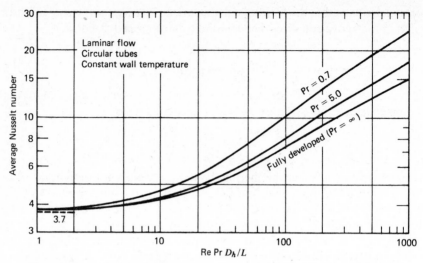

Figure 3.14.1   Average Nusselt numbers in short tubes for various Prandtl numbers.

Prandtl numbers of 0.7, 5, and ∞ are shown in Figure 3.14.1. The data of this figure can also be represented by an equation of the form of Equation 3.14.5 but with values of a, b, m, n and Nu_∞ given in Table 3.14.2.

Example 3.14.1

What is the heat transfer coefficient inside the tubes of a solar collector in which the tubes are 10 mm in diameter and separated by a distance 100 mm. The collector is 1.5 m wide of 3 m long and has total flow rate of water of 0.075 kg/s. The water is at 80 C.

Solution

The collector has 15 tubes so that the flow rate per tube is 0.005 kg/s. The Reynolds number is

$$\frac{VD}{v} = \frac{4\dot{m}}{\pi D \mu} = \frac{4 \times 0.005}{\pi \times 0.01 \times 3.6 \times 10^{-4}} = 1800$$

which indicates laminar flow. The Prandtl number is 2.2 so that

$$Re\,Pr\,\frac{D}{L} = 1800 \times 2.2 \times \frac{0.01}{3} = 13$$

From Figure 3.14.1 the average Nusselt number is 4.6 so that the average heat transfer coefficient is

$$h = Nu\,\frac{k}{D} = 4.6 \times \frac{0.67}{0.01} = 310 \ \text{W/m}^2\text{C} \qquad \blacksquare$$

Table 3.14.2   Constants for Equation 3.14.5 for Circular Tubes with Constant Wall Temperature. Average Nusselt Numbers

Prandtl Number	a	b	m	n
0.7	0.0791	0.0331	1.15	0.82
5	0.0534	0.0335	1.15	0.82
∞	0.0461	0.0316	1.15	0.84
	$Nu_\infty = 3.7$			

In the study of solar air heaters and collector-storage walls it is necessary to know the forced convection heat transfer coefficient between two flat plates. For air, the following correlation can be derived from Kay's (1966) data for fully developed turbulent flow with one side heated and the other side insulated

$$Nu = 0.0158 Re^{0.8} \tag{3.14.6}$$

where the characteristic length is the hydraulic diameter (twice the plate spacing). For flow situations in which L/D_h is 10, Kays indicates that the average Nusselt number is approximately 16% higher than that given by Equation 3.14.6. At L/D_h equal to 30 Equation 3.14.6 still underpredicts by 5%. At L/D_h equal to 100, the effect of the entrance region has largely disappeared.

Tan and Charters (1970) have experimentally studied flow of air between parallel plates with small aspect ratios for use in solar air heaters. Their results give higher heat transfer coefficients by about 10% than those given by Kays with an infinite aspect ratio.

The local Nusselt number for laminar flow between two flat plates with one side insulated and the other subjected to a constant heat flux has been obtained by Heaton et al. (1964). The results have been correlated into the form of Equation 3.14.5 with the constants given in Table 3.14.3.

For the case of parallel plates with constant temperature on one side and insulated on the other side, Mercer et al. (1967) obtained the average Nusselt

Table 3.14.3   Constants for Equation 3.14.5 for Infinite Flat Plates, One Side Insulated and constant Heat Flux on Other Side. Local Nusselt Numbers

Prandtl Number	a	b	m	n
0.7	0.00190	0.00563	1.71	1.17
10	0.00041	0.00156	2.12	1.59
∞	0.00021	0.00060	2.24	1.77
	$Nu_\infty = 5.4$			

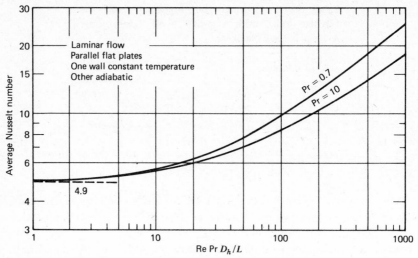

Figure 3.14.2   Average Nusselt numbers in short ducts with one side insulated and one side at constant wall temperature for various Prandtl numbers.

numbers shown in Figure 3.14.2. They also correlated this data into the form of Equation 3.14.7 for $0.1 < \mathrm{Pr} < 10$.

$$\mathrm{Nu} = 4.9 + \frac{0.0606(\mathrm{Re}\ \mathrm{Pr}\ D_h/L)^{1.2}}{1 + 0.0909(\mathrm{Re}\ \mathrm{Pr}\ D_h/L)^{0.7}\ \mathrm{Pr}^{0.17}} \qquad (3.14.7)$$

The results of Sparrow (1955) indicate that for $\mathrm{Re}\ \mathrm{Pr}\ D_h/L < 1000$, the $\mathrm{Pr} = 10$ Nusselt numbers are essentially the same as for the case when the hydrodynamic profile is fully developed (i.e., $\mathrm{Pr} = \infty$).

Example 3.14.2

(A) Determine the convective heat transfer coefficient for air flow in a 1 m wide by 2 m long channel. The channel thickness is 15 mm and the air flow rate is 0.03 kg/s. The average air temperature is 35 C. (B) If the channel thickness is halved, what is the heat transfer coefficient? (C) If the flow rate is halved, what is the heat transfer coefficient?

Solution

(A) At a temperature of 35 C the viscosity is 1.88×10^{-5} m^2/s and the thermal conductivity is 0.0268 W/m C. The hydraulic diameter, D_h, is twice the plate spacing, t, and the Reynolds number can be expressed in terms of the flow rate per unit width, $\dot{m}/W$. The Reynolds number is then

$$\mathrm{Re} = \frac{\rho V D_h}{\mu} = \frac{2\rho V t W}{W\mu} = \frac{2\dot{m}}{W\mu} = \frac{2 \times 0.03}{1 \times 1.88 \times 10^{-5}} = 3200$$

so that the flow is turbulent. From Equation 3.14.6 the Nusselt number is

$$\mathrm{Nu} = 0.0158(3200)^{0.8} = 10.1$$

and the heat transfer coefficient is $h = \mathrm{Nu}\ k/D_h = 9$ W/m^2 C. Since D_h/L is less than 100, 9 W/m^2 C is probably a few percent too low.

(B) If the channel thickness is halved, the Reynolds number remains the same but the heat transfer coefficient will double to 18 W/m^2 C.

(C) If the flow rate is halved, the Reynolds number will be 1600, which indicates laminar flow, and Equation 3.14.7 or Figure 13.14.2 should be used. The value of Re Pr D_h/L is $1600 \times 0.7 \times 0.03/2 = 16.8$ so that the Nusselt number is 6.0 and the heat transfer coefficient is the 6.2 W/m^2 C. ∎

3.15   HEAT TRANSFER DUE TO WIND

The heat loss from flat plates exposed to outside winds is important in the study of solar collectors. Sparrow et al. (1979) did wind tunnel studies on rectangular plates at various orientations and found the following correlation over the Reynolds number range of 2×10^4 to 9×10^4

$$\mathrm{Nu} = 0.86\ \mathrm{Re}^{1/2}\ \mathrm{Pr}^{1/3} \tag{3.15.1}$$

where the characteristic length is four times the plate area divided by the plate perimeter. For laminar flow (i.e., Reynolds numbers less than 10^6) over a very wide flat plate at zero angle of attack, the analysis of Pohlhausen [see Kays (1966)] yields a coefficient for Equation 3.15.1 of 0.94.*

This agreement at low Reynolds numbers suggests that Equation 3.15.1 may be valid at Reynolds numbers up to 10^6 where direct experimental evidence is lacking. This extrapolation is necessary since a 2 m by 5 m solar collector array has a characteristic length of 2.9 m and a Reynolds number of 9.4×10^5 in a 5 m/s wind. From Equation 3.15.1 the heat transfer coefficient under these conditions is approximately 7 W/m^2 C.

McAdams (1954) reports the data of Jurges for a 0.5 m^2 plate in which the convection coefficient is given by the dimensional equation

$$h = 5.7 + 3.8V \tag{3.15.2}$$

where V is the wind speed in m/s and h is in W/m^2 C. It is probable that in this equation the effects of free convection and radiation are included. For this reason Watmuff et al. (1977) report that this equation should be

$$h = 2.8 + 3.0V \tag{3.15.3}$$

For a 0.5 m^2 plate, Equation 3.15.3 yields a heat transfer coefficient of 16 W/m^2 C at a 5 m/s velocity and a temperature of 25 C. Equation 3.15.1 yields

---

* To be consistent with Equation 3.15.1, the characteristic length in the Pohlhausen solution must be changed to twice the plate length, which changes the familar coefficient of 0.664 to 0.94.

a value of 18 W/m^2 C at these conditions. Thus, there is agreement between the two at a characteristic length of 0.5 m. It is not reasonable to assume that Equation 3.15.3 is valid at other plate lengths.

The flow over a collector mounted on a house is not always well represented by wind tunnel tests of isolated plates. The collectors will sometimes be exposed directly to the wind and other times will be in the wake region. The roof itself will certainly influence the flow patterns. Also, nearby trees and buildings will greatly effect local flow conditions. Mitchell (1976) investigated the heat transfer from various shapes (actually animal shapes) and showed that many shapes were well represented by a sphere when the equivalent sphere diameter is the cube root of the volume. The heat transfer obtained in this manner is an average that includes stagnation regions and wake regions. A similar situation might be anticipated to occur in solar systems. Mitchell suggests that the wind tunnel results of these animal tests should be increased by approximately 25% for outdoor conditions. Thus, assuming a house to be a sphere, the heat transfer can be expressed as

$$\text{Nu} = 0.42\text{Re}^{0.6} \qquad (3.15.4)$$

where the characteristic length is the cube root of the house volume.

When the wind speed is very low, free convection conditions may dominate. Free convection data for hot inclined flat plates facing upwards are not available. However, results are available for horizontal and vertical flat plates. For hot horizontal flat plates with aspect ratios up to 7 : 1, Lloyd and Moran (1974) give the following equations

$$\text{Nu} = 0.76\text{Ra}^{1/4}; 2.6 \times 10^4 < \text{Ra} < 10^7 \qquad (3.15.5)$$

$$\text{Nu} = 0.15\text{Ra}^{1/3}; 10^7 < \text{Ra} < 3 \times 10^{10} \qquad (3.15.6)$$

where the characteristic length is four times the area divided by the perimeter (the original reference used A/P). For vertical plates, McAdams (1954) gives

$$\text{Nu} = 0.59\,\text{Ra}^{1/4}; 10^4 < \text{Ra} < 10^9 \qquad (3.15.7)$$

$$\text{Nu} = 0.13\,\text{Ra}^{1/3}; 10^9 < \text{Ra} < 10^{12} \qquad (3.15.8)$$

where the characteristic length is the plate height.

For large Rayleigh numbers, as would be found in most solar collector systems, Equations 3.15.6 and 3.15.8 apply and the characteristic length drops out of the calculation of the heat transfer coefficient. The heat transfer coefficients from these two equations are nearly the same since the coefficient on the Rayleigh numbers differ only slightly. This means that horizontal and vertical collectors have a minimum heat transfer coefficient (i.e., under free convection conditions) of about 5 W/m^2 C for a 25 C temperature difference and a value of about 4 W/m^2 C at a temperature difference of 10 C.

From the preceding discussion it is apparent that the calculation of wind-induced heat transfer coefficients is not well established. Until additional experimental evidence becomes available, the following guidelines are recommended.

When free and forced convection occur simultaneously, McAdams (1954) recommends that both values be calculated and the larger value used in calculations. Consequently, it appears that a minimum value of approximately 5 W/m^2 C occurs in solar collectors under still air conditions. For forced convection conditions over buildings the results of Mitchell can be expressed as

$$h = \frac{8.6V^{0.6}}{L^{0.4}} \qquad (3.15.9)$$

The heat transfer coefficient (in W/m^2 C) for roof mounted collectors can then be expressed as

$$h_{wind} = \max\left[5, \frac{8.6V^{0.6}}{L^{0.4}}\right] \qquad (3.15.10)$$

where V is in meters per second and L is the cube root of the house volume in meters. At a wind speed of 5 m/s (which is close to the world average wind speed) and a characteristic length of 8 m, Equation 3.15.10 yields a heat transfer coefficient of 10 W/m^2 C.

For flow of air across a single tube, McAdams (1954) recommends

$$Nu = 0.32 + 0.43(Re)^{0.52} \qquad (3.15.11)$$

for $0.1 < Re < 1000$ and

$$Nu = 0.24(Re)^{0.6} \qquad (3.15.12)$$

for $1000 < Re < 50{,}000$. The values calculated from these two equations should be increased by approximately 25% for outdoor conditions.

3.16   HEAT TRANSFER AND PRESSURE DROP IN PACKED BEDS

In solar air heating systems the usual energy storage media is a packed bed of small rocks or crushed gravel. The heat transfer and pressure drop characteristics of these storage devices are of considerable interest and have been extensively reviewed by Shewen et al. (1978). Although many correlations were found for both heat transfer coefficients and friction factors in packed beds, none of the correlations were entirely satisfactory in predicting the measured performance of their experimental packed bed. The following relationships are based on the recommendations of Shewen et al.

The physical characteristic of pebbles vary widely between samples. Three quantities have been used to describe pebbles, the average particle diameter D, the void fraction ε, and the surface area shape factor α. The void fraction can be determined by weighing pebbles placed in a container of volume V before and after it is filled with water. The void fraction is then equal to

$$\varepsilon = \frac{m_w/\rho_w}{V} \qquad (3.16.1)$$

Where m_w is the mass of water and ρ_w is the density of water. The density of the rock material is then

$$\rho_r = \frac{m}{V(1 - \varepsilon)} \tag{3.16.2}$$

where m is the mass of the rocks alone. The average particle diameter is the diameter of a spherical particle having the same volume and can be calculated from

$$D = \left(\frac{6m}{\pi \rho_r N}\right)^{1/3} \tag{3.16.3}$$

where N is the number of pebbles in the sample. The surface area shape factor, α, is the ratio of the surface area of the pebble to the surface area of the equivalent sphere and is difficult to evaluate. For smooth river gravel, α appears to be independent of pebble size and approximately equal to 1.5. For crushed gravel α varies with the pebble size and decreases linearly from approximately 2.5 at very small sizes to approximately 1.5 at 50-mm-diameter particles. However, large scatter is observed.

The three pebble bed parameters D, ε, and α do not fully take into account all the observed behavior of packed bed storage devices. However, exact predictions are not needed since the performance of a solar system is not a strong function of the storage unit design as long as certain criteria are met.* When measurements of the void fraction, ε, and the surface area shape factor, α, are available the pressure drop relationship recommended by Shewen et al. is due to McCorquodale et al. (1978)

$$\Delta P = \frac{LG_0^2}{\rho_{air} D} \frac{(1 - \varepsilon)\alpha}{\varepsilon^{3/2}} \left[4.74 + 166 \frac{(1 - \varepsilon)\alpha}{\varepsilon^{3/2}} \frac{\mu}{G_0 D}\right] \tag{3.16.4}$$

where G_0 is the mass velocity of the air (air mass flow rate divided by the bed frontal area). When measurements of α and ε are not available, Shewen et al. recommend the equation of Dunkle and Ellul (1972)

$$\Delta P = \frac{LG_0^2}{\rho_{air} D} \left(21 + 1750 \frac{\mu}{G_0 D}\right) \tag{3.16.5}$$

For heat transfer Shewen et al. recommend the Löf and Hawley (1948) equation

$$h_v = 650 \left(\frac{G_0}{D}\right)^{0.7} \tag{3.16.6}$$

where h_v is the volumetric heat transfer coefficient in W/m³ C, G_0 is the mass velocity in kg/s m² and D is the particle diameter in meters. The relationship

* See Table 13.2.1 and Section 9.4.

between volumetric heat transfer coefficient, h_v, and area heat transfer coefficient, h, is

$$h_v = 6h(1 - \varepsilon) \frac{\alpha}{D} \tag{3.16.7}$$

REFERENCES

Arnold, J. N., D. K. Edwards, and P. S. Wu, ASME Paper No. 78-WA/HT-5 (1978), "Effect of Cell Size on Natural Convection in High L/D Tilted Rectangular Cells Heated and Cooled on Opposite Faces."

Arnold, J. N., D. K. Edwards, and I. Catton, *Transactions of the American Society of Mechanical Engineers, Journal of Heat Transfer*, **99**, 120 (1977). "Effect of Tilt and Horizontal Aspect Ratio on Natural Convection in a Rectangular Honeycomb."

Beckman, W. A., *Solar Energy*, **13**, 3 (1971). "The Solution of Heat Transfer Problems on a Digital Computer."

Bliss, R. W., *Solar Energy*, **5**, 103 (1961). "Atmospheric Radiation Near the Surface of the Ground."

Brunt, D., *Quarterly Journal of the Royal Meteorological Society*, **58**, 389–420 (1932). "Notes on Radiation in the Atmosphere."

Buchberg, H., I. Catton, and D. K. Edwards, *Transactions of the American Society of Mechanical Engineers, Journal of Heat Transfer*, **98**, 2, 182 (1976). "Natural Convection in Enclosed Spaces: A Review of Application to Solar Energy Collection."

Charters, W. W. S. and L. J. Peterson, *Solar Energy*, **13**, 4 (1972). "Free Convection Suppression Using Honeycomb Cellular Materials."

DeGraff, J. and E. Van Der Held, *Applied Science Research*, Sec. A, **3** (1952). "The Relation Between the Heat Transfer and the Convection Phenomena in Enclosed Plane Air layers."

Dunkle, R. V., *Transactions of the American Society of Mechanical Engineers*, **76** (1954). "Thermal Radiation Tables and Applications."

Dunkle, R. V. and W. H. J. Ellul, *Transactions of the Institution of Engineers (Australia)*, MC8, 117 (1972). "Randomly Packed Particulate Bed Regenerators and Evaporative Coolers."

Edwards, D. K., *Transactions of the American Society of Mechanical Engineers, Journal of Heat Transfer*, **91**, 145 (1969). "Suppression of Cellular Convection by Lateral Walls."

Elsherbiny, S. M., K. G. T. Hollands, and G. D. Raithby, in *Heat Transfer in Solar Energy Systems*, J. R. Howell and T. Min (eds.), American Society of Mechanical Engineers, New York (1977). "Free Convection Across Inclined Air Layers With One Surface V-Corrugated."

Heaton, H. S., W. C. Reynolds, and W. M. Kays, *International Journal of Heat and Mass Transfer*, **7**, 763–781 (1964). "Heat Transfer in Annular Passages. Simultaneous Development of Velocity and Temperature Fields in Laminar Flow."

Hollands, K. G. T., *Solar Energy*, **9**, 159 (1965). "Honeycomb Devices in Flat-Plate Solar Collectors."

Hollands, K. G. T., T. E. Unny, G. D. Raithby, and L. Konicek, *Transactions of the American Society of Mechanical Engineers, Journal of Heat Transfer*, **98**, 189–193 (1976). "Free Convection Heat Transfer Across Inclined Air Layers."

Hollands, K. G. T., K. N. Marshall, and R. K. Wedel, *Solar Energy*, **21**, 231 (1978). "An Approximate Equation For Predicting The Solar Transmittance of Transparent Honeycombs."

Home Finance Agency Report #32, U.S. Government Printing Office (1954). "The Thermal Insulating Value of Airspaces."

Hottel, H. C. and A. F. Sarofim, *Radiative Transfer*, McGraw-Hill, New York (1967).

Kays, W. M., *Convective Heat and Mass Transfer*, McGraw-Hill, New York (1966).

Kreith, F., *Principles of Heat Transfer*, 3rd ed., International Textbook Co.. Scranton, PA, (1973).

Lloyd, J. R., and W. P. Moran, *Transactions of the American Society of Mechanical Engineers, Journal of Heat Transfer*, **96**, 443 (1974). "Natural Convection Adjacent to Horizontal Surface of Various Planforms."

Lof, G. O. G., and Hawley, R. W., *Industrial and Engineering Chemistry*, **40**, 1061 (1948). "Unsteady State Heat Transfer Between Air and Loose Solids."

McAdams. W. H., *Heat Transmission*, 3rd ed., McGraw-Hill, New York (1954).

McCorquodale et al., *Journal of Hydraulic Research*, **16** (2), pt. 2 (1978). "Hydraulic Conductivity of Rock Fill."

Mercer, W. E., W. M. Pearce, and J. E. Hitchcock, *Journal of Heat Transfer*, **89**, 251–257 (1967). "Laminar Forced Convection in the Entrance Region Between Parallel Flat Plates."

Meyer, B. A., M. M. El-Wakil, and J. W. Mitchell, *Thermal Storage and Heat Transfer in Solar Energy Systems*, F. Kreith, R. Boehm, J. Mitchell, and R. Bannerot (eds.), American Society of Mechanical Engineers, New York (1978), "Natural Convection Heat Transfer in Small and Moderate Aspect Ratio Enclosures—An Application to Flat-Plate Collectors."

Mitchell, J. W., *Biophysical Journal*, **16**, 561 (1976). "Heat Transfer From Spheres and Other Animal Forms."

Pivovonsky, M. and M. R. Nagel, *Tables of Blackbody Radiation Properties*, Macmillan, New York (1961).

Randall, K. R., J. W. Mitchell, and M. M. El-Wakil, *Heat Transfer in Solar Energy Systems*, J. R. Howell and T. Min (eds.), American Society of Mechanical Engineers, New York (1977). "Natural Convection Characteristics of Flat-Plate Collectors."

Randall, K. R., Ph.D. Thesis, University of Wisconsin (1978). "An Interferometric Study of Natural Convection Heat Transfer in Flat-Plate and Vee-Corrugated Enclosures."

Raithby, G. D., K. G. T. Hollands, and T. R. Unny, *Transactions of the American Society of Mechanical Engineers, Journal of Heat Transfer*, **99**, 287 (1977). "Analysis of Heat Transfer by Natural Convection Across Vertical Fluid Layers."

Richtmyer, F. K. and E. H. Kennard, *Introduction to Modern Physics*, 4th ed., McGraw-Hill, New York (1947).

Rohsenow, W. M., and H. Choi, *Heat Mass and Momentum Transfer*, Prentice-Hall, Englewood Cliffs, N.J. (1961).

Shewen, E. C., H. F. Sullivan, K. G. T. Hollands, and A. R. Balakrishnan, Report STOR-6, Waterloo Research Institute, University of Waterloo (August, 1978), "A Heat Storage Subsystem for Solar Energy."

Sargent, S. L., *Bulletin of the American Meteorological Society*, **53**, 360 (April 1972). "A Compact Table of Blackbody Radiation Functions."

Siegel, R. and J. R. Howell, *Thermal Radiation Heat Transfer*, McGraw-Hill, New York (1972).

Sparrow, E. M., NACA TN 3331 (1955). "Analysis of Laminar Forced-Convection Heat Transfer in Entrance Regions of Flat Rectangular Ducts."

Sparrow, E. M., J. W. Ramsey, and E. A. Mass, *Transactions of the American Society of Mechanical Engineers, Journal of Heat Transfer*, **101**, 2 (1979). "Effect of Finite Width on Heat Transfer and Fluid Flow About An Inclined Rectangular Plate."

Swinbank, W. C., *Quarterly Journal of the Royal Meteorological Society*, **89** (1963). "Long-Wave Radiation From Clear Skies."

Tabor, H., *Bulletin Research Council of Israel*, **6C**, 155 (1958). "Radiation Convection and Conduction Coefficients in Solar Collectors."

Tan, H. M. and W. W. S. Charters, *Solar Energy*, **13**, 121 (1970). "Experimental Investigation of Forced-Convection Heat Transfer."

Watmuff, J. H., W. W. S. Charters, and D. Proctor, *Comples*, **2** (1977). "Solar and Wind Induced External Coefficients for Solar Collectors."

Webb, R. L., *Warme-und Stoffubertragung*, **4**, 197 (1971), "A Critical Evaluation of Analytical Solutions and Reynolds Analogy Equations for Turbulent Heat and Mass Transfer in Smooth Tubes."

CHAPTER 4

Radiation Characteristics of Opaque Materials

This chapter begins with a detailed discussion of a large number of radiation characteristics of surfaces. For many solar energy calculations only two of these quantities are required, often referred to as solar absorptance and long-wave or infrared emittance, or just absorptance and emittance. Although values for these two quantities are often quoted, other radiation properties may be the only available information on a particular material. Since relationships exist between the various characteristics, it may be possible to calculate a desired quantity from available data. Consequently, it is necessary to understand exactly what is meant by the radiation terms found in the literature, to be familiar with the type of information available, and to know how to manipulate these data. The most common type of data manipulation is illustrated in the examples and readers may wish to go directly to Example 4.4.1.

The names used for the radiation surface characteristics were chosen as the most descriptive of the many names found in the literature. In many cases, the names will seem to be cumbersome, but they are necessary to distinguish one characteristic from another. For example, both a monochromatic angular-hemispherical reflectance and a monochromatic hemispherical-angular reflectance will be defined. Under certain circumstances, these two quantities are identical, but in general they are different, and it is necessary to distinguish between them.

Both the name and the symbol should be an aid for understanding the significance of the particular characteristic. The monochromatic directional absorptance, $\alpha_\lambda(\mu, \phi)$, is the fraction of the incident energy from the direction μ, ϕ at the wavelength λ that is absorbed.* The directional absorbance $\alpha(\mu, \phi)$, includes all wavelengths and the hemispherical absorptance, α, includes all directions as well all wavelengths. We will also have a monochromatic hemispherical absorptance, α_λ, which is the fraction of the energy incident from all directions at a particular wavelength that is absorbed. Thus by careful study of the name the definition should be clear.

---

* The angles θ and ϕ are shown in Figure 3.7.2. μ is equal to $\cos \theta$.

4.1 ABSORPTANCE AND EMITTANCE

The monochromatic directional absorptance is a property of a surface and is defined as the fraction of the incident radiation of wavelength λ from the direction μ, ϕ (where μ is the cosine of the polar angle and ϕ is the azimuthal angle) that is absorbed by the surface. In equation form

$$\alpha_\lambda(\mu, \phi) = \frac{I_{\lambda,\,a}(\mu, \phi)}{I_{\lambda,\,i}(\mu, \phi)} \tag{4.1.1}$$

where the subscripts a and i represent absorbed and incident.

The fraction of all the radiation (i.e., over all wavelengths) from the direction μ, ϕ that is absorbed by a surface is called the directional absorptance and is defined by the following equation:

$$\alpha(\mu, \phi) = \frac{\int_0^\infty \alpha_\lambda(\mu, \phi)I_{\lambda,\,i}(\mu, \phi)\, d\lambda}{\int_0^\infty I_{\lambda,\,i}(\mu, \phi)\, d\lambda} = \frac{1}{I_i(\mu, \phi)} \int_0^\infty \alpha_\lambda(\mu, \phi)I_{\lambda,\,i}(\mu, \phi)\, d\lambda \tag{4.1.2}$$

Unlike the monochromatic directional absorptance, the directional absorptance is not a property but a function of the wavelength distribution of the incident radiation.*

The monochromatic directional emittance of a surface is defined as the ratio of the monochromatic intensity emitted by a surface in a particular direction to the monochromatic intensity that would be emitted by a blackbody at the same temperature.

$$\varepsilon_\lambda(\mu, \phi) = \frac{I_\lambda(\mu, \phi)}{I_{\lambda b}} \tag{4.1.3}$$

The monochromatic directional emittance is a property of a surface, as is the directional emittance, defined by†

$$\varepsilon(\mu, \phi) = \frac{\int_0^\infty \varepsilon_\lambda(\mu, \phi)I_{\lambda b}\, d\lambda}{\int_0^\infty I_{\lambda b}\, d\lambda} = \frac{1}{I_b} \int_0^\infty \varepsilon_\lambda(\mu, \phi)I_{\lambda b}\, d\lambda \tag{4.1.4}$$

In words, the directional emittance is defined as the ratio of the emitted total intensity in the direction μ, ϕ to the blackbody intensity. Note that $\varepsilon(\mu, \phi)$ is a property as its definition contains the intensity $I_{b\lambda}$, which is specified when the surface temperature is known. In contrast, the definition of $\alpha(\mu, \phi)$ contains the unspecified function $I_{\lambda,\,i}(\mu, \phi)$ and is therefore not a property. An important point to remember is that these four quantities and the four to follow are all functions of surface conditions such as temperature, roughness, cleanliness, etc.

---

* Although $\alpha(\mu, \phi)$ and some other absorptances are not properties, in that they depend upon the wavelength distribution of the incoming radiation, we can consider them as properties if the incoming spectral distribution is known. As the spectral distribution of solar radiation is essentially fixed, we can consider solar absorptance as a property.

† Both the numerator and denominator could be multiplied by π so that the definition of $\varepsilon(\mu, \phi)$ could have been in terms of e_b and $e_{\lambda b}$.

From the definitions of the directional absorptance and emittance of a surface, the corresponding hemispherical properties can be defined. The monochromatic hemispherical absorptance and emittance are obtained by integrating over the enclosing hemisphere as was done in Section 3.7.

$$\alpha_\lambda = \frac{\int_0^{2\pi} \int_0^1 \alpha_\lambda(\mu, \phi) I_{\lambda, i}(\mu, \phi) \mu \, d\mu \, d\phi}{\int_0^{2\pi} \int_0^1 I_{\lambda, i} \mu \, d\mu \, d\phi} \tag{4.1.5}$$

$$\varepsilon_\lambda = \frac{\int_0^{2\pi} \int_0^1 \varepsilon_\lambda(\mu, \phi) I_{\lambda b} \mu \, d\mu \, d\phi}{\int_0^{2\pi} \int_0^1 I_{\lambda b} \mu \, d\mu \, d\phi} = \frac{1}{\pi} \int_0^{2\pi} \int_0^1 \varepsilon_\lambda(\mu, \phi) \mu \, d\mu \, d\phi \tag{4.1.6}$$

The monochromatic hemispherical emittance is seen to be a property but the monochromatic hemispherical absorptance is not a property but a function of the incident intensity.

The hemispherical absorptance and emittance are obtained by integrating over all wavelengths and are defined by

$$\alpha = \frac{\int_0^\infty \int_0^{2\pi} \int_0^1 \alpha_\lambda(\mu, \phi) I_{\lambda, i}(\mu, \phi) \mu \, d\mu \, d\phi \, d\lambda}{\int_0^\infty \int_0^{2\pi} \int_0^1 I_{\lambda, i}(\mu, \phi) \mu \, d\mu \, d\phi \, d\lambda} \tag{4.1.7}$$

$$\varepsilon = \frac{\int_0^\infty \int_0^{2\pi} \int_0^1 \varepsilon_\lambda(\mu, \phi) I_{b\lambda} \mu \, d\mu \, d\phi \, d\lambda}{\int_0^\infty \int_0^{2\pi} \int_0^1 I_{b\lambda} \mu \, d\mu \, d\phi \, d\lambda} = \frac{1}{E_b} \int_0^\infty \varepsilon_\lambda E_{b\lambda} \, d\lambda \tag{4.1.8}$$

Again the absorptance, in this case the hemispherical absorptance, is a function of the incident intensity whereas the hemispherical emittance is a surface property.

If the monochromatic directional absorptance is independent of direction [i.e., $\alpha_\lambda(\mu, \phi) = \alpha_\lambda$], then Equation 4.1.7 can be simplified by integrating over the hemisphere to yield

$$\alpha = \frac{\int_0^\infty \alpha_\lambda q_{\lambda, i} \, d\lambda}{\int_0^\infty q_{\lambda, i} \, d\lambda} \tag{4.1.9}$$

where $q_{\lambda, i}$ is the incident monochromatic radiant energy. If the incident radiation in either Equation 4.1.7 or 4.1.9 is radiation from the sun, then the calculated absorptance is called the solar absorptance.

4.2 KIRCHOFF'S LAW

A proof of Kirchoff's law is beyond the scope of this book. (See Siegel and Howell (1972) for a complete discussion.) However, a satisfactory understanding can be obtained without a proof. Consider an evacuated isothermal enclosure at temperature T. If the enclosure is isolated from the surroundings then the enclosure and any substance within the enclosure will be in thermodynamic equilibrium. In addition, the radiation field within the enclosure must be homogeneous and isotropic. If this were not so, we could have a directed flow of radiant energy at some location within the enclosure, but this is impossible since we could then extract work from an isolated and isothermal system.

If we now consider an arbitrary body within the enclosure, the body must absorb the same amount of energy as it emits. An energy balance on an element of the surface of the arbitrary body yields

$$\alpha q = \varepsilon E_b \tag{4.2.1}$$

If we place a second body with different surface properties in the enclosure, the same energy balance must apply and the ratio q/E_b must be constant.

$$\frac{q}{E_b} = \frac{\varepsilon_1}{\alpha_1} = \frac{\varepsilon_2}{\alpha_2} \tag{4.2.2}$$

Since this must also apply to a blackbody in which $\varepsilon = 1$, the ratio of ε to α for any body in thermal equilibrium must be equal to unity. Therefore, for conditions of thermal equilibrium

$$\varepsilon = \alpha \tag{4.2.3}$$

It must now be remembered that α is not a property, and since this equation was developed for the condition of thermal equilibrium, it may not be valid if the incident radiation comes from a source at a different temperature (e.g., if the source of radiation is the sun). This distinction is very important in the performance of solar collectors.

Equation 4.2.3 is sometimes referred to as Kirchoff's law, but his law is much more general. Within an enclosure the radiant flux is everywhere uniform and isotropic. The absorptance of a surface within the enclosure is then given by Equation 4.1.7 with $I_{\lambda, i}(\mu, \phi)$ replaced by $I_{\lambda b}$, and the emittance is given by Equation 4.1.8. Since the hemispherical absorptance and emittance are equal under conditions of thermal equilibrium, we can equate Equations 4.1.7 and 4.1.8 to obtain

$$\int_0^\infty I_{\lambda b} \int_0^{2\pi} \int_0^1 [\alpha_\lambda(\mu, \phi) - \varepsilon_\lambda(\mu, \phi)]\mu \, d\mu \, d\phi \, d\lambda = 0 \tag{4.2.4}$$

It is mathematically possible to have this integral equal to zero without $\alpha_\lambda(\mu, \phi)$ being identical to $\varepsilon_\lambda(\mu, \phi)$, but this is a very unlikely situation in view of the very irregular behavior of $\alpha_\lambda(\mu, \phi)$ exhibited by some substances. Thus we can say

$$\alpha_\lambda(\mu, \phi) = \varepsilon_\lambda(\mu, \phi) \tag{4.2.5}$$

This result is true for all conditions, not just thermal equilibrium, since both $\alpha_\lambda(\mu, \phi)$ and $\varepsilon_\lambda(\mu, \phi)$ are properties.*

If the surface does not exhibit a dependence on the azimuthal angle, then Equation 4.2.5 reduces to

$$\alpha_\lambda(\mu) = \varepsilon_\lambda(\mu) \tag{4.2.6}$$

---

* Kirchoff's law actually applies to each component of polarization and not to the sum of the two components as implied by Equation 4.2.5.

and if the dependence on polar angle can also be neglected, then Kirchoff's Law further reduces to

$$\alpha_\lambda = \varepsilon_\lambda \qquad (4.2.7)$$

Finally, if the surface does not exhibit a wavelength dependency, then the absorptance α is equal to the emittance ε. This is the same result obtained for any surface when in thermal equilibrium as given by Equation 4.2.3.

4.3 REFLECTION FROM SURFACES

Consider the spatial distribution of radiation reflected by a surface. When the incident radiation is in the form of a narrow "pencil" (i.e., contained within a small solid angle), two limiting distributions of the reflected radiation exist. These two cases are called *specular* and *diffuse*. Specular reflection is mirrorlike, that is, the incident polar angle is equal to the reflected polar angle and the azimuthal angles differ by 180°. On the other hand, diffuse reflection obliterates all directional characteristics of the incident radiation by distributing the radiation uniformly in all directions. In practice, the reflection from a surface is neither all specular nor all diffuse. The general case along with the two limiting situations is shown in Figure 4.3.1.

In general, the magnitude of the reflected intensity in a particular direction for a given surface is a function of the wavelength and the spatial distribution of the incident radiation. The biangular reflectance or reflection function is used to relate the intensity of reflected radiation in a particular direction by the following equation:

$$\rho_\lambda(\mu_r, \phi_r, \mu_i, \phi_i) = \lim_{\Delta\omega_i \to 0} \frac{\pi I_{\lambda, r}(\mu_r, \phi_r)}{I_{\lambda, i}\mu_i \Delta\omega_i} \qquad (4.3.1)$$

The numerator is π times the intensity reflected in the direction μ_r, ϕ_r when an energy flux of amount $I_{\lambda, i}\mu_i \Delta\omega_i$ is incident on the surface from the direction μ_i, ϕ_i. The π has been included so that the numerator "looks like" an energy flux. The physical situation is shown schematically in Figure 4.3.2.

Since the energy incident in the solid angle $\Delta\omega_i$ may be reflected in all directions, the reflected intensity in the direction μ_r, ϕ_r will be of infinitesimal size

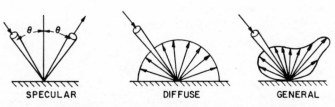

Figure 4.3.1   Reflection from surfaces. From Duffie and Beckman (1974).

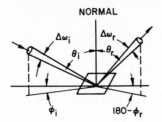

Figure 4.3.2 Coordinate system for the reflection function.

compared to the incident intensity. By multiplying the incident intensity by its solid angle (which must be finite in any real experiment) and the cosine of the polar angle, we obtain the incident radiant flux which will have values on the same order of magnitude as the reflected intensity. The biangular reflectance can have numerical values between zero and infinity; its values do not lie only between zero and one.

From an experimental point of view, it is not practical to use the scheme depicted in Figure 4.3.2 since all the radiation quantities would be extremely small. An equivalent experiment is to irradiate the surface with a nearly monodirectional flux (i.e., with a small solid angle $\Delta\omega_i$) as shown in Figure 4.3.3. The reflected energy in each direction is measured. This measured energy divided by the measurement instruments solid angle ($\Delta\omega_r$) will be approximately equal to the reflected intensity. The incident flux will be on the same order and can be easily measured.

Two types of hemispherical reflectances exist. The angular-hemispherical reflectance is found when a narrow pencil of radiation is incident on a surface and all the reflected radiation is collected. The hemispherical-angular reflectance results from collecting reflected radiation in a particular direction when the surface is irradiated from all directions.

The monochromatic angular-hemispherical reflectance will be designated by $\rho_\lambda(\mu_i, \phi_i)$ where the subscript i indicates that the incident radiation has a specified direction. This reflectance is defined as the ratio of the monochromatic radiant energy reflected in all directions to the incident radiant flux within small solid angle $\Delta\omega_i$. The incident energy ($I_{\lambda, i}\mu_i\Delta\omega_i$) that is reflected in all directions can be found using the reflection function:

$$q_{\lambda, r} = \frac{1}{\pi} \int_0^{2\pi} \int_0^1 \rho_\lambda(\mu_r, \phi_r, \mu_i, \phi_i) I_{\lambda, i}\mu_i\Delta\omega_i\mu_r \, d\mu_r \, d\phi_r \qquad (4.3.2)$$

The monochromatic angular-hemispherical reflectance can then be expressed as

$$\rho_\lambda(\mu_i, \phi_i) = \frac{q_{\lambda, r}}{I_{\lambda, i}\mu_i\Delta\omega_i} = \frac{1}{\pi} \int_0^{2\pi} \int_0^1 \rho_\lambda(\mu_r, \phi_r, \mu_i, \phi_i)\mu_r \, d\mu_r \, d\phi_r \quad (4.3.3)$$

Examination of Equation 4.3.3 shows that $\rho_\lambda(\mu_i, \phi_i)$ is a property of the surface. The angular-hemispherical reflectance, $\rho(\mu_i, \phi_i)$, can be found by integrating

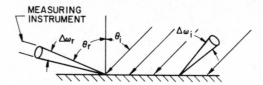

Figure 4.3.3 Schematic representation of an experiment for measuring the reflection function.

the incident and reflected fluxes over all wavelengths, but it is not a property as it depends upon the wavelength distribution of the incoming radiation.

The monochromatic hemispherical-angular reflectance is defined as the ratio of the reflected monochromatic intensity in the direction μ_r, ϕ_r to the monochromatic energy from all directions divided by π (which then "looks like" an intensity). The incident energy can be written in terms of the incident intensity integrated over the hemisphere:

$$q_{\lambda, i} = \int_0^{2\pi} \int_0^1 I_{\lambda, i} \mu_i \, d\mu_i \, d\phi_i \tag{4.3.4}$$

and the monochromatic hemispherical-angular reflectance is then

$$\rho_\lambda(\mu_r, \phi_r) = \frac{I_{\lambda, r}(\mu_r, \phi_r)}{q_{\lambda i}/\pi} \tag{4.3.5}$$

where the subscripts r in $\rho_\lambda(\mu_r, \phi_r)$ are used to specify the reflected radiation as being in a specified direction. In terms of the reflectance function, Equation 4.3.5 can be written as

$$\rho_\lambda(\mu_r, \phi_r) = \frac{\int_0^{2\pi} \int_0^1 \rho_\lambda(\mu_r, \phi_r, \mu_i \phi_i) I_{\lambda, i} \mu_i \, d\mu_i \, d\phi_i}{\int_0^{2\pi} \int_0^1 I_{\lambda, i} \mu_i \, d\mu_i \, d\phi_i} \tag{4.3.6}$$

Since $\rho_\lambda(\mu_r, \phi_r)$ is dependent upon the angular distribution of the incident intensity, it is not a surface property. For the special case when the incident radiation is diffuse, the monochromatic hemispherical-angular reflectance is identical to the monochromatic angular-hemispherical reflectance. To prove the equality of $\rho_\lambda(\mu_r, \phi_r)$ and $\rho_\lambda(\mu_i, \phi_i)$ under the condition of constant $I_{\lambda, i}$, it is necessary to use the symmetry of the reflection function as given by

$$\rho_\lambda(\mu_i, \phi_i, \mu_r, \phi_r) = \rho_\lambda(\mu_r, \phi_r, \mu_i, \phi_i) \tag{4.3.7}$$

and compare Equation 4.3.6 (with $I_{\lambda, i}$ independent of incident direction) and Equation 4.3.3. The proof of Equation 4.3.7 is beyond the scope of this book [see Siegel and Howell (1972)].

The equality of $\rho_\lambda(\mu_i, \phi_i)$ and $\rho_\lambda(\mu_r, \phi_r)$ when $I_{\lambda, i}$ is uniform is of great importance since the measurement of $\rho_\lambda(\mu_r, \phi_r)$ is much easier than $\rho_\lambda(\mu_i, \phi_i)$. This is discussed in Section 4.5.

Both $\rho_\lambda(\mu_i, \phi_i)$ and $\rho_\lambda(\mu_r, \phi_r)$ can be considered on a total basis by integration over all wavelengths. For the case of the angular-hemispherical reflectance, we have

$$\rho(\mu_i, \phi_i) = \frac{\int_0^\infty q_{\lambda r}\, d\lambda}{\int_0^\infty I_{\lambda, i}\mu_i\Delta\omega_i\, d\lambda} = \frac{1}{\pi I_i}\int_0^\infty \int_0^{2\pi} \int_0^1 \rho_\lambda(\mu_i, \phi_i, \mu_r, \phi_r) I_{\lambda, i}\mu_r\, d\mu_r\, d\phi_r\, d\lambda$$

$$(4.3.8)$$

which, unlike the monochromatic angular-hemispherical reflectance, is not a property.

When a surface element is irradiated from all directions and all the reflected radiation is collected, we characterize this by the monochromatic hemispherical reflectance defined as

$$\rho_\lambda = \frac{q_{\lambda, r}}{q_{\lambda, i}} \qquad (4.3.9)$$

The reflected monochromatic energy $q_{\lambda, r}$ can be expressed in terms of the reflection function and the incident intensity by

$$q_{\lambda, r} = \int_0^{2\pi} \int_0^1 \left[\int_0^{2\pi} \int_0^1 \frac{\rho_\lambda(\mu_r, \phi_r, \mu_i, \phi_i)}{\pi} I_{\lambda, i}\mu_i\, d\mu_i\, d\phi_i \right]\mu_r\, d\mu_r\, d\phi_r \quad (4.3.10)$$

The incident energy, expressed in terms of the incident intensity, is

$$q_{\lambda, i} = \int_0^{2\pi} \int_0^1 I_{\lambda, i}\mu_i\, d\mu_i\, d\phi_i \qquad (4.3.11)$$

Division of Equation 4.3.10 by 4.3.11 yields the monochromatic hemispherical reflectance. For the special case of a diffuse surface (i.e., the reflection function is a constant), the monochromatic hemispherical reflectance is numerically equal to the reflection function and is independent of the spatial distribution of the incident intensity.

The hemispherical reflectance is found by integration of Equations 4.3.10 and 4.3.11 over all wavelengths and finding the ratio

$$\rho = \frac{q_r}{q_i} = \frac{\int_0^\infty q_{\lambda, r}\, d\lambda}{\int_0^\infty q_{\lambda, i}\, d\lambda} \qquad (4.3.12)$$

The hemispherical reflectance depends on both the angular distribution and wavelength distribution of the incident radiation.

For low-temperature applications that do not include solar radiation, a special form of the hemispherical reflectance (often the name is shortened to "reflectance") will be found to be the most useful. The special form is Equation 4.3.12, which is based on the assumption that the reflection function is independent of direction (diffuse approximation) and wavelength (gray approximation). The diffuse approximation for the hemispherical reflectance has already been discussed and ρ_λ was found to be equal to $\rho_\lambda(\mu_i, \phi_i, \mu_r, \phi_r)$. When the gray approximation is made in addition to the diffuse approximation, the surface reflectance becomes independent of everything except possibly the temperature of the surface, and even this is usually neglected.

4.4 RELATIONSHIPS AMONG ABSORPTANCE, EMITTANCE, AND REFLECTANCE

It is now possible to show that it is necessary to know only one property, the monochromatic angular-hemispherical reflectance, and all absorptance and emittance properties for opaque surfaces can be found.

Consider a surface located in an isothermal enclosure maintained at temperature T. The monochromatic intensity in a direction μ, ϕ from an infinitesimal area of the surface consists of emitted and reflected radiation and must be equal to $I_{b\lambda}(T)$.

$$I_{b\lambda} = I_\lambda(\mu, \phi)|_{\text{emitted}} + I_\lambda(\mu, \phi)|_{\text{reflected}} \tag{4.4.1}$$

The emitted and reflected intensities are

$$I_\lambda(\mu, \phi)|_{\text{emitted}} = \varepsilon_\lambda(\mu, \phi)I_{b\lambda} \tag{4.4.2}$$

$$I_\lambda(\mu, \phi)|_{\text{reflected}} = I_{b\lambda}\rho_\lambda(\mu_r, \phi_r) \tag{4.4.3}$$

but $\rho_\lambda(\mu_r, \phi_r)$ is equal to the monochromatic angular-hemispherical reflectance, $\rho_\lambda(\mu_i, \phi_i)$, since the incident intensity is diffuse. Since $I_{b\lambda}$ can be cancelled from each term, we have

$$\varepsilon_\lambda(\mu, \phi) = 1 - \rho_\lambda(\mu_i, \phi_i) \tag{4.4.4}$$

But from Kirchoff's law

$$\varepsilon_\lambda(\mu, \phi) = \alpha_\lambda(\mu, \phi) = 1 - \rho_\lambda(\mu_i, \phi_i) \tag{4.4.5}$$

Thus, the monochromatic directional emittance and the monochromatic directional absorptance can both be calculated from knowledge of the monochromatic angular-hemispherical reflectance. Also, all emittance properties (Equations 4.1.4, 4.1.6, and 4.1.8) can be found once $\rho_\lambda(\mu_i, \phi_i)$ is known. The absorptances (Equations 4.1.2, 4.1.5, and 4.1.7) can be found if the incident intensity is specified.

The relationship between the reflectance and absorptance* of Equation 4.4.5 can be considered as a statement of conservation of energy. The incident monochromatic energy from any direction is either reflected or absorbed. Similar arguments can be used to relate other absorptances to reflectances.

---

* There are no generally accepted names used in the literature except for the simple "absorptance," "emittance," and "reflectance" which, for clarity, were prefixed with the name hemispherical. In the remainder of this book, the modifier hemispherical will generally be omitted since most available data are hemispherical. If it is necessary to distinguish directional quantities then the full name will be used.

For example, for an opaque surface, energy from all directions either monochromatic or total, is either absorbed or reflected so that

$$\rho_\lambda + \alpha_\lambda = 1 \qquad (4.4.6)$$

and

$$\rho + \alpha = 1 \qquad (4.4.7)$$

Example 4.4.1

Calculate the emittance, reflectance and absorptance for a surface at 500 C when the incident radiation is from a blackbody source at 100 C. The monochromatic reflectance is shown in the figure and is independent of surface temperature and incidence angle.

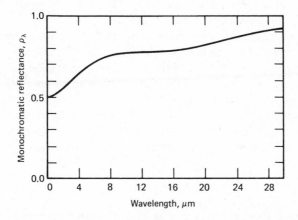

Solution

The monochromatic absorptance can be found from Equation 4.4.6 and the monochromatic emittance from Equation 4.2.7. The emittance is found by integrating Equation 4.1.8 where E_b and $E_{b\lambda}$ are evaluated at 500 C. Table 3.6.1 is useful in numerically performing this integration. Equation 4.1.8 can be expressed in terms of the fraction of the blackbody radiation between wavelengths j and $j + 1$ as

$$\varepsilon = \sum_{j=1}^{n} \Delta f_j \varepsilon_{\lambda j}$$

where Δf_j is $f_{0-\lambda T}$ at λ_{j+1} minus $f_{0-\lambda T}$ at λ_j and $\varepsilon_{\lambda j}$ is the average value of ε within this band and often evaluated at the midpoint of the interval. It is convenient to choose equal energy increments so that Δf_j is a constant. Thus, with n intervals, all Δf_j's are equal to $1/n$. The integration then reduces to summing up n values of ε_λ at particular wavelengths and dividing the sum by n. For a 5

$f_{0-\lambda T}$	λT	λ at 773 K	ρ_λ	ε_λ	λ at 373 K	ρ_λ	α_λ
0.10	2200	2.8	0.60	0.40	5.9	0.70	0.30
0.30	3120	4.0	0.64	0.36	8.4	0.76	0.24
0.50	4110	5.3	0.69	0.31	11.0	0.77	0.23
0.70	5590	7.2	0.72	0.28	15.0	0.77	0.23
0.90	9380	12.1	0.79	0.21	25.1	0.87	0.13
			Average	0.31		Average	0.23

point integration, the appropriate values of λT are found in Table 3.6.1 at $f_{0-\lambda T} = 0.10, 0.30, 0.50, 0.70,$ and 0.90. These are shown in the table above. The appropriate wavelengths are then found for the surface temperature of 773 K from $\lambda T/773$. The monochromatic emittances are found from the figure from $1 - \rho_\lambda$. The emittance at 773 K is then 0.31.

The absorptance for radiation from a source at 373 K is found by integrating Equation 4.1.7 over all angles and wavelengths. Since the absorptance is independent of angle, and the incident radiation is from a blackbody source, Equation 4.1.7 reduces to Equation 4.1.8. However, the temperature for $E_{b\lambda}$ and E_b is 373 K, the source temperature and not 773 K, the surface temperature. The same values of λT apply for a 5 point integration but the wavelengths are now 5.9, 8.4,..., as shown in the table. The absorptance for blackbody radiation from a source at 373 K is 0.23. If the surface temperature were 373 K (instead of 773 K) then the emittance at this temperature would be also 0.23. ■

4.5   MEASUREMENTS OF SURFACE RADIATION PROPERTIES

In the preceding discussion many radiation surface properties have been defined. Unfortunately, in much of the literature the exact nature of the surface being reported is not clearly specified. This situation requires that caution be exercised.

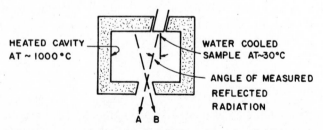

Figure 4.5.1 Schematic of a hohlraum for measurement of monochromatic hemispherical-angular reflectance. Radiation A is blackbody radiation reflected from the sample. Radiation B is blackbody radiation from the cavity. The ratio A_λ/B_λ is $\rho_\lambda(\mu, \phi)$.

Many of the reflectance data reported in the literature have been measured by a method devised by Gier et al. (1954). In this method a cool sample is exposed to blackbody radiation from a high-temperature source (a hohlraum) and the monochromatic radiation reflected from the surface is compared to mono-chromatic blackbody radiation from the cavity. The data are thus hemispherical-angular monochromatic reflectances (or angular-hemispherical monochromatic reflectances, since they are equal for diffuse incident radiation). The method is shown schematically in Figure 4.5.1. In many systems the angle between the surface normal and the measured radiation is often fixed at a small value so that measurements can be made at only one angle (approximately normal). In some of the more recent designs the sample can be rotated so that all angles can be measured. With measurements of this type, emittance and absorptance values can be found from Equation 4.4.5.

Table 4.5.1 gives data on surface properties for a few common materials. The data are total hemispherical or total normal emittances at various temperatures and normal solar absorptance at room temperature. Most of these data were calculated from monochromatic data as was done in Example 4.4.1. Table 4.5.1 was compiled from *Thermophysical Properties of Matter* by Touloukian et al., Vols. 7, 8, and 9 (1970, 1972, and 1972). These three volumes are the most complete reference to radiation properties available today. In addition to total hemispherical and normal emittance, such properties as angular spectral re-flectance, angular total reflectance, angular solar absorptance, and others, are given in this extensive compilation.

4.6 SELECTIVE SURFACES

Solar collectors must have high absorptance for radiation in the solar energy spectrum. At the same time, they lose energy by a combination of mechanisms* including thermal radiation from the absorbing surface, and it is desirable to have the long-wave emittance of the surface as low as possible, to reduce losses. The temperature of this surface in most flat-plate collectors is less than 200 C (473 K), while the effective surface temperature of the sun is approximately 6000 K. Thus, the wavelength range of the emitted radiation overlaps only slightly the solar spectrum (98 percent of the extraterrestrial solar radiation is at wavelengths less than 3.0 μm, whereas less than 1 percent of the black body radiation from a 200 C surface is at wavelengths less than 3.0 μm). Under these circumstances, it is possible to devise surfaces having high solar absorptance and low long-wave emittance, that is, selective surfaces.

The concept of a selective surface is illustrated in Figure 4.6.1. This idealized surface is called a semi-gray surface, since it can be considered gray in the solar spectrum (i.e., at wavelengths less than approximately 2.5 or 3.0 μm) and also gray, but with different properties, in the infrared spectrum (i.e., at wavelengths

---

* This will be discussed in detail in Chapter 6.

Table 4.5.1 Radiation Properties

Material		Emittance/Temperature, K	Absorptance[b]
Aluminum, pure	H[a]	$\dfrac{0.102}{573}, \dfrac{0.130}{773}, \dfrac{0.113}{873}$	0.09–0.10
Aluminum, Anodized	H	$\dfrac{0.842}{296}, \dfrac{0.720}{484}, \dfrac{0.669}{574}$	0.12–0.16
Aluminum with SiO_2 Coating	H	$\dfrac{0.366}{263}, \dfrac{0.384}{293}, \dfrac{0.378}{324}$	0.11
Carbon Black in Acrylic Binder	H	$\dfrac{0.83}{278}$	0.94
Chromium	N	$\dfrac{0.290}{722}, \dfrac{0.355}{905}, \dfrac{0.435}{1072}$	0.415
Copper, polished	H	$\dfrac{0.041}{338}, \dfrac{0.036}{463}, \dfrac{0.039}{803}$	0.35
Gold	H	$\dfrac{0.025}{275}, \dfrac{0.040}{468}, \dfrac{0.048}{668}$	0.20–0.23
Iron	H	$\dfrac{0.071}{199}, \dfrac{0.110}{468}, \dfrac{0.175}{668}$	0.44
Lampblack in Epoxy	N	$\dfrac{0.89}{298}$	0.96
Magnesium Oxide	H	$\dfrac{0.73}{380}, \dfrac{0.68}{491}, \dfrac{0.53}{755}$	0.14
Nickel	H	$\dfrac{0.10}{310}, \dfrac{0.10}{468}, \dfrac{0.12}{668}$	0.36–0.43
Paint			
Parsons Black	H	$\dfrac{0.981}{240}, \dfrac{0.981}{462}$	0.98
Acrylic White	H	$\dfrac{0.90}{298}$	0.26
White (ZnO)	H	$\dfrac{0.929}{295}, \dfrac{0.926}{478}, \dfrac{0.889}{646}$	0.12–0.18

From Duffie and Beckman (1974).

[a] H is total hemispherical emittance; N is total normal emittance.

[b] Normal solar absorptance.

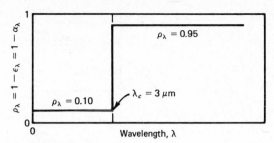

Figure 4.6.1   A hypothetical selective surface with the cutoff wavelength at 3 μm.

greater than approximately 2.5 or 3.0 μm). For this idealized surface, the reflectance below the cut-off wavelength is very low. For an opaque surface $\alpha_\lambda = 1 - \rho_\lambda$, so in this range α_λ is very high. At wavelengths greater than λ_c the reflectance is nearly unity, and since $\varepsilon_\lambda = \alpha_\lambda = 1 - \rho_\lambda$, the emittance in this range is low.

The absorptance for solar energy and emittance for long-wave radiation are determined from the monochromatic reflectance data by integration over the appropriate spectral range. The absorptance for solar radiation, usually designated in the solar energy literature simply as α, can be calculated by dividing the spectrum into portions over which ρ_λ is approximately constant and using the spectral distribution of the incident solar radiation to numerically integrate Equation 4.1.9. The emittance, usually designated simply as ε, depends on the temperature of the surface and is found by numerically integrating·Equation 4.1.8. For normal operation of flat-plate solar collectors, the temperatures will be low enough so that essentially all energy will be emitted at wavelengths greater than 3 μm.

Example 4.6.1

For the surface shown in Figure 4.6.1, calculate the absorptance for blackbody radiation from a source at 5762 K and the emittance at surface temperatures of 150 and 500 C.

Solution

The absorptance for blackbody radiation from a source at 5762 K is found by evaluating Equation 4.1.9 with the incident radiation $q_{\lambda, i}$, given by Planck's law, Equation 3.4.1. The denominator of Equation 4.1.9 is then σT^4 and the calculation of α reduces to evaluating the following

$$\alpha = \frac{C_1}{\sigma} \int_0^\infty \frac{\alpha_\lambda \, d(\lambda T)}{(\lambda T)^5 (e^{c_2/\lambda T} - 1)}$$

For this problem, α_λ has two values, α_s in the short wavelengths below λ_c and α_L in the long wavelengths. The integral then reduces to

$$\alpha = \alpha_s f_{0-\lambda T} + \alpha_L (1 - f_{0-\lambda T})$$

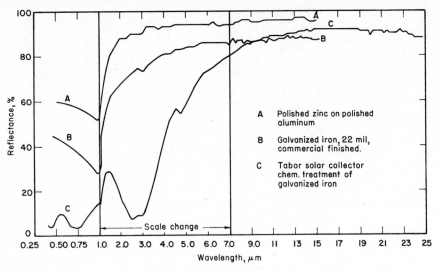

Figure 4.6.2 Spectral reflectance of several surfaces. From Edwards et al. (1960).

where $f_{0-\lambda T}$ is the fraction of the incident blackbody radiation below the critical wavelength and is found from Table 3.6.1 at $\lambda T = 3 \times 5762 = 17286$. Therefore, the absorptance is

$$\alpha = (1 - 0.10)(0.979) + (1 - 0.95)(1 - 0.979) = 0.88$$

The emittances at 150 and 500 C are found by integrating Equation 4.1.8. Again Table 3.6.1 is useful in performing this integration. Equation 4.1.8 reduces to the following

$$\varepsilon = \varepsilon_s f_{0-\lambda T} + \varepsilon_L (1 - f_{0-\lambda T})$$

where $f_{0-\lambda T}$ is now the fraction of the blackbody energy that is below the critical wavelength but at the surface temperature rather than the source temperature as was used in calculating the absorptance. For a surface temperature of 150 C (423 K), λT is 1269 and $f_{0-\lambda T}$ is 0.004. The emittance at 150 C is then

$$\varepsilon_{150} = (1 - 0.10)(0.004) + (1 - 0.95)(0.996) = 0.05$$

at a surface temperature of 500 C, $f_{0-\lambda T}$ is 0.124 and the emittance at 500 C is

$$\varepsilon_{500} = (1 - 0.10)(0.124) + (1 - 0.95)(0.876) = 0.16 \qquad \blacksquare$$

In practice, the wavelength dependence of ρ_λ does not closely approach the ideal curve of Figure 4.6.1. Examples of ρ_λ versus λ for several real surfaces are shown in Figures 4.6.2 and 4.6.3. The real selective surfaces do not have a well-defined critical wavelength, λ_c, and uniform properties in the short- and long-

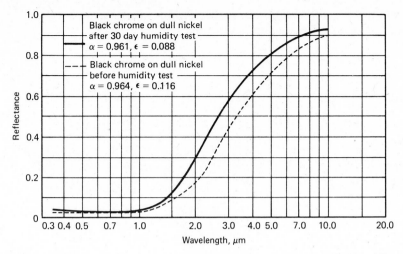

Figure 4.6.3 The spectral reflectance of black chrome on nickel, before and after humidity tests. From Lin (1977).

wavelength ranges, and values of emittance will be more sensitive to surface temperature than those of the ideal semigray surface of Figure 4.6.1. The integration procedure is the same as in Example 4.6.1, but smaller spectral ranges must be used.

Example 4.6.2

Calculate the solar absorptance and the emittance at 100 C for the surface shown in curve C of Figure 4.6.2.

Solution

The solar absorptance, α, should be calculated from Equation 4.1.9 with the incident radiation, $q_{\lambda, i}$, having the spectral distribution of solar radiation at the collector surface. Assume that the spectral distribution of Table 2.6.1, for air mass 2, represents the distribution of solar radiation. The table below gives the midpoints of the spectral bands that each contain 10% of the terrestrial solar radiation. The monochromatic absorptances of the selective surface corresponding to these wavelengths are also shown. These values are assumed to hold over their wavelength intervals, and since the intervals are all the same the solar absorptance is the sum of these values divided by the number of intervals. The reflectance is found from $1 - \alpha$ and is equal to $1 - 0.89 = 0.11$.

The emittance at a temperature of 100 C is found in the same manner as described in Example 4.4.1 and is equal to 0.15. Here, we have used a 10-interval integration rather than 5 intervals as in Example 4.4.1. The reflectance for incident radiation from a source at 373 C would be $1 - 0.15 = 0.85$. If the source temperature was not too far from 373 K, assuming the reflectance equal to 0.85 would be a good approximation.

Fract.	λ	$1 - \rho_\lambda = \alpha_\lambda$	λT	λ at 373 K	$1 - \rho_\lambda = \varepsilon_\lambda$
0.05	0.43	0.95	1880	5.0	0.43
0.15	0.52	0.93	2450	6.6	0.24
0.25	0.60	0.91	2900	7.8	0.16
0.35	0.67	0.96	3350	9.0	0.14
0.45	0.75	0.96	3830	10.3	0.11
0.55	0.85	0.93	4410	11.8	0.10
0.65	0.98	0.86	5130	13.8	0.09
0.75	1.10	0.78	6150	16.5	0.08
0.85	1.31	0.72	7850	21.3	0.10
0.95	2.05	0.90	12500	33.5	0.10^a
	Average	0.89		Average	0.15

^{a} Estimated

The potential utility of selective surfaces in solar collectors was inferred by Hottel and Woertz (1942) and noted by Gier and Dunkle (1958) and Tabor (1956, 1964, 1967). Interest in designing surfaces with a variety of ρ_λ versus λ characteristics for applications to space vehicles and to solar energy applications resulted in considerable research and compilation of data [e.g., Martin and Bell (1961), Edwards, et al. (1960), Schmidt, et al. (1964), and others]. Tabor (1967, 1977) reviewed selective surfaces and presents several methods for their preparation. Several selective surfaces are now in commercial use.

Several methods of preparing selective surfaces have been developed, which depend on various mechanisms to achieve selectivity.

1 Coatings that have high absorptance for solar radiation and high transmittance for long-wave radiation can be applied to substrates with low emittance. The coating absorbs solar energy, and the substrate is the (poor) emitter of long-wave radiation. Coatings may be homogeneous or have particulate structure; their properties are then either the inherent optical properties of the coating material or of the material properties and the coating structure. Many of the coating materials used are metal oxides and the substrates are metals. Examples are copper oxide on aluminum [e.g., Hottel and Unger (1959)], and copper oxide on copper [e.g., Close (1962)]. A nickel-zinc sulfide coating can be applied to galvanized iron [Tabor (1956)]. Four of these surfaces are in commercial use.

Selective surfaces have been in use on Israeli solar water heaters since about 1950. The nickel black surfaces are applied to galvanized iron base. The base is carefully cleaned and the black coating is applied by immersion of the plate as the cathode in an aqueous electroplating bath containing nickel sulfate, zinc sulfate, ammonium sulfate, ammonium thiocyanate, and citric acid. More details of this process are provided by Tabor (1967).

Copper oxide on copper selective blacks are formed on carefully degreased copper plates by treating the plates for various times in hot (140 C) solutions of sodium hydroxide and sodium chlorite, as described by Close (1962). Similar proprietary blackening processes are used in the United States under the name Ebanol.

The most recent commercial selective surfaces, which are being widely adopted in the United States, are black chromes. The substrate is usually nickel plating on a steel or other base. The coatings are formed by electroplating in a bath of chromic acid and other agents.* Surfaces of black chrome on dull nickel can have very desirable combination of properties. In laboratory specimens, absorptances of 0.95–0.96 and emittances of 0.08–0.14 were obtained, while the average properties of samples of production run collector plates were $\alpha = 0.94$ and $\varepsilon = 0.08$ [Moore (1976)]. Reflectance properties of these surfaces are described by McDonald (1974, 1975) and others. The surfaces appear to have good durability on exposure to humid atmospheres, as shown in Figure 4.6.3. Many references are available on preparation of chrome black surfaces, for example, Benning (1976), Pettit and Sowell (1976), and Sowell and Mattox (1976).

2 The absorptance of coatings can be enhanced by taking advantage of interference phenomena. Some coatings used on highly reflective (low ε) substrates are semiconductors, which have high absorptance in the solar energy spectrum but which have high transmittance for long-wave radiation. Many of these materials also have a high index of refraction, and thus reflect incident solar energy. This reflection loss can be reduced by secondary antireflective coatings. It has been shown by Martin and Bell (1960) that three-layer coatings such as SiO_2—Al—SiO_2 on substrates such as aluminum could have absorptances for solar energy greater than 0.90 and long-wave emittances less than 0.10. Seraphim (1975) and Meinel et al. (1973) have demonstrated the selectivity of surfaces using silicon and germanium with antireflecting coatings.

3 The surface structure of a metal of high reflectance can be designed to enhance its absorptance for solar radiation by grooving or pitting the surface to create cavities of dimensions near the desired cut-off wavelength of the surface. The surface acts as an array of cavity absorbers for solar radiation, thus having reduced reflectance in this part of the spectrum. The surface radiates as a flat surface in the long-wave spectrum, and thus shows its usual low emittance. Desirable surface structures have been made by forming tungsten dendritic crystals in substrates by reduction of tungsten hexafluoride with hydrogen [Cuomo et al. (1975)] or by chemical

---

* Plating baths are available from duPont Company (Durimir BK Black chromium process) and Harshaw Chemical Co. (Chromonyx Black Chromium Process).

vapor deposition of dendritic nickel crystals from nickel carbonyl [Grimmer et al (1976)]. Intermetallic compounds, such as Fe_2Al_5, can be formed with highly porous structures and show some selectivity (Santala (1975)). The degree of selectivity obtainable by this method is limited, and the emittances obtained to date have been 0.5 or more. However, roughening the substrate over which oxide (or other) coatings are applied can result in improved absorptance.

4   Directional selectivity can be obtained by proper arrangement of the surface on a large scale. Surfaces of deep V-grooves, large relative to all wavelengths of radiation concerned, can be arranged so that radiation from near-normal directions to the overall surface will be reflected several times in the grooves, each time absorbing a fraction of the beam. This multiple absorption gives an increase in the solar absorptance, but at the same time increases the long-wavelength emittance. However, as shown by Hollands (1963), a partially selective surface can have its effective properties substantially improved by proper configuration. For example, a surface having nominal properties of $\alpha = 0.60$ and $\varepsilon = 0.05$, used in a fixed optimally oriented flat-plate collector over a year, with 55° grooves, will have an average effective α of 0.90 and an equivalent ε of 0.10. Figure 4.6.4, from Trombe et al. (1964), illustrates the multiple absorptions obtained for various angles of incidence of solar radiation on a 30° grooved surface. Figure 4.6.5, from Hollands (1963), shows the variation of average yearly solar absorptance as a function of angle of the grooves and the absorptance of the plane surface.

5   The physical structure of coatings on reflective substrates will affect the reflectance of the surface. Williams et al. (1963) showed that the reflectance of coatings of lead sulfide is a function of the structure of the coating, and that finely divided particulate coatings of large void fraction have a low

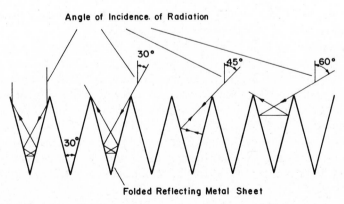

Figure 4.6.4 Absorption of radiation by successive reflections on folded metal sheets. Adapted from Trombe et al (1964).

effective refractive index and a low reflectance in the solar spectrum. This phenomenon is the basis for experimental studies of selective paints, in which binders transparent (insofar as is possible) to solar radiation are used to provide physical strength to the coatings. For example, PbS coatings of void fraction 0.8 to 0.9 on polished pure (99.99) aluminum substrates showed $\alpha = 0.8$ to 0.9 and $\varepsilon = 0.2$ to 0.3 without a binder and $\varepsilon = 0.37$ with a silicone binder. More recently, Lin (1977) has reported studies of a range of pigments (mostly metal oxides) and binders on aluminum substrates, and notes the best laboratory results obtained for an iron-manganese-copper oxide paint with a silicone binder to be $\alpha = 0.92$ and $\varepsilon = 0.13$. Many problems remain to be solved before these selective paints become practical for applications.

In addition to the methods noted above, other techniques have been experimentally applied for collectors other than flat plates. For example, vacuum sputtering processes for selective surfaces in vacuum tube collectors have been studied by Harding (1976) and Harding et al. (1976). The sputtering can be done in inert atmosphere (argon) to make metal coatings or in reactive atmospheres (argon plus 1–2% methane) to produce metal and metal carbide coatings. These coatings reportedly have extremely low emittances ($\varepsilon = 0.03$) but moderate absorptance ($\alpha = \sim 0.8$); higher absorptances are needed in most applications.

A critical consideration in the use of selective surfaces is their durability. Solar collectors must be designed to operate essentially without maintenance for many years, and the coatings and substrates must retain useful properties in

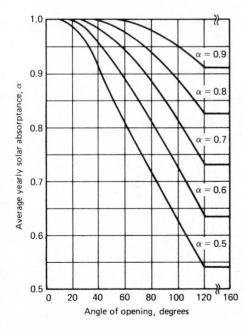

Figure 4.6.5 Average yearly solar absorptance vs groove angle for several values of absorptance of plane surfaces. From Hollands (1963).

Table 4.6.1   Properties of Some Selective Surfaces for Solar Energy Application

Surface	α	ε	Reference
"Nickel Black" on galvanized iron	0.81	0.16–0.18	Tabor et al. (1964)
"Cu Black" on Cu, by treating Cu with solution of NaOH and NaClO$_2$	0.89	0.17	Close (1962)
Ebanol C on Cu; commercial Cu-blackening treatment giving coatings largely CuO	0.90	0.16	Edwards et al. (1962)
Black-chrome plated on Ni plated steel	0.95	0.09	Mar et al. (1976), and others
CuO on Al; by spraying dilute Cu (NO$_3$)$_2$ solution on hot Al plate and baking	0.93	0.11	Hottel and Unger (1959)
CuO on Ni; made by electrode deposition of Cu and subsequent oxidation	0.81	0.17	Kokoropoulos et al. (1959)
Al$_2$O$_3$—Mo—Al$_2$O$_3$—Mo—Al$_2$O$_3$Mo—Al$_2$O$_3$ interference layers on Mo (ε measured at 500°F)	0.91	0.085	Schmidt et al. (1964)
PbS crystals on Al	0.89	0.20	Williams et al. (1963)
"Nickel Black," two layers on electroplated Ni on mild steel (α and ε after 6-hr immersion in boiling water)	0.94	0.07	Schmidt (1974)

humid, oxidizing atmospheres and at elevated temperatures. Some collector manufacturers have resisted conversion from flat black surfaces of known durability and high emittance to selective surfaces of unknown durability but low emittance. Essentially all of the data cited above and in Table 4.6.1 are for newly prepared surfaces. Data from Lin (1977) and Mar et al. (1976) and from other sources, plus experience with chrome black in other kinds of applications, suggest that this surface will retain its selective properties in a satisfactory way. Years of experience with the Israeli nickel black and Australian copper oxide on copper coatings have shown that these coatings can be durable.

Table 4.6.1 shows absorptance for solar radiation and emittance for long-wave radiation at temperatures typical of those encountered in operation of flat plate collectors. The first four are representative of commercial processes. The others are more accurately described as experimental. Data are for freshly prepared surfaces, unless otherwise noted.

In flat-plate collectors, it is generally more critical to have high absorptance than low emittance.* It is a characteristic of many surfaces that there is a relationship between α and ε as typified by data shown in Figure 4.6.6. In the case of the chrome black surface, the optimum plating time (coating thickness) is

---

* This will become evident from Chapter 6.

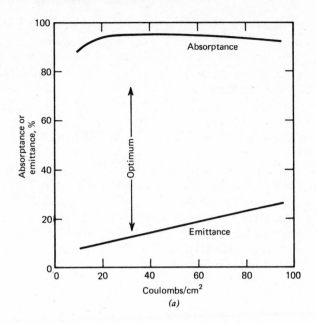

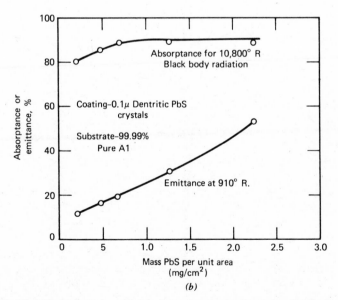

Figure 4.6.6 Variations of α and ε with (*a*) plating current density × time for chrome black, and (*b*) mass per unit area of PbS dendritic crystal coating (*a*) Adapted from Sowell and Mattox (1976). (*b*) Adapted from Williams (1961).

obvious. In the case of the PbS coating, the optimum mass per unit area is not immediately obvious. The best combination must ultimately be selected on the basis of the effects of the two properties on the annual operation of the complete solar energy system.* But the generalization can be made that α should be near its maximum for best performance.

4.7 ANGULAR DEPENDENCE OF SOLAR ABSORPTANCE

The angular dependence of solar absorptance of most surfaces used for solar collectors is not available. The directional absorptance of ordinary blackened surfaces (such as are used for solar collectors) for solar radiation is a function of the angle of incidence of the radiation on the surface. An example of dependence of absorptance on angle of incidence is shown in Figure 4.7.1. The limited data available suggests that selective surfaces may exhibit similar behavior [Pettit and Sowell (1976)].

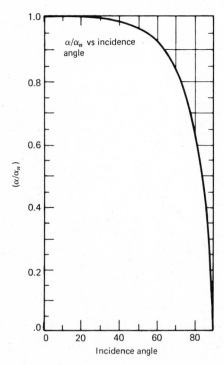

Figure 4.7.1 Ratio of solar absorptance to absorptance at normal incidence for a flat black surface. From Beckman et al. (1977).

* Methods for this evaluation are in Chapter 14.

4.8 SPECULARLY REFLECTING SURFACES

Concentrating solar collectors require the use of reflecting materials (or possibly refracting materials) to direct the beam component of solar radiation onto a target. This requires surfaces of high specular reflectance for radiation in the solar spectrum.

Specular surfaces are usually metals or metallic coatings on smooth substrates. Opaque substrates must be front-surfaced. Examples are anodized aluminum and rhodium-plated copper. The specular reflectivity of such surfaces is a function of the quality of the substrate and the plating.

Specular surfaces can also be applied to transparent substrates, including glass or plastic. If front-surface coatings are used, the nature of the substrate, other than its smoothness and stability, is unimportant. If back-surface coatings are applied, the transparency of the substrate will also be a factor, as the radiation will pass through the equivalent of twice the thickness of the substrate and twice through the front surface-air interface. (See Chapter 5 for discussion of radiation transmission through partially transparent media.)

Specular reflectance is, in general, wavelength dependent, and in principle, monochromatic reflectances should be integrated for the particular spectral distribution of incident energy. Thus, the monochromatic specular reflectance is defined as

$$\rho_{s,\lambda} = \frac{I_{\lambda,\rho s}}{I_{\lambda,i}} \qquad (4.8.1)$$

where $I_{\lambda,\rho s}$ is the monochromatic intensity specularly reflected, and $I_{\lambda,i}$ is the incident monochromatic intensity. Then, the specular reflectivity is

$$\rho_s = \frac{\int_0^\infty \rho_{s,\lambda} I_{\lambda,i}\, d\lambda}{\int_0^\infty I_{\lambda,i}\, d\lambda} \qquad (4.8.2)$$

Typical values of specular reflectance of surfaces of solar radiation are shown in Table 4.8.1.

The table includes data on front-surface and second-surface reflectors. The aluminized acrylic film is one of a number of aluminized polymeric films that have been evaluated for durability in weather, and it appears to be the best of those reported by the University of Minnesota and Honeywell (1973). Many other such materials have short lifetimes (on the order of weeks or months) under practical operating conditions.

The maintenance of high specular reflectance presents practical problems. Front surface reflectors are subject to degradation by oxidation, abrasion, dirt, and so on. Back surface reflectors may lose reflectance because of dirt or degradation of the overlying transparent medium.

Front surface reflectors may be covered by thin layers of protective materials to increase their durability. For example, anodized aluminum is coated with a thin stable layer of aluminum oxide deposited by electrochemical means and

Table 4.8.1   Normal Specular Solar Reflectances of
Surfaces

Surface	ρ
Electroplated silver, new	0.96
High-purity Al, new clean	0.91
Sputtered aluminum optical reflector	0.89
Brytal processed aluminum, high purity	0.89
Back-silvered water white plate glass, new, clean	0.88
Al, SiO coating, clean	0.87
Aluminum foil, 99.5% pure	0.86
Back-aluminized 3M acrylic, new	0.86
Back-aluminized 3M acrylic	0.85[a]
Commercial Alzac process aluminum	0.85
Aluminized Type C Mylar (from Mylar Side)	0.76

[a] Exposed to equivalent of 1 yr solar radiation.

silicon monoxide has been applied to front surface aluminum films. In general, each coating reduces the initial value of specular reflectance, but may result in more satisfactory levels of reflectance over long periods of time.

REFERENCES

Beckman, W. A., S. A. Klein, and J. A. Duffie, *Solar Heating Design*, Wiley, New York (1977).

Benning, A. C., paper in *AES Coatings for Solar Collectors Symposium*, American Electroplaters' Society, Winter Park, FL, p. 57 (1976). "Black Chromium—A Solar Selective Coating."

Close, D. J., Report E.D.7, Engineering Section, Commonwealth Scientific and Industrial Research Organization, Melbourne, Australia (1962). "Flat Plate Solar Absorbers: The Production and Testing of a Selective Surface for Copper Absorber Plates."

Cuomo, J. J., J. M. Woodall, and T. W. DiStefano, paper in *AES Coatings for Solar Collectors Symposium*, Electroplaters' Society Winter Park, FL, (1976). "Dendritic Tungsten for Solar Thermal Conversion."

Duffie, J. A. and W. A. Beckman, *Solar Energy Thermal Processes*, Wiley, New York (1974).

Edwards, D. K., K. E. Nelson, R. D. Roddick, and J. T. Gier, Report No. 60-93, Dept. of Engineering, University of California at Los Angeles (October 1960). "Basic Studies on the Use and Control of Solar Energy."

Edwards, D. K., J. T. Gier, K. E. Nelson, and R. Roddick, *Solar Energy*, 6, 1 (1962). "Spectral and Directional Thermal Radiation Characteristics of Selective Surfaces for Solar Collectors."

Gier, J. T. and R. V. Dunkle, *Transactions of the Conference of the Use of Solar Energy*, Vol. 2, Part I, p. 41, University of Arizona Press, Tucson (1958). "Selective Spectral Characteristics as an Important Factor in the Efficiency of Solar Collectors."

Gier, J. T., R. V. Dunkle, and J. T. Bevans, *Journal of the Optical Society of America*, 44, 558 (1954). "Measurements of Absolute Spectral Reflectivity from 1.0 to 15 Microns."

Grimmer, C. P., K. C. Herr, and W. V. McCreary, paper in *AES Coatings for Solar Collectors Symposium*, American Electroplaters Society, Winter Park, FL, p. 79 (1976). "A Possible Selective Solar Photothermal Absorber: Ni Dendrites Formed on Al Surfaces by the CVD of $NiCO_4$."

Harding, G. L., *Journal of Vacuum Science and Technology*, **13**, 1070 (1976). "Sputtered Metal Carbide Solar-Selective Absorbing Surfaces."

Harding, G. L., D. L. McKenzie, and B. Window, *Journal of Vacuum Science and Technology*, **13**, 1073 (1976). "The dc Sputter Coating of Solar-Selective Surfaces Onto Tubes."

Hollands, K. G. T., *Solar Energy*, **7**, 108 (1963). "Directional Selectivity, Emittance and Absorptance Properties of Vee Corrugated Specular Surfaces."

Hottel, H. C. and T. A. Unger, *Solar Energy*, **3**(3), 10 (1959). "The Properties of a Copper Oxide-Aluminum Selective Black Surface Absorber of Solar Energy."

Hottel, H. C. and B. B. Woertz, *Transactions of the American Society of Mechanical Engineers*, **14**, 91 (1942). "Performance of Flat-Plate Solar-Heat Exchangers."

Kokoropoulos, P., E. Salam, and F. Daniels, *Solar Energy*, **3**(4), 19 (1959). "Selective Radiation Coatings—Preparation and High Temperature Stability."

Lin, R. J. H., Report C00/2930-4 to ERDA (Jan. 1977). "Optimization of Coatings for Flat Plate Solar Collectors."

Mar, H. Y. B., R. E. Peterson, and P. B. Zimmer, *Thin Solid Films*, **39**, 95 (1976). "Low Cost Coatings for Flat Plate Solar Collectors."

Martin, D. C. and R. Bell, *Proceedings of the Conference on Coatings for the Aerospace Environment*, WADD-TR-60-TB, Wright Air Development Division, Dayton, Ohio (November 1960). "The Use of Optical Interference to Obtain Selective Energy Absorption."

McDonald, G. E., *Solar Energy* **17**, 119 (1975). "Spectral Reflectance Properties of Black Chrome for Use as a Solar Selective Coating."

McDonald, G. E., NASA Technical Memo NASA TMX 0171596 (1974). "Spectral Reflectance Properties of Black Chrome for Use as a Solar Selective Coating."

Meinel, A. B., M. P. Meinel, C. B. McDenney, and B. O. Seraphin, paper E13 at Paris Meeting of International Solar Energy Society (1973). "Photothermal Conversion of Solar Energy for Large-Scale Electrical Power Production."

Moore, S. W., paper in *AES Coatings for Solar Collectors Symposium*, American Electroplaters' Society, Winter Park, FL, (1976). "Results Obtained from Black Chrome Production Run of Steel Collectors."

Pettit, R. B. and R. P. Sowell, *Journal of Vacuum Science and Technology*, **13** (1976), "Solar Absorptance and Emittance Properties."

Santala, T., paper in *Workshop in Solar Collectors for Heating and Cooling of Buildings*, Report, National Science Foundation RA-N-75-919, p. 233 (May 1975). "Intermetallic Absorption Surface-Material Systems for Collector Plates."

Schmidt, R. N., K. C. Park, and E. Janssen, Tech. Doc. Report No. ML-TDR-64-250 from Honeywell Research Center to Air Force Materials Laboratory (Sept. 1964). "High Temperature Solar Absorber Coatings, Part II."

Schmidt, R. N., Honeywell Corp. private communication (1974).

Seraphin, B. O., Reports on National Science Foundation Grant GI-36731X of the Optical Sciences Center, University of Arizona (1975).

Siegel, R. and J. R. Howell, *Thermal Radiation Heat Transfer*, McGraw-Hill, New York (1972).

Sowell, R. R. and D. M. Mattox, paper in *AES Coatings for Solar Collectors Symposium*, American Electroplaters' Society, Winter Park, FL, p. 21 (1976), "Properties and Composition of Electroplated Black Chrome."

Tabor, H., *Bulletin of the Research Council of Israel*, **5A** (2), 119 (Jan. 1956). "Selective Radiation."

Tabor, H., *Low Temperature Engineering Applications of Solar Energy*, ASHRAE, New York (1967). "Selective Surfaces for Solar Collectors."

Tabor, H., J. Harris, H. Weinberger, and B. Doron, *Proceedings UN Conference on New Sources of Energy*, **4**, 618 (1964). "Further Studies on Selective Black Coatings."

Tabor, H., paper in *Applications of Solar Energy for Heating and Cooling of Buildings*, ASHRAE GRP-170, New York (1977). "Selective Surfaces for Solar Collectors."

Touloukian, Y. S., et al. *Thermophysical Properties of Matter*, Plenum Data Corporation, New York, Vol. 7 (1970): "Thermal Radiative Properties—Metallic Elements and Alloys;" Vol. 8 (1972) "Thermal Radiative Properties—Nonmetallic Solids;" Vol. 9 (1972): "Thermal Radiative Properties—Coatings."

Trombe, F., M. Foex, and M. LePhat Vinh, *Proceedings of the UN Conference on New Sources of Energy*, **4**, 625, 638 (1964). "Research on Selective Surfaces for Air Conditioning Dwellings."

University of Minnesota and Honeywell Corp. Progress Report No. 2 to the National Science Foundation, NSF/RANN/SE/GI-34871/PR/73/2 (1973). "Research Applied to Solar-Thermal Power Systems."

Williams, D. A., T. A. Lappin, and J. A. Duffie, *Journal of Transactions of the American Society of Mechanical Engineers, Engineering for Power*, **85A**, 213 (1963). "Selective Radiation Properties of Particulate Coatings."

Williams, D. A., Ph.D. Thesis, University of Wisconsin (1961). "Selective Radiation Properties of Particulate Semiconductor Coatings on Metal Substrates."

CHAPTER 5

Radiation Transmission Through Covers and Absorption by Collectors

The transmission, reflection, and absorption of solar radiation by the various parts of a solar collector are important in determining collector performance. The transmittance, reflectance, and absorptance are functions of the incoming radiation, and the thickness, refractive index, and extinction coefficient of the material. Generally, the refractive index, n, and the extinction coefficient, K, are functions of the wavelength of the radiation. However, in this chapter, all properties initially will be assumed to be independent of wavelength. This is an excellent assumption for glass, the most common solar collector cover material. Some cover materials have significant optical property variations with wavelength, and spectral dependence of properties is considered in Section 5.6. Incident solar radiation is unpolarized (or only slightly polarized). However, polarization considerations are important as radiation becomes partially polarized as it passes through collector covers.

The last sections of this chapter treat the absorption of solar radiation by collectors, windows, and collector-storage walls on an hourly and on a monthly average basis.

Reviews of important considerations of transmission of solar radiation have been presented by Dietz (1954, 1963).

5.1 REFLECTION OF RADIATION

For smooth surfaces Fresnel has derived expressions for the reflection of unpolarized radiation on passing from a medium 1 with a refractive index, n_1, to medium 2 with refractive index, n_2.

$$r_\perp = \frac{\sin^2(\theta_2 - \theta_1)}{\sin^2(\theta_2 + \theta_1)} \qquad (5.1.1)$$

$$r_\parallel = \frac{\tan^2(\theta_2 - \theta_1)}{\tan^2(\theta_2 + \theta_1)} \qquad (5.1.2)$$

$$r = \frac{I_r}{I_i} = \frac{1}{2}[r_\perp + r_\parallel] \qquad (5.1.3)$$

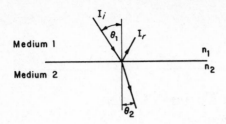

Figure 5.1.1 Angles of incidence and refraction in media having refractive indices n_1 and n_2.

where θ_1 and θ_2 are the angles of incidence and refraction, as shown in Figure 5.1.1. Equation 5.1.1 represents the perpendicular component of unpolarized radiation, $r_\perp$, and Equation 5.1.2 represents the parallel component of unpolarized radiation, $r_\parallel$. (Parallel and perpendicular refer to the plane defined by the incident beam and the surface normal.) Equation 5.1.3, then, gives the reflection of unpolarized radiation as the average of the two components. The angles θ_1 and θ_2 are related to the indices of refraction by Snell's law,

$$\frac{n_1}{n_2} = \frac{\sin \theta_2}{\sin \theta_1} \tag{5.1.4}$$

Thus, if the angle of incidence and refractive indices are known, Equations 5.1.1 through 5.1.4 are sufficient to calculate the reflectance of the single interface.

For radiation at normal incidence, both θ_1 and θ_2 are 0, and Equations 5.1.3 and 5.1.4 can be combined to yield

$$r(0) = \frac{I_r}{I_i} = \left[\frac{(n_1 - n_2)}{(n_1 + n_2)}\right]^2 \tag{5.1.5}$$

If one medium is air (i.e., a refractive index of nearly unity), Equation 5.1.5 becomes

$$r(0) = \frac{I_r}{I_i} = \left[\frac{(n - 1)}{(n + 1)}\right]^2 \tag{5.1.6}$$

Example 5.1.1

Calculate the reflectance of one surface of glass at normal incidence and at 60°. The average index of refraction of glass for the solar spectrum is 1.526.

Solution

At normal incidence, Equation 5.1.6 can be used:

$$r(0) = \left[\frac{0.526}{2.526}\right]^2 = 0.0434$$

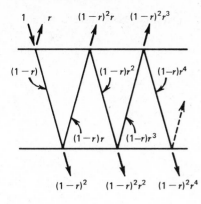

Figure 5.1.2   Transmission through one nonabsorbing cover.

At an incidence angle of 60°, Equation 5.1.4 gives the refraction angle, θ_2.

$$\theta_2 = \sin^{-1}\left[\frac{\sin 60}{1.526}\right] = 34.58$$

From Equation 5.1.3, the reflectance is

$$r(60) = \frac{1}{2}\left[\frac{\sin^2(-25.42)}{\sin^2(94.58)} + \frac{\tan^2(-25.42)}{\tan^2(94.58)}\right]$$

$$= \tfrac{1}{2}[0.185 + 0.001] = 0.093 \qquad\blacksquare$$

In solar applications, the transmission of radiation is through a slab or film of material so there are two interfaces per cover to cause reflection losses. At off-normal incidence, the radiation reflected at an interface is different for each component of polarization, so the transmitted and reflected radiation becomes partially polarized. Consequently, it is necessary to treat each component of polarization separately.

Neglecting absorption in the slab shown in Figure 5.1.2, and considering for the moment only the perpendicular component of polarization of the incoming radiation, $(1 - r_\perp)$ of the incident beam reaches the second interface. Of this, $(1 - r_\perp)^2$ passes through the interface and $r_\perp(1 - r_\perp)$ is reflected back to the first, and so on. Summing up the transmitted terms, the transmittance for the perpendicular component of polarization is

$$\tau_\perp = (1 - r_\perp)^2 \sum_{n=0}^{\infty} r_\perp^{2n} = \frac{(1 - r_\perp)^2}{(1 - r_\perp^2)} = \frac{1 - r_\perp}{1 + r_\perp} \tag{5.1.7}$$

Exactly the same expansion results when the parallel component of polarization is considered. $r_\perp$ and $r_\parallel$ are not equal (except at normal incidence) and the transmittance of initially unpolarized radiation is the average transmittance of the two components.

$$\tau_r = \frac{1}{2}\left[\frac{1 - r_\parallel}{1 + r_\parallel} + \frac{1 - r_\perp}{1 + r_\perp}\right] \tag{5.1.8}$$

where the subscript r is a reminder that only reflection losses have been considered.

For a system of N covers, all of the same materials, a similar analysis yields

$$\tau_{rN} = \frac{1}{2}\left[\frac{1 - r_\perp}{1 + (2N - 1)r_\perp} + \frac{1 - r_\|}{1 + (2N - 1)r_\|}\right] \qquad (5.1.9)$$

Example 5.1.2

Calculate the transmittance of two covers of nonabsorbing glass at normal incidence and at 60°.

Solution

At normal incidence, the reflectance of one interface from Example 5.1.1 is 0.0434. From Equation 5.1.9, with both polarization components equal, the transmittance is

$$\tau_r(0) = \frac{1 - 0.0434}{1 + 3(0.0434)} = 0.85$$

Also from Example 5.1.1 but at a 60° incidence angle, the reflectances of one interface for each component of polarization are 0.185 and 0.001. From Equation 5.1.9, the transmittance is

$$\tau_r(60) = \frac{1}{2}\left[\frac{1 - 0.185}{1 + 3(0.185)} + \frac{1 - 0.001}{1 + 3(0.001)}\right] = 0.76 \qquad \blacksquare$$

The solar transmittance of nonabsorbing glass, having an average refractive index of 1.526 in the solar spectrum, has been calculated for all incidence angles

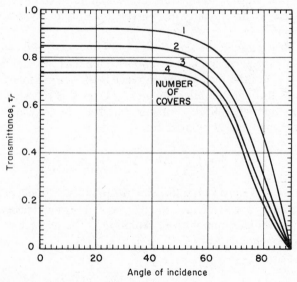

Figure 5.1.3   Transmittance of 1, 2, 3, and 4 nonabsorbing covers having an index of refraction of 1.526.

Table 5.1.1   Average Refractive Index in Solar Spectrum
of Some Cover Materials

Cover Material	Average Refractive Index
Glass	1.526
Polymethyl methacrylate	1.49
Polyvinylfluoride	1.45
Polyfluorinated ethylene propylene	1.34
Polytetrafluoroethylene	1.37
Polycarbonate	1.60

in the same manner illustrated in Examples 5.1.1 and 5.1.2. The results for from one to four glass covers are given in Figure 5.1.3. This figure is a recalculation of the results presented by Hottel and Woertz (1942).

The index of refraction of materials that have been considered for solar collector covers are given in Table 5.1.1. The values correspond to the solar spectrum and can be used to calculate the angular dependence of reflection losses similar to Figure 5.1.3.

5.2   ABSORPTION OF RADIATION

The absorption of radiation in a partially transparent medium is described by Bouguer's law, which is based on the assumption that the absorbed radiation is proportional to the local intensity in the medium and the distance the radiation travels in the medium, x:

$$dI = -IK\ dx \qquad (5.2.1)$$

where K is the proportionality constant, called the extinction coefficient, and is assumed to be a constant in the solar spectrum. Integrating along the actual pathlength in the medium (i.e., from 0 to $L/\cos\theta_2$) yields

$$\tau_a = \frac{I_\tau}{I_0} = e^{-KL/\cos\theta_2} \qquad (5.2.2)$$

where the subscript, a, is a reminder that only absorption losses have been considered. For glass, the value of K varies from approximately 4 m^{-1} for "water white" glass (which appears white when viewed on the edge) to approximately 32 m^{-1} for poor (greenish cast of edge) glass.

5.3   OPTICAL PROPERTIES OF COVER SYSTEMS

The transmittance, reflectance and absorptance of a single cover, allowing for both reflection and absorption losses, can be determined by ray-tracing techniques similar to that used to derive Equation 5.1.7. For the parallel component of polarization, the transmittance, $\tau_{\|}$, reflectance, $\rho_{\|}$, and absorptance, $\alpha_{\|}$, of the cover are:

$$\tau_{\|} = \frac{\tau_a(1 - r_{\|})^2}{1 - (r_{\|}\tau_a)^2} = \tau_a\left(\frac{1 - r_{\|}}{1 + r_{\|}}\right)\left(\frac{1 - r_{\|}^2}{1 - (r_{\|}\tau_a)^2}\right) \tag{5.3.1}$$

$$\rho_{\|} = r_{\|} + \frac{(1 - r_{\|})^2\tau_a^2 r_{\|}}{1 - (r_{\|}\tau_a)^2} = r_{\|}(1 + \tau_a\tau_{\|}) \tag{5.3.2}$$

$$\alpha_{\|} = (1 - \tau_a)\left(\frac{1 - r_{\|}}{1 - r_{\|}\tau_a}\right) \tag{5.3.3}$$

Similar results are found for the perpendicular component of polarization. For incident unpolarized radiation, the optical properties are found by the average of the two components.

The equation for the transmittance of a collector cover can be simplified by noting that the last term in Equation 5.3.1 (and its equivalent for the perpendicular component of polarization) is nearly unity, since τ_a is seldom less than 0.9 and r is on the order of 0.1 for practical collector covers. With this simplification, and with Equation 5.1.8, the transmittance of a single cover becomes

$$\tau \cong \tau_a\tau_r \tag{5.3.4}$$

This is a satisfactory relationship for solar collectors with cover materials and angles of practical interest.

The absorptance of a solar collector cover can be approximated by letting the last term in Equation 5.3.3 be unity so that

$$\alpha \cong 1 - \tau_a \tag{5.3.5}$$

Although the neglected term in Equation 5.3.3 is larger than the neglected term in Equation 5.3.1, the absorptance is much smaller than the transmittances so that the overall accuracy of the two approximations is essentially the same.

The reflectance of a single cover is then found from $\rho = 1 - \alpha - \tau$ so that:

$$\rho \cong \tau_a(1 - \tau_r) = \tau_a - \tau \tag{5.3.6}$$

The advantage of Equations 5.3.4 through 5.3.6 over Equations 5.3.1 through 5.3.3 is that polarization is accounted for in the approximate equations through the single term, τ_r, rather than by the more complicated expressions for each individual optical property. Example 5.3.1 shows the solution for transmittance by the exact equations and also by the approximate equations.

Example 5.3.1

Calculate the transmittance, reflectance, and absorptance of a single glass cover 2.3 mm thick at an angle of 60°. The extinction coefficient of the glass is 32 m^{-1}.

Solution

At an incidence angle of 60°, the optical pathlength is

$$\frac{KL}{\cos \theta_2} = 32 \times \frac{0.0023}{\cos 34.58} = 0.0894$$

where 34.58 is the refraction angle calculated in Example 5.1.1. The transmittance, τ_a, from Equation 5.2.2 is then

$$\tau_a = e^{-0.0894} = 0.915$$

Using the results of Example 5.1.1 and Equation 5.3.1 the transmittance is found by averaging the transmittances for the parallel and perpendicular components of polarization,

$$\tau = \frac{0.915}{2} \left[\left(\frac{1 - 0.185}{1 + 0.185}\right)\left(\frac{1 - 0.185^2}{1 - (0.915 \times 0.185)^2}\right) \right.$$

$$\left. + \left(\frac{1 - 0.001}{1 + 0.001}\right)\left(\frac{1 - 0.001^2}{1 - (0.915 \times 0.001)^2}\right) \right]$$

$$\tau = \tfrac{1}{2}[0.624 + 0.912] = 0.768$$

The reflectance is found using Equation 5.3.2 for each component of polarization

$$\rho = \tfrac{1}{2}[0.185(1 + 0.915 \times 0.624) + 0.001(1 + 0.915 \times 0.912)]$$
$$\rho = \tfrac{1}{2}(0.291 + 0.003) = 0.147$$

The absorptance is found from Equation 5.3.3

$$\alpha = \frac{(1 - 0.915)}{2} \left[\frac{1 - 0.185}{1 - 0.185 \times 0.915} + \frac{1 - 0.001}{1 - 0.001 \times 0.915} \right]$$

$$\alpha = \frac{0.085}{2} [0.981 + 1.000] = 0.085$$

Alternate solution

The approximate equations can also be used to find these properties. From Equation 5.3.4 and Equation 5.1.8, the transmittance is

$$\tau = \frac{0.915}{2} \times \left[\frac{1 - 0.185}{1 + 0.185} + \frac{1 - 0.001}{1 + 0.001} \right] = 0.771$$

From Equation 5.3.5, the absorptance is

$$\alpha = 1 - 0.915 = 0.085$$

and the reflectance is then

$$\rho = 1 - 0.771 - 0.085 = 0.144$$

Note that even though the incidence angle was large and poor quality glass was used in this example so that the approximate equations tend to be less accurate, the approximate method and the exact method are essentially in agreement.   ■

Although Equations 5.3.4 through 5.3.6 were derived for a single cover, they apply equally well to identical multiple covers. The quantity τ_r should be evaluated from Equation 5.1.9 and the quantity τ_a from Equation 5.2.2 with L equal to the total cover system thickness.

Example 5.3.2

Calculate the solar transmittance at incidence angles of zero and $60°$ for two glass covers, each 2.3 mm thick. The extinction coefficient of the glass is 16.1 m^{-1}, and the refractive index is 1.526.

Solution

For one sheet at normal incidence, the KL product is

$$KL = 16.1 \times 2.3/1000 = 0.0370$$

The transmittance, τ_a, is then

$$\tau_a(0) = e^{-2(0.0370)} = 0.93$$

The transmittance accounting for reflection, from Example 5.1.2, is 0.85. The total transmittance is then found from Equation 5.3.4:

$$\tau(0) = \tau_r(0)\tau_a(0) = 0.85(0.93) = 0.79$$

From Example 5.1.1, when $\theta_1 = 60°$, $\theta_2 = 34.57$, and

$$\tau_a(60) = e^{-2(0.0370)/\cos 34.57} = 0.91$$

and the total transmittance (with $\tau_r = 0.76$ from Example 5.1.2) becomes

$$\tau(60) = \tau_r(60)\,\tau_a(60) = 0.76(0.91) = 0.69 \qquad ■$$

Figure 5.3.1 gives curves of transmittance as a function of angle of incidence for systems of one to four identical covers, for three different kinds of glass. These curves were calculated from Equation 5.3.4 and have been checked by experiments [Hottel and Woertz (1942)].

In a multicover system, the ray-tracing technique used to develop Equation 5.1.7 can be used to derive the appropriate equations. Whillier (1953) has generalized the ray-tracing method to any number or type of covers, and modern radiation heat transfer calculation methods have also been applied to these complicated situations [e.g., Edwards (1977)]. If the covers are identical, the approximate method illustrated in Example 5.3.2 is recommended, although the following equations can also be used.

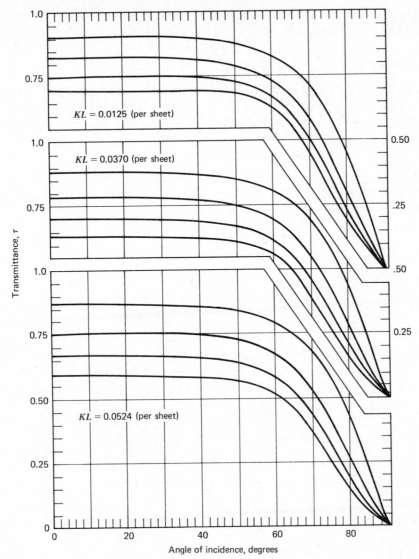

Figure 5.3.1 Transmittance (considering absorption and reflection) of 1, 2, 3 and 4 covers for three types of glass.

For a two-cover system, with covers not necessarily identical, ray tracing yields the following equations for transmittance and reflectance, where subscript 1 refers to the outer cover and subscript 2 to the inner cover.

$$\tau = \frac{1}{2}\left[\left(\frac{\tau_1 \tau_2}{1 - \rho_1 \rho_2}\right)_\perp + \left(\frac{\tau_1 \tau_2}{1 - \rho_1 \rho_2}\right)_\parallel\right] \tag{5.3.7}$$

$$\rho = \frac{1}{2}\left[\left(\rho_1 + \frac{\tau \rho_2 \tau_1}{\tau_2}\right)_\perp + \left(\rho_1 + \frac{\tau \rho_2 \tau_1}{\tau_2}\right)_\parallel\right] \tag{5.3.8}$$

Note that the reflectance from the cover system depends upon which cover is first to intercept radiation.

Example 5.3.3

Calculate the optical properties of a two-cover solar collector at an angle of 60°. The outer cover is glass with $K = 16.1$ m^{-1} and thickness of 2.3 mm. The inner cover is polyvinyl fluoride with refractive index equal to 1.45. The plastic film is thin enough so that absorption within the plastic can be neglected.

Solution

The optical properties of the glass and plastic covers alone, as calculated from Equations 5.3.1 through 5.3.3, are

$$
\begin{aligned}
\text{glass: } \tau_{\parallel} &= 0.961 & \tau_{\perp} &= 0.660 \\
\rho_{\parallel} &= 0.002 & \rho_{\perp} &= 0.304 \\
\alpha_{\parallel} &= 0.037 & \alpha_{\perp} &= 0.036
\end{aligned}
$$

$$
\begin{aligned}
\text{plastic: } \tau_{\parallel} &= 0.995 & \tau_{\perp} &= 0.726 \\
\rho_{\parallel} &= 0.005 & \rho_{\perp} &= 0.274 \\
\alpha_{\parallel} &= 0 & \alpha_{\perp} &= 0
\end{aligned}
$$

(Note that Equations 5.3.4 through 5.3.5 could have been used with each component of polarization to simplify the calculation of the above properties.)

The transmittance of the combination is found from Equation 5.3.7:

$$
\tau = \frac{1}{2}\left[\frac{0.961 \times 0.995}{1 - 0.002 \times 0.005} + \frac{0.660 \times 0.726}{1 - 0.304 \times 0.274}\right]
$$

$$
= \tfrac{1}{2}[0.956 + 0.523] = 0.739
$$

The reflectance, with the glass first, is found from Equation 5.3.8.

$$
\rho = \frac{1}{2}\left[0.002 + \frac{0.956 \times 0.005 \times 0.961}{0.995} + 0.304 + \frac{0.523 \times 0.274 \times 0.660}{0.726}\right]
$$

$$
= \tfrac{1}{2}[0.007 + 0.434] = 0.221
$$

The absorptance is then

$$
\alpha = 1 - 0.221 - 0.739 = 0.040. \qquad \blacksquare
$$

Equations 5.3.7 and 5.3.8 can be used to calculate the transmittance of any number of covers by repeated application. If subscript 1 refers to the properties of a cover system and subscript 2 to the properties of an additional cover placed under the stack, then these equations yield the appropriate transmittance and reflectance of the new system. The reflectance, ρ_1, in Equation 5.3.7 is the reflectance of the original cover system from the bottom side. If any of the covers exhibit strong wavelength dependent properties, integration over the wavelength spectrum is necessary (see Section 5.6).

5.4 TRANSMITTANCE FOR DIFFUSE RADIATION

The preceding analysis only applied to the beam component of solar radiation. Radiation incident on a collector also consists of scattered solar radiation from the sky and possibly reflected solar radiation from the ground. In principle, the amount of this radiation that passes through a cover system can be calculated by integrating the transmitted radiation over all angles. However, the angular distribution of this radiation is generally unknown.

If the incident radiation is isotropic (i.e., independent of angle), then the integration can be performed. The presentation of the results can be simplified

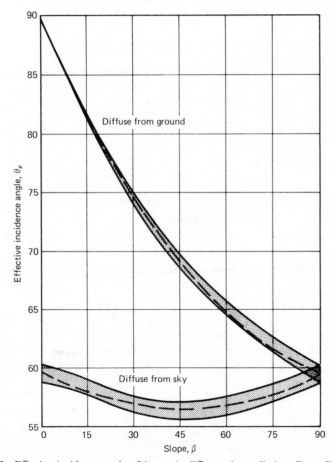

Figure 5.4.1 Effective incidence angle of isotropic diffuse solar radiation. From Brandemuehl and Beckman (1980).

by defining an equivalent angle for beam radiation that gives the same transmittance as for diffuse radiation. For a wide range of conditions encountered in solar collector applications, this equivalent angle is essentially 60°. In other words, beam radiation incident at an angle of 60° has the same transmittance as isotropic diffuse radiation.

Solar collectors are usually orientated so that they "see" both the sky and the ground. If the diffuse radiation from the sky and ground are both isotropic, then the transmittance of the glazing systems can be found by integrating the beam transmittance over the appropriate incidence angles. This integration has been done by Brandemuehl and Beckman (1980) and the results are presented in Figure 5.4.1 in terms of the effective beam radiation incidence angle. The cross-hatched region includes a wide range of glazings. The upper curve is for a one cover polyflorinated ethylene propylene glazing with no internal absorption, whereas the lower curve represents a two cover glass glazing with extinction length (KL) of 0.0524. All one and two cover systems with index of refraction between 1.34 and 1.526 and extinction lengths less than 0.0524 lie in the cross-hatched region.

The dashed lines shown on Figure 5.4.1 are given by

$$\theta_e = 90 - 0.5788\beta + 0.002693\beta^2 \qquad (5.4.1)$$

for diffuse ground radiation and

$$\theta_e = 59.68 - 0.1388\beta + 0.001497\beta^2 \qquad (5.4.2)$$

for diffuse sky radiation.

5.5 TRANSMITTANCE-ABSORPTANCE PRODUCT

To use the analysis of the next chapter, it is necessary to evaluate the transmittance-absorptance product, $(\tau\alpha)$. Of the radiation passing through the cover system and striking the plate, some is reflected back to the cover system. However, all this radiation is not lost since some of it is, in turn, reflected back to the plate.

The situation is illustrated in Figure 5.5.1, where τ is the transmittance of the cover system at the desired angle and α is the angular absorptance of the absorber plate. Of the incident energy, $\tau\alpha$ is absorbed by the absorber plate and $(1 - \alpha)\tau$ is reflected back to the cover system. The reflection from the absorber plate is assumed to be diffuse (and unpolarized) so that the fraction $(1 - \alpha)\tau$ that strikes the cover plate is diffuse radiation and $(1 - \alpha)\tau\rho_d$ is reflected back to the absorber plate. The quantity ρ_d refers to the reflectance of the cover system for diffuse radiation incident from the bottom side and can be estimated from Equation 5.3.6 at an angle of 60°. If the cover system consists of two (or more) covers of dissimilar materials, ρ_d will be different (slightly) from the diffuse reflectance of the incident solar radiation (see Equation 5.3.8). The multiple

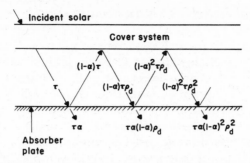

Figure 5.5.1   Adsorption of solar radiation by absorber plate.

reflection of diffuse radiation continues so that the energy ultimately absorbed is*

$$(\tau\alpha) = \tau\alpha \sum_{n=0}^{\infty} [(1-\alpha)\rho_d]^n = \frac{\tau\alpha}{1-(1-\alpha)\rho_d} \qquad (5.5.1)$$

Example 5.4.1

For a two-cover collector using glass with $KL = 0.0370$ per plate and an absorber plate with $\alpha = 0.90$ (independent of direction), find the transmittance-absorptance product at an angle of 50°.

Solution

From Figure 5.3.1, $\tau(50°) = 0.75$ and $\tau(60°) = 0.69$. From Equation 5.2.2 with $\theta_2 = 34.58°$, τ_a is 0.91. From Equation 5.3.6, ρ_d is $0.91 - 0.69 = 0.22$. From Equation 5.5.1

$$(\tau\alpha) = \frac{(0.75)(0.90)}{1-(1-0.90)(0.22)} = 0.69 \qquad \blacksquare$$

Note that the value of $(\tau\alpha)$ in this example is very nearly equal to 1.01 times the product of τ times α. This is a reasonable approximation for most practical solar collectors. Thus,

$$(\tau\alpha) \approx 1.01\tau\alpha \qquad (5.5.2)$$

can be used as an estimate of $(\tau\alpha)$ in place of Equation 5.5.1.

---

* The absorptance, α, of the absorber plate for the reflected radiation should be the absorptance for diffuse radiation. Also, the reflected radiation may not all be diffuse, and it may be partially polarized. However, the resulting errors should be negligible in that the difference between $\tau\alpha$ and $(\tau\alpha)$ is small.

5.6 SPECTRAL DEPENDENCE OF TRANSMITTANCE

Most transparent media transmit selectively; that is, transmittance is a function of wavelength of the incident radiation. Glass, the material most commonly used as a cover material in solar collectors, may absorb little of the solar energy spectrum if its Fe_2O_3 content is low. If its Fe_2O_3 content is high, it will absorb in the infrared portion of the solar spectrum. The transmittance of several glasses of varying iron content is shown in Figure 5.6.1. These show clearly that "water white" (low iron) glass has the best transmission; glasses with high Fe_2O_3 content have a greenish appearance and are relatively poor transmitters. Note that the transmission is not a strong function of wavelength in the solar spectrum except for the "heat absorbing" glass. Glass becomes substantially opaque at wavelengths longer than approximately 3 μm and can be considered as opaque to long-wave radiation.

Some collector cover materials may have transmittances that are more wavelength dependent than glass, and it may be necessary to obtain their transmittance for monochromatic radiation and then integrating over the entire spectrum. The total transmittance at angle θ then becomes

$$\tau(\theta) = \frac{\int_0^\infty \tau_\lambda(\theta) I_{\lambda, i}(\theta) \, d\lambda}{\int_0^\infty I_{\lambda, i}(\theta) \, d\lambda} \qquad (5.6.1)$$

where $\tau_\lambda(\theta)$ is calculated by the equations of the preceding section, using monochromatic values of the index of refraction and absorption coefficient. $I_{\lambda, i}(\theta)$

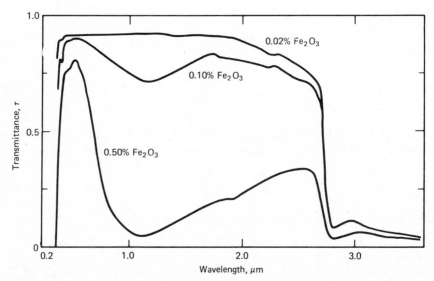

Figure 5.6.1 Spectral transmittance of 6 mm thick glass with various iron oxide contents. Data from Dietz (1954).

is the incident monochromatic intensity arriving at the cover system from angle θ.

If the absorptance of solar radiation by an absorber plate is independent of wavelength, then Equation 5.5.1 can be used to find the transmittance-absorptance product with the transmittance as calculated from Equation 5.6.1. However, if both the solar transmittance of the cover system and the solar absorptance of the absorber plate are functions of wavelength, the fraction absorbed by an absorber plate is given by:

$$\tau\alpha(\theta) = \frac{\int_0^\infty \tau_\lambda(\theta)\alpha_\lambda(\theta)I_{\lambda,\,i}(\theta)\,d\lambda}{\int_0^\infty I_{\lambda,\,i}(\theta)\,d\lambda} \qquad (5.6.2)$$

To account for multiple reflections in a manner analogous to Equation 5.5.1, it would be necessary to evaluate the spectral distribution of each reflection and integrate over all wavelengths. It is unlikely that such a calculation would ever be necessary for solar collector systems, since the error involved by directly using Equation 5.6.2 with Equation 5.5.1 would be small if α is near unity.

In a multicover system in which the covers have significant wavelength-dependent properties, the spectral distribution of the solar radiation changes as it passes through each cover. Consequently, if all covers are identical, the transmittance of individual covers increases in the direction of propagation of the incoming radiation. If the covers are not all identical, the transmittance of a particular cover may be greater or less than other similar covers in the system. Equations 5.6.1 and 5.6.2 properly account for this phenomena.

If a cover system has one cover with wavelength-independent properties (e.g., glass) and one cover with wavelength-dependent properties (e.g., some plastics), then a simplified procedure can be used. The transmittance and reflectance of each cover can be obtained separately, and the combined system transmittance and reflectance can be obtained from Equations 5.3.7 and 5.3.8.

For most plastics, the transmittance will also be significant in the infrared spectrum at $\lambda > 3$ μm. Figure 5.6.2 shows a transmittance curve for a polyvinyl

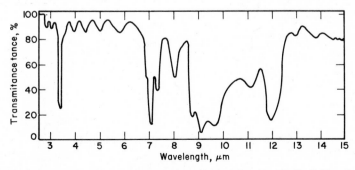

Figure 5.6.2 Infrared spectral transmittance of a polyvinyl fluoride "Tedlar" film. Courtesy of DuPont.

fluoride ("Tedlar") film for wavelengths longer than 2.5 μm. Whillier (1963) calculated the transmittance of a similar "Tedlar" film using Equation 5.6.1. The incident radiation, $I_{\lambda, i}$, was for radiation from blackbody sources at temperatures from 0 to 200 C. He found that transmittance was 0.32 for radiation from the blackbody source at 0 C, 0.29 for the source at 100 C, and 0.32 for the source at 200 C.

5.7   EFFECTS OF SURFACE LAYERS ON TRANSMITTANCE

If a film of low refractive index is deposited at an optical thickness of $\lambda/4$ onto a transparent slab, radiation of wavelength λ reflected from the upper and lower surfaces of the film will have a phase difference of π and will cancel. The reflectance will be decreased, and the transmittance will be increased relative to the uncoated material. This is the principal type of antireflection coatings used on camera lenses, binoculars, and other expensive optical equipment.

Inexpensive and durable processes have been developed for treating glass to reduce its reflectance by the addition of films having a refractive index between that of air and the transparent medium [e.g., Thomsen (1951)]. The solar reflectance of a single pane of untreated glass is approximately 8 percent. Surface treatment, by dipping glass in a silica-saturated fluosilic acid solution, can reduce the reflection losses to 2 percent, and a double layer coating can, as shown by Mar et al. (1975), reduce reflection losses to 0.9 percent. Such an

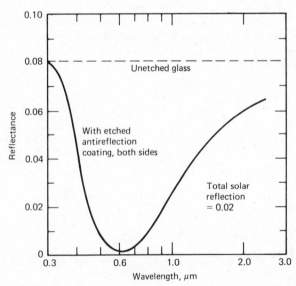

Figure 5.7.1   Monochromatic reflectance of one sheet of etched and unetched glass. From Mar et al. (1975).

Table 5.7.1 Solar Transmittance for Etched and Unetched Glass as a Function of Incidence Angle

Type of Glass	Incidence Angle						
	0°	20°	40°	50°	60°	70°	80°
Etched	0.941	0.947	0.945	0.938	0.916	0.808	0.562
Unetched	0.888	0.894	0.903	0.886	0.854	0.736	0.468

From Mar et al. (1975).

increase in solar transmittance can make a very significant improvement in the thermal performance of flat plate collectors. Figure 5.7.1 shows typical monochromatic reflectance data before and after etching. Note that unlike unetched glass, it is necessary to integrate monochromatic reflectance over the solar spectrum to obtain the reflectance to solar radiation.

Experimental values for the angular dependence of solar transmission for unetched and etched glass is given in Table 5.7.1. The etched sample not only exhibits higher transmittance than the unetched sample at all incidence angles, but the transmittance degrades less at high incidence angles.

5.8 ABSORBED SOLAR RADIATION

The prediction of collector performance requires information on the solar energy absorbed by the collector absorber plate. The solar energy incident on a tilted collector can be found by the methods of Chapter 2. This incident radiation has three different spatial distributions: beam radiation, diffuse sky radiation, and diffuse ground-reflected radiation, and each must be treated separately. On an hourly basis the absorbed radiation, S, is

$$S = I_b R_b (\tau\alpha)_b + I_d (\tau\alpha)_d \frac{(1 + \cos \beta)}{2}$$

$$+ \rho_g (I_b + I_d)(\tau\alpha)_g \frac{(1 - \cos \beta)}{2} \qquad (5.8.1)$$

where $(1 + \cos \beta)/2$ and $(1 - \cos \beta)/2$ are the view factors from the collector to the sky and from the collector to the ground, respectively.* The subscripts

---

* The product of sky area times view factor from sky to collector is identical to and has been replaced by the product of collector area times view factor from collector area to sky. A similar substitution is made for ground reflected radiation. The collector area then drops out of Equation 5.8.1.

b, d, and g represent beam, diffuse, and ground. For a given collector tilt, the effective beam incidence angle of Figure 5.4.1 can be used to find the proper transmittance values for diffuse sky and ground radiation. Equation 5.5.1 or Equation 5.5.2 can then be used to find $(\tau\alpha)_d$ and $(\tau\alpha)_g$. The angle θ, which is needed in evaluating R_b, is used to find $(\tau\alpha)_b$.

The results of the preceding sections are summarized in the following example in which the solar radiation absorbed by a solar collector is calculated.

Example 5.8.1

Calculate the radiation absorbed by the absorber plate of a single glass covered collector exposed to radiation as shown in the table. The glass has a KL product of 0.0370. The collector is tilted at an angle of 60° and the ground reflectance is 0.7. The plate solar absorptance is 0.95 and independent of direction.

Solution

The table below gives the beam and diffuse radiation on a horizontal surface, R_b, and beam incidence angle. The transmittance for beam radiation is found from Equation 5.3.4 or Figure 5.3.1. From Figure 5.4.1 the transmittances for diffuse sky and ground reflected radiation are found at angles of 57 and 65°. To find the transmittance-absorptance product it is necessary to find the reflectance at 60°. From Equation 5.2.2, τ_a at 60° is 0.96 and from Figure 5.3.1, τ at 60° is 0.82. Consequently, from Equation 5.3.6, ρ at 60° is $0.96 - 0.82 = 0.14$. The quantity $(\tau\alpha)$ for each of the three components is found from Equation 5.5.1.

$$(\tau\alpha) = \tau(\theta)\left[\frac{0.95}{1 - (1 - 0.95)(0.14)}\right] = 0.96\tau(\theta)$$

The same result is obtained from Equation 5.5.2. The quantity $(\tau\alpha)_b$ is listed in the table. For the hour 10 to 11, the absorbed beam radiation is then

$$I_b R_b (\tau\alpha)_b = 1.54 \times 2.21 \times 0.84 = 2.86 \text{ MJ/m}^2$$

The transmittance-absorptance product for diffuse sky radiation is then

$$(\tau\alpha)_d = 0.96\tau(57°) = 0.96 \times 0.82 = 0.79$$

and for diffuse ground reflected radiation is

$$(\tau\alpha)_g = 0.96\tau(65°) = 0.96 \times 0.76 = 0.73$$

The absorbed diffuse sky radiation is then

$$(\tau\alpha)_d I_d \frac{(1 + \cos\beta)}{2} = 0.79 \times 0.27 \times 0.75 = 0.16 \text{ MJ/m}^2$$

and the absorbed ground reflected radiation is

$$\rho_g(\tau\alpha)_g(I_b + I_d)\frac{(1 - \cos\beta)}{2} = 0.7 \times 0.73 \times 1.81 \times 0.25 = 0.23 \text{ MJ/m}^2$$

The table gives intermediate results for all hours.

| | | | | | | Absorbed | | | |
Hour	I_b, MJ/m^2	I_d, MJ/m^2	R_b	θ	$(\tau\alpha)_b$	Beam, MJ/m^2	Diffuse, MJ/m^2	Ground, MJ/m^2	S, MJ/m^2
7–8	0	0.02	—	63	0.75	0	0.01	—	0.01
8–9	0.05	0.27	3.34	49	0.82	0.14	0.16	0.04	0.34
9–10	0.19	0.53	2.49	35	0.84	0.40	0.31	0.09	0.80
10–11	1.54	0.27	2.21	21	0.84	2.86	0.16	0.23	3.25
11–12	1.38	0.41	2.11	7	0.85	2.48	0.24	0.23	2.95
12–1	1.64	0.29	2.11	7	0.85	2.94	0.17	0.25	3.36
1–2	1.49	0.26	2.21	21	0.84	2.77	0.15	0.22	3.14
2–3	0.59	0.42	2.49	35	0.84	1.23	0.25	0.13	1.61
3–4	0.28	0.24	3.34	49	0.82	0.77	0.14	0.07	0.98
4–5	0	0.06	—	63	0.75	0	0.04	0.01	0.05

■

In the previous example the radiation on the collector was divided into three components and each treated separately. Sometimes it is convenient to define an average transmittance-absorptance as the ratio of the absorbed solar radiation, S, to the incident solar radiation, I_T. Thus,

$$S = (\tau\alpha)_{\text{ave}} I_T \qquad (5.8.2)$$

This is especially convenient when direct measurements are available for I_T. In Example 5.8.1 the solar radiation incident on the collector for the hour 10 to 11 is 3.92 MJ/m^2. The average transmittance-absorptance product for this hour is then 0.83 which is slightly less than the beam value of $(\tau\alpha)$. When the diffuse radiation is low, using the value of $(\tau\alpha)_b$ for the average value of $(\tau\alpha)$ is often a reasonable assumption.

One further approximation is often made in estimating S. As will be seen in Chapter 6, collectors work best when the beam radiation is high. The contribution of the diffuse sky and ground reflected radiation will then be low so that $(\tau\alpha)_d$ and $(\tau\alpha)_g$ can be approximated by $(\tau\alpha)_b$ with little error in S.

5.9 MONTHLY AVERAGE ABSORBED SOLAR RADIATION

Methods for the evaluation of long-term solar system performance (see Chapter 14) require that the average radiation absorbed by a collector be evaluated for monthly periods. The solar transmittance and absorptance are both functions of the angle at which solar radiation strikes the collector. Example 5.8.1 illustrated how to calculate the absorbed solar radiation throughout a day as the incidence angle changes. This calculation can be repeated for each day of the month from which the monthly average absorbed solar radiation can be found. Klein (1979) calculated the monthly average absorbed solar radiation in this manner using many years of data. He defined a monthly average transmittance-absorptance product, which when multiplied by the monthly average radiation incident on a collector yields the monthly average absorbed radiation, $\bar{S}$.

$$\bar{S} = \bar{H}\bar{R}(\overline{\tau\alpha}) \qquad (5.9.1)$$

The following equation, analogous to the hourly evaluation of S, can be used to find $\bar{S}$.

$$\bar{S} = \bar{H}_b\bar{R}_b(\overline{\tau\alpha})_b + \bar{H}_d(\overline{\tau\alpha})_d \frac{(1 + \cos\beta)}{2} + \bar{H}\rho_g(\overline{\tau\alpha})_g \frac{(1 - \cos\beta)}{2} \qquad (5.9.2)$$

For the two diffuse terms, $(\overline{\tau\alpha})_d$ and $(\overline{\tau\alpha})_g$ can be evaluated using the effective beam radiation incidence angles given in Figure 5.4.1. For the monthly average beam radiation, the results of Klein are given in Figures 5.9.1a–f. In these figures the monthly average beam incident angle, $\bar{\theta}_b$, is given as a function of collector slope, month, latitude, and azimuth angle. These values of $\bar{\theta}_b$ were evaluated using the angular distribution of $(\tau\alpha)/(\tau\alpha)_n$ shown in Figure 5.9.2. Although this figure was derived by multiplying the angular distribution of absorptance given on Figure 4.7.1 by the angular distribution of transmission of glass covers with $KL = 0.04$, the results are not sensitive to the exact shape of the curve.

The calculations are illustrated in the following example for a passive solar heating system.

Example 5.9.1

Estimate $\bar{S}$ for a vertical collector-storage wall at a 40°N latitude location in January oriented 30° west of south. The monthly average daily beam and diffuse radiation on a horizontal surface are 3.97 MJ/m² and 2.44 MJ/m² respectively. The monthly average value of $\bar{R}_b$ is 2.29. The ground is snow covered with $\rho_g = 0.7$. The angular distribution of $(\tau\alpha)$ for the two cover glazing is as given on Figure 5.9.2 with $(\tau\alpha)_n = 0.76$.

Solution

For the vertical collector the effective beam radiation incidence angle is 59° from Figure 5.4.1 for both diffuse sky radiation and reflected ground radiation. From Figure 5.9.2 at 59°, $(\overline{\tau\alpha})/(\tau\alpha)_n$ is 0.82 so that $(\overline{\tau\alpha})_d = 1.01 \times 0.82 \times 0.76 = 0.63$.

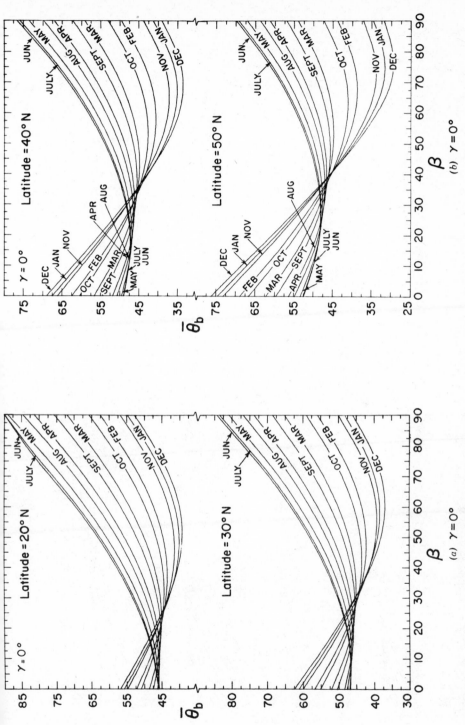

Figure 5.9.1 (a-f) Monthly average beam incidence angle for various surface locations and orientations. For Southern Hemisphere interchange months as shown on Figure 1.7.2. From Klein (1979).

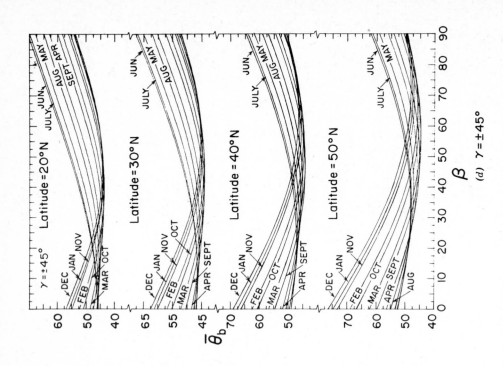

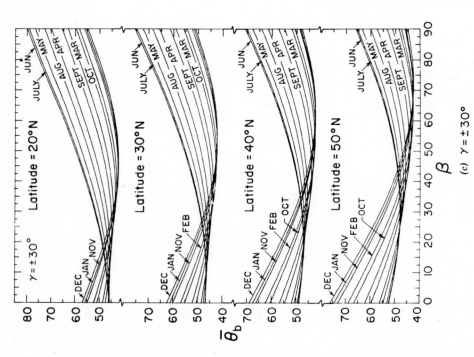

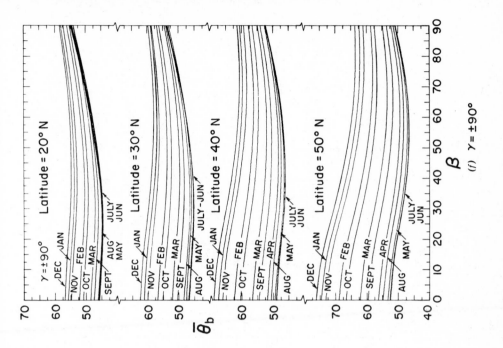

(e) Υ = ±60°

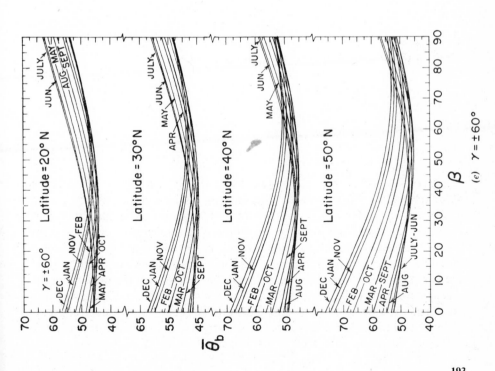

(f) Υ = ±90°

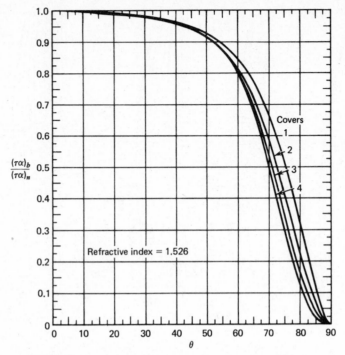

Figure 5.9.2   Typical $(\tau\alpha)_b/(\tau\alpha)_n$ curves for 1 to 4 glass covers.

From Figure 5.9.1c at $\beta = 90$, latitude = 40°N and January, $\theta_b = 48°$ so that $(\overline{\tau\alpha})_b = 1.01 \times 0.93 \times 0.76 = 0.71$. From Equation 5.9.2

$$\bar{S} = 3.97 \times 2.29 \times 0.71 + 2.44 \times 0.63 \frac{(1 + \cos 90)}{2}$$

$$+ 0.7(3.97 + 2.44)0.63 \frac{(1 - \cos 90)}{2} = 8.64 \text{ MJ/m}^2 \qquad \blacksquare$$

For collectors that face the equator, Klein (1976) found that $(\overline{\tau\alpha})_b$ could be approximated by $(\overline{\tau\alpha})$ evaluated at the incidence angle that occurs 2.5 hr from solar noon on the average day of the month. This angle can be calculated from Equation 1.6.5 (or Equation 1.6.6 for southern hemisphere) or obtained from Figure 5.9.3. This rule leads to acceptable results for solar space heating systems for which it was derived, but inaccurate results are obtained for other types of systems. Klein also found that the value of $(\overline{\tau\alpha})/(\tau\alpha)_n$ during the winter months is nearly constant and equal to 0.96 for a one-cover collector and suggests using this constant value for collectors tilted toward the equator with a slope approximately equal to the latitude plus 15° in heating system analysis. For two cover collectors a constant value of 0.94 was suggested.

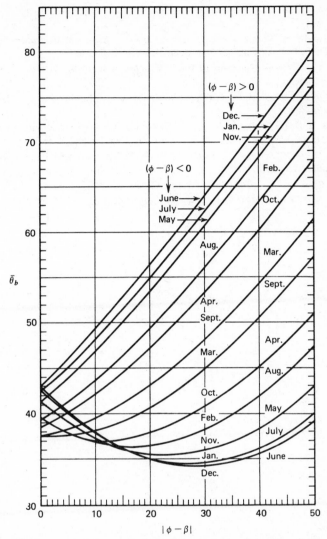

Figure 5.9.3 Monthly mean incidence angle for beam radiation for surfaces facing the equator. For the southern hemisphere, interchange the two inequality signs. From Klein (1979).

REFERENCES

Brandemuehl, M.J. and W.A. Beckman, *Solar Energy,* **24,** 511 (1980). "Transmission of Diffuse Radiation. Through CPC and Flat-Plate Collector Glazings."

Dietz, A. G. H., in *Space Heating with Solar Energy*, R. W. Hamilton (ed.), Massachusetts Institute of Technology Press, Cambridge, MA. (1954). "Diathermanous Materials and Properties of Surfaces."

Dietz, A. G. H., in *Introduction to the Utilization of Solar Energy*, A. M. Zarem, and D. D. Erway (eds.), McGraw-Hill, New York (1963). "Diathermanous Materials and Properties of Surfaces."

Edwards, D. K., *Solar Energy*, **19**, 401 (1977). "Solar Absorption by Each Element in an Absorber-Coverglass Array."

Hottel, H. C. and B. B. Woertz, *Transactions of the American Society of Mechanical Engineers*, **64**, 91 (1942). "The Performance of Flat-Plate Solar-Heat Collectors."

Klein, S. A., Ph.D. Thesis, University of Wisconsin—Madison (1976). "A Design Procedure for Solar Heating Systems."

Klein, S. A., *Solar Energy*, **23**, 547 (1979). "Calculation of the Monthly-Average Transmittance-Absorptance Product."

Mar, H. Y. B., J. H. Lin, P. P. Zimmer, R. E. Peterson, and J. S. Gross, Report to ERDA under Contract No.NSF-C-957 (September 1975), "Optical Coatings for Flat-Plate Solar Collectors."

Thomsen, S. M., *RCA Review*, **12**, 143 (1951). "Low-Reflection Films Produced on Glass in a Liquid Fluosilicic Acid Bath."

Whillier, A., *Solar Energy*, **7**, 148 (1963). "Plastic Covers for Solar Collectors."

Whillier, A., ScD. Thesis, MIT, 1953, "Solar Energy Collection and its Utilization for House Heating."

CHAPTER 6

Theory of Flat-Plate Collectors

A solar collector is a special kind of heat exchanger that transforms solar radiant energy into heat. A solar collector differs in several respects from more conventional heat exchangers. The latter usually accomplish a fluid-to-fluid exchange with high heat transfer rates and with radiation as an unimportant factor. In the solar collector, energy transfer is from a distant source of radiant energy to a fluid. Without optical concentration, the flux of incident radiation is, at best, approximately 1100 W/m² and variable. The wavelength range is from 0.29 to 2.5 μm, which is considerably shorter than that of the emitted radiation from most energy-absorbing surfaces. Thus, the analysis of solar collectors presents unique problems of low and variable energy fluxes and the relatively large importance of radiation.

Flat-plate collectors can be designed for applications requiring energy delivery at moderate temperatures, up to perhaps 100C above ambient temperature. They use both beam and diffuse solar radiation, do not require tracking of the sun, and they require little maintenance. They are mechanically simpler than concentrating collectors. The major applications of these units currently are in solar water heating and building heating, whereas potential uses include building air conditioning and industrial process heat. Passively heated buildings can be viewed as special cases of flat-plate collectors with the room or storage wall as the absorber. Passive systems are discussed in Chapter 15.

The importance of flat-plate collectors in thermal processes is such that their thermal performance is treated here in considerable detail. This is done to develop an understanding of how the component functions. In many practical cases of design calculations, the formulations of collector performance are reduced to relatively simple form.

Collector testing and costs of collectors are not considered in this chapter. This chapter deals primarily with the theory of flat-plate collectors. Chapter 7 treats testing of collectors, the use of test data, and some practical aspects of manufacture and use of these heat exchangers. Costs will be considered in chapters on applications.

6.1 GENERAL DESCRIPTION OF FLAT-PLATE COLLECTORS

The important parts of a typical liquid heating flat-plate solar collector, as shown in Figure 6.1.1, are the "black" solar energy-absorbing surface, with means for transferring the absorbed energy to a fluid; envelopes transparent to solar radiation over the solar absorber surface that reduce convection and radiation losses to the atmosphere; and back insulation to reduce conduction losses as the geometry of the system permits. Although Figure 6.1.1 depicts a water heater, and most of the analysis of this chapter is concerned with this geometry, air heaters are fundamentally the same except that the fluid tubes are replaced by a duct.

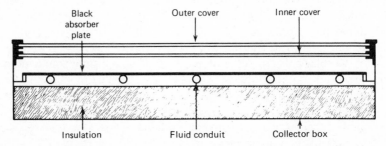

Figure 6.1.1   Cross section of a basic flat plate solar collector.

Flat-plate collectors are almost always mounted in a stationary position (e.g., as an integral part of a wall or roof structure) with an orientation optimized for the particular location in question for the time of year in which the solar device is intended to operate. In their most common forms, they are air or liquid heaters or low-pressure steam generators.

6.2 THE BASIC FLAT-PLATE ENERGY BALANCE EQUATION

In steady state, the performance of a solar collector is described by an energy balance that indicates the distribution of incident solar energy into useful energy gain, thermal losses, and optical losses. The solar radiation absorbed by a collector, S, is equal to the difference between the incident solar radiation and the optical losses and is defined by Equation 5.8.1. The thermal energy lost from the collector to the surroundings by conduction, convection, and infrared radiation can be represented by a heat transfer coefficient, U_L, times the difference between the mean absorber plate temperature, $T_{p,m}$ and the ambient temperature, T_a. In steady state the useful energy output of a collector is then the difference between the absorbed solar radiation and the thermal loss:

$$Q_u = A_c[S - U_L(T_{p,m} - T_a)] \tag{6.2.1}$$

The problem with this equation is that the mean absorber plate temperature is difficult to calculate or measure since it is a function of the collector design, the incident solar radiation, and the entering fluid conditions. This chapter is devoted to reformulating Equation 6.2.1 so that the useful energy gain can be expressed in terms of the inlet fluid temperature and a parameter, called the collector heat removal factor, which can be evaluated analytically from basic principles or measured experimentally.

Equation 6.2.1 is an energy rate equation and, in SI units, yields the useful energy gain in watts (J/s) when S is expressed in W/m^2 and U_L in W/m^2 C. Often, the most convenient time base for solar radiation is hours rather than seconds since this is the normal period for reporting meteorological data. For example, Table 2.5.2 gives solar radiation in J/m^2 for a 1-hr time period. This is the time basis for S in Equation 5.8.1 since the meaning of I is hourly J/m^2. We can consider S to be an average energy rate over a 1-hr period with units of J/m^2 hr in which case the thermal loss term, $U_L(T_{p,m} - T_a)$ must be multiplied by 3600 s/hr to obtain numerical values of the useful energy gain in J/hr. The hour time base is not a proper use of SI units but this interpretation is sometimes convenient. Alternatively, we can integrate Equation 6.2.1 over a 1-hr period. Since we seldom have data over time periods less than 1 hr, this integration can be performed only by assuming that S, $T_{p,m}$, and T_a remain constant over the hour. The resulting form of Equation 6.2.1 is unchanged except that both sides are multiplied by 3600 s/hr. To avoid including this constant in expressions for useful energy gain on an hourly basis we could have used different symbols for rates and for hourly integrated quantities (e.g., $\dot{Q}_u$ and Q_u). However, the intended meaning is always clear from the use of either G or I in the evaluation of S and we have found it unnecessary to use different symbols for collector useful energy gain on an instantaneous basis, or an hourly integrated basis. From a calculation standpoint the 3600 must still be included since S will invariably be known for an hour time period but the loss coefficient will be in SI units, which means on a second basis.

A measure of collector performance is the collection efficiency, defined as the ratio of the useful gain over some specified time period to the incident solar energy over the same time period.

$$\eta = \frac{\int Q_u \, d\tau}{A_c \int G_T \, d\tau} \tag{6.2.2}$$

The design of a solar energy system is concerned with obtaining minimum cost energy. Thus, it may be desirable to design a collector with an efficiency lower than is technologically possible if the cost is significantly reduced. In any event, it is necessary to predict the performance of a collector and that is the basic aim of this chapter.

6.3 TEMPERATURE DISTRIBUTIONS IN FLAT-PLATE SOLAR COLLECTORS

The detailed analysis of a solar collector is a complicated problem. Fortunately, a relatively simple analysis will yield very useful results. These results show the

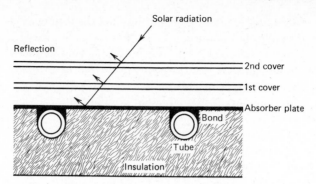

Figure 6.3.1   Sheet and tube solar collector.

important variables, how they are related, and how they affect the performance of a solar collector. To illustrate these basic principles, a liquid heating collector, as shown in Figure 6.3.1, will be examined first. The analysis presented follows the basic derivation by Whillier (1953, 1977) and Hottel and Whillier (1958).

To appreciate the development that follows, it is desirable to have an understanding of the temperature distribution that exists in a solar collector constructed as shown in Figure 6.3.1. Figure 6.3.2 shows a region between two tubes. Some of the solar energy absorbed by the plate must be conducted along the plate to the region of the tubes. Thus, the temperature midway between the tubes will be higher than the temperature in the vicinity of the tubes. The temperature above the tubes will be nearly uniform because of the presence of the tube and weld metal.

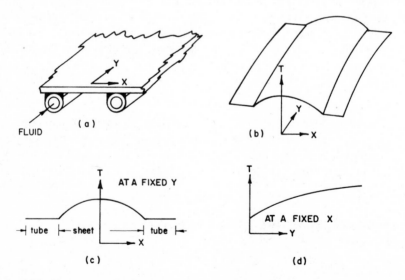

Figure 6.3.2   Temperature distribution on an absorber plate. From Duffie and Beckman (1974).

The energy transferred to the fluid will heat up the fluid causing a temperature gradient to exist in the direction of flow. Since in any region of the collector the general temperature level is governed by the local temperature level of the fluid, a situation as shown in Figure 6.3.2b is expected. At any location y, the general temperature distribution in the x direction is as shown in Figure 6.3.2c, and at any location x, the temperature distribution in the y direction will look like Figure 6.3.2d.

To model the situation shown in Figure 6.3.2, a number of simplifying assumptions can be made to lay the foundations without obscuring the basic physical situation. These important assumptions are as follows:

1   Performance is steady-state.
2   Construction is of sheet and parallel tube type.
3   The headers cover a small area of collector and can be neglected.
4   The headers provide uniform flow to tubes.
5   There is no absorption of solar energy by covers insofar as it affects losses from the collector.
6   There is one-dimensional heat flow through covers.
7   There is a negligible temperature drop through a cover.
8   The covers are opaque to infrared radiation.
9   There is one-dimensional heat flow through back-insulation.
10 The sky can be considered as a blackbody for long-wavelength radiation at an equivalent sky temperature.
11 Temperature gradients around tubes can be neglected.
12 The temperature gradients in the direction of flow and between the tubes can be treated independently.
13 Properties are independent of temperature.
14 Loss through front and back are to the same ambient temperature.
15 Dust and dirt on the collector are negligible.
16 Shading of the collector absorber plate is negligible.

In later sections of this chapter many of the above assumptions will be relaxed.

6.4   COLLECTOR OVERALL HEAT LOSS COEFFICIENT

It is useful to develop the concept of an overall loss coefficient for a solar collector to simplify the mathematics. Consider the thermal network for a two-cover system shown in Figure 6.4.1. At some typical location on the plate where the temperature is T_p, solar energy of amount S is absorbed by the plate; S is equal to the incident solar radiation, reduced by optical losses as shown in Section 5.8. This absorbed energy S is distributed to thermal losses through the top and bottom, and to useful energy gain. The purpose of this section is to convert the thermal network of Figure 6.4.1 to the thermal network of Figure 6.4.2.

The energy loss through the top is the result of convection and radiation between parallel plates. The energy transfer between the plate at T_p and the

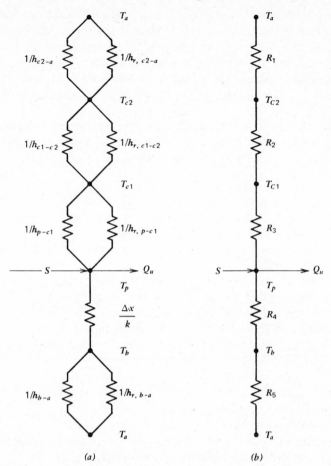

Figure 6.4.1 Thermal network for a two-cover flat-plate collector. (a) in terms of conduction, convection, and radiation resistances, (b) in terms of resistances between plates.

first cover at T_{c1} is the same as between any other two adjacent covers and is also equal to the energy lost to the surroundings from the top cover (under the assumptions listed). The loss through the top per unit area is then equal to the heat transfer from the absorber plate to the first cover

$$q_{\text{loss, top}} = h_{p-c1}(T_p - T_{c1}) + \frac{\sigma(T_p^4 - T_{c1}^4)}{\dfrac{1}{\varepsilon_p} + \dfrac{1}{\varepsilon_c} - 1} \tag{6.4.1}$$

where h_{p-c1} is the heat transfer coefficient between two inclined parallel plates from Chapter 3. If the radiation term is linearized, the radiation heat transfer coefficient can be used and the heat loss becomes

$$q_{\text{loss, top}} = (h_{p-c1} + h_{r, p-c1})(T_p - T_{c1}) \tag{6.4.2}$$

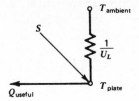

Figure 6.4.2 Equivalent thermal network for flat-plate solar collector.

where

$$h_{r,\,p-c1} = \frac{\sigma(T_p + T_{c1})(T_p^2 + T_{c1}^2)}{\dfrac{1}{\varepsilon_p} + \dfrac{1}{\varepsilon_c} - 1}$$

The resistance, R_3, can then be expressed as

$$R_3 = \frac{1}{h_{p-c1} + h_{r,\,p-c1}} \qquad (6.4.3)$$

A similar expression can be written for R_2, the resistance between the cover plates. In general, we can have as many cover plates as desired but the practical limit seems to be three, and most collectors use one or two.

The resistance from the top cover to the surroundings has the same form as Equation 6.4.3, but the convection heat transfer coefficient is for wind blowing over the collector. An equation for the wind heat transfer coefficient, h_w, is given in Section 3.15. The radiation resistance from the top cover accounts for radiation exchange with the sky at T_s. For convenience, we reference this resistance to the ambient temperature, T_a, so that the radiation heat transfer coefficient can be written as

$$h_{r,\,c2-a} = \varepsilon_c\,\frac{\sigma(T_{c2} + T_s)(T_{c2}^2 + T_s^2)(T_{c2} - T_s)}{T_{c2} - T_a} \qquad (6.4.4)$$

The resistance to the surroundings is then given by

$$R_1 = \frac{1}{h_w + h_{r,\,c2-a}}$$

For this two-cover system, the top loss coefficient from the collector plate to the ambient is

$$U_t = \frac{1}{R_1 + R_2 + R_3} \qquad (6.4.5)$$

The procedure for solving for the top loss coefficient is necessarily an iterative process. First, a guess is made of the unknown cover temperatures, from which the convective and radiative heat transfer coefficients between parallel plates are calculated. With these estimates, Equation 6.4.5 can be solved for the top loss coefficient. The top heat loss is the top loss coefficient times the overall temperature difference and since the energy exchange between plates must be equal to the overall heat loss, a new set of plate temperatures can be calculated. Beginning at the absorber plate, a new temperature is calculated for the first cover. This new first cover temperature is used to find the next cover temperature and so on. For any two adjacent plates, the new temperature of plate j can be expressed in terms of the temperature of plate i as

$$T_j = T_i - \frac{U_t(T_p - T_a)}{h_{i-j} + h_{r,\,i-j}} \qquad (6.4.6)$$

The process is repeated until the cover temperatures no longer change between each iteration. The following example illustrates the process.

Example 6.4.1

Calculate the top loss coefficient for a single glass cover with the following specifications:

Plate to cover spacing	25 mm
Plate emittance	0.95
Ambient air and sky temperature	10 C
Wind heat transfer coefficient	10 W/m^2 C
Mean plate temperature	100 C
Collector tilt	45°
Glass emittance	0.88

Solution

For this single glass cover system, Equation 6.4.5 becomes

$$U_t = \left(\frac{1}{h_{p-c} + h_{r,\,p-c}} + \frac{1}{h_w + h_{r,\,c-a}} \right)^{-1}$$

The convection coefficient between the plate and the cover, h_{p-c}, can be found using the methods of Section 3.11. The radiation coefficient from the plate to the glass, $h_{r,\,p-c}$, is

$$h_{r,\,p-c} = \frac{\sigma(T_p^2 + T_c^2)(T_p + T_c)}{\dfrac{1}{\varepsilon_p} + \dfrac{1}{\varepsilon_c} - 1}$$

and the radiation coefficient from the cover to the air, $h_{r,\,c-a}$, is

$$h_{r,\,c-a} = \varepsilon_c \sigma(T_c^2 + T_s^2)(T_c + T_s)$$

The equation for the cover glass temperature is found from Equation 6.4.6.

$$T_c = T_p - \frac{U_t(T_p - T_a)}{h_{p-c} + h_{r,\,p-c}}$$

The procedure is to guess the cover temperature from which h_{p-c}, $h_{r,p-c}$, and $h_{r,c-a}$ are calculated. With these heat transfer coefficients and h_w, the top loss coefficient is calculated. These results are then used to calculate T_c from the above equation. If T_c is close to the initial guess, no further calculations are necessary. Otherwise, the newly calculated T_c is used and the process is repeated.

With an assumed value of the cover temperature of 35 C, the two radiation coefficients become

$$h_{r,\,p-c} = 7.60 \text{ W/m}^2 \text{ C}$$
$$h_{r,\,c-a} = 5.16 \text{ W/m}^2 \text{ C}$$

Equation 3.11.4 can be used to calculate the convection coefficient between the plate and the glass. The mean temperature between the plate and the cover is 67.5 C so that the air properties are $v = 1.96 \times 10^{-5} \text{ m}^2/\text{s}$, $k = 0.0293 \text{ W/m C}$, $T = 340.5 \text{ K}$, $\text{Pr} = 0.7$. The Rayleigh number is

$$\text{Ra} = \frac{9.81(65)(0.025)^3(0.7)}{340.5(1.96 \times 10^{-5})^2} = 5.33 \times 10^4$$

and from Equation 3.11.4 the Nusselt number is 3.19. The convective heat transfer coefficient is

$$h = \text{Nu}\,\frac{k}{L} = 3.19 \times \frac{0.0293}{0.025} = 3.73 \text{ W/m}^2 \text{ C.}$$

(The same result is obtained from Figures 3.11.1 and 3.11.2. From Figure 3.11.2, $F_1 = 0.46$ and $F_2 = 0.84$. $F_1 \Delta T l^3$ is 4.7×10^5 and from Figure 3.11.1, $F_2 h l$ is 78. h is then 3.7 W/m² C.) The first estimate of U_t is then

$$U_t = \left[\frac{1}{3.73 + 7.60} + \frac{1}{5.16 + 10.0}\right]^{-1} = 6.49 \text{ W/m}^2 \text{ C}$$

The cover temperature is

$$T_c = 100 - \frac{6.49(90)}{3.73 + 7.60} = 48.5\text{C}$$

With this new estimate of the cover temperature, the various heat transfer coefficients become

$$h_{r,\,p-c} = 8.03 \text{ W/m}^2 \text{ C}$$
$$h_{r,\,c-a} = 5.53 \text{ W/m}^2 \text{ C}$$
$$h_{p-c} = 3.52 \text{ W/m}^2 \text{ C}$$

and the second estimate of U_t is

$$U_t = 6.62 \text{ W/m}^2 \text{ C}$$

When the cover glass temperature is calculated with this new top loss coefficient it is found to be 48.4 C, which is essentially equal to the estimate of 48.5 C. ∎

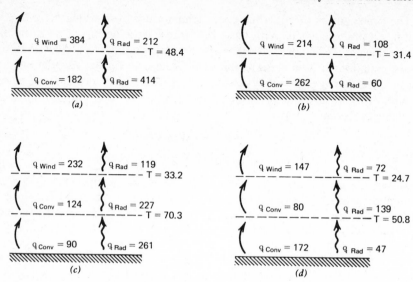

Figure 6.4.3 Cover temperature and upward heat loss for flat-plate collectors operating at 100 C with ambient and sky temperatures of 10 C, plate spacing of 25 mm, tilt of 45°, and wind heat transfer coefficient of 10 W/m² °C. (All heat flux terms in W/m².) (a) one cover, plate emittance = 0.95, U_t = 6.6 W/m²C; (b) one cover, plate emittance = 0.10, U_t = 3.6 W/m² C; (c) two covers, plate emittance = 0.95, U_t = 3.9 W/m²C; (d) two covers, plate emittance = 0.10; U_t = 2.4 W/m²C.

The results of heat loss calculations for four different solar collectors are given in Figure 6.4.3. The cover temperatures and the heat flux by convection and radiation are shown for one and two glass covers and for selective and non-selective absorber plates. Note that radiation between plates is the dominant mode of heat transfer in the absence of a selective surface. When a selective surface having an emittance of 0.10 is used, convection is the dominant heat transfer mode between the selective surface and the cover, but radiation is still the largest term between the two cover glasses in the two-cover system.

The use of a blackbody radiation sky temperature that is not equal to the air temperature will not significantly affect the top loss coefficient or the top heat loss. For example, the top loss coefficient based on the plate to ambient temperature difference for condition a of Figure 6.4.3 is increased from 6.62 to 6.76 W/m² C when the sky temperature is reduced from 10 C to 0 C. For condition b the top loss coefficient is increased from 3.58 to 3.67 W/m² C.

As illustrated by Example 6.4.1, the calculation of the top loss coefficient is a tedious process. To simplify the calculation of collector performance, Figures 6.4.4a through 6.4.4f have been prepared. These figures give the top loss coefficient for one, two, and three glass covers spaced 25 mm apart; for ambient temperatures of 40, 10, and −20 C; for wind heat transfer coefficients of 5, 10, and 20 W/m² C; for plates having an emittance of 0.95 and 0.10; for a slope of 45°; and for a range of plate temperatures.

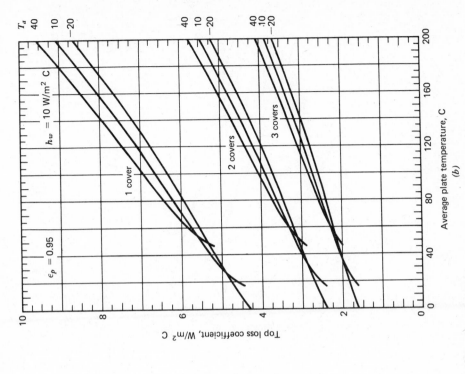

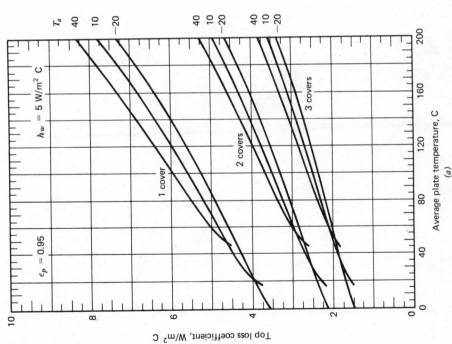

Figure 6.4.4(a–f) Top loss coefficient for slope of 45°.

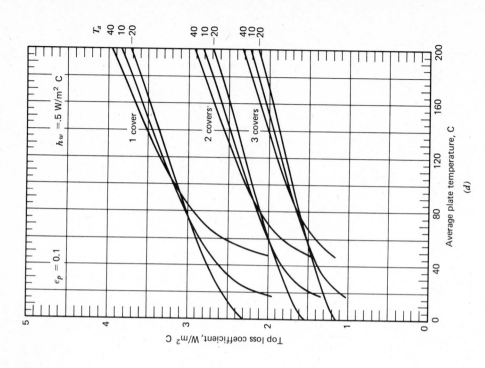

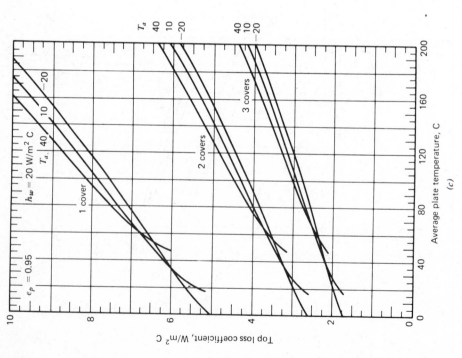

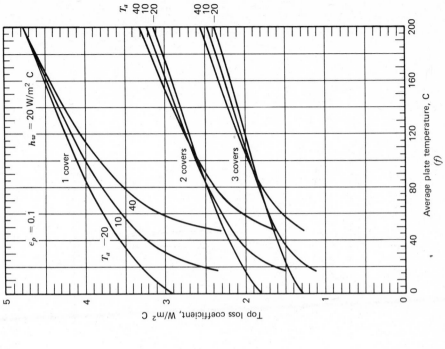

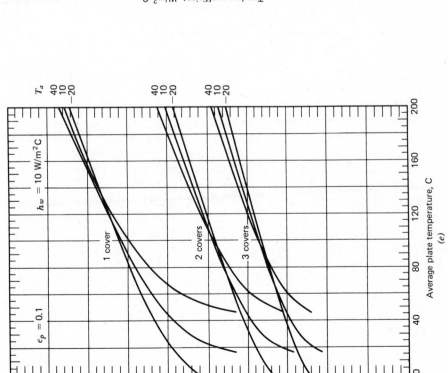

209

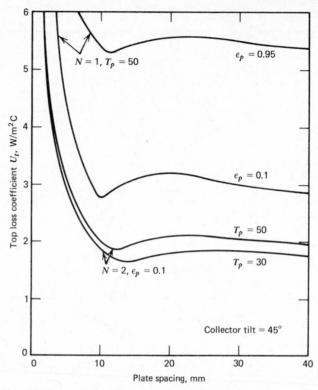

Figure 6.4.5   Typical variation of top loss coefficient with plate spacing.

Even though the top loss coefficients of Figures 6.4.4 are for a plate spacing of 25 mm they can be used for other plate spacings with little error as long as the spacing is greater than about 15 mm. Figure 6.4.5 illustrates the dependence of the top loss coefficient on plate spacing for selective and nonselective one and two cover collectors. For very small plate spacings convection is suppressed and the heat transfer mechanism through the gap is by conduction and radiation. In this range the top loss coefficient decreases rapidly as the plate spacing increases until a minimum is reached at about 10- to 15-mm plate spacing. When fluid motion first begins to contribute to the heat transfer process the top loss coefficient increases until a maximum is reached at approximately 20 mm. Further increase in the plate spacing causes a small reduction in the top loss coefficient. Similar behavior occurs at other conditions and collector designs.

Figure 6.4.4a through 6.4.4f were prepared using a tilt of 45°. In Figure 6.4.6 the ratio of the top loss coefficient at any tilt angle to that of 45° has been plotted as a function of tilt.

The graphs for U_t are convenient for hand calculations but they are difficult to use in computer simulations. The solution to the set of equations, as was done in Example 6.4.1, is time consuming even on a high-speed digital computer since many thousands of solutions may be required. An empirical equation for

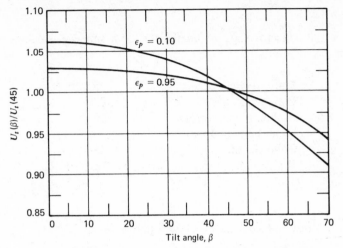

Figure 6.4.6 Dependence of top loss coefficient on slope.

U_t was developed by Klein (1979) following the basic procedure of Hottel and Woertz (1942) and Klein (1975). This new relationship fits the graphs for U_t for plate temperatures between ambient and 200 C to within ± 0.3 W/m² C.

$$U_t = \left\{ \frac{N}{\dfrac{C}{T_{p,m}} \left[\dfrac{(T_{p,m} - T_a)}{(N + f)} \right]^e} + \frac{1}{h_w} \right\}^{-1}$$

$$+ \frac{\sigma(T_{p,m} + T_a)(T_{p,m}^2 + T_a^2)}{(\varepsilon_p + 0.00591 N h_w)^{-1} + \dfrac{2N + f - 1 + 0.133\varepsilon_p}{\varepsilon_g} - N} \qquad (6.4.7)$$

where N = number of glass covers;
$\quad f = (1 + 0.089 h_w - 0.1166 h_w \varepsilon_p)(1 + 0.07866 N)$
$\quad C = 520(1 - 0.000051 \beta^2)$ for $0° < \beta < 70°$. For $70° < \beta < 90°$, use $\beta = 70°$,
$\quad e = 0.43(1 - 100/T_{p,m})$
$\quad \beta$ = collector tilt (degrees)
$\quad \varepsilon_g$ = emittance of glass (0.88)
$\quad \varepsilon_p$ = emittance of plate
$\quad T_a$ = ambient temperature (K)
$\quad T_{p,m}$ = mean plate temperature (K)
$\quad h_w$ = wind heat transfer coefficient (W/m² C)

Equation 6.4.7 is convenient both for hand calculations and for calculations on a digital computer. To use this empirical relationship or the more complicated exact equations to find U_t, it is necessary to know the mean plate temperature, $T_{p,m}$.*

---

* A method for estimating $T_{p,m}$ is given in Section 6.8.

Example 6.4.2

Determine the collector top loss coefficient for a single glass cover with the following specifications:

Plate to cover spacing	25 mm
Plate emittance	0.95
Ambient temperature	10 C
Mean plate temperature	100 C
Collector tilt	45°
Wind heat transfer coefficient	10 W/m² C

Solution

From the definition of f, C, and e below Equation 6.4.7

$$f = [1.0 + 0.0892(10) - 0.1166(10)(0.95)](1 + 0.07866) = 0.846$$
$$C = 520[1 - 0.000051(45)^2] = 466$$
$$e = 0.4299(1 - 100/373) = 0.315$$

From Equation 6.4.7

$$U_t = 2.98 + 3.65 = 6.6 \text{ W/m}^2 \text{ C}$$

which is the same as found in Example 6.4.1. ∎

The energy loss through the bottom of the collector is represented by two series resistors, R_4 and R_5, in Figure 6.4.1. R_4 represents the resistance to heat flow through the insulation and R_5 respresents the convection and radiation resistance to the environment. The magnitudes of R_4 and R_5 are such that it is usually possible to assume R_5 is zero and all resistance to heat flow is due to the insulation. Thus, the back loss coefficient, U_b, is approximately*

$$U_b = \frac{1}{R_4} = \frac{k}{L} \qquad (6.4.8)$$

where k and L are the insulation thermal conductivity and thickness, respectively.

For most collectors the evaluation of edge losses is very complicated. However, in a well-designed system the edge loss should be small so that it is not necessary to predict it with great accuracy. Tabor (1958) recommends edge insulation of about the same thickness as bottom insulation. The edge losses are then estimated by assuming one-dimensional sideways heat flow around the perimeter of the collector system. The losses through the edge should be

---

* The back losses may not be to the same temperature as the top losses.

referenced to the collector area. If the edge loss coefficient-area product is $(UA)_{edge}$ then the edge loss coefficient, based on the collector area, A_c, is

$$U_e = \frac{(UA)_{edge}}{A_c} \qquad (6.4.9)$$

The collector overall loss coefficient, U_L, is then the sum of the top, bottom, and edge loss coefficients

$$U_L = U_t + U_b + U_e \qquad (6.4.10)$$

Example 6.4.3

For the collector of Example 6.4.2 with a top loss coefficient of 6.6 W/m^2 C, calculate the overall loss coefficient with the following additional specifications:

Back insulation thickness	50 mm
Insulation conductivity	0.045 W/mC
Collector bank length	10 m
Collector bank width	3 m
Collector thickness	75 mm
Edge insulation thickness	25 mm

Solution

The bottom loss coefficient is found from Equation 6.4.8

$$U_b = \frac{k}{L} = \frac{0.045}{0.050} = 0.9 \ W/m^2 \ C$$

The edge loss coefficient for the 26-m perimeter as found from Equation 6.4.9

$$U_e = \frac{(0.045/0.025)(26)(0.075)}{30} = 0.12$$

The collector overall loss coefficient is then

$$U_L = 6.6 + 0.9 + 0.1 = 7.6 \ W/m^2 \ C \qquad \blacksquare$$

The edge loss for this 30-m^2 collector system is a little over 1 percent of the total losses. Note, however, that if this collector were 1 m by 2 m the edge losses would increase to over 5 percent. Thus, edge losses for well constructed large collector arrays are usually negligible but for small collectors the edge losses may be significant. Also note that only the exterior perimeter was used to estimate edge losses. If the individual collectors are not tightly packed together, heat loss may occur from the edge of each module.

The preceding discussion of top loss coefficients was based on covers that are opaque to long-wavelength radiation. If a plastic material is used to replace one or more covers, the equation for U_t must be modified to account for some infrared radiation passing directly through the cover. For a single cover that is partially transparent to infrared radiation, the net radiant energy transfer directly between the collector plate and the sky is

$$q_{r,\,p-s} = \tau \varepsilon_p \sigma (T_p^4 - T_s^4)/(1 - \rho_c \rho_p) \qquad (6.4.11)$$

where τ is the transmittance of the cover for radiation from T_p and from T_s (i.e., we have assumed that the transmittance is independent of source temperature). The top loss coefficient then becomes

$$U_t = \frac{q_{r,\,p-s}}{(T_p - T_a)} + \left(\frac{1}{h_{p-c} + h_{r,\,p-c}} + \frac{1}{h_w + h_{r,\,c-s}} \right)^{-1} \qquad (6.4.12)$$

The evaluation of the radiation heat transfer coefficients in Equation 6.4.12 must take into account that the cover is partially transparent. The net radiation between opaque surface 1 and partially transparent surface 2 is given by

$$q = \frac{\varepsilon_1 \varepsilon_2 \sigma [T_1^4 - T_2^4]}{1 - \rho_1 \rho_2} \qquad (6.4.13)$$

The radiation heat transfer coefficient between the plate and cover is just the net heat transfer divided by the temperature difference.

$$h_{r,\,p-c} = \frac{\varepsilon_c \varepsilon_p \sigma [T_p^4 - T_c^4]}{(1 - \rho_p \rho_c)(T_p - T_c)} \qquad (6.4.14)$$

In addition to Equation 6.4.12, Whillier (1977) presents top loss coefficients for collector cover systems of one glass cover over one plastic cover, two plastic covers, and one glass cover over two plastic covers.

6.5 TEMPERATURE DISTRIBUTION BETWEEN TUBES AND THE COLLECTOR EFFICIENCY FACTOR

The temperature distribution between two tubes can be derived if we temporarily assume the temperature gradient in the flow direction is negligible. Consider the sheet-tube configuration as shown in Figure 6.5.1. The distance between the

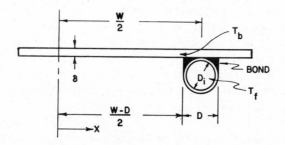

Figure 6.5.1 Sheet and tube dimensions.

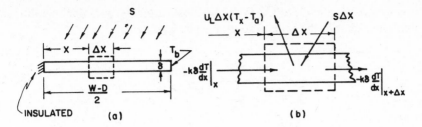

Figure 6.5.2 Energy balance on fin element.

tubes is W, the tube diameter is D, and the sheet is thin with a thickness δ. Because the sheet material is a good conductor, the temperature gradient through the sheet is negligible. We will assume the sheet above the bond is at some local base temperature, T_b. The region between the centerline separating the tubes and the tube base can then be considered as a classical fin problem.

The fin, shown in Figure 6.5.2a, is of length $(W - D)/2$. An elemental region of width Δx and unit length in the flow direction is shown in Figure 6.5.2b. An energy balance on this element yields

$$S\Delta x + U_L\Delta x(T_a - T) + \left(-k\delta\frac{dT}{dx}\right)\bigg|_x - \left(-k\delta\frac{dT}{dx}\right)\bigg|_{x+\Delta x} = 0 \quad (6.5.1)$$

where S is the absorbed solar energy defined by Equation 5.8.1. Dividing through by Δx and finding the limit as Δx approaches zero yields

$$\frac{d^2T}{dx^2} = \frac{U_L}{k\delta}\left(T - T_a - \frac{S}{U_L}\right) \quad (6.5.2)$$

The two boundary conditions necessary to solve this second-order differential equation are symmetry at the centerline and known root temperature;

$$\frac{dT}{dx}\bigg|_{x=0} = 0 \qquad T\bigg|_{x=(W-D)/2} = T_b \quad (6.5.3)$$

If we define $m^2 = U_L/k\delta$ and $\Psi = T - T_a - S/U_L$, Equation 6.5.2 becomes

$$\frac{d^2\Psi}{dx^2} - m^2\Psi = 0 \quad (6.5.4)$$

which has the boundary conditions

$$\frac{d\Psi}{dx}\bigg|_{x=0} = 0, \quad \Psi\bigg|_{x=(W-D)/2} = T_b - T_a - \frac{S}{U_L} \quad (6.5.5)$$

The general solution is then

$$\Psi = C_1 \sinh mx + C_2 \cosh mx \quad (6.5.6)$$

The constants C_1 and C_2 can be found by substituting the boundary conditions, Equation 6.5.5, into the general solution. The result is

$$\frac{T - T_a - \dfrac{S}{U_L}}{T_b - T_a - \dfrac{S}{U_L}} = \frac{\cosh mx}{\cosh m(W - D)/2} \tag{6.5.7}$$

The energy conducted to the region of the tube per unit of length in the flow direction can now be found by evaluating Fourier's law at the fin base

$$q'_{\text{fin}} = -k\delta \frac{dT}{dx}\bigg|_{x=(W-D)/2}$$

$$= \frac{k\delta m}{U_L} [S - U_L(T_b - T_a)] \tanh m \frac{W - D}{2} \tag{6.5.8}$$

but $k\delta m/U_L$ is just $1/m$. Equation 6.5.8 accounts for the energy collected on only one side of a tube; for both sides, the energy collection is

$$q'_{\text{fin}} = (W - D)[S - U_L(T_b - T_a)] \frac{\tanh m(W - D)/2}{m(W - D)/2} \tag{6.5.9}$$

It is convenient to use the concept of a fin efficiency to rewrite Equation 6.5.9 as

$$q'_{\text{fin}} = (W - D)F[S - U_L(T_b - T_a)] \tag{6.5.10}$$

where

$$F = \frac{\tanh[m(W - D)/2]}{m(W - D)/2} \tag{6.5.11}$$

The function F is the standard fin efficiency for straight fins with rectangular profile and is plotted in Figure 6.5.3.

The useful gain of the collector also includes the energy collected above the tube region. The energy gain for this region is

$$q'_{\text{tube}} = D[S - U_L(T_b - T_a)] \tag{6.5.12}$$

and the useful gain for the collector per unit of length in the flow direction is the sum of Equations 6.5.10 and 6.5.12.

$$q'_u = [(W - D)F + D][S - U_L(T_b - T_a)] \tag{6.5.13}$$

Ultimately, the useful gain from Equation 6.5.13 must be transferred to the fluid. The resistance to heat flow to the fluid results from the bond and the fluid to tube resistance. The useful gain can be expressed in terms of these two resistances as

$$q'_u = \frac{T_b - T_f}{\dfrac{1}{(h_{f,i}\pi D_i)} + \dfrac{1}{C_b}} \tag{6.5.14}$$

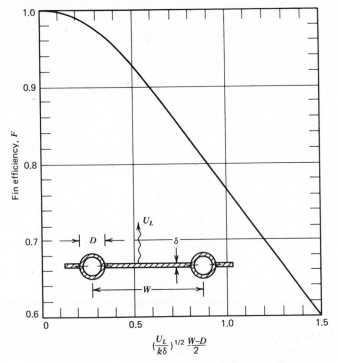

Figure 6.5.3 Fin efficiency for tube and sheet solar collectors.

where D_i is the inside tube diameter and $h_{f,i}$ is the heat transfer coefficient between the fluid and the tube wall. The bond conductance, C_b, can be estimated from knowledge of the bond thermal conductivity, k, the bond average thickness, γ, and the bond width, b. On a per unit length basis,

$$C_b = \frac{k_b b}{\gamma} \qquad (6.5.15)$$

The bond conductance can be very important in accurately describing collector performance. Whillier and Saluja (1965) have shown by experiments that simple wiring or clamping of the tubes to the sheet results in a significant loss of performance. They conclude that it is necessary to have good metal-to-metal contact so that the bond conductance is greater than 30 W/m C.

We now wish to eliminate T_b from consideration and obtain an expression for the useful gain in terms of known dimensions, physical parameters, and the local fluid temperature. Solving Equation 6.5.14 for T_b, substituting it into Equation 6.5.13, and solving the result for the useful gain, we obtain

$$q_u' = WF'[S - U_L(T_f - T_a)] \qquad (6.5.16)$$

where F', the collector efficiency factor, is

$$F' = \frac{\dfrac{1}{U_L}}{W\left[\dfrac{1}{U_L[D + (W - D)F]} + \dfrac{1}{C_b} + \dfrac{1}{\pi D_i h_{f,i}}\right]} \tag{6.5.17}$$

A physical interpretation for F' results from examining Equation 6.5.16. At a particular location, F' represents the ratio of the actual useful energy gain to the useful energy gain that would result if the collector absorbing surface had been at the local fluid temperature. For this and most, but not all geometries, another interpretation for the parameter F' becomes clear when it is recognized that the denominator of Equation 6.5.17 is the heat transfer resistance from the fluid to the ambient air. This resistance will be given the symbol $1/U_0$. The numerator is the heat transfer resistance from the absorber plate to the ambient air. F' is thus the ratio of these two heat transfer coefficients

$$F' = \frac{U_0}{U_L} \tag{6.5.18}$$

The collector efficiency factor is essentially a constant for any collector design and fluid flow rate. The ratio of U_L to C_b, the ratio of U_L to $h_{f,i}$, and the film efficiency parameter F are the only variables appearing in Equation 6.5.17 that may be functions of temperature. For most collector designs F is the most important of these variables in determining F', but it is not a strong function of temperature.

The evaluation of F' is not a difficult task. However, to illustrate the effects of various design parameters on the magnitude of F', Figure 6.5.4 has been prepared. Three values of the overall heat transfer coefficient U_L were chosen (2, 4, and 8 W/m^2 C), which cover the range of collector designs from a one cover nonselective to a two cover selective. (See Figure 6.4.4 for other combinations that will yield these same overall loss coefficients.) Instead of selecting various plate materials, the curves were prepared for various values of $k\delta$, the product of the plate thermal conductivity, and plate thickness. For a copper plate 1 mm thick, $k\delta$ is equal to 0.4 W/C; for a steel plate 0.1 mm thick, $k\delta$ is equal to 0.005 W/C. Thus, the probable range of $k\delta$ is from 0.005 to 0.4. The bond conductance was assumed to be very large (i.e., $1/C_b = 0$) and the tube diameter was selected as 0.01 m. Three values were chosen for the heat transfer coefficient inside the tube to cover the range from laminar flow to highly turbulent flow: 100, 300, and 1000 W/m^2 C. Note that increasing $h_{f,i}$ beyond 1000 W/m^2 C for this diameter tube does not result in significant increases in F'. As expected, the collector efficiency factor decreases with increased tube center to center distances and increases with increases in both material thicknesses and thermal conductivity. Increasing the overall loss coefficient decreases F' while increasing the fluid to tube heat transfer coefficient increases F'.

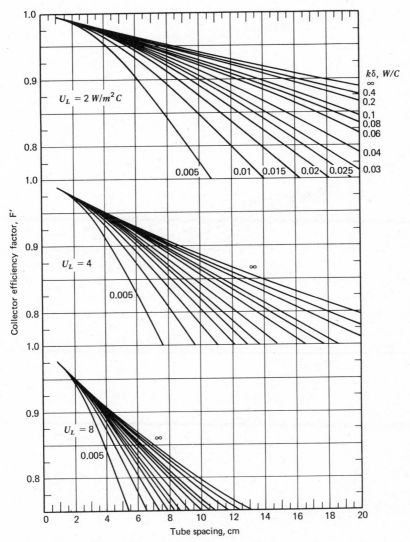

Figure 6.5.4 Collector efficiency factor F' versus tube spacing for 10 mm diameter tubes. (*a*) $h_{f,i} = 100 \ W/m^2C$.

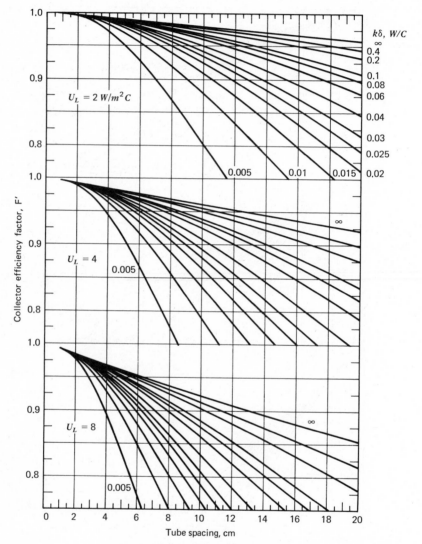

Figure 6.5.4(b) $h_{fi} = 300 \text{ W/m}^2 \text{ C}$.

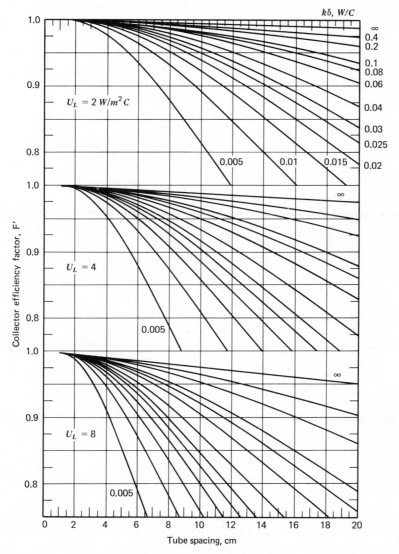

Figure 6.5.4(c) $h_{fi} = 1000$ W/m^2 C.

Example 6.5.1

Calculate the collector efficiency factor for the following specifications:

Overall loss coefficient	8.0 W/m² C
Tube spacing	150 mm
Tube diameter (inside)	10 mm
Plate thickness	0.5 mm
Plate thermal conductivity (copper)	384 W/m C
Heat transfer coefficient inside tubes	300 W/m² C
Bond resistance	0

Solution

The fin efficiency factor, F, from Equation 6.5.11 is

$$m = \left[\frac{8}{385 \times 5 \times 10^{-4}}\right]^{1/2} = 6.45$$

$$F = \frac{\tanh 6.45 \dfrac{0.15 - 0.01}{2}}{6.45 \dfrac{0.15 - 0.01}{2}} = 0.937$$

The collector efficiency factor, F', is found from Equation 6.5.17

$$F' = \frac{1/8}{0.15\left[\dfrac{1}{8[0.01 + (0.15 - 0.01)0.937]} + \dfrac{1}{\pi(0.01)300}\right]} = 0.841$$

The same result is obtained from Figure 6.5.4b.   ■

6.6 TEMPERATURE DISTRIBUTION IN FLOW DIRECTION

The useful gain per unit of flow length as calculated from Equation 6.5.16 is ultimately transferred to the fluid. The fluid enters the collector at temperature $T_{f,i}$ and increases in temperature until at the exit it is $T_{f,o}$. Referring to Figure 6.6.1, we can express an energy balance on the fluid flowing through a single tube of length Δy as

$$\left(\frac{\dot{m}}{n}\right)C_p T_f \bigg|_y - \left(\frac{\dot{m}}{n}\right)C_p T_f \bigg|_{y+\Delta y} + \Delta y q'_u = 0 \qquad (6.6.1)$$

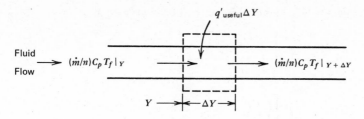

Figure 6.6.1   Energy balance on fluid element.

where $\dot{m}$ is the total collector flow rate and n is the number of parallel tubes. Dividing through by Δy, finding the limit as Δy approaches zero, and substituting Equation 6.5.16 for q_u', we obtain

$$\dot{m}C_p \frac{dT_f}{dy} - nWF'[S - U_L(T_f - T_a)] = 0 \tag{6.6.2}$$

If we assume that F' and U_L are independent of position,* then the solution for the temperature at any position y (subject to the condition that the inlet fluid temperature is $T_{f,i}$) is

$$\frac{T_f - T_a - \dfrac{S}{U_L}}{T_{f,i} - T_a - \dfrac{S}{U_L}} = e^{-[U_L nWF'y/\dot{m}C_p]} \tag{6.6.3}$$

If the collector has a length L in the flow direction, then the outlet fluid temperature, $T_{f,o}$ is found by substituting L for y in Equation 6.6.3. The quantity nWL is the collector area so that

$$\frac{T_{f,o} - T_a - \dfrac{S}{U_L}}{T_{f,i} - T_a - \dfrac{S}{U_L}} = e^{-(A_c U_L F'/\dot{m}C_p)} \tag{6.6.4}$$

6.7   COLLECTOR HEAT REMOVAL FACTOR AND FLOW FACTOR

It is convenient to define a quantity that relates the actual useful energy gain of a collector to the useful gain if the whole collector surface were at the fluid inlet temperature. This quantity is called the collector heat removal factor, F_R, and mathematically is given by

$$F_R = \frac{\dot{m}C_p(T_{f,o} - T_{f,i})}{A_c[S - U_L(T_{f,i} - T_a)]} \tag{6.7.1}$$

* Dunkle and Cooper (1975) have assumed U_L is a linear function of $T_f - T_a$.

The collector heat removal factor can be expressed as

$$F_R = \frac{\dot{m}C_p}{A_c U_L} \left[\frac{T_{f,o} - T_{f,i}}{\dfrac{S}{U_L} - (T_{f,i} - T_a)} \right]$$

$$= \frac{\dot{m}C_p}{A_c U_L} \left[\frac{\left(T_{f,o} - T_a - \dfrac{S}{U_L} \right) - \left(T_{f,i} - T_a - \dfrac{S}{U_L} \right)}{\dfrac{S}{U_L} - (T_{f,i} - T_a)} \right] \qquad (6.7.2)$$

or

$$F_R = \frac{\dot{m}C_p}{A_c U_L} \left[1 - \frac{\dfrac{S}{U_L} - (T_{f,o} - T_a)}{\dfrac{S}{U_L} - (T_{f,i} - T_a)} \right] \qquad (6.7.3)$$

which, from Equation 6.6.4 can be expressed as

$$F_R = \frac{\dot{m}C_p}{A_c U_L} [1 - e^{-(A_c U_L F'/\dot{m}C_p)}] \qquad (6.7.4)$$

To present Equation 6.7.4 graphically, it is convenient to define the collector flow factor, F'', as the ratio of F_R to F'. Thus

$$F'' = \frac{F_R}{F'} = \frac{\dot{m}C_p}{A_c U_L F'} [1 - e^{-(A_c U_L F'/\dot{m}C_p)}] \qquad (6.7.5)$$

This collector flow factor is a function of the single variable, the dimensionless collector capacitance rate, $\dot{m}C_p/A_c U_L F'$ and is shown in Figure 6.7.1.

The quantity F_R is equivalent to a conventional heat exchanger effectiveness, which is defined as the ratio of the actual heat transfer to the maximum possible heat transfer. The maximum possible useful energy gain (heat transfer) in a solar collector occurs when the whole collector is at the inlet fluid temperature; heat losses to the surroundings are then at a minimum. The collector heat removal factor times this maximum possible useful energy gain is equal to the actual useful energy gain, Q_u

$$\boxed{Q_u = A_c F_R [S - U_L (T_i - T_a)]}^* \qquad (6.7.6)$$

---

* This is the most important equation in the book. The subscript f on the fluid inlet temperature has been dropped. Whenever the meaning is not clear, this subscript will be reintroduced.

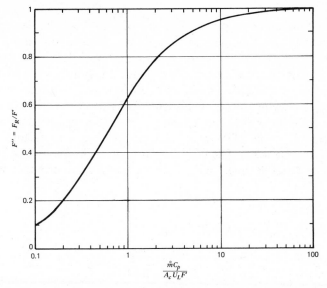

Figure 6.7.1 Collector flow factor F'' as a function of $\dot{m}C_p/A_c U_L F'$.

This is an extremely useful equation and applies to most flat-plate collectors. With it, the useful energy gain is calculated as a function of the inlet fluid temperature. This is a convenient representation when analyzing solar energy systems, since the inlet fluid temperature is usually known. However, losses based on the inlet fluid temperature are too small since losses occur all along the collector from the plate and the plate has an ever increasing temperature in the flow direction. The effect of the multiplier, F_R, is to reduce the useful energy gain from what it would have been had the whole collector absorber plate been at the inlet fluid temperature to what actually occurs. As the mass flow rate through the collector increases, the temperature rise through the collector decreases. This causes lower losses since the average collector temperature is lower and a corresponding increase in the useful energy gain. This increase is reflected by an increase in the collector heat removal factor F_R as the mass flow rate increases. Note that the heat removal factor F_R can never exceed the collector efficiency factor F'. As the flow rate becomes very large, the temperature rise from inlet to outlet decreases toward zero but the temperature of the absorbing surface will still be higher than the fluid temperature. This temperature difference is accounted for by the collector efficiency factor, F'.

Many of the equations of Sections 6.6 and 6.7 contain the ratio of the collector mass flow rate to collector area. This ratio is a convenient way to express flow rate when collector area is a design variable since increasing both in proportion will maintain a nearly constant value of F_R.

Example 6.7.1

Calculate the daily useful gain and efficiency of a bank of 10 solar collectors installed in parallel near Boulder, Colorado at a slope of 60° and a surface azimuth of 0°. The hourly radiation on the plane of the collector, I_T, the hourly radiation absorbed by the absorber plate, S, and the hourly ambient temperature, T_a, are given in the table. The methods of Sections 2.15 and 5.8 can be used to find I_T and S knowing the hourly horizontal radiation, the collector orientation and the collector optical properties. For the collector assume the overall loss coefficient, U_L, to be 8.0 W/m² C and the plate efficiency factor, F', to be 0.841 (from Example 6.5.1). The flow rate through each 1 m by 2 m collector panel is 0.03 kg/s and the inlet fluid temperature remains constant at 40 C.

Solution

The dimensionless collector mass flow rate is

$$\frac{\dot{m}C_p}{A_c U_L F'} = \frac{0.03 \times 4190}{2 \times 8 \times 0.84} = 9.35$$

so that the collector flow factor, from Equation 6.7.5 (or Figure 6.7.1), is

$$F'' = 9.35[1 - e^{-1/9.35}] = 0.948$$

and the heat removal factor is

$$F_R = F'F'' = 0.841 \times 0.948 = 0.797$$

For the hour 10 to 11, the losses, based on an inlet temperature of 40 C, are

$$U_L(T_i - T_a) = 8(40 - 2) \times 3600 = 1.09 \text{ MJ/m}^2$$

and the useful energy gain, per unit of collector area, is

$$q_u = \frac{Q_u}{A_c} = F_R[S - U_L(T_i - T_a)]$$

$$= 0.797[3.29 - 1.09] \times 10^6 = 1.76 \text{ MJ/m}^2$$

The collector efficiency, for this hour, is found from Equation 6.2.2

$$\eta = \frac{Q_u}{A_c I_T} = \frac{q_u}{I_T} = \frac{1.76}{3.92} = 0.45$$

and the day-long collector efficiency is

$$\eta_{\text{day}} = \frac{\sum q_u}{\sum I_T} = \frac{7.57}{19.79} = 0.38$$

The daily useful energy gain of 10, 1m by 2m collectors is

$$Q_u = 10 \times 2 \times 7.57 \times 10^6 = 150 \text{ MJ}$$

Time	T_{amb}, C	I_T, MJ/m^2	S, MJ/m^2	$U_L(T_i - T_a)$, MJ/m^2	q_u, MJ/m^2	η
7–8	−11	0.02	—	—	—	—
8–9	−8	0.43	0.35	1.38	0	0
9–10	−2	0.99	0.82	1.21	0	0
10–11	2	3.92	3.29	1.09	1.76	0.45
11–12	3	3.36	2.84	1.07	1.42	0.42
12–1	6	4.01	3.39	0.98	1.93	0.48
1–2	7	3.84	3.21	0.95	1.81	0.47
2–3	8	1.96	1.63	0.92	0.57	0.29
3–4	9	1.21	0.99	0.89	0.08	0.07
4–5	7	0.05	—	0.95	—	—
Sum		19.79			7.57	

$$\eta_{day} = \frac{\sum q_u}{\sum I_T} = \frac{\sum q_u}{H_T} = 0.38$$

A number of general observations can be made from the results of Example 6.7.1. The estimated performance is typical of a one cover nonselective collector, although in most systems the inlet temperatures will vary throughout the day.* The losses are both thermal and optical and during the early morning and late afternoon the radiation level was not sufficient to overcome the losses. The collector should not be allowed to operate during these periods.

Daily efficiency may also be based on the period while the collector is operating. The efficiency calculated in this manner is 7.57/18.39 or 41 percent. Reporting in this manner gives an inflated value for collector efficiency. As the collector inlet temperature is reduced, these two day-long efficiencies will approach one another. Collector efficiency is a single index that combines collector and system characteristics and as such is not always reliable for making comparisons.

The average collector outlet temperature rise varies from a high of 8.5 C between 12 and 1 to a low of 2.5 C between 2 and 3 (calculated from $\Delta T = Q_u/\dot{m}C_p$). This relatively modest temperature rise is typical of liquid heating collectors. The temperature rise can be increased by reducing the flow rate but this will reduce the useful energy gain. If the flow rate were halved, and if F' remained the same (in fact, $h_{f,i}$ would decrease which would reduce F'), then F_R would decrease to 0.76 and the temperature rise during the hour 12 to 1 would be 16.2 C which is less than twice the original temperature rise. The efficiency during this hour would be reduced from 48 to 46 percent.

---

* Temperature fluctuations are considered in Chapter 10.

6.8   MEAN FLUID AND PLATE TEMPERATURES

To evaluate collector performance, it is necessary to know the overall loss coefficient and the internal fluid heat transfer coefficients. However, both U_L and $h_{f,i}$ are functions of temperature. The mean fluid temperature can be found by integrating Equation 6.6.3 from zero to L

$$T_{f,m} = \frac{1}{L} \int_0^L T_{f,y} \, dy \qquad (6.8.1)$$

Performing this integration and substituting F_R from Equation 6.7.4 and Q_u from Equation 6.7.6, the mean fluid temperature was shown by Klein et al. (1974) to be

$$T_{f,m} = T_{f,i} + \frac{Q_u/A_c}{U_L F_R} (1 - F'') \qquad (6.8.2)$$

This is the proper temperature to evaluate fluid properties.

The mean plate temperature will always be greater than the mean fluid temperature due to the heat transfer resistance between the absorbing surface and the fluid. This temperature difference is usually small for liquid systems but may be significant for air systems.

The mean plate temperature can be used to calculate the useful gain of a collector

$$Q_u = A_c[S - U_L(T_{p,m} - T_a)] \qquad (6.8.3)$$

If we equate Equation 6.8.3 to Equation 6.7.6, and solve for the mean plate temperature, we have

$$T_{p,m} = T_{f,i} + \frac{Q_u/A_c}{U_L F_R} (1 - F_R) \qquad (6.8.4)$$

Equation 6.8.4 can be solved in an iterative manner with Equation 6.4.7. First an estimate of the mean plate temperature is made from which U_L is calculated. With approximate values of F_R, F'', and Q_u, a new mean plate temperature is obtained from Equation 6.8.4 and used to find a new value of the top loss coefficient. The new value of U_L is used to refine F_R, and F'', and the process is repeated. With a reasonable initial guess, this iterative process is seldom necessary.

Example 6.8.1

Find the mean fluid and plate temperature for Example 6.7.1.

Solution

With U_L equal to 8.0 W/m^2 C, $F'' = 0.948$, $F_R = 0.797$, and $q_u = 1.5$ MJ/m^2. We have from Equation 6.8.2.

$$T_{f,m} = 40 + \frac{(1.5 \times 10^6)/3600}{8 \times 0.797} (1 - 0.948) = 43 \text{ C}$$

The value for Q_u/A (1.5 MJ/m^2) used in the calculation for $T_{f,m}$ is approximately the average useful energy gain while the collector is operating. The mean plate temperature is found from Equation 6.8.4

$$T_{p,m} = 40 + \frac{1.5 \times 10^6/3600}{8 \times 0.797}(1 - 0.797) = 53 \text{ C} \qquad \blacksquare$$

6.9 EFFECTIVE TRANSMITTANCE-ABSORPTANCE PRODUCT

In Section 5.5, the product of cover transmittance times plate solar absorptance was discussed. In Section 6.4 the expressions for U_L were derived assuming that the glazing did not absorb solar radiation. To maintain the simplicity of Equation 6.7.6 and account for the reduced losses due to absorption of solar radiation by the glass, an effective transmittance-absorptance product, $(\tau\alpha)_e$, will be introduced. It will be shown that $(\tau\alpha)_e$ is slightly greater than $(\tau\alpha)$.

All of the solar radiation that is absorbed by a cover system is not lost, since this absorbed energy tends to increase the cover temperature and consequently reduce the losses from the plate. Consider the thermal network of a single-cover system as shown in Figure 6.9.1. The solar energy absorbed by the cover is $I_T(1 - \tau_a)$, where τ_a is the transmittance considering only absorption from Equation 5.2.2 and I_T is the hourly incident radiation. The loss for (a), without absorption, is $U_{p-c}(T_p - T_c)$ and the loss for (b), with absorption, is $U_{p-c}(T_p - T_c')$. Here we have assumed that the small amount of absorption in the cover and consequent increased cover temperature does not change the values of U_{p-c} and U_{c-a}. The difference, D, in the two loss terms is

$$D = U_{p-c}[(T_p - T_c) - (T_p - T_c')] \qquad (6.9.1)$$

The temperature difference $(T_p - T_c)$ can be expressed as

$$(T_p - T_c) = \frac{(T_p - T_a)U_t}{U_{p-c}} \qquad (6.9.2)$$

where U_t is the top loss coefficient and is equal to $U_{p-c}U_{c-a}/(U_{p-c} + U_{c-a})$.

The temperature difference $(T_p - T_c')$ can be expressed as

$$(T_p - T_c') = \frac{U_{c-a}(T_p - T_a) - I_T(1 - \tau_a)}{U_{p-c} + U_{c-a}} \qquad (6.9.3)$$

Therefore

$$D = (T_p - T_a)U_t - \frac{U_{p-c}U_{c-a}(T_p - T_a)}{U_{p-c} + U_{c-a}} + \frac{I_T(1 - \tau_a)U_{p-c}}{U_{p-c} + U_{c-a}} \qquad (6.9.4)$$

or

$$D = I_T(1 - \tau_a)\frac{U_t}{U_{c-a}} \qquad (6.9.5)$$

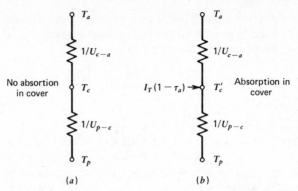

Figure 6.9.1 Thermal network for top losses for a single-cover collector with and without absorption in the cover.

The quantity D represents the reduction in collector losses due to absorption in the cover but can be considered an additional input in the collector equation. The useful gain of a collector is then

$$q_u = F_R \left[S + I_T(1 - \tau_a) \frac{U_t}{U_{c-a}} - U_L(T_i - T_a) \right] \qquad (6.9.6)$$

The quantity I_T has, in general, three components, beam, diffuse sky, and ground-reflected radiation. Each of these terms is multiplied by a separate value of $(\tau\alpha)$ to determine S (i.e., $(\tau\alpha)_b$, $(\tau\alpha)_d$, or $(\tau\alpha)_g$ as shown in Section 5.8). We can divide the radiation absorbed in the cover into these same three components. By defining the quantity $(\tau\alpha) + (1 - \tau_a)U_T/U_{c-a}$ as the effective transmittance-absorptance product for each of the three components, the simplicity of Equation 6.7.6 can be maintained. For this one-cover system

$$(\tau\alpha)_e = (\tau\alpha) + (1 - \tau_a) \frac{U_t}{U_{c-a}} \qquad (6.9.7)$$

When evaluating S, the appropriate value of $(\tau\alpha)_e$ should be used in place of $(\tau\alpha)$. As noted below, $(\tau\alpha)_e$ is on the order of 1 percent greater than $(\tau\alpha)$ for a typical single cover collector with normal glass. For a collector with low iron (water-white) glass, $(\tau\alpha)_e$ and $(\tau\alpha)$ are nearly identical.

A general analysis for a cover system of n identical plates yields

$$(\tau\alpha)_e = (\tau\alpha) + (1 - \tau_a) \sum_{i=1}^{n} a_i \tau^{i-1} \qquad (6.9.8)$$

where a_i is the ratio of the top loss coefficient to the loss coefficient from the ith cover to the surroundings and τ_a is the transmittance of a single cover from Equation 5.2.2. This equation was derived assuming that the transmittance to the ith cover could be approximated by the transmittance of a single cover raised to the $i - 1$ power.

Effects of Dust and Shading

Table 6.9.1 Constants for Use in Equations 6.9.8 and 6.9.9

Covers	a_i	$\varepsilon_p = 0.95$	$\varepsilon_p = 0.50$	$\varepsilon_p = 0.10$
1	a_1	0.27	0.21	0.13
2	a_1	0.15	0.12	0.09
	a_2	0.62	0.53	0.40
3	a_1	0.14	0.08	0.06
	a_2	0.45	0.40	0.31
	a_3	0.75	0.67	0.53

For a cover system composed of different materials (e.g., a combination of glass and plastic) the effective transmittance-absorptance product is

$$(\tau\alpha)_e = (\tau\alpha) + (1 - \tau_{a,1})a_1 + (1 - \tau_{a,2})a_2\tau_1 + (1 - \tau_{a,3})a_3\tau_2 + \cdots \quad (6.9.9)$$

where τ_i is the transmittance of the cover system above the $i + 1$ cover and $\tau_{a,i}$ is the transmittance due to absorption of the ith cover. The angular dependence of $(\tau\alpha)_e$ can be evaluated using the proper angular dependency of $(\tau\alpha)$, $\tau\alpha$, and τ_a.

The values of a_i actually depend upon the plate temperature, ambient temperature, plate emittance, and wind speed. Table 6.9.1 gives values of a_i for one, two, and three covers and for plate emittances of 0.95, 0.50, and 0.10. These values were calculated using a wind heat transfer coefficient of 24 $W/m^2 C$, a plate temperature of 100 C, and an ambient air and sky temperature of 10 C. The dependence of a_i on wind speed may be significant. However, lower wind heat transfer coefficients will increase the a_is leading to slightly higher useful energy gains. Since the total amount absorbed by the glass is small, relatively large errors in the a_is will not cause a significant error in the calculation of Q_u.

Although the value of $(\tau\alpha)_e$ can be calculated from Equation 6.9.9 with some precision, $(\tau\alpha)_e$ is usually only 1 to 2 percent greater than $(\tau\alpha)$. For a one-cover nonselective collector, a_1 is 0.27. If the cover absorbs 4 percent of the incident radiation, then $(\tau\alpha)_e$ is 1 percent greater than $(\tau\alpha)$. For a one-cover selective the two differ by approximately 0.5 percent. For a two-cover nonselective system $(\tau\alpha)_e$ is almost 2 percent greater than $(\tau\alpha)$. As discussed in Section 5.5, $(\tau\alpha)$ is approximately 1 percent greater than the product of τ times α. Since surface radiation properties are seldom known to within 1 percent, the equivalent transmittance-absorptance product can be approximated by

$$(\tau\alpha)_e = 1.02\tau\alpha \quad (6.9.10)$$

6.10 EFFECTS OF DUST AND SHADING

The effects of dust and shading are difficult to generalize. The data of Dietz (1963) shows that at the angles of interest (0 to 50°) the influence of dust can be as high as 5 percent. From long-term experiments on collectors in the Boston area, Hottel and Woertz (1942) found that collector performance decreased approximately 1 percent due to dirty glass. In a rainless 30-day experiment in India, Garg (1974) found that dust reduced the transmittance by an average of 8 percent for glass tilted at 45°. For design purposes without extensive tests, it is suggested that radiation absorbed by the plate be reduced by 2 percent to account for dust.

Shading effects can also be significant. Whenever the angle of incidence is not normal, some of the structure will intercept solar radiation. Some of this radiation will be reflected to the absorbing plate if the sidewalls are of a high-reflectance material. Hottel and Woertz recommend that the radiation absorbed by the plate be reduced by 3 percent to account for shading effects, if the net (unobstructed) glass area is used in all calculations. The net glass area accounts for the blockage by the supports for the glass. These supports may be a significant contribution to shading losses in some designs, and it may be necessary to do a more elaborate analysis.

Example 6.10.1

In Example 6.7.1, the effects of dust, shading, and absorption by the cover were all neglected. Reevaluate the daily performance taking these quantities into account.

Solution

This glass ($KL = 0.037$) absorbs approximately 4 percent of the incident radiation and, according to Table 6.9.1 and Equation 6.9.8, 27 percent of this is not lost. Thus $(\tau\alpha)_e$ is 1.01 times $(\tau\alpha)$. The effects of dust and shading reduce the absorbed radiation by 2 and 3 percent, respectively. The net effect is to decrease S by 4 percent. The table below gives new values for S and the hourly energy gains with $F_R = 0.8$ and $U_L = 8.0$ W/m^2 C. The daily efficiency is reduced from 38 to 36 percent.

Time	I_T, MJ/m^2	S, MJ/m^2	T, C	q_u, MJ/m^2	η
7–8	0.02	—	−11	—	—
8–9	0.43	0.34	−8	0	0
9–10	0.99	0.79	−2	0	0
10–11	3.92	3.16	2	1.65	0.42
11–12	3.36	2.73	3	1.33	0.40
12–1	4.01	3.25	6	1.82	0.45
1–2	3.84	3.08	7	1.70	0.44
2–3	1.96	1.56	8	0.51	0.26
3–4	1.21	0.95	9	0.05	0.04
4–5	0.05	—	7	—	—
	19.79			7.06	

$$\eta_{day} = \frac{7.06}{19.79} = 0.36$$

■

6.11   HEAT CAPACITY EFFECTS IN FLAT-PLATE COLLECTORS

The operation of most solar energy systems is inherently transient; there is no such thing as steady-state operation when one considers the transient nature of the driving forces. This observation has led to numerical studies by Klein et al. (1974), Wijeysundera (1978), and others into the effects of collector heat capacity on collector performance. The effects can be regarded in two distinct parts. One part is due to the heating of the collector from its early morning low temperature to its final operating temperature in the afternoon. The second part is due to intermittent behavior during the day whenever the driving forces such as solar radiation and wind change rapidly.

Klein et al. (1974) showed that the daily morning heating of the collector results in a loss that can be significant but is negligible for many situations. For example, the radiation on the collector of Example 6.10.1 before 10 a.m. was 1.44 MJ/m^2. The calculated losses exceeded this value during this time period because these calculated losses assumed that the fluid entering the collector was at 40 C. In reality, no fluid would be circulating and the absorbed solar energy would heat the collector without reducing the useful energy gain.

The amount of preheating that will occur in a given collector can be estimated by solving the transient energy balance equations for the various parts of the collector. Even though these equations can be developed to almost any desired degree of accuracy, the driving forces such as solar radiation, wind speed, and ambient temperature are usually known only at hour intervals. This means that any predicted transient behavior between the hourly intervals can only be approximate, even with detailed analysis. Consequently, a simplified analysis is warranted to determine if more detailed analysis is desirable.

To illustrate the method, consider a single-cover collector. We assume the absorber plate, water in the tubes, and one half of the back-insulation are all at the same temperature. We also assume that the cover is at a single temperature, but different from the plate. An energy balance on the collector absorber plate, water, and back-insulation yields

$$(mC)_p \frac{dT_p}{d\tau} = A_c[S - U_{p-c}(T_p - T_c)] \qquad (6.11.1)$$

when the subscripts c and p represent cover and plate and U_{p-c} is the loss coefficient from the plate to the cover. An energy balance on the cover yields

$$(mC)_c \frac{dT_c}{d\tau} = A_c[U_{p-c}(T_p - T_c) + U_{c-a}(T_a - T_c)] \qquad (6.11.2)$$

where U_{c-a} is the loss coefficient from the cover to the ambient air and T_a is the ambient temperature. It is possible to solve these two equations simultaneously; however, a great simplification occurs if we assume $(T_c - T_a)/(T_p - T_a)$ remains constant at its steady-state value. In other words, we must assume that the following relationship holds*

$$U_{c-a}(T_c - T_a) = U_L(T_p - T_a) \qquad (6.11.3)$$

* The back and edge losses are assumed to be small.

Differentiating Equation 6.11.3, assuming T_a is a constant, we have

$$\frac{dT_c}{d\tau} = \frac{U_L}{U_{c-a}} \frac{dT_p}{d\tau} \tag{6.11.4}$$

If we now add Equation 6.11.1 to 6.11.2 and use Equation 6.11.4, we obtain the following differential equation for the plate temperature

$$\left[(mC)_p + \frac{U_L}{U_{c-a}} (mC)_c \right] \frac{dT_p}{d\tau} = A_c[S - U_L(T_p - T_a)] \tag{6.11.5}$$

The term in the square brackets on the left-hand side represents an effective heat capacity of the collector and is written as $(mC)_e$. By the same reasoning, the effective heat capacity of a collector with n covers would be

$$(mC)_e = (mC)_p + \sum_{i=1}^{n} a_i(mC)_{c,i} \tag{6.11.6}$$

where a_i is the ratio of overall loss coefficient to the loss coefficient from the cover in question to the surroundings. This is the same quantity as presented in Table 6.9.1.

If we assume that S and T_a remain constant for some period τ, the solution to Equation 6.11.5 is

$$\frac{S - U_L(T_p - T_a)}{S - U_L(T_{p,\text{initial}} - T_a)} = e^{-(A_c U_L \tau/(mC)_e)} \tag{6.11.7}$$

The simplification introduced through the use of Equation 6.11.3 is significant in that the problem of determining heat capacity effects has been reduced to solving one differential equation. The error introduced by this simplification is difficult to assess for all conditions without solving the set of differential equations. Wijeysundera (1978) compared this one-node approximation against a two-node solution and experimental data and found good agreement for single cover collectors. For two and three cover collectors the predicted fractional temperature rise was less than 15 percent in error.

The collector plate temperature T_p can be evaluated at the end of each time period by knowing S, U_L, T_a, and the collector plate temperature at the beginning of the time period. Repeated application of Equation 6.11.7 for each hour before the collector actually operates serves to estimate the collector temperature as a function of time. An estimate of the reduction in useful gain can then be obtained by multiplying the collector effective heat capacity by the temperature rise necessary to bring the collector to its initial operating temperature.

A similar loss occurs due to collector heat capacity whenever the final average collector temperature in the afternoon exceeds the initial average temperature. This loss can be easily estimated by multiplying collector effective heat capacity times this temperature difference.

Finally, Klein et al. (1974) showed that the effects of intermittent sunshine, wind speed and ambient air temperature were always negligible for normal collector construction.

Example 6.11.1

For the collector described in Example 6.10.1, estimate the reduction in useful energy gain due to heat capacity effects. The collector has the following specifications:

Plate thickness (copper)	0.5 mm
Tube i.d. (copper)	10.0 mm
Tube wall thickness (copper)	1.2 mm
Tube spacing	150.0 mm
Glass cover thickness	3.5 mm
Back insulation thickness	50.0 mm

The collector materials have the following properties:

	C, kJ/kg C	ρ, kg/m^3
Copper	0.48	8800
Glass	0.80	2500
Insulation	0.80	50

Solution

Since the collector operates with a constant inlet temperature, only the early morning heating will influence the useful gain.

The collector heat capacity includes the glass, plate, tubes, water in tubes and insulation. The capacity of the glass is

$$0.0035 \text{ m} \times 2500 \text{ kg/m}^3 \times 0.8 \text{ kJ/kg C} = 7 \text{ kJ/m}^2 \text{ C}$$

For the plate, tubes, water in tubes, and insulation the heat capacities are 2, 1, 2, and 2 kJ/m^2 C, respectively. The insulation exposed to the ambient remains near ambient temperature so that the effective insulation heat capacitance is $\frac{1}{2}$ of its actual value. The effective collector capacity is the $2 + 1 + 2 + 1 + 0.27 \times 7 = 8$ kJ/m^2 C. From equation 6.11.7, the collector temperature at the end of the period from 8 to 9, with an initial collector temperature equal to the ambient temperature, is

$$T_p^+ = T_a + \frac{S}{U_L}\left[1 - \exp\left(-\frac{A_c U_L \tau}{(mC)_e}\right)\right]$$

$$= -8 + \frac{0.34 \times 10^6/3600}{8}\left[1 - \exp\left(-\frac{8 \times 3600}{8000}\right)\right]$$

$$= 3 \text{ C}$$

For the second hour period, the initial temperature is 3 C and the temperature at 10:00 a.m. becomes

$$T_p^+ = T_a + \frac{S}{U_L} - \left[\frac{S}{U_L} - (T_i - T_a)\right]\exp\left(-\frac{AU_L\tau}{(mC)_e}\right)$$

$$= -2 + \frac{0.79 \times 10^6/3600}{8}$$

$$- \left[\frac{0.79 \times 10^6/3600}{8} - (3+2)\right]\exp\left(-\frac{8 \times 3600}{8000}\right)$$

$$= 25\ \text{C}$$

By 10:00 a.m., the collector has been heated to within 15 C of its operating temperature of 40 C. The reduction in useful gain is the energy required to heat the collector this last 15 C or 120 kJ/m². Thus, the useful energy gain from 10 to 11 should be reduced from 1.65 to 1.53 MJ/m². Note that this collector responds quickly to the various changes as the exponential term in the preceding calculation was small. (The collector "time constant" is $(mC)_e/A_c U_L$, which is approximately 15 min. The time constant with liquid flowing is on the order of 1 to 2 min., as shown in Section 7.2.) ∎

6.12 OTHER COLLECTOR GEOMETRIES

In the preceding sections, we have considered only one basic collector design: a sheet and tube solar water heater. There are many different designs of flat-plate collectors, but fortunately it is not necessary to develop a completely new analysis for each situation. Hottel and Whillier (1958), Whillier (1977), and Bliss (1959) have shown that the generalized relationships developed for the tube and sheet case apply to most collector designs. It is necessary to derive the appropriate form of the collector efficiency factor, F', and Equations 6.7.5 and 6.7.6 then can be used to predict the thermal performance. Under some circumstances, the loss coefficient U_L will have to be modified slightly.

Figure 6.12.1 shows a number of different liquid and air collector designs. Also on this figure are equations for the collector efficiency factors that have been derived for these geometries. For (h) and (i), the Löf overlapped glass plates and the matrix air heater, the analyses to date have not put the results in a generalized form. For these two situations, it is necessary to resort to numerical techniques for analysis. Selcuk (1971) has analyzed the overlapped glass plate system, and Hamid and Beckman (1971), and Chiou et al. (1965) have studied the matrix-type air heaters.

A somewhat unconventional design is (j), the Speyer (1965) collector, which uses an evacuated glass tube for the cover. Because of the circular geometry, it is possible to evacuate the system and consequently reduce the convection heat loss. With a selective absorbing surface and low emittance

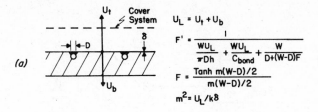

(a)

$$U_L = U_t + U_b$$

$$F' = \frac{1}{\dfrac{WU_L}{\pi Dh} + \dfrac{WU_L}{C_{bond}} + \dfrac{W}{D+(W-D)F}}$$

$$F = \frac{\text{Tanh } m(W-D)/2}{m(W-D)/2}$$

$$m^2 = U_L/k\delta$$

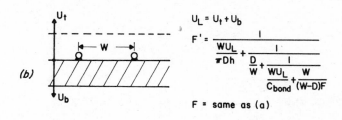

(b)

$$U_L = U_t + U_b$$

$$F' = \frac{1}{\dfrac{WU_L}{\pi Dh} + \dfrac{1}{\dfrac{D}{W} + \dfrac{1}{\dfrac{WU_L}{C_{bond}} + \dfrac{W}{(W-D)F}}}}$$

$$F = \text{same as (a)}$$

(c)

$$U_L = U_t + U_b$$

$$F' = \frac{1}{\dfrac{WU_L}{\pi Dh} + \dfrac{W}{D+(W-D)F}}$$

$$F = \text{same as (a)}$$

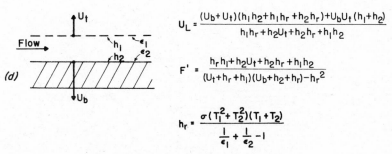

(d)

$$U_L = \frac{(U_b + U_t)(h_1 h_2 + h_1 h_r + h_2 h_r) + U_b U_t (h_1 + h_2)}{h_1 h_r + h_2 U_t + h_2 h_r + h_1 h_2}$$

$$F' = \frac{h_r h_1 + h_2 U_t + h_2 h_r + h_1 h_2}{(U_t + h_r + h_1)(U_b + h_2 + h_r) - h_r^2}$$

$$h_r = \frac{\sigma(T_1^2 + T_2^2)(T_1 + T_2)}{\dfrac{1}{\epsilon_1} + \dfrac{1}{\epsilon_2} - 1}$$

Figure 6.12.1(a–l) Collector efficiency factors.

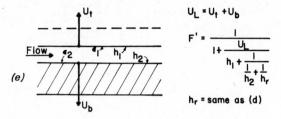

$$U_L = U_t + U_b$$

$$F' = \cfrac{1}{1+\cfrac{U_L}{h_1+\cfrac{1}{\cfrac{1}{h_2}+\cfrac{1}{h_r}}}}$$

h_r = same as (d)

(e)

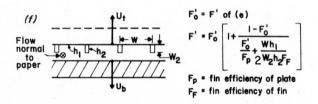

(f)

$F_o' = F'$ of (e)

$$F' = F_o'\left[1+\cfrac{1-F_o'}{\cfrac{F_o'}{F_p}+\cfrac{Wh_1}{2\,W_2h_2F_F}}\right]$$

F_p = fin efficiency of plate

F_F = fin efficiency of fin

(g)

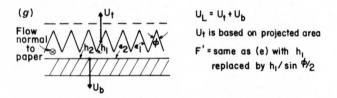

$$U_L = U_t + U_b$$

U_t is based on projected area

F' = same as (e) with h_1 replaced by $h_1/\sin\phi/2$

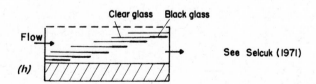

See Selcuk (1971)

(h)

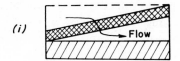

(i) Flow   See Hamid & Beckman(1971)

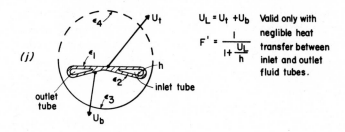

(j)

outlet tube

inlet tube

$U_L = U_t + U_b$

$F' = \dfrac{1}{1 + \dfrac{U_L}{h}}$

Valid only with neglible heat transfer between inlet and outlet fluid tubes.

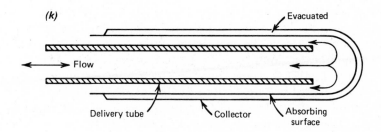

(k)

Evacuated

Flow

Delivery tube   Collector   Absorbing surface

see Eberlein (1976)

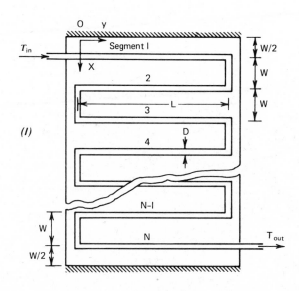

(l)

T_{in}

O   y

Segment I

X

2

3

L

D

4

N-I

N

T_{out}

W/2

W

W

W

W/2

see text

239

surfaces on the back side, the radiation loss can be reduced to relatively small values. The fluid tubes in this type of collector are usually sealed to the glass envelope at only one end to minimize thermal stress problems. This means that the fluid must flow out and return in close proximity. If the thermal resistance between the two streams is low, the plate temperature will rise until losses approach the solar gain all along the tube and the collector efficiency will be very low. This phenomena has led to designs that cut the absorber surface into two halves, one for the supply and one for the return [Ortabasi and Buehl (1975)]. The performance of a large close-packed array of these evacuated collectors is similar to a conventional flat plate collector except that the loss coefficient is much lower and the magnitude and angular dependence of the transmittance-absorptance product is somewhat different.

Another evacuated tubular design is shown in Figure 6.12.1*k* which uses three concentric tubes where the space between the outer two tubes is evacuated. The selective absorbing surface is on the outside of the middle tube and thus is protected from the elements. The collector and inner or feeder tubes are manifolded at one end so fluid travels down the feeder tube, turns around at the closed end and returns in the annular space. This basic design has been used both as a liquid heater [Mather and Beekley (1976)] and as an air heater [Grunes (1976), Eberlein (1976)]. These collectors, because of the circular geometry, are not close packed but are separated by a distance of about one diameter. The radiation passing between adjacent tubes reflects off a diffusely reflecting backing surface, and much of this radiation is absorbed on the back side

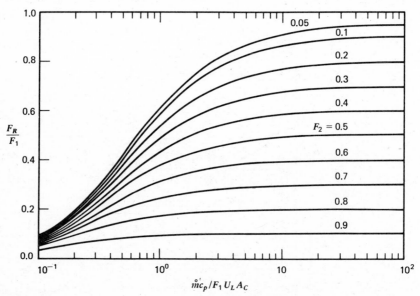

Figure 6.12.2   Heat removal factor, F_R, for the collector plate of Figure 6.12.1*l*. From Abdel-Khalik (1976).

of the tubes. An analysis of the optical characteristics and the thermal performance is given by Eberlein (1976).

Other collector geometries exist for which F' and F_R cannot easily be expressed in simple form but which are important. One such case is the serpentine
tube arrangement as shown in Figure 6.12.1*l*. Abdel-Khalik (1976) analytically
solved the case of a single bend and showed that this solution was within
5 percent of numerical solutions of the general case with any number of bends.
The solution for F_R is given graphically in Figure 6.12.2 in terms of three dimensionless parameters, F_1, F_2, and $\dot{m}C_p/U_L A_c$. The parameters F_1 and F_2
are given by

$$F_1 = \frac{\kappa}{U_L W} \frac{\kappa R(1+\gamma)^2 - 1 - \gamma - \kappa R}{[\kappa R(1+\gamma) - 1]^2 - (\kappa R)^2} \tag{6.12.1}$$

$$F_2 = \frac{1}{\kappa R(1+\gamma)^2 - 1 - \gamma - \kappa R} \tag{6.12.2}$$

where

$$\kappa = \frac{(k\delta U_L)^{1/2}}{\sinh[(W-D)(U_L/k\delta)^{1/2}]} \tag{6.12.3}$$

$$\gamma = -2\cosh\left[(W-D)\left(\frac{U_L}{k\delta}\right)^{1/2}\right] - \frac{DU_L}{\kappa} \tag{6.12.4}$$

$$R = \frac{1}{C_b} + \frac{1}{(\pi D_i h_{f,i})} \tag{6.12.5}$$

Example 6.12.1

Determine the heat removal factor for a collector with serpentine tube having
the following specifications (see Figure 6.12.1*l*):

Length of one serpentine segment, L	1.2 m
Distance between tubes, W	0.1 m
Number of segments, N	6
Plate thickness, δ	0.0015 m
Tube o.d., D	0.0075 m
Tube i.d., D_i	0.0065 m
Plate thermal conductivity, k	211 W/m C
Overall loss coefficient, U_L	5 W/m^2 C
Fluid mass flow rate, $\dot{m}$	0.014 kg/S
Fluid specific heat, C_p	3352 J/kg C
Fluid-to-tube heat trans. coefficient, $h_{f,i}$	1500 W/m^2 C

Solution

From Equation 6.12.3

$$\kappa = \frac{(211 \times 0.0015 \times 5)^{1/2}}{\sinh[(0.1 - 0.0075)(5/211 \times 0.0015)^{1/2}]}$$

$$= 3.346 \text{ W/m C}$$

From Equation 6.12.4

$$\gamma = -2 \times \cosh\left[(0.1 - 0.0075)\left(\frac{5}{211 \times 0.0015}\right)^{1/2}\right] - \frac{0.0075 \times 5}{3.346} = -2.148$$

From Equation 6.12.5

$$R = \frac{1}{(\pi \times 0.0065 \times 1500)} = 0.0326 \text{ m C/W}$$

from which $\kappa R = 3.346 \times 0.0326 = 0.1092$. From Equation 6.12.1

$$F_1 = \frac{3.346}{5 \times 0.1} \frac{(0.1029)(1 - 2.148)^2 - 1 + 2.148 - 0.1092}{[0.1029(1 - 2.148) - 1]^2 - (0.1029)^2}$$

$$= 6.692 \times \frac{1.1827}{1.2545} = 6.309$$

From Equation 6.12.2

$$F_2 = \frac{1}{1.1827} = 0.845$$

The collector area is $NWL = 6 \times 0.1 \times 1.2 = 0.72 \text{ m}^2$ and the dimensionless capacitance rate is

$$\frac{\dot{m}C_p}{F_1 U_L A_c} = \frac{0.014 \times 3352}{6.309 \times 5 \times 0.72} = 2.066$$

Therefore from Figure 6.12.2 we get

$$\frac{F_R}{F_1} = 0.145$$

and finally

$$F_R = 0.145 \times 6.309 = 0.92 \qquad \blacksquare$$

To illustrate the procedure for deriving F' and U_L for an air heater, we derive the equation for type d of Figure 6.12.1. Although type e is the more common design for an air heater, type d is somewhat more complicated to analyze. Also, type d is similar to a collector-storage wall in a passive heating system. A schematic of the collector and thermal network are shown in Figure 6.12.3. This derivation follows that suggested by Jones (1979).

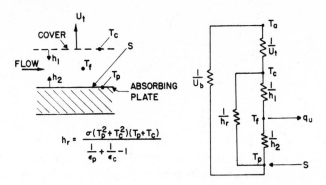

Figure 6.12.3   Type d solar air heater and thermal network.

At some location along the flow direction the absorbed solar energy heats up the plate to a temperature T_p. Energy is transferred from the plate to the ambient air through the back loss coefficient U_b, to the fluid at T_f through the convection heat transfer coefficient h_2 and by radiation to the bottom of the cover glass through the linearized radiation heat transfer coefficient h_r. Energy is transferred to the cover glass from the fluid through heat transfer coefficient h_1 and, finally, energy is lost to the ambient air through the combined convection and radiation coefficient U_t. Note that U_t can account for multiple covers.

Energy balances on the cover, the plate, and the fluid yield the following equations:

$$U_t(T_a - T_c) + h_r(T_p - T_c) + h_1(T_f - T_c) = 0 \qquad (6.12.6)$$

$$S + U_b(T_a - T_p) + h_2(T_f - T_p) + h_r(T_c - T_p) = 0 \qquad (6.12.7)$$

$$h_1(T_c - T_f) + h_2(T_p - T_f) = q_u \qquad (6.12.8)$$

The above three equations need to be solved so that the useful gain is expressed as a function of U_t, h_1, h_2, h_r, T_f, and T_a. In other words, T_p and T_c must be eliminated. The algebra is somewhat tedious and only a few intermediate steps will be given. Solving the first two equations for $(T_p - T_f)$ and $(T_c - T_f)$, we obtain

$$(T_p - T_f) = \frac{S(U_t + h_r + h_1) - (T_f - T_a)(U_t h_r + U_t U_b + U_b h_r + U_b h_1)}{(U_t + h_r + h_1)(U_b + h_2 + h_r) - h_r^2}$$

$$(6.12.9)$$

and

$$(T_c - T_f) = \frac{h_r S - (T_f - T_a)(U_t U_b + U_t h_2 + U_t h_r + U_b h_r)}{(U_t + h_r + h_1)(U_b + h_2 + h_r) - h_r^2} \qquad (6.12.10)$$

Substituting these into the equation for q_u and rearranging, we obtain

$$q_u = F'[S - U_L(T_f - T_a)] \qquad (6.12.11)$$

where

$$F' = \frac{h_r h_1 + h_2 U_t + h_2 h_r + h_1 h_2}{(U_t + h_r + h_1)(U_b + h_2 + h_r) - h_r^2} \qquad (6.12.12)$$

and

$$U_L = \frac{(U_b + U_t)(h_1 h_2 + h_1 h_r + h_2 h_r) + U_b U_t(h_1 + h_2)}{h_1 h_r + h_2 U_t + h_2 h_r + h_1 h_2} \qquad (6.12.13)$$

Note that U_L for this collector is not just the top loss coefficient in the absence of back losses but also accounts for heat transfer between the absorbing surface and the bottom of the cover. Whenever the heat removal fluid is in contact with a transparent cover, U_L will be modified in a similar fashion.

The equations for type e air heaters are derived in a similar fashion but the working fluid does not contact the cover system. For simplicity, back losses are assumed to occur from the absorber plate temperature. The following example shows performance of a type e collector.

Example 6.12.2

Calculate the outlet temperature and efficiency of a single-cover type e air heater of Figure 6.12.1 at a 45° slope when the radiation incident on the collector is 900 W/m², the plate to cover spacing is 20 mm and the air channel depth is 10 mm, the absorber plate is selective with an emittance of 0.1, the effective transmittance-absorptance product is 0.82, the inlet air temperature is 60 C, the ambient air temperature is 10 C, the air mass flow rate is 0.056 kg/s, the collector is 1 m wide by 4 m long, the wind heat transfer coefficient is 10 W/m² C, and the sum of the back and edge loss coefficients is 1.0 W/m² C (see Example 6.4.3).

Solution

From Figure 6.4.4e with an assumed average plate temperature of 70 C, the top loss coefficient is 3.3 W/m² C, and with the back and edge loss coefficient of 1.0 W/m² C, the overall loss coefficient is 4.3 W/m² C. The radiation coefficient between the two air duct surfaces is estimated by assuming a mean radiant temperature equal to the mean fluid temperature. With an estimated mean fluid temperature of 70 C, we have from equations 3.10.2 and 3.10.3

$$h_r = \frac{4\sigma \bar{T}^3}{(1/\varepsilon_1) + (1/\varepsilon_2) - 1} = \frac{4 \times 5.67 \times 10^{-8} \times 343^3}{(2/0.95) - 1}$$

$$= 8.3 \text{ W/m}^2 \text{ C}$$

The heat transfer coefficients between the air and two duct walls will be assumed to be equal. The characteristic length is the hydraulic diameter, which for flat plates is twice the plate spacing. The Reynolds number, at an assumed average fluid temperature of 70 C, is

$$Re_e = \frac{\rho V D_h}{\mu} = \frac{\dot{m} D_h}{A_f \mu} = \frac{0.056 \times (2 \times 0.01)}{(0.01 \times 1) \times 2.04 \times 10^{-5}} = 5480$$

The length to diameter ratio is

$$\frac{L}{D_h} = \frac{4}{2 \times 0.01} = 200$$

Since Re is greater than 2100 and L/D_h is large, the flow is turbulent and fully developed and from Equation 3.14.6

$$Nu = 0.0158(5480)^{0.8} = 15.5$$

The heat transfer coefficient is then

$$h = 15.5 \times \frac{k}{D_h} = \frac{15.5 \times 0.029}{2 \times 0.01} = 22 \text{ W/m}^2 \text{ C}$$

From Figure 6.12.1e with $h_1 = h_2$

$$F' = \left[1 + \frac{U_L}{h + [(1/h) + (1/h_r)]^{-1}}\right]^{-1} = 0.87$$

The dimensionless capacitance rate is

$$\frac{\dot{m} C_p}{A_c U_L F'} = \frac{0.056 \times 1009}{4 \times 4.3 \times 0.87} = 3.78$$

and from Equation 6.7.5 or Figure 6.7.1

$$F'' = 3.78[1 - e^{-1/3.78}] = 0.88$$

or

$$F_R = F'' F' = 0.88 \times 0.87 = 0.77$$

The useful gain is then from Equation 6.7.6

$$Q_u = 4 \times 0.77[900 \times 0.82 - 4.3(60 - 10)] = 1610 \text{ W}$$

The efficiency is

$$\eta = \frac{Q_u}{A G_T} = \frac{1610}{4 \times 900} = 0.45$$

and the outlet temperature is

$$T_0 = T_i + \frac{Q_u}{\dot{m} C_p} = 60 + \frac{1610}{0.056 \times 1009} = 89 \text{ C}$$

It is now necessary to check the assumed mean fluid and absorber plate temperatures. The mean temperature is found from Equation 6.8.4

$$T_{p,m} = 60 + \frac{1610/4}{4.3 \times 0.77}(1 - 0.77) = 88.C$$

and the mean fluid temperature is found from Equation 6.8.2

$$T_{f,m} = 60 + \frac{1610/4}{4.3 \times 0.77}(1 - 0.88) = 74\,C$$

since the initial guess of the plate and fluid temperatures was 70 C, another iteration is necessary. With a new assumed average plate temperature of 88 C, U_t is 3.4 W/m² C and U_L is 4.4 W/m² C. The radiation heat transfer coefficient between the two air duct surfaces is 8.7 W/m² C ($\overline{T} = 348$ K), the Reynolds number is 5400 and the heat transfer coefficient is 23 W/m² C. F', F'', and F_R are unchanged so that the useful energy gain remains at 1610 W. Note that even though the first iteration used an estimate of the plate temperature that was 18 C in error, only minor changes resulted from the second iteration.   ■

6.13   COLLECTOR PERFORMANCE

Instantaneous collector performance is shown on Figure 6.13.1 for four different liquid heating collector designs: one and two covers, selective and nonselective absorber plate. The top loss coefficient was calculated by solving the system of algebraic equations as was done in Example 6.4.1. Back and edge losses were assumed to be 1.0 W/m² C. For the one cover collectors, $(\tau\alpha)_e$ was assumed to be 0.86 and 0.81 for the nonselective and selective collectors, respectively. For the two-cover collectors, $(\tau\alpha)_e$ was 0.81 and 0.77. In all cases, F_R was held constant at 0.95.

If a constant value of U_L had been used to produce the figure all the curves would have been straight lines. An appropriate straight line can be used to fit any of the four curves to within 5 percent; this fact will be used extensively in discussing collector test results. These curves were generated with the incident radiation on the collector equal to 1000 W/m². This is a high level of radiation; if a lower radiation level had been used the curves would be even more nearly linear.

Another important observation can be made from this figure. The selection of the "best" collector depends upon the application. If low-temperature water is supplied to these collectors, the least expensive collector, the one-cover, nonselective, has the highest efficiency. As the temperature supplied to the collector increases, the performance of the one-cover, nonselective collector falls off faster than the other collectors. When the inlet temperature exceeds 100 C, the most expensive collector, the two-cover, selective collector has the highest performance.

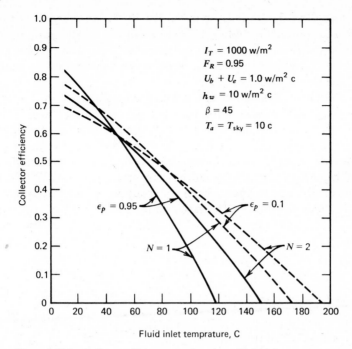

Figure 6.13.1 Collector efficiency as a function of fluid inlet temperature.

An implication of the preceding paragraph is that a fluid delivery temperature is known for a particular application. This is seldom the case. Most solar systems operate with a variable inlet temperature so that the selection of the best collector from a thermal performance viewpoint is not always possible. Even if the thermal performance of one collector exceeds another collector over a wide range of temperature, cost of the two systems must be considered before a rational judgment is possible.

6.14 SUMMARY

This chapter has been concerned with predicting instantaneous collector performance. The analysis has been detailed but the result is the simple collector model given by Equation 6.7.6. The purpose of developing these results in detail is to give collector designers the analytical tools to evaluate design options. It is much simpler and less costly to determine the effect of, for example, tube spacing and plate thickness analytically than it is to do it experimentally.

The results of the analysis of this chapter are two parameters that dictate collector performance: $F_R(\tau\alpha)_e$ and $F_R U_L$. Although we derived expressions for F_R, $(\tau\alpha)_e$ and U_L separately, F_R does not appear alone but always as a multiplier of both $(\tau\alpha)_e$ and U_L. F_R is a weak function of temperature through the temperature dependence of U_L but can usually be considered as a constant for a given design and fluid flow rate. The overall loss coefficient U_L is a function of temperature. However, for many applications it is possible to select a single value of U_L to characterize thermal losses. $(\tau\alpha)_e$ is insensitive to temperature but is a function of incidence angle.

Since collectors work best at near normal incidence, and $(\tau\alpha)_e$ is nearly independent of angle for most collectors at angles less than about 40 or 50°, a single value of $(\tau\alpha)_e$ at normal incidence will suffice to characterize the optical properties of most collectors. As a result, two numbers, $F_R(\tau\alpha)_e$ and $F_R U_L$, are widely used to report the performance of collectors.* These simplifications are vital when performing long-term system analysis to reduce computation costs. As will be seen in later chapters, a significant amount of system design and analysis can be performed with the simple collector model.

REFERENCES

Abdel-Khalik, S. I., *Solar Energy*, **18**, 59 (1976). "Heat Removal Factor for a Flat Plate Solar Collector With a Serpentine Tube."

Bliss, R. W., *Solar Energy*, **3**(4), 55 (1959). "The Derivations of Several 'Plate Efficiency Factors' Useful in the Design of Flat-Plate Solar-Heat Collectors."

Chiou, J. P., M. M. El-Wakil and J. A. Duffie, *Solar Energy*, **9**, 73 (1965). "A Slit-and-Expanded Aluminum-Foil Matrix Solar Collector."

Dietz, A. G. H., *Introduction to the Utilization of Solar Energy*, A. M. Zarem, and D. D. Erway (eds.), McGraw-Hill, New York, 1963. "Diathermanous Materials and Properties of Materials."

Dunkle, R. V. and P. I. Cooper, paper at the Los Angeles meeting of the International Solar Energy Society (1975). "A Proposed Method For The Evaluation of Performance Parameters of Flat-Plate Solar Collectors."

Eberlein, M. B., M.S. Thesis, University of Wisconsin-Madison (1976). "Analysis and Performance Predictions of Evacuated Tubular Solar Collectors Using Air as the Working Fluid."

Garg, H. P., *Solar Energy*, **15**, 299 (1974). "Effect of Dirt on Transparent Covers in Flat-Plate Solar Energy Collectors."

Grunes, H., M.S. Thesis, University of Wisconsin-Madison (1976). "Utilization and Operational Characteristics of Evacuated Tubular Solar Collectors Using Air as the Working Fluid."

Hamid, Y. H. and W. A. Beckman, *Transactions of the American Society Mechanical Engineers, Journal of Engineering and Power*, **93**, 221 (1971). "Performance of Air-Cooled, Radiatively Heated Screen Matricies."

Hottel, H. C. and B. B. Woertz, *Transactions of the American Society Mechanical Engineers*, **64**, 91 (1942). "Performance of Flat-Plate Solar-Heat Collectors."

Hottel, H. C. and A. Whillier, *Transactions of the Conference on the Use of Solar Energy*, **2**, Part I, 74, University of Arizona Press (1958). "Evaluation of Flat-Plate Collector Performance."

Jones, D. E., Personal Communication (1979).

Klein, S. A., *Solar Energy*, **17**, 79 (1975). "Calculation of Flat-Plate Loss Coefficients."

* The subscript e on $(\tau\alpha)_e$ was used in this chapter for "effective" which was a small correction to $(\tau\alpha)$. In much of the remainder of this book the subscript e will be dropped. The subscript n will be introduced to represent normal incidence.

Klein, S. A., personal communication (1979).

Klein, S. A., J. A. Duffie, and W. A. Beckman, *Journal of Engineering for Power*, **96A**, 109 (1974). "Transient Considerations of Flat-Plate Solar Collectors."

Mather, G. R., and D. C. Beekley, *Proceedings of the ISES and AS of ISES Joint Conference*, Winnepeg, Canada, **2**, 64 (1976). "Performance of an Evacuated Tubular Collector Using Non-Imaging Reflectors."

Ortabasi, U., and W. M. Buehl, *Proceedings of ISES Conference*, Los Angeles paper 32/11, 222 (1975), "Analysis and Performance of an Evacuated Tubular Collector."

Selcuk, K., *Solar Energy*, **13**, 165 (1971). "Thermal and Economic Analysis of the Overlapped-Glass Plate Solar-Air Heaters."

Speyer, E., *Journal of Engineering for Power*, **86**, 270 (1965). "Solar Energy Collection with Evacuated Tubes."

Tabor, H., *Bulletin of the Research Council of Israel*, **6C**, 155 (1958). "Radiation, Convection, and Conduction Coefficients in Solar Collectors."

Whillier, A., *Applications of Solar Energy for Heating and Cooling of Buildings*, New York, ASHRAE, 1977. "Prediction of Performance of Solar Collectors."

Whillier, A., and Saluja, G., *Solar Energy*, **9**, 21 (1965). "Effects of Materials and of Construction Details on the Thermal Performance of Solar Water Heaters."

Whillier, A., ScD. Thesis, MIT, 1953. "Solar Energy Collection and Its Utilization for House Heating."

Wijeysundera, N. E., *Solar Energy*, **21**, 517 (1978). "Comparison of Transient Heat Transfer Models for Flat Plate Collectors."

CHAPTER 7

Flat-Plate Collector Performance

Chapter 6 outlines in detail the theory of flat plate collectors. Many of the phenomena considered in the theory have been studied experimentally, such as convection coefficients, radiant exchange, and transmission of radiation through covers. The next questions we consider are: How can measurements be made of collector performance? What measurements have been made? How do they compare with the theory outlined in the previous chapter? How can they be presented in such a way as to be useful in the analysis and design of systems? Finally, the last section in this chapter outlines some practical considerations in the construction of flat plate collectors.

7.1 EARLY RESULTS

The first detailed study of the performance of flat plate collectors, by Hottel and Woertz (1942), was based on energy balance measurements on an array of collectors on an experimental solar heated building. The analysis was basically similar to that of Chapter 6, but with performance calculations based on mean plate temperature rather than on inlet temperature and F_R. They developed a correlation for thermal losses which was a forerunner of Equation 6.4.7. Their experimental data were for time periods of many days, and calculated and measured performance agreed within approximately 13 percent before effects of dust and shading were taken into account.

Tabor (1958) modified the Hottel and Woertz loss calculation by use of new correlations for convection transfer between parallel planes and values of emittance of glass lower than those used by Hottel and Woertz. These modifications permitted estimation of loss coefficients for collectors with selective surfaces where the original method did not give satisfactory results.

He found equilibrium (no fluid flow) temperatures from experiment and theory for a particular collector operation to be 172 and 176 C, indicating satisfactory agreement. He also recalculated Hottel and Woertz results using his modified heat loss coefficients and found calculated and measured losses for two sets of conditions to be 326 versus 329 W/m^2 C and 264 versus 262 W/m^2 C, again indicating excellent agreement. Moore et al. (1974) made extensive comparisons of the performance of a flat plate liquid heating collector with results predicted by use of the original Hottel and Woertz method. The operating conditions were similar to those of Hottel and Woertz, and agreement is good.

As shown by these examples, there is substantial experimental evidence that the energy balance calculation methods of Chapter 6, which are based on the earlier theoretical papers of Hottel, Woertz and Tabor, are a very satisfactory representation of the performance of many flat plate collectors.

7.2 COLLECTOR PERFORMANCE TESTS

In the mid 1970s many new collector designs appeared on the commercial market. A need developed for standard tests to produce data of the type required in process design (for example, for design methods for solar heating systems outlined in Chapter 14). In response to this need, the National Bureau of Standards [Hill and Kusuda (1974)] devised a test procedure [see also Hill and Streed (1976 and 1977)], which has been modified by ASHRAE (1977). The ASHRAE 93-77 standard procedure is the basis for this section; experiments in support of the development of the standard procedures are described by Hill et al (1979).

The basic method of measuring collector performance is to expose the operating collector to solar radiation and measure the fluid temperatures and the fluid flow rate. The useful gain is

$$Q_u = \dot{m}C_p(T_o - T_i) \qquad (7.2.1)$$

In addition, radiation on the collector, ambient temperature and wind speed are also recorded. Thus two types of information are available: data on the thermal output, and data on the conditions producing that thermal performance. These data permit the characterization of a collector by parameters that indicate how the collector absorbs energy and how it loses energy to the surroundings. More detailed information (for example, on separating the effects of transmittance and absorptance, measurement of bond conductance, etc.) must depend on measurements made in addition to those outlined in this section.

Equation 6.7.6 describing the thermal performance of a collector operating under steady conditions can be written as*

$$Q_u = A_c F_R[G_T(\tau\alpha) - U_L(T_i - T_a)] \qquad (7.2.2)$$

Here we have combined the terms that make up S into a single term, and $(\tau\alpha)$ is thus a transmittance-absorptance product that is weighted according to the proportions of beam, diffuse, and ground-reflected radiation. This is the same as $(\tau\alpha)_{ave}$ in Equation 5.8.2; it is customary to omit the subscript. This $(\tau\alpha)$ is a number which (in combination with F_R and U_L), can be used to characterize

---

* We could use $(\tau\alpha)_e$ rather than $(\tau\alpha)$. However, for brevity and because of convention in the solar heating literature we will use $(\tau\alpha)$ throughout this discussion.

a collector. It is determined under test conditions which are generally similar to conditions under which the collectors will provide most of their useful output. Collector tests are done when radiation is high and most of the incident radiation is beam, and $(\tau\alpha)$ in Equation 7.2.2 is thus nearly that for beam radiation.

Equation 7.2.2 can be written in terms of an "instantaneous" efficiency as

$$\eta_i = \frac{Q_u}{A_c G_T} = F_R(\tau\alpha) - F_R U_L \frac{(T_i - T_a)}{G_T} \qquad (7.2.3)$$

and

$$\eta_i = \frac{\dot{m} C_p (T_o - T_i)}{A_c G_T} \qquad (7.2.4)$$

Equation 7.2.3 is the basis of the standard test methods outlined in this section. Other equations are sometimes used (see Section 7.5).

Standard test procedures have been specified for liquid heating and air heating collectors, which can be applied to collectors having low heat capacity. Three methods are available for liquid heaters: a closed loop method is shown in Figure 7.2.1; and open loop methods are shown in Figure 7.2.2 and 7.2.3. Although details differ, the essential features of the methods can be summarized as follows:

1 Means are provided to feed the collector with liquid at a controlled inlet temperature; tests are made over a range of T_i.
2 Solar radiation is measured by a pyranometer on the plane of the collector.
3 Means of measuring flow rate, inlet and outlet fluid temperatures, and ambient conditions are provided.
4 Means are provided for measurements of pressure and pressure drop across the collector.

The ASHRAE standard includes a test method for air collectors, as indicated in Figure 7.2.4. The essential features are the same as for the liquid test methods, with the addition of detailed specifications of conditions relating to air flow, air mixing, air temperature measurements, and pressure drop measurements.

Measurements may be made either outdoors or indoors. Indoor tests are made using a solar simulator, that is, a source producing radiant energy that has spectral distribution, intensity, uniformity in intensity and direction closely resembling that of solar radiation. Means must also be provided to move air to produce wind. Figure 7.2.5 shows an indoor test facility that has been described by Vernon and Simon (1974) and Simon (1976). There are not many test facilities of this kind available, the results are not always comparable to those of outdoor tests, and most collector tests are done outdoors.

The general test procedure is to operate the collector in the test facility under nearly steady conditions, measure the data to determine Q_u from Equation 7.2.1 and measure G_T, T_i, and T_a, which are needed for analysis based on

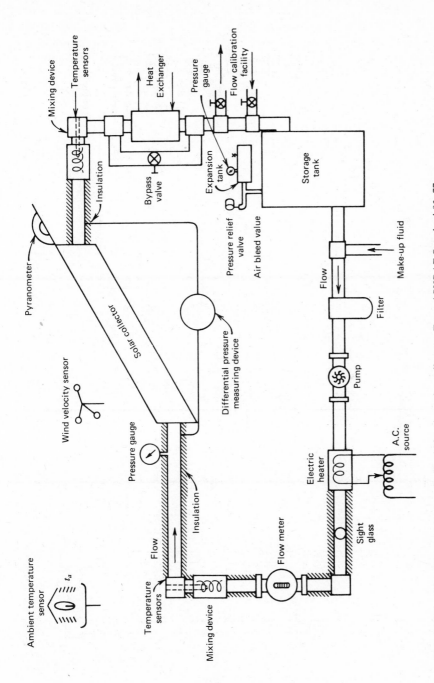

Figure 7.2.1 Closed-loop test set up for liquid heating solar collectors. From ASHRAE Standard 93–77 (1977), with permission.

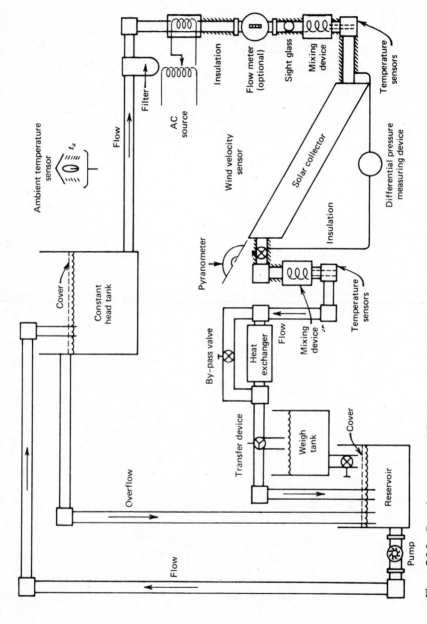

Figure 7.2.2 Open-loop test set up for liquid heating solar collectors. From ASHRAE Standard 93–77 (1977), with permission.

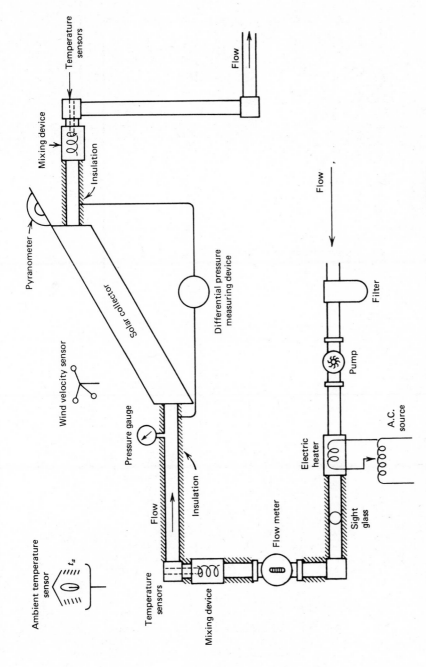

Figure 7.2.3 Open-loop through flow test set up for liquid heating solar collectors. From ASHRAE Standard 93–77 (1977), with permission.

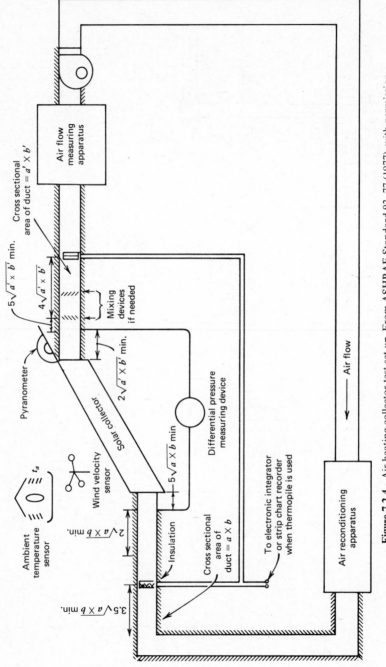

Figure 7.2.4 Air heating collector test set up. From ASHRAE Standard 93–77 (1977), with permission.

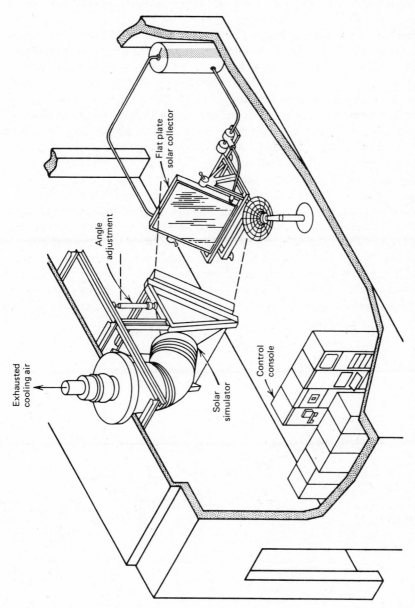

Exhausted
cooling air

Solar
simulator

Angle
adjustment

Flat plate
solar collector

Control
console

Figure 7.2.5 The NASA-Lewis indoor collector test facility. From Vernon and Simon (1974).

257

Equation 7.2.3. Of necessity, this usually means outdoor tests are done in the mid-day hours, on clear days, and usually with the beam solar radiation nearly normal to the collector. Thus the transmittance-absorptance product for these test conditions is approximately the normal incidence value, and is written as $(\tau\alpha)_n$.

Tests are made with a range of inlet temperature conditions. To minimize effects of heat capacity of collectors, tests are usually made in nearly symmetrical pairs, one before and one after solar noon, with results of the pairs averaged. "Instantaneous" efficiencies are determined from $\eta_i = Q_u/A_c G_T$ for the averaged pairs, and are plotted as a function of $(T_i - T_a)/G_T$. A sample plot of data taken at five test sites under conditions meeting ASHRAE 93-77 specifications, from Streed et al. (1979), is shown in Figure 7.2.6. A similar plot for two air heaters, from Gupta and Garg (1967), is shown in Figure 7.2.7.

If U_L, F_R and $(\tau\alpha)_n$ were all constant, the plots of η_i versus $(T_i - T_a)/G_T$ would be straight lines with intercept $F_R(\tau\alpha)_n$ and slope $-F_R U_L$. However, they are not, and the data scatter. From Chapter 6 it is clear that U_L is a function of temperatures and wind speed, with decreasing dependence as the number of covers increases. Also, F_R is a weak function of U_L. And, some variations of the relative proportions of beam, diffuse and ground-reflected components of solar radiation will occur. Thus scatter in the data are to be expected, because of temperature dependence, wind effects, and angle of incidence variations. In spite of these difficulties, for purposes of estimating long-time performance of many solar heating systems, collectors can be characterized by the intercept and slope [i.e., by $F_R(\tau\alpha)_n$ and $F_R U_L$].

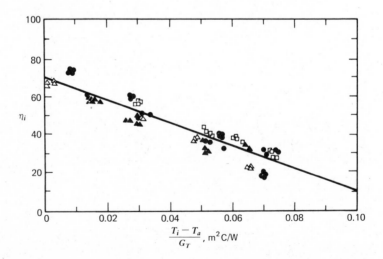

Figure 7.2.6 Experimental collector efficiency data measured for a type of liquid heating collector with one cover and a selective absorber. Sixteen points are shown for each of five test sites. The curve represents the theoretical characteristic derived from points calculated for the test conditions. Adapted from Streed et al. (1979).

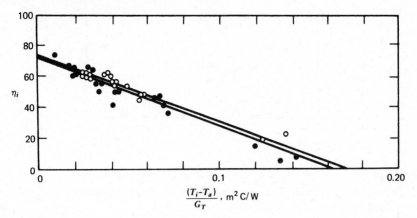

Figure 7.2.7 Experimental thermal efficiency curves for two air heaters operated outdoors. Absorbing surface was flat black paint. Adapted from Gupta and Garg (1967).

Example 7.2.1

A water heating collector with an aperture area of 4.10 m² is tested by the ASHRAE method, with beam radiation nearly normal to the plane of the collector. The following information comes from the test:

Q_u	G_T	T_i	T_a
9.05 MJ/hr	864 W/m²	18.2 C	10.0 C
1.98	894	84.1	10.0

What are $F_R(\tau\alpha)_n$ and $F_R U_L$ for this collector, based on its aperture area?

Solution

For the first data set,

$$\eta_i = \frac{9.05 \times 1000}{864 \times 3.6 \times 4.10} = 0.71$$

and

$$\frac{(T_i - T_a)}{G_T} = \frac{(18.2 - 10.0)}{864} = 0.0095 \text{ m}^2\text{C}/\text{W}$$

For the second data set, $\eta_i = 0.15$ and $(T_i - T_a)/G_T = 0.083$ m²C/W. These two points can be plotted. The slope is

$$\frac{(0.71 - 0.15)}{(0.0095 - 0.083)} = -7.62 \text{ W/m}^2 \text{ C}$$

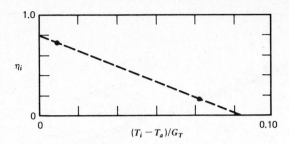

Then $F_R U_L = -$slope $= 7.62$ W/m^2 C. The intercept of the line on the η axis is 0.78, which is $F_R(\tau\alpha)_n$. (Tests produce multiple data points, and in practice a least squares fit would be used to find the best line.) ∎

The collector area appears in the denominator of the definition of η_i. Various areas have been used; gross area, glass area, aperture area, unshaded absorber plate area, etc. Gross collector area is defined as the total area occupied by a collector module, that is, the total area of a collector array divided by the number of modules in the array. Aperture area is defined as the unobstructed cover area or the total cover area less the area of cover supports. The unshaded absorber plate area is sometimes used, and in at least one design is smaller than the aperture area. ASHRAE 93-77 uses gross collector area, a satisfactory choice. It is necessary that the area used be specified clearly so that the same area basis can be used in subsequent design calculations based on the results of the collector tests.

Example 7.2.2

The collector of Example 7.2.1 has a gross area of 4.37 m^2. What are $F_R(\tau\alpha)_n$ and $F_R U_L$ based on the gross collector area?

Solution

The efficiencies for the two data points will be reduced to $0.71 \times 4.10/4.37 = 0.67$ and $0.15 \times 4.10/4.37 = 0.14$. The values of $(T_i - T_a)/G_T$ are unchanged. Thus $F_R U_L = -(0.67 - 0.14)/(0.0095 - 0.083) = 7.21$ W/m^2 C and $F_R(\tau\alpha)_n = 0.74$. ∎

7.3 COLLECTOR TIME CONSTANT

The time constant of a collector is the time required for a fluid leaving a collector to attain change through 0.632 of the total change from its initial to its ultimate steady value after a step change in incident radiation or inlet fluid

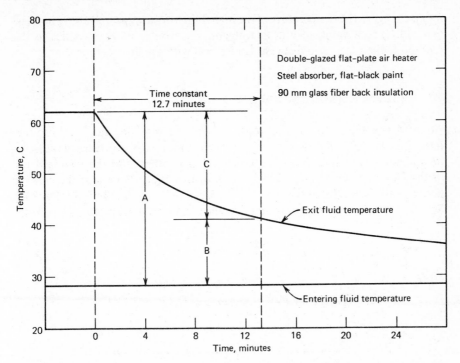

Figure 7.3.1 A time-temperature plot for a flat plate air heater showing temperature drop on sudden interruption of the solar radiation on the collector. The time constant is the time for the temperature to drop to $1/e$ of the total potential drop, that is, for B/A to reach 0.368. Adapted from Hill et al. (1979).

temperature.* The ASHRAE standard test procedure outlines two procedures for estimating the time constant of a collector. The first is to operate a collector at nearly steady conditions with inlet fluid temperature controlled at or very near ambient temperature. The solar radiation is abruptly shut off by shading or repositioning the collector and the decrease in outlet temperature (with the pump running) is noted as a function of time. The time, τ, at which the equality of Equation 7.3.1 is reached is the time constant of the collector.

$$\frac{T_{o,\tau} - T_i}{T_{o,i} - T_i} = \frac{1}{e} = 0.368 \qquad (7.3.1)$$

where $T_{o,\tau}$ is the outlet temperature at time, τ, and $T_{o,i}$ is the outlet temperature when the solar radiation is interrupted. A typical time-temperature history is shown in Figure 7.3.1.

An alternative method for measuring a collector time constant is to test a collector not exposed to radiation (i.e., at night, indoor, or shaded) and impose

---

* Other definitions of time constant are occasionally used. For example, Simon (1976) uses an approach of 0.99 to the ultimate steady value.

a step change in the temperature of the inlet fluid from a value well above ambient (e.g., 30 C) to a value very near ambient. Equation 7.3.1 also applies to this method (which may not give results identical with the first method).

7.4 INCIDENCE ANGLE MODIFIER

An incidence angle modifier, or angle of incidence modifier, can be introduced into Equation 7.2.2. The dependence of $(\tau\alpha)$ on the angle of incidence of radiation on the collector varies from one collector to another, and the standard test methods include experimental estimation of this effect. Note that the concern here is with the weighted transmittance-absorptance product in Equation 7.2.2. Souka and Safwat (1966) suggested that the incidence angle modifier be written

$$K_{\tau\alpha} = \frac{(\tau\alpha)}{(\tau\alpha)_n} \tag{7.4.1}$$

Then

$$Q_u = A_c F_R[G_T K_{\tau\alpha}(\tau\alpha)_n - U_L(T_i - T_a)] \tag{7.4.2}$$

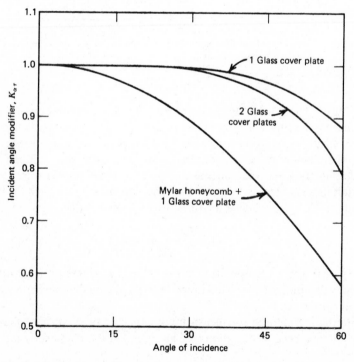

Figure 7.4.1 Incidence angle modifier, $K_{\tau\alpha}$, as a function of angle of incidence for three collectors. Adapted from ASHRAE Standard 93–77 (1977).

Variation in incidence angle modifier for three collectors is shown in Figure 7.4.1 as a function of angle of incidence. A general expression has been suggested for angular dependence of $K_{\tau\alpha}$ as

$$K_{\tau\alpha} = \frac{(\tau\alpha)}{(\tau\alpha)_n} = 1 + b_0\left(\frac{1}{\cos\theta} - 1\right) \qquad (7.4.3)$$

where b_0 is a constant, an *incidence angle modifier coefficient*. (Note that with the equation written this way, b_0 is generally a negative number. This follows the ASHRAE 93-77 convention.)

Incidence angle modifier is shown in Figure 7.4.2 for the same three collectors as a function of $[(1/\cos\theta) - 1]$. Out to values of the abscissa of 1, corresponding to an incidence angle of $60°$, the experiments show straight lines, which is to be expected if Equation 7.4.3 is useful. (At larger angles of incidence, the linear relationship no longer applies.)

Strictly speaking, the incident radiation should be considered as consisting of beam, diffuse, and ground-reflected radiation, with the three components treated separately. The beam component would be treated as outlined here, and the diffuse and ground-reflected radiation would be considered to be at the appropriate fixed angles of incidence (from Figure 5.4.1). However, collector

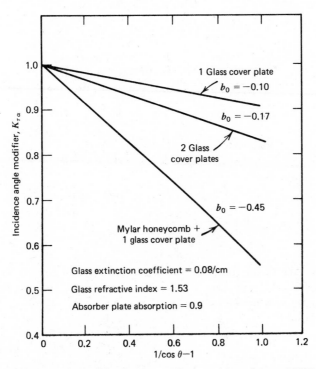

Figure 7.4.2 Incidence angle modifier for three collectors. Adapted from ASHRAE Standard 93–77 (1977).

264   **Flat-Plate Collector Performance**

tests are made under clear sky conditions when the fraction of beam is high, so the data indicate the effects of the angle of incidence of beam radiation.

ASHRAE 93-77 recommends experimental determination of $K_{\tau\alpha}$ by positioning a collector in indoor tests so that θ is 0, 30, 45, and 60°. For outdoor tests it is recommended that pairs of tests symmetrical about solar noon be done early and late in the day, when angles of incidence of beam radiation are approximately 30, 45, and 60°.

7.5 THERMAL TEST DATA CONVERSION

Data from collector tests are usually shown in the form indicated by Figures 7.2.6 and 7.2.7, a format useful in evaluating flat plate collectors for many applications. However, the data are not always presented in this form, and measurements may have been made at flow rates other than the desired rate. In this section we present methods for converting test data to the desired conditions and format.

Thermal test data may be presented in forms other than plots of η_i versus $(T_i - T_a)/G_T$. Data are sometimes plotted as η_i versus $(T_{av} - T_a)/G_T$, where T_{av} is taken as the arithmetic average of inlet and outlet fluid temperatures. In this case, the equation of the straight line is

$$\eta_i = F_{av}(\tau\alpha)_n - F_{av}U_L \frac{(T_{av} - T_a)}{G_T} \tag{7.5.1}$$

F_{av} is approximately the same as F'. If the flow rate of the fluid is known, the intercept $F_{av}(\tau\alpha)_n$ and slope $F_{av}U_L$ of the curve of Equation 7.5.1 can be corrected to obtain $F_R U_L$ and $F_R(\tau\alpha)_n$ by the following (from Beckman et al. (1977)).

$$F_R(\tau\alpha)_n = F_{av}(\tau\alpha)_n \left[\frac{\frac{\dot{m}C_p}{A_c}}{\frac{\dot{m}C_p}{A_c} + \frac{F_{av}U_L}{2}} \right] \tag{7.5.2}$$

$$F_R U_L = F_{av}U_L \left[\frac{\frac{\dot{m}C_p}{A_c}}{\frac{\dot{m}C_p}{A_c} + \frac{F_{av}U_L}{2}} \right] \tag{7.5.3}$$

Air heater test data are sometimes presented as plots of η_i versus $(T_o - T_a)/G_T$. The intercept $F_o(\tau\alpha)_n$ and slope $F_o U_L$ of these curves can be converted to the standard parameters by the following.

$$F_R(\tau\alpha)_n = F_o(\tau\alpha)_n \left[\frac{\frac{\dot{m}C_p}{A_c}}{\frac{\dot{m}C_p}{A_c} + F_o U_L} \right] \tag{7.5.4}$$

$$F_R U_L = F_o U_L \left[\frac{\dfrac{\dot{m}C_p}{A_c}}{\dfrac{\dot{m}C_p}{A_c} + F_o U_L} \right] \qquad (7.5.5)$$

Example 7.5.1

What are $F_R(\tau\alpha)_n$ and $F_R U_L$ for the two-cover air heater having performance characteristics shown in the figure below? The mass flow rate per unit area is 10.1 liter/m² s.

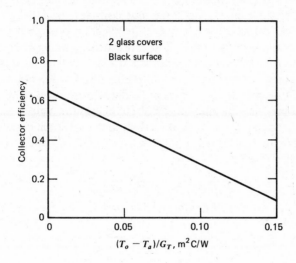

Solution

The performance curve is plotted as η_i versus $(T_o - T_a)/G_T$. For this plot:

$$F_o U_L = 3.7 \text{ W/m}^2 \text{ C}$$

$$F_o(\tau\alpha)_n = 0.64$$

For air at 20 C, $C_p = 1006$ J/kg C and = density 1.204 kg/m³. Then $\dot{m}C_p/A_c$ is

$$\frac{\dot{m}C_p}{A_c} = 0.001204 \times 10.1 \times 1006 = 12.2 \text{ W/m}^2 \text{ C}$$

Then

$$F_R U_L = (3.7)\left[\frac{12.2}{12.2 - (-3.7)} \right] = 2.84 \text{ W/m}^2 \text{ C}$$

$$F_R(\tau\alpha)_n = 0.64\left[\frac{12.2}{12.2 - (-3.7)} \right] = 0.49$$

Additional formats for data presentation may be encountered. For example, Cooper (1976) recommends a linear temperature dependence of overall loss coefficient

$$U_L = a + b(T_{av} - T_a) \tag{7.5.6}$$

and the formulation of the instantaneous collector efficiency by a second order fit in terms of T_{av}

$$\eta_i = \eta_o - a \frac{(T_{av} - T_a)}{G_T} - b\left[\frac{(T_{av} - T_a)}{G_T}\right]^2 \tag{7.5.7}$$

Here η_o is the efficiency when $T_{av} = T_a$. This formulation considers the temperature dependence of U_L but does not consider its dependence on wind speed, which can dominate the temperature effect for one-cover collectors.

Test data should be measured at flow rates corresponding to those to be used in applications. If a collector is to be used at a flow rate other than that of the test conditions, an approximate analytical correction to $F_R(\tau\alpha)_n$ and $F_R U_L$ can be obtained from the ratios of values of F_R determined by use of Equation 6.7.4 or 6.7.5. Assume that the only effect of changing flow rate is to change the temperature gradient in the flow direction, and neglect changes in F' due to changes of h_{fi} with flow rate. The ratio, r, by which $F_R U_L$ and $F_R(\tau\alpha)_n$ are to be corrected is then given by

$$r = \frac{F_R U_L|_{use}}{F_R U_L|_{test}} = \frac{F_R(\tau\alpha)_n|_{use}}{F_R(\tau\alpha)_n|_{test}} \tag{7.5.8}$$

$$= \frac{\dfrac{\dot{m}C_p}{A_c F' U_L}(1 - e^{-A_c F' U_L/\dot{m}C_p})|_{use}}{\dfrac{\dot{m}C_p}{A_c F' U_L}(1 - e^{-A_c F' U_L/\dot{m}C_p})|_{test}} \tag{7.5.9}$$

To use this equation, it is necessary to know or estimate $F'U_L$. For the test conditions, it can be calculated from $F_R U_L$. Rearranging Equation 6.7.4 and solving for $F'U_L$

$$F'U_L = -\frac{\dot{m}C_p}{A_c}\ln\left(1 - \frac{F_R U_L A_c}{\dot{m}C_p}\right) \tag{7.5.10}$$

For liquid collectors, $F'U_L$ calculated for the test conditions can be used in both numerator and denominator of Equation 7.5.9 to get an estimate of r.

Example 7.5.2

The water heating collector of Example 7.2.1 is to be used at a flow rate of 0.020 kg/m²s, rather than at 0.040 kg/m²s at which the test data were obtained. Estimate the effect of reducing the flow rate on $F_R(\tau\alpha)_n$ and $F_R U_L$.

Solution

First, calculate $F'U_L$ for the test conditions:

$$F'U_L = -\frac{0.040 \times 4187}{4.10} \ln\left(1 - \frac{7.62 \times 4.10}{0.040 \times 4187}\right) = 8.43 \text{ W/m}^2 \text{ C}$$

and $F_R/F' = 7.21/7.93 = 0.91$. Then r is obtained with Equation 7.5.9. For use conditions

$$\frac{\dot{m}C_p}{A_c F'U_L} = \frac{0.020 \times 4187}{4.10 \times 8.43} = 2.42$$

For test conditions

$$\frac{\dot{m}C_p}{A_c F'U_L} = \frac{0.040 \times 4187}{4.10 \times 8.43} = 4.85$$

Then

$$r = \frac{2.42(1 - e^{-1/2.47})}{4.85(1 - e^{-1/4.85})} = 0.91$$

Then at the reduced flow rate

$$F_R(\tau\alpha)_n = 0.78 \times 0.91 = 0.71$$

$$F_R U_L = 7.62 \times 0.91 = 6.90 \text{ W/m}^2 \text{ C} \qquad \blacksquare$$

For air heating collectors, or for liquid heating collectors where there is a strong dependence of F' on flow rate, the procedure used in Example 7.5.2 cannot be used. The defining equation for F' (Equation 6.5.17 or its equivalent from Figure 6.12.1), must be used at both use and test flow rates to estimate the effect of flow rate on F'. The collector tests do not provide adequate information with which to do this calculation, and theory must be used to estimate F' at use and test conditions.

Example 7.5.3

Test evaluations of the air heater of Example 6.12.2 give $F_R(\tau\alpha)_n = 0.63$ and $F_R U_L = 3.20 \text{ W/m}^2 \text{ C}$ when the flow rate is 0.056 kg/s through the 4.00 m^2 collector. The conditions of the tests and of anticipated use (other than flow rate) are as outlined in Example 6.12.2. What will be $F_R(\tau\alpha)_n$ and $F_R U_L$ if the flow rate through the collector is reduced to 0.028 kg/s?

Solution

The effects of flow rate on F' need to be checked for this situation, since the convection coefficient from duct to air will change significantly. This requires calculation of the convection and radiation coefficients h_{fi} and h_r as was done in Example 6.12.2, and recalculation of h for the reduced flow rate. The details of that example are not repeated here. It will be assumed that h_r and U_L are not greatly changed from the test to the use conditions, so $h_r = 8.1$ W/m^2 C and $U_L = 4.3$ W/m^2 C.

When the flow rate is reduced, the Reynolds number becomes

$$\text{Re} = \frac{0.028(2 \times 0.01)}{(0.01 \times 1) \times 2.04 \times 10^{-5}} = 2740$$

This is still in the turbulent range, so

$$\text{Nu} = 0.0158(2740)^{0.8} = 8.9$$

and

$$h_{fi} = \frac{8.9 \times 0.029}{2 \times 0.01} = 12.9 \text{ W/m}^2 \text{ C}$$

The value of F' can now be calculated at the reduced flow rate in the same way it was calculated in Example 6.12.2 for the test conditions

$$F' = \left[1 + \frac{4.3}{12.9 + \left(\dfrac{1}{12.9} + \dfrac{1}{8.1} \right)^{-1}} \right]^{-1} = 0.81$$

with the test condition calculation of $F' = 0.89$ and the F' at use conditions, we can go to Equation 7.5.9. For test conditions

$$\frac{\dot{m}C_p}{A_c F' U_L} = 3.69$$

For use conditions

$$\frac{\dot{m}C_p}{A_c F' U_L} = \frac{0.028 \times 1009}{4.00 \times 0.81 \times 4.3} = 2.03$$

So

$$r = \frac{2.03(1 - e^{-1/2.03})}{3.69(1 - e^{-1/3.69})} = 0.90$$

Then for use conditions

$$F_R(\tau\alpha)_n = 0.90 \times 0.63 = 0.57$$

$$F_R U_L = 0.90 \times 3.20 = 2.89 \text{ W/m}^2 \text{ C} \qquad \blacksquare$$

7.6 COLLECTORS IN SERIES

Collector modules in arrays may be connected in parallel, in series, or in combinations. Two situations arise in which data from tests on single collector modules must be modified before they are applied to arrays. First, it may be desirable to put collectors in series that are not identical (e.g., with the first having one cover and the second having two covers). Second, if like collectors are arranged in series *with the same flow velocity* through the modules as in the tests (e.g., with two modules in series and a flow rate per unit area of half that of the tests) the performance of the second modules will be different from the first, as the inlet temperature to the second will be the outlet from the first. For these situations it is convenient to define array characteristics for the series-connected modules that are exactly analogous to $F_R(\tau\alpha)$ and $F_R U_L$ for single modules.

If two collectors, 1 and 2, not necessarily of the same design or size, are in series then the useful output of the combination is

$$
\begin{aligned}
Q_{u,1} + Q_{u,2} = A_1 F_{R1}[(\tau\alpha)_1 I_T - U_{L1}(T_i - T_a)] \\
+ A_2 F_{R2}[(\tau\alpha)_2 I_T - U_{L2}(T_{o,1} - T_a)]
\end{aligned}
\tag{7.6.1}
$$

where T_i is the inlet fluid temperature to the pair of collectors and $T_{o,1}$ is the inlet temperature to the second collector, which is found from the outlet of the first collector:

$$
T_{o,1} = T_i + \frac{Q_{u,1}}{(\dot{m}C_p)}
\tag{7.6.2}
$$

The values of $F_R(\tau\alpha)$ and $F_R U_L$ for each collector must be the values corresponding to the actual fluid flow rate through the pair. By eliminating $T_{o,1}$ from these two equations the useful output of the combination can be expressed as

$$
\begin{aligned}
Q_{u,1+2} = [A_1 F_{R1}(\tau\alpha)_1(1 - K) + A_2 F_{R2}(\tau\alpha)_2]I_T \\
- [A_1 F_{R1} U_{L1}(1 - K) + A_2 F_{R2} U_{L2}](T_i - T_a)
\end{aligned}
\tag{7.6.3}
$$

where K is given by

$$
K = \frac{A_2 F_{R2} U_{L2}}{\dot{m}C_p}
\tag{7.6.4}
$$

The form of Equation 7.6.3 suggests that the combination of the two collectors can be considered as a single collector with the following characteristics:

$$
A = A_1 + A_2
\tag{7.6.5}
$$

$$
F_R(\tau\alpha) = \frac{A_1 F_{R1}(\tau\alpha)_1(1 - K) + A_2 F_{R2}(\tau\alpha)_2}{A}
\tag{7.6.6}
$$

$$
F_R U_L = \frac{A_1 F_{R1} U_{L1}(1 - K) + A_2 F_{R2} U_{L2}}{A}
\tag{7.6.7}
$$

If three or more collectors are placed in series, then these equations can be used for the first two collectors to define a new equivalent first collector. The equations are applied again with this equivalent first collector and the third collector becoming the second collector. The process can be repeated for as many collectors as desired.

If the two collectors are identical, then Equations 7.6.6 and 7.6.7 reduce to the following

$$F_R(\tau\alpha) = F_{R1}(\tau\alpha)_1\left(1 - \frac{K}{2}\right) \qquad (7.6.8)$$

$$F_R U_L = F_{R1} U_{L1}\left(1 - \frac{K}{2}\right) \qquad (7.6.9)$$

For N identical collectors in series, Oonk et al. (1979) have shown that repeated applications of Equation 7.6.6 and 7.6.7 yield

$$F_R(\tau\alpha) = F_{R1}(\tau\alpha)_1\left[\frac{1 - (1 - K)^N}{NK}\right] \qquad (7.6.10)$$

$$F_R U_L = F_{R1} U_{L1}\left[\frac{1 - (1 - K)^N}{NK}\right] \qquad (7.6.11)$$

Example 7.6.1

What are $F_R(\tau\alpha)$ and $F_R U_L$ for the combination of two air heating collectors connected in series with a flow rate of 0.056 kg/s. The characteristics of a single air heater are $F_R(\tau\alpha) = 0.67$ and $F_R U_L = 3.63$ W/m^2 C at a flow rate of 0.056 kg/s. Each collector is 1.00 m wide by 2.00 m long.

Solution

From Equation 7.6.4, K is

$$K = \frac{3.63 \times 2.00}{0.056 \times 1008} = 0.129$$

From Equations 7.6.8 and 7.6.9

$$F_R(\tau\alpha) = 0.67\left(1 - \frac{0.129}{2}\right) = 0.63$$

$$F_R U_L = 3.6\left(1 - \frac{0.129}{2}\right) = 3.4 \text{ W/m}^2 \text{ C} \qquad \blacksquare$$

7.7 TEST DATA FOR SEVERAL COLLECTORS

For purposes of illustrating the kinds of data that are available and the differences that can exist among collectors, we show here test results obtained by Hill et al. (1978) using the ASHRAE 93-77 standard procedures. For design of a solar process, data on the specific collector of interest should be used; the data shown here are for illustrative purposes and are not intended to be used for design.

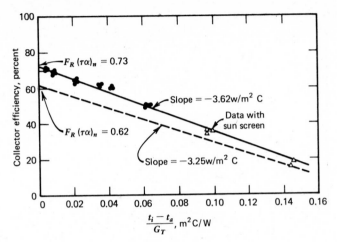

Figure 7.7.1 Test points for a liquid heater, based on collector aperture area. Also shown is the correlation based on gross collector area. The collector is double glazed with antireflective coatings on 3 glass surfaces and has a black chrome selective surface. Gross collector area is 1.66 m^2 and aperture area is 1.40 m^2. Adapted from Hill et al. (1979).

Figure 7.7.1 shows test points and correlations for a double glazed liquid heater. The points and solid curve are based on the aperture area of the collector, that is, the unobstructed glass area. The dashed curve shows the correlation based on the gross area of the collector. In some of the tests the intensity of incident radiation G_T was reduced by a shading screen to obtain points over a range of values of $(T_i - T_a)/G_T$. Figure 7.7.2 shows a set of curves for several collectors based on gross collector area, including results for three liquid based heaters and one air heater; the air heater characteristics are shown for two air flow rates.

There are obvious differences in these collector characteristics. The air collector curve lies below the liquid heaters. It is characteristic of most air heaters that they are operated at relatively low capacitance rates, have low heat transfer coefficients between absorber surface and fluid, and thus show relatively low values of F_R. The low flow rates are used to minimize pressure drop in the collectors. (Also, low air flow rates through pebble bed storage units used with air heaters in space heating applications result in high stratification in beds, which has operating advantages—see Chapters 9 and 13.) The fact that one collector may be more efficient than another in part or all of its range of operating conditions does not mean that it necessarily is a better collector in an economic context; as will be pointed out later, collectors must be evaluated in terms of their performance in systems and ultimately in terms of their costs.

Hill et al. have measured the effects of angle of incidence of radiation on the collector of Figure 7.7.1; this is shown in Figure 7.7.3. They have also measured the time constants of the four collectors shown in Figure 7.7.2; these are indicated in Table 7.7.1, together with information on the collectors. Data of the type of

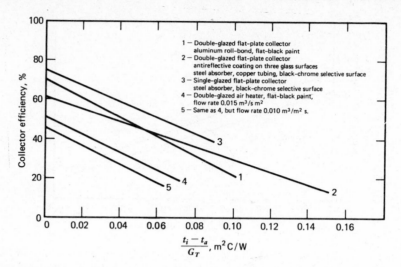

Figure 7.7.2 Collector efficiency curves for three liquid heaters and an air heater at two flow rates. Adapted from Hill et al. (1979).

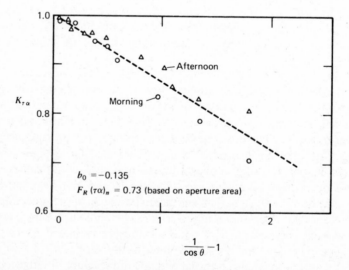

Figure 7.7.3 Data on incidence angle modifier for a double glazed water heating collector of the type indicated in Figure 7.4.1 and 7.4.2. The data were taken with the collector tilted 25° to south, with wind speed 4.5 m/s, ambient temperature 34C, and with insolation 230–830 W/m² of which an estimated 20 percent was diffuse. Adapted from Hill et al. (1979).

Table 7.7.1 Characteristics of the Collectors shown in Figure 7.7.2 and the Time Constants for the Collectors

| Collector No. (Type) | Areas. m^2 | | Plate | Glazing | Back Insulation | Time Constant, min |
	Aperture	Gross				
1 (Liquid)	1.61	1.68	aluminum roll-bond, black paint	double-glass	7.6-cm glass fiber	1.7
2 (Liquid)	1.40	1.66	steel plate with copper tubing, black chrome selective surface	double-glass, antireflective coating on three glass surfaces	8.5-cm semirigid fiber board	1.6
3 (Liquid)	1.79	1.96	steel, black-chrome selective surface	single-glass	7.0-cm glass fiber	1.8
4 (Air)	7.25	6.25	steel, black paint. Air flow is in 1.6-cm-deep channel behind plate.	double-glass	9-cm glass fiber	12.7

Figures 7.7.1 to 7.7.3 are becoming available for many collectors, as a result of programs sponsored by the U.S. Department of Energy to provide a data base for solar process design purposes.

7.8 *IN SITU* COLLECTOR PERFORMANCE

There is no fundamental reason why a collector would not operate as well installed in an application as it does on a test stand. However, there are some practical considerations that can influence measured performance on site. Differences between predicted and measured performance may arise from several sources.

1 Flow of fluid through the collector may not be uniform through all parts of the collector array. Parts of a collector array receiving reduced fluid flow will have lower F_R and poorer performance, resulting in degradation of array performance. This is discussed in the next section.

2 Flow rates may not be those at which collectors were tested. F_R is a function of flow rate for both liquid and air heaters (particularly so for air heating collectors), and changes in flow rate can make significant differences in collector performance.

3 Leaks in air collectors may also introduce differences between predicted and measured performance. Collectors are usually operated at slightly less than atmospheric pressure, resulting in leakage of cool ambient air in and reduced collector outlet temperatures [see Close and Yusoff (1978)].

4 Edge and back losses may be different in tests and applications. Edge losses may be reduced in large arrays, resulting in reduction of U_L from measured values for single collector modules.

7.9 FLOW DISTRIBUTION IN COLLECTORS

Performance calculations of collectors are based on an implicit assumption of uniform flow distribution in all of the risers in single or multiple collector units. If flow is not uniform, the parts of the collectors with low flow through risers will have lower F_R than parts with higher flow rates. Thus the design of both headers and manifolds is important in obtaining good collector performance. This problem has been studied analytically and experimentally by Dunkle and Davey (1970). It is of particular significance in large forced-circulation systems; natural circulation systems tend to be self-correcting and the problem is not as critical.

Based on the assumptions that flow is turbulent in headers and laminar in risers (assumptions logical for Australian and many other water heaters), the analysis by Dunkle and Davey shows pressure drop along the headers for the

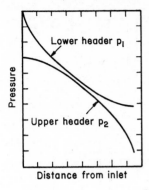

Figure 7.9.1 Pressure distribution in headers of an iso-thermal absorber bank. Adapted from Dunkle and Davey (1970).

common situation of water entering the bottom header at one side of the collector and leaving the top header at the other side. Under these circumstances, calculated pressure distributions in the top and bottom headers can be as shown in Figure 7.9.1. The implications of these pressure distributions are obvious; the pressure drops from bottom to top are greater at the ends than the center portion, leading to high flows in the end risers and low flows in the center risers.

This situation is found experimentally. Temperatures of absorber plates are measures of how effectively energy is removed, and thus differences among temperatures measured at the same relative location on individual collectors in banks is a measure of the lack of uniformity of flow in risers. Figure 7.9.2,

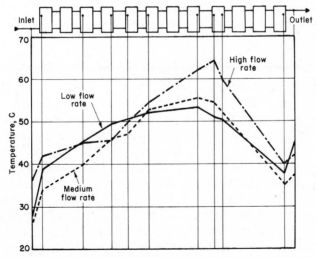

Figure 7.9.2 Experimental temperature measurements on plates in a bank of collectors connected in parallel. Adapted from Dunkle and Davey (1970).

A

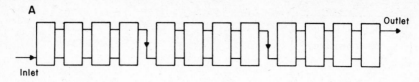

B

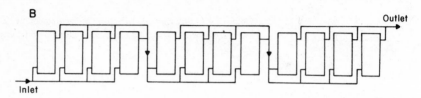

Figure 7.9.3 Examples of alternative methods of connecting collector arrays in (*a*) series-parallel and (*b*) parallel-series arrangements, as recommended by Dunkle and Davey (1970).

from Dunkle and Davey, shows measured temperatures for a bank of twelve collectors connected in parallel. The data show temperature differentials of 22°C from center to ends that are significant differences. Connecting the units in a series parallel or multiple parallel arrangements such as shown on Figure 7.9.3 results in more uniform flow distribution and temperatures.

Manufacturers of collectors have worked out recommended practices for piping collectors to avoid maldistribution of flow and minimize pressure drops, and the designer of a system should refer to those recommendations. Otherwise standard references on design of piping networks should be consulted.

7.10 PRACTICAL CONSIDERATION FOR FLAT-PLATE COLLECTORS

In this chapter we have discussed the thermal performance of collectors in tests and installed in systems. There are many other practical considerations in the design, manufacture, shipment, installation and long-term use of flat plate collectors. In this section we briefly illustrate some of these considerations and show two typical designs that are in commercial manufacture (one liquid and one air). The industry has advanced to the point that manufacturers have developed installation and service manuals; reference is made to these manuals for more details. The discussion here is based largely on the concept of factory-manufactured, modular collectors that are assembled in large arrays on a job; the same practical considerations hold for site-built collectors. Economic considerations are held for later chapters.

Equilibrium temperatures, encountered under conditions of high radiation with no fluid flowing through the collector, are substantially higher than ordinary operating temperatures, and collectors must be designed to withstand these temperatures. It is inevitable that at some time power failure, control problems, servicing, summer shut-down, or other causes will lead to no-flow conditions. (They probably will first be encountered during installation of the collectors.) Maximum plate temperatures can be estimated by evaluating the fluid temperature in Equation 6.7.6 with Q_u equal to zero. The fluid and plate temperatures are the same for the no flow condition. The equilibrium temperatures of other parts of the collector can be estimated from the ratios of thermal resistances between those parts and ambient to that of the plate to ambient. These maximum equilibrium temperatures place constraints on the material, which must retain their important properties during and after exposure to these temperatures, and on mechanical design to accommodate thermal expansion.

Extremely low temperatures must also be considered. This is particularly important for liquid heating collectors where freeze protection must be provided. This may be done by use of antifreeze fluids, or by arranging the system so that the collector will drain during periods when it is not operating. Corrosion is an important consideration with these collectors. Methods of freeze protection are in part system considerations, and will be treated in later chapters.

Materials of wide variety are used in collectors, including structural materials (such as metal used in boxes), glazing (which is usually glass), insulation, caulking materials, etc. Skoda and Masters (1977) have compiled a survey of practical experience with a range of materials and illustrate many problems that can arise if materials are not carefully selected for withstanding weather, temperature extremes, temperature cycling, and compatibility. A series of articles in Solar Age (1977) deal specifically with fabrication and corrosion prevention problems in use of copper, steel, and aluminum absorber plates.

Covers and absorbers are particularly critical; their properties determine $(\tau\alpha)$ and ε_p, and thus strongly affect thermal performance. Degradation of these properties can seriously affect long-term performance, and materials should be selected that have stable properties. Condensation of moisture sometimes occurs under covers or between covers; this imparts an increased reflectance to covers and requires energy to evaporate the condensate on start up and so can diminish collector performance, although the effect on long-term performance may be small. Some collectors use sealed spaces under the top cover, and some use breathing tubes containing desiccants to dry the air in the space under the top cover. Covers and supports should be designed so that dirt will not get under the cover. Materials used in the collector should not contain volatiles that can be evaporated during periods of high collector temperatures; these volatiles condense on the under side of covers and reduce transmittance.

Mechanical design affects thermal performance; it also is important in that structural strength of collectors must be adequate to withstand handling and installation, and also conditions to be expected over the lifetime of the units. Collectors must be water tight or provided with drains to avoid rain damage and in northern climates should be mounted to allow snow to slide off. Covers must be designed to withstand hail. Wind loads may be high and tie-downs must be adequate to withstand these loads. Some collectors may be designed to provide structural strength or serve as the water-tight envelope of a building; these functions may impose additional requirements on collector design and manufacture.

Installation costs can be an important item of cost of solar collectors. These are largely determined by three factors: handling, tie-down, and manifolding. Moving and positioning of collector modules may be done by hand if modules are light enough or may be done by machine for heavy modules or inaccessible mountings. Tie downs must be adequate to withstand wind loads; tie-down and weather proofing should be accomplished with a minimum of labor. Manifolding, that is, connecting the inlets and outlets of many collectors in an array to obtain the desired flow distribution, can be a very time-consuming and expensive operation, and module design should be such as to minimize the labor and materials needed for this operation.

Safety must also be considered in two contexts. First, it must be possible to handle and install modules with a minimum of hazards. Second, materials used in collectors must be capable of withstanding the maximum equilibrium temperatures without hazard of fire or without evolution of toxic or flammable gases.

All of these considerations are important and must be viewed in the light of the many years of low-maintenance service that should be expected of a well-

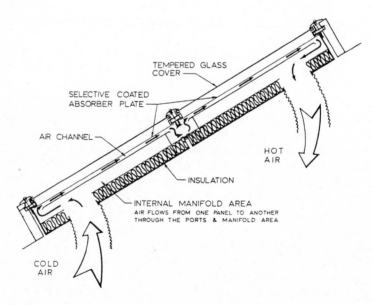

Figure 7.10.1   Cross-section view of two modules of a solar air heater. Courtesy Solaron Corp.

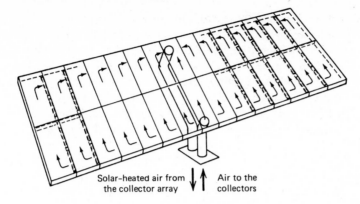

Solar-heated air from
the collector array

Air to the
collectors

Figure 7.10.2 Arrangement of collector array for a solar air heater. Courtesy Solaron Corp.

designed collector. In future chapters on economics and applications, it will be noted that lifetimes of decades are used in economic calculations; collectors in most applications should be designed to last 10 to 30 years with minimum degradation in thermal performance or mechanical properties.

Here we illustrate two collectors, both commercially manufactured. These collectors are selected as typical of good design; there are many others that are

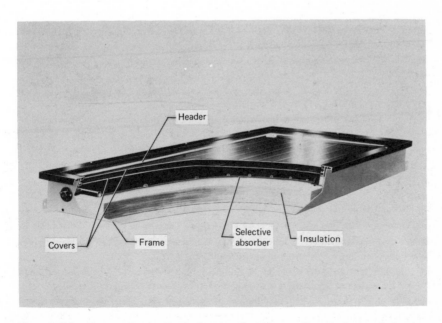

Header

Covers

Frame

Selective
absorber

Insulation

Figure 7.10.3 Construction details of a flat-plate liquid heater. Courtesy Lennox Industries.

good, and the inclusion of these two is not meant to be an endorsement of these particular collectors. Figure 7.10.1 shows a cross-sectional view of two modules of an air heating collector (the collector having the characteristics shown on curves 4 and 5 of Figure 7.7.2). Manifolds are built into the structure of the collectors, and matching openings are provided on the sides of the manifold to interconnect units placed side by side in an array. Figure 7.10.2 shows a schematic of a typical collector array assembly; one pair of openings in the back of the array will serve as entrance and exit ports for multiple modules.

Figure 7.10.3 shows details of a liquid heating collector. The absorber plate is steel, with bright nickel plating and a chrome black absorbing surface. Its design is close to that for which data are shown on Figure 7.2.6.

7.11   SUMMARY

In Chapter 7 we have shown performance measurements on flat-plate collectors, methods of testing these units, and standard methods of presenting the thermal characteristics. The thermal performance data taken on a test stand provides valuable information that will be used later in process design calculations. Performance of a collector in an application will depend not only on the design of the particular installation, but also on the skill with which the equipment is assembled and installed.

REFERENCES

ASHRAE, Standard 93-77, American Society of Heating, Refrigeration, and Air Conditioning Engineers, New York (1977). "Methods of Testing to Determine the Thermal Performance of Solar Collectors."

Beckman, W. A., S. A. Klein, and J. A. Duffie, *Solar Heating Design by the f-Chart Method*, Wiley, New York (1977).

Close, D. J. and M. B. Yusoff, *Solar Energy*, **20**, 459 (1978). "The Effects of Air Leaks on Solar Air Collector Behaviour."

Cooper, P. I., paper presented at Institution of Engineers, Australia meeting, Townsville, Qld. (1976). "The Testing of Flat-Plate Solar Collectors."

Dunkle, R. V. and E. T. Davey, Paper presented at Melbourne International Solar Energy Society Conference (1970). "Flow Distribution in Absorber Banks."

Gupta, C. L., and H. P. Garg, *Solar Energy*, **11**, 25 (1967). "Performance Studies of Solar Air Heaters."

Hill, J. E. and E. R. Streed, paper in *Applications of Solar Energy for Heating and Cooling of Buildings*, Jordan and Liu, Eds., ASHRAE GRP 170 (1977). "Testing and Rating of Solar Collectors."

Hill, J. E., J. P. Jenkins, and D. E. Jones, *ASHRAE Transactions*, **84**, Part 2 (1978). "Testing of Solar Collectors According to ASHRAE Standard 93-77."

Hill, J. E., J. P. Jenkins, and D. E. Jones, NBS Building Science Series 117, U.S. Dept. of Commerce (1979). "Experimental Verification of a Standard Test Procedure for Solar Collectors."

Hill, J. E. and T. Kusuda, National Bureau of Standards Interim Report NBSIR 74-635 to NSF/ERDA (1974). "Proposed Standard Method of Testing For Rating Solar Collectors Based on Thermal Performance."

Hill, J. E. and E. R. Streed, *Solar Energy*, **18**, 421 (1976). "A Method of Testing for Rating Solar Collectors Based on Thermal Performance."

Hottel, H. C., and B. B. Woertz, *Transactions of the American Society of Mechanical Engineers*, **64**, 91 (1942). "Performance of Flat-Plate Solar Heat Collectors."

Moore, S. W., J. D. Balcomb, and J. C. Hedstrom, paper presented at Ft. Collins ISES meeting, August 1974. "Design and Testing of a Structurally Integrated Steel Solar Collector Unit Based on Expanded Flat Metal Plates."

Oonk, R. L., D. E. Jones, and B. E. Cole-Appel, *Solar Energy*, **23**, 535 (1979). "Calculation of Performance of N Collectors in Series from Test Data on a Single Collector."

Simon, F. F., *Solar Energy*, **18**, 451 (1976). "Flat-Plate Solar Collector Performance Evaluation with a Solar Simulation as a Basis for Collector Selection and Performance Prediction."

Skoda, L. F. and L. W. Masters, Report NBSIR 77-1314 of the National Bureau of Standards (1977). "Solar Energy Systems—Survey of Materials Performance."

Solar Age, **2**, 5 (1977), a series of articles including: Butt, S. H. and J. W. Popplewell, "Absorbers, Questions and Answers"; Lyman, W. S. and P. Anderson, "Absorbers: Copper"; Kruger, P., "Absorbers: Steel"; Byrne, S. C., "Absorbers: Aluminum."

Souka, A. F. and H. H. Safwat, *Solar Energy*, **10**, 170 (1966). "Optimum Orientations for the Double Exposure Flat-Plate Collector and its Reflectors."

Streed, E. R., J. E. Hill, W. C. Thomas, A. G. Dawson, and B. D. Wood, *Solar Energy*, **22**, 235 (1979). "Results and Analysis of a Round Robin Test Program for Liquid-Heating Flat Plate Solar Collectors."

Tabor, H., *Bulletin of the Research Council of Israel*, **6C**, 155 (1958). "Radiation, Convection, and Conduction Coefficients in Solar Collectors."

Vernon, R. W. and F. F. Simon, paper presented at Ft. Collins ISES meeting, August 1974, and NASA TMX-71602. "Flat-Plate Collector Performance Determined Experimentally with a Solar Simulator."

CHAPTER 8

Concentrating Collectors

For many applications it is desirable to deliver energy at temperatures higher than those possible with ordinary flat plate collectors. Energy delivery temperatures can be increased by decreasing the area from which heat losses occur. This is done by interposing an optical device between the source of radiation and the energy absorbing surface, to raise the level of incident radiation on the relatively small absorber. In this chapter we discuss two related approaches to this question, the use of nonimaging concentrators and imaging concentrators.

Many designs have been set forth for concentrating collectors. Concentrators can be reflectors or refractors, can be cylindrical or surfaces of revolution, and can be continuous or segmented. Receivers can be convex, flat, or concave and can be covered or uncovered. Many modes of tracking are possible. Concentration ratios (the factors by which radiation flux on the energy absorbing surface is increased) can vary over four orders of magnitude. With this wide range of designs, it is difficult to develop general analyses applicable to concentrators. Thus we treat nonimaging collectors with low concentration ratios, and linear imaging collectors with intermediate concentration ratios. We also note some basic considerations of multifaceted three-dimensional concentrators.

Concentrators can have concentration ratios from low values of 1.5 or 2 to high values of the order of 10,000. Increasing ratios mean increasing temperatures at which energy can be delivered, and increasing requirements for precision in optical quality and positioning of the optical system. Thus the cost of delivered energy from a concentrating collector is a function of the temperature at which it is available. At the highest range of concentration and correspondingly highest precision of optics, concentrating collectors are termed *solar furnaces*, and are laboratory tools for study of properties of materials at high temperatures and similar purposes. Laszlo (1956) and the proceedings of a Solar Furnace Symposium (1957) include extensive discussions of solar furnaces. Our concern in this chapter is with energy delivery systems operating at low or intermediate concentrations.

From an engineering point of view, concentrating collectors present problems in addition to those of flat plat collectors. They must (except at the very low end of the concentration scale) be oriented to "track" the sun so that beam radiation will be directed onto the absorbing surface. However, the designer has open to him a range of configurations of the system that allow new sets of design parameters to be manipulated. There are also new requirements for maintenance, particularly to retain the quality of optical systems for long periods of time in the presence of dirt, weather, and oxidizing or other corrosive atmospheric components. The combination of operating problems and collector cost has restricted the utility of concentrating collectors. New materials and better engineering of systems may make them of practical importance.

To avoid confusion of terminology, the word *collector* will be applied to the total system including the receiver and the concentrator. The *receiver* is that element of the system where the radiation is absorbed and converted to some other energy form, and includes the *absorber*, associated covers and insulation. The *concentrator*, or *optical system* is the part of the collector that directs radiation onto the receiver.

The first two sections in this chapter deal with general information on optical principles and heat transfer that are important in concentrating collectors. The next two treat the performance of collectors using nonimaging concentrators. The balance of the chapter is concerned with imaging collectors and discusses linear collectors of types being used in experimental industrial process heat and pumping applications and multiple heliostat collectors now under study for "power tower" systems for generation of electrical from solar energy.

8.1   COLLECTOR CONFIGURATIONS

Many concentrator types are possible for increasing the flux of radiation on receivers. They can be reflectors or refractors. They can be cylindrical to focus on a "line," or circular to focus on a "point." Receivers can be concave, flat or convex. Examples of four configurations are shown in Figure 8.1.1.

The first is a plane receiver with plane reflectors at the edges to reflect additional radiation onto the receiver. The concentration ratios of this type are relatively low, with a maximum value of less than four. Some of the diffuse component of radiation incident on the reflectors would be absorbed at the receiver. These collectors can be viewed as flat plate collectors with augmented radiation. Analyses of these concentrators have been presented by Hollands (1971), Selcuk (1979), and others. Figure 8.1.1b shows a reflector of parabolic section, which could be a cylindrical surface (with a tubular receiver) or a surface of revolution (with a spherical or hemispherical receiver). Cylindrical collectors of this type have been studied in some detail and are being used in experimental applications.

The continuous parabolic reflector can be replaced by a Fresnel reflector, a set of flat reflectors on a moving array, as shown in Figure 8.1.1c, or by its refracting equivalent. The facets of the reflector can also be individually mounted and adjusted in position, as shown in Figure 8.1.1d. Large arrays of heliostats of this type, with receivers mounted in a tower are the basis of designs of "power tower" collectors.

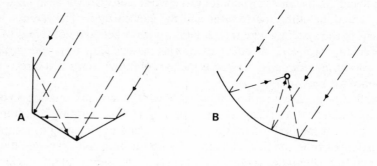

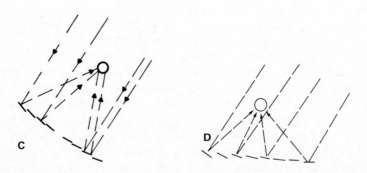

Figure 8.1.1 Possible concentrating collector configurations: (*a*) plane receiver with plane reflectors; (*b*) parabolic concentrator; (*c*) Fresnel reflector; (*d*) array of heliostats.

Flat receivers may be used for any concentrators shown on Figure 8.1.1. Cylindrical, hemispherical, or other convex shapes may also be possible or appropriate for any of the systems illustrated, and cavity receivers may be used with types *b*, *c*, or *d*.

In general, concentrators with receivers much smaller than the aperture are effective only on beam radiation. It is evident also that the orientation of the beam radiation on the concentrator is important, and that sun tracking will be required for these collectors. A variety of orienting mechanisms has been designed to move focusing collectors so that the incident beam radiation will be reflected to the receiver. The motions required to accomplish tracking vary with the design of the optical system, and a particular resultant motion may be accomplished by more than one system of component motions.

Linear (cylindrical) optical systems will focus beam radiation to the receiver if the sun is in the central plane of the concentrator, that is, the plane including the focal axis and the vertex line of the reflector. For this type of system, it is possible to rotate the optics about a single axis to meet this requirement. This axis of rotation may be north-south, east-west, or inclined and parallel to the earth's axis (in which case the rate of rotation is 15°/hr). (There are significant differences in both quantity of incident beam radiation and image quality obtained with these three modes of orientation.)

Reflectors that are surfaces of revolution (circular concentrators) generally must be oriented so that the focus, vertex, and sun are in line, and thus must be able to move about two axes. These axes may, for example, be horizontal and vertical, or one axis of rotation may be inclined so that it is parallel to the earth's axis of rotation (i.e., a polar axis) and the other perpendicular to it. The angle of incidence of beam radiation on a moving plane is indicated for the most probable modes of orientation of that plane by Equations 1.6.9 to 1.6.13.

Orientation systems can provide continuous or nearly continuous adjustments, with movement of the collector to compensate for the changing hourly position of the sun. For some low-concentration linear collectors it is possible to adjust their position intermittently, with weekly, monthly or seasonal changes possible for some designs. Continuous orientation systems may be based on manual or mechanized operation. Manual systems depend on the observations of operators and their skill at making the necessary corrections and may be adequate for some purposes if concentration ratios are not too high and if labor costs are not prohibitive; they have been suggested for use in areas of very low labor cost.

Mechanized orienting systems can be *sun-seeking systems* or *programmed systems*. Sun-seeking systems use detectors to determine system misalignment and through controls make the necessary corrections to realign the assembly. Programmed systems, on the other hand, cause the collector to be moved in a predetermined manner (e.g., 15°/hr about a polar axis as noted above) and may need only occasional checking to assure alignment. It may also be advantageous to use a combination of these tracking methods; for example, by superimposing small corrections by a sun-seeking mechanism on a programmed "rough-positioning" system. Any mechanized system must have the capability of adjusting the position of the collector from end-of-day position to that for operation early the next day, adjusting for intermittent clouds, and adjusting to a position where it can best withstand very high winds without damage.

8.2 CONCENTRATION RATIO

The most common definition of concentration ratio, and that used here, is an *area concentration ratio*,* and is the ratio of the area of aperture to the area of the receiver. (A *flux concentration ratio* is defined as the ratio of the average energy flux on the receiver to that on the aperture, but has limited utility as in many cases there are substantial variations in energy flux over the surface of a receiver.)

The area concentration ratio, in equation form, is

$$C = \frac{A_a}{A_r} \tag{8.2.1}$$

This ratio has an upper limit that depends on whether the concentration is a three-dimensional (circular) concentrator such as a paraboloid, or a two-dimensional (linear) concentrator such as a cylindrical parabolic concentrator. The following development of the maximum concentration ratio, from Rabl (1976a) is based on the second law of thermodynamics, which applies to radiative heat exchange between the sun and the receiver. Consider the circular concentrator with aperture area A_a and receiver area A_r viewing the sun of radius r at distance R, as shown in Figure 8.2.1. θ_s is the half-angle subtended by the sun. (The receiver is shown beyond the aperture for clarity; the argument is the same if it is on the same side of the aperture as the sun.)

If the concentrator is perfect, the radiation from the sun on the aperture (and thus also on the receiver) is the fraction of the radiation emitted by the sun which is intercepted by the aperture

$$Q_{s \to r} = A_a \frac{r^2}{R^2} \sigma T_s^4 \tag{8.2.2}$$

A perfect receiver (i.e., black body) radiates energy equal to $A_r T_r^4$, and a fraction of this, E_{r-s}, reaches the sun:†

$$Q_{r \to s} = A_r \sigma T_r^4 E_{r-s} \tag{8.2.3}$$

When T_r and T_s are the same, the second law of thermodynamics requires the $Q_{s \to r}$ be equal to $Q_{r \to s}$, so from Equations 8.2.2 and 8.2.3:

$$\frac{A_a}{A_r} = \frac{R^2}{r^2} E_{r-s} \tag{8.2.4}$$

---

* Usually termed simply *concentration ratio*.

† E_{r-s} is an exchange factor as used in Equation 3.8.2.

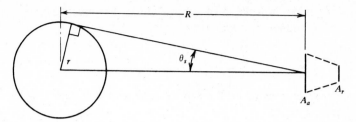

Figure 8.2.1 Schematic of sun at T_s at distance R from a concentrator with aperture area A_u and receiver area A_r. Adapted from Rabl, (1976a).

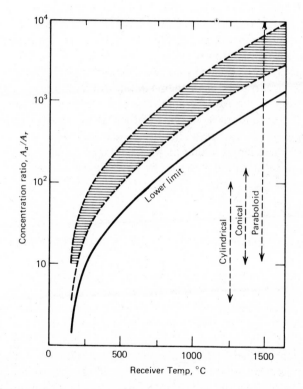

Figure 8.2.2 Relationships between concentration ratio and temperature of receiver operation. The "lower limit" curve represents concentration ratios at which the thermal losses will equal the absorbed energy; higher ratios will then result in useful gain. The shaded range corresponds to collection efficiencies of 40 to 60%, and represents the usual range of operation. Also shown are approximate ranges in which several types of reflectors might be used. (Note: This figure is not intended to be used for design. It is based on an assumed set of conditions determining the absorbed radiation and the thermal losses, and on reasonable design practice at various temperatures. The positions of these curves would shift somewhat under conditions other than those assumed.) From Duffie and Löf (1962).

and since the maximum value of E_{r-s} is unity, the maximum concentration ratio for circular concentrators is:

$$\left(\frac{A_a}{A_r}\right)_{max,c} = \frac{R^2}{r^2} = \frac{1}{\sin^2 \theta_s} \tag{8.2.5}$$

A similar development, outlined by Kreith and Kreider (1978) for linear concentrators, leads to

$$\left(\frac{A_a}{A_r}\right)_{max,l} = \frac{1}{\sin \theta_s} \tag{8.2.6}$$

Thus with circular concentrators and $\theta_s = 0.27°$, the maximum possible concentration ratio is 45,000 and for linear concentrators the maximum is 212.

The higher the temperature at which energy is to be delivered, the higher must be the concentration ratio and the more precise must be the optics of both the concentrator and the orientation system. Figure 8.2.2 shows practical ranges of concentration ratios and types of optical systems needed to deliver energy at various temperatures.

Concentrators can be divided into two categories: nonimaging and imaging. Nonimaging concentrators, as the name implies, do not produce clearly defined images of the sun on the absorber, but rather distribute radiation from all parts of the solar disc onto all parts of the absorber. The concentration ratios are in the low range, generally below 10. Imaging concentrators, in contrast, are analogous to camera lenses in that they form images (usually of very low quality by ordinary optical standards) on the absorber. Concentration ratios can range, in theory, to very high values.

8.3   THERMAL PERFORMANCE OF CONCENTRATING COLLECTORS

Calculation of the performance of concentrating collectors follows the same general outlines as for flat-plate collectors. The absorbed radiation per unit area of aperture, S, must be estimated from the optical characteristics of the concentrator and receiver. Estimation of S for several types of collectors is discussed in sections to follow. Thermal losses from the receiver must be estimated, usually in terms of a loss coefficient U_L, which is based on the area of the receiver. In principle, temperature gradients on the receiver can be accounted for by a flow-factor F_R, to allow the use of inlet fluid temperatures in energy balance calculations. This section is concerned with the estimation of U_L and F_R.

The methods for calculating thermal losses from receivers are not as easily summarized as in the case of flat-plate exchangers. The shapes are widely variable, the temperatures are higher, the edge effects are more significant, conduction terms may be quite high, and the problems may be compounded by non-uniformity of radiation flux on receivers. Thus, substantial temperature gradients may exist across the energy-absorbing surfaces. It is difficult to present a single general method of estimating thermal losses, and ultimately each receiver geometry must be analyzed as a special case.

The nature of the thermal losses for receivers of focusing-type collectors is the same as for flat-plate exchangers. Receivers may have covers transparent to solar radiation. If so, the outward losses from the absorber by convection and radiation to the atmosphere are correspondingly modified and equations similar to those of Chapter 6 can be used to estimate their magnitude. As with flat-plate systems, the losses can be estimated as being independent of the intensity of incident radiation, although this may not be strictly true, particularly if a transparent cover absorbs appreciable solar radiation. In any event, an effective transmittance-absorptance product can also be defined for focusing systems. Furthermore, with focusing systems the radiation flux at the receiver is generally such that only cover materials with very low absorptance for solar radiation can be used without thermal damage to the cover. Conduction losses occur through the supporting structure and through insulation on nonirradiated parts of the receiver.

The generalized thermal analyses of a concentrating collector is similar to that of a flat-plate collector. It is necessary to derive appropriate expressions for the collector efficiency factor F', the loss coefficient U_L, and the collector heat removal factor F_R. With F_R and U_L known, the collector useful gain can be calculated from an expression that is similar to that for a flat plate collector.

As an example of calculation of thermal loss coefficient U_L, consider an uncovered cylindrical absorbing tube that might be used as a receiver with a linear concentrator. Assume that there are no temperature gradients around the receiver tube. The loss coefficient considering convection and radiation from the surface and conduction through the support structure is

$$U_L = h_w + h_r + U_{cond} \qquad (8.3.1)$$

The linearized radiation coefficient can be calculated from

$$h_r = 4\sigma\varepsilon\overline{T}^3 \qquad (8.3.2)$$

where $\overline{T}$ is the mean temperature for radiation and ε is the emittance of the absorbing surface. If a single value of h_r is not acceptable due to large temperature gradients in the flow direction, the collector can be considered as divided into two or more segments each with constant h_r. The estimation of h_w for cylinders is noted in Section 3.15. Estimation of conductive losses must be based on knowledge of the details of construction or on measurements on a particular collector.

If the cylindrical receiver is in a cylindrical envelope (i.e., a tubular cover) the convection and radiation from absorbing surface to cover and cover to ambient must be combined in the same way as for flat plate collectors. It may be necessary to add conduction by supports for the cover to convection and radiation from absorber to cover. It may also be necessary to account for absorption of radiation by the cover.

Example 8.3.1

Calculate the overall loss coefficient for a 60-mm cylindrical receiver at 200 C. The absorber surface has an emittance of 0.91. The absorber is covered by a glass tubular cover 90 mm in diameter and the space between the two is evacuated. The wind speed is 5 m/s and the sky and air temperatures are 10 C.

Solution

Assume the cover temperature is 50 C. To estimate the wind heat transfer coefficient, it is necessary to find the Reynolds number for an average air temperature of 30 C.

$$\text{Re} = \frac{\rho V D}{\mu} = \frac{1.16 \times 5 \times 0.090}{1.86 \times 10^{-5}} = 28,100$$

The Nusselt number is then found from Equation 3.15.12 with Nu increased by 25% for outdoor conditions:

$$\text{Nu} = 0.24 \times 1.25(28100)^{0.6} = 140$$

$$h_w = 140 \times \frac{0.0265}{0.090} = 41 \text{ W/m}^2 \text{ C}$$

The radiation coefficient from the cover tube to ambient is $h_{r,c-a} = 0.88 \times 5.67 \times 10^{-8} \times 4 \times 303^3 = 5.55 \text{ W/m}^2 \text{ C}$. The radiation heat transfer coefficient between the receiver tube and cover tubes is found from Equation 3.10.2

$$h_{r,r-c} = \frac{5.67 \times 10^{-8}(473^2 + 323^2)(473 + 323)}{\dfrac{1 - 0.91}{0.91} + 1 + \dfrac{1 - 0.88}{0.88} \times \dfrac{0.06}{0.09}} = 12.45$$

There is no convection transfer from receiver tube to cover through the evacuated space, so the overall loss coefficient, U_L, based on the absorber area is

$$U_L = \left[\frac{A_r}{(h_w + h_{r,c-a})A_c} + \frac{1}{h_{r,r-c}} \right]^{-1} = \left[\frac{0.06}{(41 + 5.5)0.09} + \frac{1}{12.45} \right]^{-1}$$

$$= 10.6 \text{ W/m}^2 \text{ C}$$

With this first estimate of U_L, it is necessary to check the assumed cover temperature of 50 C. The heat transfer from absorber to cover must be equal to the heat transfer from cover to ambient.

$$A_c(h_{r,c-a} + h_w)(T_c - T_a) = A_r h_{r,r-c}(T_r - T_c)$$

Solving for the cover temperature

$$T_c = \frac{A_r h_{r,r-c} T_r + A_c(h_{r,c-a} + h_w)T_a}{A_r h_{r,r-c} + A_c(h_{r,c-a} + h_w)}$$

so that

$$T_c = \frac{12.45 \times 200 + \dfrac{0.09}{0.06}(5.5 + 41)10}{12.45 + \dfrac{0.09}{0.06}(5.5 + 41)} = 39\ C$$

A second iteration with the cover temperature equal to 39 does not significantly change the value of U_L. Therefore $U_L = 10.6\ W/m^2\ C$. ∎

Since the heat flux in a concentrating system may be high, the heat transfer resistance from the outer surface of the receiving tube to the fluid should include the tube wall. The overall heat transfer coefficient (based on the outside tube diameter) from the surroundings to the fluid is

$$U_o = \left(\frac{1}{U_L} + \frac{D_o}{h_{fi}D_i} + \frac{D_o \ln D_o/D_i}{2k} \right)^{-1} \tag{8.3.3}$$

where D_i and D_o are the inside and outside tube diameters, h_{fi} is the heat transfer coefficient inside the tube, and k is the tube thermal conductivity.

The useful energy gain per unit of collector length, q'_u, expressed in terms of the local receiver temperature, T_r, and S, the absorbed solar radiation per unit of aperture area is

$$q'_u = \frac{A_a S}{L} - \frac{A_r U_L}{L}(T_r - T_a) \tag{8.3.4}$$

where A_a is the unshaded area of the aperture of the concentrator and A_r is the area of the receiver ($= \pi D_o$ for the cylindrical absorber). In terms of the energy transfer to the fluid at local fluid temperature T_f, q'_u is

$$q'_u = \frac{(A_r/L)(T_r - T_f)}{\dfrac{D_o}{h_{fi}D_i} + \left(\dfrac{D_o}{2k} \ln \dfrac{D_o}{D_i} \right)} \tag{8.3.5}$$

If T_r is eliminated from Equations 8.3.4 and 8.3.5, we have

$$q'_u = F' \frac{A_a}{L} \left\{ S - \frac{A_r}{A_a} U_L(T_f - T_a) \right\} \tag{8.3.6}$$

where the collector efficiency factor, F', is

$$F' = \frac{1/U_L}{\dfrac{1}{U_L} + \dfrac{D_o}{h_i D_i} + \dfrac{D_o \ln(D_o/D_i)}{2k}} \tag{8.3.7}$$

or

$$F' = \frac{U_o}{U_L} \tag{8.3.8}$$

The form of Equations 8.3.6 to 8.3.8 is identical to Equations 6.5.16 to 6.5.18. If the same procedure is followed as was used to derive Equation 6.7.6, the following equation results:

$$Q_u = A_a F_R \left\{ S - \frac{A_r U_L}{A_a} (T_i - T_a) \right\} \tag{8.3.9}$$

In a manner analogous to that for a flat plate collector, the collector flow factor F'' is

$$F'' = \frac{F_R}{F'} = \frac{\dot{m} C_p}{A_r U_L F'} \{ 1 - e^{-A_r U_L F' / \dot{m} C_p} \} \tag{8.3.10}$$

The only difference between covered and uncovered receivers is in the calculation of S and U_L.

If a receiver of the type discussed above serves as a boiler, F' is the same as is given by Equation 8.3.7, but F_R is then identically equal to F' as there is no temperature gradient in the flow direction. If a part of the receiver serves as a boiler and other parts as fluid heaters, the two or three segments of the receiver must be treated separately.

Example 8.3.2

A cylindrical parabolic concentrator with width of 2.5 m and length 10 m has an estimated absorbed radiation per unit area of aperture of 430 W/m². The receiver is a cylinder, painted flat black, and surrounded by an evacuated glass cylindrical envelope. The absorbing cylinder has a diameter of 60 mm and the transparent envelope has a diameter of 90 mm. The collector is designed to heat a fluid entering the absorber at 200 C, at a flow rate of 0.139 kg/s. The fluid has $C_p = 1.26$ kJ/kg C. The heat transfer coefficient inside the tube is 300 W/m² C and the overall loss coefficient is 10.6 W/m² C (from Example 8.3.1). The tube is made of stainless steel ($k = 16$ W/m C) with a wall thickness of 5 mm. If the ambient temperature is 25 C, calculate the useful gain and exit fluid temperature.

Solution

The solution is based on Equation 8.3.9. The area of the receiver is:

$$A_r = \pi D L = \pi \times 0.06 \times 10 = 1.88 \text{ m}^2$$

Taking into account shading of the central part of the collector by the receiver,

$$A_a = (2.5 - 0.09)10 = 24.1 \text{ m}^2$$

To calculate F_R, we first calculate F' for this situation from Equation 8.3.7:

$$F' = \frac{1/10.6}{\dfrac{1}{10.6} + \dfrac{0.06}{300 \times 0.05} + \dfrac{0.06 \ln(0.06/0.05)}{2 \times 16}} = 0.96$$

Then F_R, from Equation 8.3.10

$$\frac{\dot{m}C_p}{A_r U_L F'} = \frac{0.139 \times 1260}{1.88 \times 10.6 \times 0.96} = 9.15$$

$$F'' = 9.15(1 - e^{-1/9.15}) = 0.95$$

$$F_R = F'' \times F' = 0.95 \times 0.96 = 0.91$$

Then the useful gain is

$$Q_u = 24.1 \times 0.91\left[430 - \frac{1.88 \times 10.6}{24.1}(200 - 25)\right] = 6260 \ W$$

The exit fluid temperature is

$$t_o = t_i + \frac{Q_u}{\dot{m}C_p} = 200 + \frac{6260}{0.139 \times 1260} = 236 \ C \quad \blacksquare$$

The analysis of other receiver geometries may not be as neat as that for cylinrical receivers, but the principles are the same. If concentration ratios are high, there may be temperature gradients across the receiver caused by nonuniform distribution of absorbed energy. In these circumstances it may be necessary to divide the receiver area into zones and write energy balances on each of the zones. It is unlikely that covers would be used under these conditions because of excessive cover temperatures.

8.4 OPTICAL CHARACTERISTICS OF NONIMAGING CONCENTRATORS

It is desirable to consider concentrating collectors that can function seasonally or annually with minimum requirements for tracking (with its attendant mechanical complications). The concentrators considered in this section have the capability of reflecting to the receiver all of the incident radiation on the aperture over ranges of incidence angles within wide limits. The limits define the *acceptance angle* of the concentrator. As all radiation incident within the acceptance angle is reflected to the receiver, the diffuse radiation within these angles is also useful input to the collector.

Most of this section is devoted to *compound parabolic concentrators* (designated CPC). These concentrators had their origins in instruments for detection of Cherenkov radiation in high-energy physics experiments, a development noted by Hinterberger and Winston (1966). An independent and parallel development occurred in USSR [Baranov and Melnikov (1966)]. Their potential as concentrators for solar energy collectors was pointed out by Winston (1974), and they have been the basis of detailed study since then by Welford and Winston (1978), Rabl (1976a, b) and others.

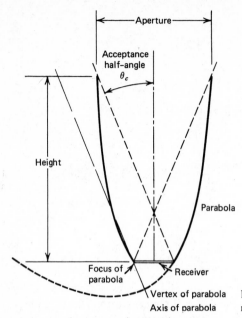

Figure 8.4.1 Cross section of a symmetrical nontruncated CPC.

The basic concept of the compound parabolic concentrator is shown in Figure 8.4.1. These concentrators are potentially most useful as linear or trough-type concentrators (although the analysis has also been done for three-dimensional concentrators), and the following is based on the two-dimensional CPC. Each side of the CPC is a parabola; the focus and axis of only the right-hand parabola are indicated. Each parabola extends until its surface is parallel with the CPC axis. The angle between the axis of the CPC and the line connecting the focus of one of the parabolas with the opposite edge of the aperture is the

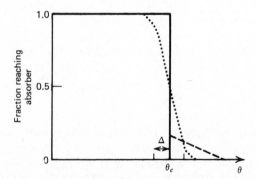

Figure 8.4.2 Fraction of radiation incident on the aperture of a CPC at angle θ which reaches the absorber surface if $\rho = 1$. θ_c is the acceptance half-angle and Δ is an angular surface error. Full CPC with no surface errors, ————; truncated CPC with no surface errors, — — — —; full CPC with surface error Δ, · · · ·. Adapted from Rabl (1976).

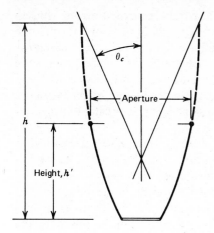

Figure 8.4.3 A CPC truncated so its height/aperture ratio is about one half of the full CPC.

acceptance half-angle, θ_c. If the reflector is perfect, any radiation entering the aperture at angles between $\pm\theta_c$ will be reflected to a receiver at the base of the concentrator by specularly reflecting parabolic reflectors.

Concentrators of the type of Figure 8.4.1 have area concentration ratios which are functions of the acceptance half angle θ_c. The relationship for an ideal two-dimensional system is*

$$C_i = \frac{1}{\sin \theta_c} \tag{8.4.1}$$

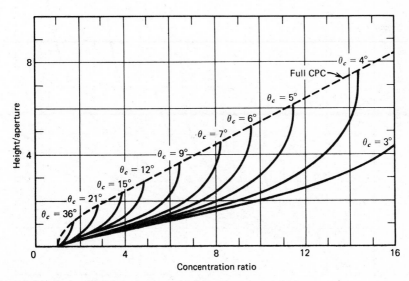

Figure 8.4.4 Ratio of height to aperture for full and truncated CPCs as a function of C and θ_c. Adapted from Rabl (1976b).

* This can be shown by the same arguments that lead to Equations 8.2.6.

An ideal CPC in this context is one which has parabolas with no errors. Thus an ideal CPC with an acceptance half angle of 23.5° will have C_i of 2.51, and one with an acceptance half-angle of 11.75° will have C_i of 4.91. Figure 8.4.2 shows the fraction of radiation incident on the aperture at angle θ which reaches the absorber, as a function of θ. For the ideal CPC, the fraction is unity out to θ_c and zero beyond. For a CPC with a surface error Δ, some radiation incident at angles less than θ_c does not reach the absorber, and some at angles greater than θ_c does reach it.

At the upper end points of the parabolas in a CPC the surfaces are parallel to the central plane of symmetry of the concentrator. The upper ends of the reflectors contribute little to the radiation reaching the absorber, and the CPC can be truncated to reduce its height from h to h' with a resulting saving in reflector area and little sacrifice in performance. A truncated CPC is shown in Figure 8.4.3. The dashed plot on Figure 8.4.2 shows the spread of the image for the

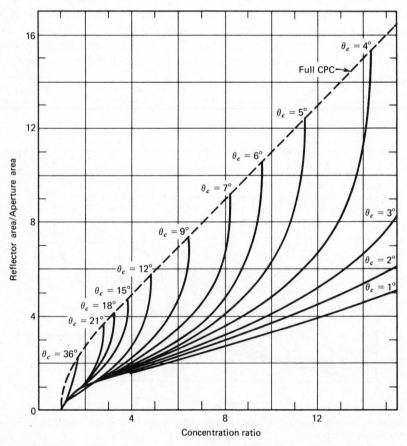

Figure 8.4.5 Ratio of reflector area to aperture area for full and truncated CPCs. Adapted from Rabl (1976b).

truncated concentrator. Truncation does not affect the acceptance angle, but it does change the height to aperture ratio, the concentration ratio and the average number of reflections undergone by radiation before it reaches the absorber surface. The effects of truncation are shown for otherwise ideal CPCs in Figure 8.4.4 to 8.4.6. Figure 8.4.4 shows the height to aperture ratio (see Figure 8.4.1) and Figure 8.4.5 shows the ratio of reflector area to aperture area. Figure 8.4.6 shows the average number of reflections undergone by radiation entering the aperture before it reaches the absorber. If the truncation is such that the average number of reflections is below the $(N)_{min}$ curve, that average number is at least $1 - 1/C$. Rabl (1976b) gives equations for all of these quantities.

The use of these plots can be illustrated as follows. An ideal full CPC has an acceptance half-angle θ_c of 12°. From Figure 8.4.4 the height/aperture ratio is 2.8 and the concentration ratio is 4.8. From Figure 8.4.5, the area of reflector required is 5.6 times the aperture. The average number of reflections undergone by radiation before reaching the absorber surface is 0.97 from Figure 8.4.6. If this CPC is truncated so that its height to aperture ratio is 1.4, from Figure 8.4.4

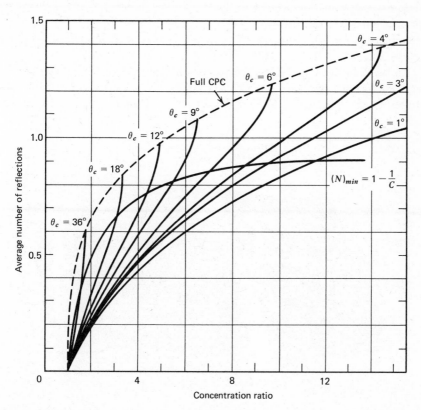

Figure 8.4.6 The average number of reflections undergone by radiation within the acceptance angle reaching the absorber surface of full and truncated CPCs. Adapted from Rabl (1976b).

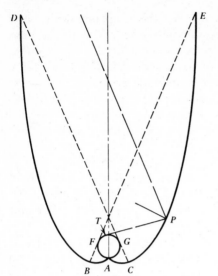

Figure 8.4.7 A CPC for a tubular receiver. Adapted from Rabl (1978).

the concentration ratio will drop to 4.2. Then from Figure 8.4.5 the reflector area/aperture ratio is 3.0 and from Figure 8.4.6 the average number of reflections will be at least $1 - 1/4.2 = 0.76$.

The preceding discussion has been based on flat receivers occupying the plane between the two foci (Figure 8.4.3). Other receiver shapes are possible, and Hinterberger and Winston (1975) showed that a CPC can be developed with

Figure 8.4.8 An array of truncated CPC reflectors with evacuated tubular receivers, with a glass cover over the array. Courtesy Energy Design Corporation.

aperture width l which will concentrate incident radiation with incidence angles between $\pm \theta_c$ onto any convex absorber with circumference $l \sin \theta_c$. The method of generation of the shape of the CPC is illustrated by Figure 8.4.7 which shows a special case of interest, a cylindrical absorber. Parts AB and AC of the reflector are convolutes of parts AF and AG of the absorber. The requirement for the rest of the reflector is that at any point P the normal to the reflector must bisect the angle between the tangent to the absorber, PT, and the line incident on P and at an angle θ_c to the axis of the CPC. This CPC is used with evacuated tubular receivers, and an example is shown in Figure 8.4.8. They can be truncated in the same way as other CPCs.

This method can be used to generate a reflector for any convex receiver shape. Thus a set of CPC-type concentrators (not necessarily parabolas) can be evolved that permit a range of choices of receiver shape. CPCs can be used in series, that is, the receiver for a primary concentrator can be the aperture of a secondary concentrator. The concentrators need not be symmetrical.

8.5 ORIENTATION AND ABSORBED ENERGY FOR CPC COLLECTORS

The requirements for orientation of a CPC collector are related to its acceptance angle, $2\theta_c$. A logical orientation for such a collector is along a horizontal east-west axis, sloped toward the equator and more or less adjustable about that axis. The CPC is arranged so that the pseudo incidence angle* of beam radiation (the projection of the angle of incidence in the north-south vertical plane) lies within the limits $\pm \theta_c$ during the times when output is needed from the collector. In practice compromises are necessary between frequency of movement of the collector and concentration ratio, with high ratios associated with small acceptance angles and relatively frequent positioning.

To estimate the radiation absorbed by the receiver of a CPC, it is necessary to determine if the angle of incidence of the beam radiation is within the acceptance angle, and then estimate the contributions of both beam and diffuse radiation. The absorbed radiation can be written†

$$S = (G_{b,\,\text{CPC}} \tau_{cb} \alpha_b + G_{d,\,\text{CPC}} \tau_{cd} \alpha_d) \tau_{\text{CPC}} \qquad (8.5.1)$$

where $G_{b,\,\text{CPC}}$ and $G_{d,\,\text{CPC}}$ are the effective beam and diffuse radiation on the aperture (effective means incident at angles within the acceptance angle), $\tau_{cb}\alpha_b$ and $\tau_{cd}\alpha_d$ are the products of transmittance for beam and diffuse of any cover which may be placed over the concentrator array and the absorptances of the absorber, and τ_{CPC} is a "transmittance" of the CPC which accounts for reflection losses.

---

* This angle is referred to in architectural literature as the solar profile angle, and by Hollands (1971) as the East-West Vertical (EWV) angle.

† Ground-reflected radiation could also be included, but usually it is not incident on a CPC within the acceptance angle. This will be further discussed below.

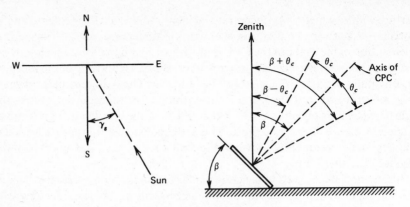

Figure 8.5.1 (a) Plan view showing solar azimuth angle, γ_s. (b) Projection on a north-south plane of CPC acceptance angles and slope, with CPC on east-west axis.

It is convenient to utilize the *solar azimuth angle*, γ_s, to describe the position of the sun. It is the angular displacement of a vertical plane including the observer and the sun from a vertical north-south plane including the observer, with angles east of south negative and west of south positive. This is shown in Figure 1.6.1 and plan view on Figure 8.5.1a. This angle can be calculated from the declination, latitude, hour angle, and zenith angle by

$$\sin \gamma_s = \frac{\cos \delta \sin \omega}{\sin \theta_z} \tag{8.5.2}$$

or

$$\tan \gamma_s = \frac{\sin \omega}{\sin \phi \cos \omega - \cos \phi \tan \delta} \tag{8.5.3}$$

Figure 8.5.1b shows the projections of the acceptance angles of a CPC on a vertical north-south plane perpendicular to a CPC oriented east-west. Two angles, $\beta - \theta_c$ and $\beta + \theta_c$, are the angles from the vertical in this plane to the two limits describing the acceptance angle. Mitchell (1979) has shown that the following condition must be met in order for the beam radiation to be useful

$$(\beta - \theta_c) \le \tan^{-1}(\tan \theta_z \cos \gamma_s) \le (\beta + \theta_c) \tag{8.5.4}$$

It is convenient to introduce a *control function*, F, which is 1 if the criterion of Equation 8.5.4 is met and 0 otherwise. If beam radiation is incident on the aperture within the acceptance angle, $F = 1$, and its contribution to energy reaching the absorber surface can be calculated. The effective incident beam energy per unit area of aperture for the CPC is

$$G_{b,\text{CPC}} = FG_{bn} \cos \theta \tag{8.5.5}$$

The angle of incidence on the aperture can be determined from the appropriate equation of Section 1.6. Two limiting cases may be encountered. If the collector is fixed, cos θ is given by Equation 1.6.2 or 1.6.5. It is conceivable (but unlikely) that a CPC would be rotated continuously about its axis; this is the other limiting case, and Equation 1.6.10 could be used to calculate cos θ. The more likely case is the first. It is also possible that occasional adjustments could be made, and in this case the collector can be treated as fixed over periods of time.

Example 8.5.1

A CPC collector array is mounted on a horizontal east-west axis, oriented at a slope of 25° from the horizontal at a location of latitude 35°N. The acceptance angle, $2\theta_c$, is 24°. At 10 a.m. on August 1 the beam normal radiation G_{bn} is 805 W/m². Estimate the effective contribution of incident beam radiation on the CPC array if the aperture area is 10 m².

Solution

The first step is to see if the criterion of Equation 8.5.4 is met, that is, whether F is 0 or 1. For this date, $n = 213$ and $\delta = 17.9°$. From Equation 1.6.4, $\cos\theta_z = 0.851$ and $\theta_z = 31.6°$. From Equation 8.5.2 the solar azimuth angle is

$$\gamma_s = \sin^{-1}\left[\frac{\cos 17.9 \sin(-30)}{\sin 31.6}\right] = -65.2°$$

For this collector, $(\beta - \theta_c) = 25 - 12 = 13$, and $(\beta + \theta_c) = 25 + 12 = 37°$. Then $\tan^{-1}(\tan 31.6 \cos(-65.2)) = 14.5°$. Since 14.5° lies between 13° and 37°, F is 1. The angle of incidence of the beam radiation on the aperture can be obtained with Equation 1.6.5:

$$\cos\theta = \cos(35 - 25)\cos 17.9 \cos 30 + \sin(35 - 25)\sin 17.9$$
$$= 0.865$$

The effective beam radiation incident on the plane of the CPC is

$$G_{b,\text{CPC}} = 805 \times 1 \times 0.865 = 0.70 \text{ kW/m}^2$$

For the 10 m² aperture, the effective beam radiation is 7.0 kW. ▪

The effective diffuse radiation on a CPC is that incident within the acceptance angle. It is a function of the acceptance angle, and can be estimated by methods analogous to those of Section 2.11 if the diffuse radiation is assumed to be isotropic.* Two situations can arise, depending on whether $\beta + \theta_c$ is greater or less than 90° (i.e., if some of the diffuse radiation within the acceptance angle $2\theta_c$ is ground-reflected radiation).

---

* If the diffuse is not isotropic and is received from the circumsolar sky, Equations 8.5.6 and 8.5.7 will underestimate the diffuse contribution at times when F is 1.

If $\beta + \theta_c$ is less than 90°, there will be no ground-reflected radiation entering the CPC, and the effective diffuse radiation within the acceptance angle can be estimated by

$$G_{d,\text{CPC}} = \frac{G_d}{C} \qquad (8.5.6)$$

For a full CPC, C can be replaced by $1/\sin \theta_c$.

If $\beta + \theta_c$ is greater than 90°, both diffuse from the sky and ground-reflected radiation should be included. This can be estimated by

$$G_{d,\text{CPC}} = G_d\left[\frac{1/C + \cos \beta}{2}\right] + G\rho\left[\frac{1/C - \cos \beta}{2}\right] \qquad (8.5.7)$$

The useful diffuse on the CPC, $G_{d,\text{CPC}}$, can then be treated in the same way as the beam radiation in estimation of the radiation reaching the absorber.

A CPC collector will probably have a transparent cover over the array of reflectors. This serves both to protect the reflecting and absorbing surfaces and to reduce thermal losses from the absorber. If a cover is used, the beam and diffuse radiation effectively entering the CPC are reduced by the transmittance of the cover. For beam radiation, the transmittance is calculated by the methods of Section 5.3. Only part of the incident diffuse radiation effectively enters the CPC, and that part is a function of the acceptance angle. A relationship between the mean angle of incidence of effective diffuse radiation and the acceptance

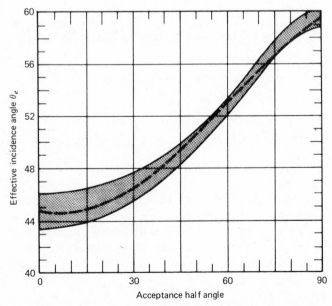

Figure 8.5.2 Equivalent incidence angle for isotropic diffuse radiation for a CPC as a function of acceptance half angle. From Brandemuehl and Beckman (1980).

half-angle θ_c is shown in Figure 8.5.2. The relationship depends on the nature of the cover system, and the figure shows a band of solutions including one and two covers, refractive indices from 1.34 to 1.526, and KL per cover up to 0.0524. An equation for the equivalent angle of incidence, θ_e (the dashed line) is

$$\theta_e = 44.86 - 0.0716\theta_c + 0.005120\theta_c^2 - 0.00002798\theta_c^3 \qquad (8.5.8)$$

Thus for a CPC with $\theta_c = 20°$, the mean angle of incidence of effective diffuse radiation is $45°$ and the transmittance of the cover for this radiation is that of beam radiation at $45°$.

The radiation reaching the absorber in the CPC, neglecting end effects (or assuming the ends to be highly reflective) also depends on the specular reflectance of the reflector, ρ, and the average number of reflections, n_r. An "effective transmittance" of the CPC can be expressed as

$$\tau_{CPC} = \rho^{n_r} \qquad (8.5.9)$$

Example 8.5.2

At the location and time of Example 8.5.1 the diffuse radiation on a horizontal surface is 320 W/m^2. The CPC is truncated and has a concentration ratio of 4.5. The average number of reflections is 0.75 (from Figure 8.4.6) and the reflectance of the CPC material is 0.88. Estimate the effective diffuse radiation on the aperture. Also, estimate the total radiation reaching the absorber, if the CPC array has a cover with $KL = 0.0125$.

Solution

For these circumstances, $\beta + \theta_c = 37°$, so there is no contribution of ground-reflected radiation and Equation 8.5.6 can be used to estimate the effective diffuse radiation:

$$G_{d,\,CPC} = 320/4.5 = 71 \ W/m^2$$

and the effective diffuse on the 10 m^2 aperture is 0.71 kW. The angle of incidence of the beam radiation on the cover is $\cos^{-1} 0.865$, or $30°$. The transmittance of the cover for this radiation, from Figure 5.3.1, is 0.90. The mean angle of incidence for the diffuse radiation, from Figure 8.5.2, is $45°$. From Figure 5.3.1 the transmittance for this radiation is 0.88. The radiation reaching the absorber for the 10 m^2 aperture is then the contribution of the beam and diffuse components, or

$$(0.90 \times 7.0 + 0.88 \times 0.71)0.88^{0.75} = 6.3 \ kW \qquad \blacksquare$$

Example 8.5.3

What would the useful incident diffuse radiation be for a full CPC collector with an acceptance half-angle of $18°$ sloped at $80°$ to south at latitude $35°$? The horizontal total radiation is 530 W/m^2, the diffuse is 175 W/m^2, and the ground reflectance is 0.7.

Solution

In this case the ground reflected radiation must be considered, and Equation 8.5.7 is used with C replaced by $1/\sin\theta_c$ for this full CPC.

$$G_{d,\,CPC} = 175\left[\frac{\sin 18 + \cos 80}{2}\right] + 530 \times 0.7\left[\frac{\sin 18 - \cos 80}{2}\right]$$

$$= 42.2 + 25.1 = 67 \text{ W/m}^2 \qquad \blacksquare$$

The absorptance of the energy absorbing surface depends on the angle of incidence of the radiation on the surface, as noted in Section 4.7. The angle of incidence varies depending on the angle of incidence of the radiation on the aperture of the CPC, the number of reflections undergone by the radiation, and the shape of the receiving surface. If the average number of reflections is small, as it will be for large acceptance angles, then as a first approximation the absorptance can be taken as that corresponding to the angle of incidence of the effective radiation on the aperture of the CPC.

Example 8.5.4

The collector of Example 8.5.1 and 8.5.2 has a flat absorbing surface with the α/α_n versus θ_i characteristics shown in Figure 4.7.1 and $\alpha_n = 0.96$. The array is covered with a single glass cover with $KL = 0.0125$. What is the absorbed energy per unit aperture area? For the 10 m^2 array?

Solution

Each of the terms in the last equation in Example 8.5.2 is to be multiplied by the appropriate absorptance for the beam and diffuse radiation. The angle of incidence of the beam radiation on the cover is $\cos^{-1} 0.86$, or $30°$. From Figure 4.7.1, at $30°$, $\alpha/\alpha_n = 0.98$, so the absorptance of the black surface is $0.98 \times 0.96 = 0.94$.

The mean angle of incidence of the diffuse radiation on the cover and black surface is $45°$, and the absorptance is estimated as 0.92. Thus the absorbed radiation, per unit area of aperture, is the sum of the beam and diffuse contributions.

$$S = (0.70 \times 0.90 \times 0.94 + 0.071 \times 0.88 \times 0.92)0.88^{0.75} = 0.59 \text{ kW/m}^2$$

and for the 10 m^2 total aperture the estimated absorbed radiation is 5.9 kW.

$\qquad \blacksquare$

There are a variety of CPC geometries that may require special treatment. For example, an evacuated tubular receiver with cylindrical absorbing surface may be used, interposing an additional cover that must be considered. Evacuated tubes with flat receivers can also be used, but the absorbing surface width will be less than the full width of the receiver, an arrangement that results in transmission losses due to the cylindrical cover and failure to intercept part of the radiation by the absorbing surfaces.

With the procedures outlined in this section, it is possible to estimate the energy absorbed by a CPC collector. In the next section, this is combined with the analysis of thermal performance of Section 8.2 to show how the collector performance can be estimated for any set of operating conditions.

8.6 PERFORMANCE OF CPC COLLECTORS

The basic equation summarizing the performance of a CPC collector is Equation 8.3.9, where S is calculated by Equation 8.5.1. The remaining question is the calculation of the appropriate thermal loss coefficient, U_L. As CPC collectors can have many configurations of both concentrator and absorber, it is difficult to generalize on how the losses should be calculated. The basic principles are the same as those outlined in Chapter 6 for flat plate collectors and in Section 8.3. The calculation must include radiation, convection, and conduction losses, and should be based on a mean absorber temperature, T_p. Rabl (1976b) presents a discussion of calculation of loss coefficients for a CPC collector geometry using a flat absorber. Figure 8.6.1 shows his estimations of overall loss coefficients, shown on the basis of unit area of absorber, for use in Equation 8.3.8. The coefficients are functions of concentration ratio; the method used in their estimation assumed fixed conduction losses independent of concentration ratio, and both radiation and convection from plate to cover are to some degree functions of C. Estimates are shown for two plate emittances and two plate temperatures.

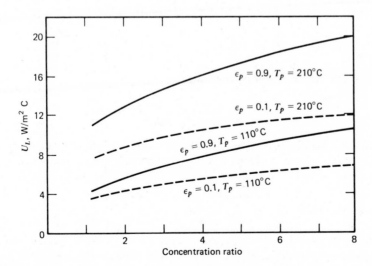

Figure 8.6.1 Estimated loss coefficients for CPC collectors with flat absorbers. Data are from Rabl (1976b).

With estimates of U_L from Figure 8.6.1, from measurements, or from calculations, and with a knowledge of the meteorological conditions and inlet fluid temperature, the procedure for calculation of output of the collector is the same as for a flat plate collector. F_R is calculated from U_L, the fluid flow rate, and F' by Equation 8.3.10. Q_u is calculated with Equation 8.3.9 and the average plate temperature is the basis for a check of the assumed T_p (and U_L).

Example 8.6.1

The CPC collector of the examples in Section 8.5 has a flat receiver with $\varepsilon_p = 0.10$. Under the conditions of its use, it is expected that F' will be 0.87. The inlet fluid temperature is 130 C and the ambient temperature is 28 C. The collector array of 10 m^2 total aperture area has 10 concentrators each with an aperture width of 0.30 m, with flow in parallel through the 10 receivers. The total flow rate is 0.135 kg/s and the fluid heat capacity is 3560 J/kg C. Estimate the useful gain from the collector, when the absorbed radiation S is 0.59 W/m^2, from Example 8.5.4. Base U_L on Rabl's estimates as indicated in Figure 8.6.1.

Solution

Assume the absorber temperature, T_p, to be 140 C. From Figure 8.6.1 the overall loss coefficient U_L is approximately 6.1 W/m^2 C for this CPC with a concentration ratio of 4.5. The flow rate per unit area of aperture, $\dot{m}/A_a = 0.135/10 = 0.0135$ kg/m^2 s. F_R can now be calculated from Equation 8.3.10

$$F_R = \frac{0.0135 \times 3560}{6.1}\left[1 - \exp\left(-\frac{6.1 \times 0.87}{0.0135 \times 3560}\right)\right] = 0.82$$

The total useful gain can be estimated from Equation 8.3.9

$$Q_u = 10 \times 0.82\left[0.59 - \frac{6.1(130 - 28)}{4.5 \times 1000}\right] = 3.7 \text{ kW}$$

The temperature rise through the collector is

$$\Delta T = \frac{3700}{0.135 \times 3560} = 7.9 \text{ C}$$

Equation 6.8.4 could be used to estimate the mean receiver temperature of 134 C, and a new U_L estimated based on the revised mean absorber temperature, T_p. It will be close to the 140 C assumed, and it is not possible to significantly improve on the estimate already made. Thus the estimated output of this CPC under the stated conditions is 3.7 kW.   ∎

Another geometry of particular interest is a CPC with an evacuated tubular receiver having a cylindrical absorber. For these receivers, an additional transmittance loss occurs, because of the outer glass cover over the absorber, but the loss coefficients for the receiver are of the order of 1 W/m^2 C.

8.7 COLLECTORS WITH LINEAR IMAGING CONCENTRATORS

Cylindrical or linear concentrators with parabolic section have been studied extensively both analytically and experimentally, and have been proposed for applications requiring intermediate concentration ratios and temperatures in the range of 100 to 500 C. Figure 8.7.1 shows an experimental installation of a collector of this type, which supplies steam to operate an engine driving an irrigation water pump. Receivers used with these concentrators may be flat (as in Figure 8.7.1) or cylindrical.

Cross sections of a linear parabolic concentrator are shown in Figure 8.7.2. Several key factors are illustrated on the diagrams. The incident beam of solar radiation has a total angular width of 0.53°, (i.e., 32′), and is incident on the

Figure 8.7.1 Collector with linear parabolic concentrator used in an experimental water pumping application. Courtesy Northwestern Mutual Life Insurance Company and Battelle Memorial Institute.

concentrator in a direction parallel to the central plane of the parabola (the plane described by the axis and focus of the parabola). The beam in the diagram is shown as normal to the aperture. (The effects of the component of angle of incidence in the axial plane perpendicular to the section will be noted below.) Radiation is shown in Figure 8.7.2b incident on the reflector at B at the rim where the "mirror radius," is a maximum at r_r. The angle ϕ_r is the *rim angle*, and is described by AFB.

For specular parabolic reflectors of perfect shape and alignment, the size of the receiver to intercept all of the solar image can be shown from these figures. For a cylindrical receiver the diameter D of this receiver is

$$D = 2r_r \sin 16' \tag{8.7.1}$$

For a receiver that is planar and normal to the axis of the parabola, as shown in Figure 8.7.2b, the width W is

$$W = \frac{2r_r \sin 16'}{\cos(\phi_r + 16')} \tag{8.7.2}$$

The distance r from a point on the reflector to the focus can be derived for the particular reflector shape. For a parabolic reflector, the focal length, f, is a constant in the equation of the surface

$$y^2 = 4fx \tag{8.7.3}$$

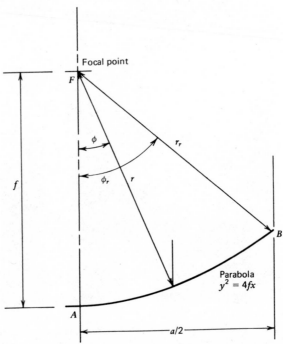

Figure 8.7.2(a) Sections of a linear parabolic concentrator showing major dimensions.

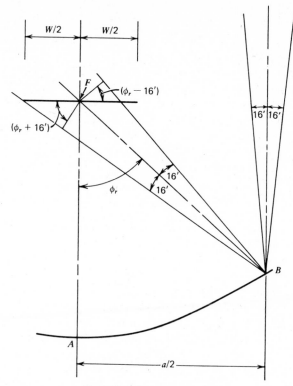

Figure 8.7.2(b) Image dimensions.

and r is given by

$$r = \frac{2f}{1 + \cos \phi} \qquad (8.7.4)$$

ϕ is the angle between the axis and a reflected beam at the focus as shown in Figure 8.7.2a. As ϕ varies from zero to ϕ_r, r increases from f to r_r, and the theoretical image size increases from $W'|_{r=f}$ to $W|_{r=r_r}$ (i.e., from $2f \sin 16'$ to $2r_r \sin 16'/\cos(\phi_r + 16')$. Thus, there is a finite image size and spreading of the image even for geometrically perfect systems. Figure 8.7.3 shows a section of an ideal solar image on a plane normal to the axis of a parabola.

The *aperture* is the unshaded opening or projected area of the optical system. For surfaces of revolution, the aperture is usually characterized by the diameter of the reflector, or for cylindrical systems, by the width, that is, the area per unit length. The focal length is a determining factor in image size as shown by Equations 8.7.2 and 8.7.4, and the aperture, a, is the determining factor in total energy; thus the image brightness or energy flux concentration at the receiver of a focusing system will be a function of the ratio a/f.

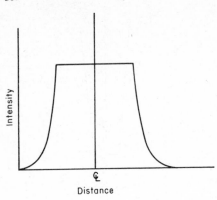

Figure 8.7.3 Cross section of a theoretical solar image on a surface normal to the axis of a parabolic reflector, assuming a uniform solar disk.

Radiation is not received uniformly from the total solar disk. As noted by de la Rue et al. (1957), the center of the sun is brightest, with the limbs (i.e., the edges of the sun) darkest. This nonuniformity tends to distort the theoretical image of Figure 8.7.3.

This discussion refers to theoretical images and applies only to very precise optical systems. Most solar reflectors of potential practical interest are not precise optical instruments, and produce images substantially larger than the theoretical; it is with these larger images that we are concerned.

Images of linear parabolic concentrators are enlarged because of several factors. First, if parabolic sections are not perfect in shape a dispersion of the image will result. This is illustrated in Figure 8.7.4, where the reflected beam has an angular width of $(\delta + 0.53)$ degrees. The increase in image size is proportional to the mirror radius, r_r, and focal length, f. The effect of the *dispersion angle*, δ, can be considered as an addition to the solar intercept angle of 32'. The image produced will be similar in shape to that shown in Figure 8.7.3, but will be larger than the theoretical. Figure 8.7.2b can be modified by replacing the 16' half width of the reflected beam by $(\delta/2 + 16')$, and Equations 8.7.1 and 8.7.2 can be rewritten to include the more dispersed reflected beam. For the cylindrical receiver

$$D = 2r_r \sin(\delta/2 + 16') \qquad (8.7.5)$$

and for the planar receiver normal to the axis of the parabolia

$$W = \frac{2r_r \sin(\delta/2 + 16')}{\cos(\phi_r + \delta/2 + 16')} \qquad (8.7.6)$$

A second reason for enlarged images is due to the orientation of the concentrator. Linear parabolic concentrators may be oriented in several ways to track the beam radiation. The most likely modes of orientation are adjustment about horizontal axes, aligned either east-west or north-south. In these situations, the beam radiation incident on the aperture of the collector will be parallel to the

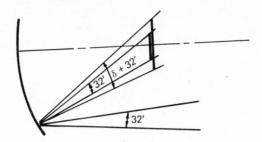

Figure 8.7.4 Schematic of a portion of a concentrator, with a dispersion angle, δ, added to the 32′ (0.53°) solar intercept angle.

central plane of the reflector but at an angle of incidence on the aperture described either by Equation 1.6.10 or by 1.6.11. A theoretical image from any point on the reflector will be enlarged by a factor $1/\cos\theta$. In effect, the focal length, f, of the parabola and its mirror radius, r, are increased by $1/\cos\theta$. Thus an east-west oriented linear parabolic concentrator would have images on the receiver very much enlarged in the early and late hours of a day and an image of minimum size at noon.

There also can be errors in orientation of the collector and errors in positioning of the receiver relative to the concentrator which result in distortion, enlargement, and displacement of the image.

It is possible to construct an image if δ is known. The reflector is divided into strip segments parallel to the axis. Each segment produces a rectangular image of a width W (Figure 8.7.2a) in the focal plane. The images from all of the segments are summed to give a total image similar in appearance to Figure 8.7.3, but with the image spread out. This procedure was used by Löf and Duffie (1963) in a study of the optical and thermal properties of collectors with linear and circular concentrators.

Example 8.7.1

A linear parabolic concentrator has a focal length of 1.00 m and an aperture width of 1.00 m. δ is 0.50°, ρ is 0.90, and the beam radiation perpendicular to the aperture is 800 W/m². The concentrator is perfectly aligned. Estimate the flux distribution on the focal plane (basing the estimate on division of the aperture into five equal segments).

Solution

A cross section of the concentrator is shown in the diagram divided into five linear segments each with a projected width of 200 mm. The image width for the midpoint of each segment is calculated, and the total energy from the segment is assumed to be uniformly distributed across the segment. The parabola has the equation $y^2 = 4fx$, and the slope of the parabola is $dx/dy = 2y/4f$.

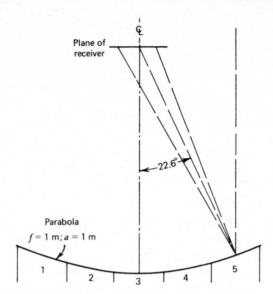

For segment 5, $y = 0.40$ m, $x = 0.040$ m and $dx/dy = 0.20$ at the midpoint of the segment. Thus the angle of incidence of radiation on the segment is $\tan^{-1} 0.20 = 11.3°$. The angle of incidence of the reflected beam on the focal plane is then $22.6°$. From Equation 8.7.4, r for this segment is $2 \times 1/(1 + \cos 22.6) = 1.04$ m. For this geometry, $(\delta/2 + 0.267)$ is $0.52°$. The image width for this segment from Equation 8.7.6 is $2 \times 1.04 \sin 0.52/\cos(22.6 + 0.52) = 0.0204$ m.

The total energy from the segment per unit length of concentrator is $0.20 \times 0.90 \times 800$ or 144 W, or for segments 1 and 5, 288 W. This energy can be considered as contributing a portion of the image uniformly distributed over a width of 0.0204 m.

For segment 2, the same procedure indicates an image width of 0.0186 m for radiation reflected from the concentrator at $y = 0.20$. The energy from segments 2 and 4 is again 288 W and is considered to be uniformly distributed across 0.0186 m.

The contribution of the center segment 3 has a width $W' = 2 \tan 0.52 = 0.018$ m. The energy in this width per unit of concentrator length is $0.20 \times 0.90 \times 800 = 144$ W. (Note that shading by a receiver has not been considered. It could be accounted for if total receiver width is known, by considering only the unshaded segments of the parabola.)

These three contributions are then added to obtain the image. This is shown in two steps on the drawing: the first is the addition of the three contributions and the second is the smoothed curve showing the shape of the image if a large number of segments had been used. (The total width of the image is found from r, and is 0.0217 m). Note that the example has been worked out on the basis of energy in the image per unit length of concentrator. ∎

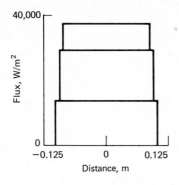

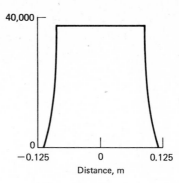

This method of analysis produces typical "hat shaped" distributions, with the shape a function of δ and a/f. An alternative representation that has been used is a Gaussian distribution, which must be based on experimental measurements.

The ratio a/W is the ratio of the area of the aperture to the total area of the receiver when the receiver is just large enough to intercept all of the specularly reflected radiation. From a/W we can get C_m, the maximum concentration ratio that leads to interception of the total image. For a concentrator producing an image with well defined boundaries and without pointing errors or mispositioning of the receiver, a/W can be determined from the rim angle ϕ_r, and the angle $(\delta + 0.53)$. From Rabl (1976a), for a tubular receiver:

$$\frac{a}{W} = \frac{\sin \phi_r}{\pi \sin(\delta/2 + 0.267)} \tag{8.7.7}$$

For a flat receiver in the focal plane, neglecting shading of the reflector by the receiver,

$$\frac{a}{W} = \frac{\sin \phi_r \cos(\phi_r + \delta/2 + 0.267)}{\sin(\delta/2 + 0.267)} \tag{8.7.8}$$

The maximum concentration ratio, if a width W of the reflector is shaded by the receiver, is $(a - W)/W$. Then

$$C_m = \frac{\sin \phi_r \cos(\phi_r + \delta/2 + 0.267)}{\sin(\delta/2 + 0.267)} - 1 \tag{8.7.9}$$

Example 8.7.2

A linear parabolic reflector has a focal length of 1.00 m and an aperture of 2.50 m. The optical quality is such that the dispersion angle, δ, is 1.10°. The reflector is oriented on an east-west axis with continuous adjustment so that the central plane of the concentrator includes the sun. If the latitude is 35°N, what would be the width of the solar image on 20 April on a flat receiver (a) at noon, and (b) at 3 p.m.?

Solution

On April 20, $n = 110$ and the declination is $11.2°$. To determine the rim angle ϕ_r, the x coordinate of the rim is calculated from the equation for the parabola, $y^2 = 4fx$

$$x = \frac{(1.25)^2}{4} = 0.391$$

Then $\tan \phi_r = 1.25/(1 - 0.39) = 2.049$, and $\phi_r = 64°$. At noon for this mode of orientation $\cos \theta = 1$ and $\theta = 0°$. Equation 8.7.8 can be used to calculate the image size at noon. The ratio is

$$\frac{a}{W} = \frac{\sin 64 \cos(64 + 0.55 + 0.267)}{\sin(0.55 + 0.267)} = 26.8$$

Then the image width for all of the specularly reflected radiation $= 2.50/26.8 = 0.093$ m. At 3 p.m., $\omega = 45°$ and for this mode of orientation, from Equation 1.6.10

$$\cos \theta = (1 - \cos^2 11.2 \sin^2 45)^{1/2} = 0.720; \theta = 43.9$$

Then at 3 p.m. the image width is $0.093/\cos 43.9 = 0.129$ m. ∎

In practice, images often do not have well-defined boundaries, and it will probably be best to use a receiver that will intercept less than all of the specularly reflected radiation. A trade-off between increasing thermal losses with increasing area and increasing optical losses with decreasing area is necessary to optimize long-term collector performance. This optimization problem has been studied by Löf et al. (1962) and Löf and Duffie (1963), with a result that for a wide range of conditions the optimum size receiver will intercept 90 to 95 % of the possible radiation. Thus an optical loss, often in a range of 5 to 10 percent, will be incurred in this type of collector. This has been expressed in terms of an *intercept factor*, γ, the fraction of the specularly reflected radiation which is intercepted by the receiver. For an image such as the arbitrary one shown in Figure 8.7.5, the total

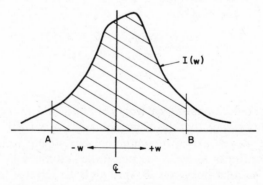

Figure 8.7.5   An arbitrary flux distribution for a cylindrical reflector; the receiver size is AB.

area under the distribution curve is the total energy reflected to the focal plane. If a receiver occupies the width A to B, it will intercept energy represented by the shaded area. Thus the definition of γ, the fraction intercepted, can be written as

$$\gamma = \frac{\int_A^B I(w)dw}{\int_{-\infty}^{\infty} I(w)dw} \tag{8.7.10}$$

where w is the distance from the center of the focal area. Similar considerations hold for concentrators that are surfaces of revolution. Tracking (pointing) errors and dispersion errors δ lead to enlarged or shifted images and consequently affect γ.

Optical losses also result from lack of perfect reflectance of the reflector, from imperfect transmittance of any cover system on the receiver, and from imperfect absorptance of the receiver. The absorbed radiation per unit area of unshaded aperture* can be written

$$S = G_{b,a}\rho\gamma\tau\alpha \tag{8.7.11}$$

where $G_{b,a}$ is the beam irradiance on the plane of the aperture, ρ is the specular reflectance of the reflector (see Table 4.8.1), τ is the transmittance of any cover system, α is the absorptance for the incident solar radiation on the receiver, and γ is the intercept factor. Both τ and α may have to be suitably integrated to account for angle of incidence of the reflected radiation on the cover and absorbing surface. A transmittance-absorptance product can be used, as noted in Section 5.5.

In order to estimate the useful output of the collector, it is necessary to estimate F_R and U_L. Methods for calculating F_R are basically the same as for flat-plate collectors, except that fin and bond conductance terms generally will not appear in F'; these are outlined in Section 8.3.

U_L may be difficult to determine. [For examples of calculation methods, for uncovered cylinders and cylinders with cylindrical covers, see Section 8.3 and Tabor (1955 and 1958).] An additional complication arises in that conduction losses through supporting structures are highly variable with design and are dependent on temperature. Some experimental data are available on loss coefficients for several receiver designs. Table 8.7.1 indicates a brief description of three receivers, and Figure 8.7.6 shows measured loss coefficients as a function of the difference between average receiver temperature and ambient temperature. These loss coefficients are based on receiver area. The table also shows area concentration ratios and concentrator dimensions.

Example 8.7.3

The linear parabolic concentrator of Example 8.7.2 is to be fitted with a liquid-heating receiver very similar to Type C in Table 8.7.1. The unit is 10.5 m long,

---

* A similar expression can be written based on the total aperture area by multiplying by a shading factor. An optical efficiency for the collector can be defined as the product $\rho\gamma\tau\alpha$ and a shading factor.

Table 8.7.1   Receiver Design and Collector Characteristics[a]

| | | Concentrator | | |
Type	Description	C	Width m	f-length m
A. Hexcel	Steel pipe, black chrome coated. Back side insulated with bulk insulation in steel jacket. On absorbing side cover is glass semicylinder. Space not evacuated.	67	2.6	0.914
B. Solar kinetics	Steel tube, black chrome plated, inside a borosilicate glass tube, not evacuated.	41	1.3	0.267
C. Suntec systems	Two parallel steel pipes, black chrome plated, with down-and back flow path. Back insulated, front covered with flat glass.	35	3.5	3.05

[a] From Leonard (1978).

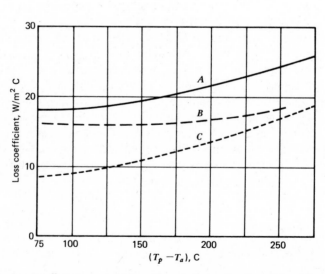

Figure 8.7.6 Loss coefficients per unit area of receiver as a function of the difference between average receiver temperature and ambient temperature for three receivers described in Table 8.6.1. Data from Leonard (1978).

and its aperture is 2.5 m. A strip of the reflector 0.21 m wide is shaded by the receiver. The receiver is designed to be just large enough to intercept all of the specularly reflected beam radiation when it is normal to the aperture, and under those conditions the distribution of radiation in the focal plane can be approximated as shown by the solid line on the figure below.

The normal beam radiation G_{bn} at noon on 20 April is 975 W/m² and at 3 p.m. it is 810 W/m². $(\tau\alpha)$ for the receiver is estimated at 0.78 with radiation normal to the aperture. ρ is 0.86. The inlet fluid temperature is 170 C, and ambient temperature is 25 C. F_R is estimated to be 0.85.

1 What will be the output of the collector at noon, and what will be its optical efficiency?
2 Estimate the output and optical efficiency at 3 p.m.

Solution

The basic equation to be used for the estimation of useful gain is Equation 8.3.9. The first step is to estimate an optical efficiency, then S then Q_u. At noon, the radiation will be normal to the aperture and, according to the design specifications, $\gamma = 1$. A fraction of the reflector of 0.21/2.50, or 0.084, is shaded by the receiver, so $(1 - 0.084)$ or 0.916 of the reflector is effective. An optical efficiency is then the product of $\rho\gamma\tau\alpha$ and this fraction:

$$= 0.86 \times 1.00 \times 0.78 \times 0.916 = 0.61$$

Based on the area of the unshaded aperture

$$S = 975 \times 0.61 = 600 \text{ W/m}^2$$

At an estimated mean receiver surface temperature of 200 C, U_L from Figure 8.7.6 is 13.7 W/m² C. Then from Equation 8.3.9, at a concentration ratio of $(2.5 - 0.21)/0.21 = 10.9$,

$$Q_u = 10.5(2.5 - 0.21) \times 0.85\left[600 - \frac{13.7}{10.9}(170 - 25)\right] = 8.5 \text{ kW}$$

At 3 p.m. the angle of incidence of beam radiation on the aperture is 43.9° (from Example 8.7.2). This leads to significant image spread and changes in γ, τ, and α. The dashed line in the figure below shows how the image will spread, with its

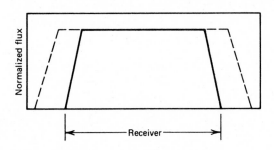

width (1/cos 43.9) greater than at noon. An approximate integration between the limits of the receiver dimensions indicates $\gamma = 0.80$. From Figures 4.7.1 and 5.3.1, α and τ will be reduced by approximately 3 % each, so an approximation to $(\tau\alpha)$ is $0.78(0.97)^2 = 0.73$. Then the optical efficiency is

$$= 0.86 \times 0.80 \times 0.73 \times 0.916 = 0.46$$

From Example 8.7.2, $\theta = 43.9°$ at 3 p.m. Based on the unshaded aperture:

$$S = 810 \times 0.46 \cos 43.9 = 268 \text{ W/m}^2$$

The useful gain is

$$Q_u = 10.5(2.5 - 0.21) \times 0.85 \left[268 - \frac{13.7}{10.9} (170 - 25) \right] = 1.7 \text{ kW} \qquad \blacksquare$$

In this example some short cuts have been shown. The mean angles of incidence of the radiation on the receiver are not known when γ is significantly less than unity, and only simple corrections accounting for incidence angles in the central plane of the receiver have been made. The integration to find γ is approximate. More detailed knowledge of the optical characteristics of reflector orienting system and receiver would have to be combined with ray-tracing techniques to improve on these calculations.

A detailed experimental study of energy balances on a linear parabolic collector has been reported by Löf et al. (1962). While it is a special case, and the concentrator f/a was smaller than optimum, it provides a useful illustration of the importance of several factors affecting performance of collectors of this type. The collector consisted of a parabolic cylinder reflector of aperture 1.89 m, length 3.66 m, and focal length 0.305 m, with bare tubular receivers of several

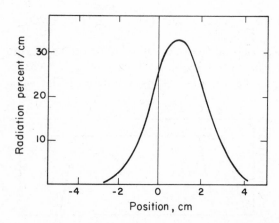

Figure 8.7.7 Experimental mean flux distribution for a parabolic cylinder reflector. From Löf et al. (1962).

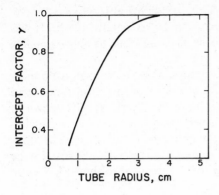

Figure 8.7.8 Intercept factors for tubes centered at position 0 of the reflector of Figure 8.7.7. From Löf et al. (1962).

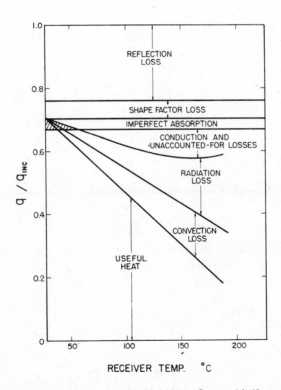

Figure 8.7.9 Distribution of incident energy for the 1.89 m reflector with 60 mm diameter receiver as a function of receiver temperature. From Löf et al. (1962).

sizes coated with a nonselective black paint having an absorptance of 0.95. The system was mounted so as to rotate on a polar axis at 15° per hr. It was operated over a range of temperatures from near ambient to approximately 180 C.

The intercept factors for various receiver sizes were determined from measurements of the flux distribution at many locations on the focal tube: these distributions were averaged and the resulting mean distribution is shown in Figure 8.7.7. This particular distribution is very similar to a Gaussian distribution curve, but is displaced from the position of the theoretical focus. The intercept factors that resulted from this distribution, with the receiver tubes fixed at the theoretical focus by the mechanical design of the system, are shown in Figure 8.7.8.

The results of many energy balance measurements are summarized in Figures 8.7.9 and 8.7.10, which show the distribution of incident beam solar energy (during operation at steady state in clear weather) into useful gain and various losses for two receiver tube sizes, the first for a 60 mm diameter tube and the second for a 27 mm diameter tube. The relative magnitudes of the losses are evident.

From these figures, it is possible to estimate the effects of design changes. For example, for this collector the use of selective surface of emittance 0.2 would

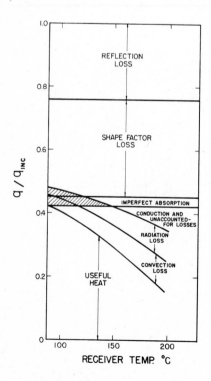

Figure 8.7.10 Distribution of incident energy for the 1.89 m reflector with 27 mm diameter receiver. From Löf et al. (1962).

reduce the radiation loss by 79 percent of the value shown at any temperature. However, radiation loss is not the dominant loss [a generalization made by Edwards and Nelson (1961)]. The most obvious initial improvements for this exchanger would be in reduction of optical losses by using surfaces of higher reflectance, and by intercept-factor improvements (the latter particularly for the smaller receiver).

In this study, the reflector and receiver tubes were supported by plates at each end; these result in heat loss by conduction from the tubes. These losses were estimated from temperature measurements along the supporting plates. Although not shown in the figures, they were estimated at 3, 6, and 10 percent of the incident clear sky radiation for receiver-surface temperatures of 100, 135, and 175 C, respectively, for the conditions of these experiments.

8.8 PARABOLOIDAL CONCENTRATORS

The previous sections outlined methods of calculation of absorbed radiation for collectors with linear parabolic concentrators. A similar analysis can be done for collectors with three-dimensional parabolic reflectors, that is, reflectors that are surfaces of revolution. Rabl (1976a) summarizes important optical aspects of these collectors. In a section through the axis of the paraboloid the collectors are represented by Figure 8.7.2, and the rim angle ϕ_r and "mirror radius" r are analogous to those for the linear concentrator. Dispersion also occurs in paraboloidal concentrators, and equations analogous to Equation 8.7.9 can be written for collectors without tracking errors. For spherical receivers (allowing for minimum shading by the receiver)

$$C_m = \frac{\sin^2 \phi_r}{4 \sin^2(\delta/2 + 0.267)} - 1 \qquad (8.8.1)$$

For flat receivers

$$C_m = \frac{\sin^2 \phi_r \cos^2(\phi_r + \delta/2 + 0.267)}{\sin^2(\delta/2 + 0.267)} - 1 \qquad (8.8.2)$$

This concentration ratio, C_m is again the maximum that can be obtained, based on interception of all of the specularly reflected radiation which is within the cone of angular width $(\delta + 0.53)$.

Cavity receivers may be used with paraboloidal concentrators, to increase absorptance and (possibly) to reduce convective losses from the absorbing surface. The equations for optical properties of systems with flat receivers will also apply to cavity receivers.

Absorbed energy for a paraboloidal collector depends on properties in the same way as linear parabolic collectors. However, at the higher concentration ratios achieved by these collectors, any absorption of solar radiation in the cover material would lead to heating of the cover and as a result covers would probably not be used. The absorbed energy for the unshaded aperture area of the collector is then

$$S = G_a A_a \rho \gamma \alpha \qquad (8.8.3)$$

where the factors have the same meaning as in Equation 8.6.8 for a linear concentrator.

Calculation of thermal losses is very much a function of receiver geometry, and may be complicated by the existence of temperature gradients on the surface of a receiver. For examples of these estimations, see reports of Martin Marietta, Boeing, and McDonnell-Douglas on the receiver subsystems of central receiver solar thermal power systems.

8.9 CENTRAL RECEIVER COLLECTORS

The "power tower" or "central receiver" concept for generation of electrical energy from solar energy is based on the use of very large concentrating collectors. The optical system consists of a field of a large number of heliostats, each reflecting beam radiation onto a central receiver. The result is a Fresnel-type concentrator, a parabolic reflector broken up into small segments, as shown in Figure 8.9.1. Several additional optical phenomena must be taken into account. Shading and blocking can occur (shading of incident beam radiation from a heliostat by another heliostat and blocking of reflected radiation from a heliostat by another which prevents that radiation from reaching the receiver). As a result of these considerations, the heliostats are spaced apart and only a fraction of the ground area, ψ, is covered by mirrors. A ψ of about 0.3 to 0.5 has been suggested as a practical value.

The maximum concentration ratio for a three-dimensional concentrator system with radiation incident at an angle θ_i on the plane of the heliostat array, ($\theta_i = \theta_z$ for a horizontal array), for a rim angle of ϕ_r and a dispersion angle of ($\delta + 0.53$), if all reflected beam radiation is to be intercepted by a spherical receiver, is

$$C_m = \frac{\psi \sin^2 \theta_r}{4 \sin^2(\delta/2 + 0.267)} - 1 \qquad (8.9.1)$$

For a flat receiver, the concentration ratio is

$$C_m = \psi \left[\frac{\cos(\phi_r + \delta/2 + 0.267)\sin \phi_r}{\sin(\delta/2 + 0.267)} \right]^2 - 1 \qquad (8.9.2)$$

As with linear concentrators, the optimum performance may be obtained with intercept factors less than unity. The designer of these collectors has additional considerations to take into account. The heliostat fields need not be

Figure 8.9.1 An experimental central receiver collector system. The receiver panel is one of an array which would constitute a complete receiver of a full scale system. Photo courtesy Sandia Laboratories.

symmetrical, the ground cover, ψ, does not have to be uniform, and the heliostat array is not necessarily all in one plane. These collectors would probably operate with high concentration ratios and at relatively high receiver temperatures.

There has been considerable work done on the design and performance of collectors of this type in the 1970s, and a large body of literature has become available. For example, see the University of Houston-Sandia Proceedings of ERDA Workshops on central receiver systems (1977). Also, many papers have appeared in journals, and as a starting point the reader could go to Hildebrandt and Vant-Hull (1977) or Vant-Hull (1977).

REFERENCES

Baranov, V. K. and G. K. Melnikov, *Soviet Journal of Optical Technology*, **33**, 408 (1966).

Brandemuehl, M.J. and W.A. Beckman, *Solar Energy,* **24**, 511 (1980) "Transmission of Diffuse Radiation Through CPC and Flat-Plate Collector Glazings."

DelaRue, R., E. Lob, J. L. Brenner, and N. K. Hiester, *Solar Energy*, **1**, No. 2–3, 94 (1957). "Flux Distribution Neat the Focal Plane."

Duffie, J. A. and G. O. G. Löf, Paper 207 III.7/5, World Power Conference, Melbourne (1962). "Focusing Solar Collectors for Power Generation."

Edwards, D. E. and K. E. Nelson, paper 61-WA-158, New York ASME Meeting (1961). "Radiation Characteristics in the Optimization of Solar-Heat Power Conversion Systems."

Hildebrandt, A. F. and L. L. Vant-Hull, *Science*, **198**, 1139 (1977). "Power with Heliostats."

Hinterberger, H. and R. Winston, *Review of Scientific Instruments*, **37**, 1094 (1966).

Hollands, K. G. T., *Solar Energy*, **13**, 149 (1971). "A Concentrator for Thin-Film Solar Cells."

Kreith, F. and J. F. Kreider, *Principles of Solar Engineering*, McGraw-Hill, New York (1978).

Laszlo, T. S., *Image Furnace Techniques*, Interscience, New York (1965). *Proceedings of the 1957 Solar Furnace Symposium, Solar Energy*, **1** (2), 3 (1957).

Leonard, J. A., paper at Solar Thermal Concentrating Collector Technology Symposium, June (1978). Denver "Linear Concentrating Solar Collectors—Current Technology and Applications."

Löf, G. O. G., D. A. Fester, and J. A. Duffie, *Journal of Engineering for Power*, **84A**, 24 (1962). "Energy Balance on a Parabolic Cylinder Solar Reflector."

Löf, G. O. G., and J. A. Duffie, *Journal of Engineering for Power*, **85A**, 221 (1963). "Optimization of Focusing Solar-Collector Design."

Mitchell, J. C., Personal Communication (1979).

Rabl, A., *Solar Energy*, **18**, 93 (1976a). "Comparison of Solar Concentrators."

Rabl, A., *Solar Energy*, **18**, 497 (1976b). "Optical and Thermal Properties of Compound Parabolic Concentrators."

Rabl, A., Paper at Solar Thermal Concentrating Collector Technology Symposium, June (1978). "Optical and Thermal Analysis of Concentrators."

Reports of Boeing Engineering and Construction Co. (Contract No. EY-76-C-03-1111), Martin Marietta Corp. (Contract EY-77-C-03-1110), and McDonnell Douglas Astronautics Co. (Contract EY-76-C-03-1108), to the U.S. Department of Energy (1977) on "Central Receiver Solar Thermal Power System."

Selcuk, M. K., *Solar Energy*, **22**, 413 (1979), "Analysis, Development and Testing of a Fixed Tilt Solar Collector Employing Reversible Vee-trough Reflectors and Vacuum Tube Receivers."

Tabor, H., *Bulletin of the Research Council of Israel*, **5C**, 5 (1955). "Solar Energy Collector Design."

Tabor, H., *Solar Energy*, **2**, (1), 3 (1958). "Solar Energy Research: Program in the New Desert Research Institute in Beersheba."

University of Houston-Sandia Laboratories, *Proceedings of the ERDA Solar Workshop on Methods for Optical Analysis of Central Receiver Systems* (1977). Available from National Technical Information Services, U. S. Dept. of Commerce.

Vant-Hull, L. L., *Optical Engineering*, **16**, 497 (1977). "An Educated Ray-Trace Approach to Solar Tower Optics."

Welford, W. T. and R. Winston, *The Optics of Nonimaging Concentrators*. Academic Press, New York (1978).

Winston, R., *Solar Energy*, **16**, 89 (1974). "Solar Concentrations of Novel Design."

Winston, R. and H. Hinterberger, *Solar Energy*, **17**, 255 (1975). "Principles of Cylindrical Concentrators for Solar Energy."

CHAPTER 9

Energy Storage

Solar energy is a time-dependent energy resource, and energy needs for a very wide variety of applications are also time-dependent, but in a different fashion than the solar energy supply. Consequently, the storage of energy or other product of the solar process is necessary, if solar energy is to meet substantial portions of these energy needs.

Energy (or product) storage must be considered in the light of a solar process system, the major components of which are the solar collector, storage units, conversion devices (such as air conditioners or engines), loads, auxiliary or supplemental energy supplies, and the control systems. The performance of each of these components is related to that of the others. The dependence of the collector performance on temperature makes the whole system performance sensitive to temperature. For example, in a solar-thermal power system, a thermal energy storage system which is characterized by high drop in temperature between input and output will lead to unnecessarily high collector temperature and/or low heat engine inlet temperature, both of which lead to poor system performance.

In passive solar heating, collector and storage components are integrated into the building structure. The performance of storage walls in passive heating systems is so interdependent with the absorption of energy that we reserve discussion of this aspect of solar energy storage for Chapter 15.

The optimum capacity of an energy storage system depends on the expected time-dependence of solar radiation availability, the nature of loads to be expected on the process, the degree of reliability needed for the process, the manner in which auxiliary energy is supplied, and an economic analysis that determines how much of the annual load should be carried by solar and how much by the auxiliary energy source.

In this chapter we set forth the principles of several energy storage methods and show how their capacities and rates of energy input and output can be calculated. In the example problems, as in the collector examples, we arbitrarily assume temperatures or energy quantities. In reality these must be found by simultaneous solutions of the equations representing all of the system components. These matters are taken up in Chapter 10; we will see there that the differential equations for storage are the key equations for most systems, with time the independent variable and storage (and other) temperatures the dependent variables.

9.1 PROCESS LOADS AND SOLAR COLLECTOR OUTPUTS

Consider a solar process in which the time-dependence of the load L and gain from the collector Q_u, are as shown in Figure 9.1.1a. During part of the time, available energy exceeds the load on the process, and at other times it is less. A storage system can be added to store the excess collector output and return it when needed. Figure 9.1.1b shows the energy stored as a function of time. Energy storage is clearly important in determining system output. If there were no storage, the useful solar gain would be reduced on the first and third days by the amount of energy added to storage on those days. This would represent a major drop in solar contribution.

It is usually not practical to meet all of the loads on a process from solar energy over long periods of time, and an auxiliary energy source must be used. Where auxiliary energy is used, information such as that illustrated by Figure 9.1.1a shows the time dependence of auxiliary input.

It is also useful to show the integrated values of the major parameters Q_u, L, and A. Examples of these are shown in Figure 9.1.1c. A major objective of system

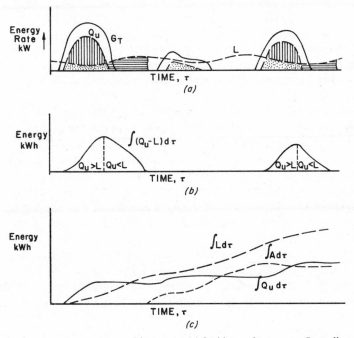

Figure 9.1.1 A solar energy process with storage. (a) Incident solar energy, G_T, collector useful gain, Q_u, and loads, L, as functions of time for a 3-day period. Vertical shaded areas show times of excess energy to be added to storage. Horizontal shaded areas show energy withdrawn from storage to meet loads. Dotted areas show energy supplied to load from collector during collector operation. (b) Energy added to or removed from storage, taking time $\tau = 0$ as a base. (c) Integrated values of useful gain from the collector, $\int Q_u \, d\tau$; load, $\int L \, d\tau$; and auxiliary energy, $\int A \, d\tau$, for the same 3-day period. In this example solar energy collected is slightly more than half the integrated load.

performance analysis is a determination of long-term values of Q_u and A; these are needed to assess the cost of delivering energy or product from the solar energy process, and to estimate the fraction of total energy or product needs met from solar and auxiliary energy sources. In practice, these integrations must be done over long periods (typically a year), and both collector area and storage capacity are variables to be considered.

9.2   ENERGY STORAGE IN SOLAR PROCESS SYSTEMS

Energy storage may be in the form of sensible heat of a solid or liquid medium, as heat of fusion in chemical systems, or as chemical energy of products in a reversible chemical reaction. Mechanical energy can be converted to potential energy and stored in elevated fluids. Products of solar processes other than energy may be stored; for example, distilled water from a solar still may be stored in tanks until needed.

The choice of storage media depends on the nature of the process. For water heating, energy storage as sensible heat of stored water is logical. If air heating collectors are used, storage in sensible or latent heat effects in particulate storage units is indicated, such as sensible heat in a pebble bed heat exchanger. In passive heating, storage is provided as sensible heat in building elements. If photovoltaic or photochemical processes are used, storage is logically in the form of chemical energy.

The major characteristics of a thermal energy storage system are (a) its capacity, per unit volume, or weight; (b) the temperature range over which it operates, that is, the temperature at which heat is added to and removed from the system; (c) the means of addition or removal of heat and the temperature differences associated therewith; (d) temperature stratification in the storage unit; (e) the power requirements for addition or removal of heat; (f) the containers, tanks, or other structural elements associated with the storage system; (g) the means of controlling-thermal losses from the storage system; and (h) its cost.

Of particular significance in any storage system are those factors affecting the operation of the solar collector. The useful gain from a collector decreases as its average plate temperature increases. A relationship between the average collector temperature and the temperature at which heat is delivered can be written as

$$
\begin{aligned}
T(\text{collector}) - T(\text{delivery}) = {} & \Delta T(\text{transport from collector to storage}) \\
& + \Delta T(\text{into storage}) \\
& + \Delta T(\text{storage loss}) + \Delta T(\text{out of storage}) \\
& + \Delta T(\text{transport from storage to application}) \\
& + \Delta T(\text{into application})
\end{aligned}
$$

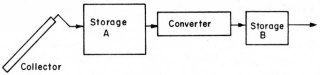

Figure 9.2.1   Schematic of alternative storage location, at *A* or *B*.

Thus, the temperature of the collector, which determines its useful gain, is higher than the temperature at which the heat is finally used by the sum of a series of temperature-difference driving forces. An objective of system design, and particularly of storage unit design, is to minimize or eliminate these temperature drops within economic constraints.

A solar process designer may have alternatives in locating the energy storage component. As an example, consider a process in which a heat engine converts solar energy into electrical energy. In such a system storage can be provided as thermal storage between the solar collector and the engine, as mechanical storage between the engine and the generator, or as chemical storage in a battery between the generator and the end application. Solar cooling with an absorption air conditioner provides another example. Thermal energy can be stored from the collector to be used by the air conditioner when needed or alternatively, the "cooling" produced by the air conditioner can be stored in a low-temperature (below ambient) thermal storage unit. These are illustrated in Figure 9.2.1.

These two alternatives are not equivalent in capacity, costs, or effects on overall system design and performance. The storage capacity required of a storage unit in position *B* is less than that required in position *A* by (approximately) the efficiency of the intervening converter. Thus the capacity of *B* must be only approximately 25 per cent of the capacity of *A* if the conversion process is operating at 25 percent efficiency. Thermal energy storage at *A* has the advantage that the converter can be designed to operate at more nearly constant rate, leading to better conversion efficiency and higher use factor on the converter; it can lower converter capacity requirements by removing the need for operation at peak capacities corresponding to direct solar input. The choice between energy storage at *A* or at *B* may have very different effects on the operating temperature of the solar collector, collector size, and ultimately on cost. These arguments may be substantially modified by requirements for use of auxiliary energy.

9.3 WATER STORAGE

For many solar systems water is the ideal material in which to store usable heat. Energy is added to and removed from this type of storage unit by transport of the storage medium itself, thus eliminating the temperature drop between transport fluid and storage medium. The typical systems in which water tanks

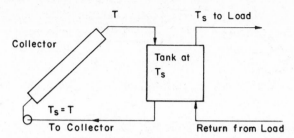

Figure 9.3.1 A typical system using water tank storage, with water circulation through collector to add energy and through the load to remove energy.

are used can be represented by the water heating system shown in Figure 9.3.1. A forced circulation system is shown, but it could be natural circulation. Implicit in the following discussion is the idea that flow rates into and out of the tanks, to collector and load, can be determined.

The energy storage capacity of a water (or other liquid) storage unit at uniform temperature operating over a finite temperature difference is given by

$$Q_s = (mC_p)_s \Delta T_s \tag{9.3.1}$$

where Q_s is the total heat capacity for a cycle operating through the temperature range ΔT_s, with m kilograms of water in the unit. The temperature range over which such a unit can operate is limited at the lower extreme for most applications by the requirements of the process, and at the upper limit by the process, the vapor pressure of the liquid, or the collector heat loss.

For a nonstratified tank, as shown in Figure 9.3.2, an energy balance on the tank yields

$$(mC_p)_s \frac{dT_s}{d\tau} = Q_u - L - (UA)_s(T_s - T_a') \tag{9.3.2}$$

where Q_u and L are rates of addition or removal of energy from the collector and to the load.

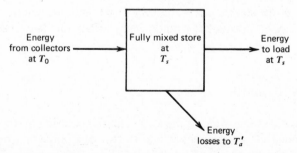

Figure 9.3.2 Unstratified storage of mass m, operating at time-dependent temperature T_s in ambient temperature T_a'.

Example 9.3.1 illustrates how the energy balances on a storage tank can be used to predict its temperature as a function of time. In this example, loads and input from the collector are given as functions of time, and the dependence of Q_u on storage-tank temperature is not shown. Examples in later chapters will illustrate this dependence.

Example 9.3.1

A fully mixed water tank storage containing 1500 kg of water has a loss co-efficient-area product of 11.1 W/°C. The tank starts a particular 24-hr period at 45 C and is in a room at a constant temperature of 20 C. Energy Q_u is added to the tank from a solar collector, and energy is extracted from the tank to meet a load L. The first column of the table indicates the time at the end of an hourly period; the second and third columns indicate values of Q_u and L for those hours. Calculate, the temperature of the tank through the 24-hour period using Euler integration.

Solution

The energy balance on the tank is represented by Equation 9.3.2, which can be rewritten in finite difference form and solved for the tank temperature at the new time, T_s^+.

$$T_s^+ = T_s + \frac{\Delta\tau}{(mC)_s} [Q_u - L - (UA)_s(T_s - T_a')]$$

Inserting the appropriate constants, with a time increment of 1 hr,

$$T_s^+ = T_s + \frac{1}{1500 \times 4190} [Q_u - L - 11.1 \times 3600(T_s - 20)]$$

With this approximation, the temperature of the tank at the end of an hour is calculated from its temperature at the beginning of that hour, from the known inputs and outputs, and assuming that the loss term can be assumed constant throughout that hour. This has been done in the following table, where column 4 is the old temperature and column 5 is the new temperature for the end of that hour calculated from the equation.

Hour	Q_u, MJ	L, MJ	T_s, C	T_s^+, C
1	0	12	45	42.9
2	0	12	42.9	40.9
3	0	11	40.9	39.0
4	0	11	39.0	37.1
5	0	13	37.1	35.0
6	0	14	35.0	32.6

(continued)

Hour	Q_u, MJ	L, MJ	T_s, C	T_s^+, C
7	0	18	32.6	29.7
8	0	21	29.7	26.3
9	21	20	26.3	26.4
10	41	20	26.4	29.7
11	60	18	29.7	36.3
12	75	16	36.3	45.6
13	77	14	45.6	55.5
14	68	14	55.5	63.8
15	48	13	63.8	69.1
16	25	18	69.1	69.9
17	2	22	69.9	66.4
18	0	24	66.4	62.3
19	0	18	62.3	59.2
20	0	20	59.2	55.7
21	0	15	55.7	53.1
22	0	11	53.1	51.2
23	0	10	51.2	49.4
24	0	9	49.4	47.8

■

Water tanks may operate with significant degrees of stratification, that is, with the top of the tank hotter than the bottom. In this case, a tank can be modeled as being divided into N nodes (sections), with energy balances written for each section of the tank. The result is a set of N differential equations that can be solved for the temperatures of the N nodes as functions of time.

To formulate these equations, it is necessary to make assumptions about how the water entering the tank is distributed to the various nodes. For example, for the five node tank shown in Figure 9.3.3, water from the collector enters at a temperature T_0 which lies between $T_{s,2}$ and $T_{s,3}$. It can be assumed that it all finds its way down inside the tank to node 3, where its density nearly matches that of the water in the tank. Alternatively, it can be assumed that the incoming water distributes itself in some way to nodes 1, 2, and 3. Unfortunately it is not now possible to state with certainty what model is best, as the actual flow will depend on the design of a particular tank, the size, location, and design of the inlets and outlets, and flow rates of entering and leaving streams. In the following discussion, a model is developed that represents a high degree of stratification; it is assumed that the water in Figure 9.3.3 finds its way into node 3. With this highly stratified model and with the fully mixed (the one node) model, it is possible to bracket the range of possible degrees of stratification.

Stratification is a quantity that is difficult to evaluate without considering the end use. If the load can use energy at the same efficiency without regard to its temperature level (that is, thermodynamic availability), then maximum stratification would provide the lowest possible temperature near the bottom of the tank and this would maximize collector output. On the other hand, if the

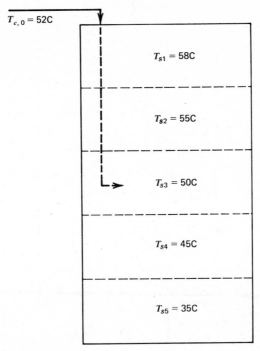

$T_{c,0} = 52C$

$T_{s1} = 58C$

$T_{s2} = 55C$

$T_{s3} = 50C$

$T_{s4} = 45C$

$T_{s5} = 35C$

Figure 9.3.3 A hypothetical five-node tank with $T_{s,2} > T_{c,0} > T_{s,3}$. Water can be considered to enter at node 3, or to be distributed among nodes 1, 2, and 3.

quality of the energy to the load is important, then minimizing the destruction of available energy may be the proper criteria for defining maximum stratification (although all parts of the system should be considered simultaneously in such an analysis). The following analysis is intended to provide a limiting case in which the bottom of the tank is maintained at a minimum temperature, but other criteria could be used.

For the three node tank, as shown in Figure 9.3.4 the flow to the collector always leaves from the bottom, node 3, and the flow to the load always leaves from the top, node 1. The flow returning from the collector will return to the node that is closest to, but less than, the collector outlet temperature. Suppose the three node temperatures are 75, 50, and 25 C, with, of course, the hottest at the top. Return water from the collector lower than 50 C will go to node 3 and between 50 and 74 C would go to node 2.

A collector control function, F_i^c, can be defined to determine which node receives collector return water:

$$F_i^c = \begin{cases} 1 & \text{if } i = 1 \text{ and } T_{c,0} > T_{s,i} \\ 1 & \text{if } T_{s,i-1} \geq T_{c,0} > T_{s,i} \\ 0 & \text{if } i = 0 \text{ or } i = N + i \\ 0 & \text{otherwise} \end{cases} \qquad (9.3.3)$$

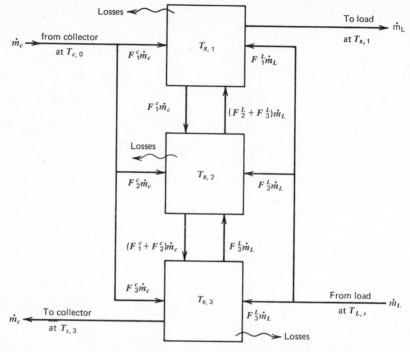

Figure 9.3.4   Three node stratified liquid storage tank.

Note that if the collector is operating, then one and only one control function can be nonzero. Also a fictitious temperature, $T_{s,0}$, of the nonexistent node zero, is assumed to be a large number. The three branches of the collector return water are controlled by this collector return control function as shown in Figure 9.3.4. For the three node tank of the previous paragraph with return water between 75 and 50, $F_1^c = 0$, $F_2^c = 1$, $F_3^c = 0$.

The liquid returning from the load can be controlled in a similar manner with a load return control function, F_i^L.

$$F_i^L = \begin{cases} 1 & \text{if } T_{s,\,i-1} \geq T_{L,r} > T_{s,\,i} \\ 1 & \text{if } i = N \text{ and } T_{L,\,r} < T_{s,\,N} \\ 0 & \text{if } i = 0 \text{ or } i = N + 1 \\ 0 & \text{otherwise} \end{cases} \tag{9.3.4}$$

The net flow between nodes can be either up or down depending upon the magnitudes of the collector and load flow rates and the values of the two control functions at any particular instant. It is convenient to define a mixed flow rate that represents the net flow into node i from node $i - 1$, excluding the effects of flow, if any, directly into the node from the load

$$\dot{m}_{m, 1} = 0$$

$$\dot{m}_{m, i} = \dot{m}_c \sum_{j=1}^{i-1} F_j^c - \dot{m}_L \sum_{j=i+1}^{N} F_j^L \qquad (9.3.5)$$

$$\dot{m}_{m, N+1} = 0$$

With the control functions defined above, an energy balance on node i can be expressed as

$$m_i \frac{dT_{s, i}}{d\tau} = \left(\frac{UA}{C_p}\right)_i (T_a' - T_{s, i}) + F_i^c \dot{m}_c (T_0 - T_{s, i}) + F_i^L \dot{m}_L (T_{L, r} - T_{s, i})$$

$$+ \begin{cases} \dot{m}_{m, i}(T_{s, i-1} - T_{s, i}) & \text{if } \dot{m}_{m, i} > 0 \\ \dot{m}_{m, i+1}(T_{s, i} - T_{s, i+1}) & \text{if } \dot{m}_{m, i+1} < 0 \end{cases} \qquad (9.3.6)$$

Where a term has been added to account for losses from node i to an environment at T_a'.

With a large number of nodes the tank model given by Equation 9.3.6 represents a high level of stratification that may not be achievable in actual experiments. There is very little experimental evidence to support the use of this model to represent high stratification, but the model is based on first principles and can be used for limiting case calculations.

As a practical matter, many tanks show some degree of stratification, and it is suggested that three nodes may represent a reasonable compromise between conservative design (represented by systems with one-node tanks) and the limiting situation of carefully maintained high degrees of stratification.

Two other factors may be significant. First, stratified tanks will have some tendency to destratify over time due to diffusion and wall conduction. This has been experimentally studied by Lavan and Thompson (1977). Second, some tanks have sources of energy in addition to that in fluids pumped into or out of the tank. If, for example, an auxiliary heater coil were present in one of the nodes, an additional term could be added to Equation 9.3.6 to account for its effect.

Numerical integration of Equations 9.3.2 or 9.3.6 can be accomplished by a number of techniques that are discussed in texts on numerical methods. The explicit Euler, the implicit Crank-Nicolson, predictor-corrector, and Runge-Kutta methods are the most common. Because of the complicated nature of the tank equations when coupled to a load and a collector, particularly the stratified tank model, it is probable that a digital computer will be used to obtain solutions.

9.4 PACKED BED STORAGE

A packed bed (pebble bed or rock pile) storage unit uses the heat capacity of a bed of loosely packed particulate material to store energy. A fluid, usually air, is circulated through the bed to add or remove energy. A variety of solids may be used, rock being the most widely used material.

A cut away drawing of a packed bed storage unit is shown in Figure 9.4.1. Essential features include a container, a screen to support the bed, support for the screen, and inlet and outlet ducts. In operation, flow is maintained through the bed in one direction during addition of heat (usually downward), and in the opposite direction during removal of heat. Note that heat cannot be added and removed at the same time; this is in contrast to water storage systems where simultaneous addition to and removal from storage is possible.

Well-designed packed beds using rocks have several characteristics that are desirable for solar energy applications: the heat transfer coefficient between the air and solid is high, which promotes thermal stratification; the costs of storage material and container are low; the conductivity of the bed is low when there is no air flow; and the pressure drop through the bed is low.

The thermal stratification of a pebble bed during heating from a constant temperature air stream is shown in Figure 9.4.2. The high heat transfer coefficient-area product between the air and pebbles means that high temperature air entering the bed quickly loses its energy to the pebbles. The pebbles near the entrance are heated but the temperature of the pebbles near the exit remains unchanged and the exit air temperature remains very close to the initial bed temperature. As time progresses a temperature front passes through the bed.

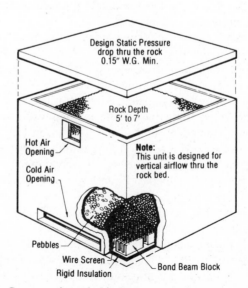

Figure 9.4.1 Cut-away of a packed bed storage unit. Courtesy of Solaron Corp.

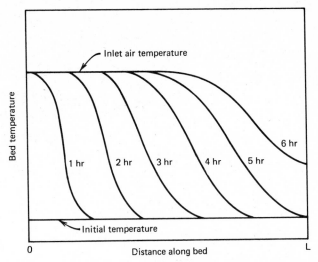

Figure 9.4.2 Temperature distribution in a pebble bed while charging with inlet air at constant temperature.

By hour 5 the front reaches the end of the bed and the exit air temperature begins to rise.

When the bed is fully charged its temperature is uniform, and reversing the flow results in a constant outlet temperature for 5 hours and then a steadily decreasing temperature until the bed is fully discharged.

If the heat transfer coefficient between air and pebbles had been infinitely large, the temperature front during charging and discharging would have been square. The finite heat transfer coefficient produces a "smeared" front that becomes more smeared as time progresses.

A packed bed in a solar heating system does not operate with constant inlet temperature. During the day the variable solar radiation, ambient temperature, collector inlet temperature, load requirements, and other time-dependent conditions result in a variable collector outlet temperature. A set of typical measured temperature profiles is shown in Figure 9.4.3.

Many studies are available on the heating and cooling of packed beds. The first analytical study was by Schumann (1929) and the following equations describing a packed bed are often referred to as the Schumann model. The basic assumptions leading to the Schumann model are: one dimensional "plug" flow; no axial conduction or dispersion; constant properties; no mass transfer; no heat loss to the environment; and no temperature gradients within the solid particles. The differential equations for the fluid and bed temperatures are

$$(\rho C_p)_f \varepsilon \frac{\partial T_f}{\partial \tau} = -\frac{(\dot{m}C_p)_f}{A} \frac{\partial T_f}{\partial x} + h_v(T_b - T_f) \tag{9.4.1}$$

$$(\rho C_p)_b (1 - \varepsilon) \frac{\partial T_b}{\partial \tau} = h_v(T_f - T_b) \tag{9.4.2}$$

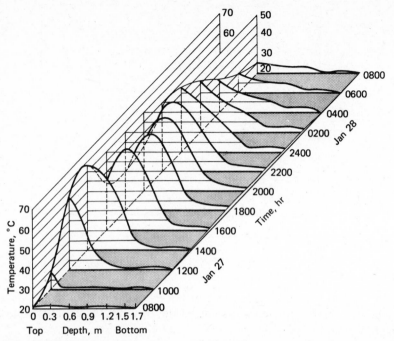

Figure 9.4.3 Temperature profiles in the Colorado State University House II pebble bed during charging and discharging. From Karaki et al. (1977).

where ε is the bed void fraction, h_v is the volumetric heat transfer coefficient between the bed and the fluid (i.e., the usual area heat transfer coefficient times the bed particulate surface area per unit bed volume) and other terms have their usual meanings. A correlation relating h_v to the bed characteristics and to the fluid flow conditions is given in Section 3.16.

For an air based system, the first term of Equation 9.4.1 can be neglected so that the equations can be written as

$$\frac{\partial T_f}{\partial (X/L)} = \text{NTU}(T_b - T_f) \tag{9.4.3}$$

$$\frac{\partial T_b}{\partial \theta} = \text{NTU}(T_f - T_b) \tag{9.4.4}$$

where NTU is the number of transfer units, $(h_v AL)/(\dot{m}C_p)_f$; θ is a dimensionless time, $\tau(\dot{m}C_p)_f/[(\rho C_p)_f(1 - \varepsilon)AL]$; A is bed cross sectional area; and L is bed length. Analytical solutions to these equations exist for a step change in inlet conditions and for cyclic operation.

For the long-term study of solar energy systems, these analytical solutions are not useful and numerical techniques must be employed. Kuhn et al. (1978)

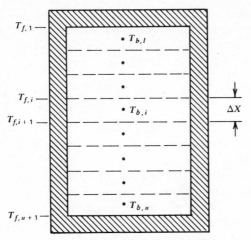

Figure 9.4.4 Packed bed divided into N segments.

investigated a large number of finite difference schemes to numerically solve Equations 9.4.3 and 9.4.4 and concluded that the "complicated effectiveness—NTU" method of Hughes (1975) was best suited for solar system simulation. The following development follows the simpler "effectiveness-NTU" method of Hughes which is the same as the method proposed by Mumma and Marvin (1976). For practical designs the two Hughes models give essentially the same results and so the simpler one is recommended.

Over a length of bed ΔX as shown in Figure 9.4.4 we can consider the bed temperature to be uniform. (The more complicated Hughes equations assumed the bed temperature to have a linear variation with distance.) The air temperature has an exponential profile and the air temperature leaving bed element i is found from

$$\frac{T_{f,i+1} - T_{b,i}}{T_{f,i} - T_{b,i}} = e^{-\text{NTU}(\Delta X/L)} \tag{9.4.5}$$

This equation is exactly analogous to a heat exchanger operating as an evaporator. The energy removed from the air and transferred to the bed in length ΔX is then

$$(\dot{m}C_p)_f(T_{f,i} - T_{f,i+1}) = (\dot{m}C_p)_f(T_{f,i} - T_{b,i})(1 - e^{-\text{NTU}/N}) \tag{9.4.6}$$

where $N = L/\Delta X$.

With Equation 9.4.6, an energy balance on the rock within region ΔX can then be expressed as

$$\frac{dT_{b,i}}{d\theta} = \eta N(T_{f,i} - T_{b,i}) \tag{9.4.7}$$

where η is a constant and equal to $1 - e^{-NTU/N}$. Equation 9.4.7 represents N ordinary differential equations for the N bed temperatures. Fluid temperatures are found from Equation 9.4.6. An extension to Equation 9.4.7 permits energy loss to an environment at T'_a to be included. Then,

$$\frac{dT_{b,i}}{d\theta} = \eta N(T_{f,i} - T_{b,i}) + \frac{(U\Delta A)_i}{(\dot{m}C_p)_f}(T'_a - T_{b,i}) \qquad (9.4.8)$$

Where $(U\Delta A)_i$ is the loss area-loss coefficient product for node i.

Hughes suggests that a Crank-Nicolson approach be used to solve Equation 9.4.8. The time derivative is replaced by $(T^+_{b,i} - T_{b,i})/\Delta\theta$ and the bed temperatures on the right-hand side of Equation 9.4.8 are replaced by $(T^+_{b,i} + T_{b,i})/2$. With all bed temperatures known, the process starts at node 1 so that the inlet fluid temperature is known. A new bed temperature is calculated from Equation 9.4.8 and an outlet fluid temperature from Equation 9.4.6. This new fluid temperature becomes the inlet fluid temperature for node 2.

The repetitive solution of the Schumann model, even in the form of Equation 9.4.8, is time consuming for year-long solar process calculations. This observation led Hughes et al. (1976) to investigate an infinite NTU model. When the complete Shumann equations are solved for various values of NTU, the long-term performance of a solar air heating system with NTU equal to 25 is virtually the same as that with NTU equal to infinity.

For infinite NTU, Equations 9.4.3 and 9.4.4 can be combined into a single partial differential equation since the bed and fluid temperatures are everywhere equal. With the addition of a container heat loss term the result is

$$\frac{\partial T}{\partial \theta} = -L\frac{\partial T}{\partial X} + \frac{(UA)_b}{(\dot{m}C_p)_f}(T'_a - T) \qquad (9.4.9)$$

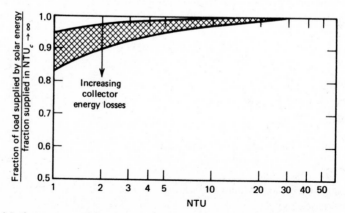

Figure 9.4.5 Long-term performance on a pebble bed as a function of NTU. The band represents different values of collector and bed characteristics for solar heating systems. From Hughes et al. (1976).

This is a single partial differential equation instead of the two coupled partial differential equations of the Schumann model. The single equation represents both the bed and air temperatures since the infinite NTU model assumes the two are identical.

Figure 9.4.5 shows the ratio of the predicted long-term fraction of a heating load by solar energy for a system with a bed having a finite NTU to the fraction by solar energy for a system with an infinite NTU bed. At NTU greater than 25 the ratio is unity but at values as low as 10, the ratio is above 0.95. Consequently, the infinite NTU model can be used for long-term performance predictions even if NTU is as low as 10. Short-term predictions of bed temperature profiles should be based on the full Schumann equations unless the bed NTU is large.

The Schumann model is based on the assumption that the temperature gradients within the particles of the packed bed are not significant; this can be relaxed by defining a corrected value of NTU. Jeffreson (1972) has shown that temperature gradients within the rocks can be accommodated by defining a corrected NTU

$$NTU_c = \frac{NTU}{(1 + Bi/5)} \tag{9.4.10}$$

Where Bi is the Biot number defined as hR/k where R is an equivalent spherical rock radius, k the rock conductivity, and h the fluid to rock heat transfer coefficient. The NTU_c can be used in any one of the equations of this section instead of NTU to include the effects of temperature gradients in the rock. If the Biot number is less than 0.1, the temperature gradients within the pebbles can be neglected.

Example 9.4.1

A pebble bed has the following characteristics: Length in flow direction is 1.80 m; cross sectional area, 14.8 m²; air velocity, 0.053 m/s; equivalent diameter of pebbles, 12.5 mm; void fraction, 0.47; density of pebble material, 1350 kg/m³; specific heat of pebbles, 0.90 kJ/kg C; thermal conductivity of pebble material, 0.85 W/m C; surface area of pebbles per unit volume, 255 m²/m³. For this pebble bed, calculate the Biot number and NTU. Will there be significant temperature gradients in the pebbles? Can the infinite NTU model be used to calculate the performance of this storage unit?

Solution

The temperature of the air is unknown. If we use a low temperature to evaluate properties (e.g. 20 C), the Biot number and NTU criteria will be more severe than if a high temperature had been used. The volumetric heat transfer coefficient from fluid to pebbles is estimated from the Löf and Hawley equation of Section 3.16.

$$h_v = 650(G/D)^{0.7}$$

$$G = 0.053 \text{ m/s} \times 1.204 \text{ kg/m}^3 = 0.0638 \text{ kg/m}^2 \text{ s}.$$

so

$$h_v = 650(0.0638/0.0125)^{0.7} = 2030 \text{ W/m}^3 \text{ C}$$

and

$$h = 2030/255 = 8.0 \text{ W/m}^2 \text{ C}$$

$$\text{Bi} = hR/k = \frac{8.0 \times 0.0125}{2 \times 0.85} = 0.059$$

Since the Biot number is less than 0.1, NTU_c is nearly the same as NTU. From the definition of NTU below Equation 9.4.4.,

$$\text{NTU} = h_v AL/(\dot{m}C_p)_f$$

$$= \frac{2030 \text{ W/m}^3 \text{ C} \times 14.8 \text{ m}^2 \times 1.80\text{m}}{0.053 \text{ m/s} \times 14.8 \text{ m}^2 \times 1.204 \text{ kg/m}^3 \times 1010 \text{ J/kg C}}$$

$$= 57$$

Thus the number of transfer units is much larger than 10, and the infinite NTU model is appropriate. ∎

9.5 PHASE-CHANGE ENERGY STORAGE

Materials that undergo a change of phase in a suitable temperature range may be useful for energy storage if several criteria can be satisfied. The phase change must be accompanied by a high latent heat effect and it must be reversible over a very large number of cycles without serious degradation. The phase change must occur with limited supercooling or superheating and means must be available to contain the material and transfer heat into it and out of it. Finally, the cost of the material and its containers must be reasonable. If these criteria can be met, phase-change energy storage systems can have high capacities (relative to energy storage in specific-heat-type systems) when operated over small temperature ranges, with relatively low volume and weight.

The storage capacity of a phase change material heated from T_1 to T_2, if it undergoes a phase transition at T^*, is the sum of the sensible heat change of the solid (the lower temperature phase) from T_1 to T^*, the latent heat at T^* and the sensible heat of the liquid (the melt, or higher temperature phase) from T^* to T_2

$$Q_s = m[(T^* - T_1)C_s + \lambda + (T_2 - T^*)C_l] \qquad (9.5.1)$$

where m is the mass of material, C_s and C_l are the heat capacities of the solid and liquid phases, and λ is the latent heat of phase transition. Thus a kilogram of $Na_2SO_4 \cdot 10H_2O$ which has C_l of approximately 1950 J/kg C, λ of 2.43×10^5 J/kg and C_l of about 3550 J/kg C on being heated from 25 to 50 C would store $(34 - 25)1950 + 2.43 \times 10^5 + (50 - 34)3350$ or 0.315 MJ.

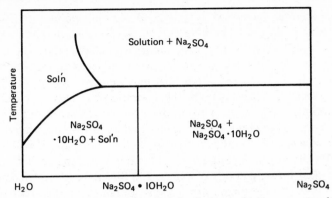

Figure 9.5.1   Phase diagram for sodium sulfate decahydrate, with incongruent melting point.

$Na_2SO_4 \cdot 10H_2O$ (Glauber's salt) was the earliest phase-change storage material to be studied experimentally for house heating applications [Telkes (1955)]. On heating at approximately 32 C it gives a solution and solid Na_2SO_4

$$Na_2SO_4 \cdot 10H_2O + Energy \;\rightleftharpoons\; Na_2SO_4 + 10H_2O$$

Energy storage is accomplished by the reaction proceeding from left to right on addition of heat. The total energy added depends on the temperature range over which the material is heated since it will include sensible heat to heat the salt to the transition temperature, heat of fusion to cause the phase change, and sensible heat to heat the Na_2SO_4 and solution to the final temperature. Energy extraction from storage is the reverse procedure, with the reaction proceeding from right to left and the thermal effects reversed.

Practical difficulties have been encountered with this material. It has been found that performance degrades on repeated cycling, with the thermal capacity of the system reduced. As shown in Figure 9.5.1, $Na_2SO_4 \cdot 10H_2O$ has an incongruent melting point, and as its temperature increases beyond the melting point it separates into a liquid (solution) phase and solid Na_2SO_4. Since the density of the salt is higher than the density of the solution, phase separation occurs. Many studies have been done to develop means to avoid phase separation, through the use of containers of thin cross section, through the use of gels or other agents, or by mechanically agitating the melt mix. Storage units based on $Na_2SO_4 \cdot 10H_2O$ can now be purchased for space heating applications.

Other possibilities for phase change storage media for temperatures in ranges of interest in comfort heating include, for example, $Na_2HPO_4 \cdot 12H_2O$ and $Fe(NO_3)_2 \cdot 6H_2O$ (which has a congruent melting point). Paraffin waxes have also been studied for this use; they can be obtained with various melting points, and as mixture with ranges of melting points. Eutectic mixtures have also been studied [see Kauffman and Gruntfest (1973)], such as $CaCl_2 \cdot MgCl_2 \cdot H_2O$, $urea \cdot NH_4NO_3$, and others. Phase change materials studied for application at higher temperatures include $AlSO_4 \cdot 10 H_2O$ (melting point 112 C), $MgCl_2 \cdot 6H_2O$ (melting point 115 C) and $NaNO_3 + NaOH$ (melting point 245 C).

A further consideration with phase-change storage materials lies in the possibility of supercooling on energy recovery. If the material supercools, the latent heat of fusion may not be recovered or it may be recovered at a temperature significantly below the melting point. This question has been approached from three standpoints, by selection of materials that do not have a strong tendency to supercool, by addition of nucleating agents, and by ultrasonic means of nucleation. These considerations are reviewed by Belton and Ajami (1973), who note that the viscosity at the melting point of a material is a major factor in determining the glass-forming ability of a melt, and thus its tendency to supercool.

Phase-change storage media are usually contained in small containers or shallow trays to provide a large heat transfer area. The general principles of packed bed storage units outlined in Section 9.4 will apply to these units. The heat transfer fluid is usually air, although water has been considered. Two additional phenomena must be considered. First, the latent heat must be taken into account which is in effect a high specific heat over a very small temperature range. Second, the thermal resistance to heat transfer within the material is variable with the degree of solidification and whether heating or cooling of the material is occurring. [Heat transfer in situations of this type have been studied, for example, by Hodgins and Hoffman (1955) and Murray and Landis (1959).] As heat is extracted from a phase-change material, crystallization will occur at the walls and then progressively inward into the material; at the end of the crystallization, heat must be transferred across layers of solid to the container walls. As a solidified material is heated, melting occurs first at the walls and then inward toward the center of the container. These effects can be minimized by design of the containers to give very short pathlengths for internal heat transfer, and the following development assumes the internal gradients to be small.

Morrison and Abdel-Khalik (1978) have developed a model applicable to phase change materials in such containers, where the length in flow direction is L, the cross-sectional area of the material is A, and the wetted perimeter is P. The heat transfer fluid passes through the storage unit in the x direction at rate $\dot{m}$ and with inlet temperature T_{fi} as shown in Figure 9.5.2.

A model can be based on three assumptions: (1) during flow, axial conduction in the fluid is negligible; (2) the Biot number is low enough that temperature gradients normal to the flow can be neglected; (3) heat losses from the bed are negligible. Then an energy balance on the material gives

$$\frac{\partial u}{\partial \tau} = \frac{k}{\rho} \frac{\partial^2 T}{\partial x^2} + \frac{UP}{\rho A} (T_f - T) \qquad (9.5.2)$$

where u, T, k, and ρ are the specific internal energy, temperature, thermal conductivity, and density of the phase change material, T_f and U are the circulating fluid temperature and overall heat transfer coefficient between the fluid and phase change material, and τ is time.

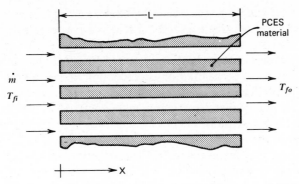

Figure 9.5.2 Schematic of a phase change storage unit. From Morrison and Abdel-Khalik (1978).

An energy balance on the fluid is

$$\frac{\partial T_f}{\partial \tau} + \frac{\dot{m}}{\rho_f A_f}\frac{\partial T_f}{\partial x} = \frac{UP}{\rho_f A_f C_f}(T - T_f) \tag{9.5.3}$$

where ρ_f, A_f, and C_f are the density, flow area, and specific heat of the fluid.

The specific internal energy is related to temperature, T, liquid fraction, χ, and the specific heats of the liquid and solid phases, C_l and C_s, by

$$u = C_s(T - T_{\text{ref}}) \qquad\qquad\qquad (T < T^*)$$

$$u = C_s(T^* - T_{\text{ref}}) + \chi\lambda \qquad\qquad T = T^*$$

$$u = C_s(T^* - T_{\text{ref}}) + \lambda + C_l(T - T^*) \qquad T > T^*$$

where T^* is the melting temperature and T_{ref} is a reference temperature where the internal energy is zero.

The equations and boundary conditions for phase change energy storage can be simplified for particular cases [see Morrison and Abdel-Khalik (1978) for details]. It has been shown that axial conduction during flow is negligible and if the fluid capacitance is small, Equations 9.5.2 and 9.5.3 become

$$\frac{\partial u}{\partial \tau} = \frac{UP}{\rho A}(T_f - T) \tag{9.5.5}$$

$$\frac{\partial T_f}{\partial x} = \frac{UP}{(\dot{m}C_p)_f}(T - T_f) \tag{9.5.6}$$

These two equations can be rewritten in terms of NTU

$$\frac{1}{\bar{C}}\frac{\partial u}{\partial \theta} = \text{NTU}(T_f - T) \tag{9.5.7}$$

$$\frac{\partial T_f}{\partial (X/L)} = \text{NTU}(T - T_f) \tag{9.5.8}$$

where $\theta = \tau(mC_p)_f/\rho AL\bar{C}$ and $\text{NTU} = UPL/(\dot{m}C_p)_f$.

Equations 9.5.7 and 9.5.8 are essentially the same as Equations 9.4.2 and 9.4.3 for air and rock beds. Morrison and Abdel-Khalik have developed these models, explored the effects of assumptions made in their development and used them in simulation studies of solar heating systems with ideal phase-change energy storage that has no superheating or supercooling and complete reversibility (see Section 13.10). They showed that the infinite NTU model is a good approximation for practical systems as it is in the rock bed storage unit.

Jurinak and Abdel-Khalik (1979) have shown that for building heating operations where the melting point is between an optimum value (usually 5 to 10 C above room temperature) and 50 C, an empirical equation for equivalent heat capacity can be written

$$(mC_p)_e = m\left[\frac{18.3}{(T^*)^2}\lambda + \left(1 - \frac{18.3}{(T^*)^2}\right)C_s + \frac{18.3}{(T^*)^2}C_l\right] \qquad (9.5.9)$$

where all temperatures are in degrees Celsius. This empirical equation has utility in design of house heating systems with phase-change storage (see Section 14.5). λ is in kJ/Kg. C_s is in $kJ/kg\ C$.

Other practical considerations include corrosion, reactions of the phase-change material with containers or other side reactions, vapor pressure, and toxicity. Cost is a major factor in many applications, and rules out all but the least expensive materials.

Comparisons of sensible heat storage and phase-change storage can only be made in the context of systems in particular applications. The system analysis methods outlined in following chapters can be used to evaluate the relative merits of the several possible methods. In doing so, it should be kept in mind that the formulations for sensible heat storage are reliable (in the sense that temperature changes of sensible heat media are very predictably related to quantities of heat added to or removed), whereas those for phase change materials may be more uncertain because of unpredictable superheating, supercooling, and lack of complete reversibility of the phase change.

9.6 CHEMICAL ENERGY STORAGE

The previous section dealt with storage of thermal energy in materials undergoing phase transitions. In this section we consider chemical reactions that might be used to store energy. None of these possibilities have yet been used in practical applications, and on all of them both technical and economic questions are yet to be answered. A review of chemical storage is presented by Offenhartz (1976).

An ideal thermochemical reaction for energy storage is an endothermic reaction in which the reaction products are easily separable and do not undergo further reactions. For example, decomposition reactions of the type

$$AB + \text{heat} \longrightarrow A + B$$

are candidates, if the reaction can be reversed to permit recovery of the stored energy. The products A and B can be stored separately, and thermal losses from the storage units are restricted to sensible heat effects, which are usually small compared to heats of reaction. Unfortunately, there are not very obvious candidates of this type useful for low temperature solar applications. Reactions in which water vapor is a product have the difficulty that its heat of condensation is usually lost. Reactions in which products like atomic chlorine are produced lose energy on formation of the dimer Cl_2.

Thermal decomposition of metal oxides for energy storage has been considered by Simmons (1976). These reactions may have the advantage that the oxygen evolved can be used for other purposes or discarded, and oxygen from the atmosphere used in the reverse reactions. Two examples are the decomposition of potassium oxide

$$4KO_2 \rightleftharpoons 2K_2O + 3O_2$$

which occurs over a temperature range of 300 to 800 C with a heat of decomposition of 2.1 MJ/kg, and lead oxide

$$2PbO_2 \rightleftharpoons 2PbO + O_2$$

which occurs over a temperature range of 300 to 350 C with a heat of decomposition of 0.26 MJ/kg. There are many practical problems yet to be faced in the use of these reactions.

Other reactions of possible interest in higher temperature ranges are the dehydration of mixtures of $MgO/Mg(OH)_2$, or $CaO/Ca(OH)_2$ [Bauerle et al. (1976)]. The storage reactions proceed at temperatures of 350 to 550 C. The reverse reactions occur at lower temperatures on the addition of water or steam. The system sulfuric acid-water has also been proposed [e.g., by Huxtable and Poole (1976)], with condensed water and acid stored until needed. There are obvious practical problems with these systems.

An example of a photochemical decomposition reaction is the decomposition of nitrosyl chloride, which can be written

$$NOCl + photons \longrightarrow NO + Cl$$

The atomic chlorine produced forms chlorine gas, Cl_2, with the release of a substantial part of the energy added to the NOCl in decomposition. Thus the overall reaction is:

$$2NOCl + photons \longrightarrow 2NO + Cl_2$$

The reverse reaction can be carried out to recover part of the energy of the photons entering the reaction [see Marcus and Wohlers (1960, 1961, 1964)].

Processes that produce electrical energy may have storage provided as chemical energy in electrical storage batteries or their equivalent. Several types of battery systems can be considered for these applications, including lead-acid, nickel-iron, and nickel-cadmium batteries. For low discharge rates and moderate charge rates the efficiencies of these systems range from 60 to 80% (ratio of watt-hour output to watt-hour input), depending on the battery. It is also possible to electrolyze water with solar-generated electrical energy, to store oxygen and hydrogen, and to recombine in a fuel cell to regain electrical energy [see Bacon (1964)]. These storage systems are characterized by the relatively high cost per kilowatt-hour of storage capacity, and can now be considered for low-capacity special applications such as auxiliary power supply for space vehicles, isolated telephone repeater power supplies, instrument power supplies, and so on.

REFERENCES

Bacon, F. T., *Proceedings of the UN Conference on New Sources of Energy*, 1, 174 (1964). "Energy Storage Based on Electrolyzers and Hydrogen-Oxygen Fuel Cells."

Bauerle, G., D. Chung, G. Ervin, J. Guon, and T. Springer, *Proceedings of the ISES Meeting*, Winnipeg, 8, 192 (1976). "Storage of Solar Energy by Inorganic Oxide Hydrides."

Belton, G. and F. Ajami, Report NSF/RANN/SE/GI27979 TR/73/4 of the University of Pennsylvania National Center for Energy Management and Power to NSF (1973). "Thermochemistry of Salt Hydrates."

Hodgins, J. W. and T. W. Hoffman, *Canadian Journal of Technology*, 33, 293 (1955). "The Storage and Transfer of Low Potential Heat."

Hughes, P. J., M.S. Thesis in Mechanical Engineering. University of Wisconsin-Madison (1975). "The Design and Predicted Performance of Arlington House."

Hughes, P. J., S. A. Klein, and D. J. Close, *Journal of Heat Transfer*, 98, 336 (1976). "Packed bed Thermal Storage Models for Solar Air Heating and Cooling Systems."

Huxtable, D. D. and D. R. Poole, *Proceedings of the ISES Meeting*, Winnipeg (1976). "Thermal Energy Storage by the Sulfuric Acid-Water System."

Jeffreson, C. P., *American Institute of Chemical Engineers Journal*, 18(2), 409 (1972). "Prediction of Breakthrough Curves in Packed Beds."

Jurinak, J. J. and S. I. Abdel-Khalik, *Energy*, 4, 503 (1979). "On the Performance of Solar Heating Systems Utilizing Phase Change Energy Storage."

Karaki, S., P. R. Armstrong, and T. N. Bechtel, Report COO-2868-3 from Colorado-State University to the U.S. Department of Energy (1977). "Evaluation of a Residential Solar Air Heating and Nocturnal Cooling System."

Kauffman, K. and I. Gruntfest, Report NCEMP-20 of the University of Pennsylvania National Center for Energy Management and Power, to NSF (1973). "Congruently Melting Materials for Thermal Energy Storage."

Kuhn, J. K., G. F. Von Fuchs, A. W. Warren, and A. P. Zob, Technical Progress Report for March 1, 1978–August 31, 1978, contract No. EG-77-C-02-4482, Boeing Computer Services Company, Seattle, Washington, September (1978). "Developing and Upgrading of Solar System Thermal Energy Storage Simulation Models."

Lavan, Z. and T. Thompson, *Solar Energy*, 19, 519 (1977). "Experimental Study of Thermally Stratified Hot Water Storage Tanks."

Marcus, R. J. and H. C. Wohlers, *Proceedings of the UN Conference on New Sources of Energy*, 1 (1964). "Chemical Conversion and Storage of Concentrated Solar Energy." See also *Solar Energy*, 5, 121 (1961); 5, 44 (1961); 4 (2), 1 (1960).

Morrison, D. J. and S. I. Abdel-Khalik, *Solar Energy*, 20, 57 (1978). "Effects of Phase-Change Energy Storage on the Performance of Air-Based and Liquid-Based Solar Heating Systems."

Mumma, S. D. and W. C. Marvin, ASME paper 76-HT-73 (1976). "A Method of Simulating The Performance of a Pebble Bed Thermal Energy Storage and Recovery System."

Murray, W. D. and F. Landis, *Journal Heat Transfer*, **81C**, 107 (1959). "Numerical and Machine Solutions of Transient Heat Conduction Problems Involving Melting or Freezing."

Offenhartz, P. O., *Proceedings of the ISES Meeting*, Winnipeg, **8**, 48 (1976). "Chemical Methods of Storing Thermal Energy."

Schumann, T. E. W., *Journal of the Franklin Institute*, **208**, 405 (1929). "Heat Transfer: A Liquid Flowing Through a Porous Prism."

Simmons, J. A., *Proceedings of the ISES Meeting*, Winnipeg, **8**, 219 (1976). "Reversible Oxidation of Metal Oxides for Thermal Energy Storage."

Telkes, M., *Solar Energy Research*, University of Wisconsin Press, Madison (1955). "Solar Heat Storage."

CHAPTER 10

System Thermal Calculations and Experiments

In Chapters 6 through 9 we have developed mathematical models for two of the key components in solar energy systems, that is, collectors and storage units. In this chapter we show how other components can be modeled and how the component models can be combined into system models. With information on the magnitude and time distribution of the system loads and the weather, it is possible to simultaneously solve the set of equations to estimate the thermal performance of a solar process over any time period. These estimates (simulations) are usually done numerically with computers, and provide information on the expected dynamic behavior of the system and long-term integrated performance.

The collector performance is a function of the temperature of the fluid entering the collector. This temperature, neglecting heat loss from the connecting pipes, is the same as the temperature in the exit portion of the storage unit. The outlet temperature from the collector becomes the inlet temperature to the storage unit. In these equations, time is the independent variable and the solution is in the form of temperatures as functions of time.

Once the temperatures are known, energy rates can be determined. It is then possible to integrate the energy quantities over time to develop information such as that in Figure 9.1.1c, and thus assess the annual performance of a system. This simulation approach can be used to estimate, for any process application, the amount of energy delivered from the solar collector to meet a load and the amount of auxiliary energy required. The simulation also can indicate whether the temperature variations for a particular system design are reasonable, for example, whether a collector temperature would rise above the boiling point of the liquid being heated.

Simulations are numerical experiments and can give the same kinds of thermal performance information as can physical experiments. They are, however, relatively quick and inexpensive, and can produce information on effects of design-variable changes on system performance by series of experiments all using exactly the same loads and weather. These design variables could include selectivity of the absorbing surface, number of covers on the collector, collector area, and so on. With cost data and appropriate economic analysis, simulation results can be used to find least-cost systems.

Simulations are not substitutes for physical experiments. Component scale experiments are necessary to understand component behavior and lend confidence to the corresponding mathematical models. System scale experiments are necessary to bring to light the many practical problems inherent in any complicated system that simulations cannot model. Careful comparisons of experiments and simulations lead to improved understanding of each. Once simulations have been verified with experiments, new systems can be designed with confidence using simulation methods.

The use of simulation methods in the study of solar processes is a recent development. Sheridan et al. (1967) used an analog computer in simulation studies of operation of solar water heaters. Gupta and Garg (1968) developed a model for thermal performance of a natural circulation solar water heater with no load, represented solar radiation and ambient temperature by Fourier series, and were able to predict a day's performance in a manner that agreed substantially with experiments. Close (1967) used numerical modeling and a factorial design method to determine which water heater system design factors are most important. Gupta (1971) used a response factor method that is amendable to hand calculation for short-term process operation. Buchberg and Roulet (1968) developed a thermal model of a house heating system, simulated its operation with a year's hourly meteorological data, and applied a pattern search optimization procedure in finding optimum designs. Other process simulations have been done by Löf and Tybout (1973), Butz et al. (1973) and Oonk et al. (1974). Since these publications, solar process simulation programs have come into widespread use.

In this chapter we provide a brief review of collector and storage models, and then show how controls, heat exchangers, and pipe and duct losses in collector circuits can be formulated. We indicate the needs for meteorological data and information on the energy loads a system is to supply. We show examples of the kinds of information that simulations and experiments can provide, and illustrate the use of a simulation program to determine dynamic system behavior and integrated performance. These simulation methods are the basis for parts of the discussions of applications to heating and cooling in later chapters.

10.1 COMPONENT MODELS

Chapters 6, 7, and 8 present a development of collector models, performance, measurements, and data, and Chapter 9 discusses storage unit models. For flat-plate collectors, Equation 6.7.6 is appropriate, whereas for focusing collectors, Equation 8.3.9 or its equivalent can be used. The flat-plate equation can be written:

$$Q_u = A_c F_R [S - U_L(T_i - T_a)]^+ \tag{10.1.1}$$

where the + sign implies a controller is present and only positive values should be used. Operation of a forced circulation collector will not be carried out when

$Q_u < 0$ (or when $Q_u < Q_{min}$, where Q_{min} is a minimum level of energy gain to justify pumping the fluid through the system). In real systems, this is accomplished by comparing the temperature of the fluid leaving the collector (i.e., in the top header) with the temperature of the fluid in the exit portion of the storage tank and running the pump only when the difference in temperatures is positive and energy can be collected. Useful gain is also given by

$$Q_u = (\dot{m} C_p)_c (T_o - T_i) \tag{10.1.2}$$

where $\dot{m}$ is the output of the pump circulating fluid through the collector.

If the storage unit is a fully mixed sensible heat unit, its performance is given by Equation 9.3.2.

$$Q_u - L - (UA)_s (T_s - T'_a) = (mC_p)_s \frac{dT_s}{d\tau} \tag{10.1.3}$$

The equivalent equations for stratified water tank storage systems, pebble bed exchangers, or heat of fusion systems are used in lieu of Equation 10.1.3, as appropriate. These equations are the basic equations to be solved in the analysis of systems such as a simple solar water heater with collector, pump and controller and storage tank. L is a time dependent rate of removal of energy to meet a load. S and T_a are also time dependent.

The performance models discussed so far have been based on the fundamental equations describing the behavior of the equipment. Models may also be expressed as empirical or stochastic representations of operating data from particular items of equipment. An example is the model of a LiBr—H$_2$O absorption cooler (see Section 16.2), which relates cooling capacity to temperatures of the fluid streams entering the machine. These empirical relations may be in the form of equations, graphs, or tabular data. In whatever form the models are expressed, they must represent component performance over the full range of operating conditions to be encountered in the solar operation.

There is very often a heat exchanger between the collector and the storage tank (the collector heat exchanger) when antifreeze is used in collectors, and if piping or ducting to and from a collector is not well insulated, the losses from the piping or ducting may have to be taken into account. Each of these can be accounted for by simple modifications of Equation 10.1.1, as noted in the next two sections.

10.2 COLLECTOR HEAT EXCHANGER FACTOR

Collectors are often used in combination with a heat exchanger between collector and storage allowing the use of antifreeze solutions in the collector loop. A common circuit of this type is shown in Figure 10.2.1.

A useful analytical combination of the equations for the collector and the heat exchanger has been derived by deWinter (1975). In this development the collector equation and the heat exchanger equation will be combined into a

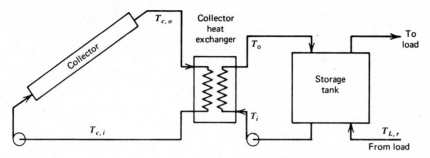

Figure 10.2.1 Schematic of liquid system using a collector heat exchanger between collector and tank.

single expression that has the same form as the collector equation alone. We will show that the combination of a collector and a heat exchanger perform exactly like a collector alone but with a reduced value of F_R. The useful gain of the collector is represented by Equations 10.1.1 and 10.1.2 written in terms of $T_{c,i}$ and $T_{c,o}$. The heat exchanger performance is expressed in terms of effectiveness [see Kays and London (1964)] by

$$Q_{HX} = (\dot{m}C_p)_{min} \varepsilon (T_{c,o} - T_i) \qquad (10.2.1)$$

where $(\dot{m}C_p)_{min}$ is the smaller of the fluid capacitance rates (flow rate, $\dot{m}$, times fluid heat capacity, C_p) on the two sides of the heat exchanger, $T_{c,o}$ is the outlet fluid temperature from the collector, and T_i inlet water temperature to the heat exchanger (the temperature in the bottom of the tank). For counterflow, the most common configuration, the heat exchanger effectiveness, ε, is given by

$$\varepsilon = \frac{1 - e^{-NTU(1-C^*)}}{1 - C^* e^{-NTU(1-C^*)}} \qquad (10.2.2)$$

where NTU is the number of transfer units defined by $[(UA)_{HX}/(\dot{m}C_p)_{min}]$ and C^* is the ratio $(\dot{m}C_p)_{min}/(\dot{m}C_p)_{max}$.

Combining Equations 10.1.1, 10.1.2, and 10.2.1, we obtain

$$Q_u = A_c F_R'[S - U_L(T_i - T_a)]^+ \qquad (10.2.3)$$

where the modified collector heat removal factor F_R' accounts for the presence of the heat exchanger and is given by

$$\frac{F_R'}{F_R} = \frac{1}{1 + \left(\dfrac{A_c F_R U_L}{(\dot{m}C_p)_c}\right)\left(\dfrac{(\dot{m}C_p)_c}{\varepsilon(\dot{m}C_p)_{min}} - 1\right)} \qquad (10.2.4)$$

The factor F_R'/F_R is an indication of the penalty in collector performance incurred because the heat exchanger causes it to operate at higher temperatures than it otherwise would. Another way of looking at the penalty is to consider the ratio F_R/F_R' as the fractional increase in collector area required for the system

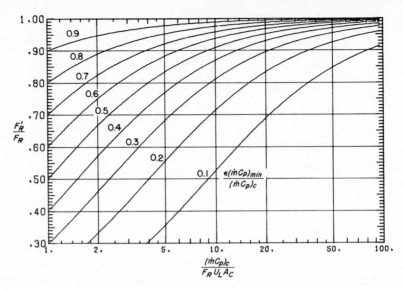

Figure 10.2.2   Collector heat exchanger correction factor as a function of $\varepsilon(\dot{m}C_p)_{min}/(\dot{m}C_p)_c$ and $(\dot{m}C_p)_c/F_R U_L A_C$. Adapted from Beckman et al. (1977).

with the heat exchanger to give the same energy output as without the heat exchanger. Equation 10.2.3 now represents the performance of a subsystem including the collector and the heat exchanger (and, implicitly, the controller and two pumps), and is of the same form as that for the collector only. F_R'/F_R can be calculated from Equation 10.2.4 or determined from Figure 10.2.2 which shows F_R'/F_R as a function of $\varepsilon(\dot{m}C_p)_{min}/(\dot{m}C_p)_c$ and $(\dot{m}C_p)_c/A_c F_R U_L$.

Example 10.2.1

A collector to be used in a solar heating system like that of Figure 10.2.1 heats antifreeze; heat is transferred to water through a collector heat exchanger. The collector $F_R U_L$ is 3.75 W/m² C. Flow rates through both sides of the heat exchanger are 0.0139 kg/s per square meter of collector. The fluid on the collector side is a glycol solution having $C_p = 3350$ J/kg C. The effectiveness of the heat exchanger is 0.7. What is F_R'/F_R?

Solution

The capacitance rate on the water side of the exchanger is $0.0139 \times 4190 = 58.2$ W/m² C. On the glycol side the capacitance rate is $0.0139 \times 3350 = 46.6$ W/m² C. Thus the minimum $(\dot{m}C_p)_c$ product is that of the glycol. The flow rates are given on a per unit area basis.

$$\frac{(\dot{m}C_p)_c}{F_R U_L A_c} = \frac{0.0139 \times 3350}{3.75} = 12.4$$

and, since the minimum capacitance rate is that through the collector,

$$\frac{(\dot{m}C_p)_{min}}{(\dot{m}C_p)_c} = 1 \quad \text{and} \quad \frac{\varepsilon(\dot{m}C_p)_{min}}{(\dot{m}C_p)_c} = \varepsilon = 0.7.$$

Then, from Figure 10.2.2 or Equation 10.2.4, $F'_R/F_R = 0.97$.

10.3 DUCT AND PIPE LOSS FACTORS

The energy lost from ducts and pipes leading to and returning from the collector in a solar energy system can be significant. Beckman (1978) has shown that the combination of pipes or ducts plus solar collector is equivalent in thermal performance to solar collector with different values of U_L and $(\tau\alpha)$. (For simplicity in terminology, the term duct will be used, but the same analysis holds for pipes. Losses from ducts are more likely to be a problem than those from pipes.)

Consider the fluid temperature distribution as shown in Figure 10.3.1. Fluid enters the portion of the duct where losses occur* at temperature T_i. Due to heat losses to the ambient at temperature T_a, the fluid is reduced in temperature by an amount ΔT_i before it enters the solar collector. The fluid passes through the collector and is heated to the collector outlet temperature. This temperature is then reduced to T_o as the fluid loses heat to the ambient while passing through the outlet ducts.

From energy balance considerations the useful energy gain of this collector-duct combination is equal to

$$Q_u = (\dot{m}C_p)_c(T_o - T_i) \tag{10.3.1}$$

This energy gain can also be related to the energy gain of the collector minus the duct losses by the following rate equation

$$Q_u = A_c F_R[G_T(\tau\alpha) - U_L(T_i - \Delta T_i - T_a)] - \text{Losses} \tag{10.3.2}$$

The duct losses are equal to the integrated losses over the inlet and outlet ducts and are given by

$$\text{Losses} = U_d \int (T - T_a)dA \tag{10.3.3}$$

where U_d is the loss coefficient from the duct. It is possible to integrate Equation 10.3.3, but in any well-designed system the losses from ducts must be small and the integral can be approximated to an adequate degree of accuracy in terms of the inlet and outlet temperatures

$$\text{Losses} \approx U_d A_i(T_i - T_a) + U_d A_o(T_o - T_a). \tag{10.3.4}$$

---

* In a house heating system, duct and pipe losses inside the heated space are not net losses but are uncontrolled gains. These gains may not be desirable during warm weather.

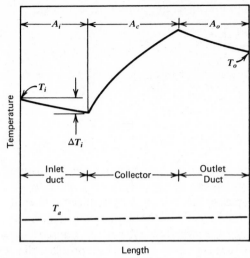

Figure 10.3.1 Temperature distribution through a duct-collector system. From Beckman (1978).

where A_i and A_o are the areas for heat loss of the inlet and outlet ducts. Upon rearranging Equations 10.3.1 and 10.3.4, the losses can be expressed in terms of the useful energy gain and the inlet fluid temperature as

$$\text{Losses} = U_d(A_i + A_o)(T_i - T_a) + \frac{U_d A_o Q_u}{(\dot{m}C_p)_c} \tag{10.3.5}$$

The decrease in temperature, ΔT_i, due to heat losses on the inlet side of the collector can be approximated by

$$\Delta T_i \approx \frac{U_d A_i (T_i - T_a)}{(\dot{m}C_p)_c} \tag{10.3.6}$$

Substituting Equations 10.3.5 and 10.3.6 into 10.3.2 and rearranging, the useful energy gain of the collector and duct system can be expressed as

$$Q_u = \frac{A_c F_R\left[G_T(\tau\alpha) - U_L\left(1 - \frac{U_d A_i}{(\dot{m}C_p)_c} + \frac{U_d(A_i + A_o)}{A_c F_R U_L}\right)(T_i - T_a)\right]}{1 + \frac{U_d A_o}{(\dot{m}C_p)_c}} \tag{10.3.7}$$

Equation 10.3.7 can be made to look like the usual collector equation by defining modified values of $(\tau\alpha)$ and U_L so that

$$Q_u = A_c F_R[G_T(\tau\alpha)' - U_L'(T_i - T_a)] \tag{10.3.8}$$

where

$$\frac{(\tau\alpha)'}{(\tau\alpha)} = \frac{1}{\left(1 + \dfrac{U_d A_o}{(\dot{m}C_p)_c}\right)} \tag{10.3.9}$$

$$\frac{U_L'}{U_L} = \frac{1 - \dfrac{U_d A_i}{(\dot{m}C_p)_c} + \dfrac{U_d(A_i + A_o)}{A_c F_R U_L}}{1 + \dfrac{U_d A_o}{(\dot{m}C_p)_c}} \tag{10.3.10}$$

Example 10.3.1

Compare the performance equations of an air collector system with ducts insulated and not insulated. The collector has an area of 50 m², an $F_R U_L$ product of 3.0 W/m² C and $F_R(\tau\alpha)$ of 0.60. The mass flow rate-specific heat product of the air through the collector is 500 W/C. The area of the inlet duct is 10 m², as is the area of the outlet duct. Insulation 33 mm thick is available, with a conductivity of 0.033 W/m C. The heat transfer coefficient outside the duct is 10 W/m² C, and that inside the duct is assumed to be large, leading to a loss coefficient, U_d for the insulated duct of 1.0 W/m² C and for the un-insulated duct of 10 W/m² C.

Solution

For the insulated duct, the modified transmittance-absorptance product is found using Equation 10.3.9:

$$\frac{(\tau\alpha)'}{(\tau\alpha)} = \frac{1}{1 + \dfrac{1.0 \times 10}{500}} = 0.98$$

and

$$F_R(\tau\alpha)' = 0.60 \times 0.98 = 0.59$$

The modified loss coefficient is obtained with Equation 10.3.10:

$$\frac{U_L'}{U_L} = \frac{1 - \dfrac{1.0 \times 10}{500} + \dfrac{1.0(10 + 10)}{50 \times 3.0}}{1 + \dfrac{1.0 \times 10}{500}} = 1.09$$

and

$$F_R U_L' = 3.27 \text{ W/m}^2 \text{ C}$$

For the system with insulated ducts the collector performance equation becomes

$$Q_u = 50[0.59 G_T - 3.27(T_i - T_a)]$$

If the ducts are not insulated, U_d is 10 W/m^2 C, $F_R(\tau\alpha)'$ is 0.50 and $F_R U_L'$ is 5.3 W/m^2 C. The performance equation is then

$$Q_u = 50[0.50G_T - 5.3(T_i - T_a)]$$

The addition of insulation has a very substantial effect on the useful gain to be expected from the system.   ■

In a liquid system when both pipe losses and a heat exchanger are present, the collector characteristics should first be modified to account for pipe losses and these modified collector characteristics used in Equation 10.2.4 to account for the heat exchanger. This assumes the heat exchanger is near the tank, which is the usual case.

Close and Yusoff (1978) have developed an analysis of the effects of leaks into air heating collectors. Their results can also be expressed in terms of modified $F_R(\tau\alpha)$ and $F_R U_L$.

10.4 CONTROLS

Two types of control schemes are commonly used on solar collectors: on-off and proportional. With an on-off controller, a decision is made to turn the circulating pumps on or off depending on whether or not useful output is available from the collectors. With a proportional controller, the pump speed is varied in an attempt to maintain a specified temperature level at the collector outlet. Both systems have advantages and disadvantages, largely depending on the ultimate use of the collected energy.

The most common control scheme requires two temperature sensors, one in the bottom of the storage unit and one on the absorber plate at the exit of a collector (or on the pipe near the plate). Assume the collector has low heat capacity. When fluid is flowing, the collector transducer senses the exit fluid temperature. When fluid is not flowing, the mean plate temperature, T_p is measured. A controller receives this temperature and the temperature at the bottom of the storage unit. This storage temperature will be called T_i; when the pump turns on, the temperature at the bottom of storage will equal the inlet fluid temperature if the connecting pipes are lossless. Whenever the plate temperature at no flow conditions exceeds T_i by a specific amount ΔT_{on}, the pump is turned on.

When the pump is on and the measured temperature difference falls below a specified amount ΔT_{off}, the controller turns the pump off. Care must be exercised when choosing both ΔT_{on} and ΔT_{off} since instabilities can occur.

When the collector pump is off, the useful output is zero and the absorber plate reaches an equilibrium temperature given by

$$[S - U_L(T_p - T_a)] = 0 \tag{10.4.1}$$

The value of S when T_p is equal to $T_i + \Delta T_{on}$ is

$$S_{on} = U_L(T_i + \Delta T_{on} - T_a) \qquad (10.4.2)$$

When the pump does turn on, the useful gain is then

$$Q_u = A_c F_R[S_{on} - U_L(T_i - T_a)] \qquad (10.4.3)$$

which when Equation 10.4.2 is substituted for S_{on} becomes

$$Q_u = A_c F_R U_L \Delta T_{on} \qquad (10.4.4)$$

The outlet temperature under these conditions is found from

$$Q_u = (\dot{m}C_p)(T_o - T_i) \qquad (10.4.5)$$

The temperature difference, $T_o - T_i$, is the temperature measured by the controller after flow begins. Consequently, the turn-off criterion must satisfy the following inequality or the system will be unstable (until the solar radiation increases to levels significantly higher than that corresponding to S_{on}).

$$\Delta T_{off} \leq \frac{A_c F_R U_L}{\dot{m}C_p} \Delta T_{on} \qquad (10.4.6)$$

Example 10.4.1

For a water collector with $A_c = 2 \text{ m}^2$, $F_R U_L = 3 \text{ W/m}^2 \text{ C}$, and $\dot{m} = 0.030 \text{ kg/s}$, what is the ratio of turn-on criterion to turn-off criterion?

Solution

From Equation 10.4.6

$$\frac{\Delta T_{on}}{\Delta T_{off}} \geq \frac{0.030 \times 4190}{2 \times 3} = 21 \qquad \blacksquare$$

From the preceding example, the turn-on criterion must be significantly greater than the turn-off criterion. Another way of looking at this situation is to assume a 5 C turn-on setting. Then, the controller of the example will have to be sensitive to temperature differences of 0.25 C or oscillations may result if the radiation stays near S_{on}. This is a small temperature difference to detect with inexpensive controllers. Raising the turn on setting to 20 C or more will not significantly reduce the useful energy collection. In fact, the pump should not be operated until the value of the useful energy collected exceeds the cost of pumping.

Even if the criterion of Equation 10.4.6 is satisfied, oscillations may still be present particularly in the morning. The fluid in the pipes or ducts between the storage unit and collector will be colder than the temperature at the bottom of storage. Consequently, when the pump first turns on cold fluid will enter the collector resulting in lower temperatures detected by the outlet sensor than expected. The controller may turn the pump off until the fluid in the collector is heated to the proper temperature for the pump to again turn on. Other than wear-and-tear on the pump and motor this is an efficient way to heat the fluid in the collector and inlet ducts up to the proper temperature.

Proportional controllers have been used to maintain either the collector outlet temperature or the temperature rise through the collector at or near a predetermined value. In such a control system the temperature sensors are used to control the pump speed. Although higher outlet temperatures can be obtained with a proportional controller than with an on-off controller the useful energy collected by an on-off system with a fully mixed tank, with both turn-off and turn-on criteria equal to zero will be greater than a proportional controller having a maximum flow rate equal to the on-off controller flow rate. This fact becomes clear when the basic collector equation is examined, as the highest value of Q_u is obtained when F_R is a maximum.

There is a growing body of literature on controls. Winn and Hull (1979) have determined the flow rate that leads to the maximum difference between energy collected and energy required for pumping fluid through a collector. Kovarik and Lesse (1976), and more recently others, have considered optimization of flow rate for systems in which reduced m leads to stratification in a tank and thus to lower T_i and improved Q_u. It has been shown that if a tank is highly stratified the optimum flow rates may be substantially lower than those usually used.

10.5 METEOROLOGICAL DATA; LOADS

Weather conditions and system loads can be thought of as forcing functions operating on the equations describing performance. Loads may be a function of weather for building air conditioning processes, or they may be determined for other kinds of applications by factors unrelated to weather. Meteorological data, including solar radiation, wind speed and ambient temperature, influence collector performance, and in general information such as that in Table 2.5.2 is needed to calculate system performance over time.

There is no alternative to the use of existing meteorological data, and it is often necessary to select a data set for use in simulations. For studies of process dynamics, data for a few days or weeks may be adequate if they include the range of conditions of interest. For design purposes it is best to use a full year's data, or a full season's data if the process is a seasonal one, and as data are available for many years for many stations it is necessary to select the best set. Klein (1976), in a study of solar heating systems, developed the concept of a *design year*. Using 8 years of heating season data, for each of the 9 months of the season the month of the eight was selected which had the radiation closest to the eight-year average. Monthly average temperatures were used as a secondary criterion where necessary. The set of months so selected constituted the design year. Discontinuities between months normally cause negligible difficulties. The selection of the design year was evaluated by simulations. Table 10.5.1 shows results of this procedure, and indicates the annual fraction of heating loads carried by solar energy for two collector areas on a particular building, for the

Table 10.5.1   The Design (Heating Season) Year for Madison, Based on 8 Year's Data (1948–49 to 1955–56)[a]

Month	From Year	Monthly Fractions by Solar	
		Area 1	Area 2
Sept.	51–52	0.97	1.00
Oct.	55–56	0.79	0.98
Nov.	49–50	0.35	0.63
Dec.	49–50	0.26	0.49
Jan.	53–54	0.23	0.43
Feb.	54–55	0.36	0.66
Mar.	53–54	0.53	0.85
Apr.	55–56	0.72	0.96
May	52–53	0.77	0.98
Annual Contribution		0.47	0.69
Annual Contribution for Eight Years		0.47	0.67

[a] From Klein (1976).

design year and for the 8 years. The design year results appear to provide a good representation of the results for the eight year period, at least for purposes of simulating solar heating systems.

A related but more detailed study by Hall et al. (1978) of 23 years of data for each of the 26 primary stations in the United States solar radiation network (as indicated in Figure 2.5.7) has led to the generation of *typical meteorological year* (TMY) tapes for these locations.* The TMY data for Madison has been used in heating system simulations and the results compared with those for the full 23 years; the TMY indicated a solar contribution of 0.60, and the value for the full 23 year simulation was 0.58.

Caution must be exercised in the use of the design year or typical meteorological year if the process is designed to provide high fractions by solar energy. Years that are far from the average year tend to include unusual sequences or extremes of weather, and the results of simulations based on the typical years may be significantly different from the long-term average. Also, year-to-year variation in weather will lead to significant year-to-year variation in a solar process output.

Another approach to the use of existing meteorological data is to synthesize from it data for short periods, which when used in simulations will provide information on longer periods of operation. For example, Anand et al. (1978)

* Available from *National Oceanographic and Atmospheric Administration.*

have worked out a two-step procedure for condensing data. First, they statistically manipulate the data (insolation and dry bulb temperature) to place it in "bins" or ranges of data pairs. Then, for these data-pairs, they curve fit expressions for diurnal variation of insolation (using a combined cosine and exponential function) and temperature (using a sine function) to get a small number of synthesized days which are representative of a large number of actual days. The method has not been extended to cover other meteorological variables.

Information must also be available on loads, L, the time-dependent rate of energy removal from the system. The determination of L is a relatively simple matter if a load is a quantity of water delivered at a minimum temperature for a service hot water requirement (with the hot water removed from the tank replaced by cold water from mains). A load may be more difficult to determine if it is the energy transferred into the generator of an absorption air conditioner, with energy requirements fixed by characteristics of the air conditioner and the building. In this case, ambient wet bulb temperature and infiltration rates will also be needed to calculate latent loads on an air conditioner. We do not go into detail on calculation of heating and air conditioning loads in buildings; reference is made to ASHRAE and other publications for information on these topics.* The general approach to determination of L is the same as that for other components, that is, to develop the set of equations or numerical data that relate energy rates and temperatures to time that can be solved simultaneously with the other solar system component equations.

The usual situation is to have the loads on a nonsolar system equal to the loads on the solar system designed for the same task. However, there are situations where the amount of energy to be supplied is influenced by the solar systems. For example, solar water heaters with larger tanks than conventional water heater tanks could have larger losses and consequently increased loads. A more extreme example is a passively heated house when the substitution of a passive element for part of a conventional wall can substantially change the energy to be supplied.

10.6 SYSTEM MODELS

System models are the assemblies of appropriate component models. The net effect of this assembly is to produce a set of coupled algebraic and differential equations, having time as the independent variable. These equations include meteorological data as forcing functions that operate on the collector, and possibly also on the load, depending on the application. These equations can be manipulated and combined algebraically or they can be solved simultaneously without formal combinations. Each procedure has some advantages in solar process simulation. If the equations are all linear (and if there are not too many

* The degree day concept for calculation of heating loads is briefly described in Chapter 14.

to manipulate) the algebraic equations can be solved and substituted into the differential equations, which can then be solved by standard methods [e.g., Hamming (1962)]. If the algebraic equations are nonlinear, or if there are a large number of them coupled so that they are difficult to solve, it may be advantageous to leave them separated and solve the set of combined algebraic and differential equations numerically.

As an example of a simple system that yields a single differential equation, consider a solar water heater with an unstratified storage unit supplying a load at a fixed flow rate and returning water back to the tank at a constant temperature, $T_{L,r}$. This is illustrated in Figure 10.2.1. Equation 10.1.1 for the collector can be combined with Equation 10.1.3 for the tank to give

$$(mC_p)_s \frac{dT_s}{d\tau} = A_c F_R[S - U_L(T_s - T_a)]^+ - (UA)_s(T_s - T_a')$$

$$- (\dot{m}C_p)_L(T_s - T_{L,r}) \tag{10.6.1}$$

Once the collector parameters, the storage size and loss coefficient, the magnitude of the load, and the meteorological data are specified, then the storage tank temperature can be calculated as a function of time. Also, gain from the collector, losses from storage, and energy to load can be determined for any desired period of time by integration of the appropriate rate quantities.

Various methods are available to numerically integrate equations like Equation 10.6.1.* For example, simple Euler integration can be used, the same technique used in Example 9.3.1. Using simple Euler, we express the temperature derivative $dT_s/d\tau$ as $(T_s^+ - T_s)/\Delta\tau$ and obtain an expression for the change in storage tank temperature for the time period in terms of known quantities. Equation 10.6.1 then becomes

$$T_s^+ = T_s + \frac{\Delta\tau}{(mC_p)_s} \{A_c F_R[S - U_L(T_s - T_a)]^+$$

$$- (UA)_s(T_s - T_a') - (\dot{m}C_p)_L(T_s - T_{L,r})\} \tag{10.6.2}$$

Integration schemes must be used with care to insure that they are stable for the desired time step and that reasonably accurate solutions are being attained. When performing hand calculations, both stability and accuracy can be problems. However, most computer facilities have subroutines that will solve systems of differential equations to a specified accuracy and automatically take care of stability problems.

Example 10.6.1

The performance of the collector of Example 6.10.1 was based upon a constant supply temperature of 40 C to the collector. Assume the collector area is 4 m². $F_R = 0.80$ and $U_L = 8.0$ W/m² C. The collector is connected to a water storage

---

* Equation 10.6.1 can be integrated analytically by making various simplifying assumptions.

tank containing 150 kg of water initially at 40 C. The storage tank loss coefficient-area product is 1.70 W/C and the tank is located in a room at 25 C. Assume water is withdrawn to meet a load at a constant rate of 10 kg/hr and is replenished from the mains at a temperature of 15 C. Calculate the performance of this system for the period from 7 a.m. to 5 p.m. using the collector and meteorological data from Example 6.10.1. Check the energy balance of the tank.

Solution

Equation 10.6.2 is used to calculate hourly temperatures of the tank. For this problem a time step of one hour is sufficient to guarantee stability

$$T_s^+ = T_s + \frac{1}{150 \times 4190} \{4 \times 0.80[S - 8.0 \times 3.6 \times 10^3(T_s - T_a)]^+$$
$$- 1.70 \times 3.6 \times 10^3(T_s - 25)$$
$$- 10 \times 4190(T_s - 15)\}$$

The first term in the bracket is the useful output of the collector, and can have only positive values. The second term is the thermal loss from the tank. The third term is the energy delivered to the load, which is specified to be a fixed mass of water per hour. This equation becomes

$$T_s^+ = T_s + 1.59\{3.20[S - 0.0288(T_s - T_a)]^+$$
$$- 0.00612(T_s - 25) - 0.0419(T_s - 15)\}$$

when S is in MJ/m^2. The data in the first four columns are from Example 6.10.1. T_s^+ is the new tank temperature calculated for the end of the hour. Q_u, loss and load are the three terms in the bracket of the equation, assuming T_s to be fixed

Time, hr	I_T, MJ/m^2	S, MJ/m^2	T_a, °C	T_s^+, °C	Q_u, MJ	Loss, MJ	Load, MJ
Start				40.0			
7–8	0.02	—	−11	38.2	0	0.09	1.05
8–9	0.43	0.34	−8	36.5	0	0.08	0.97
9–10	0.99	0.79	−2	35.0	0	0.07	0.90
10–11	3.92	3.16	2	44.8	7.07	0.06	0.84
11–12	3.36	2.73	3	50.4	4.88	0.12	1.25
12–1	4.01	3.25	6	57.8	6.31	0.16	1.48
1–2	3.84	3.08	7	62.9	5.17	0.20	1.79
2–3	1.96	1.56	8	59.3	0	0.23	2.01
3–4	1.21	0.95	9	56.0	0	0.21	1.86
4–5	0.05	—	7	53.0	0	0.19	1.72
Totals	19.79				23.43	1.41	13.87

for the hour at the initial value. The change in internal energy of the water should equal $\sum Q_U - \sum \text{Losses} - \sum \text{Loads}$

$$150 \times 4190(53.0 - 40.0) \overset{?}{=} (23.43 - 1.41 - 13.87) \times 10^6$$
$$8.17 \text{ MJ} = 8.15 \text{ MJ},$$

a satisfactory check. (The calculations shown in the table were carried out to 0.01 MJ to facilitate checking the energy balance. The result is certainly no better than ± 0.1 MJ.) The day's efficiency is

$$\frac{23.4}{19.8 \times 4} = 0.30, \text{ or } 30\% \qquad \blacksquare$$

The preceding example was simple enough that hand calculation was possible to simulate a few hours of real time. Most problems in solar simulation are not so easy, and we are usually interested in more than just a few hours of simulated data. In general, it is necessary to use a computer to obtain useful solutions.

10.7 INFORMATION FROM SIMULATIONS

In principle, all of the physical parameters for collectors, storage, and other components are the variables that need to be taken into account in the design of solar processes. The number of parameters that must be considered may be quite small, as there is a backlog of experience which indicates that the sensitivity of long-term performance or process economics to many parameters is small. For application to heating of a building for example, the primary system design variable is collector area, with storage capacity and other design variables being of secondary importance provided they are within reasonable bounds of good design practice.* Simulations can provide information on effects of collector area (or other variables) which is essentially impossible to get by other means.

For example, simulations have been done of a solar heating system for a residence in the Madison, Wisconsin, climate. The building has 150 m² floor area, is well insulated, and is to have a solar heating and hot water system to supply part of the heating load. Collector area is the primary design variable. Values of other design parameters were selected to represent good design practice, including collector characteristics and ratio of storage capacity to collector area. Simulations at several collector areas give information on the solar fraction $\mathscr{F}$, the fraction of annual heating and hot water loads carried by solar energy, that is, on the amount of energy usefully delivered to the building from solar energy. This is shown in Figure 10.7.1. A similar curve for water heating only is shown in Figure 10.7.2, which is more nearly linear over a wide range of collector areas.

* These matters are discussed in detail in Chapter 13.

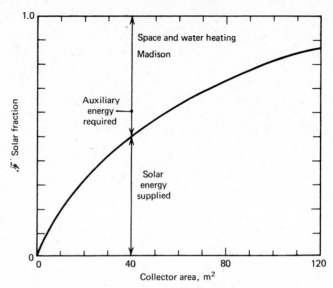

Figure 10.7.1 The thermal performance of a solar space and water heating system, showing the fraction of annual heating and hot water loads carried by solar energy as a function of collector area.

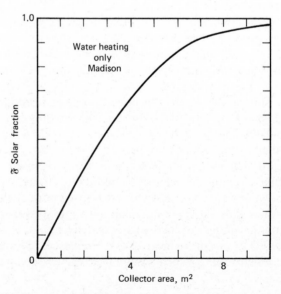

Figure 10.7.2 The thermal performance of a solar water heating system, showing the fraction of the annual hot water loads carried by solar energy as a function of collector area.

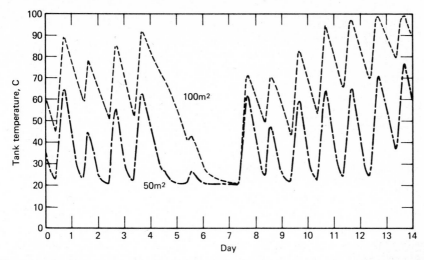

Figure 10.7.3 Tank temperature vs time for a two-week period in March for a heating and hot water system for Madison, WI, assuming a fully mixed tank, for two collector areas and fixed storage to collector area ratio.

Simulations also provide useful information on other aspects of system operation. For example, storage tank temperatures and collector outlet temperatures can be traced, providing information on the range of temperatures to be expected and the needs for energy relief valves to avoid overheating. An example is shown in Figure 10.7.3, which indicates tank temperatures for a selected two week period for two collector areas for a space and water heating system.

Other kinds of information can also be obtained. It is possible to estimate the times at which auxiliary energy is needed, and the relationship of those times to meteorological conditions. It is possible to estimate storage unit losses. In general, any variable that appears in the set of system equations can be investigated.

10.8 SIMULATIONS AND EXPERIMENTS

The extent to which simulations represent the operation of real physical systems depends on the level of detail included in the numerical experiment. Component models can vary in complexity, as can systems. However, there may be factors in system operation which are difficult to simulate, such as leaks in air systems, and operation of real systems may be less ideal than the simulation indicates.

Many component experiments have been carried out, for example, the kind noted in Chapter 7 on flat-plate collectors. A few comparisons of simulations and experiments of building heating have been made of both short term process

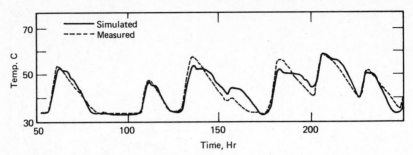

Figure 10.8.1 Comparisons of measured and predicted storage tank temperatures for Colorado State University House I. From Pawelski (1976).

dynamics and long term integrated performance. For example, Pawelski (1976) has used tank temperature as an indicator of process operation, and compared measured and computed tank temperatures over several ten-day periods from CSU House I. An example of one of these comparisons is shown in Figure 10.8.1.

Time-integrated energy quantities have also been compared by Pawelski. Comparisons of the simulated and measured useful energy gain from the collector, and solar and auxiliary contribution are shown in Table 10.8.1. In this study, the simulation was very carefully done to closely represent the physical situation; by doing so, it was possible to obtain agreements that are comparable to the accuracy of the measurements on the physical system.

Table 10.8.1 Comparison of Simulated and Measured Energy Quantities from CSU House I for Three Different Time Periods

	Period	Measured	Simulation	Difference (%)
Collected Solar	I	2388	2577	7.0
Energy (MJ)	II	2419	2292	5.2
	III	2086	2012	3.5
Air Heater Heat	I	2076	2041	1.4
Flow (MJ)	II	3243	3238	0.0
	III	1810	1736	3.6
Delivered Solar	I	—	—	—
Energy (MJ)	II	1952	2025	3.1
	III	1517	1573	2.7
Auxiliary (MJ)	I	0	0	0.0
	II	1291	1213	3.3
	III	238	162	3.7
Preheater Heat	I	398	303	3.8
Flow (MJ)	II	132	116	0.7
	III	204	132	3.5

[a] From Mitchell et al. (1979).

10.9 A SIMULATION PROGRAM

TRNSYS [Klein et al. (1975, 1976, 1979)], is a widely used, modular solar process simulation program. Subroutines are available that represent the components in typical solar energy systems. A list of the components and combinations of components in the TRNSYS library (as of 1979) is shown in Table 10.9.1. By a

Table 10.9.1 Components in the Library of the TRNSYS Solar Process Simulation Program[a]

Flat-plate solar collector
On/off differential controller with hysteresis
Pump or fan
Stratified fluid storage tank
Heat exchanger
On/off auxiliary heater
Absorption air conditioner
Three-stage room thermostat
Data reader (for input of meteorological and other data)
Rock bed thermal storage
Tee piece, flow diverter, and flow mixer
Energy/(degree-hour) space heating
Relief valve
Cyclic time-dependent forcing function
Algebraic operator
Solar radiation data processor
Wall
Roof and attic
Room and basement
Heat pump
Liquid collector-storage subsystem
Air collector-storage subsystem
Domestic hot water subsystem
Quantity integrator
Printer
Plotter
Histogram plotter
Pipe and duct
Simulation summarizer
CPC collector
Cooling coil
Psychrometric data processor
Shading overhang and wingwalls
Window
Collector-storage wall

[a] From Klein et al. (1979).

simple language, the components are "connected" together in a manner analogous to piping, ducting, and wiring in a physical system. The programmer also supplies values for all of the parameters describing the components he wishes to use. The program does the necessary simultaneous solutions of the algebraic and differential equations which represent the components and organizes the input and output. Varying levels of complexity can be used in the calculation. For example, a flat-plate collector can be represented by constant values of U_L and $(\tau\alpha)$, or it can be represented by values of U_L and $(\tau\alpha)$ which are calculated for the conditions that exist through time as the simulation proceeds. The user of the program must determine how detailed he wishes his simulation to be; the more detail the higher the cost. Most of the simulation studies reported in following chapters on specific solar applications were done with TRNSYS.

The integration algorithm selected for TRNSYS is the Modified-Euler method. It is essentially a first-order predictor-corrector algorithm using Euler's method for the predicting step and the trapezoid rule for the correcting step. The advantage of a predictor-corrector integration algorithm for solving simultaneous algebraic and differential equations is that the iterative calculations occurring during a single time step are performed at a constant value of time. (This is not the case for the Runge-Kutta algorithms.) As a result, the solutions to the algebraic equations of the system converge, by successive substitution, as the iteration required to solve the differential equations progresses. The calculation scheme can be described in the following manner.

At time τ, the values of the dependent variables, T^p, are predicted using their values and the values of their derivatives, $(dT/d\tau)_0$, from the previous time step:

$$T^p = T_0 + (\Delta\tau)\left(\frac{dT}{d\tau}\right)_0 \tag{10.9.1}$$

where T^p is the predicted value of all of the dependent variables at time τ (note that this prediction step is exactly the method used to integrate Example 10.3.1); T_0 is the value of the dependent variables at time $(\tau - \Delta\tau)$; $(\Delta\tau)$ is the time step interval at which solutions to the equations of the system model will be obtained; and $(dT/d\tau)_0$ is the value of the derivative of the dependent variables at time $(\tau - \Delta\tau)$.

The predicted values of the dependent variables, T^p, are then used to determine corrected values, T^c, by evaluating their derivatives, $(dT/d\tau)$, as a function of τ, T^p, and the solutions to the algebraic equations of the model.

$$\frac{dT}{d\tau} = f(\tau, T^p, \text{algebraic solutions}) \tag{10.9.2}$$

The corrected values of the dependent variables, T^c, are obtained by applying the trapezoidal rule:

$$T^c = T_0 + \frac{\Delta\tau}{2}\left[\left(\frac{dT}{d\tau}\right)_0 + \frac{dT}{d\tau}\right] \tag{10.9.3}$$

If

$$\frac{2(T^c - T^p)}{(T^c + T^p)} > \varepsilon \tag{10.9.4}$$

where ε is an error tolerance, the T^p is set equal to T^c and Equations 10.9.2 and 10.9.3 are repeated. When the error tolerance is satisfied, the solution for that time step is complete and the whole process is repeated for the next time step.

As an illustration of the use of the general program TRNSYS for simulating a solar energy system, and the nature of the results that can be obtained by simulations, consider the following example, shown schematically in Figure 10.6.1.

Example 10.9.1

A hot water load of 3000 kg water/day at a minimum temperature of 60 C is evenly distributed between the hours of 0700 and 2100. This load is to be met in substantial part by a solar collector assembly of total effective area 65.0 m². The two-cover collector has the following characteristics:

> Tilt $= \beta = 40°$ to south
> $U_L = 4.0$ W/m² C
> $(\tau\alpha) = 0.77$
> $F' = 0.95$
> Flow rate through collector $= \dot{m}_c = 0.903$ kg/s

The tank has the following characteristics:

> $V = 3.9$ m³
> Height/diameter ratio $= 3$
> Loss coefficient $U_L = 0.40$ W/m² C
> Ambient temperature at tank $= 21$ C
> The supply water to the tank $= 15$ C

The auxiliary heater is controlled such that if the temperature of the water from the tank is less than 60 C, it will heat the water from the storage tank temperature to 60 C. If T_s exceeds 60 C, the water delivered to meet the load will be at a temperature above 60 C.

The system shown in the diagram below, is to be operated in Boulder, Colorado, latitude 40°N, for one week in January. Hourly values of solar radiation

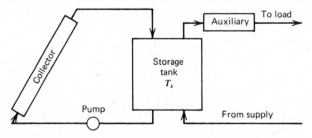

and ambient temperature are as shown in the plots below, and are the same data as given in Table 2.5.2. Assuming that the initial tank temperature at the beginning of the week is 60 C, compute the percentage of the load that is carried by solar energy.

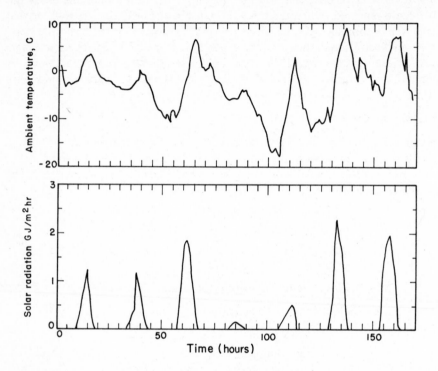

Solution

The solution to this problem was obtained using TRNSYS and a summary of the results are given in the table below. Two solutions are presented, one for an unstratified storage tank and one for a three-section approximation to a stratified storage tank. The load requirement is a fixed amount of water of at least 60 C; the two total loads are slightly different, as the stratified tank delivers water at slightly higher temperature. The minimum total load is 3.960 GJ and both systems slightly exceed this value, since the delivery temperature sometimes exceeds 60 C. The percentage of the load carried by solar energy for the two cases are

$$\text{Case 1:} \frac{2.74}{4.04} = 68\%$$

$$\text{Case 2:} \frac{3.03}{4.10} = 74\%$$

Plots of integrated energy quantities versus time for the week are shown in the figure for the case of the fully mixed tank. The points for the totals at the end of each day are connected by straight lines; smooth curves could have been obtained by using small time intervals. Below is a summary of results from Example 10.6.1. Entries are integrated energy quantities.

Case 1   Unstratified Storage Tank[a]

End of Day	Cumulative Energy			Change of Energy of Tank	Cumulative Energy		
	Incident Solar	Useful Gain	Tank Loss		Supplied from Tank	Supplied from Auxiliary	Total Load
1	0.71	0.21	0.02	−0.27	0.46	0.10	0.56
2	1.34	0.43	0.03	−0.38	0.78	0.35	1.13
3	2.84	1.17	0.04	−0.11	1.24	0.47	1.71
4	2.95	1.17	0.05	−0.45	1.57	0.71	2.28
5	3.34	1.28	0.05	−0.53	1.76	1.00	2.76
6	4.87	2.09	0.07	−0.13	2.15	1.26	3.41
7	6.43	2.81	0.08	−0.01	2.74	1.30	4.04

[a] Energy quantities are in GJ.

Case 2   Three-Section Storage Tank[a]

End of Day	Cumulative Energy			Change of Energy of Tank	Cumulative Energy		
	Incident Solar	Useful Gain	Tank Loss		Supplied from Tank	Supplied from Auxiliary	Total Load
1	0.71	0.24	0.02	−0.30	0.52	0.05	0.57
2	1.34	0.50	0.03	−0.42	0.89	0.25	1.14
3	2.84	1.28	0.04	−0.15	1.39	0.34	1.73
4	2.95	1.28	0.05	−0.58	1.81	0.50	2.31
5	3.34	1.42	0.05	−0.60	1.97	0.90	2.87
6	4.87	2.27	0.06	−0.18	2.39	1.06	3.45
7	6.43	3.05	0.08	−0.06	3.03	1.07	4.10

[a] Energy quantities are in GJ.

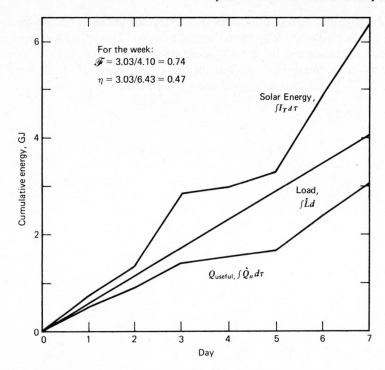

10.10 SUMMARY

In this chapter we have started with component models and shown how they can be combined to represent systems. System thermal performance can be determined by simulations, the solution of the governing equations using detailed weather data as forcing functions. Simulations are shown to compare well with experiments, and are useful in studies of system dynamics and design. For many processes, collector area is the primary design variable, and simulations can give information on fractions of loads carried by solar energy as a function of collector area. This information is used in subsequent chapters dealing with particular applications.

REFERENCES

Anand, D. K., I. N. Dief, and R. W. Allen, Paper 78-WA/Sol-16 presented at San Francisco meeting of ASME (1978). "Stochastic Predictions of Solar Cooling System Performance."

Beckman, W. A., *Solar Energy*, **21**, 531 (1978). "Duct and Pipe Losses in Solar Energy Systems."

Beckman, W. A., S. A. Klein, and J. A. Duffie, *Solar Heating Design by the f-Chart Method*, Wiley, New York (1977).

Buchberg, H. and J. R. Roulet, *Solar Energy*, **12**, 31 (1976). "Simulation and Optimization of Solar Collection and Storage for House Heating."

Butz, L. W., W. A. Beckman, and J. A. Duffie, *Solar Energy*, **16**, 129 (1974). "Simulation of a Solar Heating and Cooling System."

Close, D. J., *Solar Energy*, **11**, 112 (1967). "A Design Approach for Solar Processes."

Close, D. J., and M. B. Yusoff, *Solar Energy*, **20**, 459 (1978). "The Effects of Air Leaks on Solar Air Collector Behaviour."

deWinter, F., *Solar Energy*, **17**, 335 (1975). "Heat Exchanger Penalties in Double-Loop Solar Water Heating Systems."

Gupta, C. L., *Solar Energy*, **13**, 301 (1971). "On Generalizing the Dynamic Performance of Solar Energy Systems."

Gupta, C. L., and H. P. Garg, *Solar Energy*, **12**, 163 (1968). "System Design in Solar Water Heaters with Natural Circulation."

Hall, I. J., R. R. Prairie, H. E. Anderson, and E. C. Boes, *Proceedings of 1978 Annual Meeting, American Section of the ISES*, Denver, **2**(2), 669 (1978). "Generation of a Typical Meteorological Year."

Hamming, R. W., *Numerical Methods for Scientists and Engineers*, McGraw-Hill, New York (1962).

Kays, W. M. and A. L. London, *Compact Heat Exchangers*, McGraw-Hill, New York (1964).

Klein, S. A., Ph.D. Thesis, University of Wisconsin-Madison (1976). "A Design Procedure for Solar Heating Systems."

Klein, S. A., et al., University of Wisconsin-Madison, Engineering Experiment Station Report 38-10 (1979). "TRNSYS—A Transient System Simulation User's Manual."

Klein, S. A. and W. A. Beckman, *ASHRAE Transactions*, **82**, 623 (1976). "TRNSYS—A Transient Simulation Program."

Klein, S. A., P. I. Cooper, T. L. Freeman, D. M. Beekman, W. A. Beckman, and J. A. Duffie, *Solar Energy*, **17**, 29 (1975). "A Method of Simulation of Solar Processes and Its Application."

Kovarik, M. and P. F. Lesse, *Solar Energy*, **18**, 431 (1976). "Optimal Control of Flow in Low Temperature Solar Heat Collectors."

Löf, G. O. G. and R. A. Tybout, *Solar Energy*, **14**, 253 (1973). "Cost of House Heating with Solar Energy."

Mitchell, J. W., W. A. Beckman, and M. J. Pawelski, *Trans ASME, J. Solar Energy Engr*, **102**, 192 (1980). "Comparisons of Measured and Simulated Performance for CSU Solar House I."

Oonk, R. L., W. A. Beckman, and J. A. Duffie, *Solar Energy*, **17**, 21 (1975). "Modeling of the CSU Heating/Cooling System."

Pawelski, M. J., M.S. Thesis in Mechanical Engineering, University of Wisconsin-Madison (1976). "Development of Transfer Function Load Models and Their Use in Modeling the CSU House I."

Sheridan, N. R., K. J. Bullock, and J. A. Duffie, *Solar Energy*, **11**, 69 (1967). "Study of Solar Processes by Analog Computer."

Winn, C. B. and D. E. Hull, *Proceedings of the 1978 Annual Meeting of the American Section of ISES, Denver*, **2**(2), 493 (1978). "Optimal Controllers of the Second Kind."

CHAPTER 11

Solar Process Economics

In the first 10 chapters, we have discussed in some detail the thermal performance of components and systems and showed how the long-term thermal performance can be estimated in terms of the design parameters of the components. The value of a solar process ultimately must be assessed in economic terms; given the performance, we need methods for making economic evaluations.

Solar processes are generally characterized by high first cost and low operating costs. Thus the basic economic problem is one of comparing an initial known investment with estimated future operating costs. Most solar energy processes require an auxiliary (i.e., conventional) energy source so that the system includes both solar and conventional equipment and the annual loads are met by a combination of the sources. In essence, solar energy equipment is bought today to reduce tomorrow's fuel bill.

The cost of any energy delivery process includes all of the items of hardware and labor that are involved in installing the equipment, plus the operating expenses. Factors which may need to be taken into account include interest on money borrowed, property and income taxes, resale of equipment, maintenance, insurance, fuel and other operating expenses. The objective of the economic analysis can be viewed as the determination of the least cost method of meeting the energy need, considering both solar and non solar alternatives. For solar energy processes, the problem is to determine the size of the solar energy system that gives the lowest cost combination of solar and auxiliary energy.

In this chapter we outline the life cycle savings method of doing economic analyses. This method takes into account the time value of money and allows detailed consideration of the complete range of costs. It is introduced by an outline of cost considerations, note of economic figures of merit (design criteria) and comments on design variables which are important in determining system economics. For additional discussion of economic analyses, see Riggs (1968), De Garmo and Canada (1973), Ruegg (1975), and White et al. (1977).

Section 11.8 of the chapter describes the P_1, P_2 method of doing life cycle savings analyses. This is a quick and convenient way of carrying out the computations described in detail in earlier sections, and is the method used in economic analyses of particular processes in following chapters.

376

11.1 COSTS OF SOLAR PROCESS SYSTEMS

Investments in buying and installing solar energy equipment are important factors in solar process economics. These include the delivered price of equipment such as collectors, storage unit, pumps and blowers, controls, pipes and ducts, heat exchangers and all other equipment associated with the solar installation. Costs of installing this equipment must also be considered, as these can match or exceed the purchase price. Also to be included are costs of structures to support collectors and other alterations made necessary by the solar energy equipment. Under some circumstances credits may be taken for the solar process if its installation results in reduction of costs; for example, a collector may serve as part of the weatherproof envelope of building eliminating the need for some of the conventional siding or roofing.

Installed costs of solar equipment can be shown as the sum of two terms, one proportional to collector area and the other independent of collector area:

$$C_S = C_A A_C + C_E \tag{11.1.1}$$

where C_S = total cost of installed solar energy equipment ($)
$\quad C_A$ = total area dependent costs, ($/m^2)
$\quad A_C$ = collector area (m^2)
$\quad C_E$ = total cost of equipment which is independent of collector area ($)

The area dependent costs, C_A, include such items as the purchase and installation of the collector, and a portion of storage costs. The area-independent costs, C_E, include such items as controls and bringing construction or erection equipment to the site, which do not depend on collector area.

Operating costs are associated with a solar process. These continuing costs include: cost of auxiliary energy, energy costs for operation of pumps and blowers (this energy is often termed *parasitic* energy, and should be minimized by careful design); extra insurance costs on the solar equipment; maintenance costs (which should be low for a well-constructed system); extra real estate taxes imposed on the basis of additional assessed value of a building or facility; interest charges on any funds borrowed to purchase the equipment; and others.

There may be *income tax* implications in the purchase of solar equipment. In the United States, interest paid on a mortgage* for its purchase and extra property tax on an increased assessment due to solar equipment are both deductible from income for tax purposes, if the owner itemizes his deductions. States may allow similar deductions. The income tax reduction associated with these payments depends on the tax bracket of the owner and serves to reduce the cost of the solar process.†

---

* In this chapter we use the term "mortgage," commonly applied to funds borrowed for building projects, for any loan for purchase of solar energy equipment.

† This chapter is written with United States tax law in mind. For other countries, treatment of tax implications will have to be modified.

Equipment purchased by businesses have other tax implications. Income-producing property and equipment may be depreciated, resulting in reduced taxable income and thus reduced income tax. But, the value of fuel saved by the use of solar equipment is effectively reduced because a business already deducts the cost of fuel from its income for tax purposes. If the equipment is for purposes other than building heating or air conditioning, there may be investment tax credits available in the first year. Further, the equipment may have salvage or resale value which may result in a capital gains tax. Finally, federal and state governments may allow special tax credits to encourage the use of solar energy. (Federal and state tax laws relating to solar energy are often changed and current law should be used in any economic analysis.)

In equation form, the annual costs for both solar and nonsolar systems to meet an energy need can be expressed as:

$$
\begin{aligned}
\text{Yearly cost} = \ &\text{Mortgage payment} + \text{Fuel expense} \\
&+ \text{Maintenance and insurance} \\
&+ \text{Parasitic energy cost} \\
&+ \text{Property taxes} - \text{Income tax savings} \qquad (11.1.2)
\end{aligned}
$$

The mortgage payment includes interest and principal payment on funds borrowed to install the system. Fuel expense is for energy purchase for auxiliary or for the conventional (nonsolar) system. Maintenance and insurance are recurring costs to keep a system in operating condition and protected against fire or other losses. Parasitic energy costs are for blowing air or pumping liquids and other electrical or mechanical energy uses in a system. Property taxes are levied on many installations. Income tax savings for a nonincome-producing system (such as a home heating system) can be expressed as:

$$
\frac{\text{Income tax}}{\text{savings}} = \frac{\text{Effective}}{\text{tax rate}} \times (\text{Interest payment} + \text{Property tax}) \quad (11.1.3)
$$

If the system is an income-producing installation;

$$
\frac{\text{Income tax}}{\text{savings}} = \frac{\text{Effective}}{\text{tax rate}} \times
\left(
\begin{aligned}
&\text{Interest payments} \\
&+ \text{Property tax} \\
&+ \text{Fuel expense} \\
&+ \text{Maintenance and insurance} \\
&+ \text{Parasitic energy cost} \\
&+ \text{Depreciation}
\end{aligned}
\right) \quad (11.1.4)
$$

State income taxes are deductible from income for federal tax purposes. Where federal taxes are not deductible for state tax purposes, the effective tax rate is given by:

$$
\frac{\text{Effective}}{\text{tax rate}} = \frac{\text{Federal}}{\text{tax rate}} + \frac{\text{State}}{\text{tax rate}} - \left(\frac{\text{Federal}}{\text{tax rate}} \times \frac{\text{State}}{\text{tax rate}} \right) \quad (11.1.5)
$$

The concept of *solar savings*, as outlined by Beckman et al. (1977), is a useful one. Solar savings are the difference between the cost of a conventional system and a solar system. (Savings can be negative; they are then losses.) In equation form it is simply:

$$\text{Solar savings} = \text{Costs of conventional energy} - \text{Costs of solar energy} \tag{11.1.6}$$

In this equation it is not necessary to evaluate costs that are common to both the solar and the nonsolar system. For example, the auxiliary furnace and much of the ductwork or plumbing in a solar heating system are often the same as would be installed in a nonsolar system. With the savings concept, it is only necessary to estimate the incremental cost of installing a solar system. If the furnaces or other equipment in the two systems are different, the difference in their costs can be included as an increment or decrement to the cost of installing a solar system. Solar savings can be written:

$$
\begin{aligned}
\text{Solar savings} = \text{Fuel savings} \\
& - \text{Incremental mortgage payment} \\
& - \text{Incremental insurance and maintenance} \\
& - \text{Incremental parasitic energy cost} \\
& - \text{Incremental property tax} \\
& + \text{Income tax savings}
\end{aligned} \tag{11.1.7}
$$

The significance of the terms is the same as for Equation 11.1.2, except that here they refer to the increments in the various costs, that is, the differences between the costs for the solar energy system compared to a nonsolar system. Equations analogous to 11.1.3 and 11.1.4 can be written for the last term in Equation 11.1.7. They are, for a nonincome-producing system:

$$\text{Income tax savings} = \text{Effective tax rate} \times \left(\text{Incremental interest payment} + \text{Incremental property tax} \right) \tag{11.1.8}$$

and for an income producing system

$$\text{Income tax savings} = \text{Effective tax rate} \times \left(
\begin{aligned}
& \text{Incremental interest payment} \\
& + \text{Incremental property tax} \\
& + \text{Incremental maintenance and insurance} \\
& + \text{Incremental parasitic energy cost} \\
& + \text{Incremental depreciation} \\
& - \text{Value of fuel saved}
\end{aligned}
\right) \tag{11.1.9}$$

Fuel saved is a negative tax deduction since a business already deducts fuel expenses; the value of fuel saved is effectively taxable income.

11.2 DESIGN VARIABLES

The economic problem in solar process design is to find the lowest cost system. In principle, the problem is a multivariable one, with all of the components in the system and the system configuration having some effect on thermal performance and thus on costs. The design of the load system (the building, the industrial process using energy, or other load) must also be considered in the search for optimum design. Barley (1978) has investigated the economic trade-off between building insulation and solar heating. In practice, the problem often resolves to a simpler one of determining the size of a solar energy system for a known load, with storage capacity and other parameters fixed in relationship to collector area. Given a load that is some function of time through a year, a type of collector and a system configuration, the primary design variable is collector area. System performance is much more sensitive to collector area than to any other variable.

Three examples of the dependence of annual thermal performance on collector area are shown in Figure 11.2.1, for a solar heating operation. Curve A is for a system with a two-cover, selective-surface collector, while B is for a one-cover, nonselective collector. Curve C is for double the storage capacity but the same collector as for Curve A. Figure 11.2.2 shows the dependence of annual thermal performance on storage capacity for 60 m^2 of type A collectors.

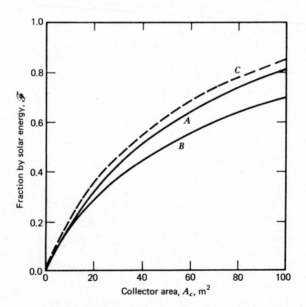

Figure 11.2.1   An example of annual fraction of heating loads carried by solar energy, for a building in Madison, WI. Curve A is for a system with a two-cover, selective-surface collector, B is for one with a one-cover nonselective collector, and C is the same system as A but with twice the storage capacity.

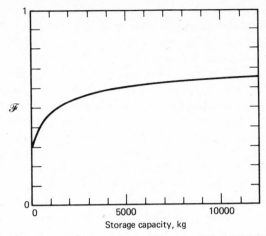

Figure 11.2.2 Annul solar fraction for the collector of type A of Figure 11.2.1, for a collector area of $60\,m^2$, as a function of storage capacity.

For this solar heating example, relative sensitivity of annual performance to collector area and relative insensitivity to collector type are apparent. The solar contribution is relatively insensitive to storage capacity (within the range shown in Figure 11.2.2) once a critical storage capacity is exceeded. Similar results are found for solar heating for a wide range of North American locations. It is difficult to generalize this experience with solar heating systems to other applications. The general procedure for determining which variables are most critical to thermal performance is a sensitivity analysis, such as that of Close (1967). The economic analysis for solar heating to meet a given load is simplified by the fact that collector area is the primary design parameter; economic analyses for other solar processes must take into account the possibility that other design variables might be of comparable importance.

11.3 ECONOMIC FIGURES OF MERIT

Several economic criteria have been proposed and used for evaluating and optimizing solar energy systems. This section outlines some of these figures of merit; two of them are discussed in more detail in following sections, and one of these, maximum life cycle savings, is applied in later chapters on solar energy applications.

Least-cost solar energy is a reasonable figure of merit for systems in which solar energy is the only energy resource. The system yielding least cost can be defined as that showing minimum owning and operating cost over the life of the system, considering solar energy only. However, the optimum design of a combined solar plus auxiliary energy system based on minimum total cost of delivering energy will generally be different from that based on least-cost solar energy, and the use of this concept is not recommended for systems using solar in combination with other energy sources.

Life cycle cost (*LCC*) is the sum of all the costs associated with an energy delivery system over its lifetime or over a selected period of analysis, in today's dollars, and takes into account the time value of money. The basic idea of life cycle costs is that anticipated future costs are brought back to present cost (discounted) by calculating how much would have to be invested at a market discount rate* to have the funds available when they will be needed. A life cycle cost analysis includes inflation when estimating future expenses. This method can include only major cost items or as many details as may be significant. *Life cycle savings* (*LCS*) or *net present worth* is defined as the difference between the life cycle costs of a conventional fuel-only system and the life cycle cost of the solar plus auxiliary energy system. Life cycle savings analysis is outlined in Sections 11.6 and 11.8 and is applied in later chapters to solar processes.

A special case of life cycle savings is based on cash flow, that is, the sum of the items on the right-hand side of Equation 11.1.2. Cash flow may be an important consideration in residential solar heating applications where the willingness of lending institutions to provide mortgage funds may be dependent on a borrower's ability to meet periodic obligations. The main impacts of solar heating will be increased mortgage payments and decreased fuel costs.

Annualized life cycle cost (*ALCC*) is the average yearly outflow of money (cash flow). The actual flow varies with year, but the sum over the period of an economic analysis can be converted to a series of equal payments in today's dollars that are equivalent to the varying series. The same ideas apply to an *annualized life cycle savings* (*ALCS*).

Payback time is defined in many ways. Below are listed several which may be encountered; these are illustrated in Figure 11.3.1.

A The time needed for the yearly cash flow to become positive.
B The time needed for the cumulative fuel savings to equal the total initial investment, that is, how long it takes to get an investment back by savings in fuel. The common way to calculate this payback time is without discounting the fuel savings. It can also be calculated using discounted fuel savings.
C The time needed for the cumulative savings to reach zero.
D The time needed for the cumulative savings to equal the down payment on the solar energy system.
E The time needed for the cumulative solar savings to equal the remaining debt principal on the solar energy system.

The most common definition of payback time is **B**, while **D** is least often encountered. Each of these payback times may have significance in view of economic objectives of various solar process users. Calculation of payback periods can be done including only major items or including many details.

* Market discount rate is the rate of return on the best alternative investment. See Section 11.4.

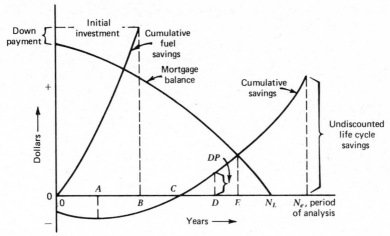

Figure 11.3.1 Changes in mortgage balance, cumulative fuel savings, and cumulative savings as a function of time through the period of a life-cycle cost analysis.

Care must be used in interpreting reported payback periods that the definition of the period and the items included in it are fully understood. Calculation of payback times is described in Section 11.8.

Return on investment (ROI) is the market discount rate which results in zero life cycle savings, that is, the discount rate that makes the present worth of solar and nonsolar alternatives equal. This is illustrated in Figure 11.3.2, which shows an example of life cycle savings as a function of market discount rate.

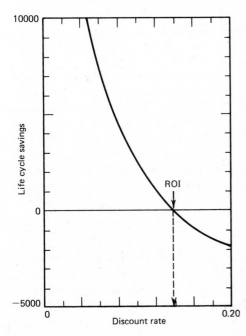

Figure 11.3.2 An example of life cycle savings as a function of market discount rate for a solar energy application.

11.4 DISCOUNTING OF FUTURE COSTS: INFLATION

The most complete approach to solar process economics is to use life cycle cost methods that take into account all future expenses. This method provides a means of comparison of future costs with today's costs. This is done by discounting all anticipated costs to the common basis of present worth (or present value), that is, what would have to be invested today, at the best alternative investment rate, to have the funds available in the future to meet all of the anticipated expenses.

Conceptually, in a life cycle cost analysis, all anticipated costs are tabulated and discounted to a present worth, and the life cycle cost is the sum of all of these present worths. As a practical matter, the calculations can be simplified. For example, the cash flow (net payment) for each year can be calculated and the life cycle cost found by discounting each annual cash flow to its present value and finding the sum of these discounted cash flows. When the present values of all future costs have been determined for each of the alternative systems under consideration, including solar and nonsolar options, the system that yields the lowest life cycle cost is selected as the most cost effective. Life cycle costing requires that all costs be projected into the future; the results obtained from analyses of this type usually depend very much on predictions of future costs.

The reason that cash flow must be discounted lies in the "time value of money." The relationship for determining the present worth of one dollar needed N periods (usually years) in the future, with a market discount rate of d (fraction per time period) is

$$PW = \frac{1}{(1 + d)^N} \tag{11.4.1}$$

Thus an expense that is anticipated to be $1.00 in five years is equivalent to an obligation of $0.681 today at a market discount rate of 8 percent. To have $1000 available in 5 years it would be necessary to make an investment of $681 today at an annual rate of return of 8 percent.

Many recurring costs can be assumed to inflate (or deflate) at a fixed percentage each period. Thus an expense of $1.00, when inflated at a rate i, will be $(1 + i)$ at the end of a time period, $(1 + i)^2$ at the end of an additional time period, etc. If a cost A is considered to be incurred as of the end of the first time period (e.g., a fuel bill is to be paid at the end of the month or year) that recurring cost at the Nth period is

$$C_N = A(1 + i)^{N-1} \tag{11.4.2}$$

Thus a cost which is $1.00 at the end of the first period and inflates at 6 percent per year will be $(1 + 0.06)^4$ or 1.26 at the end of five periods.

The progression of a series of payments which are expected to inflate at a rate i is shown in Figure 11.4.1. The first payment in the series is A, the second is

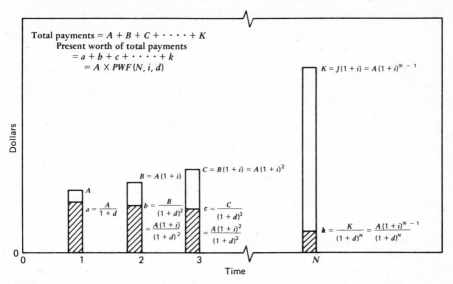

Figure 11.4.1   Present worth of a geometric series of inflating payments. Each payment is made at the end of a period. The bars show both the payment and the present worth of the payment.

B and the Nth is K. The payments are made at the ends of the periods. The shaded portions of the bars show the present worth of the anticipated payment. At the Nth period the cost is $A(1 + i)^{N-1}$, and the present worth of the cost is

$$PW_N = \frac{A(1 + i)^{N-1}}{(1 + d)^N} \tag{11.4.3}$$

This equation is useful for calculating the present worth of any one payment of a series of inflating payments. It is also useful for calculating the present worth of any one-time costs that are anticipated during the period of the analysis, and for which the present cost is known.

Example 11.4.1

(a) A fuel cost is expected to inflate at the rate of 7 percent per year, and for the first year is $400. The market discount rate is 10 percent per year. What is the present worth of the payment to be made at the end of the third year? (b) It is expected that the blower in an air heater system will need to be replaced at the end of 10 years. The cost in the first year is $300. Under the same assumptions of inflation and discount rates, what is the present worth of replacing the blower?

Solution

(a) Using Equation 11.4.3, the present worth of the third year fuel payment is

$$PW_3 = 400(1.07)^2/(1.10)^3 = \$344$$

(b) The present worth of the blower replacement after 10 years is

$$PW = 300(1.07)^9/(1.10)^{10} = \$213.$$ ■

The analyses and examples in this chapter are all based on the premise that the costs are known and payable at the end of the first time period. If a cost A' is known as of the beginning of the first time period, Equation 11.4.2 becomes

$$C_N = A'(1 + i)^N \qquad\qquad (11.4.4)$$

and Equation 11.4.3 becomes

$$PW_N = A'\left(\frac{1 + i}{1 + d}\right)^N \qquad\qquad (11.4.5)$$

11.5 PRESENT WORTH FACTOR

In the previous section we dealt with calculating the present worth of a single future payment. If an obligation recurs each year and inflates at a rate i per period, a present worth factor, PWF, of N such payments can be found by summing Equation 11.4.3 over N periods.

$$PWF(N, i, d) = \sum_{j=1}^{N} \frac{(1 + i)^{j-1}}{(1 + d)^j} = \begin{cases} \dfrac{1}{(d - i)}\left(1 - \left(\dfrac{1 + i}{1 + d}\right)^N\right) & \text{if } i \neq d \\[2ex] N/(1 + i) & \text{if } i = d \end{cases}$$

$$(11.5.1)$$

If the function of Equation 11.5.1 is multiplied by the first value of a series of payments that are made at the end of the periods, the result is the sum of N such payments, discounted to the present with a discount rate of d. This is represented by the sum of the cross-hatched bars on Figure 11.4.1. This function is tabulated in Appendix F for increments of 5 in N, from 5 to 30, for values of i from 0 to 12 percent, and for values of d from 0 to 20 percent.

Example 11.5.1

What is the present worth of a series of 20 yearly payments, the first of which is $500, which are expected to inflate at the rate of 8 percent per year, if the market discount rate is 10 percent per year?

Solution

From Equation 11.5.1

$$PW = 500 \frac{1}{0.10 - 0.08}\left[1 - \left(\frac{1.08}{1.10}\right)^{20}\right] = 500 \times 15.35892 = 7679$$

or, from Appendix F, the present worth factor for $N = 20$, $i = 0.08$, and $d = 0.10$ is 15.359, and the present worth of the series is $15.359 \times 500 = \$7680$. ■

The present worth factor defined by Equation 11.5.1 can be used to find the periodic loan payment on a mortgage which involves a series of N_L fixed payments over the lifetime of the loan. Since all mortgage payments are equal, we have a series of payments with an inflation rate of zero. The discount rate in Equation 11.5.1 becomes the mortgage interest rate. Thus the periodic loan payment* is

$$\text{Periodic payment} = \frac{M}{PWF(N_L, 0, m)} \qquad (11.5.2)$$

where m is the mortgage interest rate, N_L is the period of the mortgage, and M is the mortgage principal.

Example 11.5.2

What is the annual payment and yearly interest charge if an \$11,000 solar installation is to be financed by a 10 percent down payment with the balance borrowed at an annual interest rate of 9% for 20 years? The payments are to be made at the end of each year. The market discount rate is 8%. What is the present worth of the series of interest payments?

Solution

The present worth factor is used in Equation 11.5.2 to calculate the annual payment. The present worth of the sum of all payments is the mortgage, or $0.9 \times 11,000 = \$9900$. The yearly payment is

$$\frac{9900}{PWF(20, 0, 0.09)} = \frac{9900}{9.129} = \$1084.46$$

The interest charge varies with time, since the mortgage payment includes a principal payment and interest. In this example, the interest for the first year is $0.09 \times 9900 = 891$. The payment is \$1084.46, so the principal is reduced by \$193.46 to 9706.54. The second year's interest is $0.09 \times 9706.54 = \$873.59$, the principal is reduced by \$210.87 to \$9495.67. This progression of payments, remaining principal, interest payment, and the present worth of the interest payment is shown in the table overleaf (to the nearest dollar).

* In this and following sections the examples are based on periods of a year. The identical principles and methods hold for monthly (or other) periods. Interest, inflation and discount rates must correspond to the time period used.

Year	Mortgage Payment	Remaining Principal	Interest Payment	PW of Interest Payment
1	1084	9707	891	825
2	1084	9496	874	749
3	1084	9266	855	678
4	1084	9015	834	613
–	–	–	–	–
–	–	–	–	–
–	–	–	–	–
20	1084	0	90	19
	Total present worth of interest payments			$6730

The total present worth of all of the mortgage interest payments can be calculated as in Example 11.5.2, but it is tedious. Present worth factors can be used to obtain this total by the following equation:

$$PW_{int} = M\left[\frac{PWF(N_{min}, 0, d)}{PWF(N_L, 0, m)} + PWF(N_{min}, m, d)\left(m - \frac{1}{PWF(N_L, 0, m)}\right)\right]$$

(11.5.3)

where M is the initial mortgage principal, N_L is the term of mortgage, N_e is the term of economic analysis, N_{min} is the lesser of N_L or N_e, m is mortgage interest rate, and other terms are as previously defined. Note that it is not necessary that the term of the economic analysis coincide with the term of the mortgage.

Example 11.5.3

Calculate the total present worth of all of the mortgage interest payments in Example 11.5.2, over the term of the mortgage.

Solution

Here $N_L = N_e = 20$, $m = 0.09$, and $d = 0.08$. The PWF values can be calculated from Equation 11.5.1, or they can be obtained from Appendix F. Using Equation 11.5.3:

$$PW_{int} = 9900\left[\frac{9.818}{9.129} + 20.242\left(0.09 - \frac{1}{9.129}\right)\right] = \$6731$$

This is in agreement with the value obtained by summing the individual present worths in Example 11.5.2.

11.6 LIFE CYCLE SAVINGS METHOD

The previous two sections dealt with discounting of future costs. Now we apply these ideas in a series of examples to illustrate the principles and steps in life cycle savings analysis. The first two examples are for fuel payments for a conventional process. The next is a calculation of solar savings, the difference in present worth of a solar plus fuel design compared to the fuel only process. The last is a calculation of optimum system design based on solar savings calculations for several collector areas.*

These examples are intended to illustrate the method and the particular costs used are not intended to have significance. These costs vary widely from one location to another, with time, and as state and federal legislation is enacted which impacts the costs of solar equipment.

Example 11.6.1

For a nonsolar process using fuel only, what is the present worth of the fuel cost over 20 years if the first year's cost is $1255 (i.e., 125.5 GJ at $10.00/GJ), the market discount rate is 8 percent per year and the fuel cost inflation rate is 10 percent per year?

Solution

A tabulation of the yearly progression of fuel costs and their present worth is shown below. Each year's fuel cost is the previous year's cost multiplied by $(1 + i)$. Each item in the PW column is calculated from column 2 using Equation 11.4.1.

Year	Fuel Costs	PW of Fuel Cost
1	1255	1162
2	1381	1183
3	1519	1206
–	–	–
–	–	–
–	–	–
20	7675	1647

Total Present Worth of Fuel Costs 27822

The same result is obtained by multiplying $PWF(20, 0.10, 0.08)$ times the first year fuel cost. ∎

---

* Section 11.8 shows an additional and more convenient way of approaching these same calculations.

The next example also shows a calculation of the present worth of a series of fuel costs, but in this case the costs of fuel are expected to inflate at rates dependent on time. It illustrates the general principle that costs need not be expected to change at fixed rates. The variation can be anything from nil to completely irregular; the basic ideas of life cycle cost hold whether payments are regular or otherwise.

Example 11.6.2

For a nonsolar process, using fuel only, what is the present worth of fuel cost over 8 years if the first year's cost is $1200, it inflates at 15 percent per year for 3 years and then at 10 percent per year. The market discount rate is 8 percent per year.

Solution

A tabulation of the yearly progression of fuel costs and their present worth is given below. Each item in the present worth column is calculated from the cost column using Equation 11.4.1.

Year	Fuel Cost	PW of Fuel Cost
1	1200	1111
2	1380	1183
3	1587	1260
4	1825	1341
5	2008	1366
6	2208	1391
7	2429	1417
8	2672	1444
	$15,309	$10,513

Note that in this example, the sum of the expected payments is $15,309, about 50% more than the present worth of the payments. ∎

Alternate Solution

This example can also be solved using present worth factors. Part of the present worth is a series of four payments, the first $1200 and the next three inflating at 15 percent. From Equation 11.5.1

$$PWF(4, 0.15, 0.08) = \frac{1}{0.08 - 0.15}\left[1 - \left(\frac{1.15}{1.08}\right)^4\right]$$
$$= 4.080$$

So, the present worth of this part is $1200 \times 4.080 = \$4896$. The second part is a series starting 5 years in the future, with i of 10 percent. The PWF for this series as of the beginning of year 5 is

$$PWF(4, 0.10, 0.08) = \frac{1}{0.08 - 0.10}\left[1 - \left(\frac{1.10}{1.08}\right)^4\right]$$

$$= 3.808$$

This PWF, when multiplied by A', the 5th year payment, gives the present worth of that series as of the beginning of the 5th year. To find A', the original cost is inflated three times by 1.15 and once by 1.10

$$A' = 1200(1.15)^3(1.10) = \$2008$$

The second series is then discounted to the present by

$$PW = \frac{3.808 \times 2008}{(1.08)^4} = 5620$$

The sum of the present worths of the two series is then $5620 + 4896 = \$10,516$.
■

With either Equation 11.4.1 or 11.5.1 it is possible to discount any future cost or series of costs to a present worth. In the same way, future savings (negative costs) can be discounted to a present worth. In the following examples we apply these methods to systems using combined solar and auxiliary (conventional) sources.

Example 11.6.3

A combined solar and fuel system to meet the same energy need as in Example 11.6.1 is to be considered. The proposed collector and associated equipment will supply energy so as to reduce fuel purchase by 56 percent, will cost $11,000, and is to be 90 percent financed over 20 years at an interest rate of 9 percent. The first year's fuel cost for a system without solar would be $1255. Fuel costs are expected to rise at 10 percent per year. It is expected that the equipment will have a resale value at the end of 20 years of 40 percent of the original cost.

In the first year, extra insurance, maintenance, and parasitic energy costs are estimated to be $110; extra property tax is estimated to be $220. These are expected to rise at a general inflation rate of 6 percent per year. Extra property taxes and interest on mortgage are deductible from income for tax purposes; the effective income tax rate is expected to be 45 percent through the period of the analysis.

What is the present worth of solar savings for this process over a 20 year period if the market discount rate is 8 percent?

Solution

The table below shows the incremental yearly costs and savings. Year 1 includes the estimates of first year extra costs as outlined in the problem statement. The annual payment on the $9900 mortgage is calculated as $9900/PWF(20, 0, 0.09)$ or 1084.46. The entries in the tax savings column are calculated by Equation 11.1.3; for example, in year 1, from the interest and property taxes paid in that year:

Interest in year $1 = 0.09(9900) = 891$
Principal payment $= 1084.46 - 891 = 193.46,$
Principal balance $= 9900 - 193.46 = 9706.54$
Tax savings $= 0.45(891 + 220) = 500$
Interest in year $2 = 0.09(9706.54) = 873.59$
Tax savings $= 0.45(873.59 + 233) = 498$

Solar savings for each year are the sum of columns 2 to 6. Each year's solar savings is brought to a present worth using the market discount rate of 8 percent. The down payment is 1100; this is a negative present worth of solar savings. The resale value of $4400 in year 20 is shown as a second entry in year 20, and is positive as it contributes to savings. The sum of the last column, $4206, is the total present worth of the gains from the solar energy system compared to the fuel only system, and is termed "life cycle solar savings" or simply "solar savings."

Year	Fuel Savings	Extra Mort. Pt.	Extra Ins Maint, Energy	Extra Prop. Tax	Income Tax Savings	Solar Savings	PW of Solar Savings
0						−1100	−1100
1	703	−1084	−110	−220	500	−211	−195
2	773	−1084	−117	−233	498	−163	−140
3	850	−1084	−124	−247	496	−109	−87
4	935	−1084	−131	−262	493	−49	−36
5	1029	−1084	−139	−278	490	18	12
–	–	–	–	–	–	–	–
–	–	–	–	–	–	–	–
20	4298	−1084	−333	−666	340	2555	548
20						4400	944

Total Present Worth of Solar Savings = $4203

∎

Another approach to this problem is to do a life cycle cost analysis of both the solar and the nonsolar systems. The difference in the present worths of the two systems (that is, the difference in the life cycle costs) is the life cycle solar savings.

More information is needed to do the analyses separately if they are to be complete, as equipment common to both systems should be included. In the calculation of Example 11.6.3 common costs do not influence life cycle solar savings.

In the previous examples, collector area (i.e., system cost) and annual fraction of loads met by solar were given. In Example 11.6.4, the relationship between solar fraction $\mathscr{F}$ and the area from thermal performance calculations is given, as are C_A and C_E, and the life cycle solar savings are calculated for various sizes of solar energy systems to find the collector size (combination of solar and auxiliary) which provides the highest savings.

Example 11.6.4

A thermal analysis of the process of Example 11.6.3 indicates a relationship between collector area and solar fraction $\mathscr{F}$ as indicated in the first two columns of the table below. Area dependent costs are $200/m^2$, and fixed cost is $1000. All economic parameters are as in Example 11.6.3. What is the optimum collector area which shows the maximum life cycle solar savings relative to the fuel only system?

Solution

The costs of the solar energy system are calculated by Equation 11.1.1, with $C_A = \$200/m^2$ and $C_E = \$1000$. The third column indicates the total system cost. Calculations of the kind shown in Example 11.6.3 give column 4, life cycle solar savings as a function of the collector area. A calculation is made for a very small collector area, where the cost of the system is essentially C_E, to establish the "zero-area" life cycle savings. The solar savings of column 4 are plotted in the figure. The maximum savings are realized at a collector area of about 39 m^2, and positive savings are realized over an area range of approximately 3 to 110 m^2.

Area, m^2	Solar Fraction, $\mathscr{F}$	Installed Cost	Solar Savings
0.01	0	1000	-1036
25	0.37	6000	4088
39	0.49	8800	4531
50	0.56	11000	4203
75	0.71	21000	3204
150	0.92	31000	-6468

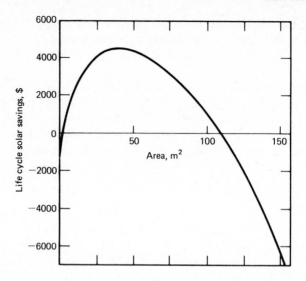

These examples show only regularly varying costs. All anticipated costs can be included, whether they are recurring, regularly varying, or however they may be incurred. The method, in essence, is to construct a table such as that in Example 11.6.3 and sum the last column. Nonrecurring items such as expected replacement covers in years hence, changes in income tax bracket, or other increments can be included in the present worth calculation, as were the down payment and resale value in the examples. Example 11.6.3 shows expenses of types associated with nonincome-producing applications. Income-producing applications require that additional terms be taken into account, for example, depreciation.

11.7 EVALUATION OF OTHER ECONOMIC INDICATORS

The principles of the preceding sections can be used to find payback times. Payback time B (Section 11.3) is the time needed for the cumulative fuel savings to equal the total initial investment in the system. Consider first the case where fuel savings are not discounted. The fuel savings in the jth year are given by $\mathscr{F}LC_{F1}(1 + i_F)^{j-1}$, where $\mathscr{F}L$ is the energy saved, C_{F1} is the first period's unit energy cost delivered from fuel,* and i_F is the fuel cost inflation rate. Summing these over N_p, the payback time and equating to the initial investment as given by Equation 11.1.1

$$\sum_{j=1}^{N_p} \mathscr{F}LC_{F1}(1 + i_F)^{j-1} = C_S \qquad (11.7.1)$$

---

* The energy delivered from a unit of fuel is the product of the fuel lower heating value times the furnace efficiency.

Summing the geometric series

$$\frac{\mathscr{F}LC_{F1}[(1 + i_F)^{N_P} - 1]}{i_F} = C_S \qquad (11.7.2)$$

This can be solved for N_p, the payback period

$$N_p = \frac{\ln\left[\dfrac{C_S i_F}{\mathscr{F}LC_{F1}} + 1\right]}{\ln(1 + i_F)} \qquad (11.7.3)$$

The present worth factors of Appendix F can also be used to find this payback period. The sum of the fuel savings is the first year's saving, $\mathscr{F}LC_{F1}$, times the PWF at zero discount rate:

$$\mathscr{F}LC_{F1} \times PWF(N_p, i_F, 0) = C_S \qquad (11.7.4)$$

By interpolating from the tables in Appendix F, the value of N_p can be found for which the PWF is $C_S/\mathscr{F}LC_{F1}$.

Example 11.7.1

What is the undiscounted payback time B for an $11,000 investment in solar energy equipment which meets 56% of an annual load of 156 GJ? The first year fuel cost is 8.00/GJ and is expected to inflate at 10% per year.

Solution

Using Equation 11.7.3

$$N_p = \frac{\ln\left[\dfrac{11,000 \times 0.1}{0.56 \times 156 \times 8} + 1\right]}{\ln 1.10} = 9.9 \text{ years}$$

Or, using present worth factors

$$\frac{C_S}{\mathscr{F}LC_F} = \frac{11,000}{0.56 \times 156 \times 8} = 15.74$$

At a discount rate of 0 percent and an inflation rate of 10 percent, interpolating between Tables F1 and F2 of Appendix F, N_p is nearly 10 years. ■

By similar procedures it is possible to equate discounted fuel costs to initial investment. If fuel costs are discounted, an equation like 11.7.4 can be written

$$PWF(N_p, i_F, d) = \frac{C_S}{\mathscr{F}LC_{F1}} \qquad (11.7.5)$$

and the appropriate value of N_p can be found by interpolating from the tables.

Example 11.7.2

Repeat Example 11.7.1 but discount future fuel costs at a rate of 8 percent per year.

Solution

In this case, $C_S/\mathscr{F}LC_{F1}$ is the same as before, but the interpolation is between tables F2 and F3 at an inflation rate of 10% and discount rate of 8%. N_p is approximately 15 years. ∎

Other payback times are defined in Section 11.3 in terms of cumulative savings. For these it is necessary to use Equation 11.1.7, including in the equation whatever terms are significant, calculating the solar savings each year and finding the year in which the cumulative savings meet whatever criteria is established for the particular payback time desired. Table 11.7.1 shows informa-

Table 11.7.1   Mortgage Balance Fuel Savings, Cumulative Fuel Savings, Solar Savings, Cumulative Solar Savings, from Example 11.6.4

Year	Mortgage Balance	Fuel Savings	Cumulative Fuel Savings	Solar Savings	Cumulative Solar Savings
				−1100	−1100
1	9707	703	703	−211	−1311
2	9496	773	1476	−163	−1474
3	9266	850	2326	−109	−1583
4	9105	935	3261	−49	−1632
5	8742	1029	4290	18	−1614
6	8445	1132	5422	94	−1520
7	8120	1245	6667	175	−1345
8	7766	1370	8037	268	−1077
9	7381	1570	9607	369	−708
10	6961	1657	11264	481	−227
11	6503	1823	13087	607	380
12	6004	2005	15092	745	1125
13	5459	2206	17298	900	2025
14	4866	2426	19724	1070	3095
15	4220	2669	22393	1260	4355
16	3515	2936	25329	1469	5824
17	2747	3229	28558	1701	7525
18	1910	3552	32110	1958	9483
19	997	3907	36017	2241	11724
20	—	4298	40315	2550	14274
		Resale Value			4400
		Undiscounted Savings Including Resale Value			18674

tion from Example 11.6.3 and illustrates the several payback periods defined in Section 11.3.

A   Solar savings become positive (and the cumulative solar savings reach a minimum) by year 5.

B   The undiscounted cumulative fuel savings exceed the total initial investment in year 10. This is in agreement with Example 11.7.1 (The time for the cumulative discounted fuel savings to reach the initial investment cannot be determined from this table, but is obtained as shown in Example 11.7.2).

C   The cumulative solar savings reach zero during year 10.

D   The cumulative solar savings exceed the remaining debt principal (mortgage balance) by the end of year 15.

E   The cumulative solar savings exceeded the down payment of $1100 by year 12.

Annual cash flow (or savings) vary with years (as shown in the next-to-last column of the tabulation of results in Example 11.6.3). These costs can be "annualized" or "levelized" by determining the equal payments that are equivalent (in present dollars) to the varying series. The annualized life cycle cost, $ALCC$, or savings, $ALCS$, are determined from

$$ALCC = \frac{LCC}{PWF(N_e, 0, d)} \qquad (11.7.6)$$

$$ALCS = \frac{LCS}{PWF(N_e, 0, d)} \qquad (11.7.7)$$

Thus the annualized life-cycle savings in Example 11.6.3 are

$$\frac{4203}{PWF(20, 0, 0.08)} = \$428 \text{ per year.}$$

The series of variable savings is equivalent to annual savings of $428 in today's dollars. (An annualized cost per unit of delivered solar energy can also be calculated. It is necessary to divide annualized life cycle cost by the total annual load.)

The return on investment, ROI, of a particular solar process may be found by determining the market discount rate, d_0, which corresponds to zero life cycle solar savings. This can be done by trial and error, by plotting LCS versus d to find d_0 at $LCS = 0$. For the 39 m^2 optimum area of Example 11.7.1, variation of d results in a plot of LCS versus d as shown in Figure 11.7.1. The return on investment under these circumstances is about 21% (Note that maximizing the return on investment does not lead to the same area as maximizing solar savings).

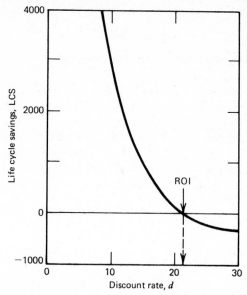

Figure 11.7.1   Solar savings as a function of discount rate for the 39 m² system of Example 11.6.4.

11.8   THE P_1, P_2 METHOD

It is possible to view the calculations of Example 11.6.3 in a different way, by obtaining the present worth of each of the columns, and summing these (with appropriate signs) to get the present worth the solar savings. The life cycle cost of insurance, maintenance and parasitic power, property taxes, and mortgage payments and the life cycle fuel savings are determined from the appropriate present worth factors as previously noted. The life cycle benefit of tax savings can be determined by multiplying the present worth factor for interest paid on the mortgage and the present worth factor for property taxes by the effective income tax rate. This view of the calculation is shown in Example 11.8.1. In this example, we use several new symbols:

MS_1 = Miscellaneous costs (maintenance, insurance, parasitic power) payable at the end of the first period

PT_1 = Property tax payable at the end of the first period

$\bar{t}$ = Effective federal-state income tax bracket, from Equation 11.1.5

Example 11.8.1

Redo Example 11.6.3, by obtaining the life cycle costs of each of the columns in the table in that example, and summing them to get the solar savings.

Solution

The present worth of fuel savings is given by

$$C_{F_1} \times L \times \mathscr{F} \times PWF(N_e, i_F, d) = 703 \times 22.1687 = 15585$$

The present worth of the miscellaneous costs is given by

$$-MS_1 \times PWF(N_e, i, d) = -110 \times 15.5957 = -1716$$

The present worth of the series of mortgage payments is

$$-M \times PWF(N_L, 0, d) = -1084 \times 9.8181 = -10643$$

The present worth of the extra property tax is

$$-PT_1 \times PWF(N_e, i, d) = -220 \times 15.5957 = -3431$$

The present worth of the income tax savings on the interest paid on the mortgage is $\bar{t}$ times the present worth of the series of interest payments (from Equation 11.5.3), ot $\bar{t}\,(PWF)_{\text{int}}$:

$$0.45 \times 9900\left[\frac{9.818}{9.129} + 20.242\left(0.09 - \frac{1}{9.129}\right)\right] = 3028$$

The present worth of the income tax saving due to property taxes is $\bar{t}$ times the present worth of the extra property tax:

$$\bar{t}[PT_1 \times PWF(N_e, i, d)] = 0.45 \times 3431 = 1544$$

The downpayment is 1100 and the present worth of the resale value is $4400/(1.08)^{20} = 944$. Thus solar savings are

$$15585 - 1716 - 10643 - 3431 + 3028 + 1544 - 1100 + 944 = \$4211$$

Within round-off errors, this is the same result as was obtained in Example 11.6.3.   ∎

An examination of the terms in the savings calculation of Example 11.8.1 suggests that a general formulation of solar savings can be developed. Two facts are apparent. First, there is one term that is directly proportional to the first year's fuel savings. Second, the remainder of the terms are all related directly to the initial investment in the system (or to the mortgage, which in turn is a fraction of the initial investment). Using these facts, Brandemuehl and Beckman (1979) have shown how the present worth factors in terms like those in Example 11.8.1 can be combined to a simple formulation for life cycle solar savings:*

$$LCS = P_1 C_{F1} L\mathscr{F} - P_2(C_A A_C + C_E) \tag{11.8.1}$$

where P_1 is the ratio of the life cycle fuel cost savings to the first year fuel cost savings and P_2 is the ratio of the life cycle expenditures incurred because of the additional capital investment to the initial investment. Other terms are as defined previously.

---

* Equation 11.8.1 is written with the implicit assumption that the loads are independent of the size of the solar energy system. To calculate life-cycle savings where there are significant differences between loads with solar and loads without solar, the first term of the equation can be written in terms of the difference in auxiliary energy required, or $P_1 C_{F1}(A_{ns} - A_s)$.

Any costs that are proportional to the first year fuel cost can be included in the analysis by appropriate determination of P_1, and any costs that are proportional to the investment can be included in P_2. Thus the full range of costs noted in the examples of Section 11.6 can be included as needed. P_1 is given by

$$P_1 = (1 - C\bar{t})PWF(N_e, i_F, d) \qquad (11.8.2)$$

where C is a flag indicating income-producing or nonincome-producing (1 or 0, respectively), i_F is the fuel inflation rate, d is the discount rate, N_e is the term of the economic analysis, and $\bar{t}$ is the effective income tax rate, from Equation 11.1.5. P_2 is given as

$$P_2 = D + (1 - D)\frac{PWF(N_{min}, 0, d)}{PWF(N_L, 0, m)} - (1 - D)\bar{t}$$

$$\times \left[PWF(N_{min}, m, d)\left(m - \frac{1}{PWF(N_L, 0, m)}\right) + \frac{PWF(N_{min}, 0, d)}{PWF(N_L, 0, m)} \right]$$

$$+ (1 - C\bar{t})M_s \times PWF(N_e, i, d) + t(1 - \bar{t})V \times PWF(N_e, i, d)$$

$$- \frac{C\bar{t}}{N_D} PWF(N'_{min}, 0, d) - \frac{(1 - c\bar{t}) R_V}{(1 + d)^{N_e}} \qquad (11.8.3)$$

where

$m =$ Annual mortgage interest rate
$i =$ General inflation rate
$N_L =$ Term of loan
$N_{min} =$ Years over which mortgage payments contribute to the analysis (usually the minimum of N_e or N_L)
$N_D =$ Depreciation lifetime in years
$N'_{min} =$ Years over which depreciation contributes to the analysis (usually the minimum of N_e or N_D)
$t =$ Property tax rate based on assessed value
$D =$ Ratio of down payment to initial investment
$M_s =$ Ratio of first year miscellaneous costs (parasitic power, insurance and maintenance) to initial investment
$V =$ Ratio of assessed valuation of the solar energy system in first year to the initial investment in the system
$R_V =$ Ratio of resale value at end of period of analysis to initial investment

In this equation the first term on the right represents the down payment; the second term represents the life cycle cost of the mortgage principal and interest; the third, income tax deductions of the interest; the fourth, miscellaneous costs such as parasitic power, insurance and maintenance; the fifth, net property tax costs (tax paid less income saved); the sixth, straight line depreciation tax deduction;* and the seventh the present worth of resale value at the end of the period of the economic analysis, all in proportion to the initial investment. Terms may be added to or deleted from an analysis as appropriate.

The contributions of loan payments to the analysis depends on N_L and N_e. If $N_L \leq N_e$, all N_L payments will contribute. If $N_L > N_e$, only N_e payments would be made during the period of the analysis. Accounting for loan payments past N_e depends on the rationale for choosing N_e. If N_e is a period over which the discounted cash flow is calculated without regard for costs outside of the period, then $N_{min} = N_e$. If N_e is the expected operating life of the system, and all payments are expected to continue as scheduled, then $N_{min} = N_L$. If N_e is chosen as the time to anticipated sale of the facility, the remaining loan principal at N_e would be repaid at that time, the life cycle loan cost would consist of the present worth of N_e loan payments plus the principal balance in year N_e. The principal balance would then be deducted from resale value.

Similar arguments can be made about the period over which depreciation deductions contribute to an analysis. The contributions will depend on the relationship of N_D and N_e.

The equations for P_1 and P_2 include only present worth factors and ratios of payments to initial investments in the system. They do not include collector area or solar fraction. As P_1 and P_2 are independent of A_c and $\mathscr{F}$, systems in which the primary design variable is A_c can be optimized by use of Equation 11.8.1. At the optimum, the derivative of the savings with respect to collector area is zero:

$$\frac{\partial(LCS)}{\partial A} = 0 = P_1 C_{F1} L \frac{\partial \mathscr{F}}{\partial A} - P_2 C_A \qquad (11.8.4)$$

Rearranging, the maximum savings are realized when the relationship between collector area and solar load fraction satisfies

$$\frac{\partial \mathscr{F}}{\partial A} = \frac{P_2 C_A}{P_1 C_{F1} L} \qquad (11.8.5)$$

The relationship of the optimum area to the annual thermal performance curve is shown in Figure 11.8.1. The optimum area occurs where the slope of the $\mathscr{F}$ versus A curve is $P_2 C_A / P_1 C_{F1} L$.

---

* Straight line depreciation is assumed in Equation 11.8.3. For $N'_{min} > N_D$, the sixth term for double declining balance or sum of digits [from Barley and Winn (1978)] may be written as

$$DB = C\bar{\imath} + \frac{2C\bar{\imath}}{N_D}\left[PWF\left(N_D - 1, \frac{-2}{N_D}, d\right) - \frac{PWF\left((N_{D-1}), \frac{-2}{N_D}, 0\right)}{(1+d)^{N_D}} \right]$$

$$SOD = \frac{2\bar{\imath}}{N_D(N_D + 1)}\left[PWF(N_D, 0, d) + \left(\frac{N_D - 1 - PWF(N_D - 1, 0, d)}{d}\right)\right]$$

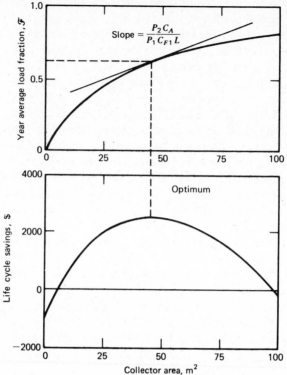

Figure 11.8.1 Optimum collector area determination from the slope of the $\mathscr{F}$ vs A_c thermal performance curve. From Brandemuehl and Beckman (1979).

Example 11.8.2

Redo Example 11.6.4 using the P_1, P_2 method.

Solution

$C = 0$, since the application is not an income-producing one. P_1 is calculated from Equation 11.8.2

$$P_1 = PWF(20, 0.10, 0.08) = 22.169$$

P_2 is calculated from Equation 11.8.3:

$$P_2 = 0.1 + 0.9 \frac{PWF(20, 0, 0.08)}{PWF(20, 0, 0.09)}$$

$$- 0.9 \times 0.45 \left[PWF(20, 0.09, 0.08) \left(0.09 - \frac{1}{PWF(20, 0, 0.09)} \right) \right.$$

$$\left. + \frac{PWF(20, 0, 0.08)}{PWF(20, 0, 0.09)} \right] + 0.01 \times PWF(20, 0.06, 0.08)$$

$$+ 0.02 \times 0.55 \times 1.0 \times PWF(20, 0.06, 0.08) - \frac{0.4}{(1.08)^{20}}$$

$$P_2 = 0.1 + 0.968 - 0.275 + 0.156 + 0.172 - 0.086 = 1.035$$

From Equation 11.8.5, with $C_A = \$200/m^2$ and $C_{F_1}L = \$1255$,

$$\frac{\partial \mathscr{F}}{\partial A} = \frac{1.035 \times 200}{22.169 \times 1255} = 0.00744$$

A plot of the $\mathscr{F}$ versus A_c data from Example 11.6.4 is shown on the figure. The collector area where the slope $= 0.0074$ is about 40 m².

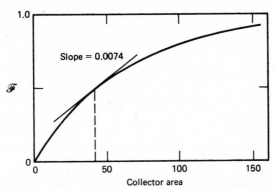

The P_1, P_2 method is not limited to regularly varying costs. The requirements are for P_1 that the fuel expenses be proportioned to the first year unit energy cost, and for P_2 that the owning costs be·proportioned to the initial investment. For the irregularly varying fuel costs of Example 11.6.2 the value of P_1 can be found from the ratio of the life cycle fuel cost to the first year fuel cost, that is, $P_1 = 10516/1200 = 8.7633$. This value of P_1 can be used with any other first year fuel cost that has the same inflation rate schedule as Example 11.6.2. Also P_1 and P_2 can be obtained from a single detailed calculation of life-cycle savings such as that in Example 11.6.3, and can then be applied to all collector areas and solar fractions. If there are highly irregular costs, this may be the easiest way to determine P_1 and P_2.

The P_1, P_2 method is quick, convenient and extremely useful. It is used in developing economic evaluations of specific applications in later chapters.

11.9   UNCERTAINTY ANALYSIS

Many assumptions and uncertainties are involved in the use of the economic analysis methods presented in this chapter. The analyst must make estimates of many economic parameters, with varying degrees of uncertainty. In particular, the projection of future energy costs is difficult, in view of unsettled international energy affairs. Thus it is desirable to determine the effects of uncertainties on the calculated values of life cycle savings (and optimum system design).

For a given set of conditions, the change in life cycle savings resulting from a change in a particular parameter, Δx_j, can be approximated by

$$\Delta LCS = \frac{\partial LCS}{\partial x_j} \Delta x_j \tag{11.9.1}$$

When there are uncertainties in more than one variable the maximum possible uncertainty is given by

$$\Delta LCS_{\max} = \sum_{j=1}^{n} \left| \frac{\partial LCS}{\partial x_j} \right| \Delta x_j \tag{11.9.2}$$

A "most probable" uncertainty in savings can be written

$$\Delta LCS_{\mathrm{prob}} = \sqrt{\sum_{j=1}^{n} \left(\frac{\partial LCS}{\partial x_j} \Delta x_j \right)^2} \tag{11.9.3}$$

From Equation 11.8.1,

$$\frac{\partial LCS}{\partial x_j} = \frac{\partial (P_1 C_{F1} L \mathcal{F})}{\partial x_j} - \frac{\partial P_2 (C_A A + C_E)}{\partial x_j} \tag{11.9.4}$$

The partial derivatives of P_1 and P_2 (from Equations 11.8.2 and 11.8.3) for selected variables are as follows:

For the fuel inflation rate:

$$\frac{\partial P_1}{\partial i_F} = (1 - C\bar{t}) \frac{\partial}{\partial i_F} PWF(N_e, i_F, d) \tag{11.9.5}$$

For the general inflation rate:

$$\frac{\partial P_2}{\partial i} = [(1 - C\bar{t})M_s + (1 - \bar{t})tV] \frac{\partial}{\partial i} PWF(N_e, i, d) \tag{11.9.6}$$

For effective income tax bracket:

$$\frac{\partial P_1}{\partial \bar{t}} = -C \, PWF(N_e, i_F, d) \tag{11.9.7}$$

$$\frac{\partial P_2}{\partial \bar{t}} = (D - 1)\left[\frac{PWF(N_{\min}, 0, d)}{PWF(N_L, 0, m)} + PWF(N_{\min}, m, d)\left(1 - \frac{1}{PWF(N_L, 0, m)}\right)\right]$$
$$- (tV)PWF(N_e, i, d) - C\left[M \, PWF(N_e, i, d) + \frac{1}{N_D} PWF(N'_{\min}, 0, d)\right] \tag{11.9.8}$$

For property tax rate:

$$\frac{\partial P_2}{\partial t} = V(1 - \bar{t})PWF(N_e, i, d) \tag{11.9.9}$$

For resale value:

$$\frac{\partial P_2}{\partial R_V} = \frac{1}{(1 + d)^{N_e}} \tag{11.9.10}$$

The partial derivative of the life cycle savings with respect to the fraction by solar is

$$\frac{\partial LCS}{\partial \mathcal{F}} = P_1 C_{F1} L \tag{11.9.11}$$

A complete set of the partial derivatives is provided by Brandemuehl and Beckman (1979).

It is also necessary to know the partial derivatives of the present worth functions. From Equation 11.5.1

$$\frac{\partial}{\partial a} PWF(a, b, c) = -\frac{1}{c-b}\left(\frac{1+b}{1+c}\right)^a \ln\left(\frac{1+b}{1+c}\right) \tag{11.9.12}$$

$$\frac{\partial}{\partial b} PWF(a, b, c) = \frac{1}{c-b}\left[PWF(a, b, c) - \frac{a}{1+b}\left(\frac{1+b}{1+c}\right)^a\right] \tag{11.9.13}$$

$$\frac{\partial}{\partial c} PWF(a, b, c) = \frac{1}{c-b}\left[\frac{a}{1+c}\left(\frac{1+b}{1+c}\right)^a - PWF(a, b, c)\right] \tag{11.9.14}$$

In the example below, the effect of uncertainty of one variable is illustrated. To estimate the effects of uncertainties of more than one variable, the same procedure is used as for one variable in determining the appropriate terms in Equation 11.9.2 and 11.9.3.

Example 11.9.1

In Example 11.8.2, the fuel inflation rate was taken as 10 percent/year. For a 50 m^2 collector area, what are the life cycle savings? What is the uncertainty in life cycle savings if the fuel inflation rate is uncertain to ± 2 percent? From the data of Example 11.6.4, the 50 m^2 collector will provide 0.56 of the annual loads. The first year's fuel cost is $1255 and the installed cost is $11,000. $N_e = 20$ years, and $d = 0.08$. From example 11.8.2, $P_1 = 22.169$ and $P_2 = 1.035$.

Solution

The life cycle savings are, from Equation 11.8.1,

$$LCS = 22.169 \times 1255 \times 0.56 - 1.035 \times 11000 = \$4195$$

The effect of fuel inflation rate is only on P_1. From Equation 11.9.5, with $C = 0$,

$$\frac{\partial P_1}{\partial i_F} = \frac{\partial}{\partial i_F} PWF(N_e, i_F, d)$$

and from Equation 11.9.13:

$$\frac{\partial}{\partial i_F} PWF(N_e, i_F, d) = \frac{1}{0.08 - 0.10}\left[PWF(20, 0.10, 0.08) - \frac{20}{1.10}\left(\frac{1.10}{1.08}\right)^{20}\right]$$

$$= 204$$

The uncertainty in LCS is obtained from Equations 11.9.2 or 11.9.3 (which lead to the same result when considering only one variable) and 11.9.4:

$$\Delta LCS = \frac{\partial LCS}{\partial i_F} \Delta i_F = C_{F_1} L \mathscr{F} \frac{\partial P_1}{\partial i_F} \Delta i_F$$
$$= 1255 \times 0.56 \times 204 \times 0.02 = \$2867$$

Thus the uncertainty in LCS due to the uncertainty of 2 percent in the fuel inflation rate is over half of the projected savings. An increase in i_F results in an increase in LCS.   ■

The effect of any one variable on the life cycle savings is largely determined by the values of the many other variables, it is not easy to find "rules of thumb," and a quantitative analysis is usually required.

11.10 SUMMARY

In this chapter we have outlined the kinds of investments and operating costs that may be expected with a solar process, and indicated that in many circumstances collector area can be considered the primary design variable, once a system configuration and collector type are established.

A variety of economic figures of merit have been proposed and used including payback times, cash flow, and life cycle savings. The life cycle costing method is the most inclusive and takes into account any level of detail the user wishes to include, including the time value of money. The use of the life cycle savings method is recommended and will be illustrated for specific solar processes in the later chapters. The P_1, P_2 method of calculating life cycle savings is particularly useful. The results of these calculations are very dependent on values assumed for N_e, i_F, and d. If $i_F > d$ (an abnormal situation that has existed in recent years) the choice of a sufficiently high value of N_e will make the life-cycle savings of solar (or other capital-intensive fuel-saving technologies) appear positive.

REFERENCES

Barley, C. D., Paper presented at the AS of ISES Denver Meeting, 2(1), 163 (1978). "Optimization of Space Heating Loads."

Barley, C. D. and C. B. Winn, *Solar Energy*, 21, 279 (1978). "Optimal Sizing of Solar Collectors by the Method of Relative Areas."

Beckman, W. A., S. A. Klein, and J. A. Duffie, *Solar Heating Design by the f-chart Method*, Wiley-Interscience, New York (1977).

Brandemuehl, M. J. and W. A. Beckman, *Solar Energy*, 23, 1 (1979). "Economic Evaluation and Optimization of Solar Heating Systems."

Close, D. J., *Solar Energy*, 11, 112 (1967). "A Design Approach for Solar Processes."

De Garmo, E. P. and J. R. Canada, *Engineering Economy*, Macmillan, New York (1973).

Riggs, J. L., *Economic Decision Models*, McGraw-Hill, New York (1968).

Ruegg, R. T., U.S. Dept. of Commerce, NBSIR 75-712 (July 1975). "Solar Heating and Cooling in Buildings: Methods of Economic Evaluation."

White, J. A., M. H. Agee, and K. E. Case, *Principles of Engineering Economic Analysis*, Wiley, New York (1977).

CHAPTER 12

Solar Water Heating

This is the first of a set of chapters on thermal energy applications. In this chapter we treat the use of solar heating for domestic or institutional hot water supplies. Descriptions of systems, components, and important design considerations are outlined. Considerations important in designing water heating systems are also basic to solar heating and cooling systems, applications that are covered in succeeding chapters. Many of the principles noted here for service hot water for buildings also apply to industrial process heat, discussed in Chapter 17.

12.1 WATER HEATER SYSTEMS

The basic elements in solar water heaters can be arranged in several system configurations. The most common of these are shown in Figure 12.1.1. Auxiliary energy is shown added in three different ways; these are interchangeable among the four methods of transferring heat from the collector to the tank.

A passive water heater, that is, a natural circulation system, is shown in Figure 12.1.1a. The tank is located above the collector, and water circulates by natural convection whenever solar energy in the collector adds energy to the water in the collector leg and so establishes a density difference. Auxiliary energy is shown added into the tank near the top to maintain a supply of hot water.

Figure 12.1.1b shows an example of a forced circulation system. A pump is required, which is usually controlled by a differential thermostat turning on the pump when the temperature at the top header is higher than the temperature of the water in the bottom of the tank by a sufficient margin to assure control stability (as outlined in Section 10.4). A check valve is needed to prevent reverse circulation and resultant nighttime thermal losses from the collector. Auxiliary energy is shown added to the water in the pipe leaving the tank to the load by a heater having no storage capacity.

In climates where freezing temperatures occur, these designs are modified. Examples of systems using nonfreezing fluids in the collector are shown in Figure 12.1.1c and d. The collector heat exchangers can be either internal or external to the tank. Auxiliary energy is shown added to the water in the storage tank in Figures 12.1.1c by a heat exchanger in the tank. The auxiliary energy

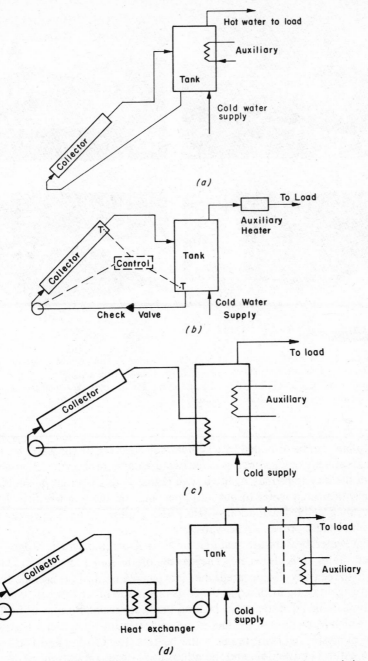

Figure 12.1.1 Schematics of common configurations of water heaters. (*a*) A natural circulation system. (*b*) One tank forced circulation system. (*c*) System with antifreeze loop and internal heat exchanger. (*d*) System with antifreeze loop and external heat exchanger. Auxiliary is shown added in the tank, in a line heater, or in a second tank; any of these auxiliary methods can be used with any of the collector-tank arrangements.

Figure 12.1.2 Solar water heaters. (*a*) Natural circulation system on hostel, Griffith, Australia. Photo courtesy of CSIRO. (*b*) Domestic system, with coupled tank and collector, Perth, Australia. (*c*) Hospital service hot water system, Madison, WI. Photo courtesy Affiliated Engineers, Inc. (*d*) Collectors for a domestic system, Madison, WI.

supply can also be provided by a standard electric, oil, or gas water heater with storage capacity of its own; this is the two-tank system shown in Figure 12.1.1*d*. Any of these systems may be fitted with tempering valves that mix cold supply water with heated water to put an upper limit on the temperature of the hot water going to the distribution system. Other equipment not shown can include surge tanks and pressure relief valves.

Solar water heaters are manufactured in Australia, Israel, Japan, Greece, United States, and elsewhere. They were common in Florida and California early in the century, disappeared when inexpensive natural gas became available, and are again being installed as the costs of gas and other fuels rise. Solar water heating has the advantage that heating loads are usually uniform through the year, which leads to high use factors on solar heating equipment. Figure 12.1.2 shows domestic and institutional water heaters. The Griffith and Perth systems utilize natural convection, and the others use forced convection.

The collectors in use in many water heating systems are similar to that shown in Figure 7.10.3, with parallel riser tubes 0.10 to 0.15 m apart. Plate materials may be copper or steel. Other plate designs are also used. For example, many are manufactured of two spot-welded, seam-welded, or roll-bonded

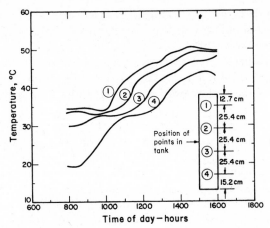

Figure 12.1.3 Temperature distribution in a vertical cylindrical water tank used in a thermal circulation water heater. From Close (1962).

plates of stainless or ordinary steel, copper, or aluminum. The fluid passages between the plates are formed by hydraulic expansion after welding. Serpentine tubes are also used. The absorber plates are mounted in a metal box, with 50 to 100 mm of insulation behind the plate and one or two glass covers over the plates. The dimensions of a typical single collector module made in the United States are typically approximately 0.9 × 2.1 m; the Australian and Israeli heater modules are typically 0.6 × 1.2 m or 1.2 × 1.2 m. The thermal performance characteristics of these and other collectors can be determined by equations given in Chapter 6.

It is advantageous to maintain stratification in the storage tanks, and the location and design of tank connections is important. The schematics in Figure 12.1.1 show approximate locations of connections in typical use. Close (1962) measured tank temperatures at various levels in an experimental natural circulation water heating system operated for a day with no hot water removal from the tank. These data are shown in Figure 12.1.3; the degree of stratification is evident. Tanks without baffles will stratify to some degree in forced circulation systems, if the entrance and exit velocities are not too high. It is also possible to design tanks with internal baffles to improve stratification but at the present time baffled tanks are not being used in commercial water heaters. Storage tanks should be well insulated, and good practice is to use 0.2 m or more of mineral wool or glass wool insulation on sides, top, and bottom. Piping connections to a tank should also be well insulated.

12.2 FREEZING AND BOILING

Solar water heaters must be designed to avoid damage from freezing or boiling. Low ambient temperatures during periods of no solar radiation can result in plate temperatures below 0 C. If no energy is withdrawn from a system, or if a circulating pump should not operate during times of high radiation, the plate temperature may exceed 100 C.

Freeze protection can be provided by draining the water from the collectors, use of nonfreezing solutions, or warming of the water in the collectors. Five approaches have been developed to protect collectors against damage by freezing.

First, antifreeze solutions can be used in the collector loop with a heat exchanger between the collector and the storage tank. As shown in Figure 12.1.1c and d, the heat exchanger can be external to the tank or it can be a coil within the tank, relying on natural circulation of the water in the tank for heat transfer. For either arrangement, the performance of the collector-heat exchanger combination can be treated by the F'_R method outlined in Section 10.2. A typical overall heat transfer coefficient for a coil in a tank is 600 W/m² C.

Common antifreeze liquids are ethylene glycol-water and propylene glycol-water solutions. Their physical properties are included in Appendix E. Ethylene glycol is toxic, as are some commonly used corrosion inhibitors, and many plumbing codes require the use of two metal interfaces between the toxic fluid and the potable water supply. This can be accomplished either by the use of two heat exchangers in series or by double-walled heat exchangers that can be either internal coils in the tank or external to the tank.

Second, air can be used as the heat transfer fluid in the collector-heat exchanger loop of Figure 12.1.1d. Air heating collectors have lower $F_R(\tau\alpha)$ and $F_R U_L$ than liquid heating collectors. However, no toxic fluids are involved, no second heat exchanger interface is needed, leakage is not critical, and boiling is not a problem.

The third method of freeze protection is to circulate warm water from the tank through the collector to keep it from freezing. Thermal losses from the system are significantly increased, and an additional control mode must be provided. This method can only be considered in climates where freezes are infrequent. In emergencies when pump power is lost the collector and piping subject to freezing temperatures must be drained.

The fourth method is based on draining water from the collectors when they are not operating. Draining systems must be arranged so that collectors and piping exposed to freezing temperatures are completely emptied, and the collectors must be vented. Draining systems may drain back into the tank or to a heat exchanger in the tank, or they may drain out of the system to waste.

The fifth method is to design the collector plate and piping so that it will withstand occasional freezing. Designs have been proposed using butyl rubber risers and headers that can expand if water freezes in them.

High collector temperatures may also be a problem, as equilibrium temperatures of good collectors can be well above the boiling point of water under conditions of no fluid circulation, high radiation, and high ambient temperature. These conditions can be expected to occur, for example, when occupants of a residence are away from home in the summer. Several factors may mitigate this problem. First, antifreeze solutions used in collector loops have elevated boiling points (see Appendix E); 50 percent ethylene glycol and propylene glycol solutions in water have boiling points at atmospheric pressure of 112 and 108 C, respectively. Second, many systems are operated at pressures of several atmospheres, which further raises the boiling point; the boiling points of the 50 percent glycol solutions at four atmospheres absolute are about 150 and 145 C, respectively. Third, collector loss coefficients rise as plate temperatures rise as shown in Figure 6.4.4. Practical systems should include pressure relief valves and vent tanks to relieve excess pressure and contain any antifreeze solution that is vented.

Example 12.2.1

A collector with $(\tau\alpha)$ equal to 0.78 has a temperature dependence of overall loss coefficient as shown on the plot below. The fluid being heated is a 50 percent solution of propylene glycol in water. The glycol loop is pressurized to a limit of 4.0 atm. On a summer day the radiation on the collector is 1.15 kW/m^2 and ambient temperature is 38 C. Will the solution in the collector boil if the circulating pump does not operate?

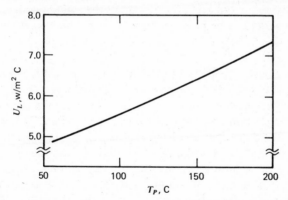

Solution

With no flow, absorbed energy equals losses

$$G_T(\tau\alpha) = U_L(T_{p,m} - T_a) = 1150 \times 0.78 = 900 \text{ W/m}^2$$

so the plate temperature is

$$T_{p,m} = T_a + \frac{900}{U_L}$$

Since the loss coefficient is a function of plate temperature, it is necessary to simultaneously solve the energy balance equation and the U_L versus T_p relationship. Assume for a first estimate that $U_L = 6.5$ W/m^2 C. Then

$$T_{p,m} = 38 + \frac{900}{6.5} = 176 \text{ C}$$

A second estimate of 6.8 W/m² C yields

$$T_{p,m} = 38 + \frac{900}{6.8} = 170\ \text{C}$$

The loss coefficient at this temperature is close to the second estimate, so $T_p = 170$ C is a reasonable estimate of the temperature of the plate and the temperature of the fluid under no flow conditions. The boiling point of a 50 percent propylene glycol solution at 4 atm is approximately 145 C and the solution would boil in the collector under these circumstances.   ∎

12.3 AUXILIARY ENERGY

The degree of reliability desired of a solar process to meet a particular load can be provided by a combination of properly sized collector and storage units and an auxiliary energy source. In a few unique areas where there seldom are clouds of significant duration, it may be practical to provide all of the loads with the solar systems. However, in most climates auxiliary energy is needed to provide high reliability and avoid gross overdesign of the solar system.

Auxiliary energy can be provided in any of the three ways shown in Figure 12.3.1.

1   Energy can be supplied to the water in the tank, at location A. Auxiliary energy is controlled by a thermostat in the top part of the tank which keeps the temperature in the top portion at or above a minimum set-point. This is the simplest and probably the least expensive method. However, it has the disadvantage that auxiliary energy will usually increase the temperature of the water in the bottom of the tank and thus the collector inlet temperature, resulting in reduction of solar gain.
2   Auxiliary energy can be supplied to the water leaving the tank, thus "topping off" the solar energy with auxiliary energy. This requires a

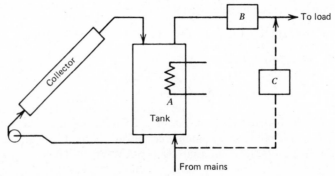

Figure 12.3.1   Schematic of alternative locations for auxiliary energy supply to a one-tank forced circulation solar water heater: A, in tank: B, in line to load; C, in a bypass around the tank.

heater separate from the solar tank. This heater may be a simple line heater, or it may be a conventional water heater with storage capacity of its own. Auxiliary energy is controlled to maintain the outlet temperature from the auxiliary heater at a desired level. This method has the advantage of using the maximum possible solar energy from the tank without driving up the collector temperature, but additional heat loss will occur from the auxiliary heater if it has storage capacity.

Table 12.3.1 $\mathscr{F}$, **Ratio of Solar Energy to Total Load for 31-Day Period for a Solar Water Heating System**

Type	Auxiliary Method		
	1	2	3
N-L1-S2-32-S	0.91	0.92	0.90
L2	0.80	0.82	0.79
L3	0.61	0.62	0.57
M-L1-S2-32-S	0.86	0.88	0.85
L2	0.73	0.76	0.73
L3	0.54	0.57	0.53
N-L1-S1-32-W	0.84	0.85	0.84
L2	0.67	0.69	0.66
L3	0.48	0.49	0.47
M-L1-S1-32-W	0.74	0.77	0.73
L2	0.59	0.60	0.58
L3	0.40	0.42	0.40
N-L1-S1-43-S	0.86	0.88	0.84
L2	0.72	0.74	0.68
L3	0.51	0.55	0.50
M-L1-S1-43-S	0.78	0.81	0.75
L2	0.63	0.67	0.60
L3	0.46	0.47	0.43
M-L1-S1-66-S	0.70	0.76	0.61
L2	0.54	0.61	0.45
L3	0.40	0.46	0.32

Adapted from Gutierrez et al. (1974).
Key: M, early morning loads. N, noon loads. Magnitude of loads, L1 = 8.5 MJ/m^2; L2 = 11.9 MJ/m^2; L3 = 17.0 MJ/m^2. Storage capacity, S1 = 0.20 MJ/m^2 C; S2 = 0.40 MJ/m^2 C. Minimum control temperature, 32, 43, or 66 C. Month, summer S, or winter W.

3 Auxiliary energy can be added directly to the incoming supply water by by-passing the tank when auxiliary energy is needed. This is a simple method, but has the disadvantage that collected solar energy in the tank which does not result in the water temperature being above the set temperature may be lost from the tank.

The relative merits of these three methods have been examined by Gutierrez et al. (1974). By simulating a month's operation of the three systems having forced circulation, fixed collector size and a partially stratified (three-node) storage tank under a range of conditions of ambient temperature, time of day and magnitude of loads, the relative portion of the loads carried by solar and by auxiliary energy were estimated. It was shown for the system and climate data used, when the set point temperature was not greatly above ambient temperatures, the method of adding auxiliary was not critical, with method 2 showing minor advantages. However, when the minimum hot water temperature was raised to set points usually used in domestic hot water systems, method 2 showed significant advantages over method 1, which in turn was better than method 3. Examples of the estimated fraction of the load carried by solar energy are shown in Table 12.3.1.

The major reasons for the changes in performance on changing the method of adding auxiliary concern the temperature at which the collector operates. Adding auxiliary energy to the top of the tank (method 1) can result in a higher mean collector temperature, poorer collector performance, and higher requirements for auxiliary energy. Method 3, which bypasses the tank when its top section is not hot enough, results in failure to use some collected solar energy. Method 2, with a modulated auxiliary heater (or its equivalent in an on-off auxiliary heater) maximizes use of the solar collector output and minimizes collector losses by operating at the lowest mean collector temperature of any of the methods. The magnitude of the differences may depend on the degree of stratification in the tank and the insulation on the tank. Losses from the auxiliary heater must be maintained at very low levels, or method 2 may not be best (see Section 12.4).

12.4 FORCED CIRCULATION SYSTEMS

Most solar water heaters in the United States are forced circulation systems, and in this section we review comparative simulation studies by Buckles et al. (1979) of the long term performance of systems of several configurations.

The simulations were based on a typical domestic hot water load pattern shown in Figure 12.4.1 [Mutch (1974)]. Hot water use is concentrated in morning and evening hours; to supply large fractions of these loads, water heated by solar energy one day must be stored for use the following day. (Minor changes in time dependence of loads do not have a major effect in long-term performance of domestic hot water systems. However, variations of larger magnitude, such

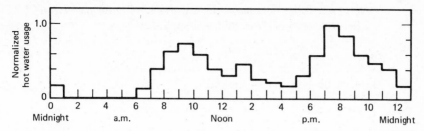

Figure 12.4.1 A normalized profile of hourly hot water use for a domestic application.

as caused by week-end closures of commercial buildings, can have a significant impact on water heater performance and design.)

In the systems studied, daily loads were kept fixed at 300 liters of water, heated from 10 to 50 C, for an annual heating load of 18.3 GJ. The collectors were well-designed single cover selective surface collectors with $F_R(\tau\alpha)_n = 0.84$ and $F_R U_L = 4.67$ W/m^2 C. Collector areas were taken as 2, 3, or 4 modules each of area 1.44 m^2.

The tanks in all cases had loss coefficients of 1.67 W/m^2 C. Single tank systems were modeled as three-node tanks with auxiliary energy supplied to the top node by immersion heaters. Two tank systems were modeled with the same total volume as the single tanks, with one third of the volume in the auxiliary tank and the preheat tank with the remaining two-thirds treated as a two-node tank. The collector heat exchangers inside the tank were assumed to have an effectiveness of 0.5, unless otherwise noted. The direct systems were modeled without heat exchangers, with circulation from the bottom of the tank to collector and return to the top of the tank.

An index of performance of these systems is the yearly quantity of solar energy delivered to heat the water. This quantity is often calculated without consideration of thermal losses from the tanks. However, the total energy required by a conventional heater would include losses from the tank. The data noted below are comparisons of solar system performance relative to a conventional system in which the total load is the energy required to heat the water, 18.3 GJ for a year, plus estimated losses from a conventional tank of 3.6 GJ, for a total of Q_{conv} of 21.6 GJ. The fraction by solar, $\mathscr{F}$, is then the ratio of the solar energy delivered to the total conventional load, or $(1 - Q_{aux}/Q_{conv})$ and is a measure of the reduction in consumption of energy relative to the conventional system.

Results of the simulations, done for a year in the Madison climate, are summarized in Table 12.4.1. The annual solar fraction is shown for five system configurations and three collector areas. Under the assumptions made in this study, that is, with a reasonable degree of stratification and with auxiliary energy supplied to the top third of a single tank or in the second tank of a two-tank system, there is very little difference between one- and two-tank systems. In some cases one-tank systems are slightly better because of less area for heat loss from the tanks. The two-tank direct systems without tempering valves have essentially the same performance as those with tempering valves. During times

Table 12.4.1   Solar Contribution to Water Heating Loads for Several System Configurations. From Buckles et al. (1979).

System, Number of Collector Modules	$\mathscr{F}$
One tank, direct, with tempering valve	
2	0.55
3	0.69
4	0.77
Two tank, direct, with tempering valve	
2	0.52
3	0.66
4	0.74
One tank, heat exchanger with $\varepsilon = 0.5$, with tempering valve	
2	0.51
3	0.64
4	0.71
Two tank, heat exchanger with $\varepsilon = 0.5$, with tempering valve	
2	0.49
3	0.62
4	0.72
Two tank, direct, without tempering valve	
2	0.52
3	0.65
4	0.73

when the system is oversized (i.e., during very good weather) and the water delivered from storage is hot enough to require tempering with cold water, the higher thermal losses negate any advantages of delivery of less heated water from the tank, and at times when the system is undersized the tempering valve does not function.

An additional set of simulations was done in which the average daily hot water load was kept at 300 liters, but which had a 15-fold variation from the largest daily draw to the smallest daily draw in each week. The percent of the week's hot water load drawn in each of the days of the week (starting with Sunday), was 5.7, 42.9, 2.8, 2.8, 14.4, 2.8, and 28.6 percent. For the one-tank, external heat exchanger system with three collector modules, the fraction of the load met by solar energy was 0.64 for the regularly recurring load and 0.58 for the irregular load. This simulation study suggests that the choice of system configuration makes less difference* in the annual output than does the kind of variation in the day-to-day loads which might be expected with a domestic system.

Sizing of domestic forced circulation systems can be done by "rule of thumb" or by thermal and economic analyses and design procedures.† Typical home

---

* See Section 14.7 for note of experimental data comparing performance of several system configurations.

† See Chapter 14 for details on a design procedure.

hot water usage in the United States is 75 liters per day per person, and typical collector areas per person are about 1.5 m² for systems which deliver approximately 0.5 to 0.8 of the annual loads by solar energy. However, these figures vary with the quality of the collector and the climate, and there are wide individual variations from average water use.

Systems for institutional, commercial, and office buildings are almost all forced circulation systems. The magnitude and time dependence of loads in these buildings may be easier to predict than that for residences. The loads may go to zero on weekends and holidays in some office and commercial buildings. Careful design of the system is warranted by the larger investments in them. Designs can be based on correlation methods (see Chapter 14) if systems and loads approximate the conditions for which the correlations were developed, or they can be based on simulations.

12.5 NATURAL CIRCULATION SYSTEMS

Circulation in solar heaters such as that shown in Figure 12.1.1*a* occurs when the collector warms up enough to establish a density difference between the leg including the collector and the leg including the tank and the feed line from tank to collector. The density difference is a function of temperature difference, and the flow rate is then a function of the useful gain of the collector which produces the temperature difference. Under these circumstances, these systems are self-adjusting, with increasing gain leading to increasing flow rates through

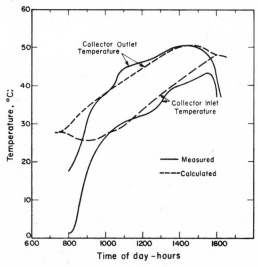

Figure 12.5.1 Collector inlet and outlet water temperatures for a natural circulation water heater. From Close (1962).

the collector. It has been observed by Löf and Close (1967) and by Cooper (1973) that under wide ranges of conditions the increase in temperature of water flowing through many collectors in natural circulations systems (particularly those of Australian design) is approximately 10 C.

Close (1962) worked out an analysis of circulation rates in natural circulations systems and compared computed and experimental inlet and outlet temperatures. His results, some of which are shown in Figure 12.5.1, confirm the suggestion that temperature increases of approximately 10 C are representative of these systems if they are well designed and without serious flow restrictions. Gupta and Garg (1968) also show inlet and outlet water temperatures for two collectors that suggest nearly constant temperature rise across the collectors.

There are two alternative methods of modeling the performance of a collector in natural circulation systems. The first is by an analysis of the temperature and density distributions and resulting flow rates based on pressure drop calculations, as outlined by Close. The second is to assume a constant temperature increase of water flowing through the collector and calculate the flow rate which will produce this temperature difference at the estimated collector gain. The basic collector equations are

$$Q_u = A_c F_R [S - U_L (T_i - T_a)] \qquad (12.5.1)$$

and

$$Q_u = \dot{m} C_p (T_o - T_i) = \dot{m} C_p \Delta T_f \qquad (12.5.2)$$

solving for the flow rate

$$\dot{m} = \frac{A_c F_R [S - U_L (T_i - T_a)]}{C_p \Delta T_f} \qquad (12.5.3)$$

This equation can be solved for $\dot{m}$ if it is assumed that F' is independent of flow rate. Substituting Equation 6.7.4 for F_R into Equation 12.5.3 and rearranging gives

$$\dot{m} = - \frac{U_L F' A_c}{C_p \ln \left[1 - \dfrac{U_L (T_o - T_i)}{S - U_L (T_i - T_a)} \right]} \qquad (12.5.4)$$

Example 12.5.1

A natural circulation water heater operates with a nearly constant increase in water temperature of 10 C. The collector has an area of 4 m^2, an overall loss coefficient 4.2 W/m^2 C, and an F' of 0.91. If the water inlet temperature is 30 C, the ambient temperature is 15 C, and the radiation absorbed by the collector plate is 780 W/m^2, what is the useful gain from the collector?

Solution

All of the information needed to estimate $\dot{m}$ using Equation 12.5.4 is available

$$\dot{m} = \frac{-4.2 \times 0.91 \times 4}{4190 \ln\left[1 - \dfrac{4.2 \times 10}{780 - 4.2(30 - 15)}\right]}$$

$$= 0.060 \text{ kg/s}$$

$$Q_u = 0.060 \times 4190 \times 10 = 2550 \text{ W}$$

We have assumed F' is fixed. If necessary, a first estimate can be made of F', a first iteration of $\dot{m}$ obtained as shown, and the assumed value of F' checked using the calculated $\dot{m}$ to see if a second iteration is needed. If the fluid temperature rise were 20 C, the calculation shows $Q_u = 2450$ W, with the reduction due to lower F_R. The useful gain is not very sensitive to ΔT_f, except at low radiation levels or with collectors with high loss coefficients. ∎

Collector operation with fluid temperature rises of approximately 10 C implies for practical systems that water circulates through the collector several times per day. Tabor (1969) suggested an alternative, that resistance to flow and ΔT_f be higher, with flow rates such that the water in the tank makes about one pass through the collector in a day. He calculated that the daily efficiency of a "one-pass" high ΔT_f system will be about the same as a system using several passes per day and lower ΔT_f. This is dependent on maintenance of good stratification in the tank. It can be inferred from this that the effects of reduced F_R are offset by reduced mean fluid inlet temperature. Many Israeli water heaters are designed to operate this way.

In Australia, where natural circulation systems are widely used, sizing of systems is based on experience [CSIRO (1964)]. They are designed to produce water at 65 C, and at a daily average usage of 45 liters per person per day. If an all-solar system is to be used, for example, in Darwin (which is characterized by almost continuous good solar weather and expensive conventional fuels), a storage capacity of 2.5 times the daily requirement is suggested. For a family of four, in Darwin, a collector area of about 4 m^2 is suggested. If an auxiliary energy source is to be used (e.g., in Melbourne, where radiation is more intermittent and conventional energy is less expensive), the recommended tank size is approximately 1.5 times the daily requirement. Most Australian solar water heaters are natural circulation systems, as freezing is not a problem in most of the country.

12.6 ECONOMICS OF SOLAR WATER HEATING

Most solar water heating installations are designed for use through the year, which results in a high use factor on the equipment and makes solar water heating more competitive with conventional methods than solar applications which are seasonal. A solar collector used in a water heater will deliver more energy per unit area per year than will a comparable installation used seasonally, such as winter space heating or harvest-time crop drying.

The economic principles outlined in Chapter 11 apply directly to solar water heating. The installed costs include the purchase and installation costs of collectors, tanks, controls, piping, and ducting. The operating costs include interest and principal payments, parasitic power for fluid circulation, insurance, maintenance, taxes, and any other costs which may be significant. The following example shows the application of life-cycle savings analysis by the P_1, P_2 method to a solar water heater installation.

Example 12.6.1

A solar water heater is proposed to be installed on a residence in Madison. The annual load is estimated to be 22.2 GJ. The collector areas and tank capacities available are such that it is not practical to consider collector area and tank size as continuous variables, and a set of discrete system designs must be considered. The following table shows collector areas (for 2, 3, or 4 collector modules), tank capacities, installed costs, and fractions of annual hot water loads carried by solar energy.

Design	Collector Area, m^2	Tank Capacity, kg	System Cost, $	Fraction by Solar, $\mathscr{F}$
A	3.66	310	1700	0.41
B	5.49	310	2100	0.55
C	5.49	454	2400	0.56
D	7.32	454	2800	0.67

The water heaters would be purchased by a 20 percent downpayment with the balance to be paid over a 10 year period with interest at 9 percent per year. The present cost of electrical energy (the auxiliary and alternative energy supply) is $14.00/GJ, and it is expected to inflate at 8.5 percent per year. Insurance, maintenance, and parasitic power costs in the first year are expected to be 1 percent of the system cost. The real estate tax increment in the first year will be 1.5 percent of the system costs. The insurance, maintenance, parasitic power, and real estate taxes are expected to rise at a general inflation rate of 6 percent per year.

The owner's effective income tax bracket is 0.45. The market discount rate is to be 8 percent per year. If the analysis is done over 15 years and the system is assumed to have no appreciable resale value at that time, which system (if any) should be bought? For the best system, how long would it take to recover the system cost in savings on the purchase of electricity?

Solution

The analysis is done by the P_1, P_2 method of Chapter 11, with life cycle savings in each of the four cases calculated as shown in Example 11.8.2. P_1 and P_2 are common to all four designs and are calculated by Equations 11.8.2 and 11.8.3.

$$P_1 = PWF(15, 0.085, 0.08) = 14.348$$

$$P_2 = 0.20 + 0.80 \frac{PWF(10, 0, 0.08)}{PWF(10, 0, 0.09)} - 0.80(0.45)$$

$$\times \left\{ PWF(10, 0.09, 0.08) \left[0.09 - \frac{1}{PWF(10, 0, 0.09)} \right] \right.$$

$$+ \left. \frac{PWF(10, 0, 0.08)}{PWF(10, 0, 0.09)} \right\}$$

$$+ 0.01 PWF(15, 0.06, 0.08) + 0.015(1 - 0.45) PWF(15, 0.06, 0.08)$$

$$= 1.112$$

Then savings are calculated from Equation 11.8.1

$$LCS = 14.348 \times 14 \times 22.2 \times \mathscr{F} - 1.112 \times \text{Cost}$$

The savings for the four designs are, under these assumptions of costs:

A	$ -62
B	$ 118
C	$-171
D	$-127

Design B, the only one showing positive savings, is the one which should be purchased. However, the others are close, and small changes in the cost assumptions could shift all of them either way. The time to recover the installed costs in undiscounted electricity savings can be determined from Equation 11.7.3

$$N_p = \frac{\ln \left[\dfrac{2100 \times 0.085}{0.55 \times 22.2 \times 14} + 1 \right]}{\ln(1 + 0.085)} = 8.8 \text{ years} \qquad \blacksquare$$

The competitive position of solar water heating is very much dependent on the costs of alternative energy and how rapidly it is expected to rise. It may also depend on the potential resale value of the equipment at the end of the period of economic analysis. Figure 12.6.1 was generated for design B, with economic assumptions the same as in Example 12.6.1. Fuel cost inflation rate is plotted against first year fuel cost for combinations of C_{F1} and i_F, which result in zero life cycle savings ("breakeven" combinations.) Any combination of C_{F1} and its inflation rate which lie above and to the right of the line results in positive solar savings and economically feasible solar water heating.

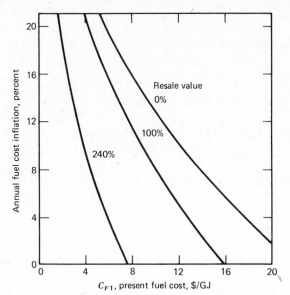

Figure 12.6.1 Examples of relations between fuel cost and its inflation rate for break-even solar water heating system operation. The three curves are for three assumptions of resale value at the end of the period of the economic analyses. Areas above curves represent combinations leading to economically feasible solar water heating under the assumptions of Example 12.6.1 in Madison, WI.

The 0 percent curve is based on the assumption of zero resale value at the end of the period of the economic analysis. Two additional lines are shown that are based on the assumptions that the system is very well built and will have a resale value equal to its first cost (the curve marked 100 percent), or at its first cost inflated at the general inflation rate (the curve marked 240 percent). If the present cost of energy from fuel is \$10/GJ and it is expected to increase at 10 percent/year, this system would show positive life cycle savings if its resale value will be 100 percent but not if it will have no resale value. Resale value results in significantly improved competitive positions for solar water heating, for short periods of economic analysis. (Note, however, that this is a specific example. A new set of curves would have to be generated for new sets of economic parameters and locations. In particular, tax credits that effectively reduce first cost and other changes in installed cost of the water heaters may substantially alter the economic feasibility.)

12.7 RETROFIT WATER HEATERS

It is generally difficult to add solar equipment to buildings not designed for them. Of all solar energy systems, water heaters are the easiest to retrofit. Collector areas of residential water heating systems are manageable, and the addition of a separate solar storage tank can result in a system configuration such as shown in Figure 12.1.1d, where the existing heater is the second tank. The solar heating system then provides preheated water to the conventional system and integration into the existing system is easy. Or, a solar tank equipped with an auxiliary heater can replace the existing heater. The major problems are likely to be collector mounting and space for pipe or duct runs. As these runs are likely to be longer than in buildings designed for solar applications, they must be carefully insulated to avoid excessive losses. The methods of Section 10.3 can be used to estimate the effects of pipe or duct losses.

It is also possible to adapt existing tanks for solar operation, as shown by Czarnecki and Read (1978), by use of special fittings that allow the needed extra connections to be made to the tanks for circulation through the collector. Single openings in a tank can serve as dual openings by use of fittings having concentric center tubes; the center tube and annulus provide two connections that can be separated by lengthening the tube. Czarnecki and Read found that the measured thermal performance of such a system was poorer than that of a similar one-tank forced circulation system using an outlet to collector at the bottom of the tank and a return from the collector located half way up the tank. Thus, reduced installation cost resulted in reduced solar output from the system. It is difficult to generalize on the merits of these various configurations, but as long as collector costs dominate the costs of the system, it is advisable to consider the use of other components with good performance characteristics which will produce the most useful gain from the collectors.

12.8 WATER HEATING IN SPACE HEATING AND COOLING SYSTEMS

The following chapters treat space heating and cooling, and most systems for these applications will include provision for water heating. The water heating subsystems in these systems often include heat exchanger to extract heat either from the fluid circulated through the collector or from the main storage tank. A "solar preheat" tank to store solar heated water and a conventional water heater (auxiliary) may be used, as shown in Figure 12.8.1, or a single-tank

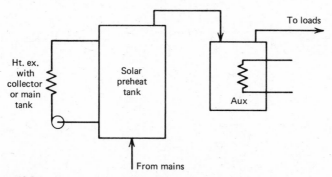

Figure 12.8.1 A two tank hot water subsystem for solar heating/cooling systems.

system including the auxiliary source can be used. The collectors for heating and cooling are usually sized for these functions, and are oversized for water heating during periods when heating or cooling loads are low. Even in northern U. S. climates, annual residential hot water loads are typically approximately one-fourth to one-fifth of annual heating loads, so significant contributions to total annual loads can be obtained by well-designed hot water facilities.

12.9   COMBINED COLLECTOR-STORAGE

Water heaters that combine the functions of collector and storage are manu-factured in Japan [Tanishita (1970)]. A cross-sectional schematic of this type is shown in Figure 12.9.1. In the morning, water is put in the tubes, which are black plastic cylinders about 0.2 m in diameter. Through the day, the water is heated by absorbed solar energy. The top cover and back insulation reduce energy losses from the cylinders. Analysis of the performance of this type of collector can be done by solving the differential equation describing the energy balance on the tank, as shown in the following example.

Example 12.9.1

A water heater that combines collector and storage is constructed of 150 mm black tubes, placed side by side in a box, insulated on its bottom to the extent that bottom losses are small, and covered in the top by a single glass cover. A section of the heater is shown in Figure 12.9.1.

The heater is filled with water at 18 C at 6 a.m. The heater is sloped toward the equator with $\beta = \phi = 40°$. Meteorological data for a 24-hr period are shown in the second and third columns of the table.

Assume that no water is drawn off at any time during the 24-hr period. Estimate the temperature history of the water through the 24-hr period.

Hour Ending	I_T, MJ/m^2	T_a, C	T_w^+, C	Hour Ending	I_T, MJ/m^2	T_a, C	T_w^+, C
7 a.m.	0.05	7	17.5	7 p.m.	—	15	43.4
8	0.32	9	17.7	8	—	14	42.0
9	1.09	14	19.3	9	—	12	40.5
10	1.23	16	21.2	10	—	10	38.9
11	2.36	16	25.0	11	—	9	37.4
12	3.95	17	31.3	12	—	9	36.0
1 p.m.	3.90	20	37.4	1 a.m.	—	9	34.7
2	3.52	20	42.5	2	—	8	33.3
3	2.55	21	45.7	3	—	6	32.0
4	1.38	22	46.9	4	—	7	30.7
5	0.46	21	46.4	5	—	7	29.5
6	0.04	16	44.9	6	—	7	28.4

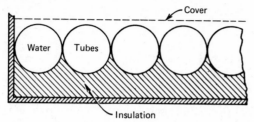

Figure 12.9.1 Schematic cross section of a solar water heater combining collector and storage.

Solution

A set of assumptions must be made to estimate the temperature history of this heater. These are as follows:

1 Absorptance = 0.95.
2 Emittance of black plastic = 0.95.
3 Solar transmittance of cover = 0.89.
4 Heat capacity of cover and other structure are negligible compared to that of the water.
5 Water and tubes are at uniform temperature.
6 Wind speed is such that h_w is 10 W/m² C.
7 $U_t = U_L$ = constant through the period.

First, estimate U_L. If the plate were flat, U_t would be approximately 5.5 W/m² C, estimated from Figure 6.4.4b. The area of the plate is $0.5\pi D/D$ larger than the cover, or

$$A_p = 1.57 A_c$$

As a first order estimate

$$U_t = 5.5\left(1 + \frac{0.57}{2}\right) = 7.0 \text{ W/m}^2$$

a value used without change in the calculation below.

The basic equation used is Equation 6.11.7. This can be written

$$T_w^+ = T_a + \frac{S}{U_L} - \left[\frac{S}{U_L} - (T_w - T_a)\right]e^{-A_c U_L t/(mc)_e}$$

For each hour,

$$S = \frac{1.02 \times 0.89 \times 0.95 I_T}{7.0 \times 3600} = 34.1 I_T$$

$$\frac{A_c U_L t}{(mc)_e} = \frac{1 \times 7.0 \times 3.6 \times 0.15}{(\pi/4)(0.15)^2 \times 4190} = 0.0511$$

and

$$e^{-0.0511} = 0.95$$

a working equation is thus

$$T_w^+ = T_a + 34.1I_T - [34.1I_T - T_w + T_a]0.95$$

The temperatures at the end of each hour, T^+, are shown in the last column in the table. These heaters are suitable for use where hot water is wanted at the end of the collection period. Thermal losses from the stored hot water are sufficiently high at night or during prolonged cloudy periods that their use for continuous hot water is usually impractical.   ■

Water heaters combining collector and storage are passive systems, in that they require no mechanical energy for operation. They are similar in many ways to combined collector-storage walls used in passive space heating (see Chapter 15).

12.10 SWIMMING POOL HEATING

Heating of swimming pools to prolong swimming seasons represents a significant consumption of energy in some areas, and a substantial expense to pool owners and operators. The temperatures required for pools are low, usually not much above average ambient temperatures. For these reasons, solar heating of pools has become an application of increasing interest. Two general approaches have been taken, the first based on covering the pool in such a way to use it as the collector, and the second based on the use of separate collector systems.

Open pools lose heat by conduction to the ground, by convection to the air, and by evaporation,* as outlined by Löf and Löf (1977). The simplest and least expensive method for heating pools is to cover them when they are not in use with a transparent plastic cover which floats on the surface of the water. The cover may fit within the edges of the pool, or it may overlap the edges and be held down on the area around the pool by weights placed on the plastic to keep it in position. Sections of pools with such covers are shown in Figure 12.10.1. As long as the top of the cover is dry, evaporation losses are effectively eliminated. The plastic cover, if largely transparent to solar radiation, admits solar energy which is absorbed in the pool. The cover also serves to keep dirt out of the pool. It is difficult to quantify the performance of such a system, as times of use, winds, humidity, etc. are highly variable. However, experiments in Denver have indicated that through the summer months an uncovered pool will have an average temperature below ambient, while a pool covered except for periods

* Section 20.2 discusses solar evaporators, an analogous situation.

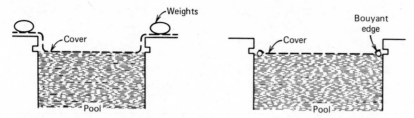

Figure 12.10.1 Schematic sections of plastic pool covers: (*a*) held on pool edges by weights; (*b*) with floating edge.

of occasional use will have average temperatures of 5 to 10 C above ambient temperature.

Either single layer or double layer covers are used. Single layers store in less volume when not on the pool, and have higher transmittance. Double-layer covers float without addition of other bouyant materials but have lower transmittance for solar radiation and do not store in quite as compact a manner as single-layer covers.

Separate collectors are also used for heating pools. They operate with water inlet temperatures near ambient, by forced circulation, without covers and with a minimum of insulation. They may be made of metal or plastic, and may be installed as a roof material for a patio or pool cabana. Swimming pool heaters have been described by Andrassy (1964) and others; deWinter (1973) has written a monograph on design and construction of pool heaters of this type, and an analysis of pool-heating collectors without covers has been presented by Farber (1978).

12.11 SUMMARY

Water heating is a practical application of solar energy in many parts of the world. Natural circulation systems are widely used in climates where freezing does not occur. Forced circulation systems using drain-down or nonfreezing fluids in collector-heat exchanger loops are used in climates where freezing is a problem. Auxiliary energy is used in essentially all systems where high reliability is wanted, and care must be taken to use auxiliary in such a way that it does not drive up collector temperatures.

REFERENCES

Andrassy, S., *Proceedings of the UN Conference on New Sources of Energy*, **5**, 20 (1964). "Solar Water Heaters."

Buckles, W. E., S. A. Klein, and J. A. Duffie, paper presented at ISES meeting, Atlanta (1979). "Analysis of Solar Water Heating Systems."

Close, D. J., *Solar Energy*, **6**, (1), 33 (1962). "The Performance of Solar Water Heaters with Natural Circulation."

Cooper, P. I., personal communication (1973).

Czarnecki, J. T. and W. R. W. Read, *Solar Energy*, **20**, 75 (1978). "Advances in Solar Water Heating for Domestic Use in Australia."

deWinter, F., publication of the Copper Development Association, Inc. (1973). "How to Design and Build a Solar-Energy Swimming Pool Heater."

Farber, J., paper in *Proceedings of the 1978 Annual Meeting of the American Section of the ISES*, Denver, **2** (1), 235 (1978). "Analysis of Low Temperature Plastic Collectors."

Gupta, C. L. and H. P. Garg, *Solar Energy*, **12**, 163 (1968). "System Design in Solar Water Heaters with Natural Circulation."

Gutierrez, G., F. Hincapie, J. A. Duffie, and W. A. Beckman, *Solar Energy*, **15**, 287 (1974). "Simulation of Forced Circulation Water Heaters; Effects of Auxiliary Energy Supply, Load Type and Storage Capacity."

Löf, G. O. G. and D. J. Close, *Low Temperature Engineering Applications of Solar Energy*, ASHRAE, New York (1967). "Solar Water Heaters."

Löf, G. O. G. and L. G. A. Löf, paper in *Proceedings of the 1977 Annual Meeting of the American Section of the ISES*, Orlando, **1**, 31-1 (1977). "Performance of a Solar Swimming Pool Heater—Transparent Cover Type."

Mutch, J. J., RAND Report R 1498 (1974). "Residential Water Heating, Fuel Consumption, Economics and Public Policy."

"Solar Water Heaters, Principles of Design, Construction and Installation," Div. of Mech. Engr. Circular No. 2, Commonwealth Scientific and Industrial Research Organization, (CSIRO), Melbourne, Australia (1964).

Tabor, H., *Bulletin, Cooperation Mediterraneene pour L'Energie Solaire* (COMPLES), No. 17, 33 (1969). "A Note on the Thermosyphon Solar Hot Water Heater."

Tanishita, I., paper presented at Melbourne International Solar Energy Society Conference (1970). "Present Situation of Commercial Solar Water Heaters in Japan."

CHAPTER 13

Solar Heating

Heat for comfort in buildings can be provided from solar energy by systems that are similar in many respects to the water heater systems described in Chapter 12. The two most common heat transfer fluids are water and air, and systems based on each of these are described in this chapter. The basic components are the collector, storage unit, and load (i.e., the house or building to be heated). In temperate climates, an auxiliary energy source must be provided and the design problem is in part the determination of the optimum combination of solar energy and auxiliary (i.e., conventional) energy.

In this and the following chapter, we deal with "active" solar heating systems, which use collectors to heat a fluid, storage units to store solar energy until needed, and distribution equipment to provide the solar energy to the heated spaces in a controlled manner. In combination with conventional heating equipment solar heating provides the same levels of comfort, temperature stability, and reliability as conventional systems.

The term "solar house" is also applied to buildings that include as integral parts of the building elements that admit, absorb, store, and release solar energy and thus reduce the needs for auxiliary energy for comfort heating. Architectural design can be used to maximize solar gains in the winter (and minimize them in the summer) to reduce heating (and cooling) loads that must be met by other means. Elements in the building (floors, walls) may be constructed to have high heat capacity to store thermal energy and reduce temperature variations. Movable insulation may be used to control losses (and gains) from windows or other architectural elements in the building. These concepts will be treated in Chapter 15 on passive heating. There is no substitute for good energy-conserving architectural design which (as far as other constraints allow) maximize the solar gains in the building itself. In this and the following chapter we are concerned with methods for meeting the energy needs for heating (and hot water) that are not eliminated by careful design.

Since 1970 there has been a tremendous surge of interest and activity in solar heating, and a small but growing industry has developed to design, develop, manufacture, sell, and install solar heating equipment and systems in the United States. As of 1977, patterns in the configurations of many air and liquid systems were emerging. These "standard" configurations, which are used with many variations, are discussed here in some detail. In addition, other system types are noted. We show operating data for several systems that are illustrative of the long-term thermal performance data needed to evaluate the economics of solar heating.

The design of solar heating systems is a two-part problem, including thermal performance analysis and economic analysis. The latter can be readily done with the P_1, P_2 method of Chapter 11. The solar heating industry had grown to the point in 1978 that methods were needed for estimating thermal performance of systems, without resorting to simulations. These methods are reviewed in Chapter 14, and one of them, the f-chart method, is presented in detail.

13.1 HISTORICAL NOTES

A gift by Godfrey L. Cabot to the Massachusetts Institute of Technology for solar energy studies in 1938 marked the beginning of modern research in solar heating. The MIT program resulted in development of methods of calculating collector performance (see Chapter 6) which, with some modification, are standard methods in use today. It also resulted in the successive development of a series of four solar heated structures, with the last at Lexington, Massachusetts. The building was equipped with a carefully engineered and instrumented system based on solar water heaters and water storage. The solar heating system was designed to carry approximately two-thirds of the total winter heating loads in the Boston area, and its performance was carefully monitored (see Section 13.5).

Telkes and Raymond (1949) described a solar house that was constructed at Dover, Massachusetts, that utilized vertical south-facing air heater collectors and energy storage in the heat of fusion of sodium sulfate decahydrate. This system was designed to carry the total heating load, having theoretical capacity in the storage system to carry the design heating loads for 5 days.

Bliss (1955) constructed and measured the performance of a fully solar heated house in the Arizona desert, using a matrix, through-flow air heater and a rock bed energy storage unit. It was noted that the system as built did not represent an economic optimum, and a smaller system using some auxiliary energy would have resulted in lower cost.

Löf designed an air heating system using overlapped glass-plate collectors and a pebble bed for energy storage and, using these concepts, built a residence near Denver, in which he and his family have lived since 1959. The performance of this system during the first years of its operation was studied and reported by Löf et al. (1963, 1964) and was again measured in 1976–78 (see Section 13.3). The data are particularly significant in that they are the only data available on a system that has been in operation over a time span of 18 years, a time period comparable to the time periods over which solar heating systems are being amortized.

Close et al. (1968) described a heating system used for partial heating of a laboratory building in Australia that has been in operation for many years.

This system uses a 56 m² vee-groove air heater and a pebble bed storage unit. Air flow through the collectors is modulated to obtain a fixed 55 C air outlet temperature.

Since 1970, many varied experimental systems have been built, and the performance of a few of them has been measured and reported. These experiments have led to commercial production of both liquid and air systems, and these are being sold and installed by the thousands.

13.2 SOLAR HEATING SYSTEMS

Figure 13.2.1 is a schematic of a basic solar heating system using air as the heat transfer fluid, with a pebble bed storage unit and auxiliary furnace. The various modes of operation are achieved by appropriate damper positioning. Most air systems are arranged so that it is not practical to combine modes by both adding energy to and removing energy from storage at the same time. The use of auxiliary energy can be combined with energy supplied to the building from collector or storage if that supply is inadequate to meet the loads. In this system configuration it is possible to by-pass the collector and storage unit when auxiliary alone is being used to provide heat.

A more detailed schematic of an air system is shown in Figure 13.2.2. Blowers, controls, means of obtaining service hot water, and more details of ducting are shown. The hot water subsystem is the same as that shown in Figure 12.8.1. Auxiliary energy for space heating is added to "top off" that available from the solar energy system. This is the air system configuration on which the design procedure of Chapter 14 is based. Table 13.2.1 shows typical design parameters for heating systems of this type.

Air systems have a number of advantages compared to those using liquid heat transfer media. Problems of freezing and boiling in the collectors are eliminated and corrosion problems are reduced. The high degree of stratification possible in the pebble bed leads to lower collector inlet fluid temperatures. The

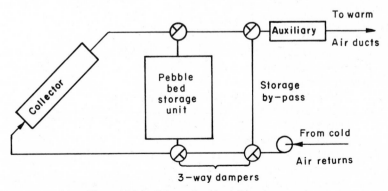

Figure 13.2.1   Schematic of basic hot air system.

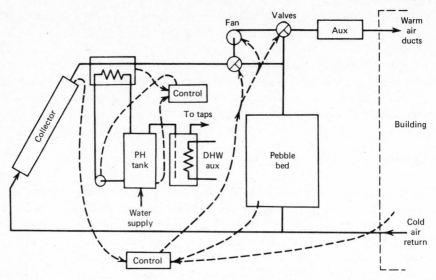

Figure 13.2.2   Detailed schematic of a solar heating system using air as the heat transfer fluid.

working fluid is air, and warm air heating systems are in common use. Control equipment is readily available that can be applied to these systems. Disadvantages include the possibility of relatively high fluid pumping costs (if the system is not carefully designed), relatively large volumes of storage, and the difficulty of adding air conditioning to the systems. Air systems are also relatively difficult to seal; leakage of solar heated air from collectors and ductwork can represent a significant energy loss from the system. Air collectors are operated at

Table 13.2.1   Typical Design Parameters for Solar Air Heating Systems

Collector air flowrate	5 to 20 liters/m² s
Collector slope	$(\phi + 15°) \pm 15°$
Collector surface azimuth angle	$0° \pm 15°$
Storage capacity	0.15 to 0.35 m³ pebbles/m²
Pebble size (graded to uniform size)	0.01 to 0.03 m
Bed length, flow direction	1.25 to 2.5 m
Pressure drops:	
Pebble Bed	55 Pa minimum
Collectors	50 to 200 Pa
Ductwork	10 Pa
Maximum entry velocity of air into pebble bed (at 55 Pa pressure drop in bed)	4 m/s
Water preheat tank capacity	1.5 × conventional water heater

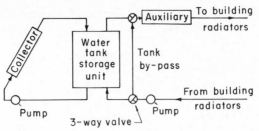

Figure 13.2.3 Schematic of basic hot water system.

lower fluid capacitance rates and thus with lower values of F_R than are liquid heating collectors.

Figure 13.2.3 is a schematic of a basic water heating system with water tank storage and auxiliary energy source. This system allows independent control of the solar collector-storage part of the system on the one hand, and storage-auxiliary-load part of the system on the other, as solar heated water can be added to storage at the same time that hot water is removed from storage to meet building loads. In the system illustrated, a bypass around the storage tank is provided to avoid heating the storage tank with auxiliary energy.

More details of a liquid based system are shown in Figure 13.2.4. A *collector heat exchanger* is shown between the collector and storage tank, allowing the

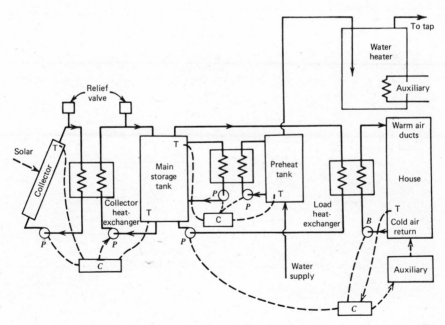

Figure 13.2.4 Detailed schematic of a liquid based solar heating system. P, pump; C controller; T, temperature sensor; B, blower.

Table 13.2.2 Typical Design Parameter Ranges for Liquid Solar Heating Systems

Collector flowrate	0.010 to 0.020 kg/s m^2
Collector slope	$(\phi + 15°) \pm 15°$
Collector azimuth	$0° \pm 15°$
Collector heat exchanger	$F_R'/F_R > 0.9$
Storage capacity	50 to 100 liters/m^2
Load heat exchanger	$1 < \varepsilon_L C_{min}/(UA)_h < 5$
Water preheat tank capacity	$1.5 \times$ capacity of conventional heater

use of antifreeze solutions in the collector. Relief valves are shown for dumping excess energy should the collector run at excessive temperatures. A *load heat exchanger* is shown to transfer energy from the tank to the heated spaces. Means of extracting energy for service hot water are indicated. Auxiliary energy for heating is added so as to "top off" that available from the solar energy system. This system is the basis of the liquid system design method discussed in Chapter 14. Typical design parameter values for systems of this type are shown in Table 13.2.2.

The load heat exchanger, the exchanger for transferring solar heat from the storage tank to the air in the building, must be adequately designed to avoid excessive temperature drop and corresponding increase in tank and collector

(a)

Figure 13.2.5 Three solar heated buildings: (*a*) The Denver House, (*b*) MIT House IV (Courtesy H. C. Hottel), (*c*) CSU House I.

temperatures. The parameter describing this exchanger is $\varepsilon_L C_{\min}/(UA)_h$, where ε_L is the effectiveness of the exchanger, $C_{\min}$ is the lower of the two fluid capacitance rates (mass flow rate times specific heat of the fluid) in the heat exchanger (usually that of the air), and $(UA)_h$ is the building overall energy loss coefficient-area product.*

An alternative to the use of antifreeze loops to transfer heat from the collector is a drain-down system. This method of freeze protection was used in MIT House IV.

Advantages of liquid heating systems include high collector F_R, smaller storage volume, and relatively easy adaptation to supply of energy to absorption air conditioners.

Use of water also involves problems. Freezing of collectors must be avoided, as outlined in Chapter 12 for water heaters. Solar heating systems using liquids will operate at lower water temperatures than conventional hydronic systems, and will require more baseboard heater area (or other heat transfer surface) to transfer heat into the building. In spring and fall, solar heaters will operate at excessively high temperatures and means must be provided to vent energy to avoid pressure build-up and boiling. Care must also be exercised to avoid corrosion problems.

It is useful to consider solar systems as having four basic modes of operation, depending on the conditions that exist in the system at a particular time:

Mode A If solar energy is available and heat is not needed in the building, energy gain from the collector is added to storage.

Mode B If solar energy is available and heat is needed in the building, energy gain from the collector is used to supply the building need.

Mode C If solar energy is not available, heat is needed in the building, and the storage unit has stored energy in it, the stored energy is used to supply the building need.

Mode D If solar energy is not available, heat is needed in the building, and the storage unit has been depleted, auxiliary energy is used to supply the building need.

Note that there is a fifth situation that will exist in practical systems. The storage unit is fully heated, there are no loads to be met, and the collector is absorbing radiation. Under these circumstances, there is no way to use or store the collected energy and this energy must be discarded. This can happen through operation of pressure relief valves or other energy dumping mechanisms, or the collector temperature will rise until the absorbed energy is dissipated by thermal losses.

Additional operational modes may also be provided, for example, to provide service hot water. It is possible with many systems to combine modes, that is, to operate in more than one mode at a time. Many systems do not allow direct

---

* The quantitative assessment of the effect of this parameter will be shown in Chapter 14.

heating from collector to building, but transfer heat from collector to storage whenever possible and from storage to load whenever needed.

To illustrate the system configurations, modes of operation and designs of air and water systems, the following four sections describe the design and performance of four residential-scale building systems. These buildings were carefully engineered, and their performance has been measured and the results documented. The buildings and systems are the Löf residence in Denver and CSU House II (air systems), and MIT House IV and CSU House I (liquid systems). Photographs of three of these buildings are shown in Figure 13.2.5; these show variations in architectural treatment. The thermal characteristics and appearance of the two CSU buildings are essentially identical.

13.3 THE DENVER HOUSE

The Denver house was constructed in 1958 to 1959, and has been described in some detail by Löf et al. (1963, 1964). The house uses air-heating collectors, pebble bed storage unit, and natural gas furnace for an auxiliary source. The house has a heated floor area of approximately 195 m² and 102 m² in the basement. The building is of contemporary design with substantial areas of windows and a flat roof. The design heat load was calculated to be 31.8 kW at an ambient temperature of − 18 C and wind speed of 3.9 m/s.

The heating system is shown in Figure 13.3.1. Collectors are in two banks, angled up from the flat roof at a slope of 45°. Each bank has a nominal area of

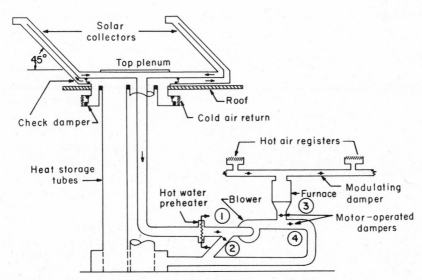

Figure 13.3.1 Schematic of the Denver house air heating system. Adapted from Löf et al. (1963) and UN Conf. Proc (1964).

27.9 m^2 for a gross collector area of 55.7 m^2. The collector aperture area, allowing for collector frames, is 49.2 m^2. Collectors are of the overlapped glass plate design shown in Figure 6.12.1h; air flows through two modules in series, the first having one cover and the second having two.

The storage medium is 10,600 kg of rock of a nominal mean size of 30 mm diameter, and a specific heat of 750 J/kg C. It is contained in two cylindrical tubes 0.91 m in diameter and 5.5 m high. A duct leads down through one of the storage tubes as a means of access between collector banks on the roof and equipment in the basement.

Some solar energy is provided for service hot water heating by an air-water heat exchanger serving as a preheater. The balance of needed heat for hot water is provided in a fuel-fired conventional heater. Other components are shown in the diagram, such as the blower, furnace, dampers, cold and warm air registers, and so on.

Solar radiation is sensed by a resistance thermometer mounted on a black plate above and parallel to one of the collectors, which is designed to have a thermal response similar to the black plate in the collector. Another resistance thermometer is located in the exit (in this case the top) of the storage units, and air to the collector is turned on or off by the detection of an appropriate temperature difference between the two resistance elements.

Room thermostats calling for heat in the house control dampers that direct air from collector or storage to the house. If, after a period of about 10 min of air flow from collector or storage, the condition called for by the thermostat is not met, the furnace turns on. When collectors are operating, heated air goes to the building if needed; otherwise it goes to storage.

The four modes of operation of the previous section (and combinations thereof) are used in this system:

1   When the house does not need heat and solar energy is high enough to justify collection, dampers 1 and 4 are open, 2 and 3 are closed. Air flow is from collector to water preheater to blower to storage unit and return to collector.

2   When house needs can be met directly by solar energy, dampers 1 and 3 are open, 2 and 4 are closed. Air flow is from collector to water preheater to blower to furnace to hot air registers, through the building to cold air returns to collector. In this system some gas heat may be supplied by the furnace if solar heat is not sufficient to meet house demands.

3   For house heating from storage, dampers 2 and 3 are open, 1 and 4 are closed. Air flow is from rooms to cold air returns, down through storage to blower to furnace to hot air registers to room.

4   Operation completely on auxiliary (gas) is the same as C, except that the furnace is on. Thus some heat from storage can be supplemented by gas or all energy may come from gas.

The results of measurements of performance of the system have been reported for the 1959 to 1960 heating season. They are based on measurements of air

Table 13.3.1   Energy Balance of Denver Solar House Winter 1959 to 1960[a]

	Energy, GJ
Total energy incidence on 55.7m² collector area	239
Total solar incidence on 55.7m² collector area when collection cycles operated	170
Useful collected heat	59
Net collector efficiency, percent	35
Solar heat to storage tubes	27
Solar heat to water preheater	4
Heat delivered by natural gas for house heating	150
Heat delivered by natural gas for water heating	22
Total heat load	230
Useful collected heat absorbed by water preheater	7%
Total water heating load supplied by solar energy	16%
House heat load supplied by solar energy (including water preheating but excluding water heating)	28%
House heat load supplied by solar energy (including both water preheating and heating)	25%

[a] From Löf et al. (1964).

temperatures, air flow rates, storage temperatures, indication of modes of operation, and gas consumption for space heating and for hot water. With these data, energy quantities in various parts of the system were integrated for months and for the total heating season. Table 13.3.1, from Löf et al., shows the year's data for the Denver house. Figure 13.3.2 shows cumulative performance data for the heating season.

The Denver house solar system is small relative to the heating loads of the building, and supplied approximately 25 percent of the season's heating and hot water loads by solar energy. Most systems carry substantially higher fractions of the annual loads by solar, usually in the range of 50 to 75 percent. The parasitic power to operate this system is significant; the shape of the energy storage tubes is such that velocities and air flow pathlengths through the pebble beds are higher than they should be. It has been estimated that pumping power required by the storage unit would be reduced to approximately 5 percent of its current value if the storage unit had the shape of a cube with the same storage capacity.

The system in this house was in routine operation from 1959 to 1977 with essentially no maintenance [Ward and Löf (1977)], indicating its durability. The performance of the system was measured again for the 1974 to 1975 heating season [Ward and Löf (1976)]. After 18 years the solar output was estimated to be 78 percent of its initial value. Most of the reduction in output is ascribed to glass breakage of the interior clear glass plates and black glass plates in the overlapped plate air heating collectors. None of the outer covers were broken.

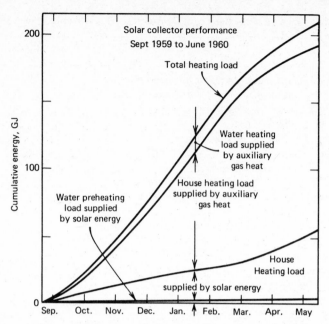

Figure 13.3.2 Cumulative total heating loads, auxiliary energy used, and solar energy supplied for the Denver house for 1959–60. Adapted from Löf et al. (1964).

Current collector designs are not subject to this difficulty. These performance measurements indicate that air systems can be expected to operate over many years with very low maintenance, if the systems are properly designed and installed.

13.4 CSU HOUSE II—AIR SYSTEM

Colorado State University House II uses an air system basically similar to that of the Denver house, but with more conventional air heating collectors and a nearly cubic pebble bed energy store. The system configuration is shown in Figure 13.4.1 and is essentially the same as is shown in Figure 13.2.2. The house is nearly identical in appearance to CSU House I, as shown in Figure 13.2.5. The system is described and its performance is summarized by Ward et al. (1977), and Karaki et al. (1977, 1978).

The building serves as an office and laboratory, and has 130 m² floor area on the first floor and an equal area on a basement level. The collectors are site-built* and are similar in design to air collectors now available on the commercial

---

* New collectors have been put on this "test-bed" building since the experiments reported here. This system also includes capability to obtain some cooling by nocturnal chilling of the pebble bed with evaporatively cooled air. This will be described in Chapter 16.

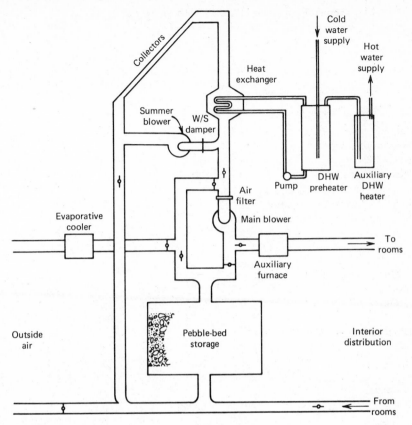

Figure 13.4.1 Schematic of the system in CSU House II. From Karaki et al. (1977).

market. They occupy an area of 68.4 m² on the south roof, and have 64.1 m² net collector area. The storage unit is 10.2 m³ of 20 to 40 mm diameter pebbles. The auxiliary energy source is a gas-fired duct furnace. Domestic water preheat is obtained by an air-to-water heat exchanger; the preheat tank capacity is 300 liters. The system configuration requires only one blower for the heating operation, but has a second blower for providing heated air to the hot water subsystem in the summer.

Controls on the collector loop during the heating season are based on temperature measurements in the top of the collectors and the bottom of the storage unit; when the difference of these two exceeds 11 C, the blower is turned on, and it remains on until the difference drops to 0.6 C. The solar heated air goes to the building if heat is needed; otherwise, it goes to storage. A thermostat in the building senses when the rooms need heat and positions the dampers to obtain heat from the collector (if available) or storage. If the room temperature drops 1.1 C below the control point a second stage contact is made and the auxiliary gas duct heater is turned on.

Table 13.4.1   Thermal Performance Data for Solar House II[a]

	1976		1977			
	Nov 24 days	Dec 10 days	Jan 26 days	Feb 20 days	Mar 27 days	April 18 days
Collector (MJ/m² day)						
Total solar insolation	13.5	15.8	15.1	15.6	17.7	14.2
Solar insolation						
while collecting	11.2	13.8	12.9	13.6	15.6	12.2
Heat collected	3.1	4.9	4.6	4.7	5.0	3.7
Efficiency (percent):						
Based on total solar insolation	23	31	30	30	28	26
Based on solar insolation						
while collecting	28	36	36	35	32	30
Space and DW Heating (MJ/day)						
Total energy required						
Space	311.6	323.2	432.9	319.9	319.0	192.3
DHW	40.6	52.3	73.0	82.9	75.4	62.8
Total	352.2	375.5	505.9	402.8	394.4	255.1
Solar contribution						
Space	198.7	257.5	231.4	239.6	245.8	166.1
DHW	21.3	30.8	63.5	75.4	67.4	51.8
Total	220.0	288.5	294.9	315.0	313.2	217.9
Solar fraction						
Space	0.64	0.80	0.53	0.75	0.77	0.86
DHW	0.52	0.59	0.87	0.91	0.89	0.82
Total	0.62	0.77	0.58	0.78	0.79	0.85

[a] From Karaki et al. (1977).

The house and its heating system have been extensively instrumented, and data for most of the winter of 1976–77 (except for some days when instrumentation was not operable) are shown in Table 13.4.1. For the 1976–77 heating season, the solar heating system carried 72 percent of the heating and domestic hot water (DHW) loads. The parasitic power for operation of the solar energy system was 7.4 percent of the delivered solar energy for space and water heating, resulting in a "coefficient of performance," the ratio of solar energy delivered to electrical energy to operate the system, of 13.5 for the heating season.

13.5   THE MIT HOUSE IV

The MIT House IV has been described in detail by Engebretson and Ashar (1960) and Engebretson (1964), and is shown in Figure 13.2.5. It was built in

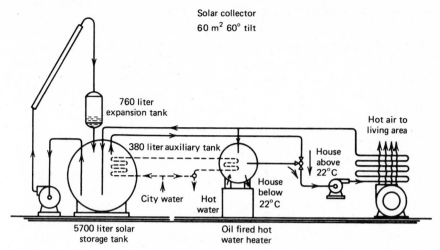

Figure 13.5.1   Schematic of the MIT House IV heating system. Adapted from Engebretson (1964).

1958–1959 and data on its operation are available for the two heating seasons of 1959–1960 and 1960–1961. The house used water in the collectors and storage tank, and an oil-fired auxiliary furnace to supply space heating and hot water needs of the two story, 135 m^2 building. The house was designed specifically to incorporate solar heating in the most effective way. The collectors were part of the envelope of the building, which had minimum surface-to-volume ratio. At the same time, the house was designed to be a pleasant and functional structure. The collector was the dominant architectural feature (in contrast to the design of the Denver house, which used a relatively smaller collector partially hidden from view).

A schematic of the system is shown in Figure 13.5.1. The collector had an area of 59.5 m^2 at a tilt of 60° to the south, using two low-iron content glass covers, black painted aluminum absorbing surface, copper tubing in risers, and headers, and both air space and fiberglass insulation on the back. For this collector, $F_R = 0.86$, $\alpha = 0.97$, and $U_L = 3.97$ W/m^2 C. The storage tank had a capacity of 5670 liters. An expansion tank of 757 liter capacity permitted draining of the collector system when it was not in use. Heat was transferred to air in the house through a water-to-air heater exchanger. The oil-fired auxiliary heater included a 380 liter tank from which water was circulated to the water-air heat exchanger. Hot water was supplied by passing supply water in series through a heating coil in the main storage tank and another coil in the auxiliary tank; the resulting hot water was mixed with mains water to obtain the desired 60 C water.

The collector pump was controlled by the difference in temperature between sensor elements applied to the collector plate and the storage tank. Since this collector was drained when not in use, the control allowed a small amount of overheating of the collector before start-up, and also allowed some delay in the form of thermal response lag in the collector sensor to avoid premature shut-down when the first cool water from the piping system entered the collector.

Space heating control was by means of a two-stage thermostat in the house. As the temperature in the room dropped, the heat exchanger pump (and circulating fan) turned on to provide heat from the storage tank. If the temperature continued to drop, that is, if the stored solar energy could not meet the load, the position of the motorized valve shifted to provide circulation of hot water from the auxiliary tank rather than the main tank.

These modes of operation were somewhat different than those for the air heating system, as the collector-storage part of the system was controlled and operated independently of the storage-load part. In the nomenclature of Section 13.2, *Mode A* operated whenever the energy incident on the collector was enough to deliver useful energy to the fluid at the bottom storage tank temperature. *Mode B* did not exist in this system. *Mode C* operated when the house needed heat and the storage tank could provide it, the first stage of operation called for by the room thermostat. *Mode D* operated on the second stage of the thermostat, with heat delivered by the auxiliary heater.

The house was instrumented with means for measuring appropriate temperatures and flow rates to permit calculation of energy balances on all components and on the system as a function of time. These energy balances were integrated to give cumulative performance through the heating seasons. The results of these measurements are shown in Figure 13.5.2 and 13.5.3 [from Engebretson (1964)], each of which shows 1959 to 1960 data (solid lines) and 1960 to 1961

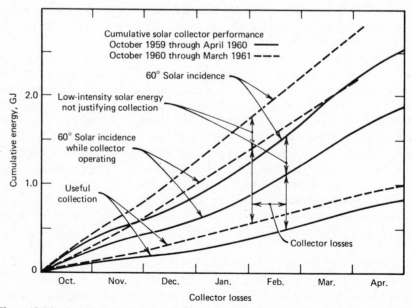

Figure 13.5.2   Cumulative collector performance for MIT House IV. From Engebretson (1964).

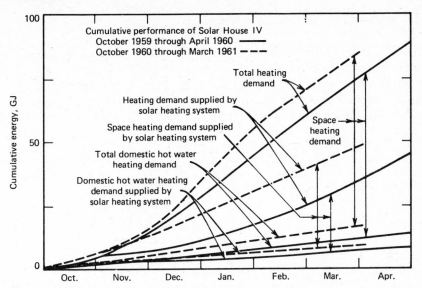

Figure 13.5.3 Cumulative system thermal performance for MIT House IV. From Engebretson (1964).

data (dashed lines). Figure 13.5.2 shows collector performance. Figure 13.5.3 shows performance of the house, indicating cumulative building needs for space heat and for domestic hot water heat, and how each of these was met by a combination of solar and auxiliary energy. These figures show that the weather was significantly better and the solar energy delivered by the system to the house was substantially higher in the second year than in the first. Table 13.5.1 summarizes some of the integrated energy quantities for the building for the 2 years.

Table 13.5.1   Integrated Performance of MIT House IV over Two Heating Seasons[a]

	GJ	
	1959–60	1960–61
Space heating demand	72.5	70.7
Space heating from solar energy	33.6	40.2
Water heating demand	14.7	17.6
Water heating from solar energy	8.4	9.7
Total heating demand	87.1	88.3
Total heating from solar energy	41.9	49.9
Percent from solar energy	48.1	56.6

[a] From Engebretson (1964).

13.6 CSU HOUSE I—LIQUID SYSTEM

The CSU House I is essentially the same as House II in its size and construction, and is shown in Figure 13.2.5. The system is a liquid-based system, very similar in configuration to that shown in Figure 13.2.4. Performance data for the system are summarized by Karaki et al. (1978); more detailed design information and early performance data are in Löf and Ward (1976). Data are shown in Table 13.6.1.

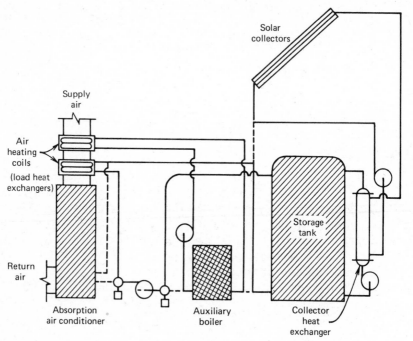

Figure 13.6.1 Schematic of the Liquid Solar System in Solar House I. Adapted from Karaki et al. (1978).

A schematic of the system is shown in Figure 13.6.1.* The collector supplies energy to the storage tank through the collector heat exchanger. The loads are met by supplying energy from the tank to the lower of two air heating coils. Auxiliary energy is normally supplied by a separate circuit through the upper air heating coil. The system is arranged to permit experiments with changing configurations; some of these connections are indicated as dashed lines on the diagram.

Two collectors have been used on this building. The first was a site-built collector shown schematically in Figure 13.6.2. For this collector, measured

---

* It includes a solar-operated absorption air conditioner; cooling by this method will be discussed in Chapter 16.

Table 13.6.1 Thermal Performance Data for CSU House I[a]

	Flat Plate Collectors 1976				Evacuated Tube Collectors 1977			
	Feb. 29 days	Mar. 4.5 days	April 23 days	May 12 days	Jan. 26 days	Feb. 20 days	Mar. 27 days	April 18 days
Collector (MJ/m² day)								
Total solar insolation	19.0	16.6	16.5	19.6	14.8	16.2	17.5	15.3
Solar insolation while collecting	9.3	8.5	6.8	9.0	13.2	14.8	16.6	12.7
Heat collected	3.5	2.9	2.3	3.1	5.6	6.1	7.7	3.2
Efficiency (percent)								
Based on total solar insolation	18	17	14	16	38	38	44	21
Based on solar insolation while collecting	38	34	34	34	43	42	46	25
Space and DW Heating (MJ/day)								
Total energy required:								
Space					544.3	384.1	300.9	147.9
DHW					121.1	102.1	105.0	189.0
Total	305.0	573.3	214.4	326.0	665.4	486.2	405.9	336.9
Solar contribution:								
Space					457.9	361.6	268.8	135.4
DHW					39.5	39.7	63.9	107.7
Total	231.5	392.7	159.9	241.2	497.4	401.3	332.6	243.1
Solar fraction								
Space					0.84	0.94	0.89	0.92
DHW					0.33	0.39	0.61	0.57
Total	0.76	0.68	0.75	0.74	0.75	0.83	0.82	0.72

[a] From Karaki et al. (1978).

449

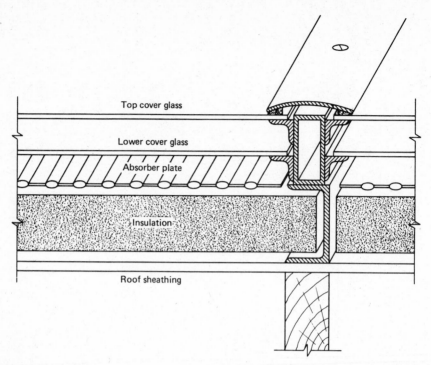

Figure 13.6.2 Collector section of site-built collector on CSU House I. From Löf and Ward (1976).

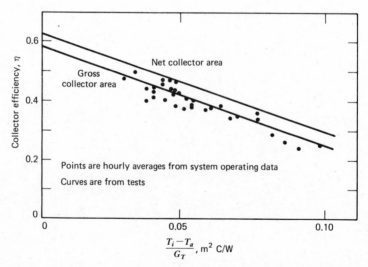

Figure 13.6.3 Performance of collectors on CSU House I. Adapted from Löf and Ward (1976).

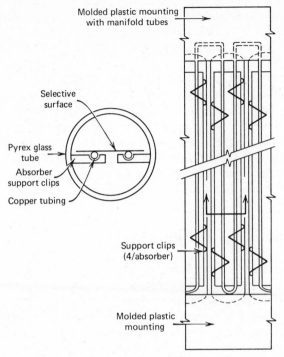

Figure 13.6.4 Cross-section and longitudinal-section schematics of the Corning evacuated tubular collector. From Karaki et al. (1978).

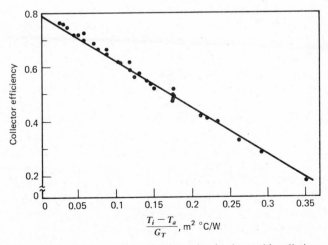

Figure 13.6.5 Evacuated collector test data taken in a solar simulator with radiation normal to the absorber and no wind, based on absorbing surface area. Adapted from Karaki et al. (1978).

451

performance data from system tests and curves from single module tests are shown in Figure 13.6.3. These indicate $F_R U_L$ of 3.3 W/m^2 C and $F_R(\tau\alpha)_n$ of 0.58, both based on the gross collector area of 71.3 m^2; wind speed was less than 0.3 m/s for the tests.

The second collector array was an assembly with gross area of 75 m^2 of tubular modules in which long, narrow flat absorbers with a selective surface were located inside evacuated tubular jackets. Sections of a unit of the collector are shown in Figure 13.6.4 and experimental data on the collector are shown in Figure 13.6.5. The energy absorbing area of the collector array was 39.9 m^2. This collector had $F_R(\tau\alpha)_n$ of 0.79, and $F_R U_L$ of 2.03 W/m^2 C, based on the absorber area. The lower loss coefficient means that this collector operated at lower radiation levels than the flat plate collector.

13.7 A HEATING SYSTEM SIMULATION STUDY

In preceding sections we discussed the design and measured performance of several solar heating systems. Of necessity, each of these experiments is with a fixed design. Simulations, on the other hand, offer the opportunity to determine the effects of changes of design parameters and system configuration on the annual performance of systems. In this section we summarize the results of a simulation study of an air system in the Madison climate, to show how collector area and storage capacity affect the monthly and annual solar fractions. The meteorological data used are hourly data for a Madison design year, as defined by Klein et al. (1976).

The system in this simulation is similar to that shown in Figure 13.2.2, and is shown in Figure 13.7.1. It is to provide space heating and hot water for a well-insulated residence of 150 m^2 floor area. Internal heat generation, infiltration, and capacitance of the building are considered. The hot water demand is 300 liters/day, assumed distributed through the day according to the load profile shown in Figure 12.4.1, with water entering at 11 C and supplied at 55 C.

The collectors are air heaters of design similar to that in Figure 6.12.1e, with $F' = 0.87$, $F_R = 0.75$, $F_R(\overline{\tau\alpha}) = 0.62$. $F_R U_L$ is assumed fixed at 3.38 W/m^2 C, and air flow rate to collectors is 0.0125 kg/m^2 s. Collector areas of 10, 30, 55, and 105 m^2 were used in the simulations. The collector slope is 58° and surface azimuth angle is zero. Ground reflectance is assumed to be 0.2 at all times.

The pebble bed is sized so that its volume is at a fixed ratio to the collector area, at 0.25 m^3/m^2. The bulk density of the pebble bed is 1600 kg/m^3 and the specific heat of the pebbles is 840 J/kg C. The loss coefficient of the pebble bed is 0.5 W/m^2 C, and losses from the bed are to a 20 C environment. To show the effects of relative pebble bed size, simulations were done at 25, 50, and 200 percent of the nominal size, for collector areas of 30 and 55 m^2. (See Sections 13.10 and 13.11 for further discussion of effects of storage size.)

The auxiliary heater has a maximum capacity of 13.9 kW. The heating system is controlled by a two stage thermostat which is designed to keep the building

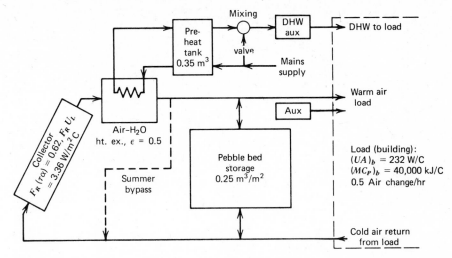

Figure 13.7.1 Schematic of the air system simulated in the Madison simulation example. Details of blowers, pumps, etc. are not shown.

Room temp, °C	Direction of change	Solar heating system	Auxiliary heating system
21.5	↓	Off	Off
	↑	Turns off[1]	Off
20.0	↓	Turns on	Off
	↑	On	Off
19.9	↓	On	Off
	↑	Turns on[2]	Turns off[1]
18.5	↓	Turns off	Turns on
	↑	Off	On

[1] Turns off if on
[2] Turns on if off

Figure 13.7.2 The control logic used in the Madison simulations. In addition to the functions shown, it is a requirement that the solar energy system must be able to supply air at a temperature above 30 C. If it cannot, it is off regardless of the room temperature.

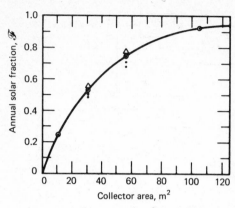

Figure 13.7.3 Solar fraction, $\mathcal{F}$, as a function of collector area, for the Madison simulations. The circled points and curves are for the normal storage size of 0.25 m³/m². The triangles show the effect of increasing storage size to 0.50 m³/m², and the dots show the effects of decreasing it to 0.125 m³/m² and 0.063 m³/m².

near 20 C. When heat is needed, solar heat is called first. If that will not hold the temperature, the solar heating system is turned off and the auxiliary is turned on. These control modes are indicated in Figure 13.7.2. In summer (June, July, and August), the pebble bed is bypassed so that heat is supplied only to the hot water preheat tank, and the collector fan operates when the preheat tank is below its control temperature and the collector can deliver energy. The volume of the water preheat tank is 0.35 m³, and its loss coefficient is 0.5 W/m² C.

Figure 13.7.3 shows the annual solar fraction, $\mathcal{F}$, as a function of the collector area, for the normal storage volume to collector area ratio, and also shows points

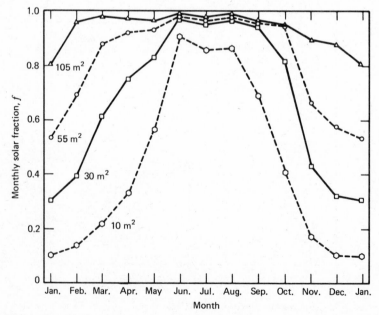

Figure 13.7.4 Monthly fraction of loads carried by solar energy for four collector areas, for the Madison simulations.

for larger and smaller storage capacities for the 30 and 55 m² collector areas. This is a typical $\mathscr{F}$ versus A_c curve for house heating, in that the slope is substantially higher for small collector areas than for large collector areas. The larger collectors are oversized a greater part of the year than are the small collectors. Thus an increment in collector area produces more useful energy at small collector areas than at large areas. The effects of change of storage size are small for the 30 m² collector, as for much of the year the bottom of the pebble bed stays near room temperature. The effects of decreasing storage size are more pronounced with the larger collector, as the bottom of the bed is at a higher average temperature, resulting in increased air temperature to the collector.

Figure 13.7.4 shows the monthly fraction of loads carried by solar for the four collector areas for the normal storage volume to collector area ratio. All of these systems will carry most of the summer hot water loads. (There are a few times during the design year when auxiliary heat is used to supply hot water during prolonged cloudy periods, and storage capacity limits solar fractions to less than unity.) The largest system meets 95 percent or more of the loads through 9 months, while the smaller systems meet relatively small fractions of the monthly loads in mid winter.

Figure 13.7.5 shows essentially the same information, but is in terms of the monthly efficiencies of the system for the four areas. System efficiency is defined

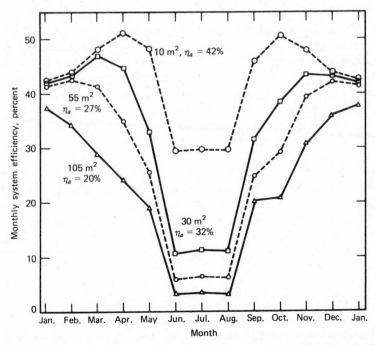

Figure 13.7.5   Monthly system efficiencies for the four collector areas, for the Madison simulations.

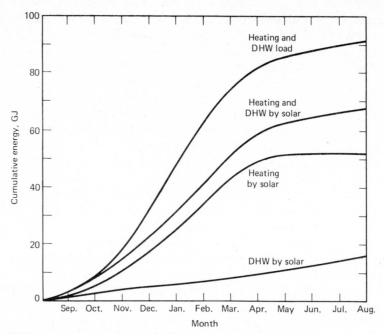

Figure 13.7.6   Cumulative loads and solar contributions for the 55 m² collector area and 0.25 m³/m² storage volume, for the Madison simulations.

as the total useful solar energy delivered to the building for the month divided by the total energy incident on the collector in the month. Annual efficiencies, η_a, are also shown for the four collector areas. It is clear that the efficiencies of this collector can vary over a very wide range, depending on the relative magnitudes of loads and insolation. The larger collectors are oversized for longer periods of the year, and operate at higher temperatures and lower efficiencies.

     Figure 13.7.6 indicates the cumulative energy quantities, starting from zero at an arbitrary time base at the beginning of the heating season, for the 55 m²

Table 13.7.1   **Summary of Annual Energy Quantities for the Madison Simulation**

Collector Area	Storage Volume	Solar Energy to Space Heating	Solar Energy to Hot Water	Total Heating, Hot Water Loads[a]	$\mathscr{F}$
10 m²	2.50 m³	11.1 GJ	10.7 GJ	86.9 GJ	0.25
30	7.50	32.7	14.5	89.3	0.53
55	13.75	51.7	15.9	91.2	0.74
105	26.25	68.3	16.9	92.7	0.92

[a] The loads are to some degree a function of system size, as larger systems keep the building at slightly higher mean temperatures.

system with normal storage capacity. It indicates the cumulative total loads and the heating and hot water loads met by solar through the year.

The cumulative energy quantities are summarized in Table 13.7.1. These are used in the discussion of economics of solar heating (see Section 13.12).

13.8   SOLAR ENERGY-HEAT PUMP SYSTEMS

Heat pumps use mechanical energy to transfer thermal energy from a source at a lower temperature to a sink at a higher temperature. Electrically driven heat pump heating systems have attracted wide interest as an alternative to electric resistance heating or expensive fuels. They have two advantages: a COP (ratio of heating capacity to electrical input) greater than unity for heating, which saves on purchase of energy, and usefulness for air conditioning in the summer. Heat pumps may use air or water sources for energy, and dual source machines are under development that can use either.

Yanagimachi (1958, 1964) and Bliss (1964) have built and operated heating and cooling systems that use uncovered collectors as daytime collectors and nighttime radiators, "hot" and "cold" water storage tanks to supply heating or cooling to the buildings, and heat pumps to assure maintenance of adequate temperature differences between the two tanks. The Yanagimachi system was applied to several houses in the Tokyo area, and the Bliss system was used on a laboratory in Tucson, Arizona.

Heat pumps have been studied by Jordan and Threlkeld (1954), and by an AEIC-EEI Heat Pump Committee (1956). An office building in Albuquerque was heated and cooled by a collector-heat pump system [Bridgers et al. (1957a,b)]. More recent systems have been built on residential buildings (e.g., Converse (1976), Kuharich (1976), and Terrell (1978)).

A schematic of an air-to-air heat pump is shown in Figure 13.8.1, operating in the heating mode. The most common type in small sizes are air-to-air units. For a discussion of the design and operation of heat pumps, see, for example, ASHRAE (1976). Typical operating characteristics of a residential scale heat pump are shown in Figure 13.8.2. As ambient air temperature (the evaporator fluid inlet temperature) increases, the COP increases, as does the capacity. As the air temperature drops, frost can form on the evaporator coils which adds heat transfer resistance and blocks air flow. Brief operation of the heat pump in the cooling mode removes the frost (a defrost cycle), and causes the ir-regularity shown in the capacity and COP curves. Figure 13.8.2 also shows a typical building heating requirement curve, which crosses the capacity curve at the balance point (BP). At ambient temperatures below this point a heat pump will have inadequate capacity to heat the building and the difference must be supplied by a supplemental source, which is often an electric resistance heater. At ambient temperatures above the balance point, the heat pump has excess capacity.

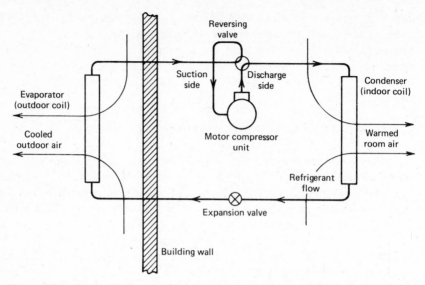

Figure 13.8.1 Schematic of a reversible heat pump system shown operating as a space heater.

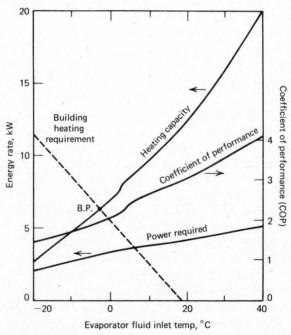

Figure 13.8.2 Operating characteristics of a typical residential scale air to air heat pump, as a function of ambient air temperatures, for delivery of energy to a building at 20°C.

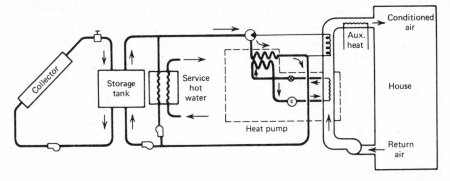

Figure 13.8.3 Schematic of a series solar energy-heat pump system, arranged so that solar heat can be supplied directly to the house when the storage tank temperature is above room temperature. From Freeman et al. (1979).

The performance of collectors is best at low temperatures, and the performance of heat pumps is best at high evaporator temperatures. This combination leads to consideration of *series systems*, in which the evaporator of the heat pump is supplied with energy from the solar system. A schematic of a series system is shown in Figure 13.8.3. This system is arranged and controlled so that solar heat can be added to the building directly from the storage unit when the storage temperature is high enough. The system shown is a liquid system, using a water-to-air heat pump.

A parallel system is shown in Figure 13.8.4, using a liquid based solar energy system with a water-to-air load heat exchanger and an air-to-air heat pump. The solar system could also be an air system with pebble bed storage. From the solar energy point of view, the operation of this system is the same as a conventional solar heating system, but with the heat pump supplying the auxiliary energy.*

A *dual source* system is shown in Figure 13.8.5. The heat pump evaporator is supplied with energy from either the storage unit (in this case a liquid) or from ambient air. Solar energy can be supplied directly to the building. Controls can be arranged to select the source leading to best heat pump COP, that is, the higher of the two source temperatures, although other control strategies may lead to better long-term system performance. An alternative system analogous to that shown in Figure 13.8.5 would be a solar system using air and an air-to-air heat pump, which could be supplied either from ambient air or solar heated air.

An integrated overall energy balance for a heat pump-solar energy heating system over a long time period includes: Q_{solar}, the energy supplied by the collector and tank; Q_{air}, the energy extracted from the source (air or water) by the heat pump; W_{elec}, the electric energy used to operate the heat pump, and Q_{aux}, the supplemental or auxiliary energy needed to assure meeting total heating loads Q_{load}. In equation form

$$Q_{solar} + Q_{air} + W_{elec} + Q_{aux} = Q_{load} \qquad (13.8.1)$$

* Design of parallel heat pump systems is discussed in Section 14.8.

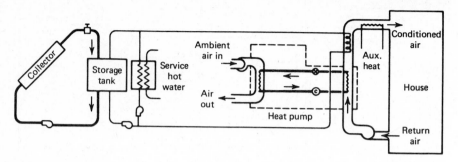

Figure 13.8.4   Schematic of a parallel solar-energy-heat pump system. From Freeman et al. (1979).

A useful index of system performance of these systems is the ratio of "non-purchased" energy to the total load, $\mathscr{F}'$

$$\mathscr{F}' = \frac{Q_{\text{solar}} + Q_{\text{air}}}{Q_{\text{load}}} \qquad (13.8.2)$$

This fraction is analogous to the solar fraction, $\mathscr{F}$, defined for conventional solar heating systems.

Simulation studies have been used to compare some of these systems. Marvin and Mumma (1976) studied air systems of four configurations. Karman et al. (1976) considered two air systems as well as the dual source system of Figure 13.8.5, and included simulations in Madison and Albuquerque climates. Mitchell et al. (1978) have computed the performance of the three systems shown in Figures 13.8.3, 13.8.4, and 13.8.5 and a standard liquid solar heating system of the type of Figure 13.2.4, using Madison meteorological data. In this study, the heat pump was modeled using the published performance data of a typical recent model commercial heat pump; the simulations were done using TRNSYS. Figure 13.8.6 summarizes the results for Madison, showing $\mathscr{F}'$ as a function of collector area for a typical residential type building. Figure 13.8.7 shows comparative information for collector area of 30 m^2 for each of the three combined systems, the conventional solar and heat pump only systems.

The results of these simulations show that with the same collector the parallel system is substantially better than the series system and slightly better than the dual source system in all collector sizes, in that it delivers a greater fraction of the loads from "nonpurchased" sources. This arises because the heat pumps in the series and dual source systems must operate to deliver all solar energy stored below 20 C. The extra electrical energy required to deliver this energy more than compensates for the combined advantages of higher collector

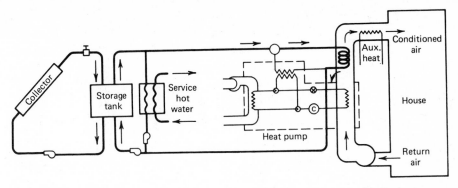

Figure 13.8.5 Schematic of a dual source solar energy-heat pump system with the heat pump source either the storage tank or ambient air. From Freeman et al. (1979).

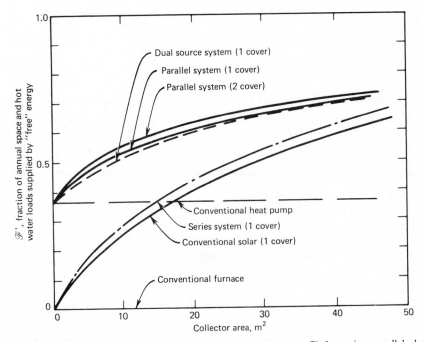

Figure 13.8.6 The fraction of energy from nonpurchased sources, $\mathscr{F}'$, for series, parallel, dual source, heat pump only, and standard solar energy systems as a function of collector area, based on simulations of the systems in the Madison climate for a residential building. From Freeman et al. (1979).

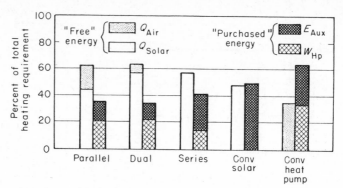

Figure 13.8.7 Sources of energy used for heating under the same circumstances as those of Figure 13.8.6. Collector areas are 30 m^2 for the solar energy systems. From Freeman et al. (1979).

efficiency and higher heat pump COP. The temperature of storage in the series system (and in the dual source system) is rarely high enough in these simulations to permit direct solar heating.

These simulations were done with the same collector for each of the system configurations. The series collector will run cooler for part of the time than will the collector for the parallel or dual source systems. Thus there is the possibility that less expensive collectors will be adequate for the series system. An economic comparison must be made which takes into account the collector operating temperature ranges for each of the systems.

The combined height of the left bars of Figure 13.8.7 for each of the systems is $\mathscr{F}'$. The systems are arranged from left to right in order of decreasing $\mathscr{F}'$, or increasing purchased energy. The data for the series and dual source heat pump systems show higher solar energy contribution, as is to be expected with reduced collector temperature, but they also show higher purchased energy. The yearly computed COP for the heat pumps for the parallel, dual source, series and heat pump only systems were 2.0, 2.5, 2.8, and 2.1, showing that the series system heat pump with the solar source does have the highest mean evaporator temperature and highest annual COP.

Mitchell et al. have explored system-design variables, including storage size, heat pump characteristics, and reduced minimum tank temperatures (i.e., antifreeze solutions in the storage tank) and found no significant changes in the results of the simulations. Marvin and Mumma (1976) also concluded that parallel operation led to best thermal performance. Many experimental systems in the series configuration have been installed, and based on economic considerations there may be circumstances for which series systems would be better than parallel systems.

Heat pumps are capital intensive, as are solar energy systems. Consideration of economics of solar energy-heat pump systems indicates that it may be difficult to justify the investment in a heat pump to improve solar energy system performance, unless the cost of the heat pump can be justified on the basis of its use for air conditioning, or unless the use of the heat pump would reduce collector costs to a small fraction of their present levels.

13.9 SOLAR AND OFF-PEAK ELECTRIC SYSTEMS

Auxiliary energy can be supplied to solar energy systems from supplies stored on site (oil or LPG) or from utilities (gas or electricity). On-site storage of auxiliary energy poses no unique problems for the distributors of oil or LPG. Supply of auxiliary by utilities, on the other hand, can cause significant peaking problems for the electric utilities if a large number of solar buildings all call for auxiliary energy during periods of bad weather. A potential solution to this problem is to supply electric auxiliary during off-peak periods when the utility has excess capacity, and store the energy in the building. (The same possible solutions may apply to systems that are nonsolar, that is, all electric.) The purpose of this section is to outline a simulation study by Hughes et al. (1977) of three off-peak electric auxiliary options in air solar heating systems. The three systems are shown in Figure 13.9.1.

The first is a conventional air system configuration that requires the delivery of auxiliary energy whenever the solar energy system cannot meet the loads; it is the same as the systems described in Sections 13.3 and 13.4. The second, Figure 13.9.1b, uses an electric furnace supplying off-peak electrical energy to a separate storage system. The utility supplies energy to charge the auxiliary storage when it has the generating capacity. The operating requirement is that at the end of each off-peak period there must be at least enough energy in the auxiliary storage unit to meet the maximum anticipated heating loads on the building until the next off-peak period. From the standpoint of solar energy systems thermal performance, this is identical to the conventional configuration of Figure 13.9.1a. It is, however, mechanically more complex.

The system of Figure 13.9.1c uses a single combined storage unit. During off-peak periods, electrical energy is added to the pebble bed to assure having adequate energy stored in the bed to last until the next off-peak period. Solar energy from the collector is then added to the bed as available. The pebble bed will operate at significantly higher temperatures with this system than with the system with two storage units; this leads to higher inlet collector temperature and reduction in collector performance. Stratification in the bed helps diminish the impact of the elevation in temperature, but it remains significant. This single storage system suffers a performance penalty compared to the two-store or conventional solar configurations, but it is mechanically simpler.

Comparisons of single storage and two storage configurations have been made using two types of collectors. The first collector is a two-cover, nonselective air heating collector of the type having characteristics shown in Figure 6.12.1e with a U_L of approximately 4.0 W/m^2 C. The second is an evacuated tubular collector of the type shown in Figure 6.12.1k, functioning as an air heater, with an effective U_L of approximately 0.8 W/m^2 C.

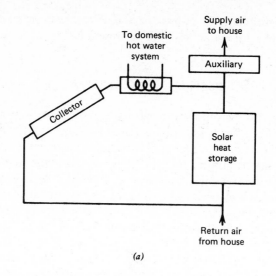

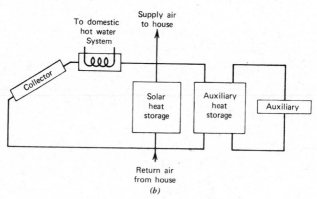

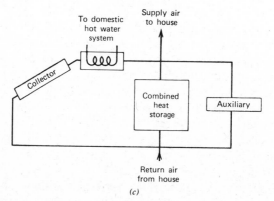

Figure 13.9.1 Schematics of three solar air heating systems: (*a*) a conventional system; (*b*) an off-peak system using two storage units, one for solar and one for auxiliary; (*c*) an off-peak system using a single storage unit for both auxiliary energy and solar energy. From Hughes et al. (1977).

Table 13.9.1   Comparisons of Calculated Performance of Systems Using One and Two Storage Units on Annual Performance, on a House at Arlington, WI for Two Collectors[a]

System	A_c, m^2	H_T, GJ	Q_u, GJ	$\mathscr{F}$
2 storage, flat-plate collectors	50.6	148	54	0.58
2 storage, evacuated tube collector	44.1	129	62	0.66
1 storage, flat-plate collector	50.6	148	46	0.49
1 storage, evacuated tube collector	44.1	129	59	0.62

[a] Adapted from Hughes et al. (1977).

The differences in thermal performance of four systems (the one and two store systems each with the two types of collector and a reasonable storage capacity for each) are shown in Table 13.9.1. The areas of the two types of collectors are different as the comparisons were made for a specific building that could accommodate different areas of the two collectors. The performance penalties of elevated collector temperatures are shown by reduced solar contribution for the single storage system and are, as expected, much smaller for the collector with the lower collector loss coefficient.

Figure 13.9.2 shows the effects of storage capacity on the thermal performance of the single storage system with each of the collector types. The computations were done using a fixed bed height and variable cross section area. Annual

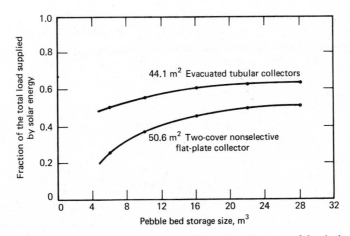

Figure 13.9.2   Effects of storage capacity on the annual thermal performance of the single storage system. From Hughes et al. (1977).

system performance is more sensitive to storage capacity for the flat-plate collector than it is for the evacuated tubular collector, again a result of the low loss coefficient for the evacuated collector and resulting insensitivity to operating temperature.

The off-peak addition of energy to a pebble bed can be controlled by comparing the average bed temperature to a set minimum average temperature, which assures adequate heat supply until the next off-peak period. The minimum average temperature may be set at a fixed level for the whole year (which is the simplest control strategy) or it may be varied to account for the variation in the anticipated maximum space heating load with time of year. Performance of systems using a single storage unit for both solar and auxiliary energy will be dependent on the storage unit temperature and thus on this control strategy. Simulations indicate that the annual fraction by solar rises from 0.60 to 0.62 in going from a fixed yearly minimum bed temperature to a minimum bed temperature set according to anticipated monthly loads, when the evacuated tubular collector is used. A system using a flat-plate collector will be much more sensitive to control strategy.

An experimental system using evacuated tubular collectors and a single pebble bed storage unit has been constructed at Arlington, Wisconsin, and is described by Hughes et al. (1976). Its performance through part of the 1977–78 heating season is summarized by Erdman and Persons (1978). The system functioned with little difficulty. Its performance was very near that predicted for March, but below predictions in January and February when the weather was unusually severe and very heavy snow interfered with collector operation. Changes in design may correct these problems.

13.10 PHASE-CHANGE STORAGE SYSTEMS

In Chapter 9 we reviewed phase-change energy storage materials, methods, and problems, and presented a model developed by Morrison and Abdel-Khalik for a phase-change energy storage unit. In this section we show the results of the use of this model in simulation studies of solar heating systems to evaluate the possible impact of successful development of a phase-change storage unit on heating system performance.

Several buildings have been built using phase-change storage. Telkes and Raymond (1949) described a solar house constructed at Dover, MA, that utilized vertical south-facing air heating collectors and energy storage in Glauber's salt, $Na_2SO_4 \cdot 10H_2O$, contained in 5-gal drums. The system was designed to have 5 days storage capacity and to carry the total heating load by solar energy. No data are available on its operation, which was terminated after a few years. Solar One, an experimental house at the University of Delaware [Böer (1973), Telkes (1975)] used phase-change storage units for energy storage at two temperature levels, in a solar energy-heat pump system. Baer (1973) reported the use of Glauber's salt in drums in a storage wall. In 1978 a number of solar

Table 13.10.1   Properties of Phase Change Storage Media and Rock[a]

Property	Paraffin Wax	$Na_2SO_4 \cdot 10\,H_2O$	Rock
C_{ps}, J/kg C	2890	1920	840
C_{pl}, J/kg C	[b]	3260	
k_s, W/m C	0.138	0.514	0.125
k_l, W/m C	[b]	[c]	
T^*, C	46.7	32	
λ, J/kg	2.09×10^5	2.51×10^5	
ρ_s, kg/m^3	786	1460	1600
ρ_l, kg/m^3		1330	

[a] From Morrison and Abdel-Khalik (1978).
[b] Assumed equal to value for the solid phase.
[c] Unknown; assumed to be 0.475 W/m C, the value for liquid $Na_2HPO_4 \cdot 12\,H_2O$.

heating systems have been built using commercially manufactured phase-change storage units. However, there are few data available on their performance.

An alternative method for study of heating systems using phase-change storage is by simulations. Morrison and Abdel-Khalik (1978), postulated an idealized phase-change operation, which was free of superheating, supercooling and property degradation, developed models for phase change storage units for incorporation in simulations of liquid and air heating systems, and made comparative simulation studies of several systems. Their infinite NTU model of the storage unit is outlined in Section 9.5. The phase-change material properties used were those of Glauber's salt and P116 wax; the properties of the materials are indicated in Table 13.10.1. Meteorological data for Madison, Wisconsin and Albuquerque, New Mexico were used.

Typical results of this study are shown in Figures 13.10.1 and 13.10.2 for air based systems and in Figures 13.10.3 and 13.10.4 for liquid systems. The air systems are the same configuration as that of Figure 13.2.2, using either the pebble bed or the phase change storage unit. The model of the phase-change storage allows for flow reversal and temperature stratification in the bed, in a manner similar to that of the pebble bed. The figures show comparisons of annual fraction of heating loads carried by solar energy for typical residential buildings for collector areas of 50 m^2 as a function of storage volume. Figures 13.10.1 and 13.10.2 for the air systems show, for the idealized phase-change unit, that the same thermal performance is obtained with smaller volumes of the phase-change materials, and that the annual performance of systems with any of the three storage media are nearly the same as long as the volume of the storage media is above the "knee" of the curve. Similar results were obtained at other collector areas, and with collectors with two different heat loss coefficients.

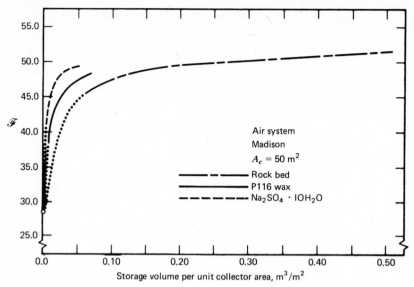

Figure 13.10.1 Comparison of annual performance of air systems with phase change and sensible heat storage, in Madison, based on simulations. From Morrison and Abdel-Khalik (1978).

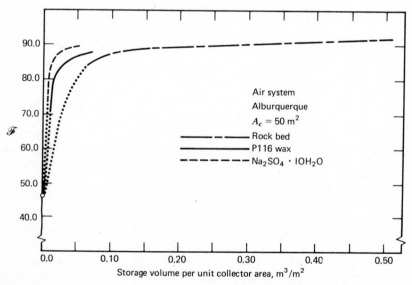

Figure 13.10.2 Comparisons similar to that of Figure 13.10.1, but for Albuquerque. From Morrison and Abdel-Khalik (1978).

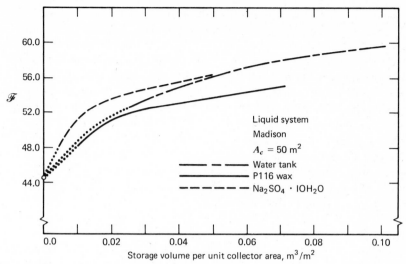

Figure 13.10.3 Comparisons of liquid systems for Madison. From Morrison and Abdel-Khalik (1978).

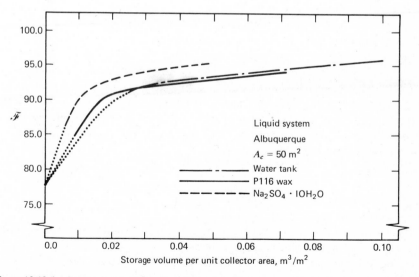

Figure 13.10.4 Comparisons of liquid systems for Albuquerque. From Morrison and Abdel-Khalik (1978).

Figures 13.10.3 and 13.10.4 show similar results for the liquid-based systems. The basic system configuration is that shown in Figure 13.2.4, with storage being either the tank or the phase change unit. These simulation results indicate that the $Na_2SO_4 \cdot 10H_2O$ unit is slightly better for the example shown in Albuquerque, but the same general conclusions can be reached as for air systems, i.e., the annual performance of heating systems using an idealized phase change storage is nearly the same as that with sensible heat storage, but the storage volumes are significantly smaller.

In all of these systems, the phase-change material operates in sensible heat modes as a solid part of the time, as a liquid part of the time, and in combined phase change and sensible heat mode part of the time. Systems carrying larger fractions of annual loads by solar tend to operate as liquid phase sensible heat stores larger fractions of the time.

These results, based on ideal phase-change storage characteristics, represent upper bounds to the performance of heating systems using Glaubers salt or P116 wax phase change storage. Any change from the ideal can be expected to result in a greater proportion of the operation being in sensible heat modes and a reduction in effective storage capacity. This will lead to a small reduction in annual system output if the storage capacity is very large. If the storage volume is at or near the knee in the curve, deviations from idealized behavior may cause large reductions in annual system output.

Jurinak and Abdel-Khalik (1978) have used the same phase-change storage model and idealized behavior, and explored the possibilities of improvements in system performance on changing the melting point, latent heat, and melting point range (of mixes of wax type materials), and found that "fine-tuning" of the properties of the materials did not lead to significant increases in long-term system performance. In another paper (1979) they developed the concept of an effective heat capacity of a phase change material, as indicated in Section 9.5. This concept permits use of quick design procedures (see Chapter 14) to estimate the performance of systems with phase change storage.

Successful application of phase-change storage in solar heating depends on availability of materials that can cycle thousands of times without significant degradation, and which can be packaged or handled without corrosion. The ultimate feasibility will depend on economic factors, that is, what the phase change material costs, the expense of packaging it, and the value of the space it saves in the building.

13.11 SEASONAL ENERGY STORAGE

The possibility of storing solar energy collected in the summer for use for building heating in the winter was considered by Speyer (1959). He concluded that increasing storage capacity to high levels results in reduction of collector area requirements but that the cost of seasonal storage is not justified in terms of increased annual output of the system. Similar conclusions have been reached

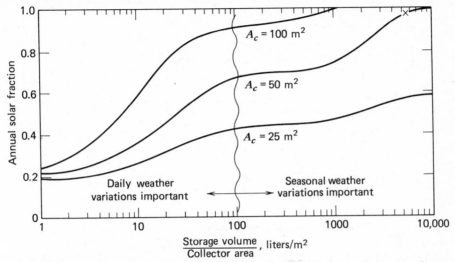

Figure 13.11.1 Fraction by solar as a function of storage capacity for three collector areas, for an example based on Madison meterological data. Adapted from Braun (1980).

by others, and most solar heating studies have concerned storage capacities equivalent to approximately 1 to 3 days design heating loads of the building. A more optimistic view of long-term storage, and a method of estimating heat losses from tanks to ground, are provided by Hooper and Attwater (1977). In this section we outline some of the factors that enter into consideration of seasonal storage and show examples of the relationships of solar fraction to storage capacity and collector area, determined by simulations.*

The plots of solar fraction versus water storage capacity shown in Figures 13.10.3 and 13.10.4 cover the usual range of storage capacities considered in building heating. Figure 13.11.1 from Braun (1980) is based on a similar analysis, but shows several orders of magnitude variation in water tank storage capacity. The simulations were done using Madison meteorological data and system parameters other than storage characteristic of systems with ordinary storage capacity (except that the tank is very heavily insulated). Three collector areas are shown, which with liquid storage of 75 liters/m² would deliver 39, 63, and 85 percent of the annual loads by solar energy. The 50 m² collector would supply 21 percent of the load with no storage.

On a linear plot solar fraction increases sharply as storage is added until approximately 30 liters/m² is reached. At this point the storage capacity is adequate to smooth out much of the diurnal solar variation. As more capacity is added it begins to provide storage for more than a day and a gradual increase in solar fraction is noted. As the storage capacity increases by two orders of magnitude, a second "knee" in the curve appears as the capacity becomes

---

* A simplified design method for systems with long-term storage is shown in Example 18.3.1.

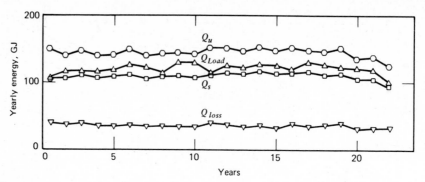

Figure 13.11.2 Year to year variation of performance of the 50 m² system with a storage capacity of 5500 liters/m². From Braun (1980).

adequate for seasonal storage, that is, of energy collected in the summer for winter use.

As the solar fraction approaches unity in a system with seasonal storage, year to year variations in weather must be considered. The 50 m² system shown at point X on Figure 13.11.1 was simulated for 21 years, with the results shown in Figure 13.11.2. Useful gain from the collector, Q_u, losses from storage, Q_{Loss}, load, L, and energy supplied by solar, Q_s, are shown as integrated quantities for each year. The difference between L and Q_s must be supplied from an auxiliary source. In four of these years very little auxiliary energy was required; in other years substantial amounts were required.

Thermal storage losses in these simulations were to the ground at a constant temperature equal to the annual average ambient temperature, with a constant

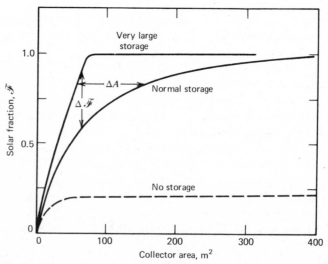

Figure 13.11.3 Fraction by solar vs collector area for systems with very large storage, standard storage (75 liters/m²), and no storage. Adapted from Braun (1980).

thermal resistance from tank to ground. The tank sizes required for seasonal storage are such that the only feasible location is in the ground. The losses shown are significant; if thermal energy is to be stored over time spans of months, energy loss rates must be low or the cumulative losses will be prohibitive.

Figure 13.11.3 shows the solar fraction as a function of collector area for a "normal" storage capacity, for very large storage capacity, and for no storage. For large storage, annual output is limited by collector area and is nearly proportional to collector area until the solar fraction approaches unity. The difference between the two curves at constant area is the possible improvement in annual system performance ($\Delta\mathscr{F}$) that can be achieved by addition of large storage capacity. At constant solar fraction, the difference in the two curves (ΔA) represents the increment in area that can be saved by going to very large storage capacity. If no storage is provided there is very little gain on increasing collector area beyond 50 m^2. (The examples shown in these figures were calculated assuming no building capacitance. At low storage capacity, including building capacitance has the effect of adding some storage, so that the solar fractions for no storage are lower than they would be if building capacitance were considered.)

13.12 SOLAR HEATING ECONOMICS

The first major economic studies of solar heating by Tybout and Lof (1970), and Lof and Tybout (1973), were based on constant annual savings without anticipation of changing future costs. They devised a thermal model for a liquid-based solar heating system to estimate annual thermal performance (based on one year's meteorological data), and developed a set of cost assumptions to calculate costs of delivered solar energy for houses of two sizes in eight U. S. locations of differing climate types. Several system design parameters were studied in addition to collector area, to establish the range of optimum values, including collector slope, number of covers, and heat storage capacity per unit collector area. Their results (which are in general agreement with the conclusions of others) can be summarized as follows:

1 The optimum tilt is in the range of the latitude plus 10° to the latitude plus 20°, and variation of 10° either way outside of this range, that is, from latitude to latitude plus 30°, has relatively little effect on the cost of delivered energy for heating.
2 The best number of (ordinary) glass covers with nonselective absorbers was found to be two for all locations except those in the warmest and least severe climates, that is, Miami and Phoenix, where one cover produces less expensive energy from the solar heating system.
3 The best storage tank capacity per unit collector area was indicated to be in the range of 50 to 75 liters/m^2. Increasing the size of storage several fold for fixed collector size had relatively small effect on the cost of delivered solar energy or on the fraction of total heating loads carried by solar.

The costs assumed by Löf and Tybout (approximately $20 and $40/m^2 of collector, installed) are unrealistically low in the light of 1978 costs ($200 to $400/m^2 of factory-built collector installed on new residential buildings). However, their conclusions on collector orientation and storage capacity have been confirmed by others. The availability of durable selective surfaces is leading to wider applicability of one-cover collectors than they predicted.

The solar process economic considerations noted in Chapter 11 apply directly to solar heating. The costs of installed solar heating equipment (the first costs) include purchase and installation of all collectors, storage units, pumps, blowers, controls, ductwork, piping, heat exchangers, etc. Operating costs include costs of auxiliary energy, parasitic power, insurance, maintenance, taxes, etc. In the following discussions, as in Chapter 11, first costs of solar heating systems are considered as the incremental costs, that is, the difference in cost between the solar heating system and a conventional heating system.

The basic method used in calculating the economics of solar heating is the P_1, P_2 life cycle savings method described in Section 11.9. In this analysis there are a large number of economic parameters that must be determined. This discussion shows how some of these parameters affect the economic viability of active solar heating, and is introduced by a brief review of their nature and significance.

A primary economic consideration is the first cost of a solar heating system. In 1978, the installed costs of many systems are in the range of $200 to $400/m^2 of collector, with fixed costs of the order of $1000 to $3000. Retrofit systems tend to be more expensive, particularly on large flat-roof buildings where new structure must be provided for collector support. Some residential scale systems cost less, if integrated into a building or constructed by the owner. Most of the tax incentives proposed or enacted by the Congress and state legislatures are tax credits, which have the net effect of reducing first costs. If a system is well built, it can have significant salvage or resale value, and if its major components are as durable as the basic building, its value may appreciate in the same way the building appreciates.

The costs of energy delivered from fuel at the time of installation, C_{F1}, are a second important consideration. They are widely variable with time, location, type of fuel, and efficiency of the fuel burning system. In 1978 in the United States, some regulated natural gas prices were at the low end of the scale with costs of delivered energy from $2 to 3/GJ. Typical costs of energy delivered from oil were $5 to 8/GJ. LPG energy costs were higher than oil. Some of the most expensive energy is electricity, which if used in resistance heating in some metropolitan areas costs more than $20/GJ. Thus 1978 energy costs varied over an order of magnitude.

In the decade 1970–1980, most energy costs have inflated at higher rates than the general inflation rate. Life cycle cost calculations require that projections be made of future energy costs. Future fuel energy costs will be dependent on gas and oil discoveries and on technological developments relating to fuel conversion and use. They will also be dependent on international political developments as many of the worlds industrialized nations are energy importers.* Fuel cost projections are uncertain, and interject corresponding uncertainties into the life cycle cost analysis.

A third category of economic factors relates to costs of operation of systems. Insurance, maintenance, and property taxes will probably increase with the general inflation rate. Parasitic power should be small in relation to other costs, and is often included with insurance and maintenance. It can also be included with the auxiliary energy costs or accounted for separately.

A fourth category of economic considerations relates to the costs of money and the time periods over which analyses are made. Interest rates and terms of initial mortgages or building improvement loans both affect the life cycle costs of the capital-intensive solar energy systems. Time periods for economic analysis are sometimes selected as the period of time over which it is expected the building will be occupied or owned, over the expected lifetime of the equipment, or over the term of the loan used to purchase the equipment. In general, longer periods of analysis will tend to improve life cycle solar savings, although costs or savings far in the future have relatively small impact on life cycle costs when discounted to present value. Life cycle costs are sensitive to discount rates, with lower discount rates generally improving life cycle savings for solar or other energy-conserving and capital-intensive measures. Market discount rates are often taken as approximately 2 percent more than the general inflation rate for individuals, approximately 4 percent more for established businesses, and may be as high as 20 percent higher than general inflation rates for fast growing industries. Most industries have developed their own discount rates.

Examples of the solar savings to a homeowner are shown in Figures 13.12.1 and 13.12.2. The configuration is the air system described in Section 13.7, on a well insulated building in Madison. The economic parameters assumed are: term of analysis, 20 years; term of mortgage, 20 years; mortgage interest rate, 9.5 percent/yr; downpayment, 20 percent; market discount rate, 9 percent/yr; general inflation rate, 7 percent/yr; real estate taxes in first year, 2 percent of system first cost; insurance, maintenance, and parasitic power in first year, 2 percent of system first cost; income tax bracket through the period of analysis, 45 percent. The system is assumed to have a resale value of 100 percent of its first cost, that is, it does not decline in value, nor does it appreciate as the building is expected to appreciate.

First year fuel costs are taken as \$4, \$8, and \$12/GJ, and these are assumed to inflate at rates of 10 and 15 percent per year. This covers a wide range of present costs and possible inflation rates.

Figure 13.12.1 is based on first costs of $C_A = \$150/m^2$ and $C_E = \$500$. These costs are representative of the low end of the range of net costs after tax incentives of commercially installed systems or of owner-built systems. Figure 13.12.2 is based on $C_A = \$300/m^2$ and $C_E = \$1000$, which is in the range of

---

* In 1978 the United States imported about $\$45 \times 10^9$ worth of petroleum products.

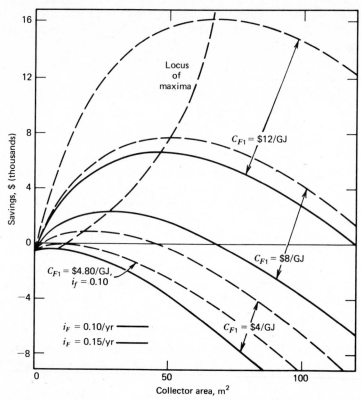

Figure 13.12.1   Life cycle solar savings for the Madison example with $C_A = \$150/m^2$ and $C_E = \$500$, for present fuel costs of \$4, 8 and \$12/GJ inflating at 0.10 and 0.15/yr. The dotted curve shows a "break-even" fuel cost of \$4.80/GJ at $i_F = 0.10$.

1978 costs of commercially installed systems on new residential buildings without tax credits.

Several trends are clear from these figures. Lower first costs of the system lead to more favorable savings, as do higher present fuel costs and higher fuel cost inflation rates. As fuel costs or inflation rates rise, the optimum collector area increases. The approximate loci of the maxima are shown. As the optimum collector area increases, the range of areas over which positive savings can be expected will also increase. In all cases, the optima are broad, and the selection of area is not critical. For example from Figure 13.12.1 at $C_{F1} = \$8/GJ$ and $i_F = 10$ percent/year, the optimum area is nearly 30 m², any area between 20 and 40 m² results in essentially the same savings, and any area between 2 and 68 m² results in positive savings.

Given a full set of economic parameters (other than fuel costs) and the thermal performance of a system, it is possible to determine combinations of present fuel cost and fuel cost inflation rate that will make solar heating just competitive with the alternative, i.e., when savings = \$0. An example is shown by the

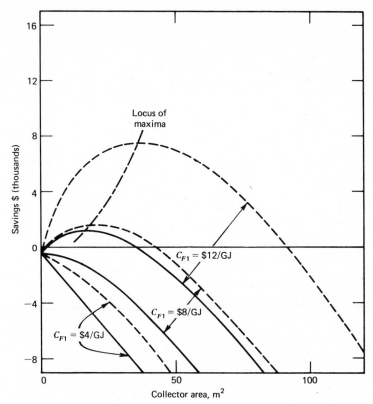

Figure 13.12.2   Life cycle solar savings for the Madison example, with $C_A = \$300/m^2$ and $C_E = \$1000$, for present fuel costs of $4, 8 and \$12/GJ inflating at 0.10 and 0.15/year.

dotted curve on Figure 13.12.1. At a fuel cost inflation rate of 0.10 per year, a C_{F1} of $4.80 is a "break-even" fuel cost. If fuel cost is above this level solar savings become positive.

The results shown in the figures are for a specific system, location, and set of economic assumptions. Many economic parameters have not been varied in the estimates that are shown on these figures. As was noted in Chapter 11, the explicit effect of any one variable on the life cycle savings is influenced by the values of many other variables. Thus "rule of thumb" sensitivity trends are not easily recognized, and quantitative sensitivity analyses such as that outlined in Section 11.10 are required.

With the economic analysis methods of Chapter 11 and the heating process optimization method to be described in the next chapter, surveys can readily be made of the relative economic feasibility of solar heating and hot water in various climates. This has been done by Duffie et al. (1977) for the continental United States and southern Canada. Results are shown in Figures 13.12.3 and 13.12.4, and illustrate the substantial differences that are found in the optimized systems from one geographical location to another.

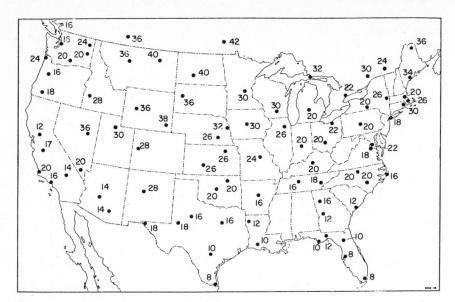

Figure 13.12.3 Optimum collector areas for various North American stations, for a 150 m²
building, for a set of economic parameters which are the same for all stations. From Duffie et al.
(1977).

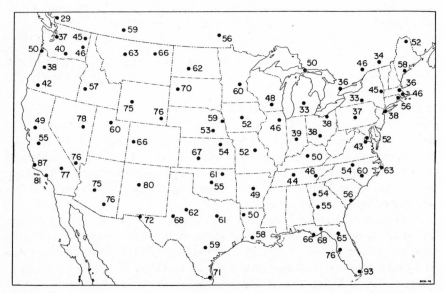

Figure 13.12.4 Solar fractions at the optimum collector area of Figure 13.12.3. From Duffie et al.
(1977).

Two basic assumptions underlie the numbers shown on these maps. First, the current cost of energy from fuel is taken as the same for all locations. This is obviously not true but permits comparisons to be made of optimized heating systems in various climates. Second, the building to be heated in each location is a residence with 150 m² floor area, insulated to ASHRAE 90-75 standards. In all cases, the collectors are south-facing with slope equal to latitude.

The two maps show optimum collector areas and the percentage of annual loads carried by solar energy at the optimum areas. Substantial variations of both collector area and solar fractions are found in nearby locations (e.g., Madison, WI and Lansing, MI, which are upwind and downwind of the western Great Lakes). Large variations are also noted in collector area from one part of the country to another (e.g., for a solar fraction near half, the areas required for Madison, WI and Little Rock, AR, are different by a factor of two). A similar map could be drawn indicating life cycle savings for these stations using a common C_{F1} for all stations; it would show that the best climates for solar heating are in the Western Mountain states where the heating season is long and where solar radiation is high. If current local fuel costs were used, the areas with high C_{F1} (such as northeastern US) would tend to appear more favorable.

13.13   ARCHITECTURAL CONSIDERATIONS

Active solar heating poses challenges to architectural design of buildings. Many approaches to these challenges have been devised; Shurcliff (1978) has compiled information on a variety of solar heated buildings. In this section we outline in general terms some of the major architectural considerations to be taken into account in solar building design. While not addressed specifically here, it is implicit throughout this discussion that any building design should be energy conserving, as solar energy and the fuels with which it competes will be expensive.

Economic studies of active solar heating indicated that the optimum fractions of total annual loads range from zero to over three-fourths. For some locations the architect must design into the building collector areas in the range up to approximately one half of the floor area of the house (depending on the collector, the climate and the degree of insulation in the building). A basic problem faced by architects and engineers is to integrate the collectors into the building design in such a way that thermal performance is satisfactory and the structure is aesthetically satisfying.

Collectors should be oriented within the slope and azimuth angle limits noted in Tables 13.2.1 and 13.2.2. Vertical collectors may be useful at high latitudes to answer problems of integration of collectors into the building and avoid snow accumulation. Space must be provided in the structure for energy storage units, piping and ducts, controls, auxiliaries, and all associated equipment. Anderson et al. (1955) have addressed these and related questions on solar house architecture. Similar considerations will apply to institutional buildings.

(a)

(b)

480

(c)

(d)

Figure 13.13.1 Four examples of solar heated buildings: (a) and (b) residences, Madison, WI; (c) McKay Environmental Center, Madison, WI; (d) Bank, Longmont, CO. Courtesy Solaron Corp.

The collector may be a part of the envelope of the building (as in MIT House IV) or separate (as in the Denver Solar House). The orientation of the collector is substantially fixed and if it is part of the envelope of the house, the collector will probably become an important or dominant architectural feature of the structure. The collector may serve as part of the weatherproof enclosure and thus allow a reduction in cost of roofing or siding; such a reduction is a credit which reduces the cost of the collector. Separate collectors, on the other hand, can permit greater flexibility in house design and allow buildings that are more conventional (contemporary or traditional) in appearance.

Figure 13.2.5 shows three solar heated residential buildings, including MIT House IV, the Denver House, and Laboratory buildings at Colorado State University. Figure 13.13.1 shows four additional solar heated buildings, illustrating a variety of architectural approaches.

Storage is usually not a major architectural problem, other than recognizing the need for appropriate space and access to it within the structure. The volume of storage per unit area of collector depends on the system used. Pebble bed storage units will usually occupy a volume of roughly 0.15 to 0.35 m^3/m^2 of collector and water tanks about 0.050 to 0.10 m^3/m^2. The volume of heat of fusion storage systems would be less than that of water systems. The most common location for storage in buildings is in basements.

Providing solar heat to larger buildings, such as apartment buildings, presents a special set of problems. It may be necessary to consider vertical mounting for collectors, which may cause a significant reduction in their performance. This possibility has been studied by Lorsch and Niyogi (1971). Otherwise, collectors may be mounted like awnings, with improvements in performance but with increased cost of installation. If the building geometry is such that the roof area is adequate, banks of collectors can be mounted on the roof with appropriate piping or ducts leading to the space to be heated.

Most solar heating studies to date have been concerned primarily with new buildings designed to include solar heating systems. Adding solar heating systems to existing buildings presents more formidable tasks. For example, Lior et al. (1978) have fitted a solar heating system to an existing town house in Philadelphia. Consideration should be given to the problems of designing buildings that can accommodate the addition of solar heating after construction.

REFERENCES

Anderson, L. B., H. C. Hottel, and A. Whillier, *Solar Energy Research*, 47, University of Wisconsin Press, Madison (1955). "Solar Heating Design Problems."

AEIC-EEI Heat Pump Committee, Edison Electric Inst. Bull, 77 (1956). "Possibilities of a Combination Solar-Heat Pump Unit." Based on a report by G. O. G. Löf.

ASHRAE *Systems Handbook*, American Society of Heating, Refrigeration, and Air Conditioning Engineers, New York (1976).

Baer, S., report in *Proceedings of the Solar Heating and Cooling for Buildings Workshop*, 186, Washington (March 21 to 23, 1973), R. Allen, ed., University of Maryland. "The Drum Wall."

Bliss, R. W., *Air Conditioning, Heating and Ventilating*, **52** (10), 92 (October 1955). "Design and Performance of the Nation's Only Fully Solar-Heated House." See also *Proceedings of the World Symposium on Applied Solar Energy*, 151, SRI, Menlo Park, California (1956).

Bliss, R. W., *Proceedings of the UN Conferences on New Sources of Energy*, **5**, 148 (1964). "The Performance of an Experimental System Using Solar Energy for Heating, and Night Radiation for Cooling a Building."

Boer, K., paper presented at ISES Congress, Paris (1973). "A Combined Solar Thermal Electrical House System."

Braun, J., M.S. Thesis, University of Wisconsin-Madison (1980). "Seasonal Storage of Energy in Solar Heating."

Bridgers, F. H., D. D. Paxton, and R. W. Haines, paper 57-SA-26 presented at ASME meeting (June 1957a). "Solar Heat for a Building."

Bridgers, F. H., D. D. Paxton, and R. W. Haines, *Heating, Piping, and Air Conditioning*, **29**, 165 (1957b). "Performance of a Solar Heated Office Building."

Close, D. J., R. V. Dunkle, and K. A. Robeson, *Mechanical and Chemical Engineering Transactions* Inst. Engrs, Australia, **MC4**, 45 (1968). "Design and Performance of a Thermal Storage Air Conditioner System."

Converse, A. O., *Proceedings of the Joint Conference of the American Section of ISES and SES of Canada*, Winnipeg, **3**, 277 (1976). "Solar Heating in Northern New England."

Duffie, J. A., W. A. Beckman, and J. G. Dekker, *Mechanical Engineering*, **99**, 36 (1977). "Solar Heating in North America."

Engebretson, C. D. and N. G. Ashar, paper 60-WA-88 presented at the New York ASME Meeting (1960). "Progress in Space Heating with Solar Energy."

Engebretson, C. D., *Proceedings of the UN Conference on New Sources of Energy*, **5**, 159 (1964). "The Use of Solar Energy for Space Heating—M.I.T. Solar House IV."

Erdman, D. A. and R. W. Persons, paper at the Department of Energy Conference on Solar Heating, Denver (1978). "The Arlington Solar House of the University of Wisconsin-Madison."

Freeman, T. L., J. W. Mitchell, and T. E. Audit, *Solar Energy*, **22**, 125 (1979). "Performance of Combined Solar-Heat Pump Systems."

Hooper, F. C. and C. R. Attwater, paper in *Heat Transfer in Solar Energy Systems*, American Society of Mechanical Engineers, New York (1977). "A Design Method for Heat Loss Calculation for In-Ground Heat Storage Tanks."

Hughes, P. J., W. A. Beckman, and J. A. Duffie, *Solar Energy*, **19**, 317 (1977). "Simulation Study of Several Solar Heating Systems with Off-Peak Auxiliary."

Hughes, P. J., T. L. Freeman, and H. Grunes, paper #76-4534 presented at the American Society Agricultural Engineers, Chicago, (December 1976). "Design of An Experimental Solar House Heating System."

Jordan, R. C. and J. L. Threlkeld, *Heating, Piping and Air Conditioning*, **28**, 122 (1954). "Design and Economics of Solar Energy Heat Pump Systems."

Jurinak, J. J. and S. I. Abdel-Khalik, *Solar Energy*, **21**, 377 (1978) "Properties Optimization for Phase-Change Energy Storage in Air-Based Solar Heating Systems."

Jurinak, J. J. and S. I. Abdel-Khalik, *Solar Energy*, **22**, 355 (1979). "Sizing Phase-Change Energy Storage Units for Air-Based Solar Heating Systems."

Karaki, S., W. S. Duff, and G. O. G. Löf, report COO-2868-4 to U. S. Department of Energy (1978). "A Performance Comparison Between Air and Liquid Residential Solar Heating Systems."

Karaki, S., P. R. Armstrong, and T. N. Bechtel, report COO-2868-3 to U. S. Department of Energy (1977). "Evaluation of a Residential Solar Air Heating and Nocturnal Cooling System."

Karman, V. D., T. L. Freeman, and J. W. Mitchell, *Proceedings of the Joint Conference of the American Section of the ISES and the SES of Canada*, Winnipeg, **3**, 324 (1976). "Simulation Study of Solar Heat Pump Systems."

Klein, S. A., W. A. Beckman, and J. A. Duffie, *Solar Energy*, **18**, 113 (1976). "A Design Procedure for Solar Heating Systems."

Kuharich, R. F., *Proceedings of the Joint Conference of the American Section of the ISES and the SES of Canada*, Winnipeg, **3**, 378 (1976). "Operational Analysis of a Solar Optimized Heat Pump."

Lior, N., S. Shore, J. A. Lepore, and G. F. Jones, paper in *Proceedings of the 1978 Annual Meeting of the American Section of the ISES*, Denver, **2** (1), 425. "Solar Heating Retrofit of an Urban Row House: Construction and Start Up."

Löf, G. O. G., M. M. El-Wakil, and J. P. Chiou, *Transactions of the ASHRAE*, **77** (October 1963). "Residential Heating with Solar Heated Air—the Colorado Solar House."

Löf, G. O. G., M. M. El-Wakil, and J. P. Chiou, *Proceedings of the UN Conference on New Sources of Energy*, **5**, 185 (1964). "Design and Performance of Domestic Heating System Employing Solar Heated Air—The Colorado House."

Löf, G. O. G. and R. A. Tybout, *Solar Energy*, **14**, 253 (1973). "Cost of House Heating with Solar Energy."

Löf, G. O. G. and D. S. Ward, report COO/2577-76/1 to NATO Committee on Challenges of Modern Society (1976). "Design, Construction and Testing of the Colorado State University Solar House I Heating and Cooling System."

Lorsch, H. G. and B. Niyogi, report NSF/RANN/SE/GI 27976/TR72/18, to NSF from University of Pennsylvania (August 1971). "Influence of Azimuthal Orientation on Collectible Energy in Vertical Solar Collector Building Walls."

Marvin, W. C. and S. A. Mumma, *Proceedings of the Joint Conference of the American Section of the ISES and the SES of Canada*, Winnipeg, **3**, 321 (1976). "Optimum Combination of Solar Energy and the Heat Pump for Residential Heating."

Mitchell, J. W., T. L. Freeman, and W. A. Beckman, *Solar Age*, **3** (7), 20 (1978). "Heat Pumps."

Morrison, D. J. and S. I. Abdel-Khalik, *Solar Energy*, **20**, 57 (1978). "Effects of Phase Change Energy Storage on the Performance of Air Based and Liquid-Based Solar Heating Systems."

Shurcliff, W. A., *Solar Heated Buildings of North America*, Brick House Publishing Co., Harrisville, NH (1978).

Speyer, E., *Solar Energy*, **3** (4), 24 (1959). "Optimum Storage of Heat with a Solar House."

Telkes, M. and E. Raymond, *Heating and Ventilating*, **80** (1949). "Storing Solar Heat in Chemicals —A Report on the Dover House."

Telkes, M., *Proceedings Workshop on Solar Energy Storage Subsystems for the Heating and Cooling of Buildings*, Charlottesville, VA, (1975), page 17. "Thermal Storage for Solar Heating and Cooling."

Terrell, R. E., paper presented at the American Section Meeting of the ISES, Denver, August (1978), "Performance of a Heat Pump Assisted Solar Heated Residence in Madison, Wisconsin."

Tybout, R. A. and G. O. G. Löf, *Natural Resources Journal*, **10**, 268 (1970). "Solar Energy Heating."

Ward, D. S., G. O. G. Löf, C. C. Smith, and L. L. Shaw, *Solar Energy*, **19**, 79 (1977). "Design of a Solar Heating and Cooling System for CSU House II."

Ward, J. C. and G. O. G. Löf, *Solar Energy*, **18**, 301 (1976). "Long-Term (18 years) Performance of a Residential Solar Heating System."

Ward, J. C. and G. O. G. Löf, *Proceedings of the Annual Meeting of the American Section of the ISES*, **1**, 10–12 (1977). "Maintenance Costs of Solar Air Heating System."

Yanagimachi, M., *Transactions of the Conference on Use of Solar Energy*, **3**, 32, University of Arizona Press (1958). "How to Combine: Solar Energy, Nocturnal Radiational Cooling, Radiant Panel System of Heating and Cooling, and Heat Pump to Make a Complete Year-Round Air Conditioning System."

Yanagimachi, M., *Proceedings of the UN Conference on New Sources of Energy*, **5**, 233 (1964). "Report on Two and a Half Years' Experimental Living in Yanagimachi Solar House II."

CHAPTER 14

Design of Solar Heating Systems

Design methods for solar thermal systems fall into two main categories. In the first are large, new, or one-of-a-kind systems which are best designed with the use of simulation methods. The detailed information that simulations provide is needed for new and unique kinds of systems, and the cost of doing simulations is small compared to the total costs involved in new or large systems. The methods outlined in Chapter 10 for doing hour by hour performance calculations are useful tools for these design problems.

The second category includes design of active systems that fit standard configurations, where adequate information on details of dynamics of performance is already available, and where the cost of the project does not warrant the expense of a simulation. Design procedures are available for many of these systems that are easy to use and provide adequate estimates of long-term thermal performance. In this chapter we briefly review some of these methods. The *f*-chart method, applicable to heating of buildings where the minimum temperature for energy delivery is approximately 20 C, is outlined in detail. Methods for designing systems delivering energy at other minimum temperatures, as are encountered in solar absorption air conditioning or industrial process heat applications, are presented in Chapter 18. Design of passive heating systems is treated in Chapter 15.

14.1 REVIEW OF DESIGN METHODS

Design methods for solar thermal processes can be put in three general categories, according to the assumptions on which they are based and the ways in which the calculations are done. They all produce estimates of the long-term useful outputs of solar processes, but do not provide detailed information on process dynamics.

The first category applies to systems in which the collector operating temperature is known, and for which critical radiation level can be established (i.e., radiation levels above which useful energy can be collected). The first of these, the utilizability method, is based on analysis of hourly weather data to obtain the fraction of the total month's radiation that is above a critical level.*

_____
* This method and a further development are outlined in Chapter 18.

Another example in this category is the heat table method of R. Morse as described by Proctor (1975). This is a straightforward tabulation of integrated collector performance as a function of collector characteristics, location, and orientation, assuming fixed fluid inlet temperatures.

The second category of design methods includes those that are correlations of the results of a large number of detailed simulations. The f-chart method of Klein et al. (1976, 1977) is an example. The results of many simulations are correlated in terms of easily calculated dimensionless variables. The results of the f-chart method have served as the basis for further correlations, e.g., by Ward (1976) who has used only January results, by Barley and Winn (1978) who used a 2 point curve fit to obtain location dependent annual results and by Lameiro and Bendt (1978) who also obtained location dependent annual results with 3 point curve fits.

Another example, in the second category is the method of Los Alamos Scientific Laboratory [Balcomb and Hedstrom (1976)], which is a correlation of the outputs of simulations for specific systems and two collector types.

The third category of design methods is based on short-cut simulations. In these methods, simulations are done using representative days of meteorological data and the results are related to longer term performance. The SOLCOST method [Connelly et al. (1976)] simulates a clear day and a cloudy day and then weighs the results according to average cloudiness to obtain a monthly estimate of system performance.

14.2   THE f-CHART METHOD

This and the following sections outline the f-chart method for estimating the annual thermal performance of active heating systems for buildings (using either liquid or air as the working fluid) where the minimum temperature of energy delivery is 20 C. The system configurations that can be evaluated by the f-chart methods are expected to be common in residential applications. This material is presented in more detail in *Solar Heating Design by the* f-*Chart Method* by Beckman et al. (1977).

The f-chart method provides a means for estimating the fraction of total heating load that will be supplied by solar energy for a given solar heating system. The primary design variable is collector area; secondary variables are collector type, storage capacity, fluid flow-rates, and load and collector heat exchanger sizes. The method is a correlation of the results of many hundreds of thermal performance simulations of solar heating systems. The conditions of the simulations were varied over appropriate ranges of parameters of practical system designs. The resulting correlations give f, the fraction of the monthly heating load (for space heating and hot water) supplied by solar energy as a function of two dimensionless parameters, one related to the ratio of collector losses to heating loads and the other related to the ratio of absorbed solar radiation to heating loads.

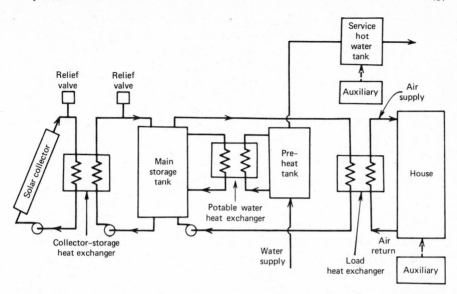

Figure 14.2.1   Schematic of the standard system configuration using liquid heat transfer and storage media.

The *f*-charts have been developed for three standard system configurations, liquid and air systems for space (and hot water) heating, and systems for service hot water only. A schematic diagram of the standard heating system using liquid heat transfer fluids is shown in Figure 14.2.1. This system normally uses an antifreeze solution in the collector loop and water as the storage medium. Collectors may be drained when energy is not being collected, in which case water is used directly in the collectors and a collector heat exchanger is not needed. A water-to-air load heat exchanger is used to transfer heat from the storage tank to the building. A liquid-to-liquid heat exchanger is used to transfer energy from the main storage tank to a domestic hot water sub system. Although Figure 14.2.1 shows a two-tank domestic hot water system, a one-tank system could be used as described in Section 12.4. An auxiliary heater is provided to supply energy for the space heating load when it cannot be met from the tank. The ranges for major design variables used in developing the correlations are given in Table 14.2.1.

The standard configuration of a solar air heating system with a pebble bed storage unit is shown in Figure 14.2.2. Other equivalent arrangements of fans and dampers can be used to provide the same modes of operation. Energy required for domestic hot water is provided by an air-to-water heat exchanger in the hot air duct leaving the collector. During summer operation, it is best not to store solar energy in the pebble bed, so a manually operated storage bypass is usually provided in this system to allow summer water heating. The ranges of design parameters used in developing the correlations for this system are shown in Table 14.2.2.

Table 14.2.1 Ranges of Design Parameters Used in Developing the f-Charts for Liquid Systems[a]

$0.6 \leq (\tau\alpha)_n$	≤ 0.9
$5 \leq F'_R A$	$\leq 120 \text{ m}^2$
$2.1 \leq U_L$	$\leq 8.3 \text{ W/m}^2\text{ C}$
$30 \leq \beta$	$\leq 90 \text{ deg}$
$83 \leq (UA)_h$	$\leq 667 \text{ W/C}$

[a] From Klein et al. (1976).

The standard configuration for a solar domestic water heating system is shown in Figure 14.2.3. The collector may heat either air or liquid. The solar energy is transferred via a heat exchanger to a domestic hot water (DHW) preheat tank, which supplies solar heated water to a conventional water heater or an in-line low capacitance "zip" heater where the water is further heated to the desired temperature if necessary. A tempering valve may be provided to maintain the tap water below a maximum temperature. These changes in the system configuration do not have major effects on the performance of the system (see Section 12.4).

Detailed simulations of these systems have been used to develop correlations

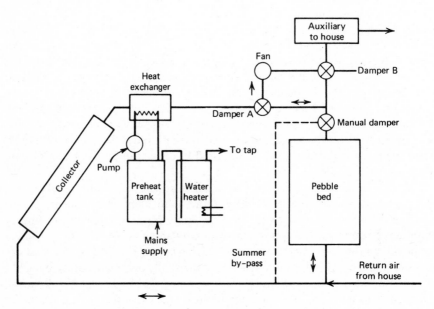

Figure 14.2.2   The standard air system configuration.

Table 14.2.2 Ranges of
Design Parameters Used in
Developing the f-Charts for
Air Systems[a]

$0.6 \le (\tau\alpha)_n$	≤ 0.9
$5 \le F_R A$	$\le 120 \text{ m}^2$
$2.1 \le U_L$	$\le 8.3 \text{ W/m}^2 \text{ C}$
$30 \le \beta$	$\le 90 \text{ deg}$
$83 \le (UA)_h$	$\le 667 \text{ W/ C}$

[a] From Klein et al. (1977).

between dimensionless variables and f, the monthly fraction of loads carried by solar energy. The two dimensionless groups are

$$X = \frac{A_c F'_R U_L (T_{\text{ref}} - \overline{T}_a)\Delta\tau}{L} \tag{14.2.1}$$

$$Y = \frac{A_c F'_R (\overline{\tau\alpha}) \overline{H}_T N}{L} \tag{14.2.2}$$

where A_c = collector area (m^2)

F'_R = collector-heat exchanger efficiency factor*

U_L = collector overall loss coefficient (W/m^2 C)

$\Delta\tau$ = total number of seconds in the month

$\overline{T}_a$ = monthly average ambient temperature (C)

T_{ref} = an empirically derived reference temperature (100 C)

L = monthly total heating load for space heating and hot water (J)

$\overline{H}_T$ = monthly average daily radiation incident on the collector surface per unit area (J/m^2)

N = days in month

$(\overline{\tau\alpha})$ = monthly average transmittance-absorptance product

The Equations 14.2.1 and 14.2.2 can be rewritten

$$X = F_R U_L \times \frac{F'_R}{F_R} \times (T_{\text{ref}} - \overline{T}_a) \times \Delta\tau \times \frac{A_c}{L} \tag{14.2.3}$$

$$Y = F_R(\tau\alpha)_n \times \frac{F'_R}{F_R} \times \frac{(\overline{\tau\alpha})}{(\tau\alpha)_n} \times \overline{H}_T N \times \frac{A_c}{L} \tag{14.2.4}$$

---

* Although we indicate only a modification to F_R to account for the collector-storage heat exchanger, both $F_R(\tau\alpha)$ and $F_R U_L$ can be modified to account for the collector-storage heat exchanger or duct losses or both (see Sections 10.2 and 10.3).

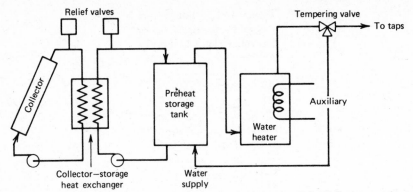

Figure 14.2.3 The standard system configuration for water heating only. Collector may heat air or water.

$F_R U_L$ and $F_R(\tau\alpha)_n$ are obtained from collector test results by the methods noted in Chapter 7. F'_R/F_R corrects for various temperature drops between the collector and the storage tank and is calculated by methods shown in Sections 10.2 and 10.3. $(\overline{\tau\alpha})/(\tau\alpha)_n$ is estimated by the methods noted in Section 5.9. $\overline{T}_a$ is obtained from meteorological records for the month and location desired. $\overline{H}_T$ is found from the monthly average daily radiation on the surface of the collector as outlined in Chapter 2. The monthly loads, L, are discussed in Section 14.3. A_c is the collector area. Thus, all of the terms in these two equations are readily determined from available information.

Example 14.2.1

A solar heating system is to be designed for Madison, WI (latitude 43°N), using one-cover collectors with $F_R(\tau\alpha)_n = 0.74$ and $F_R U_L = 4.00$ W/m² C as determined from standard collector tests. The collector is to face south with a slope of 60° from the horizontal. The average daily radiation on a 60° surface for January in Madison is 12.9 MJ/m² and the average ambient temperature is -7 C (from Appendix G). The heating load is 36.0 GJ for space and hot water. The collector-heat exchanger correction factor, F'_R/F_R, is 0.97. The ratio of the monthly average to normal incidence transmittance-absorptance product $(\overline{\tau\alpha})/(\tau\alpha)_n$, is 0.96 for the one-cover collectors with this orientation as noted in Section 5.9. Calculate X and Y for these conditions for collector areas of 25 and 50 m².

Solution

From Equations 14.2.3 and 14.2.4 with $A_c = 25$ m²,

$$X = 4.0 \text{ W/m}^2 \text{ C} \times 0.97 \times [100 - (-7)] \text{ C}$$
$$\times 31 \text{ days} \times 86400 \text{ s/day} \times 25 \text{ m}^2/36.0 \times 10^9 \text{ J}$$
$$= 0.77$$
$$Y = 0.74 \times 0.97 \times 0.96 \times 12.9 \times 10^6 \text{ J/m}^2 \text{ day}$$
$$\times 31 \text{ days} \times 25 \text{ m}^2/36.0 \times 10^9 \text{ J}$$
$$= 0.19$$

For 50 m^2, the values of X and Y are proportionally higher.

$$X = 0.77 \times 50/25 = 1.54$$
$$Y = 0.19 \times 50/25 = 0.38$$ ∎

As will be shown in later sections, the variables X and Y are used to determine f, the monthly fraction of the load supplied by solar energy. The energy contribution for the month is the product of f, and the total monthly heating load, L. The fraction of the annual heating load supplied by solar energy is the sum of the monthly solar energy contributions divided by the annual load

$$\mathscr{F} = \frac{\sum f_i L_i}{\sum L_i} \tag{14.2.5}$$

14.3   CALCULATION OF HEATING LOADS

A detailed discussion of heating and hot water loads is beyond the scope of this book, and for such a discussion the reader is referred to the ASHRAE Handbook of Fundamentals (1977). In this section we outline one method of estimating heating loads, the "degree-day" method, and a method for estimating hot water loads. The f-chart method is not dependent on the method used to calculate loads, and other means of estimating L can be used at the discretion of the designer.

The degree-day method of estimating loads is based on the principle that the energy requirement for space heating is primarily dependent on the difference in temperature between indoors and outdoors. The monthly space heating load for a building maintained at 24 C (75 F) is assumed to be proportional to the number of degree-days in the month, DD,

$$L_s = (UA)_h \text{DD} \tag{14.3.1}$$

where $(UA)_h$ is a loss coefficient-area product for the building. The number of degree-days (reported by the United States Weather Service) is the difference between 18.3 C and the mean daily ambient temperature, with positive values only included in the compilation. The difference between 24 C and 18.3 C allows for ordinary levels of internal energy generation in the building. Experience has shown that the fuel consumption for heating is approximately proportional to the number of degree-days. Long-term averages of degree-days by months are tabulated for many North American cities in Appendix G. (Recent trends toward lower thermostat settings as a means of reducing energy requirements and use of increased amounts of thermal insulation will lead to a reduction in the temperature base used in the definition of a degree-day.)

For existing structures where fuel consumption records are available, $(UA)_h$ may be calculated from

$$(UA)_h = \frac{(N_F H_F \eta_F)}{\text{DD}} \tag{14.3.2}$$

where N_F = units of fuel consumed
H_F = heating value of fuel consumed
η_F = efficiency of furnace (usually 0.5 to 0.7 for gas or oil burners)

For new structures, $(UA)_h$ can be calculated from details of building construction by methods outlined in the ASHRAE Handbook of Fundamentals. From the design heating load and design temperature difference:

$$(UA)_h = \frac{\text{Design heating load}}{\text{Design temperature difference}} \qquad (14.3.3)$$

Loads in the f-chart method may also include energy requirements for heating of hot water. The water heating loads are given by

$$L_w = C_p M(T_w - T_m) \qquad (14.3.4)$$

where C_p = specific heat of water
M = hot water requirements for the month
T_w = set temperature for delivery of hot water
T_m = mains (supply) water temperature

Heat losses from the auxiliary hot water tank should be added to L_w, unless losses from that tank contribute to meeting the space heating loads in the winter. They can be estimated from $(UA)_{\text{tank}}(T_w - T_{\text{room}})$ or from manufacturers data.

Example 14.3.1

A residence in Madison has $(UA)_h = 463$ W/C. It is expected that hot water requirements will be 450 kg/day, heated from 12 to 60 C. For the month of January, estimate the loads for space heating and hot water heating.

Solution

The Madison January degree days are 830 from Appendix G so the space heating load is

$$L_s = 463 \times 830 \times 24 \times 3600 = 33.2 \text{ GJ}$$

The water heating load is

$$L_w = 450 \times 31 \times 4190(60 - 12) = 2.8 \text{ GJ}$$

The total loads are $33.2 + 2.8 = 36.0$ GJ for the month. (Note: heat losses from the auxiliary water heater are assumed to contribute to meeting the space heating load and are thus not included in the water heating load. In a summer month these losses would be added to L_w.) ■

14.4 THE *f*-CHART FOR LIQUID SYSTEMS

The fraction, f, of the monthly total load supplied by the solar space and water heating system shown in Figure 14.2.1 is given as a function of X and Y in Figure 14.4.1. The relationship between X, Y, and f in equation form is

$$f = 1.029Y - 0.065X - 0.245Y^2 + 0.0018X^2 + 0.0215Y^3 \quad (14.4.1)$$

Because of the nature of Equation 14.4.1, it should not be used outside of the range shown by the curves of Figure 14.4.1. If a calculated point falls outside of this range, the graph can be used for extrapolation with satisfactory results.

Example 14.4.1

The solar heating system described in Example 14.2.1 is to be a liquid system. What fraction of the annual heating load will be supplied by solar energy for a collector area of 50 m²? The monthly combined loads on the system are indicated in the table below. (See Example 2.16.1 for $\overline{H}_T$ values.)

Solution

From Example 14.2.1, the values of X and Y for 50 m² are 1.54 and 0.38, respectively, in January. From Figure 14.4.1 (or Equation 14.4.1), $f = 0.26$. The total heating load for January is 36.0 GJ. Thus, the energy delivery from the solar heating system in January is

$$fL = 0.26 \times 36.0 \text{ GJ} = 9.4 \text{ GJ}$$

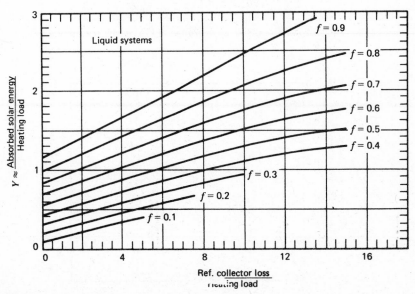

Figure 14.4.1 The *f*-chart for systems using liquid heat transfer and storage media. From Beckman *et al.* (1977).

The fraction of the annual heating load supplied by solar energy is determined by repeating the calculation of X, Y, and f for each month, and summing the results as indicated in Equation 14.2.5. The table shows the results of these calculations. From Equation 14.2.5 the annual fraction of the load supplied by solar energy is

$$\mathscr{F} = \frac{87.0}{203.2} = 0.43$$

Monthly and Annual Performance of a Liquid Heating System in Madison, for Example 14.4.1

Month	$\bar{H}_T$, MJ/m^2	$\bar{T}_a$	Load, GJ	$(\overline{\tau\alpha})/(\tau\alpha)_n$	X	Y	f	fL, GJ
Jan.	12.9	−7	36.0	0.96	1.54	0.38	0.26	9.4
Feb.	15.6	−6	30.4	0.94	1.64	0.48	0.34	10.3
Mar.	16.0	0	26.7	0.93	1.95	0.62	0.43	11.5
Apr.	14.5	7	15.7	0.92	2.99	0.92	0.57	8.9
May	15.4	13	9.2	0.90	4.92	1.68	0.86	7.9
June	16.2	19	4.1	0.89	9.96	3.79	1.00[a]	4.1
July	16.5	21	2.9	0.89	14.19	5.64	1.00[a]	2.9
Aug.	16.5	20	3.4	0.92	12.25	4.97	1.00[a]	3.4
Sept.	15.8	15	6.3	0.94	6.80	2.54	1.00	6.3
Oct.	14.9	10	13.2	0.94	3.55	1.18	0.70	9.2
Nov.	10.4	1	22.8	0.96	2.19	0.47	0.30	6.8
Dec.	9.5	−5	32.5	0.96	1.68	0.31	0.19	6.2
Total			203.2					87.0

[a] These points have coordinates outside of the range of the f-chart correlation.

■

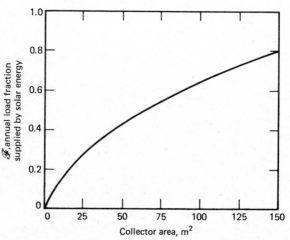

Figure 14.4.2 Annual load fraction versus collector area.

To determine the economic optimum collector area, the annual load fraction corresponding to several different collector areas must be determined. The annual load fraction is then plotted as a function of collector area, as shown in Figure 14.4.2. The information in this figure can then be used for economic calculations, as shown in Chapter 11.

For liquid systems, *f*-chart calculations can be modified to estimate changes in long-term performance due to changes in storage tank capacity and load heat exchanger characteristics. This is done by modifying the values of X or Y as described below.

Storage Capacity

Annual system performance is relatively insensitive to storage capacity, as long as capacity is more than approximately 50 liters of water per square meter of collector. When the costs of storage are considered, there are broad optima in the range of 50 to 200 liters of water per square meter of collector.

The *f*-chart was developed for a storage capacity of 75 liters of stored water per square meter of collector area. The performance of systems with storage capacities in the range of 37.5 to 300 liters/m^2 can be determined by multiplying the dimensionless group X by a storage size correction factor, X_c/X, from Figure 14.4.3 or Equation 14.4.2.

$$\frac{X_c}{X} = \left(\frac{\text{Actual storage capacity}}{\text{Standard storage capacity}}\right)^{-0.25}$$ (14.4.2)

for

$$0.5 \le \left(\frac{\text{Actual}}{\text{Standard}}\right) \le 4.0$$

where the standard storage capacity is 75 liters of water per square meter of collector area.

Example 14.4.2

For the conditions of Example 14.4.1, what would be the annual solar contribution if the storage capacity of the tank is doubled, to 150 liters/m^2?

Solution

To account for changes in storage capacity, the value of X calculated in the previous examples must be modified using Figure 14.4.3 or Equation 14.4.2. The ratio of actual storage size to standard storage size is 2.0, so

$$\frac{X_c}{X} = (2.0)^{-1/4} = 0.84$$

For January the corrected value of X is then

$$X_c = 0.84 \times 1.54 = 1.29$$

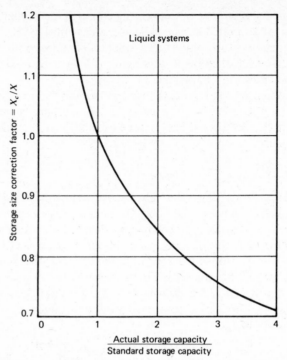

Figure 14.4.3 Storage size correction factor for liquid systems. Standard storage capacity is 75 liters/m^2.

The value of Y remains 0.38. From the f-chart, $f = 0.28$. The solar contribution for January is

$$fL = 0.28 \times 36.0 \text{ GJ} = 9.9 \text{ GJ}$$

Repeating these calculations for the remaining 11 months gives an annual solar load fraction of 0.45 (versus 0.43 for the standard storage size). ■

Load Heat Exchanger Size

As the heat exchanger used to heat the building air is reduced in size, the storage tank temperature must increase to supply the same amount of heat, resulting in higher collector temperatures and reduced collector output. A measure of the size of the heat exchanger needed for a specific building is provided by the dimensionless parameter, $\varepsilon_L C_{\min}/(UA)_h$, where ε_L is the effectiveness of the water-air load heat exchanger and $C_{\min}$ is the minimum fluid capacitance rate (mass flow rate times the specific heat of the fluid) in the heat exchanger and is generally that of the air. $(UA)_h$ is the overall energy loss coefficient-area product for the building used in the degree-day space heating load model.

From thermal considerations, the optimum value of $\varepsilon_L C_{\min}/(UA)_h$ is infinity. However, system performance is asymptotically dependent upon the value

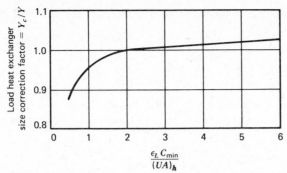

Figure 14.4.4   Load heat exchangers size correction factor.

of this parameter, and for values of $\varepsilon_L C_{min}/(UA)_h$ greater than 10, performance will be essentially the same as that for the infinitely large value. The reduction in performance due to an undersized load heat exchanger will be significant for values of $\varepsilon_L C_{min}/(UA)_h$ less than about 1. Practical values of $\varepsilon_L C_{min}/(UA)_h$ are generally between 1 and 3 when the cost of the heat exchanger is considered. See Beckman et al. (1977) for further discussion.

The f-chart for liquid systems was developed with $\varepsilon_L C_{min}/(UA)_h = 2$. The performance of systems having other values of $\varepsilon_L C_{min}/(UA)_h$ can be estimated from the f-chart by modifying Y by a load heat exchanger correction factor, Y_c/Y, as indicated in Figure 14.4.4 or Equation 14.4.3.

$$\frac{Y_c}{Y} = 0.39 + 0.65 e^{-(0.139(UA)_h/\varepsilon_L C_{min})} \qquad (14.4.3)$$

for

$$0.5 < \left(\frac{\varepsilon_L C_{min}}{(UA)_h}\right) < 50$$

Example 14.4.3

For the conditions of Example 14.4.1, what will be the solar contribution if the load heat exchanger is used under the following circumstances: air flow rate = 520 liters/s, water flow rate = 0.694 liters/s, and the heat exchanger effectiveness at these flowrates is 0.69. $(UA)_h$, the building overall energy loss coefficient-area product, is 463 W/C.

Solution

First, the value of C_{min} is determined. This is usually the capacitance rate of the air, which in this example is

$$C_{min} = 520 \text{ liters/s} \times 1.20 \text{ kg/m}^3 \times 1010 \text{ J/kg C}/1000 \text{ liters/m}^3$$
$$= 630 \text{ W/C}$$

The capacitance rate of the water is 2910 W/C so that of the air is lower. Then

$$\frac{\varepsilon_L C_{min}}{(UA)_h} = 0.69 \times \frac{(630 \text{ W/C})}{(463 \text{ W/C})}$$

$$= 0.94$$

This heat exchanger is smaller than the standard value of 2 used in developing Figure 14.4.1. The correction factor from Figure 14.4.3 or Equation 14.4.3 is

$$\frac{Y_c}{Y} = 0.95$$

$$Y_c = 0.95 \times 0.38 = 0.36$$

From Figure 14.4.1, $f = 0.24$ for January, and the solar energy contribution for the month is

$$fL = 0.24 \times 36.0 \text{ GJ} = 8.8 \text{ GJ}$$

The annual solar load fraction is found to be 0.41.   ∎

If both the storage and load heat exchanger sizes differ from the standards used to develop the f-chart, the correction factors discussed in Examples 14.4.2 and 14.4.3 would both be applied to find the appropriate values of X_c and Y_c for determination of f. Thus if the storage correction of Example 14.4.2 and the load heat exchanger correction of Example 14.4.3 are both needed, f would be determined at $X_c = 1.22$ and $Y_c = 0.33$ where $f = 0.24$.

14.5   THE f-CHART FOR AIR SYSTEMS

The monthly fraction of total heating load supplied by the solar air heating system shown in Figure 14.2.2 has been correlated with the same dimensionless parameters X and Y as were defined in Equations 14.2.1 and 14.2.2. The correlation is given in Figure 14.5.1 and Equation 14.5.1. It is used in the same manner as the f-chart for liquid-based systems. The equation is subject to the same cautions concerning the range of X and Y as the liquid system equation.

$$f = 1.040Y - 0.065X - 0.159Y^2 + 0.00187X^2 - 0.0095Y^3 \quad (14.5.1)$$

Example 14.5.1

A solar air heating system is to be designed for a building in Madison, WI with 2-cover collectors facing south at a slope of 58°. The air heating collectors have the following characteristics: $F_R U_L = 2.84 \text{ W/m}^2 \text{ C}$ and $F_R(\tau\alpha)_n = 0.49$. $(\overline{\tau\alpha})/(\tau\alpha)_n = 0.93$ for this application of the two cover collector. The total space and water heating load for January is 36.0 GJ (as in Examples 14.4.1 to 14.4.3). What fraction of the load would be supplied by solar energy with a system having a collector area of 50 m^2?

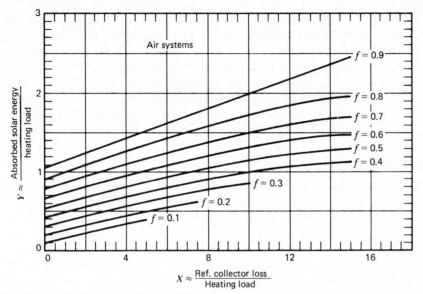

$$X \approx \frac{\text{Ref. collector loss}}{\text{Heating load}}$$

Figure 14.5.1 The *f*-chart for air systems of the configuration shown in Figure 14.4.2. From Beckman *et al.* (1977).

Solution

For air systems, there will be no correction factor for the collector heat exchanger and we will assume duct losses are small, so that $F_R'/F_R = 1$. From Equations 14.2.3 and 14.2.4

$$X = 2.84 \text{ W/m}^2 \text{ C} \times 1 \times [100 - (-7)] \text{ C} \times (31 \text{ days})$$

$$\times (86400 \text{ s/day}) \times \frac{50 \text{ m}^2}{36.0 \times 10^9 \text{ J}}$$

$$= 1.13$$

$$Y = 0.49 \times 1 \times 0.93 \times (13.2 \times 10^6 \text{ J/day m}^2)$$

$$\times (31 \text{ days}) \times \frac{50 \text{ m}^2}{36.0 \times 10^9 \text{ J}}$$

$$= 0.259$$

Then f for January, from Figure 14.5.1 or Equation 14.5.1, is 0.19. The solar energy supplied by this system in January is

$$fL = 0.19 \times 36.0 \text{ GJ} = 6.8 \text{ GJ}$$

As with the liquid systems, the annual system performance is obtained by summing the energy quantities for all months. The result of the calculation is that 37% of the annual load is supplied by solar energy. ∎

Air systems require two correction factors, one to account for effects of storage size if it is other than 0.25 m³/m², and the other to account for air flow rate that affects stratification in the pebble bed. In addition, care must be exercised to be sure that the values of $F_R(\tau\alpha)_n$ and $F_R U_L$ from collector tests are obtained for the same air flow rates as will be used in an installation. The corrections shown in Section 7.5 can be used to convert test data from one flow rate to another. The correction factors for storage capacity and air flow rate are outlined below. There is no load heat exchanger in air systems.

Air Flow Rate

An increase in air flow rate tends to improve system performance by increasing F_R, and tends to decrease performance by reducing the thermal stratification in the pebble bed. The f-chart for air systems is based on a standard collector air flow rate of 10 liters/s of air per square meter of collector area. The performance of systems having other collector air flow rates can be estimated by using the appropriate values of F_R and Y and then modifying the value of X by a collector air flow rate correction factor, X_c/X, as indicated in Figure 14.5.2 or Equation 14.5.2 to account for the degree of stratification in the pebble bed.

$$\frac{X_c}{X} = \left(\frac{\text{Actual air flow rate}}{\text{Standard air flow rate}} \right)^{0.28} \tag{14.5.2}$$

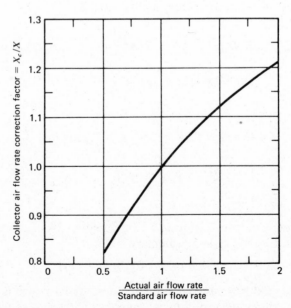

Figure 14.5.2 Correction factor for air flowrate to account for stratification in the pebble bed. The standard flowrate is 10 liters/m² s.

for

$$0.5 < \left(\frac{\text{Actual}}{\text{Standard}} \right) < 2$$

where the standard air flow rate is 10 liters/s per square meter of collector area.

Example 14.5.2

The system of Example 14.5.1 is to be designed using a collector air flow rate of 15 liters/s per square meter of collector. Estimate the change in annual performance of the system resulting from the increased air flow.

Solution

Increasing the air flow rate affects F_R and stratification in the pebble bed. The effects of air flow rate on F_R and thus on $F_R U_L$ and $F_R(\tau\alpha)_n$ must be determined either by collector tests at the correct air flow rate or estimated by the methods of Section 7.5. In this case, $F_R(\tau\alpha)_n = 0.52$ and $F_R U_L = 3.01$ W/m^2 C at 15 liters/s m^2. The corrected X to account for pebble bed stratification is found from Equation 14.5.2 or Figure 14.5.2

$$\frac{X_c}{X} = \left(\frac{15}{10} \right)^{0.28} = 1.12$$

Thus the X to be used is the value from Example 14.5.1 corrected for the modified F_R and for the air flow rate

$$X_c = 1.13 \times \frac{3.01}{2.84} \times 1.12 = 1.34$$

Correcting Y for the new value of F_R (i.e., $F_R(\tau\alpha)_n$)

$$Y_c = 0.259 \times \frac{0.52}{0.49} = 0.27$$

From the air *f*-chart, $f = 0.19$ and $fL = 6.8$ GJ for January. The calculation for the year indicates that 38 % of the annual load is supplied by solar energy. (This is a slight increase over the 37 % at the standard air flow rate; the increase must be evaluated in light of the increased fan power required at the higher air flow rate.) ∙   ■

Pebble Bed Storage Capacity

The performance of air systems is less sensitive to storage capacity than that of liquid systems. Air systems can operate in the collector-load mode, in which the storage unit is bypassed. Also, pebble beds are highly stratified and additional capacity is effectively added to the cold end of the bed, which is seldom heated and cooled to the same extent as the hot end.

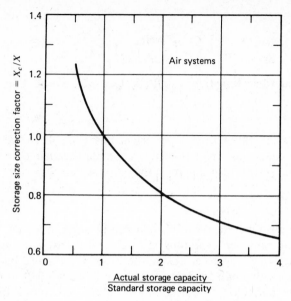

Figure 14.5.3 Storage size correction factors for air systems. The standard storage capcity is 0.25 m³/m².

The *f*-chart for air systems is for a storage capacity of 0.25 cubic meters of pebbles per square meter of collector area, which corresponds to 350 kJ/m² C for typical void fractions and rock properties. The performance of systems with other storage capacities can be determined by modifying X by a storage size correction factor, X_c/X, as indicated in Figure 14.5.3 or Equation 14.5.3.

$$\frac{X_c}{X} = \left(\frac{\text{Actual storage capacity}}{\text{Standard storage capacity}} \right)^{-0.30} \tag{14.5.3}$$

for

$$0.5 < \left(\frac{\text{Actual}}{\text{Standard}} \right) < 4.0$$

where the standard storage capacity is 0.25 m³/m².

Example 14.5.3

If the system of Example 14.5.1 has storage capacity which is 60 percent of the standard capacity, what fraction of the annual heating load would the system be expected to supply?

Solution

The storage size correction factor, from Figure 14.5.3 (or Equation 14.5.3), is 1.17. Then for January

$$\frac{X_c}{X} = 1.17$$

$$X_c = 1.17 \times 1.13 = 1.32$$

Y remains 0.26. From Figure 14.5.1 or Equation 14.5.1, $f = 0.18$ and $fL = 0.18 \times 36.0$ GJ $= 6.5$ GJ. The fraction of the annual load supplied by solar energy is 0.35 (compared to 0.37 for the standard storage size).   ∎

If both air flow rate and storage size are non standard, there will be two corrections on X to be made (in addition to any corrections due to changes in F_R) and the final X will be the product of the uncorrected value and the two correction factors.

If a phase-change energy storage unit is used in place of the rock bed, Equation 9.5.1, an empirical equation for the equivalent rock bed storage capacity, can be used to predict system performance. The properties and mass of the phase-change material are used to estimate the size of an equivalent rock bed, which is then used in the air f-chart correlations. Some phase-change material properties are given in Table 13.10.1.

14.6 SERVICE WATER HEATING SYSTEMS

Figure 14.4.1, the f-chart for liquid heating systems, can be used to estimate the performance of solar water heating systems having the configuration shown in Figure 14.2.3, by defining an additional correction factor on X. The mains water supply temperature, T_m, and the minimum acceptable hot water temperature, T_w, both affect the performance of solar water heating systems. Both T_m and T_w affect the average system operating temperature level and thus the collector energy losses. The dimensionless group X, which is related to collector energy losses, can be corrected to include these effects. If monthly values of X are multiplied by a water heating correction factor, X_c/X in Equation 14.6.1 the f-chart for liquid-based solar space and water heating systems (Equation 14.4.1 or Figure 14.4.1) can be used to estimate monthly values of f for water heating systems. All temperatures are in C.

$$\frac{X_c}{X} = \frac{(11.6 + 1.18T_w + 3.86T_m - 2.32\overline{T}_a)}{(100 - \overline{T}_a)} \tag{14.6.1}$$

This method of estimating water heater performance is based on storage capacity of 75 liters/m^2 and on the typical day's distribution of hot water use occurring each day as shown in Figure 12.4.1. If other distributions of use occur, a system may not perform as well as indicated by the f-chart. Effect of storage size is difficult to predict and will depend on use patterns.

The water heating correction factor is based on the assumption of a well-insulated solar preheat tank, and losses from an auxiliary tank were not included in the f-chart correlations. For systems supplying hot water only, loads on the system should also include losses from the auxiliary tank. (These are normally included in the energy supplied to a conventional water heater). Tank losses can be estimated from the insulation and tank area, but this frequently leads to their underestimation as losses through connections, mounting brackets, etc. can be significant. It is recommended that tank loss calculations be based on the assumption that the entire tank is at the water set temperature, T_w.

The use of a tempering valve on the supply line to mix cold supply water with solar heated water above the water set temperature has little effect on the overall output of the solar system, as noted in Section 12.4, and the method indicated here can be used for systems either with or without the tempering valve.

Example 14.6.1

A solar water heating system is to be designed for a residence in Madison, WI (latitude 43°N). The collectors considered for this purpose have two covers with $F_R'(\tau\alpha)_n = 0.64$ and $F_R'U_L = 3.64$ W/m² C. The collectors are to face south at a slope of 45°. The estimated water heating load is 400 liters/day heated from 11 C to 60 C. The storage capacity of the preheat tank is to be 75 liters of water per square meter of collector area. The auxiliary tank has a capacity of 225 liters, and has a loss coefficient of 0.62 W/m² C. The tank is a cylinder 0.50 m diameter and 1.16 m high. Estimate the fraction of the January heating load supplied by solar energy for this system with a collector area of 10 m². The radiation on the collector, $\bar{H}_T$, is 11.8 MJ/m² and $(\bar{\tau\alpha})/(\tau\alpha)_n$ is 0.94.

Solution

The monthly load is the energy required to heat the water from T_m to T_w plus the auxiliary tank losses. For January, the energy to heat the water is

400 liters/day × 1 kg/liter × 4190 J/kg C × (60 − 11) C

$$\times \ 31 \ \text{day} \times \text{GJ}/10^9 \ \text{J} = 2.55 \ \text{GJ}$$

The loss rate from the auxiliary tank is $UA(T_w - T_a')$. The tank area is 2.21 m², so the loss rate for T_a' of 20 C is

$$0.62 \times 2.21(60 - 20) = 55 \ \text{W}$$

Energy required to supply this loss for the month is

$$\frac{55 \times 31 \times 24 \times 3600}{10^9} = 0.15 \ \text{GJ}$$

The total load to be used in calculation of X and Y is then

$$2.55 + 0.15 = 2.70 \ \text{GJ}$$

We now calculate X_c and Y

$$X_c = X\left(\frac{X_c}{X}\right) = \frac{10 \times 3.64 \times [100 - (-7)] \times 31 \times 24 \times 3600}{2.70 \times 10^9}$$

$$\times \frac{[11.6 + 1.18(60) + 3.86(11) - 2.32(-7)]}{[100 - (-7)]}$$

$$= 5.10$$

$$Y = 0.64 \times 0.94 \times 11.8 \times 10^6 \times 31 \times \frac{10}{2.70 \times 10^9} = 0.81$$

From Figure 14.4.1 or Equation 14.4.1, $f = 0.41$. This process can be repeated for each of the twelve months and the annual solar contribution estimated. ∎

14.7 *f*-CHART RESULTS

In the original development of *f*-charts [Klein (1976)], it was necessary to make a number of assumptions about systems and their performance. Several of these are worth noting as they are useful in interpreting results obtained from this method.

First, all liquid storage tanks were assumed to be fully mixed, both for main storage tanks for liquid systems and preheat tanks for all water heating. This assumption, as shown in Chapter 10, tends to lead to conservative estimates of long-term performance by overestimating collector inlet temperatures. Second, for reasons of economy in simulations, all days were considered symmetrical about solar noon. This also leads to conservative estimates of system outputs. For water heating only, it has been noted above that energy in water above the set temperature is not considered useful. Thus, the computations tend to be conservative in their predictions. On the other hand, it has been assumed that there are no leaks in systems; most air systems leak to some extent, and this will tend to degrade performance below predicted levels.

There are implicit assumptions in this method; systems are well-built, flow distribution to collectors is uniform, flow rates are as assumed, and the system configurations are close to those for which the correlations were developed. If these are not true, systems can not be expected to perform as estimated by the *f*-chart method.

The results obtained with *f*-chart have been compared to results of detailed simulations for a variety of locations. Agreement is generally to within a few percent. Table 14.7.1 shows some of these results for a liquid house-heating system.

Agreement of monthly solar fractions is not nearly as good as annual fractions, and the *f*-chart method should be used to estimate annual performance only.

There are a few measured annual performance data on heating systems that can be compared to *f*-chart results. MIT House IV (Section 13.5) was 52 percent

Table 14.7.1   Comparisons of Annual Fractions
Met by Solar Energy, for *f*-Chart and Detailed
Simulations

Location	*f*-Chart	Simulation
Albuquerque, NM	0.64	0.64
Bismark, MD	0.68	0.69
Blue Hill, MA	0.71	0.74
Charleston, SC	0.71	0.70
Dodge City, KA	0.70	0.68
Madison, WI	0.65	0.67
Medford, OR	0.69	0.74
New York, NY	0.55	0.55
Phoenix, AZ	0.67	0.70
Santa Maria, CA	0.49	0.51
Seattle, WA	0.65	0.73

heated by solar energy over 2 years. *f*-chart estimates (based on measured meteorological conditions) indicates a solar fraction of 57 percent. (The system configuration is close to, but not the same as, the *f*-chart system.) CSU House II, the air system (Section 13.4) was supplied with 72 percent of its heat by solar energy; *f*-chart predicts 76 percent for the period.

A year long experimental study of solar domestic hot water systems at the U. S. National Bureau of Standards was done by Fanney (1979). The annual fraction of the total load (water draw-off plus auxiliary tank losses) supplied by solar from measurements and from *f*-chart predictions is given in Table 14.7.2. Although there are differences between measured and predicted performance, the results agree reasonably well.

Table 14.7.2   Measured and Predicted Values of Annual Solar
Fraction

Number of Tanks	Type	Heat Exchanger	Measured	Predicted
1	Liquid	External	0.36	0.37
2	Liquid	External	0.37	0.40
1	Liquid	Internal	0.45	0.43
2	Liquid	Internal	0.33	0.30
1	Air	External	0.20	0.21

14.8   PARALLEL SOLAR ENERGY-HEAT PUMP SYSTEMS

For the parallel solar energy-heat pump system shown in Figures 13.8.4 and 14.8.1, Anderson (1979) and Anderson et al. (1979) have developed a design method based on a combination of the "bin" method and the f-chart method. In the parallel mode of operation the solar system in the primary energy source and its operation is unaffected by the presence of a heat pump, that is, the heat pump system acts as the solar system auxiliary energy source. Consequently, the f-chart method can be used to determine the solar contribution to the heating load. The remaining portion of the load is met by a combination of the energy delivered by the heat pump and auxiliary energy. Although the performance of the heat pump is affected by the presence of the solar system, the Anderson et al. study observed that this interaction is small and can be neglected. This means that the only effect of the solar system on the heat pump is a reduction of the load that the heat pump will meet. The results of bin method calculations for a heat pump only system can then be modified to predict heat pump performance in the presence of a solar system.

A typical set of heat pump and load characteristics are shown in Figure 14.8.2 as a function of ambient temperature. When the ambient temperature is above the balance point, the heat pump can supply more energy, Q_{del}, than is needed by the load, Q_L. When the ambient temperature is below the balance point, auxiliary energy must be used in addition to the heat pump.

The bin method for estimating the monthly energy usage of a stand-alone heat pump system is described in ASHRAE (1976). The method uses long-term weather data to determine the number of hours in which the ambient temperatures were within 2.8 C (5 F) temperature ranges called bins. The number of hours in each temperature bin for a particular month can be used to estimate the monthly purchased energy. For example, suppose a month has 15 hr in a temperature bin centered around 10 C. To meet the load during this 15 hr, the

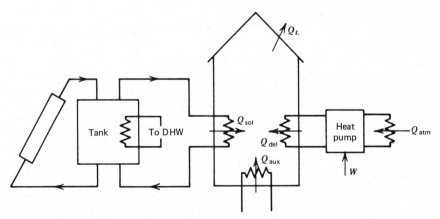

Figure 14.8.1   Parallel solar-heat pump system. From Anderson (1979).

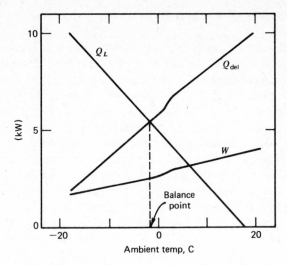

Figure 14.8.2 Typical heat pump and load characteristics as a function of ambient temperature. From Anderson (1979).

system needs to run only 15 hr times the ratio of Q_L to Q_{del} or 15 (2.2/8.1) = 4.1 hr. In this 4.1 hr the energy required by the heat pump is 4.1 × 3.4 = 13.9 kW-hr (50 MJ). This calculation must be repeated for each temperature bin above the balance point.

At temperatures below the balance point the heat pump alone cannot meet the load and auxiliary energy must be used to make up the deficit. If 12 hr are in the bin centered around −10 C, the heat pump will run continuously for the 12 hr at a rate of 2.1 kW for a total electrical requirement of 2.1 × 12 = 25.2 kW-hr (91 MJ). In addition, auxiliary energy must make up the difference between the load of 7.7 kW times 12 hr and the delivered energy of 4.6 kW times 12 hr, or 37.2 kW-hr (134 MJ). The total purchased energy for this bin is then 25.2 + 37.2 = 62.4 kW-hr (225 MJ). By repeated application of these calculations, the monthly purchased energy can be estimated from which annual values can be calculated.

For a house with a parallel solar-heat pump system in Columbia, MO the combined system performance is shown in Figure 14.8.3 as a function of collector area. For zero collector area the system is a stand-alone heat pump and the fraction of the load supplied by non purchased energy from the air is $FATM_0$. At any finite collector area, some energy is supplied by solar and some is from the ambient air. The design procedure assumes that on a monthly basis

$$FATM = FATM_0(1 - f) \tag{14.8.1}$$

where f is the monthly fraction by solar. This equation is a result of assuming that the only effect of the solar system on the heat pump performance is a reduction in the load.

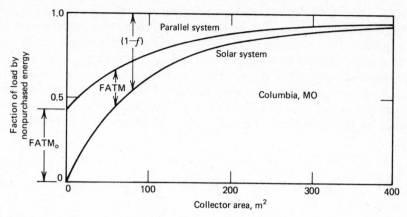

Figure 14.8.3 Parallel solar-heat pump system performance for January in Columbia, MO. From Anderson *et al.* (1979).

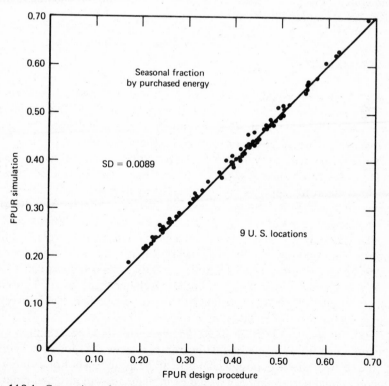

Figure 14.8.4 Comparison of purchased energy fractions as calculated by design procedure and by detailed simulations. From Anderson *et al.* (1979).

With monthly results from f-chart, a bin method calculation, and Equation 14.8.1, the monthly fraction of the load supplied by nonpurchased energy can be estimated. The remainder of the load is supplied by a combination of compressor work supplied to the heat pump and auxiliary energy. If the auxiliary energy is electricity, there is no need to separate the purchased energy into fractions by compressor work and by auxiliary. However, if the auxiliary is not electricity, it is necessary to know each of the two fractions to do an economic assessment. Equation 14.8.2 is recommended by Anderson et al. to find the work fraction

$$FW = FW_0(1 - f) \tag{14.8.2}$$

where FW_0 is the work fraction for the stand alone heat pump. The auxiliary fraction is then

$$FAUX = (1 - FATM_0 - FW_0)(1 - f) \tag{14.8.3}$$

The results of using the design procedure given by Equations 14.8.1 through 14.8.3 have been compared to detailed computer simulations and typical results are shown in Figure 14.8.4. In this figure the fraction of the total load by purchased energy calculated by the two methods compare very well. A similar conclusion can be made concerning the work fraction, the fraction from the atmosphere and the fraction by solar.

14.9 SUMMARY

The f-chart method provides a means of quickly estimating the long-term performance of solar heating systems of standard configurations. The data needed are monthly average radiation and temperature, the collector parameters available from standard collector tests, and estimates of loads.

It should be recognized that there are uncertainties in the estimates obtained from the f-chart procedure. The major uncertainties arise from several sources. First, the meteorological data can be in error by as much as 5 to 10 percent, particularly when the horizontal data are converted to radiation on the plane of the collector. Second, average data are used in the calculations, and any particular year may vary widely from that average. Third, it is extremely difficult to predict what building heating loads will be as it is dependent on the habits of the occupants. Fourth, systems must be carefully engineered and constructed, with minimal heat losses, leakage, and other mechanical and thermal problems. Finally (and probably least important), there are some differences between the f-chart correlation and individual data points.

It is difficult to quantitatively assess the impacts of these uncertainties on the results obtained from the method. However, two generalizations can be made. First, the relative effects of design changes can be established. For example, the effects on annual performance of a change in plate absorptance and emittance can be shown. The second decimal place is significant in this context. Second, the method will predict the performance of a given system, but because of the uncertainties only the first decimal should be considered as significant.

The calculation of the f-chart method can easily be done by hand, but they can be tedious. The method has been programmed, in combination with life cycle economic analysis, in an available program FCHART (1978).

REFERENCES

Anderson, J. V., M. S. thesis, University of Wisconsin-Madison (1979) "Procedures for Predicting the Performance of Air-to-Air Heat Pumps in Stand-Alone and Parallel Solar-Heat Pump Systems."

Anderson, J. V., J. W. Mitchell, and W. A. Beckman, Paper at International Solar Energy Society Meeting, Atlanta (1979). "A Design Method for Parallel Solar-Heat Pump Systems."

ASHRAE *Systems Handbook*, American Society of Heating, Refrigeration, and Air Conditioning Engineers, NY (1976).

ASHRAE *Handbook of Fundamentals*, American Society of Heating, Refrigeration, and Air-conditioning Engineers, NY (1977).

Balcomb, J. D. and J. C. Hedstrom, *Proceedings of the International Solar Energy Society Conference*, Winnipeg, 4, 281 (1976). "A Simplified Method for Sizing a Solar Collector Array for Space Heating."

Barley, C. D. and C. B. Winn, *Solar Energy*, 21, 279 (1978). "Optimal Sizing of Solar Collectors by the Method of Relative Areas."

Beckman, W. A., S. A. Klein, and J. A. Duffie, *Solar Heating Design by the f-Chart Method*, Wiley-Interscience, New York (1977).

Connelly, M., R. Giellis, G. Jenson, and R. McMordie, *Proceedings of the International Solar Energy Society Conference*, Winnipeg, 10, 220 (1976). "Solar Heating and Cooling Computer Analysis—A Simplified Sizing Method for Non-Thermal Specialists."

Fanney, A. H., Personal Communication on the National Bureau of Standards Solar Domestic Hot Water Test Facility (1979).

FCHART Users Manual, University of Wisconsin Engineering Experiment Station, Report 49 (1978).

Klein, S. A., W. A. Beckman, and J. A. Duffie, *Solar Energy*, 18, 113 (1976). "A Design Procedure for Solar Heating Systems."

Klein, S. A., W. A. Beckman, and J. A. Duffie, *Solar Energy*, 19, 509 (1977). "A Design Procedure for Solar Air Heating Systems."

Lameiro, G., and P. Bendt, paper in *Proceedings of the 1978 Meeting of the American Section of the International Solar Energy Society*, Denver, 2 (1), 113 (1978), "The GFL Method for Designing Solar Energy Space Heating and Domestic Hot Water Systems."

Proctor, D., paper at International Solar Energy Society Meeting, Los Angeles (1975). "Methods of Predicting the Heat Production of Solar Collectors for System Design."

Ward, J. C., *Proceedings of the International Solar Energy Society Conference*, Winnipeg, 4, 336 (1976). "Minimum-Cost Sizing of Solar Heating Systems."

CHAPTER 15

Passive Solar Heating

The active solar heating systems described in Chapter 13 for heating buildings are based on collectors and storage units that are separated in order that losses from the collector do not occur when the collector does not operate and that losses from storage can be controlled by insulation. Passive heating methods depend on the same principles as have been presented for active systems, except that the collectors and storage units are integrated into the building structure and often depend on movable insulation to control thermal losses.

Passive heating will depend on auxiliary energy sources for high reliability in climates where heating is critical. The level of control of temperature in passive buildings may be different than that in buildings heated with active or conventional systems. Temperature variations may be larger, and the solar contribution to the total loads may depend on how wide the tolerances are for temperature variation. Systems described as passive may require mechanical energy to move insulation, to add or remove energy from storage, or to distribute energy through the building and in that sense are not strictly "passive." In milder climates they may be entirely passive in the sense of not requiring any mechanical energy for their operation.

Passive heating poses new problems and opportunities for the architect. Passive buildings use the south* facades for large windows or storage wall elements. Shading is provided to control overheating in summer. Energy storage can be provided by heavy structural units, so arranged that solar energy is stored when excess is available and recovered when needed. Adequate insulation must be provided to control losses from the passive elements, and (as with any building) energy conservation measures should be adopted.

Solar heating can be viewed as comprising a spectrum of system types: (a) energy-conserving architecture; (b) use of solar collection and storage units which are integral parts of the structure that function without mechanical energy; (c) use of integrated collector and storage elements and mechanical means to control and move energy; (d) use of elements less closely integrated into the structure, where mechanical energy is used to move fluids rather than insulation, that is, active systems; (e) use of more complex mechanical systems

---

* This discussion is based on northern hemisphere application.

such as heat pump-solar energy systems. In this chapter we are concerned with systems b and c in this spectrum.

A survey of passive solar buildings has been prepared by AIA (1978), and many passive buildings are included in broader surveys of solar heated buildings by Szokolay (1975) and by Shurcliff (1978). Many of the concepts now being studied and developed were set forth by Olgyay (1963). Anderson (1977) reviews passive heating concepts.

15.1 CONCEPTS OF PASSIVE HEATING

Several concepts for passive solar heating have developed that are sufficiently distinct that they provide a useful basis of discussion of the principles and functions of passive systems. These are direct gain, collector-storage wall, and greenhouse.

Direct gain of energy through windows can meet part of building heating loads. The window acts as a collector and the building itself provides some storage. Overhangs, wingwalls, or other architectural devices are used to shade the windows during times when heating is not wanted. It is also necessary, in cold climates, to insulate the windows during periods of low solar radiation to prevent excessive losses. Direct gain can provide energy to the south side of a building; means may have to be provided to distribute energy to rooms not having south windows, but good architectural design can mitigate the problem.

*Collector-Storage wall** combines the functions of collector and storage into a single unit that is part of a building structure. A south wall may be single or double glazed; inside the glazing is a massive wall of masonry material or water tanks, painted black to absorb solar radiation. Heat is transferred from the storage wall to the room by radiation and convection from the room side of the wall, and by forced or natural convection of room air through the space between glazing and wall. Room air may enter this space through openings in the bottom of the wall and return to the room through openings in the top. In principle the storage unit could be part of the ceiling, in which case it would probably be an uncovered collector. Losses during nonsunny periods can be an important factor in long-time performance, and movable insulation may have to be provided in any but mild climates.

Greenhouse attachments to buildings have been used as solar collectors, with pebble bed storage units and forced air circulation to the rooms added as options to improve storage and utilization of absorbed energy. In cold climates, energy losses from the greenhouse structure can exceed the absorbed energy, and care must be used to assure that net gains accrue from such a system.

In this chapter we show methods for quantitatively estimating the performance of systems using these concepts, and show results of simulation

---

* Sometimes referred to as a Trombe wall or Trombe-Michel wall.

studies which indicate the effects of some design parameters and control methods on performance.

15.2 COMFORT CRITERIA AND HEATING LOADS

Active solar heating systems can be designed to provide the same levels of control of conditions in the heated (or cooled) spaces as conventional systems. With indoor temperatures essentially fixed at or above a minimum, load estimations can be done by conventional methods such as the degree-day method noted in Section 14.3. Passively heated buildings in many cases are not controlled within the same narrow temperature ranges, the thermal storage capacity of the building parts or contents is usually significant, and load calculation methods must be used that take into account variable internal temperatures and capacitance of building components. For simulations, the best load calculation method is probably the response factor method, as described in ASHRAE (1977). Loads and indoor temperature variations are the desired results of these calculations.

For residential buildings, there are limits in the variation of indoor conditions (temperature and humidity) considered comfortable by most people. The limits are not well defined and are to a degree, subjective. They depend on the activities of the individuals in the building and on their clothing. Air movement is also important, as are the temperatures of the interior surfaces with which an occupant exchanges radiation. An extensive discussion of comfort under various conditions of temperature, humidity, air movement, radiant exchange, activity levels, and clothing is provided by ASHRAE (1977) and Fanger (1972).

15.3 MOVABLE INSULATION AND CONTROLS

Passive elements such as direct gain and storage wall elements can lose energy at excessive rates if measures are not taken to control the losses. An example of this situation, the equivalent of a storage wall using water tanks, was shown in Example 12.9.1 to have very large night losses. Hollingsworth (1947) and Dietz and Czapek (1950) noted the same problems based on measurements on MIT solar houses. In active systems, storage is in an insulated and compact container and energy is transported to and from storage by pumping fluids. In passive heating, storage units are also energy absorbing units and thus have large relatively uninsulated areas, and other means must be taken to control losses.

Movable insulation is the evident possibility. Movable insulation can take several forms. Drapes, shades, screens, and shutters provide nominal levels of insulation. More extensive insulation can be provided by movable foam plastic or glass wool panels, or by such devices as "beadwall" panels described by Harrison (1975), in which lightweight plastic beads are pneumatically moved

Table 15.3.1   Typical Average Insulating Values of Windows without and with Movable Insulation[a]

Window, Insulation	Loss Coefficient, U
Single glazed, no insulation	$6.0 \ W/m^2 \ C$
Double glazed, 12 mm gap, no insulation	3.20
Single glazed, with tight fitting shade or drape	2.61
Double glazed, with tight fitting shade or drape	1.23
Double glazed, with 20 mm of foam insulation, tight fitting, with air gap between glazing and foam	0.91
Double glazed, with foam insulation as above but 100 mm thick	0.27

[a] Data from ASHRAE (1977).

into or out of spaces between glazings. Table 15.3.1 indicates typical insulating values for windows with movable insulation, which have been proposed for use in passive heating.

If insulation is moved by mechanical (electrical) energy, controls are needed to determine when insulation should be moved into or out of place, and detectors and control strategies must be arranged to maximize the net gains from the system while keeping indoor conditions within acceptable limits. Controls must also be provided for mechanical devices that may be used for transferring heat from storage to rooms, such as fans used to circulate air around storage walls. Controls will be discussed further in sections on direct gain and storage walls.

15.4   SHADING

Shading devices are used with passive heating to reduce gains during times when heat is not wanted in the building. They are most commonly overhangs for south-facing surfaces and may also be wingwalls for surfaces with other surface azimuth angles. As these shading devices may partially shade absorbing surfaces during periods when collection is desired, it is necessary to estimate their effect on the absorbed radiation. In this section we show how this can be formulated for any point in time, and also how monthly average effects of overhangs can be estimated. This discussion treats overhangs of finite and infinite length, and is a summary of a detailed treatment by Utzinger (1979a) and Utzinger and Klein (1979). For overhangs of infinite length (very long compared to the width of the receiver), the method of Jones (1980) can be used.

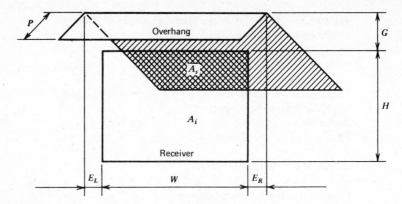

Figure 15.4.1 Diagram of the shading of a vertical receiver by a horizontal overhang. From Utzinger (1979).

A horizontal overhang shading device perpendicular to the wall is shown schematically in Figure 15.4.1. Its geometry is described by a set of dimensions: the projection, P, the gap between the top of the receiver and the overhang, G, and the left and right extensions, E_L and E_R. The receiver (the window) height is H and width is W. Using the receiver height H as a characteristic dimension, the other dimensions can be expressed in dimensionless form as ratios to the height. Thus the relative width is w, the relative projection is p, the relative gap is g, and the relative extension (with the left and right values the same) is e.

The ratio of beam radiation received by the shaded receiver to that received by the unshaded receiver, f_i, is the same as the fraction of the receiver area A_r, which is exposed to direct beam radiation:

$$f_i = \frac{A_i}{A_r} \qquad (15.4.1)$$

The value of f_i at any time depends on the dimensions of the overhang and receiver and on the angle of incidence of the beam radiation on the wall. An algorithm for computing this fraction has been developed by Sun (1975).

The area-average radiation on the shaded receiver at any time is the sum of beam, diffuse and ground-reflected radiation

$$G_r = G_b R_b f_i + G_d F_{r-s} + G\rho F_{r-g} \qquad (15.4.2)$$

The three terms have the same general significance as those in Equation 2.15.7. Diffuse and ground-reflected radiation are again assumed to be isotropic. The first term includes f_i to account for shading from beam radiation. The view factor of the receiver for radiation from the sky F_{r-s}, is reduced from its value of $(1 + \cos 90)/2$ by the overhang. Values of F_{r-s} are shown in Table 15.4.1. The third term is the ground-reflected radiation. F_{r-g} is $(1 - \cos 90)/2$, or $\frac{1}{2}$,

Table 15.4.1   Receiver Radiation View Factor of the Sky, F_{r-s}[a]

			F_{r-s} at p =								
e	g	w	0.10	0.20	0.30	0.40	0.50	0.75	1.00	1.50	2.00
0.00	0.00	1.0	0.46	0.42	0.40	0.37	0.35	0.32	0.30	0.28	0.27
		4.0	0.46	0.41	0.38	0.35	0.32	0.27	0.23	0.19	0.16
		25.0	0.45	0.41	0.37	0.34	0.31	0.25	0.21	0.15	0.12
	0.25	1.0	0.49	0.48	0.46	0.45	0.43	0.40	0.38	0.35	0.34
		4.0	0.49	0.48	0.45	0.43	0.40	0.35	0.31	0.26	0.23
		25.0	0.49	0.47	0.45	0.42	0.39	0.34	0.29	0.22	0.18
	0.50	1.0	0.50	0.49	0.49	0.48	0.47	0.44	0.42	0.40	0.38
		4.0	0.50	0.49	0.48	0.46	0.45	0.41	0.37	0.31	0.28
		25.0	0.50	0.49	0.47	0.46	0.44	0.39	0.35	0.27	0.23
	1.00	1.0	0.50	0.50	0.50	0.49	0.49	0.48	0.47	0.45	0.43
		4.0	0.50	0.50	0.49	0.49	0.48	0.46	0.43	0.39	0.35
		25.0	0.50	0.50	0.49	0.48	0.47	0.44	0.41	0.35	0.30
0.30	0.00	1.0	0.46	0.41	0.38	0.35	0.33	0.28	0.25	0.22	0.20
		4.0	0.46	0.41	0.37	0.34	0.31	0.26	0.22	0.17	0.15
		25.0	0.45	0.41	0.37	0.34	0.31	0.25	0.21	0.15	0.12
	0.25	1.0	0.49	0.48	0.46	0.43	0.41	0.37	0.34	0.30	0.28
		4.0	0.49	0.47	0.45	0.42	0.40	0.34	0.30	0.24	0.21
		25.0	0.49	0.47	0.45	0.42	0.39	0.33	0.29	0.22	0.18
	0.50	1.0	0.50	0.49	0.48	0.47	0.45	0.42	0.39	0.35	0.33
		4.0	0.50	0.49	0.48	0.46	0.44	0.40	0.36	0.30	0.26
		25.0	0.50	0.49	0.47	0.46	0.44	0.39	0.34	0.27	0.22
	1.00	1.0	0.50	0.50	0.49	0.49	0.48	0.47	0.45	0.42	0.40
		4.0	0.50	0.50	0.49	0.48	0.48	0.45	0.43	0.38	0.34
		25.0	0.50	0.50	0.49	0.48	0.47	0.44	0.41	0.35	0.30

[a] From Utzinger and Klein (1979).

for vertical receivers, neglecting secondary reflections from the under side of the overhang.

Although in principle it is possible to calculate G_r at any time from Equation 15.4.2, the determination of f_i is tedious. For design purposes, we are not normally concerned with what happens at any particular time, but rather with monthly means. A monthly mean fraction of the beam radiation received by the shaded receiver relative to that on the receiver if there were no overhang, $\bar{f_i}$, can be calculated by integrating beam radiation with and without shading over a month:

$$\bar{f_i} = \frac{\int G_b R_b f_i \, d\tau}{\int G_b R_b \, d\tau} \qquad (15.4.3)$$

Then, with $\bar{f}_i$, an equation analogous to Equation 15.4.2 can be written for the time and area average daily radiation on the shaded vertical receiving surface

$$\bar{H}_r = \bar{H}\left[\left(1 - \frac{\bar{H}_d}{\bar{H}}\right)\bar{R}_b\,\bar{f}_i + \frac{\bar{H}_d}{\bar{H}}\,F_{r-s} + \frac{\rho}{2}\right] \tag{15.4.4}$$

This is analogous to Equations 2.16.1 and 2.16.2 for monthly average radiation on a tilted, unshaded surface. The $\bar{H}_d/\bar{H}$ and $\bar{R}_b$ terms are found by the methods described in Chapter 2.

Utzinger and Klein present plots of $\bar{f}_i$ as a function of e, g, w, and p for various latitudes. Figures 15.4.2 to 15.4.4 shows three sets of the more extensive plots of Utzinger and Klein, for north latitudes 35, 45, and 55°. At a latitude of 35°, for a receiver with a width ratio, w, of 4.0, an extension ratio, e, of 0.3 on both sides, a gap ratio, g, of 0.2 and a projection ratio, p, of 0.5, from Figure 15.4.2 $\bar{f}_i$ is 0.13 in April.

The nature of these curves dictates that a particular interpolation method be used. The general procedure to obtain an $\bar{f}_i$ consists of a set of steps to account for gap, width and latitude; this procedure is modified in some cases to account for discontinuities in the relationships. The curves are shown as plots of $\bar{f}_i$ versus p for months. The first interpolation is a linear interpolation for gap, g. The second step is another linear interpolation for e. Next, interpolation for relative width, w, is done by

$$\bar{f}_i = \bar{f}_{i2} + (\bar{f}_{i1} - \bar{f}_{i2})\left[\frac{\dfrac{1}{w} - \dfrac{1}{w_2}}{\dfrac{1}{w_1} - \dfrac{1}{w_2}}\right] \tag{15.4.5}$$

Then, linear interpolation is used to account for intermediate values of latitude.

This procedure is modified by two circumstances that are clear on examination of the plots. First, if the projection is large enough, $\bar{f}_i$ is independent of p for some months. The value of p at which $\bar{f}_i$ becomes independent of p is a linear function of the relative gap. Care must be used in interpolating for gap when on or near the horizontal portion of the curves. Second, during winter months when the receiver area is not shaded at all, $\bar{f}_i$ has a limiting value of 1. The intersections of $\bar{f}_i$ curves for a given month with the $\bar{f}_i = 1$ axis is also a linear function of the relative gap. Care must be used in interpolating for gap when near these intersections.

All of the computations of $\bar{f}_i$ described above are for surface azimuth angles of zero. Utzinger and Klein have shown that for surface azimuth angles within $\pm 15°$, there is negligible difference in the estimated radiation on shaded receivers from that on a south-facing surface. When the surface azimuth angles exceed $\pm 30°$, the calculation for south-facing surfaces underpredicts summer radiation and overpredicts the winter radiation on the receiver by substantial amounts and a more detailed computation is needed. If the surface azimuth

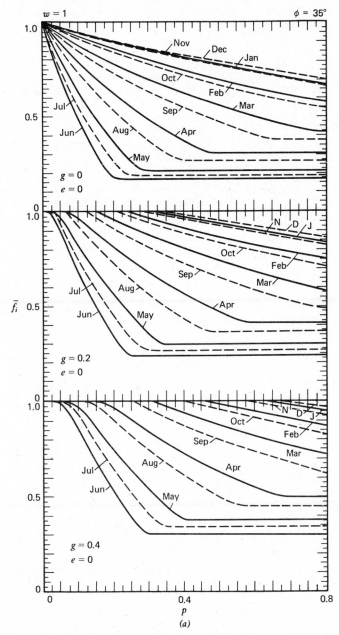

Figure 15.4.2 Monthly mean fraction of vertical receiver area receiving beam radiation as a function of relative overhang dimensions, for latitude 35°. For southern hemisphere interchange months as shown on Figure 1.7.2. From Utzinger (1979).

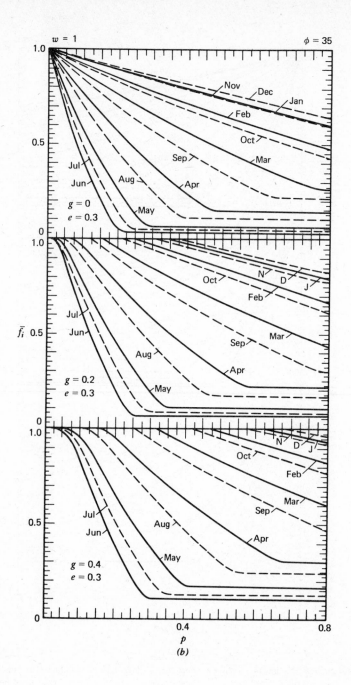

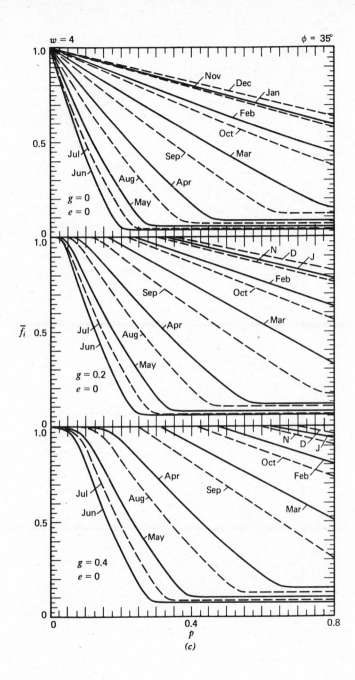

w = 4　　　　　　　　　　　φ = 35°

$\bar{f_i}$

g = 0
e = 0

Nov　Dec　Jan
Feb
Oct
Jul　Sep　Mar
Jun　Apr
Aug
May

g = 0.2
e = 0

N　D　J
Feb
Sep　Oct
Mar
Jul　Apr
Jun　Aug
May

g = 0.4
e = 0

N　D　J
Oct
Feb
Apr
Sep
Jul　Mar
Jun　Aug
May

p

(c)

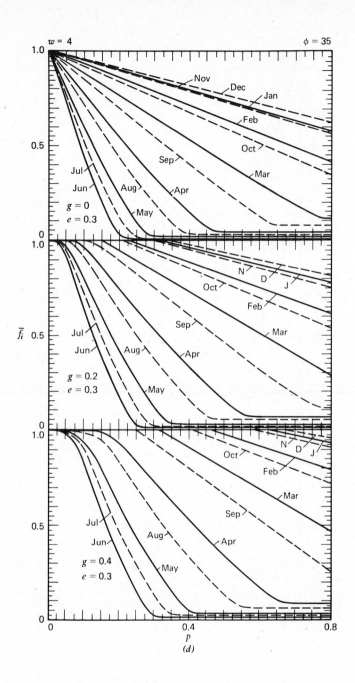

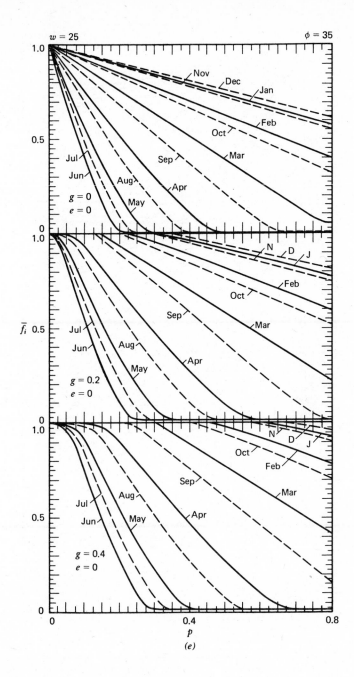

$w = 25$

$\phi = 35$

$g = 0$
$e = 0$

$g = 0.2$
$e = 0$

$g = 0.4$
$e = 0$

$\bar{f_i}$

p

(e)

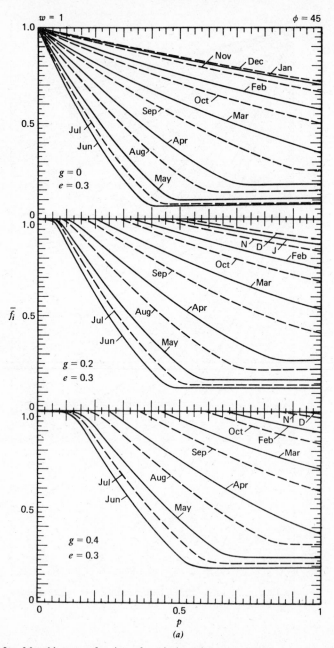

Figure 15.4.3a Monthly mean fraction of vertical receiver area receiving beam radiation as a function of relative overhang dimensions, for latitude 45°. For southern hemisphere interchange months as shown in Figure 1.7.2. From Utzinger (1979).

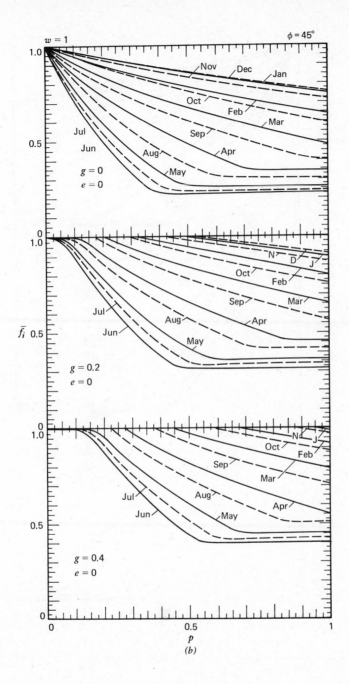

$w = 1$

$\phi = 45°$

Nov Dec Jan

Oct
Feb
Mar

Jul
Jun
Sep
Apr
Aug
May

$g = 0$
$e = 0$

$\bar{f_i}$

N D J
Oct
Feb
Sep
Mar
Jul
Aug
Apr
Jun
May

$g = 0.2$
$e = 0$

N J
Oct
Feb
Sep
Mar
Jul
Aug
Apr
Jun
May

$g = 0.4$
$e = 0$

p

(b)

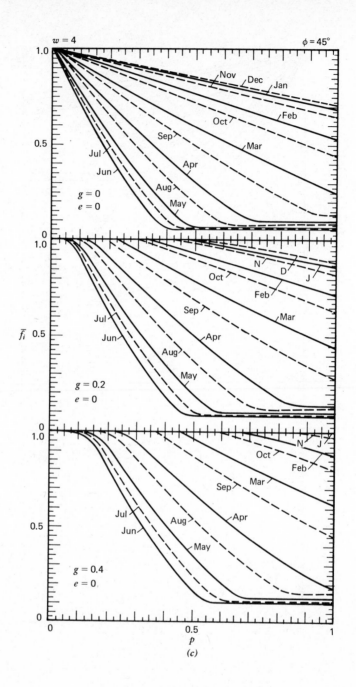

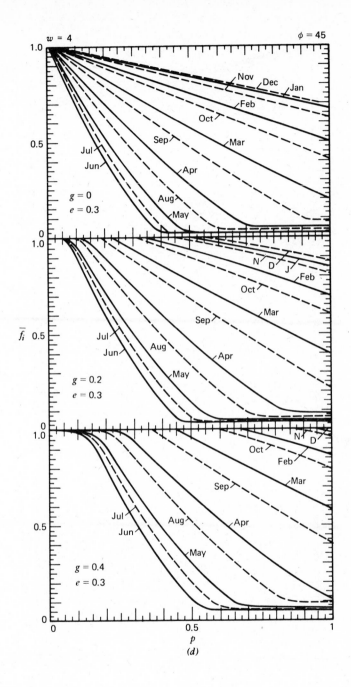

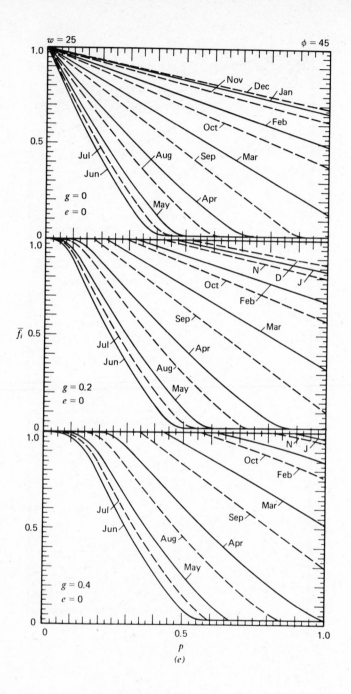

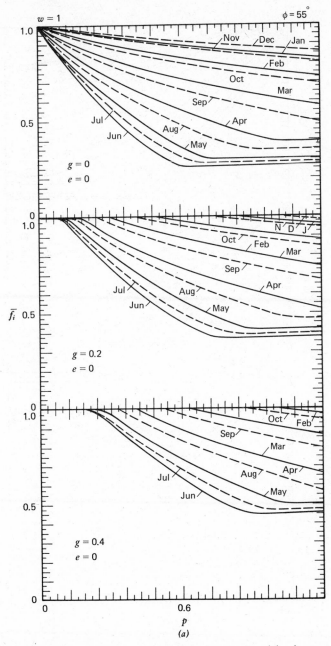

Figure 15.4.4a Monthly mean fraction of vertical receiver area receiving beam radiation as a function of relative overhang dimensions, for latitude 55°. For southern hemisphere interchange months as shown on Figure 1.7.2. From Utzinger (1979).

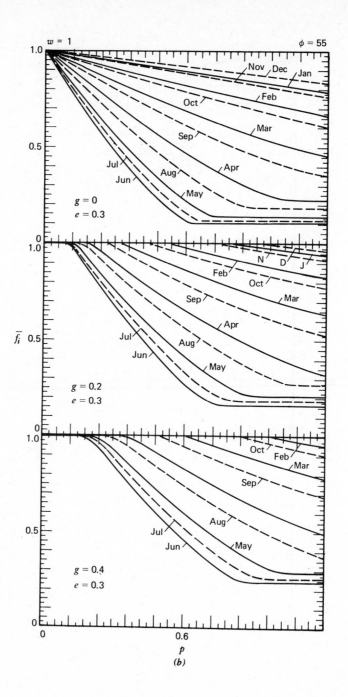

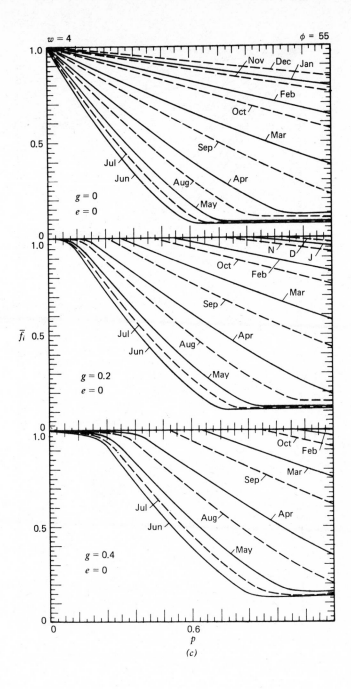

$w = 4$ $\phi = 55$

$g = 0$
$e = 0$

$g = 0.2$
$e = 0$

$g = 0.4$
$e = 0$

$\bar{f_i}$

(c)

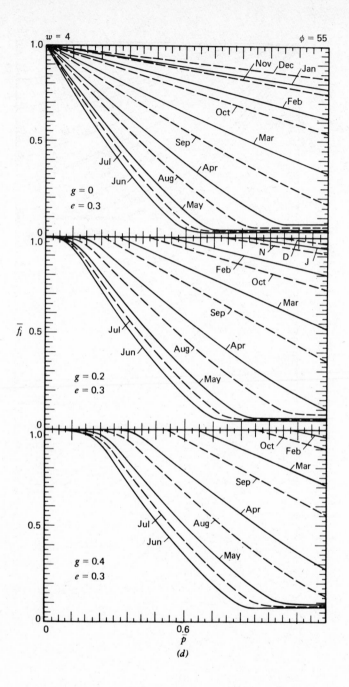

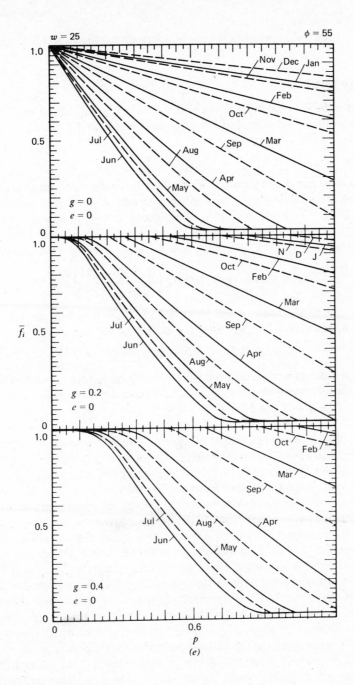

angles exceed $\pm 15°$, wingwalls may be more important in shading than overhangs.

Example 15.4.1

What is $\bar{f}_i$ for a window receiver with overhang which has $w = 12$, $g = 0.4$, $p = 0.5$ and $e = 0$ at latitude 40° for the month of March?

Solution

Interpolate w and ϕ, in that order. The values of $\bar{f}_i$ from which the interpolations are done are as follows: At $\phi = 35°$, $e = 0$, and $g = 0.4$, if $w = 4$ $\bar{f}_i = 0.80$; if $w = 25$, $\bar{f}_i = 0.78$. At $\phi = 45°$, $e = 0$, and $g = 0.40$, if $w = 4$, $\bar{f}_i = 0.96$; if $w = 25$, $\bar{f}_i = 0.96$. Equation 15.4.5 is used to interpolate for width. The inverse widths are $1/w = 0.0833$, $1/w_1 = 0.25$, and $1/w_2 = 0.04$. At a latitude of 30°

$$\bar{f}_{i,\,35} = 0.80 + (0.78 - 0.80)\left(\frac{0.0833 - 0.25}{0.04 - 0.25}\right) = 0.79$$

At a latitude of 45°, both f_{i1} and f_{i2} are 0.96, so $f_{i,45} = 0.96$. We can now linearly interpolate between 0.79 and 0.96 for a latitude of 40°

$$f_i = 0.79 + 0.5\,(0.96 - 0.79) = 0.88$$

Note that if the gap ratio had been 0.3, for which curves are not shown, the four initial values shown above would each have been the result of linear interpolations between $g = 0.2$ and 0.4. ∎

Figure 15.4.5 shows the results of month-by-month calculation of radiation on shaded and unshaded south-facing receivers at Albuquerque, NM ($\phi = 35°$) for a particular overhang geometry, calculated by Equation 15.4.4. The overhang is sufficiently large that beam radiation is essentially eliminated during the summer months, yet the total radiation on the shaded receiver in the summer is approximately two-thirds of that on the unshaded receiver. A similar computation for Minneapolis, MN ($\phi = 45°$) shows summer month radiation on the shaded receiver is more than half of that on the unshaded receiver.

Jones (1980) has devised a useful way of estimating the effects of overhangs where the extensions are large. A shading plane, which has a slope $\beta = 90 + \psi$, is defined by the outer edge of the overhang and the bottom edge of the receiver, as shown in Figure 15.4.6. Beam radiation incident on this shading plane will reach the plane of the receiver. A gap can be accounted for by defining a second shading plane between the top of the receiver and the outer edge of the overhang, and subtracting any beam radiation passing through that plane from that through the first shading plane. The monthly average radiation on the receiver is calculated using Equation 15.4.4. The value of $\bar{R}_b\,\bar{f}_i$ to be used in Equation 15.4.4 is found from $\bar{R}_b$ for the two shading planes

$$\bar{R}_b\bar{f}_i = \bar{R}_{b,SP1}\,[(1 + g)^2 + p^2]^{1/2} - \bar{R}_{b,SP2}\,[g^2 + p^2]^{1/2} \qquad (15.4.6)$$

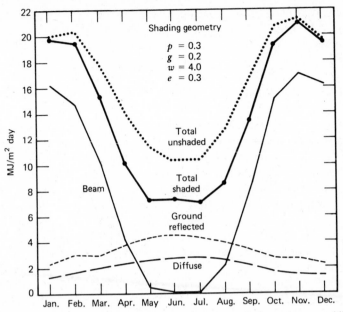

Figure 15.4.5 Monthly estimated radiation on vertical, south facing receivers at Albuquerque, shaded and unshaded. From Utzinger and Klein (1979).

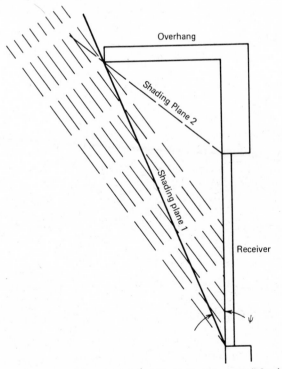

Figure 15.4.6 Overhand and receiver geometry, with shading plane, as defined by Jones (1980).

where the subscripts $SP1$ and $SP2$ refer to the two shading planes. (In Figure 15.4.6 $\bar{R}_{b,SP2}$ is zero; it can be greater than zero if the gap is large or the sun is low in the sky.)

15.5 DIRECT GAIN WINDOWS

A direct gain passive system is most often a south-facing window, with movable insulation, overhang shading to reduce incident summer solar radiation, and with the floor, walls, or furnishings behind the window arranged to absorb and store solar energy entering the structure. The window and the room are, in effect, a vertical south-facing flat-plate collector with thermal capacitance. The methods noted in Section 15.4 for estimating radiation incident on a shaded surface are modified to account for transmittance of the glazing and absorptance of the room to estimate direct gains. Loss coefficients are in principle very similar to those of flat-plate collectors, but the use of movable insulation must be taken into account.

The net rate of energy transfer across a window (receiver) of area A_r at any time can be written as

$$Q_r = A_r[\alpha F_c(G_b R_b \tau_b f_i + G_d \tau_d F_{r-s} + \rho G \tau_g/2) - U_L F_I(T_r - T_a)] \quad (15.5.1)$$

The effective absorptance, α, of the room can be approximated by

$$\alpha = \frac{1 - \rho_R}{1 - \rho_R + \rho_R \tau_d A_r/A_R} \quad (15.5.2)$$

where ρ_R is the reflectance of the room surfaces, A_r is the receiver (window) area, and A_R is the room surface area. F_c is a control function which is unity when there is no movable insulation covering the window and zero (or near zero) if movable insulation is in place. The three radiation terms in parentheses are the three terms of Equation 15.4.2, each multiplied by the transmittance determined at appropriate angles of incidence from Figure 5.3.1 or its equivalent. The last term in the bracket represents the thermal losses. U_L is the loss coefficient for the uninsulated window, and $(T_r - T_a)$ is the temperature difference between room and ambient. F_I is a special control function, which is unity when $F_c = 1$. With insulation in place, F_I is the ratio of the loss coefficient with insulation in place to the loss coefficient without the insulation. Typical window loss coefficients, U_L, are available from ASHRAE (1977) or from Table 15.3.1.

The room temperature, T_r, may be nearly constant if it is prevented from falling by the use of auxiliary energy. It will rise during periods when there is net gain through the windows, but will be prevented from dropping below a set temperature by an auxiliary energy source. The monthly average net gains (or losses) from a direct gain system can be estimated by considering the solar gain into the rooms independently of the thermal losses through the window. The calculation of both solar gain and thermal losses must account for movable insulation. Thus the gain and loss terms of Equation 15.5.1 can be written in terms of hourly irradiation (I, I_b, and I_d) and summed for the month.

$$\sum Q_r = A_r \alpha \sum F_c(f_i \tau_b I_b R_b + \tau_d I_d F_{r-s} + \rho \tau_g I/2) - A_r \sum U_L F_I(T_r - T_a)$$

(15.5.3)

These calculations can be more conveniently done using monthly averages. The f_i for the individual hours can be replaced by a single value of $\bar{f_i}$ obtained from Figures 15.4.2–15.4.4. Each of the solar gain terms includes the transmittance corresponding to the monthly mean angle of incidence determined by methods of Chapter 5. If movable insulation is used $\bar{F_c}$, a radiation-weighted value of the control function F_c must be used. If the insulation is used only when no significant radiation is incident on the receiver, $\bar{F_c}$ will be unity. If this is not the case, information must be available on which to estimate $\bar{F_c}$. The average daily radiation absorbed in the room can then be written

$$A_r \bar{S} = A_r \alpha \bar{F_c}(\bar{f_i} \bar{\tau_b} \bar{H_b} R_b + \bar{\tau_d} F_{r-s} \bar{H_d} + \rho \bar{\tau_g} \bar{H}/2)$$

(15.5.4)

The average thermal losses for a day are

$$\bar{Q_l} = 24 A_r \overline{U_L F_I}(\bar{T_r} - \bar{T_a})$$

(15.5.5)

Example 15.5.1

A direct gain system is to be used at a latitude of 40°. The south facing, double glazed window (with glass of $KL = 0.0370$), without insulation, has an overall loss coefficient (inside air to outside air) of 3.2 W/m² C.

The window is 15.0 m wide and 1.25 m high. It is shaded by a rectangular overhang 0.5 m above the top edge of the window which projects out 0.625 m and is the same width as the window. The room surface area is 440 m² and has a solar reflectance of 0.6.

For a day characterized by average March conditions, the interior of the building is kept at 20 C, and the average ambient temperature is 3 C. $\bar{H}$ is 19.93 MJ/m², $\bar{H_d}$ is 6.26 MJ/m², $\bar{H_b}$ is 13.67 MJ/m², and ground reflectance is 0.2. What will be the net gain (or loss) from the window?

Solution

The average day of the month for March is the 16th. The average transmittance of the glazing can be estimated from Figure 5.9.1. At $\phi = 40°$, $\bar{\theta_b} = 58°$, and from Figure 5.3.1, $\tau_b = 0.71$. From Figure 5.4.1 the mean angle of incidence of both diffuse and ground-reflected radiation for a vertical surface is nearly 59°, so the same value of transmittance can be used for the diffuse and ground-reflected components. (Note that Figure 5.4.1 does not give an exact equivalent angle for diffuse radiation for a shaded receiver, but uncertainties such as those due to assumption of isotropic diffuse radiation are probably greater than a correction for overhang.) $\tau_d = \tau_g = 0.70$. $\bar{R_b}$ from Figure 2.16.1d is 0.92.

For this shading overhang, $p = 0.625/1.25 = 0.5$, $g = 0.5/1.25 = 0.4$, and $e = 0$. The width is $15/1.25 = 12$. These are the dimensions assumed in Example 15.4.1, and from that example $\bar{f_i} = 0.87$. From Table 15.4.1, F_{r-s} is 0.42.

Absorptance of the room-window combination is estimated by Equation 15.5.2

$$\alpha = \frac{1 - 0.6}{1 - 0.6 + 0.6 \times 0.70 \times 18.75/440} = 0.96$$

The estimated absorbed energy in the rooms, through the window area of 18.75 m², is

$$18.75 \times 0.96 \bigg(0.87 \times 0.71 \times 13.67 \times 0.92$$

$$+ 0.70 \times 6.26 \times 0.42 + 0.2 \times 0.70 \times \frac{19.93}{2} \bigg) = 198 \text{ MJ}$$

The thermal losses from the window over a 24 hour period, when $F_I = 1$, is

$$18.75 \times 24 \times 3.2 \times (20 - 3) \times 3600 = 88 \text{ MJ}$$

The net gain is $198 - 88 = 110$ MJ per day for the month .   ∎

Thus this window gives an estimated net gain of 110 MJ/day. By varying F_I, taken as unity in this example, the effects of control strategy on this day's performance can be shown. A control method that excludes significant radiation will affect both $\bar{f}_i$ and $\bar{\tau}_b$, and hourly calculations would be needed. If very effective movable insulation were to reduce collector losses to near zero whenever there is a net loss from the wall, the direct gain system would then produce the maximum possible net input to the rooms. (The $\bar{\phi}$-methods simplify these computations—see Chapter 18.)

In a direct gain system the energy supplied by the window may exceed the building load and storage capacity, and lead to overheating. Thus it may be necessary to vent some of the energy which is indicated as useful energy by calculations such as those of Example 15.5.1. This may not be a problem in midwinter when heating loads are high, but in milder months a significant fraction of the apparent net gains may not be useful. The heat capacity and allowable temperature variation of the room have significant effect on how much energy must be vented. Monsen (1980) has studied this problem.

15.6 COLLECTOR-STORAGE WALLS AND ROOFS

A collector-storage wall is essentially a high capacitance solar collector coupled directly to the spaces to be heated. A diagram is shown in Figure 15.6.1. Solar radiation is absorbed on the outer surface of the wall. Energy is transferred from the room side of the wall to the spaces to be heated, by convection and radiation. Energy can be transferred to the room by air circulating through the gap between the wall and glazing through openings at the top and bottom of the

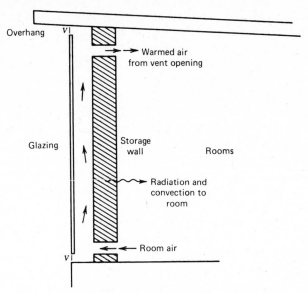

Figure 15.6.1 Section of a collector-storage wall, showing two means of transfer of heat from wall to rooms. Vent dampers v are used to vent the gap between glazing and wall in summer.

wall. Circulation can be by natural convection, controlled by dampers on the openings, or by forced circulation by fans.

Radiation transmitted by the glazing and absorbed by the wall is calculated for any time or time period by the same methods as for the direct gain component noted in Section 15.5, with the additional consideration that absorptance must be evaluated at the appropriate angles of incidence

$$A_r S = A_r [f_i F_c G_b R_b (\tau\alpha)_b + G_d (\tau\alpha)_d F_{r-s} + \rho G (\tau\alpha)_g / 2] \qquad (15.6.1)$$

and

$$A_r \bar{S} = A_r \bar{F}_c [\bar{H}_b \bar{R}_b \bar{f}_i (\overline{\tau\alpha})_b + \bar{H}_d (\overline{\tau\alpha})_d F_{r-s} + \rho \bar{H} (\overline{\tau\alpha})_g / 2] \qquad (15.6.2)$$

where $\bar{F}_c$ is an energy-weighted control function that will be less than one if movable insulation is in place during times when significant radiation is available.

If a collector-storage wall uses water to provide thermal capacitance, so that there are negligible temperature gradients in the wall, then the situation is exactly analogous to a flat-plate collector with high capacity and Equation 6.11.7 can be used to predict the wall temperature as a function of time. A proper interpretation of U_L is needed, as losses out through the glazing are indeed losses but losses out the back are gains to the space to be heated. Hourly calculations are essentially the same as Example 6.11.1, with S calculated from Equation 15.6.1, except that convection and radiation from the back must be included (and also calculated separately for use in the energy balances on the rooms).

A solid storage wall [e.g., a wall of concrete, as used by Trombe et al. (1977)] can serve both structural and thermal purposes. There will be temperature gradients through these walls, variable with time, which can be determined by considering the wall to be a set of nodes, connected together by a thermal network, each with a temperature and capacitance. The network used in TRNSYS (1979) [which is similar to that used by others, for example by Balcomb et al. (1977) and Ohanessian and Charters (1978)] is shown in Figure 15.6.2. The wall is shown divided into n nodes, with the surface nodes having half of the mass of the interior nodes. U_t is based on the inner glazing temperature, as it is in Figure 6.12.1d. Heat is transferred by convection to air flowing in the gap from the absorbing surface and from the inner glazing. (The glazing is frequently colder than the air and at those times there is convective transfer from air to glazing.)

If there is no air flow, heat transfer will be directly across the gap by both radiation and convection. It is possible that movable insulation (such as a metallized shade) could be placed between glazings or in the gap during periods of no collection; if so, the formulation of losses from the storage wall must take into account the additional resistances. The thermal circuit of Figure 15.6.2 can be modified depending on the particular mode of operation of the component.

The air flowing through the gap is usually circulated to the room, and the energy added to the space by this mechanism is $(\dot{m}C_p)_a(T_o - T_r)$. It may also be desirable to vent this air to the outside, as a means of "dumping" energy when collection is high and loads are low (such as in spring and fall months) and as means of inducing circulation through the rooms in summer.

Air may circulate through the gap to the rooms by either natural convection or forced convection, and in either case it is necessary to estimate flow rates and convection coefficients. Flow by natural convection is difficult to estimate. Trombe et al. (1977) have published measurements that indicate that most of the pressure drops are due to flow through the vents. Analyses have been done for laminar flow [e.g., see Fender and Dunn (1978)], but have not included pressure drops in the vents. It is probable that flow will be turbulent, due to entrance and exit effects.

A solution of Bernoulli's equation can be used to estimate the mean velocity in the gap, and is based on the assumption that density and air temperature in the gap vary linearly with height. The average velocity through the gap is

$$\bar{V} = \sqrt{\frac{2gh}{\left[C_1\left(\dfrac{A_g}{A_v}\right)^2 + C_2\right]} \cdot \frac{T_m - T_r}{T_m}} \qquad (15.6.3)$$

The term $[C_1(A_g/A_v)^2 + C_2]$ represents the pressure drop in the gap and vents and C_1 and C_2 are dimensionless empirical constants. The wall height is h and T_m is the mean air temperature in the gap. From data in Trombe et al. (1977)

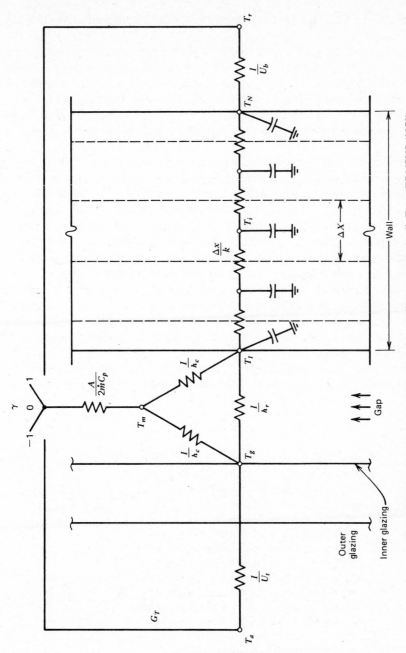

Figure 15.6.2 A thermal circuit diagram for a collector-storage wall. From TRNSYS (1979).

values of C_1 and C_2 have been determined as 8.0 and 2.0, respectively by Utzinger (1979a).

The heat transfer coefficient between air in the gap and the wall and glazing depends on whether the vents are closed or open. If the vents are closed, the loss coefficients can be calculated by the same methods as discussed in Section 3.11 for vertical collectors. If there is flow through the gap, Equation 3.14.6 can be used for flow in the turbulent region, and Equation 3.14.7 can be used for the laminar region. The loss coefficient U_t can be calculated by the standard methods of Chapter 6.

A collector-storage wall supplying energy to a room and the auxiliary energy supply to the room, may be controlled by several alternative strategies. If forced circulation through the gap is used, a thermostat must turn on the fan when the rooms need energy, or call for auxiliary energy if the storage wall is unable to provide the needed energy. If natural circulation is used, it may be desirable to control addition of energy to the rooms by means of dampers on the vents. If summer venting to the outside is used, dampers will have to be positioned; this may be a single seasonal change which can be done manually. It may be necessary to move insulation into the air gap to control losses, and controls and mechanisms for accomplishing this must be provided. The equations for both the collector-storage wall and the rooms must reflect the modes of operation at any time. An example of the results of alternative control strategies is provided by Sebald et al. (1979).

It is not feasible to calculate hour by hour performance of collector-storage wall systems by hand, but simulations can be used to provide information on

Table 15.6.1 Relative Auxiliary Energy Needs for a Building in the Madison Climate

System	Q_{aux}/Q_{conv}
A Free convection with control only by passive backdraft dampers, and no night insulation	0.81
B Forced convection at air rates comparable to those used in flat-plate air heaters when the temperature of the air in the gap is greater than room temperature, otherwise no air flow, and no night insulation.	0.81
C Free convection as in A, with night insulation of approximately 0.1 m of foam insulation placed over the glazing resulting in U of 0.5 W/m^2 C at night.	0.72
D Forced convection as in B and night insulation as in C.	0.74

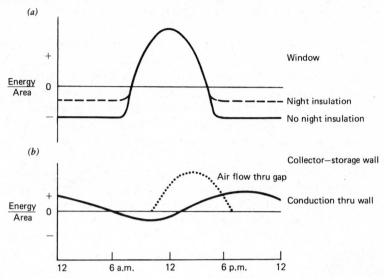

Figure 15.6.3 Energy flows through a direct gain system (top) and a collector storage wall (bottom) as a function of time of day.

short or long-term performance and assess the effects of control strategies. Results of a study by Monsen (1979) of performance of a building with a collector-storage wall in the severe climate of Madison illustrate the kind of information that can be obtained. A rectangular building was simulated with both a conventional south wall and with a south collector storage wall and several control and circulation schemes. The four systems simulated and the ratio of auxiliary required to energy required to heat the building with the conventional south wall are shown in Table 15.6.1.

The variations with time of day of energy flow into rooms from direct gain windows and collector-storage walls are quite different. Figure 15.6.3 shows this contrast for south-facing components for clear days. The energy absorbed in the rooms from the window occurs when the radiation enters the spaces (where it is absorbed by the structure and contents, from which it can later provide heating). Energy is added to the rooms from the collector-storage wall by two mechanisms, by heating of air flowing through the gap and by radiation and convection from the room side of the storage wall. The energy added by air flow lags the direct gain, and is in turn lagged by the energy flow through the wall.

This discussion has dealt with collector-storage walls. Similar concepts have been developed for flat roofs with storage capacity by Hay and Yellott (1970) and Hay (1973). The performance of a building using this method is described by Niles (1976). The principles are similar, but energy transport into the rooms is primarily by radiation from the storage ceiling. Collector-storage units are uncovered, and movable insulation is placed over them when collection is not occurring.

15.7 SOLAR LOAD RATIO DESIGN METHOD

Balcomb and McFarland (1978) have devised the solar load ratio (*SLR*) method for estimating the performance of collector-storage wall passive heating systems. Many detailed hour-by-hour simulations of four different systems were done for 29 locations and the results were correlated in terms of the ratio of absorbed solar energy to loads. The systems included a solid storage wall and a water wall, each with and without night insulation. The use of the *SLR* method is restricted to systems that have characteristics nearly like the major assumptions listed in Table 15.7.1.

The *SLR* is defined as the ratio of monthly solar energy absorbed on the storage wall surface to the monthly building load (including what would have been lost through the wall in the absence of solar gains). Average daily absorbed solar energy is calculated by Equation 15.6.2; monthly absorbed radiation is then $A_r \bar{S} N$, where N is the number of days in the month. Loads are calculated by standard techniques (such as the degree-day method) for all but the collector-storage wall. A hypothetical loss through the storage wall (that would have occurred in the absence of solar gains) must be added, based on a steady-state loss coefficient for the wall which combines the effects of wall, glazing, and insulation averaged over a 24 hour day. Balcomb and McFarland use the loss coefficients for the storage walls as shown in Table 15.7.2. Loads are calculated from

$$L = [(UA)_h + A_r U_W](\text{degree-days}) \tag{15.7.1}$$

The solar load ratio, $A_r \bar{S} N/L$, is calculated for each month and a monthly solar heating fraction (*SHF*) is determined from Figure 15.7.1 for the particular system. The monthly auxiliary energy required is $(1 - SHF)L$ where L includes the hypothetical loss through the collector-storage wall. The annual auxiliary

Table 15.7.1 Characteristics of Systems Simulated in Development of the *SLR* Design Method[a]

Storage Capacity	0.92 MJ/m^2 C
Room Temperature Range	19–24 C
Night Insulation Resistance (where used)	1.6 m^2 C/W
Night Insulation Time	5 P.M. to 8 A.M.
Wall to Room Conductance	5.68 W/m^2 C
Storage Wall Thermal Conductivity	k = 1.73 W/m C
Storage Wall Thermal Capacitance	ρC_p = 2.0 MJ/m^3 C
Double Glazed	
Building Thermal Mass Other than Storage Wall is Negligible	
Storage Wall has Vents with Backdraft Dampers	

[a] From Balcomb and McFarland (1978).

Table 15.7.2 Hypothetical 24-Hour Average Loss Co-efficients for Collector Storage Walls For Use in Calculating Loads in the *SLR* Method

	U_w (W/m^2 C)	
Type of Wall	Double Glazed, No Night Insulation	Double Glazed With R9 Night Insulation
Water wall	1.87	1.02
18 in. concrete wall	1.25	0.68

[a] From Balcomb and McFarland (1978).

energy requirement is then the sum of the monthly auxiliary requirements. (The existence of a passive system modifies the heating loads on a building, and the concept of annual solar fraction is less useful than the annual auxiliary energy required.)

Figure 15.7.1 was developed from the results of simulations of the systems described by Table 15.7.1. The curves were fit to the data so as to minimize

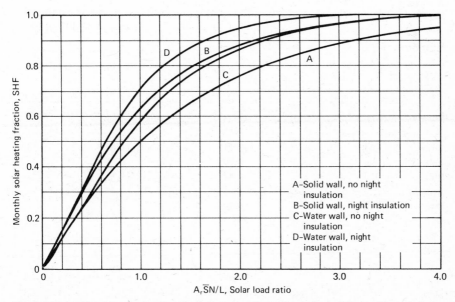

A-Solid wall, no night insulation
B-Solid wall, night insulation
C-Water wall, no night insulation
D-Water wall, night insulation

Figure 15.7.1 Monthly solar heating fraction versus solar load ratio for buildings with south-facing collector-storage wall systems. Adapted from Balcomb and McFarland, Proc. 2nd Passive Conf. (1978).

errors in annual solar heating fraction. The *SLR* must be calculated in the same way as was done in the development of the correlations. The standard deviations of annual solar fractions determined by the correlations from those of the simulations were in the range of 0.023 to 0.028 for the four systems. The standard deviations for particular locations are not known. As with the *f*-chart method for active systems, individual months can vary substantially from the results of simulations, and only annual results from the *SLR* method should be considered significant.

Example 15.7.1

A residential building in Tulsa (latitude 36°) has a south-facing concrete collector-storage wall without night insulation of 15 m^2 area. The design parameters are the same as those of Table 15.7.1. The building (exclusive of the collector-storage wall) has a UA of 200 W/C (including effects of infiltration and internal generation). The average absorbed solar radiation and degree-days are shown by months in the table; these were calculated·from the data in Appendix G using the methods of Chapters 2 and 5 to obtain $\bar{S}$. $(\tau\alpha)_n$ was taken as 0.75 for this calculation. Estimate the annual amount of auxiliary energy required to heat this building.

Solution

The detailed calculations are shown for January only. The table indicates the results for all months. First the hypothetical loss coefficient due to the collector-storage wall must be added to the building coefficient. The modified load using Equation 15.7.1 is

$$L = (200 + 15 \times 1.25)489 \times 3600 \times 24 = 9.24 \text{ GJ}$$

Month	$\bar{S}$ (MJ/m^2)	DD(C-days)	*Load* (GJ)	$A, \bar{S}N$ (GJ)	*SLR*	*SHF*	*AUX*
Jan.	8.1	489	9.24	3.76	0.41	0.24	7.0
Feb.	8.1	370	6.99	3.40	0.49	0.28	5.0
Mar.	7.3	293	5.44	3.39	0.61	0.34	3.6
Apr.	5.3	98	1.85	2.39	1.29	0.59	0.8
May	4.2	16	0.30	1.95	6.46	~1	—
June	4.0	0	—	—	—	—	—
July	4.1	0	—	—	—	—	—
Aug.	5.0	0	—	—	—	—	—
Sept.	6.7	6	—	—	26.6	~1	—
Oct.	9.0	79	1.49	4.19	2.80	0.87	0.2
Nov.	8.8	260	4.91	3.96	0.81	0.42	2.8
Dec.	7.8	434	8.20	3.63	0.44	0.25	6.2

Total Auxiliary = 25.6 GJ

The absorbed solar radiation is

$$A_r \bar{S} N = 15 \times 8.1 \times 31 = 3.76 \text{ GJ}$$

The *SLR* is

$$SLR = 3.76/9.24 = 0.41$$

From Figure 15.7.1, using curve *A*, the month's *SHF* is 0.24. The auxiliary required for January is then $(1 - 0.24) \times 9.24 = 7.0$ GJ. The table shows the results for 12 monthly calculations. The annual auxiliary energy indicated by the *SLR* method is 25.6 GJ. ∎

McFarland and Balcomb (1979) have studied the sensitivity of performance of some storage-collector wall systems to several design parameters, including wall thickness and thermal characteristics, number of glazings, vent area, and others. These studies were done by simulation methods. They were not compared directly with results from *SLR* calculations but the directions and magnitude of the changes should apply. Wray et al. (1978) have extended the *SLR* method to direct gain systems.

15.8 RESISTANCE NETWORK DESIGN METHOD

Monsen et al. (1979) have used detailed simulation models in a study of the thermal phenomena that occur in collector-storage walls. The network representing the walls is shown in Figure 15.6.2. The wall is part of a building that has allowable room temperature swings of 6 C (i.e., auxiliary is added if the space gets too cool and ventilation or air conditioning is provided if it gets too warm). Two useful observations have come from these simulations. The first is that the amount of air flowing through the gap between the wall and glazing has little effect on the total auxiliary heat required by the building. (This was also observed by Ohanessian, (1976) and by Utzinger (1979).) Increasing air flow tends to increase heat transfer into the room by convection but also results in lower temperature of the outer wall surface and less heat transfers into the room by conduction through the wall. The mechanisms and time of transfer into the room vary, but over a month the energy into the heated space is not sensitive to air flow rates.

The second observation is that the monthly average temperature profile through the storage wall is linear. An example of computed monthly average profiles is shown in Figure 15.8.1; similar results were found for a wide range of collector-storage wall designs. Thus for monthly average computations, the network representing the wall can be replaced with a single resistance that is given by the thickness to conductivity ratio.

With these observations, and the fact that over a month the net energy stored in a collector-storage wall is small relative to the building load, the

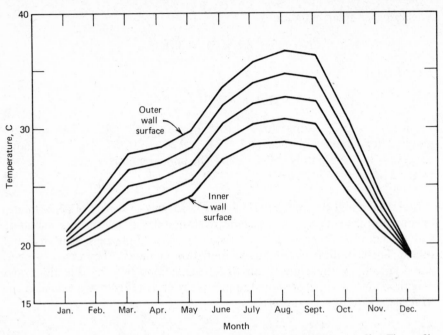

Figure 15.8.1 An example of computed monthly average wall temperature profiles. The profiles are for equally spaced positions in the wall. Adapted from Monsen (1980).

thermal network of Figure 15.6.2 reduces to the very simple network of Figure 15.8.2. On a monthly basis, the solar energy absorbed by the wall, $\bar{S}$, is distributed to losses to the ambient and to energy to the room, Q_{in}. In equation form

$$\bar{S} = U_L(\bar{T}_{w,o} - \bar{T}_a)\Delta\tau + Q_{in} \tag{15.8.1}$$

where $\Delta\tau$ is the number of seconds in a day and U_L is the monthly average loss coefficient (in W/m^2 C) between the outer wall surface at $\bar{T}_{w,o}$, and the ambient at $\bar{T}_a$ and can be found from Equation 6.4.7. The energy to the room, Q_{in}, is found from

$$Q_{in} = \frac{U_i k}{k + tU_i} (\bar{T}_{w,o} - \bar{T}_R)\Delta\tau \tag{15.8.2}$$

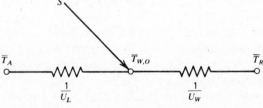

Figure 15.8.2 Simple linear network model of a collector-storage wall.

where U_i is the heat transfer coefficient between the inside wall and the heated space, t is the wall thickness, k is the wall thermal conductivity, $\overline{T}_R$ is the average room temperature.

When night insulation is used, the $U_L \Delta \tau$ product is replaced by

$$U_L \Delta \tau = U_{L,d}(\Delta \tau - \Delta \tau_n) + U_{L,n}\Delta \tau_n \tag{15.8.3}$$

where $\Delta \tau_n$ is the number of seconds during which the night insulation is in place.

Equations 15.8.1, 15.8.2, and 6.4.7 can be solved simultaneously for Q_{in}, the average daily energy entering the heated space through the collector-storage wall. Not all of this energy necessarily contributes to reduction of the auxiliary heating load. Whenever the energy from the collector-storage wall results in overheating of the building, some energy will have to be vented to maintain comfortable conditions. The amount of venting will depend on the thermal capacitance of the storage wall as well as the thermal capacitance of the remainder of the building. During winter when the building loads are high relative to Q_{in}, very little energy will have to be vented.

Example 15.8.1

For the same circumstances as in Example 15.7.1, estimate the solar contribution to space heating for January using the resistance network design method. The estimated heat transfer coefficient from wall to room is 5.68. The wall thermal conductivity is 1.73 W/m C and the thickness is 0.46 m. For Tulsa in January, $\overline{T}_a = 3$ C.

Solution

The process is to solve Equations 15.8.1 and 15.8.2 for $\overline{T}_{w,o}$ and calculate Q_{in} from Equation 15.8.2. Since U_L is to some degree a function of $\overline{T}_{w,o}$, it is necessary to assume a wall temperature, base U_L on it, calculate $\overline{T}_{w,o}$, adjust U_L if necessary, and again solve for $\overline{T}_{w,o}$. The monthly average room temperature is assumed to be 18.3 C, as solar heating will seldom drive it higher in this month. From Example 15.7.1, $\overline{S} = 8.1$ MJ/m^2. Assume that $\overline{T}_{w,o}$ is 30 C. By the methods of Section 6.4, U_L is approximated as 2.4 W/m^2 C.

$$8.1 \times 10^6 = 2.4(\overline{T}_{w,o} - 3)86,400$$

$$+ \frac{5.68 \times 1.73}{1.73 + 0.46 \times 5.68}(\overline{T}_{w,o} - 18.3)$$

$$\overline{T}_{w,o} = 30.5 \text{ C}$$

This is close to the 30 C assumed, and U_L will not change significantly at the new wall temperature. With this temperature Q_{in} can be calculated. For the 15 m^2 collector-storage wall.

$$Q_{in}A_r = 2.26 \times 15 \times 86,400 \times 31(30.5 - 18.3) = 1.1 \text{ GJ}$$

This is the net gain for the month. This cannot be compared directly to the difference between the hypothetical loads and auxiliary energy of Example 15.7.1. The auxiliary energy required for the month calculated by the SLR method is 7.0 GJ; the auxiliary energy required calculated by the resistance network method would be the load (in this case $200 \times 489 \times 3600 \times 24 \times 10^{-9}$, or 8.45 GJ) less the net solar contribution of 1.1 GJ or 7.3 GJ. ∎

Monsen (1980) has studied the problem of energy venting in passive systems consisting of either direct gain elements (windows) or collector-storage walls. He was able to correlate the vented energy in terms of the vented energy that would occur in a house with no thermal capacitance. Without thermal capacitance, the heated space quickly reaches the upper temperature limit for comfort and consequently the maximum amount of energy will have to be vented. This maximum amount was calculated using the utilizability ideas of Chapter 18 (although in this situation "unutilizability" is better terminology).

15.9 GREENHOUSES

Collectors in the form of greenhouses may be used on the south side of buildings, to provide a combination of energy gain and space which is uniquely useful. Storage may be provided only in the structure itself, or it may be provided separately, for example in the form of a pebble bed using either natural or forced circulation. Circulation of warm air from the greenhouse to the rooms of the building can be arranged, usually with the use of a fan and appropriate controls.

The utility of an attached greenhouse as a solar collector must be estimated by integrating energy balances on the structure over time. The design and construction of greenhouses vary widely, as will solar gains and thermal losses. The basic methods of direct gain systems (Section 15.5) can be used to estimate the solar energy absorbed in the greenhouse. Losses can be calculated by standard methods, for example using ASHRAE loss coefficients for windows for the glazed portions of the structure.

Many greenhouse structures have glazed areas that are significantly larger than the effective solar energy collection area, and the losses will be larger than from the equivalent flat plate collector. Also, losses from greenhouses occur continuously, although from temperatures that should be lower than those of operating collectors. Thus it is essential that the evaluation of a greenhouse as a collector be based on an analysis that estimates the temperature in the structure as a function of time and properly integrates both gains and losses. In cold climates the annual net losses may exceed the absorbed solar energy, and extensive movable insulation may have to be used to control the losses.

15.10 EXAMPLES OF PASSIVE APPLICATIONS

A great variety of passively heated buildings have been built and used. Data are available on the performance of a few of these. Some experimental structures have been instrumented and their performance measured [Grimmer et al. (1979)]. It is difficult to quantitatively evaluate most of the designs. Many inventive ideas and approaches are evolving. In this section we have chosen several interesting passive heating ideas and buildings as illustrations of the range of possibilities.

One of the best known of modern applications is the group of residential buildings at Odeillo, France. The collector-storage wall in the first of these buildings consists of double glazings, spaced 0.12 m from the concrete storage wall which is 0.60 m thick. The wall is painted black, and has openings 0.11 × 0.56 m at top and bottom which are 3.5 m apart. The collector occupies most of the south wall of the building. The auxiliary energy source is electricity. The system is estimated by Trombe et al. (1977) to have supplied between 60 and 70 percent of the energy for an average house in the Odeillo climate. Based on extensive studies of this system, a building with three apartments has been constructed that is much better insulated, has smaller collector area per cubic meter of heated space (0.1 rather than 0.16 m²/m³), and utilizes storage walls of concrete 0.37 m thick.

A residential building using a collector-storage wall, the Hunn residence in Los Alamos, NM, is shown in Figure 15.10.1a [see Hunn (1979)].

An example of a residence using direct gain is the Balcomb home in Santa Fe, NM, shown in Figure 15.10.1b. This two story house has 0.35 m thick adobe walls that provide thermal capacity, and also a pebble bed storage unit to provide additional storage capacity [see AIA (1978)]. A monastery using direct gain, with the upper row of clerestory windows providing energy for a

Figure 15.10.1 Examples of architectural designs of passively heated buildings. Courtesy of S. Sargent. (a) Residence, Los Alamos, NM.

Figure 15.10.1 (*b*) Residence, Santa Fe, NM.

Figure 15.10.1 (*c*) Monastery, Pecos, NM.

second row of rooms, is shown in Figure 15.10.1c. All windows on the south side are double glazed, and the direct gain system meets most of the heating loads on the building [see AIA (1978)].

Warehouse heating, where temperature control may not be as critical as in buildings for human occupancy, is another type of application of interest. The basic criterion may be that freezing temperatures may not be allowable. Thermal storage may be provided by the contents of the warehouse, with the system a direct gain system, or by a masonry wall in which part or all of the south wall of the building is a collector-storage wall. Applications of each of these types in New England is described by Keller et al. (1978).

Direct gain solar heating systems have several problems associated with them. The contents of the room are, by definition, subjected to solar radiation, which can cause problems such as fabric degradation and fading of dyes. If floors are to be made heavy to provide storage, they cannot be covered with rugs and furnishings that reduce the rates at which energy can be added to or removed from storage. An experimental building at MIT is being used in a study of possible answers to these problems. Beam radiation incident on the windows is deflected upwards by narrow-slat venetian blinds which have the slats shaped to be concave upwards, and which have reflective surfaces on the top of the slats. These blinds reflect the beam radiation upward to the ceiling rather than admitting it to the floor. The heat storage material is a phase-change material $Na_2SO_4 \cdot 10H_2O$ in thin layers in ceiling panels. Heat transfer from the ceiling to the rooms is primarily by radiation [see Mahone (1978)].

15.11 SUMMARY

Passive heating concepts are similar to active concepts, with major distinctions being in the manner of controlling thermal losses and the degree of integration of the collection and storage components into the building. In many cases, energy losses are controlled by moving insulation rather than fluids. The methods of estimating the performance of passive heating systems are in many respects similar to those of active systems. Many control methods are possible, and the choice can affect both the variability of conditions maintained in the rooms and the amount of auxiliary energy required.

Coupled with energy-conserving building design, passive concepts can be useful in substantially reducing the heating loads on a building. At the same time, there are interesting and innovative architectural treatments of the passive elements that lead to buildings that are often aesthetically pleasing. To date, most applications in the United States in which substantial fractions of heating loads are provided by solar are in the southwest where solar radiation is available most of the time.

Utzinger (1979b) suggests that large collector-storage wall systems are best applied in climates with high K_T. In cold, cloudy climates such as those of the north-central part of the United States, he suggests that the appropriate strategy is to decrease building losses (that is, $(UA)_h$) as far as possible, and size direct-gain windows with night insulation to provide enough gain on a clear mid-winter day to meet the loads for 24 hours. (He also notes that if a building is very well insulated, internal generation can supply large fractions of the energy needs for heating, and the standard 18.3 C (65 F) base for determining degree days is not adequate.)

There are a number of general sources of information to which the engineer interested in passive heating can refer for a wide range of descriptive and quantitative papers. Proceedings are available of three National Passive Solar Conferences (1976, 1978, and 1979). The 1979 Conference proceedings include many quantitative papers on performance of systems and on design methods. Other collections of papers are in the Proceedings of the 1978 meeting of AS/ISES, and of earlier ISES conferences.

REFERENCES

AIA Research Corporation, Report the Department of Housing and Urban Development HUD-PDR-287, Washington (1978). "A Survey of Passive Solar Buildings."

Anderson, B., *Solar Energy, Fundamentals in Buildings Design*, McGraw-Hill, New York (1977).

ASHRAE *Handbook of Fundamentals*, American Society of Heating, Refrigeration and Air Conditioning Engineers, New York (1977).

Balcomb, J. D., J. C. Hedstrom, and R. D. McFarland, *Solar Energy*, **19**, 277 (1977). "Simulation Analysis of Passive Solar Heated Buildings—Preliminary Results."

Balcomb, J. D. and R. D. McFarland, paper in *Proceedings of the 2nd National Passive Solar Conference*, **2** (2), 377 (1978). "A Simple Empirical Method for Estimating the Performance of a Passive Solar Heated Building of the Thermal Storage Wall Type."

Conference and Workshop Proceedings, Passive Solar Heating and Cooling, Albuquerque (1976) Report of Los Alamos Scientific Laboratory, available from National Technical Information Service.

Dietz, A. G. H. and E. L. Czapek, *Heating, Piping and Air Conditioning*, 118 (Mar. 1950). "Solar Heating of Houses by Vertical South Wall Storage Panels."

Fanger, P. O., *Thermal Comfort Analysis and Applications in Environmental Engineering*, McGraw-Hill, New York (1972).

Fender, D. A. and J. R. Dunn, *Proceedings of the ASME Winter Annual Meeting*, 78-WA/Sol-11, (1978). "A Theoretical Analysis of Solar Collector/Storage Panels."

Grimmer, D. P., R. D. McFarland, and J. D. Balcomb, *Solar Energy*, **22**, 351 (1979). "Initial Experimental Tests on the Use of Small Passive-Solar Test—Boxes to Model the Thermal Performance of Passively Solar-Heated Building Designs."

Harrison, D., *Solar Energy*, **17**, 317 (1975). "Beadwells."

Hay, H. R. and J. I. Yellott, Mech. Engr., **92**, 19 (1970). "A Naturally Air Conditioned Building."

Hay, H. R., *Mechanical Engineering*, **95** (11) 18, (1973). "Energy, Technology and Solarchitecture."

Hollingsworth, F. N., *Heating and Ventilating*, 76 (May 1947). "Solar Heat Test Structure at MIT."

Hunn, B. D., *ASHRAE Journal*, **21** (4) 25 (1979). "Hybrid Passive/Active Solar System: Performance and Cost."

Jones, R. E., *Solar Energy*, **24**, 305 (1980). "Effects of Overhang Shading of Windows having Arbitrary Azimuth."

Keller, S. F., A. V. Sedrick, and W. C. Johnson, paper in *Proceedings of the 2nd National Passive Solar Conference*, Philadelphia, **1**, 81 (1978). "Solar Experiments with Passive Retrofit."

Mahone, D., *Solar Age*, **3** (9), 20 (1978). "Three Solutions for Persistent Passive Problems."

McFarland, R. D. and J. D. Balcomb, Proceedings of the 3rd National Passive Solar Conference, American Section of the ISES, **3**, 54 (1979). "The Effects of Design Parameter Changes on the Performance of Thermal Storage Wall Passive Systems."

Monsen, W. A., M.S. Thesis, University of Wisconsin, Madison (1980), "A Procedure for Sizing Building-Integrated Solar Heating Components."

Monsen, W. A., S. A. Klein, W. A. Beckman, and D. M. Utzinger, *Proceedings of the 4th National Passive Solar Conference*, 119 (1979). "The Resistance Network Design Method for Passive Solar Systems."

Monsen, W. A., Personal communication (1979).

Niles, P. W. B., *Solar Energy*, **18**, 413 (1976). "Thermal Evaluation of a House Using a Movable-Insulation Heating and Cooling System."

Ohanessian, P., M.S. Thesis, University of Melbourne (1976). "Numerical Modelling of a Passive Solar Energy House Heating System."

Ohanessian, P. and W. W. S. Charters, *Solar Energy*, **20**, 275 (1978). "Thermal Simulation of a Passive Solar House using a Trombe-Michel Wall Structure."

Olgyay, V., *Design with Climate*, Princeton University Press, Princeton, NJ (1963).

Proceedings of the 2nd National Passive Solar Conference, American Section of the ISES, Volumes 2.1, 2.2., and 2.3, Philadelphia (1978).

Proceedings of the 3rd National Passive Solar Conference, American Section of the ISES, Volume 3 San Jose (1979).

Proceedings of the 1978 Annual Meeting of the American Section of the ISES, Volume 2.2, Denver (1978).

Sebald, A. V., J. R. Clinton, and F. Langenbacher, *Solar Energy*, **23**, 479 (1979). "Performance Effects of Trombe Wall Control Strategies."

Shurcliff, W. A., *Solar Heated Buildings of North America*, Brick House Publishing Co., Harrisville, NH (1978).

Sun, T. Y., *ASHRAE Task Group on Energy Requirements*, 48 (1975). "SHADOW 1, Procedure for Determining Heating and Cooling Loads for Computerizing Energy Calculations."

Szokolay, S. V., *Solar Energy and Building*, The Architectural Press, London (1975).

TRNSYS Manual, University of Wisconsin-Madison, Engineering Experiment Station Report No. 38–10 (1979).

Trombe, F., J. F. Robert, M. Cabanot, and B. Sesolis, *Solar Age*, **2**, 13 (1977). "Concrete Walls to Collect and Hold Heat."

Utzinger, D. M. and S. A. Klein, *Solar Energy*, **23**, 369 (1979). "A Method of Estimating Monthly Average Solar Radiation on Shaded Receivers."

Utzinger, D. M., M.S. Thesis, University of Wisconsin, Madison (1979a). "Analysis of Building Components Related to Direct Solar Heating of Buildings."

Utzinger, D. M., Personal Communication (1979b).

Wray, W. O., J. D. Balcomb, and R. D. McFarland, *Proceedings 3rd National Passive Solar Conference* of American Section of the ISES, **3**, 395 (1979). "A Simple Empirical Method for Estimating the Performance of Direct Gain Passive Solar Heated Buildings."

CHAPTER 16

Solar Cooling

The use of solar energy to drive cooling cycles has been considered for two different but related purposes. The first of these is to provide refrigeration for food preservation, and the second to provide comfort cooling. In Section 16.1 we briefly review some of the literature relating to both of the applications, since there is a common underlying technology. From then on, we consider problems relating to solar air conditioning. In particular, for application in the United States, we address questions of the use of flat-plate collectors for both winter heating and summer cooling.

Solar cooling of buildings is an attractive idea. Cooling is important in space conditioning of most buildings in warm climates and in large buildings in cooler climates. Cooling loads and availability of solar radiation are approximately in phase. The combination of solar cooling and heating should improve the economics, compared to heating alone. Solar air conditioning can be accomplished by three classes of systems: absorption cycles, desiccant cycles, and solar-mechanical processes. Within these classes there are many variations, using continuous or intermittent cycles, hot or cold side energy storage, various control strategies, various temperature ranges of operation, different collectors, etc. Each of these methods is reviewed in this chapter, with emphasis on solar absorption cooling since that is the process closest to commercialization.

Very substantial research and development efforts on solar cooling are in progress, and the technology is evolving. It is possible to buy absorption cooling equipment specifically designed for solar operation and further advances can be expected. Desiccant and solar-mechanical cooling, however, are in much earlier, experimental states of development.

The future of many of the methods will depend on developments beyond the cooling process. Temperature constraints in the operation of collectors limit what can be expected of solar cooling processes. If collector temperatures are pushed upward, storage may then become a critical problem. The relationship of collector and storage characteristics to cooling performance will be evident in the following discussions.

Solar cooling is expensive, as is solar heating. Reduction in cooling loads through careful building design and insulation will certainly be warranted and, within limits, will be less expensive than providing additional solar cooling. For any air conditioning and heating system, good building design and construction are needed. Our concern is with cooling loads that cannot be avoided by building design.

16.1 REVIEW OF SOLAR ABSORPTION COOLING

Two approaches have been taken to solar operation of absorption coolers. The first is to use continuous coolers, similar in construction and operation to conventional gas or steam fired units, with energy supplied to the generator from the solar collector-storage-auxiliary system whenever conditions in the building dictate the need for cooling. The second approach is to use intermittent coolers similar in concept to that of commercially manufactured food coolers used many years ago in rural areas (the Crosley "Icyball") before electrification and mechanical refrigeration were widespread. Intermittent coolers have been considered for refrigeration but most work in solar air conditioning has been based on continuous cycles.

Continuous absorption cycles can be adapted to operation from flat-plate collectors. The principles of these cooling cycles are described in ASHRAE (1975). A diagram of one possible arrangement is shown in Figure 16.1.1. The present temperature limitations of flat-plate collectors restrict consideration among commercial machines to lithium bromide-water systems. LiBr—H_2O machines require cooling water for cooling the absorber and condenser, and their use will probably require use of a cooling tower. Solar operation of ammonia-water coolers such as those now marketed for steam or gas operation is difficult because of the high temperatures required to operate the machines. Coolers based on other refrigerant-absorbent systems may be possible candidates for solar operation, but are not yet developed to the point where they can be evaluated.

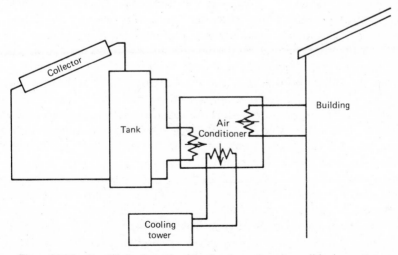

Figure 16.1.1   Simplified schematic of a solar absorption air conditioning system.

A commercial lithium bromide-water air conditioner, slightly modified to allow supplying the generator with hot water rather than steam, was operated from a flat-plate water heater by Chung et al. (1963). An analytical study of solar operation of a LiBr—H$_2$O cooler and flat-plate collector combination by Duffie and Sheridan (1965) identified critical design parameters and assessed the effects of operating conditions on integrated solar operation. Under the assumptions made in their study, design of the sensible heat exchanger between absorber and generator, cooling water temperature, and generator design are important; the latter is more critical here than in fuel-fired coolers because of the coupled performance of the collector and cooler. An experimental program was also developed at the University of Queensland, Australia, in a specially designed laboratory house [Sheridan (1970)].

From these and other experiments, it was clear that LiBr—H$_2$O absorption air conditioners could be adapted for solar operation. Without modification, the commercial machines operated at reduced capacities, but could be modified to operate at nominal capacities with energy supply to generator by hot water. Part load operation could be accomplished at little loss of the coefficient of performance (COP), with reduced dehumidification. Temperatures required for these air conditioners would be in a range suitable for flat-plate collectors (with the collectors operating at about the same temperature levels above ambient as those for winter heating operation).

If cooling requirements rather than heating loads fix collector size requirements, it may be an advantage to design coolers with improved COP. For example, double-effect evaporators can be used to decrease energy input requirements [see Whitlow and Swearingen (1959) or Chinnappa (1973)]. The conditions and constraints of solar operation lead to cooler designs different than those for fuel operation.

Intermittent absorption cooling may be an alternative to continuous systems. Most work to date on these cycles has been directed at food preservation rather than comfort cooling. These cycles may be of interest in air conditioning because they offer potential solutions to the energy storage problem. In these cycles, distillation of refrigerant from the absorbent occurs during the regeneration stage of operation, and the refrigerant is condensed and stored. During the cooling portion of the cycle, the refrigerant is evaporated and reabsorbed. A schematic of the simplest of these processes is shown in Figure 16.1.2. Thus, "storage" is in the form of separated refrigerant and absorbent. Modifications of this basically simple cycle, using pairs of evaporators, condensers, or other arrangements, may result in an essentially continuous cooling capacity and improved performance.

Refrigerant-absorbent systems used in intermittent cycles have been H$_2$SO$_4$—H$_2$O, NH$_3$—H$_2$O and NH$_3$—NaSCN. In the latter system, the absorbent is a solution of NaSCN in NH$_3$, with NH$_3$ the refrigerant. This system has been studied by Blytas and Daniels (1962) and by Sargent and Beckman (1968), and it appears to have good thermodynamic properties for cycles for ice manufacture. Williams et al. (1958) reported an experimental study of an

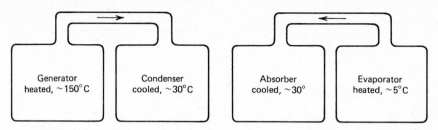

Figure 16.1.2   Schematic of an intermittent absorption cooling cycle. On the left, the regeneration cycle. On the right, the refrigeration cycle. The generator-absorber is a single vessel performing both functions, and the condenser-evaporator is also a single vessel performing both functions.

intermittent NH_3—H_2O cooler using a concentrating collector for regeneration.

Chinnappa (1961, 1962) and Swartman and Swaminathan (1970) experimentally studied the operation of intermittent NH_3—H_2O machines in which flat-plate collectors served as the energy supply. The absorber and generator are separate vessels. The generator was an integral part of the flat-plate collector, with refrigerant-absorbent solution in the tubes of the collector circulated by a combination of thermosyphon and a vapor lift (bubble) pump.

Using approximately equal cycle times for the regeneration and refrigeration steps (5 to 6 hr each), overall coefficients of performance were found to be approximately 0.06 at generator temperatures rising from ambient to approximately 99 C during regeneration. Evaporator temperatures were below 0 C. With cooling water available at approximately 30 C, the effective cooling per unit area of collector surface per day for the experimental machine was in the range of 50 to 85 kJ/m² for clear days.

This is not a complete history of developments of solar absorption cooling, but it provides an indication of the basis for interest in the combination of solar energy supply and absorption air conditioning. In the following section, we discuss some aspects of the theory of absorption cooling, particularly as it relates to solar operation, and in Sections 16.4 and 16.5 show data on solar cooling system performance.

16.2   THEORY OF ABSORPTION COOLING

Operation of absorption air conditioners with energy from flat-plate collector and storage systems is the most common approach to solar cooling. A schematic of a solar absorption cooler is shown in Figure 16.2.1; this system (or variations on it using other methods of energy storage, auxiliary energy input, multiple stage coolers, etc.) has been the basis of much of the experience to date with solar air conditioning.

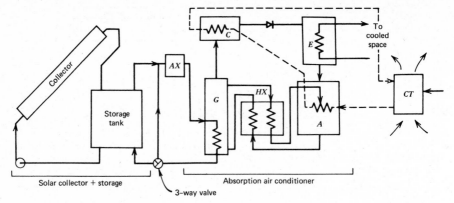

Figure 16.2.1 Schematic of a solar operated absorption air conditioner. *AX* is auxiliary energy source. The essential components of the cooler are: *G*, generator; *C*, condenser; *E*, evaporator; *A*, absorber; *HX*, heat exchanger to recover sensible heat; *CT*, cooling tower.

The coolers used in most experiments are LiBr—H_2O machines with water-cooled absorber and condenser. A pressure-temperature-concentration equilibrium diagram for LiBr and H_2O is shown in Figure 16.2.2. The idealized operation of a cycle is indicated on the diagram. The pressure in the condenser and generator is fixed by the condenser fluid coolant temperature. The pressure in the evaporator and absorber is fixed by the temperature of the cooling fluid to the absorber. The letters on the lines representing the cycle correspond to the processes occurring in the components indicated in Figure 16.2.1. The generation process is one of increasing the concentration from 55 to 60 percent while the

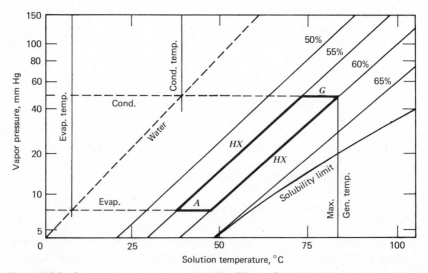

Figure 16.2.2 Pressure-temperature-concentration diagram for Li-Br-H_2O, showing an idealized cooling cycle with letters on the segments corresponding to the process components of Figure 16.2.1.

equilibrium temperature of the solution rises from 72 to 82 C at the pressure of the condenser. In the absorber, the solution concentration drops from 60 to 55 percent as the solution temperature drops from 48 to 38 C, all at the evaporator pressure. In a real cycle, some sensible heat will have to be transferred in the generator and absorber (the amount dependent on the effectiveness of exchanger HX), there will be pressure changes through the generator due to hydrostatic head, temperature differences across all heat exchangers, and other nonideal behavior. Exact pressures, temperatures, and concentrations will vary with the machine and operation conditions; the numbers used here are for illustration of the nature of the process.

The maximum solution temperature in the generator is shown on Figure 16.2.2; the temperature of the heated fluid to the generator must be above the maximum generator temperature, which is determined by the condenser pressure and the concentration of the solution leaving the generator. The generator temperatures must be kept within the limits imposed by the characteristics of flat-plate collectors. The critical design factors and operational parameters include solution concentrations, effectiveness of the heat exchangers and coolant temperature.

The pressure differences between the high- and low-pressure sides of $LiBr-H_2O$ systems is small enough that these systems can use a vapor-lift pump and gravity return from absorber to generator as an alternative to mechanical pumping to move the solution from the low pressure to the high pressure side. Early absorption machines used the vapor-lift pump but more recent designs use a mechanical pump because of improved performance.

An overall steady-state energy balance on the absorption cooler indicates that the energy supplied to the generator and to the evaporator must equal the energy removed from the machine via the coolant flowing through the absorber and condenser, plus whatever net losses may occur with the surroundings

$$Q_G + Q_E = Q_A + Q_C + Q_{\text{Losses}} \qquad (16.2.1)$$

The coefficient of performance, COP, is defined as the ratio of energy into the evaporator, Q_E, to the energy into the generator, Q_G:

$$\text{COP} = Q_E/Q_G \qquad (16.2.2)$$

Coefficient of performance is a useful index of performance in solar cooling, where collector costs (and thus costs of Q_E) are important. Many $LiBr-H_2O$ machines have nearly constant COP as the generator temperatures vary over the operating range, as long as they are above a minimum. The COP is usually in the range of 0.6 to 0.8, and the major effect of variation in the solar energy temperature to the generator is to vary Q_E, the cooling rate.

With water used as a coolant in the absorber and condenser, the generator temperatures are in the range 70–95 C. The temperature of the fluid supplied to the generator must be higher than this which means that there is a very small temperature range over which an unpressurized water storage tank can operate.

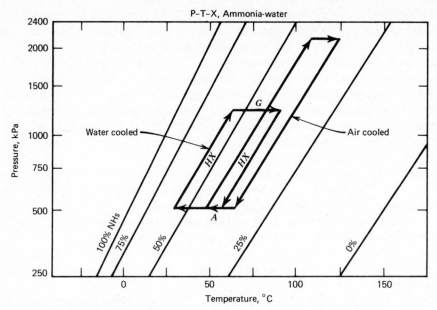

Figure 16.2.3 Pressure-temperature-concentration diagram for NH_3-H_2O, with two idealized cycles corresponding to water cooled condenser and absorber, and air cooled condenser and absorber.

Operation of most flat-plate collectors near 100 C is marginal. In addition, cooling towers are needed. These are three major problem areas in solar application of $LiBr$—H_2O coolers.

The schematic diagram of an ammonia-water cooler is very nearly the same as that of the cooler of Figure 16.2.1, except that a rectifying section must be added to the top of the generator to reduce the amount of water vapor going to the condenser. Pressure-temperature-concentration data for the ammonia-water system are shown in Figure 16.2.3. The basic solution processes are similar to those of the $LiBr$—H_2O system, but the pressures and pressure differences are much higher and mechanical pumps are needed to return solutions from the absorber to the generator. In many applications the condenser and absorber are air cooled with generator temperatures in a range of 125 to 170 C. In applications where water cooling is used, generator temperatures may be in the range of 95 to 120 C. Both air-cooled and water-cooled cycles are shown on Figure 16.2.3. The condensing temperatures for the air-cooled condenser correspond to much higher generator temperatures than those for liquid-cooled systems.

There has been relatively little done on experimental applications of NH_3—H_2O systems with solar operation. The generator temperatures required in today's commercial coolers are too high for present flat-plate collectors. Work on NH_3—H_2O systems has been directed at development of cycles using higher concentrations of NH_3 [see Farber et al. (1966) and Dao et al. (1976)] with the objective of lowering generator temperatures.

There are two approaches to the calculation of performance of absorption coolers. It is possible to write the full set of energy balances, material balances, rate equations, and equilibrium relationships, considering each component in the cooler, and simultaneously solve the set to determine the operating conditions and energy rates. This approach has been taken, for example, by Allen et al. (1973). The solution of the equations is expensive in computer time, and there is the additional difficulty that real absorption coolers may include other components such as concentration adjusters (small containers that retain volumes of refrigerant or absorbent, depending on pressures in the machine, and so change the solution concentrations), which are difficult to include in the analysis.

The second alternative is to devise empirical models based on operating data from specific machines. These are much less cumbersome and expensive to use, particularly in simulations where repetitive computations are done. An early model of a LiBr—H_2O cooler was devised by Butz (1973) and used in an early simulation of absorption air conditioning [Butz et al. (1974)]. This model has undergone several modifications, by Ward and Löf (1975) and Oonk et al. (1975) to reflect data on later machines, and by Blinn (1979) who has introduced a method of modeling transient operation of the cooler.

As a result of the on-off control strategy most commonly used in residential applications, chillers often do not operate at steady state. The room thermostat calls for cooling when the room temperature rises above a set point, and shuts off the chiller when room temperature drops below the set point. If the cooling capacity is significantly greater than the building cooling load, the thermostat will cycle and the chiller will spend part of its operating time in a transient mode.

When a cooler is started up, the lithium-bromide-water solution will begin circulating between the generator and absorber. No vapor will be evolved until the generator and all the solution held up in the generator have been heated to a temperature $T_{\min}$. This temperature is the boiling point of a lithium bromide solution whose concentration corresponds to the chiller's initial charge, at a pressure set by the condenser temperature.

If the generator, sensible heat exchanger, and absorber are all modeled as constant effectiveness heat exchangers during start up, if the generator is modeled as a single-node thermal capacitance, and if it is assumed that the absorber and sensible heat exchanger respond much more rapidly than the generator, then Blinn shows that the generator temperature during startup will vary exponentially

$$T_G = T_{G,ss} + (T_{G,o} - T_{G,ss})e^{-\tau/\tau_H} \tag{16.2.3}$$

where T_G = the generator temperature
 $T_{G,ss}$ = the steady-state generator temperature
 $T_{G,o}$ = the initial generator temperature
 τ_H = the generator time constant for start up.

Measurements of the time-temperature history of the generator of a three-ton capacity chiller (ARKLA model WF-36) located at Colorado State University Solar House I show that during startup the generator is described well by Equation 16.2.5, with a time constant, τ_H, of about 8.0 min.

During cooldown, after the chiller has been shut off, solution will drain to the lowest point of the chiller (usually the absorber) and lose heat to the surrounding air by conduction and natural convection. In that case

$$T_S = T'_a + (T_{G,o} - T'_a)e^{-\tau/\tau_c} \qquad (16.2.4)$$

where T_S = the solution temperature
$\quad T'_a$ = the temperature of the surroundings
$\quad \tau_C$ = the cooldown time constant

The cooldown time constant τ_C will tend to be much larger than the startup time constant, since τ_C depends on free rather than forced convection. The cooldown time constant measured for the three-ton chiller at CSU was about 63 min.

To complete the transient model, it is assumed that the instantaneous cooling delivered is a unique function of generator temperature, condensing water temperature (which fixes condenser and absorber temperatures) and evaporator temperature. The evaporator temperature is assumed constant, and manufacturer's performance data is used to determine cooling capacity and coefficient of performance as empirical functions of generator and condensing water temperatures.

When the generator temperature is above T_{min}, cooling and generator heat uptake are given by

$$Q_E = f_1(T_G, T_{CW}) \qquad (16.2.5)$$

$$Q_G = f_2(T_G, T_{CW}) \qquad (16.2.6)$$

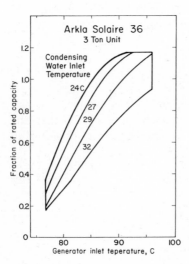

Figure 16.2.4 Manufacturer's data for 3 tan Arkla LiBr-H$_2$O water chiller, indicating capacity as a function of water inlet temperatures to generator and condenser, at chilled water outlet temperature of 7.2 C. From Blinn (1979).

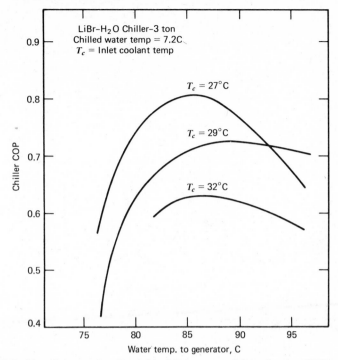

Figure 16.2.5 Manufacturers data for Arkla Solaire water chiller, showing COP as a function of water inlet temperatures to generator and condenser. From Blinn (1979).

where f_1 and f_2 are curve fits to manufacturer's data. When the generator temperature is less than T_{min}, $Q_E = 0$, and $Q_G = (UA)_G(T_s - T_G)$.

Figures 16.2.4 and 16.2.5 show the measured steady-state operation of a LiBr—H_2O chiller, indicating fraction of rated capacity and COP as functions of inlet temperature of water to the generator and the coolant temperature.

The average COP over a long term can be evaluated as

$$\overline{COP} = \frac{\int Q_E \, d\tau}{\int Q_G \, d\tau} \tag{16.2.7}$$

Simulations using meteorological data for Charleston, SC, over a cooling season indicate that the transient operation of the chiller leads to approximately 8 percent lower $\overline{COP}$ than would be expected if transients were neglected. The months with the lowest cooling loads are the months with the highest differences, as those are the months when the cooler is oversized and cycles most frequently.

16.3 COMBINED SOLAR HEATING AND COOLING

Many applications of solar air conditioning will be done in conjunction with solar heating, with the same collector, storage, and auxiliary energy system

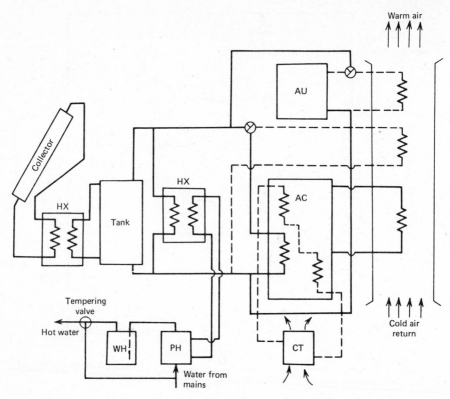

Figure 16.3.1 Schematic of a combined solar heating, air conditioning and hot water system using a LiBr-H_2O absorption cooler. Solid flow lines are for cooling, dashed lines are for heating and dotted lines are air conditioner coolant flow. AU is auxiliary energy source, AC is air conditioner, CT is cooling tower, PH is preheat tank, and WH is water heater.

serving both functions and supplying hot water. Figure 16.3.1 shows a combined heating and cooling system, based on absorption cooling for air conditioning.

An important consideration in combined heating and cooling systems is the relative importance of the summer and winter loads. Either one may dictate the needed capacity of the collector, and consequently its size and design. Climate is a major determining factor, and cooling requirements will dominate in climates like those of Phoenix and Miami. Commercial buildings are likely to have design fixed by cooling loads, even in cool climates. Also important are the building design features that can affect relative energy requirements for the two loads. These include, for example, fenestration, shading by overhangs, wingwalls and foliage, and building orientation. Less obvious is the performance of the heating and cooling system; a poor absorption cooler would require a larger collector area than one with a high COP, and thus could shift the determination of collector needs from winter heating to summer cooling.

The location of storage, whether inside or outside the building, will have an effect on heating or cooling loads. If heat is to be stored, and if the storage unit is inside the structure, heat losses from storage become uncontrolled gains during the heating season, and additional loads during the cooling season. If collectors are part of the envelope of the building, back losses from the collector will also become uncontrolled gains during heating and additional loads during cooling.

Collector orientation may be affected by which load dominates; optimum orientation is approximately $\beta = \phi + 15°$ for winter use, $\beta = \phi - 15°$ for summer use, and $\beta = \phi$ for all-year use. Heating loads are likely to be higher in the morning, suggesting that the surface azimuth angle γ should be negative, while cooling loads peaking in the afternoon suggest that γ should be positive. Simulations can be used to answer these questions [see Oonk et al. (1975) for an example]; fortunately collector orientation is usually not critical.

As with solar heating alone, the major design problem is the determination of optimum collector area, with underdesign leading to excessive use of auxiliary energy, and overdesign leading to low use factor on the capital-intensive solar energy system. Absorption air conditioners are now significantly more expensive than mechanical air conditioners. In climates where annual cooling loads are low, the use of absorption coolers will lead to higher cooling costs because of low use factors on the coolers.

16.4   SIMULATION STUDY OF SOLAR AIR CONDITIONING

Simulations provide useful information on effects of design changes on the long-term performance of solar coolers. In this section we show the results of a simulation study of the system of Figure 16.3.1, a liquid system with a LiBr—H_2O chiller modeled as described in Section 16.2. The simulations were done using meteorological data for Albuquerque, NM.

The building simulated has a floor area of 150 m^2, is well insulated with a UA of 232 W/C, has infiltration of one-half air change per hour and reasonable levels of internal heat generation and building capacitance. Latent cooling loads are estimated as 0.3 of the sensible cooling loads [ASHRAE (1977)]. The desired room temperature range is 19 to 25 C. The hot water load is 300 kg/day heated from 11 to 55 C, and a tempering valve prevents delivery of water at temperatures above 55 C. Storage capacity of the preheat tank is 0.35 m^3.

The collectors have $F_R(\tau\alpha)_n = 0.72$ and $F_R U_L = 4.94$ W/m^2 C, and are sloped toward the south at $\beta = 36°$ at the Albuquerque latitude of 35°N. F'_R/F_R is 0.94. Simulation results are shown for five collector areas, from five to fifty square meters. An antifreeze solution with $C_p = 3900$ J/kg C is circulated at $\dot{m}/A = 0.015$ kg/m^2 s. The storage tank volume is 0.10 m^3/m^2, it has a loss coefficient of 10.5 W/m^2 C, and its temperature is limited to 100 C. The minimum tank temperature for solar heating is 30 C.

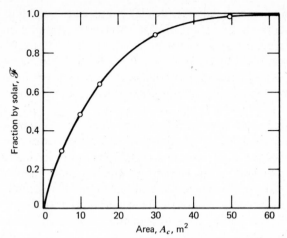

Figure 16.4.1 Annual fraction of energy supplied by solar energy for the Albuquerque simulations.

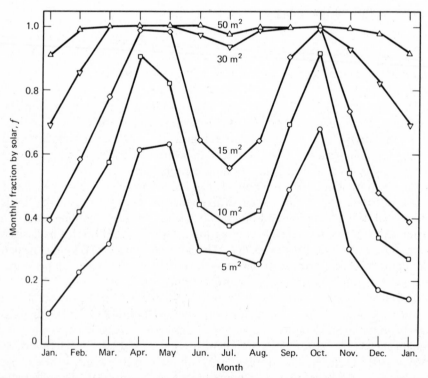

Figure 16.4.2 Monthly fractions of loads met by solar energy for five collector areas for the Albuquerque simulation.

The chiller has a nominal capacity of 4.2 kW, a startup time constant of 0.133 hr and a cool down time constant of 1.05 hr. The minimum useful source temperature is 77 C. Auxiliary in this set of simulations is provided in either of two ways. The primary method is by addition of heat to the generator of the chiller when needed. If the temperature in the building continues to be above the control temperature, parallel cooling by a separate (mechanical) cooler is computed to provide an approximation of energy required to meet any cooling loads not met by the absorption chiller.

Figure 16.4.1 shows the variation of the fraction of the annual energy needs (for heating, hot water, and cooling) met by this system as a function of collector area. The curve has the same general shape as those for space heating. In the Albuquerque climate, the solar fraction reaches 0.98 at a collector area of 50 m^2 in the particular year used in these simulations. The monthly fractions are shown in Figure 16.4.2 where the inability of smaller collectors to meet large winter heating loads or summer cooling loads is evident. As with solar heating, the larger systems show poorer integrated efficiencies as they are oversized during

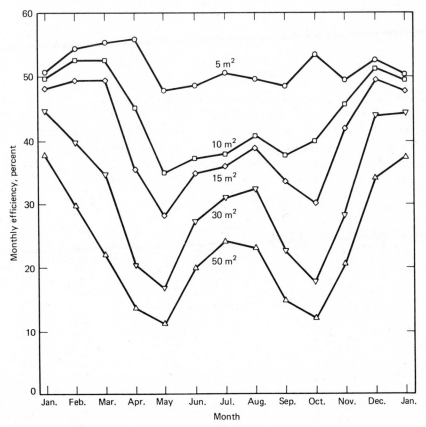

Figure 16.4.3 Monthly collector efficiencies for five collector areas for the Albuquerque simulations.

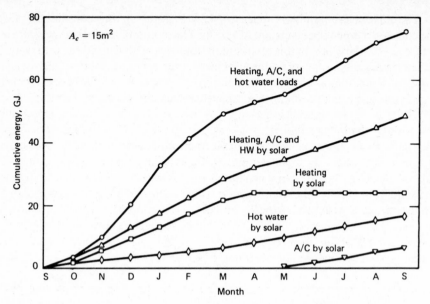

Figure 16.4.4 Cumulative loads and solar contributions for the 15 m² collector area for the Albuquerque simulations.

spring and fall. The larger systems run at higher temperatures and have more thermal losses from the collectors. Figure 16.4.3 shows a plot of monthly efficiency (the ratio of the collected energy to the total radiation incident on the collector for the month) for the five collector areas. These data are not useful for design, but show the wide range in integrated efficiencies encountered as collector area increases.

Table 16.4.1   Summary of Annual Performance for the Albuquerque Simulations of Solar Heating, Hot Water, and Cooling

Collector area, m²	5	10	15	30	50
Space heating load, GJ	38.6	40.2	41.0	42.1	43.3
Air conditioning load, GJ	13.9	13.6	13.2	14.3	16.9
DHW load, GJ	20.2	20.2	20.2	20.2	20.2
Total load, GJ	72.8	74.0	74.5	76.6	80.4
Solar to space heating, GJ	8.7	17.9	24.5	36.1	42.5
Solar to air conditioning, GJ	0.0	5.0	6.7	13.8	16.8
Solar to DHW, GJ	12.9	15.2	16.6	18.6	19.6
Total solar to load, GJ	21.6	35.6	47.8	68.4	78.9
Fraction of load met by solar	0.30	0.48	0.64	0.89	0.98
Collector efficiency	52%	44%	39%	30%	22%
Energy delivered, GJ/m²	4.32	3.56	3.19	2.28	1.58

Figure 16.4.4 shows, for a collector area of 15 m^2, the cumulative heating, cooling, and hot water loads and the contributions by solar to meeting those loads. The year starts in September, between the heating and cooling seasons.

The annual results of these simulations are summarized in Table 16.4.1. The space heating and cooling loads increase slightly with collector area, as the control scheme maintains the building at higher mean temperatures in the winter and lower mean temperatures in the summer as the size of the solar energy system increases. (Different controls would have made some difference in these loads, but the general trends and conclusions would not change.) Annual efficiencies are shown; they decrease as collector area increases. Annual energy delivered per square meter of collector is shown, as it is an approximate index of energy delivered per unit cost of the system (insofar as system cost is proportional to collector area).

16.5 OPERATING EXPERIENCE WITH SOLAR COOLING

Performance measurements of a few solar cooling systems have been published. These systems are to a substantial degree experiments, but the data provide an indication of the technical feasibility of solar absorption air conditioning.

In this section we summarize data on the performance of a residential scale installation on Colorado State University House I, as described by Duff et al. (1978). The system is essentially that shown in Figure 16.3.1, with the addition of limited cold side storage between the chiller and the spaces to be cooled. It includes a 3-ton Arkla LiBr—H$_2$O absorption chiller. The building and heating system are described in Section 13.6. The collector in these experiments was an evacuated tube type with a flat absorber shown in Figure 13.6.4, with a gross area of 75.2 m^2 and an absorber area of 39.9 m^2. For this collector, $F_R(\tau\alpha)$

Table 16.5.1 Summary of 1977 Solar Cooling Data on CSU House I^a

	August	September
Incident solar radiation	1287	1584
Solar energy delivered to generator	344	402
Auxiliary energy required for cooling	459	236
Solar fraction, cooling	0.43	0.63
Solar energy delivered to hot water	73	50
Auxiliary energy required for hot water	47	38
Solar fraction, hot water	0.61	0.57

^{a} From Duff et al. (1978). Energy units are averages of daily energy quantities, MJ.

was 0.79 and $F_R U_L$ was 2.03 W/m^2 C, based on absorber area. The Arkla cooler had a maximum capacity of about 14 kW at a COP of nearly 0.8; the minimum generator operating temperature was 66 C with a capacity of 5.6 kW when operated in the Fort Collins climate.

Data for substantial parts of two months of operation are summarized in Table 16.5.1, for August and September, 1977. During these periods the experimental energy balances closed to within 5 percent. From these experiments, Duff et al. conclude that the combination of a collector with low loss coefficients, a LiBr—H$_2$O chiller and associated equipment is effective in meeting heating, cooling and hot water needs of a residential scale building.

16.6 ECONOMICS OF SOLAR AIR CONDITIONING

In an early study of the economics of solar cooling, Löf and Tybout (1974) concluded that the combination of heating and cooling was more economical than either of them alone. Their study was based on early and unrealistically low costs of collectors and coolers, and utilized cash-flow economic analysis methods. Life cycle cost methods allow consideration of a more complete range of cost factors, and are applied here to solar absorption cooling. Similar economic considerations apply to other cooling methods to be outlined in later sections.

There are two major additional factors in solar air conditioning, in contrast to solar heating. First, there is a substantial additional item of cost for the air conditioner and its associated piping, controls, auxiliary energy supply, etc. Second, in many climates there will be a substantial annual increment in the useful energy supplied from the collector, as it will not be oversized in the cooling season as is the case for heating alone.

The simulation study of Section 16.4 provides the basis for an example of economic evaluation of solar heating and cooling. The cost and economic parameters are: term of analysis, 20 years; term of mortgage, 20 years; mortgage interest rate, 9.5 percent/yr; downpayment, 20 percent; market discount rate, 9 percent/yr; general inflation rate, 7 percent/yr; real estate taxes in first year, 2 percent of installed cost; insurance, maintenance, and parasitic power in first year, 4 percent of installed cost*; income tax bracket through the period of analysis, 45 percent; resale value, 100 percent of installed cost. Fuel costs are taken as $8 and $12/GJ, inflating at 10 and 15 percent/yr. System costs are taken as $C_A = 300$/m^2 and $C_E = 1000$ (for heating) + 3000 (an estimate for the air conditioner), or $4000 for the combined system.

Figure 16.6.1 shows life cycle savings as a function of collector area for the four combinations of fuel costs and fuel cost inflation rates. The curves are very similar to those of Figures 13.12.1 and 13.12.2, and indicate the same kind of

---

* These assumptions are the same as those of Section 13.12, except that an extra allowance is made for the consumption of parasitic power by the cooling tower.

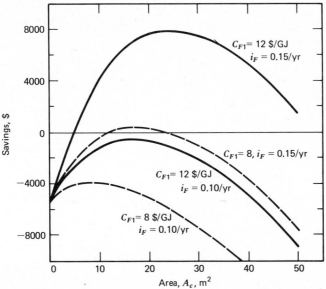

Figure 16.6.1 Life cycle savings as a function of collector area, for the Albuquerque simulations, for two first year fuel energy costs and fuel cost inflation rates.

dependence of economic merit on fuel cost and its inflation rates as for heating systems. (The numbers assumed are not necessarily representative of present or future costs, but other sets of economic parameters would have led to similar kinds of curves.)

16.7 SOLAR DESICCANT COOLING

The second class of solar air conditioners is based on open cycle dehumidification-humidification processes. These systems take in air, either from outside or from the building, dehumidify it with a solid or liquid desiccant, cool it by exchange of sensible heat, and then evaporatively cool it to the desired state. The desiccant is regenerated with solar energy. The components used include heat exchangers, heat and mass exchangers for dehumidification, and evaporative coolers. Many cycles have been proposed, and in this section we review several of these and then show results of a simulation study of one of them.

Löf (1955) suggested solar operation of a system as shown in Figure 16.7.1. In this system, the drying agent is liquid triethylene glycol. The glycol is sprayed into an absorber where it picks up moisture from the building air. It is then pumped through a sensible heat exchanger to a stripping column where it is sprayed into a stream of solar heated air. The high temperature air removes water from the glycol, which then returns to the heat exchanger and absorber.

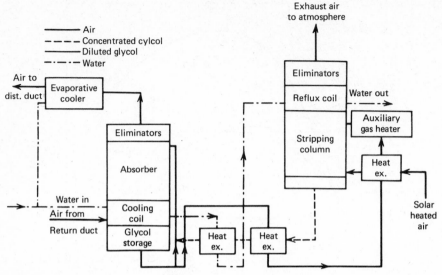

Figure 16.7.1 Schematic of a triethylene glycol open cycle air conditioning system. From Löf (1955).

Heat exchangers are provided to recover sensible heat, maximize the temperature in the stripper, and minimize the temperature in the absorber. Eliminators remove glycol spray from the air streams. This type of cycle, operated by steam, is marketed commercially and used in hospitals and other large installations. Solar operation has been studied by Lodwig et al. (1977), and a variation using the collector as the stripper is described by Collier (1979).

Dunkle (1965) showed a cycle designed for air conditioning in humid tropical or subtropical areas, which is based on use of desiccant beds such as silica gel for drying air. The desiccants are regenerated by solar heated air, and a pebble bed energy storage unit is included to allow operation during times of inadequate solar radiation. Rotary heat exchangers are provided to maximize cycle effectiveness. Figure 16.7.2 shows a schematic of Dunkle's cycle and Figure 16.7.3 shows the cycle on a psychrometric chart, with the state numbers corresponding to conditions on the cycle schematic.

Nelson et al. (1978) have simulated seasonal solar operation of a Munters Environmental Control (MEC) cycle, in the Miami climate. The process is shown in Figure 16.7.4, and is traced on the psychrometric chart of Figure 16.7.5 where the state points are numbered to correspond to the points on the process schematic. Ambient air is dried and heated by a dehumidifier, from 1 to 2, regeneratively cooled by exhaust air from 2 to 3, evaporatively cooled from 3 to 4, and introduced into the building. Exhaust air at state 5 moves in the countercurrent direction and is evaporatively cooled to 6, heated to 7 by the energy removed from the supply air in the regenerator, heated by solar or other source to 8, and then passed through the dehumidifier (desiccant) where it regenerates the desiccant.

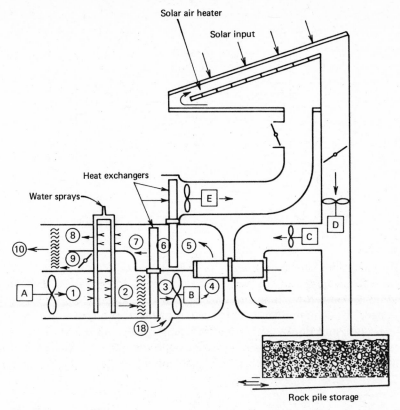

Figure 16.7.2 Schematic of open-cycle solar air conditioning system using rotary desiccant bed and heat exchangers. From Dunkle (1965).

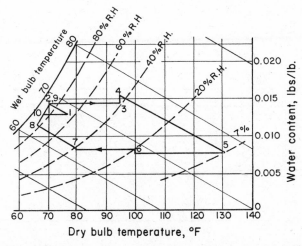

Figure 16.7.3 The cycle on a psychrometric chart. From Dunkle (1965).

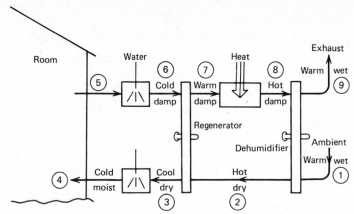

Figure 16.7.4 Schematic of a solar-MEC system. From Nelson (1976).

The conditions shown on Figure 16.7.5 are based on reasonable assumptions of the effectivenesses of the various components. In operation the ambient conditions change over time and the response to changing conditions must be estimated. The dehumidifier was modeled using the method developed by Banks (1972), Close and Banks (1972), and Maclaine-Cross and Banks (1972). Component effectiveness, control strategies and collector area were studied. A sample of the results is shown in Table 16.7.1 and Figure 16.7.6; in these simulations

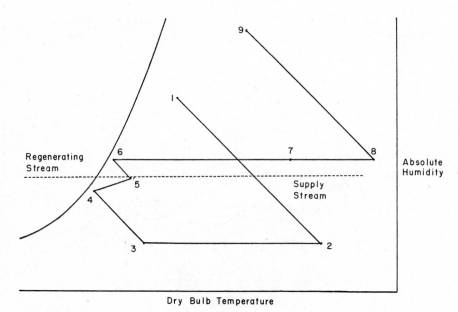

Figure 16.7.5 The solar-MEC cycle in a psychrometric chart. From Nelson (1976).

Table 16.7.1   Results of the Solar-MEC System Simulations For April Through September 1955[a]

Parameter	Collector Area, m^2			
	7.5	15	30	45
Collector mass flow rate, kg/s	0.104	0.208	0.417	0.625
Tank heat exchanger mass flow rate, kg/s	0.167	0.333	0.670	1.00
Tank volume, m^3	0.56	1.13	2.25	3.38
Sensible load met, GJ	20.8	20.8	20.8	20.8
Latent load met, GJ	15.7	15.7	15.9	16.7
Solar energy collected, GJ	12.2	23.0	41.0	54.4
Solar energy dumped,[b] GJ	0	0	0.026	2.4
Auxiliary energy supplied to system, (GJ)	33.6	24.2	11.4	4.2
Fraction of total energy requirement met by solar energy	0.27	0.49	0.78	0.93

[a] From Nelson (1978).
[b] As a result of boiling of the water-ethylene glycol mixture.

a conventional flat-plate collector [$F_R(\tau\alpha) = 0.65$ and $F_R U_L = 3.47$ W/m^2 C] supplied heat between state points 7 and 8 via a water to air heat exchanger.

The temperature of the regeneration stream (from the solar heating system) was below 65 C approximately 60 percent of the time and was above 80 C only 6 percent of the time in these simulations, regardless of collector area. This suggests that flat-plate collectors may be used to operate a cooling cycle of this type.

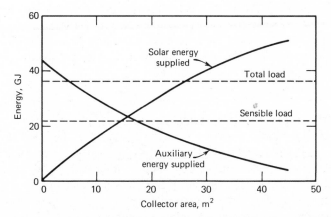

Figure 16.7.6   Performance of a solar-MEC system as a function of collector area. From Nelson (1976).

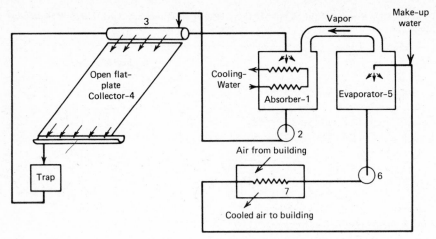

Figure 16.7.7   LiCl-H$_2$O open cycle cooling system. From Baum *et al.* (1973).

The Nelson study provides an indication of the possible performance of these cooling cycles. The seasonal COP values are comparable to those obtained with closed cycle absorption coolers. These results have not been verified by experiment and no conclusive cost studies have as yet been made. The application of these cycles to solar air conditioning is in early stages of development.

Baum et al. (1973) are working with a related system for solar cooling. Their cycle is shown schematically in Figure 16.7.7. The absorbing liquid is a solution of lithium chloride in water. Starting at the absorber 1, dilute LiCl solution is transferred by pump 2 to a heat exchanger-distributor-header 3 and then to an open flat-plate collector 4 where water evaporates. The concentrated solution returns, via the heat exchanger 3 for recovery of sensible heat, to the absorber. Water is cooled in the evaporator 5, with water vapor from the evaporator going to the absorber. The chilled water from the evaporator is moved by pump 6 to an air to water heat exchanger 7, which cools the building air (in this case, without direct contact with the absorbent solution). Means are provided to deaerate the solutions, recover sensible heat, and add makeup water. The absorber is cooled with a separate cooling coil.

16.8   SOLAR MECHANICAL COOLING

The third type of solar cooling system that has received attention in recent years couples a solar-powered Rankine cycle engine with a more or less conventional air conditioning system. The design of conventional air conditioning systems is well established; the problems associated with solar operation are basically those associated with generating mechanical energy from solar energy, and adaptation of air conditioning equipment for part load operation.

Studies on Rankine-cycle solar air conditioning systems to date have concentrated on theoretical analysis [e.g., see Teagan and Sargent (1973), Beekman (1975), and Olson (1977)]. An experimental system in a mobile laboratory [Barber (1974) and Prigmore and Barber (1974)] has been in operation for several years, but no long-term operating data are available. The National Security and Resources Study Center at Los Alamos National Laboratory is partially cooled by a 77 ton solar vapor-compression system. Biancardi and Meader (1976), Eckhard and Bond (1976), and others describe development studies of engine-compressor systems.

The problems associated with solar Rankine air conditioning systems are substantial. Generation of mechanical power from solar radiation has not yet been shown to be economical in large-scale systems, and it is difficult to envision a household scale system that will convert solar to mechanical energy at less expense than a large-scale solar system.

A simple Rankine-cycle cooling system is shown in Figure 16.8.1a. Energy from the storage tank is transferred through a heat exchanger to a heat engine. The heat engine exchanges energy with the surroundings and produces work. As shown in Figure 16.8.1b the efficiency of the solar collector decreases as the operating temperature increases, while the efficiency of the heat engine will increase as the operating temperature increases. Figure 16.8.1c shows the overall system efficiency for converting solar energy to mechanical work and indicates an optimum operating temperature exists for steady-state operation.

A schematic diagram of a Rankine heat engine is shown in Figure 16.8.2. The corresponding temperature-entropy diagram is shown in Figure 16.8.3. This cycle is somewhat different from the conventional power plant cycle using water as a working fluid. In a conventional power plant cycle superheating and

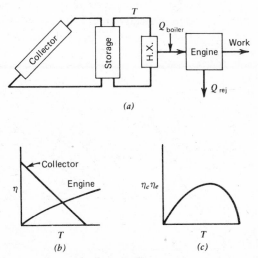

Figure 16.8.1 (a) Schematic of a solar operated Rankine cycle cooler. (b) Collector and power cycle efficiencies as a function of operating temperature. (c) Overall system efficiency.

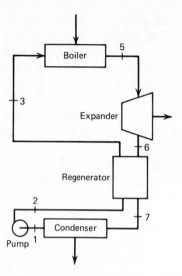

Figure 16.8.2   Rankine power cycle with regeneration.

extraction are used to increase the cycle efficiency and to prevent moisture from eroding the turbine blades. In a solar energy system super-heating is not desirable due to the increased temperature requirements of the collector, and extraction is not economical in small systems. To avoid moisture in the turbine, a proper fluid such as R 114 must be selected which has a positive slope to the saturated vapor line on a T–S diagram. With R 114, the outlet temperature of the turbine is significantly higher than the temperature of condensation, so a regenerator can be used to preheat the fluid leaving the condenser before it enters the boiler [see Olson (1977)].

The steady-state thermodynamic analysis of such a cycle is not difficult. However, the analysis under unsteady operation is difficult. The prediction

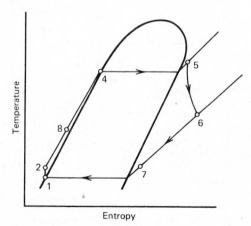

Figure 16.8.3   Temperature-entropy diagram of Rankine power cycle.

of component performance at off-design conditions and the matching of components into a complete system so that the overall performance is optimized is not easy. For example, as the storage tank temperature changes through the day, the temperature to the boiler will change. This will result in variable energy being added to the working fluid. To ensure that fluid does not enter the turbine from the two-phase region, either auxiliary energy must be added at the boiler or the circulation rate of the fluid must be reduced. Both of these options result in design problems.

The expander that drives the air conditioner can be either a turbine (which will require a gearbox speed reduction unit) or a piston engine. It is not now possible to determine which system is best for solar operation. Some designs use a single working fluid for both the expander and the air conditioner compressor.

When a Rankine heat engine is coupled with a constant-speed air conditioner, the output of the engine will seldom match the required input to the air conditioner. Consequently, a control system is needed to ensure matching of the engine and air conditioner. When the engine output is greater than needed, matching can be accomplished by throttling the energy supply to the engine, which means that the engine will work at an off-design condition and available energy will be wasted. Or, excess energy from the engine can be used to produce electrical energy for other purposes. When the engine output is less than that required by the air conditioner, auxiliary energy must be supplied. This auxiliary can be supplied in the form of heat to the Rankine cycle or the form of mechanical work to drive the air conditioner. The system can be designed to operate at variable speed. However, the operation of the air conditioner will be off-design with subsequent reduction in output.

As with all solar processes, steady-state conditions cannot be used to find optimum designs. It is necessary to evaluate design options based on estimates of integrated yearly performance. The effects of design variables are not intuitively obvious. The design and control of Rankine engine-vapor compression cooling has been studied by Olson (1977) (see Section 19.1), who found that for a given location and collector size, there are optimum sizes of both the engine and the storage tank which maximize annual solar contribution to meeting the cooling load.

16.9 SOLAR-RELATED AIR CONDITIONING

Some components of systems installed for the purpose of heating a building can be used to cool the building, but without the direct use of solar energy. In this section we note three examples: (1) night cold storage systems, (2) sky radiation systems and (3) heat pump systems. These are referred to as "solar-related" methods for cooling, and the economics of the processes are interrelated with the economics of solar heating.

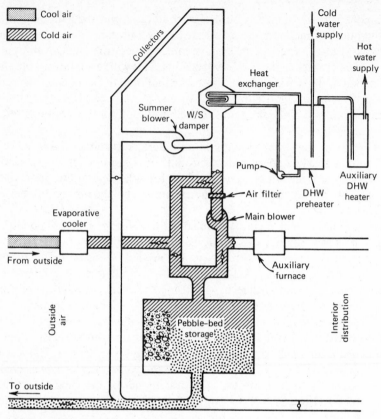

Figure 16.9.1 (*a*) The night cold storage system on CSU House II: nocturnal cooling of the bed. From Karaki *et al.* (1977).

Night chilling of pebble beds to store "cold" for use the following day can provide some cooling capacity. In climates where night temperatures and humidities are low, pebble bed storage units of solar air heating systems can be cooled by passing outside air through an evaporative cooler, through the pebble bed and to exhaust. This system was used on part of a laboratory building in Melbourne, Australia [see Close et al. (1968)]. During the air conditioning season the rock pile storage unit was cooled with evaporatively cooled air at night when, in that climate, the ambient wet-bulb temperatures were at most 15 to 20 C.

Essentially the same system is used on CSU House II [see Karaki et al. (1977)] and is shown in Figure 16.9.1. During the night, air flow is through the evaporative cooler, down through the pebble bed, and then to exhaust. When building cooling is used, flow is from the rooms, up through the pebble bed and return to rooms. The pebble bed stratifies, as it does in the heating mode, and sample profiles are shown in Figure 16.9.2. Karaki et al. found in the climate of Fort Collins there was very little charging of the bed on the warmest summer

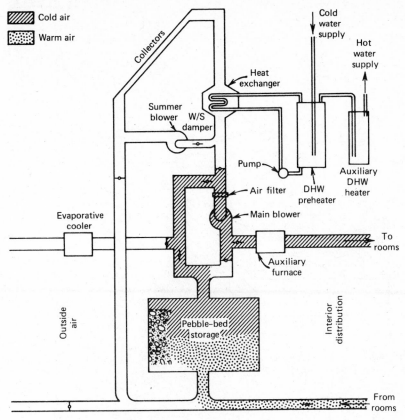

Figure 16.9.1 (*b*) Cooling from storage.

nights, and the bed was underdesigned for cooling purposes during the height of the cooling season. Early and late in the cooling season when night time temperatures are low, there is increased cooling capacity, with the bed chilled to a minimum temperature of 10 C by air entering at 7.5 C from the evaporative cooler.

Radiation to the night sky has been used to dissipate energy in several experimental systems. The nearly horizontal radiators are also used as collectors for solar heating. Bliss (1964), Yanagimachi (1958 and 1964), Hay (1973), and Hay and Yellott (1970) describe systems that use nocturnal radiation to chill water for subsequent cooling. Bliss found that on a monthly average the uncovered collectors on his laboratory at Tucson could dissipate at most approximately 4.1 MJ/m² per night. For the 93 m² collector-radiator on the Tucson laboratory, the monthly average cooling was equivalent to about 3.5 kW of continuous cooling.

Two problem areas are encountered in design and operation of systems of this type. First, the characteristics that make a good collector are not those

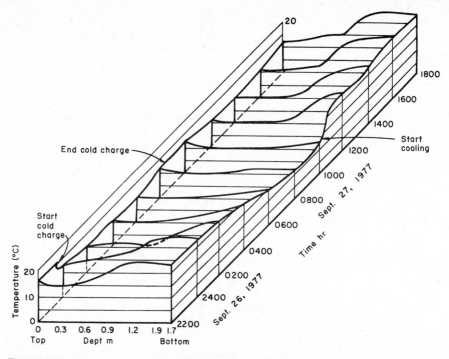

Figure 16.9.2   Temperature profiles in the pebble bed in CSU House II for a day late in the cooling season. From Karaki *et al.* (1977).

that make a good radiator. Neither covers nor selective surfaces can be used (unless movable covers are provided), as either effectively limits nocturnal radiation to very low levels. Thus collectors are limited to the kinds that can be used at temperatures close to ambient, for example, collectors that could be used as sources for series heat pump systems. The second problem is a climatic one. These systems can only be used where night sky temperatures are low, that is, where atmospheric moisture and dust content are low. Also, it is necessary that the night time wind speeds are low.

Heat pumps used as part of solar heating systems can also be used for cooling. If the cost of the heat pump can be justified by its use for air conditioning, the result will be to make the economics of the solar energy-heat pump systems described in Section 13.8 more favorable.

REFERENCES

Allen, R. W., F. H. Morse *et al.*, University of Maryland, Mech. Engr. Dept. Reports on project NSF/RANN/SE/GI 39117 (1973-76). "Optimization Study of Solar Absorption Air Conditioning Systems."

ASHRAE Handbook of Fundamentals, American Society of Heating, Refrigerating, and Air Conditioning Engineers, New York (1977).

ASHRAE Handbook and Product Directory/Equipment, American Society of Heating, Refrigerating and Air Conditioning Engineers, New York (1975).

Banks, P. J., *Chemical Engineering Science*, **27**, 1143 (1972). "Coupled Equilibrium Heat and Single Adsorbate Transfer in Fluid Through a Porous Media: Part I, Characteristic Potentials and Specific Capacity Ratios."

Barber, R. E., report from Barber-Nichols to Honeywell (October 3, 1974). "Final Report on Rankine Cycle Powered Air Conditioning System."

Baum, V. A., A. Kakabaev, A. Khandurdyev, O. Klychiaeva, and A. Rakhmanov, paper presented at International Solar Energy Congress, Paris (1973). "Utilization de L'energie Solair Dans Les Conditions Particulieres Des Regions A Climate Torride Et Aride Pour La Climatisation En Ete."

Beekman, D. M., M.S. Thesis, Mech. Engr., University of Wisconsin-Madison (1975). "The Modeling of a Rankine Cycle Engine for Use in a Residential Solar Energy Cooling System."

Biancardi, F. and M. Meader, paper in *Proceedings of the Second Workshop on the Use of Solar Energy for the Cooling of Buildings*, 137 (1976). "Demonstration of a 3 ton Rankine Cycle Powered Air Conditioner."

Blinn, J. C., M.S. Thesis (Chem E), University of Wisconsin-Madison (1979). "Simulation of Solar Absorption Air Conditioning."

Bliss, R. W., *Proceedings of the UN Conference on New Sources of Energy*, **5**, 148 (1964). "The Performance of an Experimental System Using Solar Energy for Heating, and Night Radiation for Cooling a Building."

Blytas, G. C. and F. Daniels, *Journal of the American Chemical Society*, **84**, 1075 (1962). "Concentrated Solutions of NaSCN in Liquid Ammonia: Solubility, Density, Vapor Pressure, Viscosity, Thermal Conductance, Heat of Solution, and Heat Capacity."

Butz, L. W., M.S. Thesis in Mechanical Engineering, Madison, University of Wisconsin (1973). "Use of Solar Energy for Residential Heating and Cooling."

Butz, L. W., W. A. Beckman, and J. A. Duffie, *Solar Energy*, **16**, 129 (1974). "Simulation of a Solar Heating and Cooling System."

Chinnappa, J. C. V., *Solar Energy*, **5**, 1 (1961). "Experimental Study of the Intermittent Vapour Absorption Refrigeration Cycle Employing the Refrigerant-Absorbent System of Ammonia Water and Ammonia Lithium Nitrate."

Chinnappa, J. C. V., *Solar Energy*, **6** (4), 143 (1962). "Performance of an Intermittent Refrigerator Operated by a Flat-Plate Collector."

Chinnappa, J. C. V., Paper presented at International Solar Energy Congress, Paris (1973). "Computed Year-Round Performance of Solar-Operated Multi-State Vapour Absorption Air Conditioners at Georgetown, Guyana, and Colombo, Ceylon."

Chung, R., J. A. Duffie, and G. O. G. Löf, *Mechanical Engineering*, **85**, 31 (1963). "A Study of a Solar Air Conditioner."

Close, D. J., R. V. Dunkle, and K. A. Robeson, *Mechanical and Chemical Engineering Transactions*, Inst. Engineers Australia, **MC4**, 45 (1968). "Design and Performance of a Thermal Storage Air Conditioning System."

Close, D. J. and P. J. Banks, *Chemical Engineering Sciences*, **27**, 1143 (1972). "Coupled Equilibrium Heat and Single Absorbate Transfer in Fluid Flow Through a Porous Media: Part II Predictions for Silica-Gel Air-Drier Using Characteristic Charts."

Collier, R. K., *Solar Energy*, **23**, 357 (1979). "The Analysis and Simulation of an Open Cycle Absorption Refrigeration System."

Dao, K., M. Simmons, R. Wolgast, and M. Wahlig, paper in *Proceedings of the ISES Meeting*, Winnipeg (1976). "Performance of an Air-Cooled Ammonia-Water Absorption Air Conditioner at Low Generator Temperature."

Duff, W. S., T. M. Conway, G. O. G. Löf, D. B. Meredith, and R. B. Pratt, *Proceedings of the CCMS/ISES Conference on Performance of Solar Heating and Cooling Systems*, NATO Committee on Challenges to Modern Society, CCMS Report 85, p. 217 (1978). "Performance of Residential Solar Heating and Cooling System with Flat-Plate and Evacuated Tubular Collectors: CSU House I."

Duffie, J. A. and N. R. Sheridan, *Mechanical and Chemical Engineering Transactions*, Inst. Engrs. Australia, **MC1**, 79 (1965). "Lithium Bromide-Water Refrigerators for Solar Operation."

Dunkle, R. V., *Mechanical and Chemical Engineering Transactions*, Inst. Engrs., Australia, **MC1**, 73 (1965). "A Method of Solar Air Conditioning."

Eckard, S. E. and J. A. Bond, paper in *Proceedings of the Workshop on the Use of Solar Energy for the Cooling of Buildings*, 110 (1976). "Performance Characteristics of a 3-Ton Rankine Powered Vapor-Compression Air Conditioner."

Farber, E. A., F. M. Flanigan, L. Lopez, and R. W. Polifka, *Solar Energy*, **10**, 91 (1966). "University of Florida Solar Air Conditioning System."

Hay, H. R. and J. I. Yellott, *Mechanical Engineering*, **92** (1), 19 (1970). "A Naturally Air Conditioned Building."

Hay, H. R., *Mechanical Engineering*, **95** (11), 18 (1973). "Energy Technology and Solarchitecture."

Karaki, S., P. R. Armstrong, and T. N. Bechtel, Colorado State University Report COO-2868-3 to U. S. Dept. of Energy (Dec. 1977). "Evaluation of a Residential Solar Air Heating and Noctural Cooling System."

Lodwig, E., D. A. Wilkie, and J. Bressman, paper in *Proceedings of the ISES Meeting*, Orlando, **1**, 7–1 (1977). "A Solar Powered Desiccant Air Conditioning System."

Löf, G. O. G., *Solar Energy Research*, University of Wisconsin Press, Madison, p. 33 (1955). "House Heating and Cooling with Solar Energy."

Löf, G. O. G. and R. A. Tybout, *Solar Energy*, **16**, 9 (1974). "The Design and Cost of Optimized Systems for Cooling Dwellings by Solar Energy."

Maclaine-Cross, I. L. and P. J. Banks, *International Journal of Heat and Mass Transfer*, **15**, 1225 (1972). "Coupled Heat and Mass Transfer in Regenerators-Predictions Using an Analogy with Heat Transfer."

Nelson, J. S., M. S. thesis, *Mech. Engr.*, University of Wisconsin-Madison (1976). "An Investigation of Solar Powered Open Cycle Air Conditioners."

Nelson, J. S., W. A. Beckman, J. W. Mitchell, and D. J. Close, *Solar Energy*, **21**, 273 (1978). "Simulations of the Performance of Open Cycle Desiccant Systems Using Solar Energy."

Olson, T. J., M.S. Thesis, Mech. Engr., University of Wisconsin-Madison (1977). "Solar Source Rankine Cycle Engines For Use in Residential Cooling."

Oonk, R. L., W. A. Beckman, and J. A. Duffie, *Solar Energy*, **17**, 21 (1975). "Modeling of the CSU Heating/Cooling System."

Prigmore, D. R. and R. E. Barber, 9th IECEC Conference (1974). "A Prototype Solar Powered, Rankine Cycle System Providing Residential Air Conditioning and Electricity."

Sargent, S. L. and W. A. Beckman, *Solar Energy*, **12**, 137 (1968). "Theoretical Performance of an Ammonia-Sodium Thiocyanate Intermittent Absorption Refrigeration Cycle."

Sheridan, N. R., paper presented at the ISES Conference, Melbourne (1970). "Performance of the Brisbane Solar House."

Swartman, R. K. and C. Swaminathan, paper presented at the ISES Conference, Melbourne (1970). "Further Studies on Solar-Powered Intermittent Absorption Refrigeration."

Teagen, W. P. and S. L. Sargent, ISES Paris Conference, paper EH-94-1 (1973). "A Solar Powered Combined Heating/Cooling System With the Air Conditioning Unit Driven by an Organic Rankine Cycle Engine."

Ward, D. S. and G. O. G. Löf, *Solar Energy*, **17**, 13 (1975). "Design and Construction of a Residential Solar Heating and Cooling System."

Whitlow, E. P. and J. S. Swearingen, paper presented at Southern Texas American Institute of Chemical Engineers Meeting (1959). "An improved Absorption-Refrigeration Cycle."

Williams, D. A., R. Chung, G. O. G. Löf, D. A. Fester, and J. A. Duffie, *Refrigeration Engineering*, **66**, 33 (Nov. 1958). "Cooling Systems Based on Solar Regeneration."

Yanagimachi, M., *Proceedings of the UN Conference on New Sources of Energy*, **5**, 233 (1964). "Report on Two and a Half Years' Experimental Living in Yanagimachi Solar House II."

Yanagimachi, M., *Transactions of the Conference on Use of Solar Energy*, **3**, 32, University of Arizona Press, Tucson (1958). "How to Combine: Solar Energy, Nocturnal Radiational Cooling, Radiant Panel System of Heating and Cooling, and Heat Pump to Make a Complete Year-Round Air Conditioning System."

CHAPTER 17

Solar Industrial Process Heat

Very large amounts of energy are used for low-temperature process heat in industry, for such diverse applications as drying of lumber or food, cleaning in food processing, extraction operations in metallurgical or chemical processing, cooking, curing of masonry products, paint drying, and many others. Temperatures for these applications can range from near-ambient temperatures to those corresponding to low-pressure steam, and can be provided from flat-plate collectors or concentrating collectors of low concentration ratios.

The principles of operation of components and systems outlined in earlier chapters apply directly to industrial process heat applications. The unique features of these applications lie in the system configurations and controls needed to meet industrial requirements, and the integration of the solar energy supply system with the auxiliary energy source and the industrial process itself. Process simulations similar to those outlined in Chapter 10 and chapters on heating and cooling are very useful tools in the study of industrial applications. The design methods outlined in Chapter 18 for systems delivering energy at temperatures other than 20 C are applicable to many solar industrial process heating operations.

Other than traditional crop drying techniques and solar evaporation, which have been practiced over centuries, solar applications for industrial process heat have been on a relatively small scale and are experimental in nature. In this chapter we outline some general design and economic considerations and then briefly describe several examples of experimental industrial applications to illustrate the potential utility of solar energy to industry. These are described, with others, in the *Proceedings of Solar Industrial Process Heat Symposium* (1977) and *Conference* (1978).

17.1 INTEGRATION WITH INDUSTRIAL PROCESSES

Two primary questions to be considered in a possible industrial process application concern the use to which the energy is to be put and the temperature at which it is to be delivered. If a process requires hot air for direct drying, an air heating system is probably the solar energy system best matched to the need. If steam is needed to operate an autoclave or indirect dryer, the solar energy system must be designed to produce steam and concentrating collectors will probably be required. If hot water is needed for cleaning in food processing, the solar energy system will be a liquid heater. An important factor in determination of the best system for a particular use is the temperature of the fluid to the collector. The generalizations of building heating applications (e.g., that return air to collectors in air systems is usually at or near room temperatures) do not necessarily carry over to industrial processes as the system configurations and energy uses may be quite different.

The energy may be needed at a particular temperature or over a range of temperatures. If low-pressure steam is condensed in an indirect dryer, the condensate will probably be recirculated, and the solar process system will be called on to deliver essentially all of the energy at a constant temperature level. A once-through cleaning process may call for fresh water to be heated from supply temperature to some useful minimum level, so that energy can be added to the water over a range of temperatures. A system in which a working fluid is recirculated back to the tank will probably operate over an intermediate range of temperatures.

The temperature at which energy is used in many industrial processes has not been limited by the characteristics of conventional energy supplies. The partial replacement of conventional sources by solar energy in retrofit applications is usually limited to operations in the same temperature ranges as that of the sources replaced. In new applications, since solar collectors operate more efficiently at lower temperatures, the industrial process itself should be examined to see if the temperature of energy delivery can be optimized.

Storage would usually be used in industrial processes, except where the maximum rate at which the solar energy system can deliver energy is not appreciably larger than the rate at which the process uses energy. In these cases the annual fraction of the energy needs delivered by solar energy will be small. If storage is used, the energy balances and rate equations of Chapter 9 can be used to describe the storage subsystem.

The investments in industrial processes are generally large, and the transient and intermittent characteristics of a solar energy supply are so unique, that the study of options in solar industrial applications can be done by simulation methods at costs that are very small compared to the investments.

17.2 MECHANICAL DESIGN CONSIDERATIONS

Many industrial processes use large amounts of energy in small spaces. If solar is to be considered for these applications, the location of collectors can be a problem. It may be necessary to locate collector arrays on ajoining buildings or grounds, resulting in long runs of pipe or duct. Or collector area may be limited by building roof area and orientation. Existing buildings are generally not designed or oriented to accommodate arrays of collectors, so in many cases structures to support collector arrays must be added to the existing structures.

New buildings can be readily designed, often at little or no incremental cost, to allow for collector mounting and access.

Interfacing with conventional energy supplies must be done in a way that is compatible with the process. If air to dryers is to be preheated it must be possible to get the solar preheated air into the dryer air supply. In food processing, the sanitation requirements of the plant must be met. It is not possible to generalize on these matters; the engineering of the solar energy process and the industrial process must be mutually compatible. In most of the examples in the following sections, solar heating is retrofitted, supplies a relatively small part of the plant loads, and in varying degrees the range of problems noted here have been encountered.

17.3 ECONOMICS OF INDUSTRIAL PROCESS HEAT

Economic analysis of industrial processes is the same as outlined in Chapter 11, but in contrast to nonincome-producing application, deduction of fuel costs as a business expense, investment tax credits, and depreciation must be taken into account. And, industry usually requires a higher return on investment (i.e., shorter payback time) than do individuals.

The investments in solar industrial process systems are as outlined in Section 11.1. For a retrofit system, mounting a collector array may be a major item of investment; in some experimental installations the cost of providing mountings exceeded the cost of the collectors. In new construction, provision for collector mounting may mean only nominal changes in the structure. The operating costs include the same kinds of items as for other solar applications, such as parasitic power, maintenance, insurance, and real estate taxes.

The generalization made for solar heating that collector area can be considered as the primary design variable does not necessarily extend to industrial process applications. There is no general relationship between time dependence of loads and energy supply, and both collector area and storage capacity (and possibly others) may be important design parameters. In general, the design of a solar industrial process will have to be based on a study of the appropriate ranges of variables of all of the important parameters, done in view of the characteristics of the energy using process.

As high reliability is needed for industrial applications, solar will normally be combined with a conventional energy supply. The useful energy produced by the solar energy system serves to reduce the fuel consumption of the conventional (auxiliary) energy supply. Industry buys fuels in quantities that are large compared to homeowners, and the price of that fuel is generally less than is paid by the small buyer. In addition, funds spent for purchase of fuel are business expenses, and therefore deductible from corporate income taxes. With a basic corporate income tax rate of 50 percent, the effective cost of energy is halved. As with nonindustrial applications, it is difficult to build solar heating systems that can deliver energy at a sufficiently low cost to compete with inexpensive fuels. Differences in taxation of fuels, investment tax credits, and depreciation of equipment can make very substantial difference in the economic viability of these processes and are matters of legislation.

Miscellaneous expenses for parasitic power, insurance, and maintenance are also business expenses and therefore tax deductible. Investment tax credits effectively reduce first costs, and thus encourage investments in equipment. Equipment costs can be depreciated, which has the effect of decreasing corporate taxes. Depreciation can be by straight line, declining balance, sum of digits, or other methods. Most industries will have established methods for depreciation and discount rates.

Seasonal industrial processes, which can only utilize the output of solar heating systems over part of the year, will show less favorable economics than will processes that can be operated all year. Food processing applications in temperate climates provide examples. This disadvantage can, in part, be minimized by collector orientation to maximize output over the period of use. It may also be possible to develop other off-season uses for energy, such as building heating, and one of the examples in a following section shows such a combination of uses.

17.4 AIR HEATING APPLICATION

Heated outside air is used in many industrial applications where recirculation of air is not practical because of contaminants. Examples are drying, supplying fresh air to hospitals, and paint spraying operations. Heating of ambient air is an ideal operation for a collector, as it operates very close to ambient temperature. In this case, Equation 6.7.6 becomes simply

$$Q_u = A_c F_R S \tag{17.4.1}$$

F_R may be essentially fixed if flow rate is fixed, or it may vary if, for example, system controls are arranged to provide a fixed outlet temperature with varying inlet ambient temperatures by adjusting the flow rate.

Equation 17.4.1 is evaluated for the hours of operation of the solar system. All of the methods of Chapter 2, 4, and 5 in determining S, Chapter 6 in calculating F_R and Chapter 7 in determination of $F_R(\tau\alpha)_n$ and angle of incidence effects apply directly to these calculations. For a full month's operation of such a system, the average daily useful gain can be approximated by

$$\bar{Q}_u = A_c F_R(\tau\alpha)_n \frac{(\overline{\tau\alpha})}{(\tau\alpha)_n} \bar{H}_T \tag{17.4.2}$$

where $(\overline{\tau\alpha})/(\tau\alpha)_n$ can be estimated by the methods of Section 5.8. This calculation of monthly gain will be good for systems which operate whenever the incident radiation is high enough to justify operation of the fan.

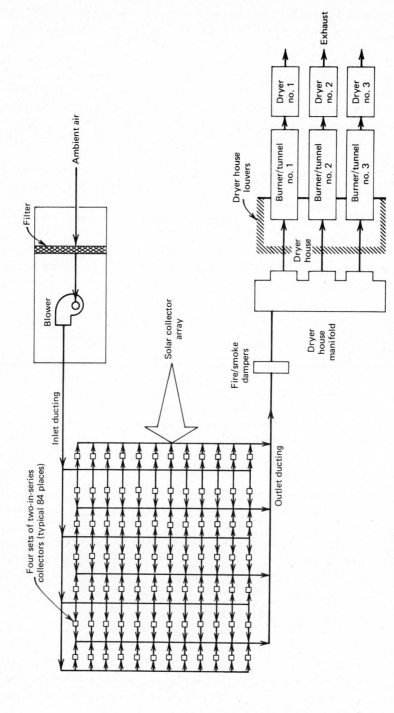

Figure 17.4.1 Layout of collector array and soybean drying installation. Adapted from Guinn (1978).

Figure 17.4.2   The 1200 m^2 air heating collector array at the Gold Kist soybean plant, Decatur, Alabama. Courtesy Solaron Co.

An experimental application of air heating to drying of soybeans at the Gold Kist, Inc. plant at Decatur, Alabama, is one of a series of demonstrations of solar industrial process heating sponsored in part by the United States Department of Energy. The system described by Guinn (1978), consists of 1200 m^2 (672 modules) of air heating collectors supplying warmed air to the dryer. The air is mixed with additional ambient air, heated by oil to the desired temperature, and used in the drying operation. Since the maximum output from the collectors is less than the energy needs of the dryer, all collected energy can be used (during the drying season) and no storage is provided. Figure 17.4.1 is a schematic of the process, and Figure 17.4.2 shows the collector array.

In this system, the filter and blower are on the upstream side of the collectors, to put the collectors under positive pressure and avoid ingestion of dusty air into the system where it might become an explosive mixture. The size of the collector array was essentially fixed by the dimensions of a parking lot over which the array was mounted. The collector slope is 15°, chosen as a compromise between performance and greater structural costs associated with larger slopes. The atmosphere in the plant area is dusty, and keeping the collectors reasonably clean has been a significant operational problem.

Systems of this type, which are designed for small contributions by solar in relation to the total loads, can be operated without energy storage. No energy will be dumped as long as the maximum output of the collector is less than the energy needs of the application at the time the collector maximum occurs. It may be that the time of collector operation would be determined by the process itself (e.g., times when paint spraying is going on, or when materials are in the dryer ready to be dried) and under these circumstances storage may be needed.

17.5   ONCE-THROUGH INDUSTRIAL WATER HEATING

Large quantities of water are used for cleaning in the food processing industries, and recycling of used water is not practical because of contaminants picked up by the water in the cleaning processes. These processes involve heating of water from main supply temperature to (or toward) a desired temperature level for the process. As with other systems, if the solar collector output is always smaller than the loads on a process, the solar heating operation can be carried out without storage, but solar contributions relative to total loads will be small. Storage should usually be used with system schematics similar in concept to the domestic water system configurations shown in Figure 12.1.1. Systems are also designed in which supply water passes through the collector once and then to storage or the process. If there is no storage or if the amount of water in storage is constant, flow through the collector is fixed by draw off at the point of use. If variable-capacity storage is provided, flow through the collector can be controlled independently of the loads on the system.

An example of this application is the experimental water heater for the Campbell soup plant at Sacramento, CA, shown in Figure 17.5.1. Water supply from a well is pumped in series through a flat plate collector and a concentrating collector, and then to an insulated storage tank. The temperature of this water is boosted as needed by an auxiliary steam heater to bring it to 85 to 90 C as required in the can washer. It is mixed with heated water from the conventional system, used for can washing, and discarded. The system, described by Vindum and Bonds (1978), was designed to provide 75 percent of the energy to one can line of 20 in the plant, so its output is small compared to the total energy requirement for hot water for the plant. This system, while it produces hot water in the same temperature range as that of flat-plate collectors used in absorption cooling, uses concentrating collectors for part of the heating process.

17.6   RECIRCULATING INDUSTRIAL WATER HEATING

An experimental solar heating system to preheat air and boiler feedwater in an onion drying plant in Gilroy, CA is based on the use of evacuated tubular collectors delivering hot water at temperatures of 70–100 C. The gas fired burner

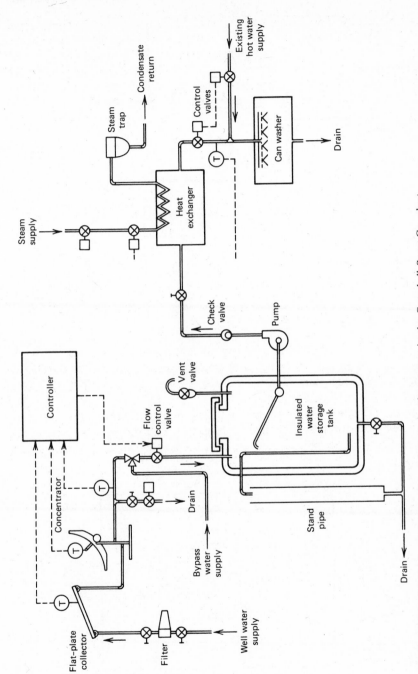

Figure 17.5.1 Schematic of the experimental water heating system in the Campbell Soup Co. plant, Sacramento, CA. Adapted from Vindum and Bonds (1978).

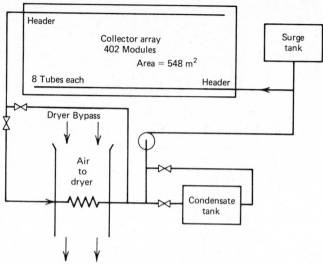

Figure 17.6.1 Schematic of the water heating system for the Gilroy Foods onion drying facility. Adapted from Graham *et al.* (1978).

receives preheated air from the solar system and delivers 93 C air to the dryer. The capacity of the collectors to deliver energy is small compared to the energy requirements of the plant, and no storage is used. A schematic of the solar heating process is shown in Figure 17.6.1.

In normal operation of the system, water is circulated through the collectors and the air-water preheat exchanger to heat drying air before it goes to the gas burners. It is also possible to heat water in the condensate tank which is part

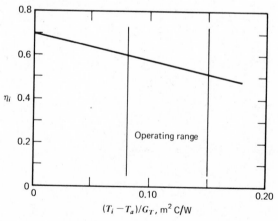

Figure 17.6.2 Characteristics of evacuated tubular collectors used in the Gilroy Foods onion dryer. Adapted from Graham *et al.* (1978).

of the plant boiler system, so during months when drying is not done output from the collector can still be used.

The collector elements are evacuated collector tubes similar to the type shown in Figure 6.12.1k and are fitted with vee-shaped specular reflectors behind the tubes. The steady-state thermal characteristics of the collectors are shown in Figure 17.6.2, with the normal range of operating conditions indicated. Tubes are assembled into modules, with each 1.36 m² module including eight tubes and reflectors with appropriate manifold and mounting units. Four hundred and two modules are used for a total collecting area of 548 m². The collector array is mounted on the roof of a warehouse near the building housing the dryers.

17.7 SHALLOW POND WATER HEATERS

Large horizontal collectors consisting of plastic envelopes to contain water, rigid covers supported on concrete curbs, and underlying foam glass and sand insulation have been studied for possible application to provide large quantities

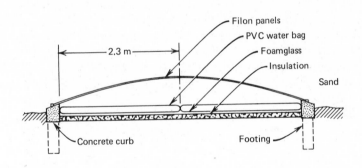

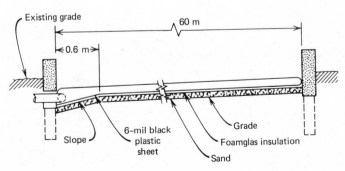

Figure 17.7.1 Schematic cross sections of a shallow pond water heater. Adapted from Guinn and Hall (1977).

of hot water for mineral leaching operations, for providing service hot water to a military base, and for heating water used for various purposes in a poultry processing plant [see Dickenson (1976) and Guinn and Hall (1977)]. A cross section of a collector module of this type is shown in Figure 17.7.1. Dimensions of single collector units are of the order of 3 or 4 m wide and 50 m long, with water depths of the order of 0.1 m. Details of construction of such a collector are provided by Casamajor and Parsons (1979).

The deeper the pond is, the less the temperature will rise and the higher the integrated efficiency. Pond depth is a function of the desired water temperature and cost of energy; it is also influenced by the practical considerations of preparation of large areas of very level ground for support of the collectors. Operation of the shallow pond collectors is usually in a batch mode. In batch operation the collectors are filled in the morning, the temperature rises through the day as energy is absorbed, and when losses exceed absorbed energy the heated water is drained to an insulated storage tank for subsequent use.

A scematic for a system proposed for preheating water for a poultry processing plant is shown in Figure 17.7.2. This plant requires a total of 150,000 liters of water per day, 40 percent of which is for boiler feedwater and 60 percent for can washing and plant cleanup operations. In this design essentially all of the water from the insulated tank is used each day and the collectors are fed with fresh water at supply temperatures each morning.

Collectors of this type have been studied experimentally at Livermore, CA and at Grants, NM, by Dickenson (1976) and Casamajor et al. (1977). Some practical problems have been encountered. The collectors must be very carefully sloped to drain, so as to diminish the amount of water left in them

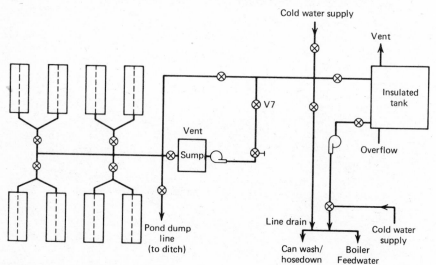

Figure 17.7.2 Schematic of intermittant shallow pond solar water heating system for a poultry processing plant. Adapted from Guinn and Hall (1977).

at night. Snow loads on the wide, flat spans of the covers have resulted in structural problems. The durability of the system is not yet established.

Collectors of this type are restricted to horizontal orientation, so there is strong latitude and seasonal variation of the incident energy on them. The monthly average radiation data from Appendix G are directly applicable to these collectors. Hour-by-hour estimates of the performance of these collectors can be made by the methods of Section 6.11, as illustrated by Example 12.11.1. Long-term performance estimates can be made by the $\bar{\phi}$ method (see Chapter 18).

17.8 SUMMARY

Use of solar energy for industrial process heat represents a range of potential applications that could replace large quantities of fossil fuels. These applications face formidable competition. Industry purchases fuel in large quantities at relatively low cost. These costs are effectively further reduced by their deduction as an operating expense from corporate income with a resulting reduction of corporate income taxes. On the other hand, investment tax credits and depreciation of equipment tend to make the economics of solar applications more favorable. In this chapter we have described several experimental applications of solar industrial process heat. Others could have been included, including low pressure steam applications.

REFERENCES

Casamajor, A. B. and R. E. Parsons, Lawrence Livermore Laboratory Report UCRL 52385 Rev 1 to U.S. Dept. of Energy (1979). "Design Guide for Shallow Solar Ponds."

Casamajor, A. B., T. A. Trautt, and J. M. Flowers, *Proceedings of the Solar Industrial Process Heat Symposium* (1977), p. 141. "Shallow Solar Pond Design Improvements."

Dickenson, W. W., Report SAN/1038-76/1 to NATO Committee on Challenges of Modern Society for U.S. Energy Research and Development Administration (1976). "Performance of the Sohio Solar Water Heating System Using Large Area Plastic Collectors."

Graham, B. J., M. Saarlas, and P. D. Sierer, *Proceedings of the Solar Industrial Process Heat Symposium* (1978), p. 79. "Application of Solar Energy to the Dehydration of Onions."

Guinn, G. R., *Proceedings of the Solar Industrial Process Heat Symposium* (1978), p. 63. "Process Drying of Soybeans Using Heat from Solar Energy."

Guinn, G. R. and B. R. Hall, *Proceedings of the Solar Industrial Process Heat Symposium* (1977), p. 161. "Solar Production of Industrial Process Hot Water Using Shallow Solar Ponds."

Proceedings of the Solar Industrial Process Heat Symposium, College Park, MD (1977). Report No. Conf. 770966. U.S. Govt. Printing Office Stock 061-000-00109-3.

Proceedings of the Solar Industrial Process Heat Conference (1978). Report No. Conf. 781015, available from National Technical Information Service.

Vindum, J. O. and L. P. Bonds, *Proceedings of the Solar Industrial Process Heat Symposium* (1978), p. 15. "Solar Energy for Industrial Process Hot Water."

CHAPTER 18

Thermal Design Methods

In Chapter 14 the f-chart method was presented as a design tool for solar systems that deliver energy to a load at minimum temperatures near 20 C. There are, however, many applications that can use energy at temperatures higher or lower than 20 C. A warehouse heating system may be required to keep the building above freezing so that all energy delivered above 0 C is useful. A solar operated absorption cycle air conditioning system may be able to use all energy above 75 C. Industrial process heat temperature requirements can be at almost any level. This chapter describes design methods for systems for which the f-chart method does not apply.

The design of active solar systems can be done with detailed computer simulations. The expense involved in this type of design can be high but should be considered in the final analysis of all large systems. Preliminary designs and designs for small systems require inexpensive methods for predicting long-term performance. The methods presented here are in this category.

The first method is monthly average hourly utilizability for flat-plate collectors (the ϕ method) as developed by Whillier (1953) and Hottel and Whillier (1958) and later generalized by Liu and Jordan (1963). The second method is due to Klein (1978) and is an extension of hourly utilizability to monthly average daily utilizability ($\bar{\phi}$ method), which significantly reduces the computational effort. Collares-Pereira and Rabl (1979a, b) independently extended hourly utilizability to daily utilizability and included concentrating collectors. Finally, Klein and Beckman (1979) combined daily utilizability with the f-chart concept to account for finite storage capacity ($\bar{\phi}$, f-chart method).

At this point it may appear that design methods are available for all solar-thermal systems, but this is not the case. The utilizability methods all require knowledge of the collector inlet fluid temperature, which is usually not known. The $\bar{\phi}$, f-chart method allows the collector inlet temperature to vary with the storage tank temperature but requires that the load be a closed loop with the fluid returned to the tank at or above a minimum temperature. Also, any device between the solar system and the load must have a conversion efficiency which is independent of the temperature level at which energy is delivered, as long as it is above the minimum temperature. This last requirement rules out solar-to-mechanical systems.

18.1   UTILIZABILITY

Utilizability is defined as the fraction of the incident solar radiation that can be converted to useful heat (i.e., utilized) by a collector having $F_R(\tau\alpha) = 1$ and operating at a fixed inlet to ambient temperature difference. Although the collector has no optical losses and has a heat removal factor of unity, the utilizability is always less than one since the collector does have thermal losses from the top, sides and through the back.*

An analytical expression for utilizability can be derived from Equation 6.7.6, expressed in terms of the hourly radiation incident on the plane of the collector

$$Q_u = A_c F_R[I_T(\tau\alpha) - U_L(T_i - T_a)]^+ \qquad (18.1.1)$$

The radiation level must exceed a critical value before useful output is produced. This critical level is found by setting Q_u in Equation 18.1.1 equal to zero.

$$I_{T,c} = \frac{F_R U_L(T_i - T_a)}{F_R(\tau\alpha)} \qquad (18.1.2)$$

The collector useful output can be expressed in terms of the critical radiation level as

$$Q_u = A_c F_R(\tau\alpha)(I_T - I_{T,c})^+ \qquad (18.1.3)$$

If the critical radiation level is constant for a particular hour (say 10 to 11) for for one month (N days), then the monthly average hourly collector output for this hour is given by

$$\bar{Q}_u = \frac{A_c F_R(\tau\alpha)}{N} \sum^N (I_T - I_{T,c})^+ \qquad (18.1.4)$$

The monthly average radiation in this particular hour is $\bar{I}_T$, so the average useful output can be expressed as

$$\bar{Q}_u = A_c F_R(\tau\alpha)\bar{I}_T \phi \qquad (18.1.5)$$

where, ϕ, the utilizability, is defined as

$$\phi = \frac{1}{N} \sum^N \frac{(I_T - I_{T,c})^+}{\bar{I}_T} \qquad (18.1.6)$$

The procedure for calculating ϕ is illustrated in Figure 18.1.1 for Blue Hill Massachusetts, where a cumulative distribution curve is plotted for the measured radiation incident on a vertical surface in January for the hour pair 11–12 and 12–1. Days are assumed symmetrical about solar noon so hour pairs are used.

* If a collector heat exchanger is present, F_R' can be used in place of F_R.

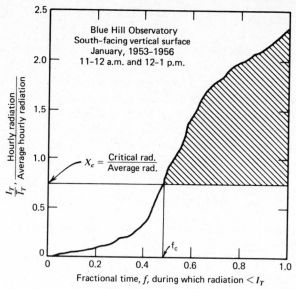

Figure 18.1.1   Cumulative distribution curve for hourly radiation on a south facing vertical surface in Blue Hill, MA. Adapted from Liu and Jordan (1963).

The shaded area represents the fraction of the total time that the radiation level is above a critical radiation ratio, X_c, defined by

$$X_c = \frac{I_{T,c}}{\bar{I}_T} = \frac{F_R U_L (T_i - T_a)}{F_R(\tau\alpha)\bar{I}_T} \tag{18.1.7}$$

In this example the critical radiation ratio is 0.75. Graphical integration of this figure for different ratios yields the result shown in Figure 18.1.2.

Whillier (1953) and later Liu and Jordan (1963) have shown that in a particular location for a 1-month period, ϕ is essentially the same for all hours. Thus, although the curve of Figure 18.1.2 was derived for the hour pair 11 to 12 and 12 to 1, it is valid for all hour pairs in Blue Hill.

The line labeled "limiting curve of identical days" of Figure 18.1.2 represents a cumulative distribution curve that is a horizontal line on Figure 18.1.1 at a value of the ordinate of 1.0. In other words, every day of the month looks like the average day. The difference between the actual ϕ curve and this limiting case represents the error that would be made by using a single average day to represent a whole month.

Example 18.1.1

Calculate the average output of a south facing vertical solar collector in Blue Hill, MA for the month of January when the collector inlet temperature remains at 50 C for the entire month. The collector $F_R(\tau\alpha)$ and $F_R U_L$ are 0.7 and 4.0

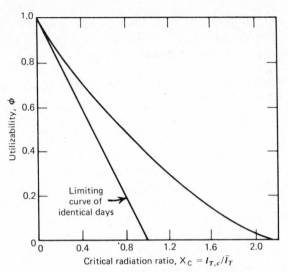

Figure 18.1.2 Utilizability curve derived by numerically integrating Figure 18.1.1 Adapted from Liu and Jordan (1963).

W/m² C, respectively. The average daytime ambient temperature during the month of January in Blue Hill is -2.0 C and the average January solar radiation on a vertical surface is 1.52, 1.15, and 0.68 MJ/m² for the hour pairs 0.5, 1.5, and 2.5 hr from solar noon.

Solution

For simplicity in this example, we will assume $F_R(\tau\alpha)$ is independent of incidence angle. If the incidence angle modifier were known for this collector, the incidence angle and then $F_R(\tau\alpha)$ could be found for each hour pair. The critical radiation level for this collector is found from Equation 18.1.2

$$I_{T,c} = \frac{4.0 \times 3600[50 - (-2)]}{0.7} = 1.07 \text{ MJ/m}^2$$

For the hour pair 11 to 12 and 12 to 1, the dimensionless critical radiation ratio X_c, from Equation 18.1.7, is

$$X_c = \frac{1.07}{1.52} = 0.70$$

and the utilizability, from Figure 18.1.2, is 0.54. The useful output of the collector during this hour is found from Equation 18.1.5.

$$\frac{\bar{Q}_u}{A_c} = 0.7 \times 1.52 \times 0.54 = 0.57 \text{ MJ/m}^2$$

For the pair 10 to 11 and 1 to 2 the value of X_c is 0.93, ϕ is 0.43 and $\bar{Q}_u/A_c$ is 0.35 MJ/m². For the hour pair 9 to 10 and 2 to 3, $\bar{Q}_u/A_c$ is 0.07 MJ/m². The average output for the month of January is then

$$\sum \left(\frac{\bar{Q}_u}{A_c}\right) N = 2(0.57 + 0.35 + 0.07) \times 31 = 61.4 \text{ MJ/m}^2 \qquad \blacksquare$$

18.2 GENERALIZED UTILIZABILITY

To solve Example 18.1.1, it was necessary to have the ϕ curve for a vertical surface in Blue Hill for the month of January. For most locations this type of data is not available, but it is possible to utilize the observed statistical nature of solar radiation to generate ϕ curves that depend only on $\bar{K}_T$, latitude, and collector slope. As mentioned earlier, ϕ curves are nearly independent of the time of day (i.e., of the particular hour-pair from solar noon). It was observed in early studies [e.g., Whillier (1953)] that ϕ curves based on daily totals of solar radiation are also nearly identical to hourly ϕ curves. It is possible to generate ϕ curves from average hourly values of radiation using the methods of Chapter 2 to break daily total radiation into hourly radiation. However, it is easier to generate ϕ curves from daily totals and this is the procedure to be described here.

The radiation data that is most generally available is monthly average daily radiation on a horizontal surface. Thus, with $\bar{K}_T$ and the long-term distribution of days having particular values of K_T from Figure 2.9.2, it is possible to generate sequences of days that represent the long-term average distribution of daily total radiation. The order of occurrence of the days is unknown, but for ϕ curves the order is irrelevant.

For each of these days, the daily total radiation on an inclined collector can be estimated by a procedure similar to that presented in Section 2.16 for monthly average radiation. For a particular day, the radiation on a tilted surface is*

$$H_T = (H - H_d)\bar{R}_b + H_d \frac{1 + \cos\beta}{2} + H\rho \frac{1 - \cos\beta}{2} \qquad (18.2.1)$$

where the monthly average conversion of daily beam radiation on a horizontal surface to daily beam radiation on an inclined surface, $\bar{R}_b$, is used rather than the value for the particular day since the exact date within the month is unknown. The value of $\bar{R}_b$ is found from Equation 2.16.3. If we divide by the monthly average extraterrestrial daily radiation, $\bar{H}_o$, and introduce K_T based on $\bar{H}_o$ (i.e., $K_T = H/\bar{H}_o$), Equation 18.2.1 becomes

$$\frac{H_T}{\bar{H}_o} = K_T \left[\left(1 - \frac{H_d}{H}\right)\bar{R}_b + \frac{H_d}{H} \frac{1 + \cos\beta}{2} + \rho \frac{1 - \cos\beta}{2} \right] \qquad (18.2.2)$$

---

* Section 2.16 is concerned with monthly average daily radiation on a tilted surface. Here we want the average radiation on an inclined surface for all days having a particular value of K_T.

The ratio H_d/H is the daily fraction of diffuse radiation and can be found from Figure 2.11.2 (or Equation 2.11.1) as a function of K_T. Therefore, for each of the days selected from the generalized distribution curve, Equation 18.2.2 can be used to estimate the radiation on a tilted surface. The average of all the days yields the long-term monthly average radiation on the tilted surface so the ratio $H_T/\overline{H}_T$ can be found for each day. The data for the whole month can then be plotted in the form of a cumulative distribution curve, as illustrated in Figure 18.1.1. The ordinate will be daily totals rather than hourly values but, as has been pointed out, the shape of the two curves are nearly the same. Finally, integration of the frequency distribution curve yields a utilizability curve as illustrated in Figure 18.1.2. The process is illustrated in the following example.

Example 18.2.1

Calculate and plot utilizability as a function of the critical radiation ratio for a collector tilted 40° to the south at a latitude of 40°. The month is February and $\overline{K}_T$ is 0.5.

Solution

Since the only radiation information available is $\overline{K}_T$, it will be necessary to generate a ϕ curve from the generalized $\overline{K}_T$ frequency distribution curves. Twenty days, each represented by a K_T from Figure 2.9.2 at $\overline{K}_T = 0.5$, are given in the table below. (Twenty days from the generalized distribution curves are sufficient to represent a month.) For any day with daily total horizontal radiation H and daily diffuse horizontal radiation H_d, the ratio of daily radiation on a south facing tilted surface to extraterrestrial horizontal radiation is found from Equation 18.2.2. For the condition of this problem, $\overline{R}_b$ is 1.79 from Equation 2.16.3. The view factors from the collector to the sky and to the ground are $(1 + \cos \beta)/2 = 0.88$, and $(1 - \cos \beta)/2 = 0.12$. The ground will be assumed to be covered with snow so that ρ is 0.7. Equation 18.2.2 reduces to

$$\frac{H_T}{\overline{H}_o} = K_T \left[1.87 - 0.91 \left(\frac{H_d}{H} \right) \right]$$

For each day in the table below, H_d/H is found from Figure 2.11.2 (or Equation 2.11.1) using the corresponding value of K_T. The results of these calculations are given in columns 2 through 4. The average of column 4 is 0.721. Column 5, the ratio of daily total radiation on a tilted surface, H_T to the monthly average value, $\overline{H}_T$, is calculated by dividing column 4 by its average value. Column 5 is plotted in the first figure below as a function of the day since the data are already in ascending order. The integration, as indicated in this figure, is used to determine the utilizability, ϕ. The area under the whole curve is 1.0. The area above a particular value of $H_T/\overline{H}_T$ is the fraction of the time that the radiation is above this level. For $H_T/\overline{H}_T$ equal to 1.2, the radiation is above this level 13 percent of the time. The utilizability is plotted in the second figure. Although daily totals were used to generate this figure, the hourly ϕ curves will have approximately the same shape. Consequently, the curve can be used in hourly calculations to determine collector performance as illustrated in Example 18.1.1.

1	2	3	4	5
day	K_T	H_d/H	$H_T/\overline{H}_0$	$H_T/\overline{H}_T$
1	0.08	0.99	0.078	0.11
2	0.15	0.99	0.145	0.20
3	0.21	0.95	0.211	0.29
4	0.26	0.92	0.269	0.37
5	0.32	0.87	0.345	0.48
6	0.36	0.82	0.405	0.56
7	0.41	0.76	0.483	0.67
8	0.46	0.68	0.576	0.80
9	0.49	0.62	0.640	0.89
10	0.53	0.55	0.726	1.01
11	0.57	0.47	0.822	1.14
12	0.59	0.43	0.872	1.21
13	0.61	0.39	0.924	1.28
14	0.63	0.36	0.972	1.35
15	0.65	0.33	1.020	1.41
16	0.67	0.30	1.070	1.48
17	0.69	0.27	1.121	1.55
18	0.72	0.24	1.189	1.65
19	0.74	0.23	1.229	1.70
20	0.79	0.21	1.326	1.84

Average = 0.721

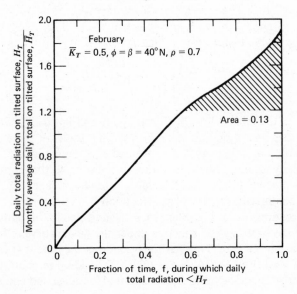

Fraction of time, f, during which daily total radiation $< H_T$

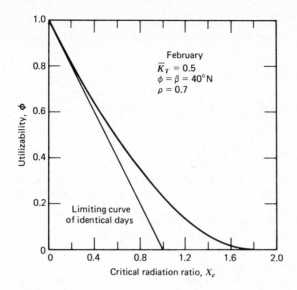

In the preceding example, a ϕ curve was generated from knowledge of the monthly average solar radiation and the known statistical behavior of solar radiation. To calculate average collector output it is necessary to know average *hourly* radiation. The method described in Section 2.13 and illustrated in Example 2.13.3 can be used to determine monthly average hourly radiation from knowledge of monthly average daily radiation (i.e., $\overline{K}_T$). In the next example, useful output is estimated for a particular collector system operating at a constant critical radiation level knowing only the monthly average radiation and ambient temperature.

Example 18.2.2

Estimate the monthly average output of a solar collector for the conditions of Example 18.2.1. The collector $F_R(\tau\alpha)_n$ is 0.7 and $F_R U_L$ is 7.0 W/m² C. The incidence angle modifier coefficient, b_0 (Equation 7.4.3) is 0.17. The average daytime temperature is -5 C and the collector inlet temperature is 30 C. The collector area is 10 m².

Solution

At a latitude of 40°N in February the monthly average daily extraterrestrial radiation is 20.1 MJ/m² and the monthly average declination is $-13.0°$. The sunset hour angle and the day length for February 16, the mean day of the month, are 78.9° and 10.5 hr, respectively. The monthly average ratio of diffuse radiation to daily total, $\bar{H}_d/\bar{H}$, is 0.39 from Figure 2.12.2. For each hour pair from solar noon, the monthly average radiation incident on the collector is given by

$$\bar{I}_T = (\bar{H}r_t - \bar{H}_d r_d)R_b + \bar{H}_d r_d\left(\frac{1 + \cos\beta}{2}\right) + \rho\bar{H}r_t\left(\frac{1 - \cos\beta}{2}\right)$$

or by dividing by $\bar{H}$ and introducing $\bar{H} = \bar{K}_T\bar{H}_0$

$$\bar{I}_T = \bar{K}_T\bar{H}_0\left[\left(r_t - \frac{\bar{H}_d}{\bar{H}}r_d\right)R_b + \frac{\bar{H}_d}{\bar{H}}r_d\left(\frac{1 + \cos\beta}{2}\right) + \rho r_t\left(\frac{1 - \cos\beta}{2}\right)\right]$$

The ratios r_t and r_d are found from Figures 2.13.1 and 2.13.2 for each hour pair and are given in the table below. For the hour pair 11 to 12 and 12 to 1 we have

$$\bar{I}_T = 0.5 \times 20.1[(0.158 - 0.39 \times 0.146)1.62 + 0.39 \times 0.146 \times 0.88$$
$$+ 0.7 \times 0.158 \times 0.12] = 2.28\,\text{MJ/m}^2$$

where R_b is found from Equation 1.7.2. The remaining hour pairs are treated in a similar manner. The results are given in the table. The average transmittance-absorptance product will be obtained using only the beam component for simplicity. From Equation 7.4.3, the ratio of $(\tau\alpha)/(\tau\alpha)_n$ for the first pair is

$$\frac{(\tau\alpha)}{(\tau\alpha)_n} = 1 - 0.17\left(\frac{1}{\cos 15} - 1\right) = 0.99$$

The results for other hour pairs are given in the table. The critical radiation ratio for the hour pair 11 to 12 and 12 to 1 is then found from Equation 18.1.7

$$X_c = \frac{F_R U_L(T_i - T_a)}{F_R(\tau\alpha)_n \times (\tau\alpha)/(\tau\alpha)_n \times \bar{I}_T} = \frac{7 \times 3600 \times 35}{0.7 \times 0.99 \times 2.28 \times 10^6} = 0.56$$

and ϕ is then 0.51 from the Figure of Example 18.2.1. The average energy output for this hour is then found from Equation 18.1.5.

$$\bar{Q}_u = A_c F_R(\tau\alpha)\bar{I}_T\phi = 10 \times 0.7 \times 0.99 \times 2.28 \times 10^6 \times 0.51 = 8.1\,\text{MJ}$$

The monthly average collector outputs for the remaining hour pairs are given in the table. Finally, the total output for the month is

$$N\sum\bar{Q}_u = 28 \times 2(8.1 + 6.4 + 3.7 + 0.5) = 1.05\,\text{GJ}$$

Hour	θ_T	R_b	r_t	r_d	$\bar{I}_T$	$\dfrac{(\tau\alpha)}{(\tau\alpha)_n}$	X_c	ϕ	Q_u
11–12, 12–1	15	1.62	0.158	0.146	2.28	0.99	0.56	0.51	8.1
10–11, 1–2	26	1.65	0.140	0.134	2.04	0.98	0.63	0.46	6.4
9–10, 2–3	39	1.73	0.109	0.111	1.62	0.95	0.82	0.34	3.7
8–9, 3–4	54	1.91	0.069	0.075	1.08	0.88	1.33	0.08	0.5
7–8, 4–5	68	2.63	0.026	0.035	0.47	0.72	3.72	0	0.0

■

Liu and Jordan (1963) have generalized the calculations of Example 18.2.1. They found that the shape of the ϕ curves was not strongly dependent on the ground reflectance or the view factors from the collector to the sky and ground. Consequently, they were able to construct a set of ϕ curves for a fixed value of $\overline{K}_T$. The effect of tilt was taken into account by using the monthly average ratio of beam radiation on a tilted surface to monthly average beam radiation on a horizontal surface, $\overline{R}_b$, as a parameter. The *generalized ϕ curves* are shown in Figure 18.2.1 for values of $\overline{K}_T$ of 0.3, 0.4, 0.5, 0.6, and 0.7. The method of constructing these curves is exactly like Example 18.2.1 except that the tilt used in their calculations was 47° and the ground reflectance was 0.2. A comparison of the ϕ curve from Example 18.2.1, in which the tilt was 40° and the ground reflectance was 0.7, with the generalized ϕ curve for $\overline{K}_T = 0.5$ and $\overline{R}_b = 1.79$ shows that the two are nearly identical.

With the generalized ϕ curves, it is possible to predict the output of any collector that operates at a constant critical level by knowing only the long-term average radiation and ambient temperature. This procedure was illustrated in Example 18.2.2. Rather than use the ϕ curve calculated in Example 18.2.1, the generalized ϕ curve could have been used. The only additional calculation is determining $\overline{R}_b$ so that the proper ϕ curve can be selected.

ϕ curves are very powerful design tools, but they can be misused. They cannot be directly applied to most conventional solar heating systems since the critical level varies considerably during the month due to the finite storage capacity. Two limiting cases of solar heating do fit into the ϕ curve restrictions; air heating systems in midwinter and annual storage systems. For the case of air systems, the inlet air temperature to the collector during much of the winter will be the return air from the house. This is because little excess energy is available for storage and what is stored is effectively stratified by the rock bed. Only in the spring and fall will the collector inlet temperature rise much above the room temperature. In the case of annual (long-term) storage, the storage tank temperature varies slowly during a month so that a monthly average tank temperature, and therefore critical level, can be found by trial and error. This is illustrated in the next section. Certain industrial process heat applications can also be analyzed with ϕ curves. The point to be remembered is that the critical level must be a constant in order to use this design method.*

The ϕ charts graphically illustrate why a single average day should not be used to predict system performance under most conditions. The line on each ϕ chart labeled "limiting curve of identical days" is the ϕ curve that would be obtained if all days in a month were identical. Only if $\overline{K}_T$ is high or if the critical level is very low do all ϕ curves approach this limit. For most situations the error in using one average day to predict performance is substantial.

---

* Actually, the critical level can vary from hour to hour as long as it is the same variation for each day of the month. It is unlikely that many systems behave in such a manner.

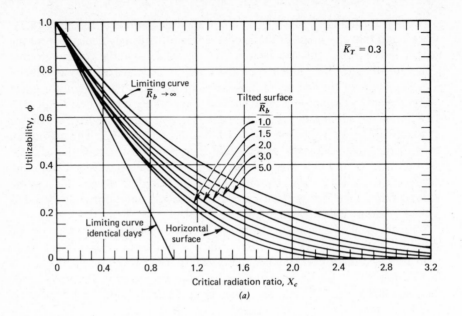

(a)

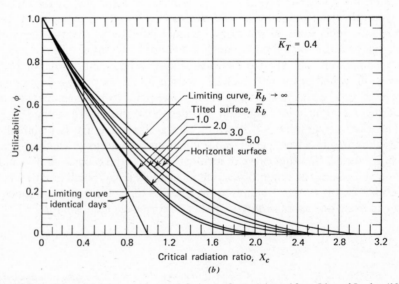

(b)

Figure 18.2.1 Generalized ϕ curves for south facing surfaces. Adapted from Liu and Jordan (1963).

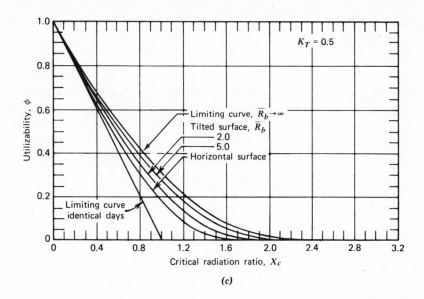

(c)

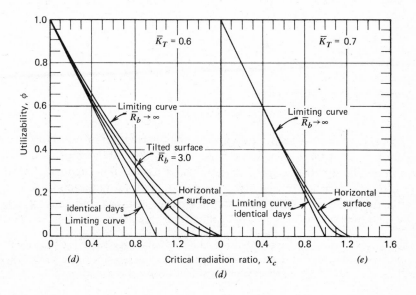

(d)

Example 18.2.3

Repeat Example 18.2.2 but use the generalized ϕ-charts.

Solution

Everything in Example 18.2.2 remains the same except that ϕ is found from one of the graphs of Figure 18.2.1 rather than the Figure of Example 18.2.1. Since $\bar{K}_T$ is 0.5, Figure 18.2.1c must be used. From Example 18.2.1, $\bar{R}_b$ is 1.79 so that values of ϕ must be obtained by interpolation between $\bar{R}_b = 2$ and the curve for a horizontal surface. Values of ϕ are essentially the same as found in Example 18.2.2.

18.3 DAILY UTILIZABILITY

Although the ϕ curves of the previous section were derived from daily totals of radiation, they must be used hourly. This means that 3 to 6 hourly calculations must be performed to obtain the useful output for a month. The amount of calculation is significant and led Klein (1978) to develop the concept of monthly average daily utilizability, $\bar{\phi}$.

Daily utilizability is defined as the sum for a month, over all hours and all days, of the radiation on a tilted surface that is above a critical level, divided by the monthly radiation. In equation form

$$\bar{\phi} = \sum_{\text{days}} \sum_{\text{hours}} \frac{(I_T - I_{T,c})^+}{\bar{H}_T N} \tag{18.3.1}$$

where the critical level, $I_{T,c}$ is similar to that defined by Equation 18.1.2 except that monthly average transmittance-absorptance must be used in place of $(\tau\alpha)$ and T_i and $\bar{T}_a$ are representative inlet and daytime temperatures for the month.

$$I_{T,c} = \frac{F_R U_L (T_i - \bar{T}_a)}{F_R \overline{(\tau\alpha)}} \tag{18.3.2}$$

The value of $\overline{(\tau\alpha)}$ can be found from Equation 5.9.2. The monthly average daily useful energy gain is then given by

$$\sum \bar{Q}_u = A_c F_R \overline{(\tau\alpha)} \bar{H}_T \bar{\phi} \tag{18.3.3}$$

The value of $\bar{\phi}$ for a month depends on the distribution of hourly values of radiation in the month. If it is assumed that all days are symmetrical about solar noon and that the hourly distributions are as shown in Figures 2.13.1 and 2.13.2, then $\bar{\phi}$ depends on the distribution of daily total radiation, that is, the relative frequency of occurrence of below average, average, and above average daily radiation values.* Figure 18.3.1 illustrates this point. The days in the top

---

* Klein assumed symmetrical days in his development of $\bar{\phi}$. It can be shown that departure from symmetry within days (e.g., if afternoons are brighter than mornings) will lead to increases in $\bar{\phi}$; thus a $\bar{\phi}$ calculated from the correlations of this section are somewhat conservative.

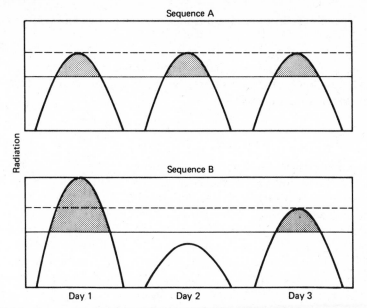

Figure 18.3.1 Two sequences of days with the same average radiation. Two critical radiation levels on the plane of the collector are indicated. From Klein (1978).

sequence are all average days; for the low critical radiation level represented by the solid horizontal line, the shaded areas show useful gain, whereas for the high critical level represented by the dotted line, there is no collection. The bottom sequence shows three days of varying radiation with the same average as before; collection for the low critical radiation level is significantly higher than before, and there is collection above the high critical radiation level. Thus the effect of increasing variability of days is to increase $\bar{\phi}$, particularly at high critical radiation levels.

The monthly distribution of daily total radiation is a unique function of $\bar{K}_T$, as shown by Figure 2.9.2. Thus the effect of daily radiation distribution on $\bar{\phi}$ is related to a single variable, $\bar{K}_T$.

Klein has developed correlations for $\bar{\phi}$ as a function of $\bar{K}_T$ and two variables, a geometric factor $\bar{R}/R_n$ and a dimensionless critical radiation level, $\bar{X}_c$.

$\bar{R}/R_n$ is the ratio of $\bar{R}$, the monthly ratio of radiation on a tilted surface to radiation on a horizontal surface, to R_n, the noon ratio of radiation on the tilted surface to that on a horizontal surface for an average day of the month. $\bar{R}$ is found from equation 2.16.1. R_n is calculated by the methods of Section 2.15 and for the hour around solar noon can be written as*

$$R_n = \left(1 - \frac{r_{d,n}H_d}{r_{t,n}H}\right)R_{b,n} + \left(\frac{r_{d,n}H_d}{r_{t,n}H}\right)\left(\frac{1 + \cos\beta}{2}\right) + \rho\left(\frac{1 - \cos\beta}{2}\right) \quad (18.3.4)$$

* R_n is for a particular day that has a daily total radiation equal to the monthly average daily total radiation. It is not the monthly average value of R at noon.

where, $r_{t,n}$ and $r_{d,n}$ can be obtained from Equations 2.13.1 and 2.13.3 evaluated at solar noon.

The monthly average critical radiation ratio, $\bar{X}_c$, is the ratio of the critical radiation, $I_{T,c}$, to the noon radiation level for a day of the month in which the total radiation for the day is the same as the monthly average. Writing this ratio for the noon hour, we have

$$\bar{X}_c = \frac{I_{T,c}}{r_{t,n}R_n\bar{H}} = \frac{F_R U_L(T_i - \bar{T}_a)/F_R(\bar{\tau\alpha})}{r_{t,n}R_n\bar{K}_T\bar{H}_o} \tag{18.3.5}$$

Klein obtained $\bar{\phi}$ as a function of $\bar{X}_c$ for various values of $\bar{R}/R_n$ by the following process. For a given $\bar{K}_T$, a sequence of daily radiation totals is determined that has the correct long-term average distribution. (This is the process illustrated in Example 18.2.1.) The radiation in each of the days in this sequence is divided into hours by the information in Section 2.13. These hourly values of beam and diffuse radiation are used to find the total hourly radiation on an inclined surface. Then, critical radiation levels are subtracted from these hourly estimates and summed according to Equation 18.3.1 to arrive at $\bar{\phi}$. $\bar{\phi}$ curves calculated in this manner are shown in Figures 18.3.2a–e, for $\bar{K}_T$ of 0.3 to 0.7. The curves can be represented by the equation

$$\bar{\phi} = e^{[a + b(R_n/\bar{R})][\bar{X}_c + c\bar{X}_c^2]} \tag{18.3.6}$$

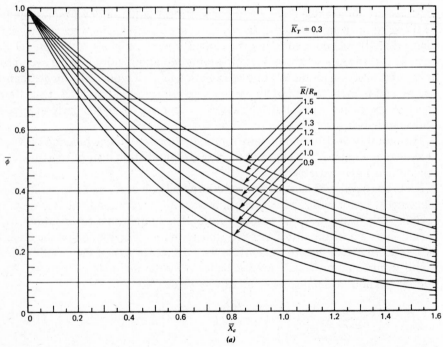

Figure 18.3.2   (a) Monthly average daily utilizability as a function of $\bar{X}_c$ and $\bar{R}/R_n$. From Klein (1978).

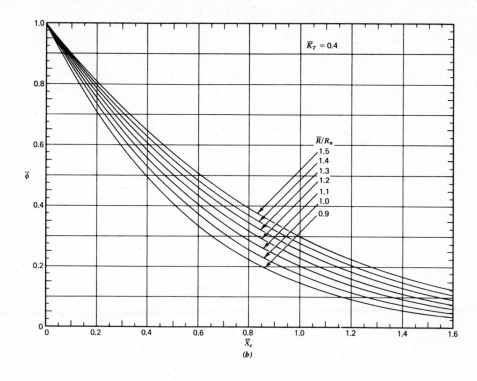

(b)

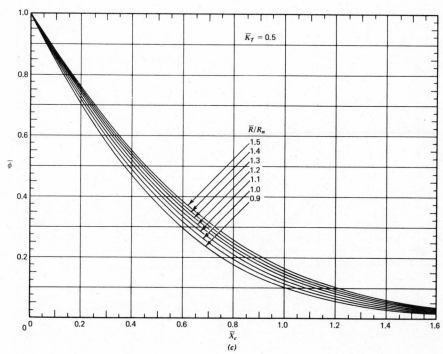

(c)

615

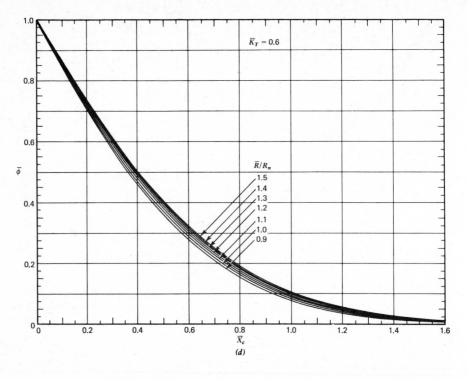

(d)

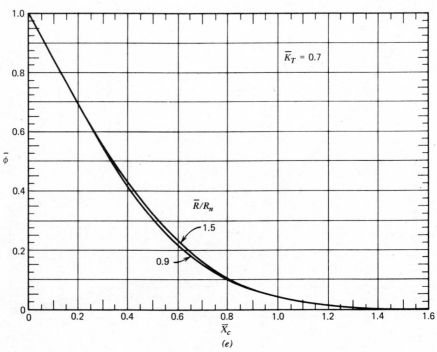

(e)

where $a = 2.943 - 9.271\overline{K}_T + 4.031\overline{K}_T^2$
  $b = -4.345 + 8.853\overline{K}_T - 3.602\overline{K}_T^2$
  $c = -0.170 - 0.306\overline{K}_T + 2.936\overline{K}_T^2$

In the next example, a problem involving seasonal storage will be used to illustrate the $\overline{\phi}$ method.

Example 18.3.1

Estimate the annual performance of a liquid solar heating system in Madison having a very large storage volume (seasonal storage). The collector area is 50 m², and is tilted 60° to the south. The storage tank contains 250,000 liters of water. The losses from the tank are to the ground at an average temperature of 7 C. The tank area-loss coefficient product is 30 W/C. The house UA is 200 W/C and the characteristics of the collector-heat exchanger-piping combination are $F_R'(\tau\alpha)_n = 0.78$, $F_R' U_L = 4.55$ W/m² C. The monthly average weather data are given in the table below.

Solution

Since the storage tank is very large, its temperature will not change significantly through each month so that an average collector inlet temperature can be assumed and the $\overline{\phi}$ method can be used. An energy balance on the tank for a one month period yields

$$(mC_p\Delta T)_{\text{tank}} = \text{Collector output} - \text{Tank losses} - \text{Energy to load}$$

The procedure is to guess a tank temperature at the beginning of March. March is used as the beginning of the year since it usually has the lowest tank temperature and is therefore easiest to guess. Once the tank temperature is known at the beginning of the month, the tank temperature at the end of the month is guessed. The average tank temperature for the month is used to calculate the collector output and tank losses. The tank energy balance equation is used to determine the tank temperature at the end of the month. This calculated temperature is compared with the assumed value and if they agree, the calculation continues with the next month. If they disagree, another monthly average temperature is estimated and a new final tank temperature is calculated. The procedure is repeated for all 12 months and the final tank temperature is compared to the initial guess made for March. If they agree, the calculations are complete; if they disagree, the calculations must be repeated. The details of the March calculations are given below and the results for all 12 months are given in the table. The house load is found from Equation 14.3.1

$$(UA) \times (\text{Degree Days}) = 200 \times 3600 \times 24 \times 599 = 10.4 \text{ GJ}$$

Month	ρ_g	$\bar{T}_a$, C	DD, C day	$\bar{H}$, MJ/m^2	$\bar{K}_T$	$\bar{R}$	R_n	$\bar{X}_c$	$\bar{\phi}$	$N \sum \bar{Q}_u$, GJ	Q_{loss}, GJ	Q_{load}, GJ	T, C
Mar.	0.4	2	599	12.89	0.500	1.24	1.13	0.26	0.66	12.2	1.6	10.4	27.4
Apr.	0.2	9	328	15.87	0.475	0.91	0.93	0.23	0.69	11.1	1.7	5.7	27.6
May	0.2	15	165	19.78	0.507	0.78	0.86	0.21	0.70	12.5	2.2	2.9	31.2
June	0.2	21	40	22.11	0.535	0.73	0.83	0.22	0.68	12.3	2.8	0.7	38.3
July	0.2	23	8	21.95	0.547	0.75	0.85	0.27	0.61	11.7	3.5	0.1	46.7
Aug.	0.2	22	22	19.38	0.545	0.85	0.93	0.34	0.54	10.3	4.0	0.4	54.4
Sept.	0.2	17	96	14.75	0.518	1.07	1.04	0.45	0.45		4.2	1.7	60.1
Oct.	0.2	12	263	10.34	0.500	1.45	1.20	0.55	0.40	6.9	4.4	4.5	62.0
Nov.	0.2	3	505	5.72	0.395	1.84	1.20	0.96	0.33	3.8	3.8	8.7	60.1
Dec.	0.4	-3	742	4.41	0.374	2.17	1.29	1.00	0.37	4.2	3.2	12.8	51.8
Jan.	0.7	-5	830	5.85	0.442	2.19	1.46	0.58	0.47	7.0	2.3	14.3	40.6
Feb.	0.7	-4	696	9.12	0.490	1.72	1.37	0.36	0.58	9.6	1.6	12.0	3.13

The tank losses depend upon the monthly average tank temperature, which is unknown. An average tank temperature of 27.5 C will be assumed (and later checked).

$$(UA)_{tank} \times (\overline{T}_{tank} - T_{ground}) = 30 \times 3600 \times 24 \times 31 \times (27.5 - 7) = 1.6 \text{ GJ}$$

The monthly average daily useful energy gain will be estimated first using the assumed 27.5 C tank temperature. The monthly average daily radiation on the collector surface is 16.0 MJ/m² from Example 2.16.1. The monthly average ratio $(\overline{\tau\alpha})/(\tau\alpha)_n$ will be assumed to be a constant at 0.96. The critical radiation level $I_{T,c}$ is found from Equation 18.3.2.

$$I_{T,c} = \frac{4.55(27.5 - 2)}{0.78 \times 0.96} = 155 \text{ W/m}^2$$

Here the daytime average ambient temperature was estimated to be 2 C higher than the 24 hr average. For the average March day (March 16) the sunset hour angle is 87.7 from Equation 1.6.7 so that $r_{t,n}$ and $r_{d,n}$ are 0.146 and 0.134 from Equations 2.13.1 and 2.13.3, respectively. The ratio H_d/H is 0.60 from Figure 2.11.2. From Equation 18.3.4 the factor R_n is then

$$R_n = \left(1 - \frac{0.134 \times 0.60}{0.146}\right)1.38 + \frac{0.134 \times 0.60}{0.146}\left(\frac{1 + \cos 60}{2}\right)$$

$$+ 0.4\left(\frac{1 - \cos 60}{2}\right) = 1.13$$

where the factor $R_{b,n}$ was found from Equation 1.7.2 evaluated at noon. $\overline{R}$, from Example 2.16.1, is 1.24 so that $\overline{R}/R_n$ is 1.24/1.13 = 1.10. From Equation 18.3.5, the dimensionless critical level is

$$\overline{X}_c = \frac{155 \times 3600}{0.146 \times 1.13 \times 12.89 \times 10^6} = 0.26$$

Using Equation 18.3.6, with $\overline{K}_T = 0.50$, $a = -0.685$, $b = -0.819$, and $c = 0.411$, $\overline{\phi}$ is 0.66. The average monthly energy gain is found from multiplying Equation 18.3.3 by the number of days in the month.

$$N \sum \overline{Q}_u = 31 \times 50 \times 0.78 \times 0.96 \times 1.24 \times 12.89 \times 10^6 \times 0.66 = 12.2 \text{ GJ}$$

The drop in tank temperature during the month of March is then

$$\Delta T = (12.2 - 1.6 - 10.4) \times \frac{10^9}{4190 \times 0.25 \times 10^6} = 0.20$$

The tank temperature started the month at 27.4 and it ends at 27.6. The average tank temperature for the month is then 27.5 C, which is the value chosen as the average tank temperature to begin the March calculations. This calculation process is continued for each month and the results are shown in the table. At the end of February the tank temperature is back to 27.4 C. Note that late in the winter when the tank temperature is low the heat exchanger between the

tank and the house may limit the useful energy that can be extracted from the
tank.   ■

The $\bar{\phi}$ depend on $\bar{R}$ and R_n, which in turn depend on the division of total
radiation into beam and diffuse components. As noted in Section 2.11, there are
substantial uncertainties in determining this division. The correlation of
$\bar{H}_d/\bar{H}$ versus $\bar{K}_T$ of Liu and Jordan (1960) were used by Klein (1978) to generate
the $\bar{\phi}$ charts. The correlation of Ruth and Chant (1977), which indicates significantly higher fractions of diffuse radiation, was also used to generate $\bar{\phi}$
charts and the results were not significantly different from those of Figures
18.3.2. A ground reflectance of 0.2 was used in generating the charts, but a value
of 0.7 was also used and it made no significant difference. Consequently, even if
the diffuse to total correlation is changed as a result of new experimental
evidence, the $\bar{\phi}$ curves will remain valid. Of course, using different correlations
will change the predictions of radiation on a tilted surface which will change the
performance estimates.

18.4   THE $\bar{\phi}$, f-CHART METHOD

The utilizability design concept is useful whenever the collector operates at a
known critical radiation level throughout a month. In a more typical situation
the collector is connected to a tank so that the monthly sequence of weather
and the load time distribution result in a fluctuating storage tank temperature
and consequently a variable critical radiation level. The f-chart method was
developed to overcome the restriction of constant critical level but it is limited
to systems which deliver energy to a load near 20 C.

In this section, the $\bar{\phi}$ concept is combined with the f-chart idea to produce
a design method for closed-loop solar systems shown in Figure 18.4.1. In these
systems energy supplied to the load must be above a specified minimum useful
temperature and it must be used at a constant coefficient of performance or
thermal efficiency so that the load on the solar system can be calculated.

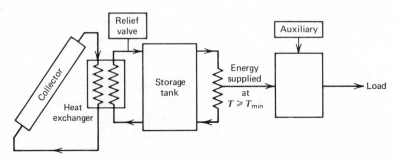

Figure 18.4.1   Schematic of a closed-loop solar energy system.

(If the load is, for example, a heat engine that performs better as the temperature increases, then the thermal load on the solar system cannot be calculated from knowledge of the heat engine output.) The storage tank is assumed to be pressurized so that energy dumping does not occur. The return temperature from the load is always at or above T_{min}. A separate auxiliary system is in parallel with the solar system and makes up any energy deficiency of the solar system.

The maximum monthly average daily energy that can be delivered by such a system is given by

$$\sum \overline{Q}_{max} = A_c F_R(\overline{\tau\alpha}) \overline{H}_T \overline{\phi}_{max} \tag{18.4.1}$$

which is similar to Equation 18.3.3 except that $\overline{\phi}$ is replaced by $\overline{\phi}_{max}$. The maximum daily utilizability is calculated from the minimum monthly average critical radiation ratio

$$\overline{X}_{c,min} = \frac{F_R U_L (T_{min} - \overline{T}_a)/F_R(\overline{\tau\alpha})}{r_{t,n} R_n \overline{H}} \tag{18.4.2}$$

The method of calculating $\overline{\phi}_{max}$ is exactly as illustrated in Example 18.3.1.

For a particular storage size to collector area ratio, Klein and Beckman (1979) correlated the results of many detailed simulations of the system of Figure 18.4.1 with two dimensionless variables. These variables are similar to the f-chart variables but are not exactly the same. The f-chart ordinate, Y, is replaced by $\overline{\phi}_{max} Y$

$$\overline{\phi}_{max} Y = \overline{\phi}_{max} \frac{A F_R(\overline{\tau\alpha}) \overline{H}_T N}{L} \tag{18.4.3}$$

and the f-chart abscissa X is replaced by a modified variable, X'

$$X' = \frac{A F_R U_L (100) \Delta\tau}{L} \tag{18.4.4}$$

The change in X is that $(100 - T_a)$ has been replaced by an empirical constant 100 (or 180 if English units are used).

Figures 18.4.2a through d present $\overline{\phi}$, f-charts for four different storage volumes to collector area ratios. The information in these figures can be represented analytically by

$$f = \overline{\phi}_{max} Y - 0.015(e^{3.85f} - 1)(1 - e^{-0.15X'})(R_s)^{0.76} \tag{18.4.5}$$

where R_s is the ratio of the standard storage heat capacity per unit of collector area of 350 kJ/m^2 C to the actual storage capacity. Although f is given implicitly by Equation 18.4.5 it is easy to solve by Newton's method or by trial and error.

The $\overline{\phi}$, f-charts are used in the same manner as the f-charts. Values of $\overline{\phi}_{max} Y$ and X' are calculated from long-term radiation data and load patterns for the location in question. The value of f is then determined from the figures or from Equation 18.4.5. The product of f times L is the average monthly contribution

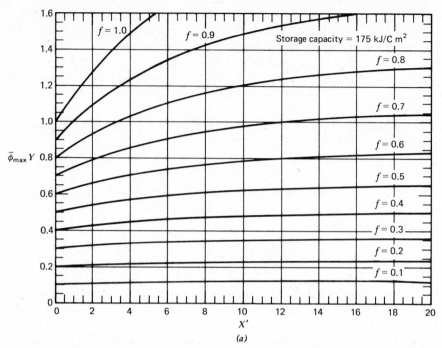

Figure 18.4.2 $\bar{\phi}, f$-charts for various storage capacities. From Klein and Beckman (1979).

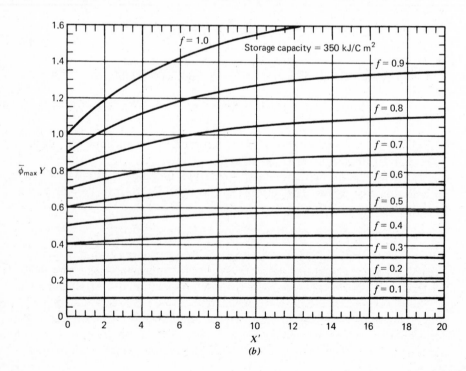

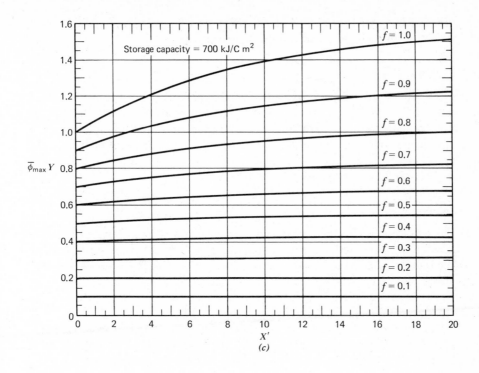

(c)

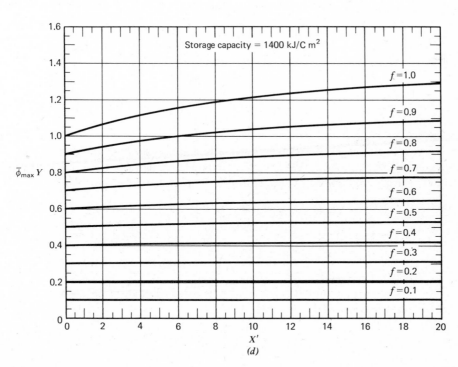

(d)

supplied by solar energy. The calculations are repeated for each month from which the annual fraction $\mathscr{F}$ can be determined.

Example 18.4.1

An industrial solar energy system in Omaha, Nebraska (latitude 41°) requires energy above 60 C at a rate of 12 KW for a 12-hr period each day. The average ambient temperature and the monthly average daily radiation on a horizontal surface are given in the table. The collector-heat exchanger characteristics are $F'_R(\alpha)_a = 0.72$, $F'_R U_L = 2.63$ W/m² C, $F_R = 0.8$. Tilt = 40°, $\gamma = 0$, area = 50 m². The storage is 4180 liters of water. What is F?

Solution

The detailed calculations for January will be illustrated. Values of intermediate calculations for all months are given in the table. The radiation incident on the tilted surface is calculated by the methods of Section 2.16. The ground reflectance will be assumed to be 0.2. $\bar{H}_0$ is 14.4 MJ/m² for January so $\bar{K}_T$ is 0.60 and $\bar{H}_d/\bar{H}$ is 0.30 from Equation 2.12.2. From Equation 2.16.1, $\bar{R}$ is then 1.91 and $r_{t,n}$ and $r_{d,n}$ are 0.178 and 0.164 from Equations 2.13.1 and 2.13.3.

The beam radiation conversion factor at noon, $R_{b,n}$, is 1.97 from Equation 1.7.2. For an average day with $K_T = 0.6$, the ratio of diffuse to total radiation H_d/H is 0.41 from Figure 2.11.2. From Equation 18.3.4, the noon ratio of radiation on a tilted surface to that on a horizontal surface for the average day of the month is

$$R_n = \left(1 - \frac{0.164 \times 0.41}{0.178}\right)1.97 + \frac{0.164 \times 0.41}{0.178}\left(\frac{1 + \cos 40}{2}\right)$$

$$+ 0.2\left(\frac{1 - \cos 40}{2}\right) = 1.59$$

The ratio $\bar{R}/R_n$ is then 1.20.

The monthly average transmittance-absorptance product divided by the normal incidence value is 0.94 from the method given in Section 5.9. The critical level at the minimum useful temperature is found from Equation 18.4.2.

$$\bar{X}_c = \frac{\dfrac{2.63[60 - (-5)]}{0.72 \times 0.94}}{0.178 \times 1.59 \times 8.6 \times \dfrac{10^6}{3600}} = 0.37$$

The value of $\bar{\phi}_{max}$ is found from Figure 18.3.2 or Equation 18.3.6 with $a = -1.17$, $b = -0.330$, and $c = 0.704$

$$\bar{\phi}_{max} = \exp\left[\left(-1.17 - \frac{0.33}{1.20}\right)(0.37 + 0.704 \times 0.37^2)\right] = 0.51$$

1 Average Day	2 $\bar{H}_o$, MJ/m²	3 $\bar{T}_a$, °C	4 $\bar{H}$, MJ/m²	5 $\bar{R}_b$	6 $\bar{K}_T$	7 $\frac{H_d}{\bar{H}}$	8 $\bar{R}$	9 $r_{t,n}$	10 $r_{d,n}$	11 $R_{b,n}$	12 R_n	13 $\bar{R}/R_n$	14 $\frac{(\bar{\tau\alpha})}{(\tau\alpha)_n}$	15 $\bar{X}_c$ (min)	16 $\bar{\phi}$ (max)	17 Y	18 $\bar{\phi}Y$ (max)	19 X'	20 f	21 L GJ	22 fL GJ
Jan. 17	14.4	−5	8.6	2.32	0.60	0.30	1.91	0.178	0.164	1.97	1.59	1.20	0.94	0.37	0.51	1.07	0.55	2.19	0.53	16.1	8.5
Feb. 16	19.8	−3	11.6	1.83	0.59	0.33	1.54	0.162	0.148	1.66	1.37	1.12	0.94	0.34	0.54	1.17	0.63	2.19	0.60	14.5	8.6
Mar. 16	26.7	3	14.9	1.40	0.56	0.37	1.23	0.146	0.134	1.38	1.18	1.04	0.93	0.31	0.58	1.18	0.69	2.19	0.65	16.1	10.4
Apr. 15	34.0	10	19.4	1.08	0.57	0.40	1.02	0.132	0.121	1.16	1.07	0.95	0.93	0.26	0.63	1.28	0.81	2.19	0.74	15.6	11.6
May 15	39.2	17	21.5	0.89	0.55	0.44	0.91	0.123	0.112	1.03	0.99	0.92	0.91	0.24	0.66	1.24	0.81	2.19	0.74	16.1	13.0
June 11	41.3	22	23.6	0.81	0.57	0.44	0.87	0.118	0.109	0.98	0.96	0.91	0.90	0.21	0.69	1.28	0.89	2.19	0.80	15.6	12.5
July 17	40.2	25	23.8	0.84	0.59	0.42	0.88	0.120	0.110	1.00	0.98	0.90	0.91	0.18	0.73	1.32	0.97	2.19	0.85	16.1	13.7
Aug. 16	36.0	23	21.8	0.99	0.61	0.38	0.97	0.128	0.117	1.10	1.05	0.92	0.92	0.18	0.73	1.35	0.99	2.19	0.87	16.1	13.9
Sept. 15	29.3	19	16.6	1.26	0.57	0.38	1.14	0.140	0.128	1.29	1.14	1.00	0.93	0.22	0.69	1.22	0.84	2.19	0.76	15.6	11.9
Oct. 15	21.8	12	12.3	1.67	0.56	0.36	1.41	0.156	0.143	1.55	1.28	1.10	0.94	0.27	0.64	1.13	0.72	2.19	0.67	16.1	10.8
Nov. 14	15.7	4	8.3	2.17	0.53	0.35	1.75	0.174	0.159	1.88	1.40	1.25	0.93	0.39	0.53	0.94	0.50	2.19	0.48	15.6	7.5
Dec. 10	13.0	−2	7.0	2.50	0.54	0.33	1.99	0.184	0.169	2.10	1.53	1.30	0.94	0.44	0.48	0.91	0.44	2.19	0.43	16.1	6.8
																			$\sum$	189.6	129.2

The value of Y is found from Equation 18.4.3

$$Y = \frac{50 \times 0.72 \times 0.94 \times 1.91 \times 8.6 \times 10^6 \times 31}{12{,}000 \times 12 \times 3600 \times 31} = 1.07$$

and $\bar{\phi}_{max} Y$ is then $0.51 \times 1.07 = 0.55$.

From Equation 18.4.4 the value of X' is

$$X' = \frac{50 \times 2.63 \times 100 \times 31 \times 24}{12{,}000 \times 12 \times 31} = 2.19$$

and remains constant for the whole year. The storage capacity per unit of collector area is

$$\frac{MC_p}{A_c} = \frac{4180 \times 4190}{50 \times 1000} = 350 \text{ kJ/C m}^2$$

so that Figure 18.4.2b can be used to find f. Alternatively, Equation 18.4.4 can be used but a trial and error solution is necessary. With X' and $\bar{\phi}_{max} Y$ equal to 2.19 and 0.55 respectively, f is 0.53. The load for the month of January supplied by solar is then $0.53 \times 16.1 = 8.5$ GJ. The other months are shown in the table. The annual fraction by solar is then the sum of column 22, the energy from solar, divided by the sum of column 21, the yearly load.

$$\mathscr{F} = \frac{129.2}{189.6} = 0.68 \qquad\qquad \blacksquare$$

The $\bar{\phi}$, f-chart calculations overestimate f due to the assumptions that there are no losses from the tank and that the load heat exchanger is infinite in size. Corrections can be applied to eliminate both of these assumptions.

The rate at which energy is lost from the storage tank to the surroundings at T'_a is given by

$$\dot{Q}_{tank} = (UA)_T (T_{tank} - T'_a) \qquad\qquad (18.4.6)$$

If T'_a and $(UA)_T$ are both constant for a month, then integration of Equation 18.4.6 over a month yields

$$Q_{tank} = (UA)_T (\overline{T}_{tank} - T'_a)\Delta\tau \qquad\qquad (18.4.7)$$

where $\overline{T}_{tank}$ is the monthly average tank temperature.

The total load on the solar system is the useful load plus the energy required to meet the tank losses. If the tank losses are modest so that the tank seldom drops below the minimum temperature, then the solar system cannot tell the difference between the energy withdrawn to supply the load and the tank losses. In equation form, the fraction of the total load supplied by solar, including the tank losses, is

$$f_{TL} = \frac{Q_s + Q_{tank}}{L + Q_{tank}} \qquad\qquad (18.4.8)$$

where Q_s is the solar energy supplied to the load. Once the Q_{tank} is known, f_{TL}, can be calculated from the $\bar{\phi}$, f-charts in exactly the same manner as illustrated in Example 18.4.1. The usual interpretation for fraction of the load supplied by solar energy is the ratio Q_s/L, the solar energy supplied to the load divided by the useful load. If we use the symbol f for this fraction, Equation 18.4.8 can be written as

$$f = f_{TL}\left(1 + \frac{Q_{tank}}{L}\right) - \frac{Q_{tank}}{L} \qquad (18.4.9)$$

The tank losses cannot be calculated exactly but two limiting values can be determined which should bracket the actual losses. A low estimate for tank losses is to assume the tank remains at T_{min} all month. An upper bound for tank losses is to assume the average tank temperature is the same as the monthly average, collector inlet temperature, $\bar{T_i}$. The actual average tank temperature will be lower than this value since the collector does not operate 24 hr a day. An estimate for $\bar{T_i}$ can be found using the $\bar{\phi}$ charts. The average daily utilizability is

$$\bar{\phi} = \frac{f_{TL}}{Y} \qquad (18.4.10)$$

With this value of $\bar{\phi}$, the monthly average operating critical level can be found from the $\bar{\phi}$ charts and $\bar{T_i}$ can then be found. Klein and Beckman (1979) recommend that the arithmetic average of T_{min} and $\bar{T_i}$ be used to evaluate tank losses from Equation 18.4.7. A more conservative estimate is to use $\bar{T_i}$ for estimating tank losses.

The process is iterative. An estimate is first made of the monthly average tank temperature from which Q_{tank} is found from Equation 18.4.7. This estimate of tank losses is used as part of the total load and the $\bar{\phi}$, f-chart method is used to estimate f_{TL}. $\bar{\phi}$ is then calculated from Equation 18.4.10 and $\bar{X}_c$ is found from the $\bar{\phi}$ charts. This critical level is used to find $I_{T,c}$ and $\bar{T_i}$ from Equations 18.3.5 and 18.3.2. The tank temperature is compared with the initial guess and the process is repeated if necessary. With the final value of Q_{tank}, Equation 18.4.9 is used to find f, the fraction of the useful load supplied by solar.

Example 18.4.2

Consider the solar system of example 18.4.1 but include the effect of tank losses. The tank has a loss coefficient-area product of 5.9 W/C and tank losses are to a temperature of 20 C.

Solution

Only the month of January will be considered in this example. For January the average tank temperature will be assumed to be 62 C so that the tank losses, from Equation 18.4.7, are

$$Q_{tank} = 5.9(62 - 20) \times 3600 \times 24 \times 31 = 0.7 \text{ GJ}$$

The total load is then $16.1 + 0.7 = 16.8$ GJ. The values of $\bar{\phi}_{max} Y$ and X' are then $16.1/16.8$ times the values from Example 18.4.1

$$\bar{\phi}_{max} Y = 0.55 \times 16.1/16.8 = 0.53$$

$$X' = 2.19 \times 16.1/16.8 = 2.10$$

From the $\bar{\phi}, f$ charts (or Equation 18.4.5) we obtain

$$f_{TL} = 0.51$$

From Equation 18.4.10, $\bar{\phi}$ is then

$$\bar{\phi} = 0.51/1.03 = 0.49$$

and $\bar{X}_c$ is 0.39 from Figure 18.3.2 or Equation 18.3.6 (by trial and error). Since the original value of $\bar{X}_c$ was 0.37, the temperature difference of 65 C must be increased by the ratio of 0.39/0.37. The estimate of $\bar{T}_i$ is then

$$\bar{T}_i = 65 \times \frac{0.39}{0.37} - 5 = 63.5$$

and the average tank temperature is estimated to be

$$\frac{63.5 + 60}{2} = 61.7$$

which is close to the initial guess so no iterations are necessary. The fraction by solar is then found from Equation 18.4.9

$$f = 0.51\left(1 + \frac{0.7}{16.1}\right) - \frac{0.7}{16.1} = 0.49$$

The tank losses have reduced the fraction of the load by solar from 51 to 49 percent and results in $0.49 \times 16.1 = 7.9$ GJ being supplied by solar rather than 8.5 GJ. ∎

The load heat exchanger adds thermal resistance between the storage tank and the load. This resistance elevates the storage tank temperature which results in reduced collector useful energy collection and increased tank losses. In the development of the $\bar{\phi}, f$-charts, the load heat exchanger was assumed to be infinite in size and consequently the value of f will be optimistic and a correction is necessary.

The average rate of solar energy supplied to the load is found by dividing Q_s by the number of seconds during the month in which the load was required, $\Delta\tau_L$. The average increase in the tank temperature necessary to supply the required energy rate is

$$\Delta T = \frac{Q_s/\Delta\tau_L}{\varepsilon_L C_{min}} = \frac{fL/\Delta\tau_L}{\varepsilon_L C_{min}} \tag{18.4.11}$$

where ε_L is the load heat exchanger effectiveness and C_{min} is the smaller of the two fluid capacity rates in the heat exchanger. This temperature difference is added to T_{min} to find the monthly average critical radiation ratio from Equation 18.4.2. Since f is unknown at the beginning, it is necessary to first estimate ΔT, then follow the procedure illustrated in Example 18.4.2 to find f. This value of f is used in Equation 18.4.11 to check the estimate of ΔT. The calculations are illustrated in the following example.

Example 18.4.3

Include the effect of a load heat exchanger on the performance of the system described in Examples 18.4.1 and 18.4.2. The heat exchanger effectiveness is 0.45 and the minimum capacitance rate is 3000 W/C.

Solution

From Example 18.4.1 for January, $R_n = 1.59$, $\bar{R}/R_n = 1.20$, $r_{t,n} = 0.178$, $\bar{K}_T = 0.6$, $\bar{H} = 8.6$ MJ/m^2, $\bar{T}_a = -5$ C, and $(\overline{\tau\alpha})/(\tau\alpha)_n = 0.94$. As a first estimate of ΔT we will use 4 C so from Equation 18.4.2 the minimum critical radiation level is

$$\bar{X}_{c,min} = \frac{\dfrac{2.63[60 + 4 - (-5)]}{0.72 \times 0.94}}{0.178 \times 1.59 \times 8.6 \times \dfrac{10^6}{3600}} = 0.40$$

From Figure 18.3.2 or Equation 18.3.6

$$\bar{\phi}_{max} = 0.48$$

Since we wish to also consider tank losses, a guess of the tank temperature is necessary to determine the total load. With a tank temperature of 66 C, $Q_{tank} = 0.7$ MJ. The total load is $16.1 + 0.7 = 16.8$ MJ and Y is then

$$Y = \frac{50 \times 0.72 \times 0.94 \times 1.91 \times 8.6 \times 10^6 \times 31}{16.8 \times 10^9} = 1.03$$

Thus $\bar{\phi}_{max} Y$ is $0.48 \times 1.03 = 0.49$. The value of X' is

$$X' = \frac{50 \times 2.63 \times 100 \times 31 \times 24 \times 3600}{16.8 \times 10^9} = 2.10$$

From Equation 18.4.5 or Figure 14.4.2b

$$f_{TL} = 0.47$$

We must now check the tank loss approximation by evaluating $\bar{X}_c$ at $\bar{\phi} = 0.47/1.03 = 0.46$. From Figure 18.3.2d or Equation 18.3.6 $\bar{X}_c$ is 0.42. From Equation 18.4.2

$$\bar{T}_i - \bar{T}_a = \frac{0.42 \times 0.178 \times 1.59 \times 8.6 \times 10^6 \times 0.72 \times 0.94}{2.63 \times 3600} = 73$$

so that $\overline{T}_i = 73 - 5 = 68$. The average temperature for tank losses is then

$$\overline{T} = \frac{64 + 68}{2} = 66$$

Since this is the same as the guess, no iteration is necessary for tank losses. From Equation 18.4.9, the fraction by solar is then

$$f = 0.47\left(1 + \frac{0.7}{16.1}\right) - \frac{0.7}{16.1} = 0.45$$

This value of f is used in Equation 18.4.11 to find ΔT

$$\Delta T = \frac{0.45 \times 12000}{0.45 \times 3000} = 4\ C$$

Since this is the same as the initial guess, the calculations for January are complete. The load by solar is then $0.45 \times 16.1 = 7.2$ GJ. ■

It is interesting to compare the results of the last three example problems for January. With no tank losses or load heat exchanger, the contribution by solar is 8.5 GJ. With tank losses considered, this was reduced to 7.9 GJ and with the addition of the load heat exchanger, the energy supplied by solar is 7.2 GJ. These are not insignificant reductions.

At present it is impossible to compare the $\overline{\phi}$, f-chart predictions with experiments since no long-term data are available. However, it is possible to make comparisons with detailed performance predictions from a simulation program such as TRNSYS.

Table18.4.1 Comparison of TRNSYS, f-chart, and $\overline{\phi}$, f-chart results for $T_{\min} = 20$ C[a]

	Space Heating Annual Solar Load Fractions		
	TRNSYS	f chart	$\overline{\phi}$, f-chart
Albuquerque, NM (1959)	0.79	0.78	0.81
Blue Hill, MA (1958)	0.49	0.50	0.52
Boulder, CO (1956)	0.67	0.68	0.72
Madison, WI (1948)	0.45	0.47	0.47
Medford, OR (1969)	0.55	0.53	0.56
Seattle, WA (1960)	0.57	0.56	0.59

[a] From Klein and Beckman (1979).

Table 18.4.2 Solar load fractions for a process heating application ($T_{min} = 60$ C)^a

	Albuquerque, NM		New York, NY	
	TRNSYS	$\bar{\phi}, f$-chart	TRNSYS	$\bar{\phi}, f$-chart
Jan.	0.89	0.86	0.77	0.76
Feb.	0.96	1.00	0.65	0.67
Mar.	0.80	0.82	0.58	0.63
Apr.	0.90	1.00	0.49	0.52
May	0.75	0.78	0.30	0.32
June	0.66	0.64	0.20	0.16
July	0.69	0.66	0.26	0.21
Aug.	0.74	0.71	0.39	0.37
Sept.	0.95	0.93	0.73	0.68
Oct.	0.92	0.89	0.58	0.56
Nov.	0.91	0.90	0.45	0.40
Dec.	0.94	0.93	0.66	0.66
Year	0.84	0.84	0.50	0.49

^{a} From Klein and Beckman (1979).

In Table 18.4.1, the results for space heating systems, which delivers energy to the load with a minimum temperature of 20 C, are compared in six climates. Also, the f-chart results are presented. The estimates from all three methods are in good agreement.

In Table 18.4.2, the monthly results of an industrial process heating example are compared for two climates. The process uses energy above a minimum temperature of 60 C at a constant rate between 6 a.m. and 6 p.m., 7 days a week. The agreement between TRNSYS and $\bar{\phi}, f$ chart results is excellent in both locations. Differences between the TRNSYS results and $\bar{\phi}, f$-chart results on a monthly basis are expected since TRNSYS results are affected by energy carry over from month to month and the actual radiation data used in the TRNSYS simulations may be too small a sample to represent the long-term statistical distribution.

The $\bar{\phi}, f$-chart method has potential for misuse and as a result its limitations need to be emphasized. The method is intended for applications in which the load can be characterized by a single temperature, T_{min}. The load must be relatively uniform on a day-to-day basis and will produce inaccurate results for processes in which the load distribution is highly irregular. The method is not applicable for systems in which energy supplied to the load is used at an efficiency or COP that depends upon temperature, such as solar-Rankine engine.

18.5 SUMMARY

In this chapter three design methods have been presented: ϕ; $\bar{\phi}$; and $\bar{\phi}$, f-chart. The ϕ method can usually be replaced by the $\bar{\phi}$ method with approximately a fourfold reduction in calculation. The hourly and daily utilizability methods require that the collector critical radiation level remain constant for a month, which means that the collector inlet temperature is fixed. In practice this requires that the storage tank is either very large or nonexistent with the collector inlet fluid from a constant temperature source. Neither of these two situations are very common. The $\bar{\phi}$, f-charts were developed for closed-loop systems with finite storage and where the load is characterized by a single minimum useful temperature. This is a common system but it does not cover all practical applications. Systems that are not covered by these methods must be designed using detailed simulations.

REFERENCES

Collares-Pereira, M. and A. Rabl, *Solar Energy*, **23**, 235 (1979a). "Simple Procedure for Predicting Long Term Average Performance of Nonconcentrating and of Concentrating Solar Collectors."

Collares-Pereira, M. and A. Rabl, *Solar Energy*, **23**, 223 (1979b). "Derivation of Method for Predicting Long Term Average Energy Delivery of Solar Collectors."

Hottel, H. C. and A. Whillier, *Transactions of the Conference on the Use of Solar Energy*, **2**, part I, 74, University of Arizona Press, (1958). "Evaluation of Flat-Plate Collector Performance."

Klein, S. A., *Solar Energy*, **21**, 393 (1978). "Calculation of Flat-Plate Collector Utilizability."

Klein, S. A. and W. A. Beckman, *Solar Energy*, **22**, 269 (1979). "A Generalized Design Method for Closed-Loop Solar Energy Systems."

Liu, B. Y. H. and R. C. Jordan, *Solar Energy*, **7**, 53 (1963). "A Rational Procedure for Predicting the Long-Term Average Performance of Flat-Plate Solar-Energy Collectors."

Ruth, D. W., and R. E. Chant, *Solar Energy*, **18**, 153 (1976). "The Relationship of Diffuse Radiation to Total Radiation in Canada."

Whillier, A., Sc.D. Thesis (Mechanical Engineering), Massachusetts Institute of Technology (1953). "Solar Energy Collection and Its Utilization for House Heating."

CHAPTER 19

Conversion to Mechanical Energy

Conversion of solar energy to mechanical energy has been the objective of experiments for at least a century. In 1872, Mouchot exhibited a steam powered printing press at the Paris Exposition; in 1902 the Pasadena Ostrich Farm system was publicly exhibited; and in 1913, a solar operated irrigation plant started its brief period of operation at Meadi, Egypt. These and other similar developments utilized concentrating collectors to supply steam to heat engines, producing output when beam solar radiation was high enough. An interesting historical review of these experiments is provided by Jordan and Ibele (1956).

In this chapter we briefly describe some of these conversion processes where the end objective is the production of mechanical energy. The processes are similar to other solar energy processes, and the principles treated in earlier chapters on radiation, collectors, storage, and systems form the basis for estimating the performance of solar thermal power systems.

Much of the early attention to solar thermal-mechanical systems was for small-scale applications, with outputs ranging up to 100 KW and with water pumping as the end objective. In the last two decades the possibilities of operation of vapor compression air conditioners have been studied analytically and experimentally (see Section 16.8), and interest has grown in much larger systems, for generation of substantial quantities of electrical energy.

Our concern is with generation of mechanical energy from solar thermal energy, which is one of three possible solar-mechanical processes. Others are photovoltaic processes with subsequent operation of electric motors, and solar-biological processes that produce fuels for operation of conventional engines or power plants.

19.1 THERMAL CONVERSION SYSTEMS

The basic thermal process for conversion of solar to mechanical energy is shown in a simplified schematic in Figure 19.1.1. Energy is collected by either flat-plate or concentrating collectors, stored (if appropriate), and then used to

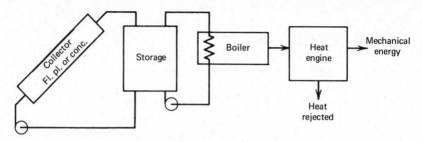

Figure 19.1.1   Schematic of a solar thermal conversion system.

operate a heat engine. Problems of these systems lie in the fact that the efficiency of a collector diminishes as its operating temperature rises, while the efficiency of the engine rises as its operating temperature rises. The maximum operating temperatures for flat-plate collectors are low relative to desirable input temperatures for heat engines, and system efficiencies are low if flat-plate collectors are used.

Hottel (1955) has pointed out that the availability of cover materials with high transmittance for solar radiation may make it possible to consider the use of flat-plate collectors to supply energy to heat engines. Recent developments in collector loss control by honeycombs, CPCs, concentrating collectors, or by use of evacuated tubular collectors may open up additional systems for thermal conversion of solar to mechanical energy.

Methods of estimation of system performance are based on analysis of the components in the system. The analysis of the collectors has been covered in Chapters 6 through 8. If arrays are large, two factors must be taken into account. First, if collector modules are arranged in series, the methods of Section 7.6 can be used to calculate the effective $F_R U_L$ and $F_R(\tau\alpha)$. Second, large arrays inevitably mean long runs of piping, and the methods of Section 10.3 can be used or adapted to account for these losses in a relatively easy fashion. If pipe runs are very long, their heat capacity and that of the fluid in them can represent an appreciable factor in the energy balances during transient operation.

Models are also needed for engines. These can be derived from basic thermodynamic relationships, or they can be graphical or analytical representations of operating data. Olson (1977) has developed a method of sizing engine components for design conditions for a reciprocating expander engine based on user-supplied design parameters. The engine parameters, plus properties of the working fluid, are then used to calculate a "performance map" for the engine that can be used to estimate engine performance for off-design conditions.

A sample of such a map for a reciprocating expander with Refrigerant 114 as the working fluid is shown in Figure 19.1.2. The design parameters are as follows.

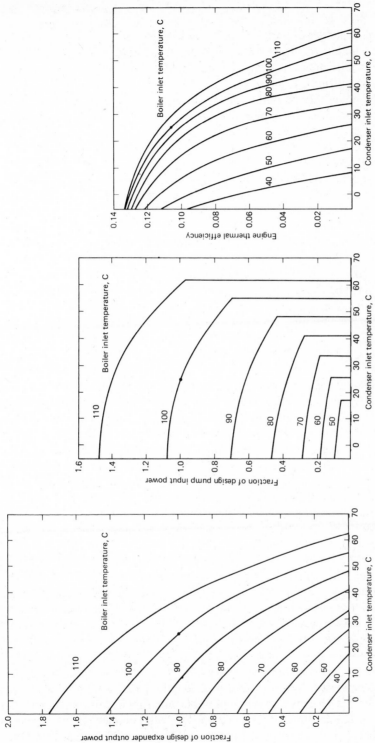

Figure 19.1.2 Performance map for a reciprocating engine using Refrigerant 114. (*a*) Expander output. (*b*) Pump energy requirements. (*c*) Engine thermal efficiency. All as functions of inlet temperatures to the boiler and condenser. From Olson (1977). Dots indicate design point.

635

1   The expander output power = 2.24 kW (3 hp).
2   The boiler water side inlet temperature = 100 C.
3   The boiler liquid to liquid section mean overall heat transfer coefficient
    = 454 W/m^2 C.
4   The boiler inlet to outlet water side temperature drop = 5.6 C.
5   The boiler outlet temperature difference = 2.8 C.
6   The condenser air inlet temperature = 27 C.
7   The condenser air to vapor section mean overall heat transfer coefficient
    = 312 W/m^2 C.
8   The condenser inlet to outlet air temperature rise = 1.7 C.
9   The condenser outlet temperature difference = 5.6 C.
10 The expander shaft rotational speed = 3600 rpm.

The diagrams indicate engine output, pump input energy and engine thermal efficiency, all as functions of the temperatures of the fluids entering the boiler and the condenser. With this information, the operation of this engine in a system can be determined by solving these relationships simultaneously with the equations for the other components.

19.2   THE GILA-BEND PUMPING SYSTEM

The Gila-Bend solar irrigation pumping system described by Alexander et al. (1979) is shown schematically in Figure 19.2.1. It is based on an array of cylindrical parabolic collectors (pictured in Figure 8.7.1) and is designed to produce hot water at 150 C (pressurized at about 8 atm). The hot water goes to a heat exchanger which serves as a Freon boiler where its temperature drops about 9 C in vaporizing the Refrigerant 113 working fluid. The hot water then goes through the preheater where it preheats the fluid returning to the boiler. The water then returns to the collector at about 135 C. There is no thermal energy storage in this system. Water is pumped into the irrigation canals whenever possible, and storage is in the form of water at the desired higher elevation.

The working fluid vapor goes from the boiler through an entrainment separator and to a 50 hp turbine. The turbine drives a 630 liters/s (10,000 gpm) pump through a gearbox. After leaving the turbine, the vapor goes through a regenerator to partially preheat the boiler feed liquid, and then to the condenser where it is condensed at approximately 32 C by part of the irrigation water. Condensate is returned by a feed pump to the boiler via the regenerator and preheater.

The collector array includes nine rows of concentrating collectors, each 24.4 m long, with an aperture of 2.45 m and focal length of 0.91 m, for a total collecting area of 537 m^2. The rows are rotated about horizontal north-south axes. The receivers are copper tubes 41 mm in diameter, with nonselective black coatings, and insulated on the side away from the concentrator. First year experiments

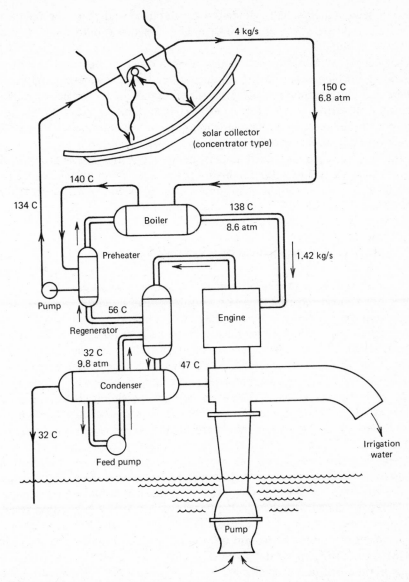

Figure 19.2.1 Schematic of the Gila-Bend solar pumping system. Design flow and conditions are shown. Adapted from Alexander *et al.* (1979).

were done with a glass half-cylinder cover, and in second year experiments the cover was removed. Each of the nine rows of collectors is individually oriented to track the sun.

The report of Alexander et al. describes a range of practical design and operating problems that have been encountered in this experimental system. These include maintenance of specular reflectance of the reflector, tracking accuracy on various kinds of days, maintenance of mechanical and electronic components, integration of the solar energy system with an electric drive system to increase the use factor on the pump, and others.

This system, even though it underwent frequent and prolonged developmental changes, operated for 323 hr in the first year, at an average capacity of 240 liters/s and a maximum of 570 liters/s. In its second year, it operated for 188 hr and delivered 1.24×10^5 m^3 of irrigation water.

19.3 CENTRAL RECEIVER POWER SYSTEMS

In Chapter 8, note was taken of some of the optical properties of large "power tower" or central receiver type concentrators. The potential applications of these collectors is in large scale conversion processes with outputs in the megawatt range. Large scale systems must collect energy from large areas. In these systems beam radiation from a large array of (relatively) small heliostats is focused on a central receiver, thus collecting by optical rather than thermal means. A photograph of an experimental system is shown in Figure 8.8.1.

An early paper on a central receiver concept, by Baum (1956), outlined a design in which heliostats were to be mounted on semicircular railway tracks, with the assemblies of mirrors moved through the day to keep beam radiation focused on the central receiver. More recent design concepts call for heliostats on mounts at fixed locations, and movable about two axes of rotation to accomplish concentration. Major research and development efforts in the United States are aimed at solving the range of optical, thermal, and mechanical problems associated with the development of electric power generation systems based on these concepts. These efforts are reported in the Proceedings of the Department of Energy Workshop on Central Solar Thermal Electric Systems (1978). A review of general design considerations is provided by Vant-Hull in a paper in the workshop proceedings.

The major components in the system are the heliostat field, the heliostat controls, the receiver, the storage system, and the heat engine which drives the generator. Heliostat design concepts were briefly outlined in Section 8.9. The objective of heliostat design is to deliver radiation to the receiver at the desired flux density at minimum cost. For an extensive discussion of optical problems, see the Proceedings of the ERDA Workshop on Optical Analysis (1977).

A range of receiver shapes has been considered, including cavity receivers and cylindrical receivers. The optimum shape is a function of intercepted radiation, thermal losses, receiver cost, and the design of the heliostat field. Vant-Hull suggests that for a large heliostat field a cylindrical receiver has advantages when used with Rankine cycle engines, particularly for radiation from heliostats at the far edges of the field. If higher temperatures are required for operation of Brayton cycle turbines, it may be necessary to use cavity receivers with larger tower height/heliostat field area ratios.

It has been observed in many solar power studies that the solar collector represents the largest cost in the system. Under these circumstances, an efficient engine is justified to obtain maximum useful conversion of the collected energy. Several possible thermodynamic cycles have been considered. First, Brayton or Stirling gas cycle engines operate at inlet temperatures of 800 to 1000 C, provide high engine efficiencies, but are limited by low gas heat transfer coefficients, by the need for recouperators, and by the practical constraints on collector design (i.e., the need for cavity receivers) imposed by the requirements of 1000 C temperatures. Second, turbines driven from steam generated in the receiver would operate at 500 to 550 C and have several advantages over the Brayton cycle. Heat transfer coefficients in the steam generator are high, allowing the use of high energy flux densities and smaller receivers with energy absorption on the outer surface. Cavity receivers are not needed, and cylindrical receivers permit larger heliostat fields to be used. Use of reheat cycles would improve steam turbine performance but entails mechanical design problems. It is also possible to use steam turbines with steam generated from an intermediate heat transfer fluid circulated through collector and boiler. The fluids could be molten salts or liquid metals, and cylindrical receivers could be operated at around 600 C with such systems. These indirect systems are the only ones that readily lend themselves to the use of storage. Low-temperature steam or organic fluid turbines may also be used, with collector temperatures around 300 C in receivers with lower solar radiation flux densities.

Electrical power generation by central receiver systems can provide daytime power with little or no storage. Until such systems can be built and operated on a very large scale they cannot be expected to replace base-load generating capacity. The major storage problem is one of providing sufficient storage capacity to keep the turbine-generator going during a few hours, to provide continuity of operation during intermittent clouds or to provide capacity during early evening peak load periods. In a system using molten salts or liquid metals, storage techniques may be analogous to those used in lower temperature operations, but with temperatures in the range of 500 to 600 C.

REFERENCES

Alexander, G., D. F. Busch, R. D. Fischer, and W. A. Smith, Battelle Columbus Laboratories report SAND 79–7009 (1979). "Final Report on the Modification and 1978 Operation of the Gila-Bend Solar-Powered Irrigation Pumping System."

Baum, V. A., paper in *Proceedings of the World Symposium on Applied Solar Energy*, Stanford Research Institute, Menlo Park, CA., p. 289 (1956). "Prospects for the Application of Solar Energy, and Some Research Results in the USSR."

Hottel, H. C., paper in *Solar Energy Research*, University of Wisconsin Press, Madison, p. 85 (1955). "Power Generation with Solar Energy."

Jordan, R. C. and W. C. Ibele, paper in *Proceedings of the World Symposium on Applied Solar Energy*, Stanford Research Institute, Menlo Park, CA, p. 81 (1956). "Mechanical Energy from Solar Energy."

Olson, T. J., M.S. Thesis, Mech. Engr., University of Wisconsin-Madison (1977). "Solar Source Rankine Cycle Engines for Use in Residential Cooling."

Proceedings of the ERDA Solar Workshop on Methods for Optical Analyses of Central Receiver Systems, University of Houston—Sandia Laboratories (1977). Available from the U.S. Dept. of Commerce.

Proceedings of the 1978 DOE Workshop on Systems Studies for Central Solar Thermal Electric, University of Houston—Sandia Laboratories (1978). Available from the U.S. Dept. of Commerce.

CHAPTER 20

Evaporative Processes and Salt-Gradient Ponds

In this final chapter we note three evaporative processes, distillation of salt water to produce fresh water, evaporation of salt brines to produce salt, and drying to remove moisture from solids. A related process, the salt gradient pond, is discussed in the final section. Evaporation for salt production and drying of crops are many centuries old and are the basis of commercial and agricultural enterprises in many parts of the world. A substantial body of know-how and design information has been developed on evaporation and drying, and we only note the principles involved. Distillation developments have occurred over the past century and are treated in some detail. In all cases our concern is with direct distillation, evaporation or drying in which solar energy is absorbed in the apparatus where evaporation occurs rather than in a separate collector.

20.1 SOLAR DISTILLATION

The first known application of solar distillation was in 1872 when a still at Las Salinas on the northern deserts of Chile started its three decades of operation to provide drinking water for animals used in nitrate mining. The still utilized a shallow black basin to hold the salt water and absorb solar radiation; water vaporized from the brine, condensed on the underside of a sloped transparent cover, ran into troughs, and was collected in tanks at the end of the still. Most stills built and studied since then have been based on the same concepts, though many variations in geometry, materials, methods of construction and operation have been employed. This section is concerned with these basin type stills. Other configurations have been used, based on evaporation from wicks or from brines cascading over weirs, to allow other than horizontal energy absorbing surfaces. A very comprehensive summary of literature on all aspects of solar distillation is provided by Talbert et al. (1970).

The basin type still is shown in section in Figure 20.1.1. A still may have many bays side by side, each of the type shown. The covers are usually glass; they may also be air-supported plastic films. The basin may be on the order of 10 to 20 mm deep (referred to as shallow basins) or they may be 100 mm or more deep (referred to as deep basins). The widths are of the order of 1 to 2 m, with lengths widely variable up to 50 to 100 m.

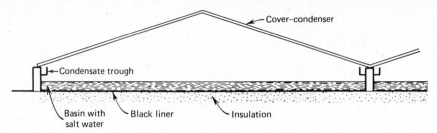

Figure 20.1.1 Schematic cross section of a basin type solar still.

In operation, solar radiation is transmitted through the cover and absorbed by the salt water and basin. The solution is heated, water evaporates and vapor rises to the cover by convection where it is condensed on the under side of the cover. Condensate flows by gravity into the collection troughs at the lower edges of the cover; the covers must be at sufficient slope that surface tension of the water will cause it to flow into the troughs without dropping back into the basin. The trough is constructed with enough pitch along its length so that condensate will flow to the lower end of the still, where it drains into a product collection system. Operation of a still may be continuous or batch. If sea water (approximately 3.5 percent salt) is used as feed, the concentration is usually allowed to double before the brine is removed, so about half of the water in the feed is distilled off.

Figure 20.1.2 shows the major energy flows in a still while it is operating. The objective of still design is to maximize Q_{evap}, the transport of absorbed solar radiation to the cover-condenser by water vapor, as this is directly proportional

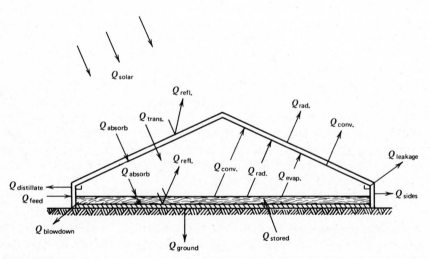

Figure 20.1.2 The major energy transport mechanisms in a basin type still. From Talbert *et al.* (1970).

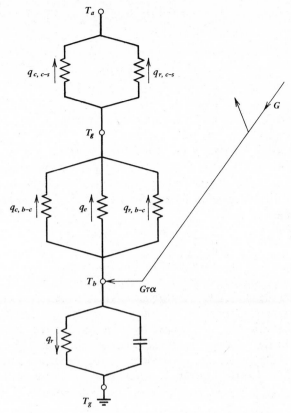

Figure 20.1.3   Basic thermal network for a basin type still.

to still productivity. All other energy transfer from basin to surroundings should be suppressed, as far as is possible. Most energy flows can be evaluated from basic principles, but terms such as leakage and edge losses are difficult to quantify and may be lumped together in a catch-all term determined experimentally for a particular still.

The basic concepts of solar still operation have been set forth by Dunkle (1961). The thermal network for a basin type still is similar in principle to Figure 6.4.1 for a flat plate collector, but with three differences. Energy transfer from basin to cover occurs by evaporation-condensation, in addition to convection and radiation. The losses from the back of the still are to the ground. The depth of the water in the still is usually such that its capacitance must be taken into account. There is no useful energy gain in the sense of that of a flat-plate collector; still output is measured by the evaporation-condensation transfer from basin to cover. A thermal network is shown in Figure 20.1.3, where the resistances correspond to the energy flows in Figure 20.1.2. Terms for leakage, edge losses, entering feedwater, and leaving brine or product are not shown.

An energy balance on the water in the basin (and the basin itself), per unit area of basin, can be written

$$G\tau\alpha = q_e + q_{r,b-g} + q_{c,b-g} + q_k + (mC_p)_b \frac{dT_b}{d\tau} \qquad (20.1.1)$$

where the subscripts $e, r, c,$ and k represent evaporation-condensation, radiation, convection, and conduction, respectively. The subscripts b and g refer to basin, and glazing (cover).

The capacitance of the glazing will normally be small compared to that of the water and basin. In most modern still designs the pitch of the covers is small, so the cover area is approximately the same as the basin area. An energy balance on the cover, neglecting its capacitance and solar energy absorbed by it, can be written as

$$q_e + q_{r,b-g} + q_{c,b-g} = q_{c,g-a} + q_{r,g-a} \qquad (20.1.2)$$

These two equations are analogous to Equations 6.11.1 and 6.11.2 for flat-plate collectors. q_e is not a linear function of the temperature difference between the plate and the glazing, and the two equations need to be solved simultaneously to find T_b, T_c, and q_e as functions of time.

Dunkle (1961) provides convenient ways of estimating the terms for internal heat transfer in the still for use in these equations. The radiation exchange between basin and cover, $q_{r,b-g}$ is calculated by Equation 3.8.3. The cover is usually glass, and during operation a thin layer of condensate forms on most of its under surface. Dunkle recommends that the term be written in the form

$$q_{r,b-g} = 0.9\sigma(T_b^4 - T_g^4) \qquad (20.1.3)$$

For estimating the convection energy transfer from basin to cover, $q_{c,b-g}$, he suggests that the normal Rayleigh number must be modified to account for bouyancy effects due to the fact that heat and mass transfer occur simultaneously. The buoyancy term in the Grashof number is modified by the density gradient caused by the composition gradient (in addition to the temperature gradient). In a horizontal enclosed air gap, a relationship between Nusselt and Rayleigh numbers is

$$\text{Nu} = 0.075(\text{Ra})^{1/3} \qquad (20.1.4)$$

where the temperature difference in the Rayleigh number, $\Delta T'$, is an equivalent temperature difference accounting for density differences due to water vapor concentration differences. For air and water

$$\Delta T' = (T_b - T_g) + \left(\frac{p_{wb} - p_{wg}}{2016 - p_{wb}}\right)T_b \qquad (20.1.5)$$

where p_{wb} and p_{wg} are the vapor pressures in mm Hg of the solution in the basin at T_b and of water at the cover temperature T_g. Temperatures are in degrees Kelvin.

From Equations 20.1.4 and 20.1.5 the convection coefficient in a still is

$$h'_c = 0.884\left[T_b - T_g + \left(\frac{p_{wb} - p_{wg}}{2016 - p_{wb}}\right)T_b\right]^{1/3} \qquad (20.1.6)$$

and the heat transfer between the basin and cover is

$$q_{c,b-g} = h'_c(T_b - T_g) \qquad (20.1.7)$$

By analogy between heat and mass transfer, the mass transfer rate can be written

$$\dot{m}_D = 9.15 \times 10^{-7}\, h'_c(p_{wb} - p_{wg}) \qquad (20.1.8)$$

and the heat transfer by evaporation-condensation is

$$q_e = 9.15 \times 10^{-7}\, h'_c(p_{wb} - p_{wg})h_{fg} \qquad (20.1.9)$$

where $\dot{m}_D$ is the mass transfer rate in kg/m² s, and h_{fg} is the latent heat of water, in J/kg.

The heat transfer terms from cover to ambient are formulated in the same way as for flat plate collectors, by the method of Sections 3.8 and 3.15. If the still has insulation under the basin, heat loss to the ground can be written

$$q_k = U_G(T_b - T_a) \qquad (20.1.10)$$

where U_G is an overall loss coefficient to ground assuming the ground to be at a temperature equal to ambient. This term will be small in a well-designed still.

If the basin is very shallow and well insulated, the heat capacity term in Equation 20.1.1 can be neglected and steady-state solutions found. However, for practical reasons, most stills will have sufficient depth that the capacitance of the basin should be considered. If the still is not well insulated, an effective ground capacitance will also have to be considered.

This set of equations, with data on radiation, temperature, and wind speed and with design parameters of the still, can be solved for T_b as a function of time, and the productivity is then calculated from Equation 20.1.8. This analysis does not include capacitance rates of feed water or leaving brine and product nor edge effects and leakage, which are difficult to formulate; these are often lumped together in a term determined experimentally as that required to make energy balances close. There is evidence for the existence of temperature gradients in still basins that make the surface temperature different from the bulk temperature of the salt water.

The efficiency of a still is defined as the ratio of the heat transfer in the still by evaporation-condensation to the radiation on the still.

$$\eta_i = \frac{q_e}{G} \qquad (20.1.11)$$

This is usually integrated over some extended period (e.g., day or month) to indicate long-term performance. If there is any loss of product water back into the still (by dripping from the cover or leakage from collecting troughs), less product would be available than is indicated by this equation. Efficiency from experimental measurements is

$$\eta_i = \frac{\dot{m}_p h_{fg}}{G} \tag{20.1.12}$$

where $\dot{m}_p$ is the rate at which distillate is produced from the still (which may be less than $\dot{m}_D$) and h_{fg} is the latent heat of vaporization.

The objective of still design is to maximize q_e (i.e., $\dot{m}_D$), which is proportional to the vapor pressure difference between basin and cover. Thus it is desirable to have the basin temperature as high as possible, which will increase the ratio of heat transfer by evaporation-condensation to that by convection and radiation. Shallow basins with small heat capacity will heat up more rapidly than will deep basins, and operate at higher mean temperatures.

Many practical considerations govern still design and operation. Shallow basins require precise leveling of large areas, which is costly. Crystals of salt build up on dry spots in basins, leading to reduced overall absorptance and reduced effective basin area. Leakage can cause problems in three ways: distillate can leak back into the basin; salt water can leak out of the basin; and humid air from inside the still can leak out through openings in the cover. Occasional flushing of still basins has been found necessary to remove accumulations of salt and organisms such as algae which grow in the brines. Growths can be controlled by additions of algicides.

A wide variety of experimental basin-type stills have been built and studied. Two design trends have evolved. Large area deep basin stills as shown in Figure 20.1.4 can be built by standard construction techniques, are durable, and are relatively inexpensive. Modular shallow basin stills as shown in Figure 20.1.5 have lower thermal capacitance, produce somewhat more water, but may be more expensive to construct.

Details of these and many other designs are included in Talbert et al. (1970). A less extensive report emphasizing the potential of distillation applications in developing economies was published by the United Nations (1970). Cooper, in a series of papers (e.g., 1973) has made a detailed study of simulation of solar still performance. Proctor (1973) has experimentally investigated the possibility of augmenting solar still output with waste heat.

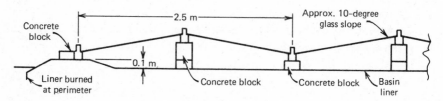

Figure 20.1.4 Section schematic of experimental deep-layer basin type still at Daytona Beach, FL. From Talbert *et al.* (1970).

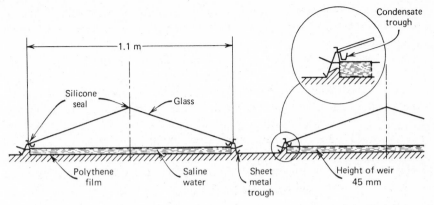

Figure 20.1.5   Section schematic of an experimental shallow basin still at Muresk, W. A. Adapted from Morse and Read (1968).

20.2   EVAPORATION

Solar evaporation to produce salts from sea water or other brines is a large-scale industrial operation, and on a world-wide basis approximately one-third of the annual salt production is from solar processes. In the United States, solar salt production facilities are concentrated in the Great Salt Lake and San Francisco Bay areas. This is a well-established technology.

A description of the history and operation of salt manufacture by solar evaporation of sea water is provided by See (1960). The process is a fractional crystallization. Shallow pans or ponds are filled with the brine to be evaporated and exposed to radiation. As long as evaporation exceeds rainfall, concentration will increase with time, and crystallization will occur when saturation is reached. Pond operation depends on the source of feed water. Sea water contains approximately 3.5 percent salts, of which approximately two thirds is NaCl (the usual desired product). Sea water evaporation can be considered to consist of three steps. The first is brine concentration to bring the solution to saturation (approximately 23 percent salts) and is the largest in area. This solution is transferred to second-stage pans, where the first salts to crystallize are calcium, magnesium and iron carbonates and sulfates. Then the solution goes to the crystallizing ponds, where NaCl crystallizes out. More water is allowed to evaporate, until the solution nears the saturation point for the magnesium and potassium chlorides and bromides and other salts (the "bittern" salts), at which point the solution is removed from the pond. The salt precipitated in the crystallizer pond is NaCl in a reasonably pure form, and is harvested mechanically.

Solar evaporation is used with feeds other than sea water. Sodium sulfate solutions are concentrated in 440,000 m^2 ponds on the Atacama desert of Chile for the manufacture of anhydrous sodium sulfate [Suhr (1970)]. The Caracol, a solar evaporating process at Mexico, D.F., concentrates brines from the Lake Texcoco area to feed a fuel fired evaporator plant producing sodium carbonate decahydrate.

Solar evaporation has been of substantial interest for salt production and also for estimation of evaporative losses from water storage reservoirs. A number of studies of the process have been published, including reports of measurements from salt pans [e.g., Bloch et al. (1951)] and lakes and reservoirs [e.g., Geological Survey (1954)]. (Salt solutions in evaporators will have lower vapor pressures than lake water; otherwise the processes are similar.)

An energy balance over a fully mixed pond is very much like Equation 20.1.1 for the basin of a solar still, but heat is transferred to the atmosphere by evaporation, convection, and radiation, rather than to a cover.

$$\alpha G = q_e + q_{r,\,p-a} + q_{c,\,p-a} + q_k + (mC_p)\frac{dT_p}{d\tau} \qquad (20.2.1)$$

where q_e, q_k, and capacitance terms have the same significance as for a still, and the subscript p refers to the pond. (For deep ponds that are thermally stratified, the heat capacity term requires more detailed treatment.)

For evaporation from open water surfaces, the ratio of energy transport from the water by convection to that by evaporation has been shown by Bowen (1926) to be nearly constant regardless of wind speed. The ratio is given by

$$\frac{q_{c,\,p-a}}{q_e} = \frac{0.46(T_p - T_a)}{p_{wp} - p_a} \times \frac{p}{760} \qquad (20.2.2)$$

where T_p and T_a are the temperatures of the brine surface and ambient, p_{wp} and p_a are the vapor pressure of the surface solution and the ambient vapor pressure of water, and p is the barometric pressure, with all pressures in mm Hg.

These equations have been applied to evaporation from lakes by Cummings (1946) and others and to evaporation from salt pans by Bloch et al. (1951) [see also Spiegler (1955)]. Bloch et al. in an experimental study of evaporation from brines measured diurnal temperature variations of solutions, temperature gradients in solutions during evaporation and at night, and effects of dyes which increase absorptance of the brines for solar radiation.

Net annual evaporation from salt pans in the San Francisco area is usually in the range of 0.9 to 1.1 m with an annual rainfall in the range of 0.25 to 0.5 m, and in dry years the net annual evaporation was 1.2 m [See (1960)]. Thus in this area a half to one cubic meter of water is evaporated per year per square meter of pond.

20.3 DIRECT SOLAR DRYING

Direct crop drying is another well established use of solar energy, in which the standard practice is to spread the materials to be dried in thin layers to expose them to radiation and wind. (Indirect drying, by supply of solar heated air to more or less conventional dryers, was noted in Chapter 17.) The practices of solar drying are based on long experience. Questions regarding further applications concern possible improvements in the process or changes that might be made to improve product quality or dry additional kinds of materials.

The equations describing a radiant drying process can be written in form similar to those for evaporation from pans. However, several additional phenomena may have to be considered. If the material is opaque to solar radiation, drying occurs at the surface, and moisture is transported to the surface by capillary action or diffusion. As with other drying processes, there may be constant rate periods when energy balances determine rates of drying (when the surface of the material is always saturated with moisture) and falling rate periods when moisture transport through the drying material to its surface controls the rate of evaporation. The determination of the vapor pressure of water at the surface of the drying material may be difficult.

In addition to field crops, drying of grapes in tiers of trays has been reported by Wilson (1965), and an analysis of this type of dryer is presented by Selcuk et al. (1974). Experiments on direct drying of an oil shale, to reduce its moisture content prior to retorting, have been described by Talwalkar et al. (1965).

20.4 SALT-GRADIENT SOLAR PONDS

Temperature inversions have been observed in natural lakes, with higher temperatures in the bottom layers than near the surface. These lakes have high concentrations of dissolved salts in the bottom layers and much more dilute solutions at the surface. This phenomenon suggested the possibility of constructing and using ponds as large-scale horizontal solar collectors. Tabor (1964), Weinberger (1964), and Tabor and Matz (1965) have reported a series of theoretical and experimental studies of these salt-gradient ponds.* Salt-gradient ponds have been proposed for use in power production, salt production, and for providing thermal energy for buildings.

Gradients of salinity and temperature in an experimental pond are shown in Figure 20.4.1. The basic criterion for solar pond operation is that the density of the concentrated solution in the bottom zone must be higher at its maximum temperature than the density of the more dilute layers above it. If solar radiation is transmitted through the upper layers and absorbed to a significant extent by the pond bottom and the lower layers, the bottom layers will become heated. The nonconvective layer provides insulation, as heat is transferred up out of the heated layer only by conduction. In practice, as observed by Nielson, there is convection in a thin surface layer and in the bottom layer; these are separated by the nonconvective zone.

---

* These are often referred to simply as "solar ponds." As there are other types of ponds called by the same name (e.g. the collector described in Section 17.7) the term salt-gradient pond avoids confusion.

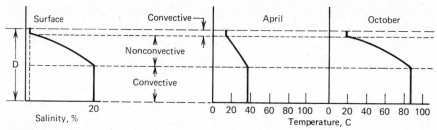

Figure 20.4.1 Salinity and temperature profiles for a salt-gradient pond. From Nielson (1978), with permission of the American Section of the International Solar Energy Society.

Absorption of radiation in an experimental pond is shown in Figure 20.4.2 for a pond of 1.0 m depth. At the time of these measurements, the zenith angle of the sun was approximately 12°, so radiation was nearly normal to the water surface. Approximately 6 percent of the incident radiation should have been lost by reflection at the water surface. In this pond there is significant radiation transmitted to the bottom layers.

Many practical problems arise in the development of solar ponds. Pond salt gradients must not undergo major disturbances, or the salinity gradients may be destroyed and the whole pond undergo convective mixing. Removal of heat from the bottom layers is difficult, but can be accomplished either by means of a large area heat exchanger submerged in the pond or by simultaneous slow removal of brine from the lower level at one side of the pond and return of (cooled) brine to the lower level on the other side. Water evaporates from the top layer of the pond, and must be replaced, either by pure water or by dilute salt solution. There will be some diffusion of salt upward that will tend to eliminate the salinity gradients; this can be overcome by addition of water at the top and removal of corresponding amounts of water (e.g., by flashing

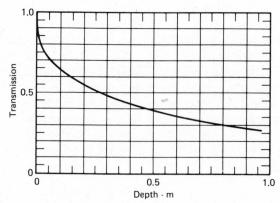

Figure 20.4.2 Measured transmission of solar radiation into a salt gradient pond one meter in depth. Adapted from Tabor and Matz (1965).

vapor from the draw-off brine) from the lower level. Cleanliness of the ponds may be a problem if they are not large in extent as contaminants can reduce transmittance; some "dirt" will float at the surface, some will sink to the bottom, and some will float at intermediate levels where its density matches that of the solution. Research is in progress to seek answers to these problems.

It is also possible to construct ponds with solutes having solubility-temperature and density-temperature dependences such that denser and more concentrated solutions can exist at higher temperatures in pond bottoms in equilibrium with less dense and less concentrated solutions near the surface. Examples of salt solutions exhibiting suitable properties are $MgCl_2$-water and borax-water. These pond are stable so that they are not subject to irreversible loss of salinity gradient on mixing. The vapor pressure of the top layer is lower than in the typical salt gradient pond resulting in less evaporation loss [see Ochs and Bradley (1979)].

20.5 SUMMARY

In this chapter we have briefly described some of the major considerations regarding distillation, evaporation, and direct drying. The discussion of salt-gradient ponds could have been included in an earlier chapter, such as Chapter 17 on industrial process heat, but as its application may be in salt production and it parallels direct evaporation, it is included here. Much of the information needed in estimating the performance of these processes can be found by methods outlined in earlier chapters. They all are horizontal collectors, some of which have integral storage, and evaporation of water is either an objective or a problem.

REFERENCES

Bloch, M. R., L. Farkas, and K. S. Spiegler, *Industrial and Engineering Chemistry*, **43**, 1544 (1951). "Solar Evaporation of Salt Brines."

Bowen, I. S., *Physical Review*, **27B**, 779 (1926). "The Ratio of Heat Loss by Conduction and by Evaporation from Any Water Surface."

Cooper, P. I., *Solar Energy*, **14**, 451 (1973). "Digital Simulation of Experimental Solar Still Data."

Cummings, N. W., *Transactions of the American Geophysical Union*, **27**, 81 (1946). "The Reliability and Usefulness of the Energy Equations for Evaporation."

Dunkle, R. V., *International Developments in Heat Transfer*, Conference at Denver, Part 5, 895 (1961). "Solar Water Distillation: The Roof Type Still and a Multiple Effect Diffusion Still."

Geological Survey Professional Paper 269, U.S. Government Printing Office (1954). "Water Loss Investigations: Lake Hefner Studies, Technical Report."

Morse, R. N. and W. R. Read, *Solar Energy*, **12**, 5 (1968). "A Rational Basis for the Engineering Development of a Solar Still."

Nielson, C. E., paper in *Proceedings of the 1978 Annual Meeting of the American Section of the ISES*, **2**(1), 932 (1978). "Equilibrium Thickness of the Stable Gradient Zone in Solar Ponds."

Ochs, T. L. and J. O. Bradley, paper presented at the Atlanta ISES meeting (1979), "The Physics of a Saturated $Na_2O \cdot 2B_2O_3 \cdot 10H_2O$ Non-Convecting Solar Pond."

Proctor, D., *Solar Energy*, **14**, 433 (1973). "The Use of Waste Heat in a Solar Still."

See, D. S., in *Sodium Chloride: The Production and Properties of Salt and Brine*, D. W. Kaufman (ed.), American Chemistry Society Monograph #145, Chapter 6 (1960). "Solar Salt."

Selcuk, M. K., O. Ersay, and M. Akyurt, *Solar Energy*, **16**, 81 (1974). "Development, Theoretical Analysis and Performance Evaluation of Shelf Type Solar Dryers."

Spiegler, K. S., paper in *Solar Energy Research*, University of Wisconsin Press (1955), p. 119. "Solar Evaporation of Salt Brines in Open Pans."

Suhr, H. B., Paper No. 4/27 presented at the International Solar Energy Society Conference, Melbourne (1970). "Energy-Balance Calculations on the Production of Anhydrous Sodium Sulfate with Solar Energy and Waste Heat."

Tabor, H., *Proceedings of the UN Conference on New Sources of Energy*, **4**, 59 (1964). "Large Area Solar Collectors (Solar Ponds) for Power Production."

Tabor, H. and R. Matz, *Solar Energy*, **9**, 177 (1965). "Solar Pond Project."

Talbert, S. G., J. A. Eibling, and G. O. G. Löf, *Manual on Solar Distillation of Saline Water*, Office of Saline Water, U.S. Dept. of Interior, Research and Development Progress Report #546 (1970).

Talwalkar, A. T., J. A. Duffie, and G. O. G. Löf, *Proceedings of the UN Conference on New Sources of Energy*, **5**, 284 (1965). "Solar Drying of Oil Shale."

United Nations Department of Economic and Social Affairs, Publication Sales No. E. 70. II.B.1, (1970), "Solar Distillation."

Weinberger, H., *Solar Energy*, **8**, 45 (1964). "The Physics of the Solar Pond."

Wilson, B. W., *Proceedings of the UN Conference on New Sources of Energy*, **5**, 296 (1965). "The Role of Solar Energy in the Drying of Vine Fruit."

APPENDIX A

Problems

The problems below are arranged by chapters. Most of them have quantitative answers; a few have descriptive answers. Time is assumed to be solar time unless otherwise stated.

A solution manual for the problems is available from the authors at the Solar Energy Laboratory, University of Wisconsin, 1500 Johnson Drive, Madison, WI 53706.

The most important equations for solution of these problems have been programmed for hand-held calculators. For information on these programs contact FCHART, P.O. Box, 5562, Madison, WI 53705.

1.1 From the diameter and effective surface temperature of the sun, estimate the rate at which it emits energy. What fraction of this emitted energy is intercepted by the earth? Estimate the solar constant, given the mean earth-sun distance.

1.2 What fraction of the extraterrestrial radiation is at wavelengths below 0.5 μm? 2 μm? What fraction is included in the wavelength range 0.5 μm to 2 μm?

1.3 Divide the extraterrestrial solar spectrum into 10 equal increments of energy. Specify a characteristic wavelength for each increment.

1.4 Calculate the angle of incidence of beam radiation at 1400 solar time on January 20 at latitude 35°N on surfaces with the following orientation:
  a Horizontal
  b Tilted to south at slope of 40°
  c At slope of 40°, but facing 25° west of south
  d Vertical, facing south
  e Vertical, facing west

1.5 Determine the sunset hour angle and day length for Madison and for Miami, for the following dates: January 1, March 22, July 1.

1.6 **a** When it is noon Pacific Standard Time in San Francisco ($L = 122°$ W) on March 1, what is the corresponding solar time?
  b What Central Daylight Time corresponds to solar noon in Madison ($L = 89°$ W) on September 30?
  c When it is 10 a.m. Mountain Standard Time in Salt Lake City ($L = 112°$ W) on January 26, what is the Solar Time?

1.7   Estimate R_b for a collector sloped 60° from horizontal, with $\gamma = 0°$, at Madison ($\phi = 43°$) at 2:30 on March 5.

1.8   What is R_b for a collector at latitude 43° sloped 45°, with surface azimuth angle of 15°, at 2:30 solar time on March 5?

1.9   At Denver (latitude 40°), on December 16, what is the extraterrestrial radiation on a horizontal surface, H_o? What is the December monthly mean extraterrestrial radiation on a horizontal surface, $\bar{H}_o$?

1.10   **a**   What is the day's extraterrestrial radiation on a horizontal surface for Madison for June 16? How does this compare to the monthly mean value?

          b   What is the H_o for Madison for January 26? What is $\bar{H}_o$ for Madison for January?

1.11   What is the angle of incidence of beam radiation on the aperture of a concentrating collector located at $\phi = 35°$, at 1500 solar time on February 15 if the collector is rotated about a single axis to minimize the angle of incidence:

          a   If the axis is horizontal and east–west?

          b   If the axis is horizontal and north–south?

          c   If the axis is parallel to the earth's axis?

1.12   Estimate the ratio of beam radiation on a collector tilted 45° toward the south to that on a horizontal surface, if located at a latitude of 40° on March 1, **a** at noon, **b** at 3:30 pm.

1.13   The plane of the collectors on the Arlington (WI) solar house is inclined 60° from the horizontal, is at a latitude of 43.3°N, and has a surface azimuth angle of 0°. What is R_b for this surface on Feb. 7 at 11:30 a.m.? What will it be for the hour from 3 to 4 p.m. on July 4? If the collector were tilted at 30°, what would be the value of R_b? (Note: Do this problem by equation and graphically.)

2.1   **a**   For the clear day illustrated in Figure 2.5.1, determine the hourly and daily solar radiation on a horizontal surface.

          b   For the cloudy day illustrated in Figure 2.5.1, determine the hourly and daily solar radiation on a horizontal surface.

2.2   **a**   For Poona, India estimate the monthly average radiation for January and July from the average hours of sunshine data of Table 2.7.2.

          b   Estimate the monthly average radiation on a horizontal surface in January and June in Madison starting with the average hours of sunshine data from Table 2.7.1. Compare your results with data from Appendix G.

2.3   Solar radiation on a horizontal surface integrated over the day of January 9th at Boulder ($\phi = 40°$) is 4.48 MJ/m². What is the clearness index, K_T, for that day? What is the estimated fraction of the day's energy which is diffuse?

2.4   The average daily solar radiation on a horizontal surface in Lander, WY ($\phi = 42.8°$) is 18.9 MJ/m² in March. How much of this is beam, and how much is diffuse?

2.5 At $\phi = 43$ on December 22, K_T was 0.63. Estimate the total radiation on a horizontal surface in the hour from 10:00 to 11:00 a.m.

2.6 Estimate the hourly beam and diffuse radiation on a horizontal surface in Madison for May 18, 1960; $H = 29.7\ \text{MJ/m}^2$. Assume symmetry about solar noon.

2.7 From Madison, 43°N, 60° tilt, January 20, the total radiation on a horizontal surface is 8.0 MJ/m².

    a Estimate the total horizontal radiation for 10 to 11.

    b Estimate the beam and diffuse for 10 to 11.

    c What is R_b for that hour?

    d If all radiation is treated as beam, what is total radiation on the tilted surface for that hour?

    e If the diffuse radiation is considered to be independent of orientation, what is I_T for that hour?

    f If the diffuse radiation is uniform over sky, and ground reflectance is considered, what is I_T for that hour? Let ρ be 0.2 (bare ground) and 0.7 (snow).

2.8 The day's radiation on a horizontal surface in Madison ($\phi = 43°$) on a Dec. 22 is 8.80 MJ/m². There is a fresh snow cover. Estimate the diffuse radiation, ground-reflected radiation and the total radiation on a south-facing vertical surface during the hour 11 to 12.

2.9 A collector is installed at Boulder ($\phi = 40°$) at a slope of 60°, facing south. What will be the hourly beam and diffuse components of solar radiation on this collector on January 13, if the total radiation on a horizontal surface for that day is 10.9 MJ/m², and ground reflectance is 0.7?

2.10 Estimate for May 11, for Madison, WI the hourly total radiation on **a** a horizontal surface, and **b** on a surface tilted toward the equator with a slope equal to the latitude. Weather bureau data show a total solar radiation for the day of 622 cal/cm².

Note: For purposes of comparison, the weather bureau hourly data are as follows:

Time	cal/cm²	Time	cal/cm²
5–6	0	12–1	82
6–7	7	1–2	76
7–8	22	2–3	53
8–9	39	3–4	37
9–10	54	4–5	19
10–11	77	5–6	4
11–12	82	6–7	0

2.11 Estimate the radiation on the plane of a collector sloped at 60°, $\gamma = 0°$, $\phi = 43.3°\text{N}$, for the hour from 11 to 12 a.m. on Feb. 7. The total day's radiation on a horizontal surface is 11.8 MJ/m².

2.12 What is the average daily radiation in March on a south-facing surface tilted at a slope of 35° for Albuquerque?

2.13 For Albuquerque, plot the monthly average daily radiation as a function of month for collectors with the following orientations:
  a Horizontal surface.
  South facing:
  b Collector tilt equal to latitude (35°).
  c Collector tilt of 50°.
  d Collector tilt of 20°.
  e Vertical collector.
  Note: let ground reflectance be 0.2 for all months.

2.14 A window faces south at a location with latitude 36°. Estimate the January average radiation on the window if $\bar{K}_T$ for the month is 0.47.

2.15 $\bar{H}$ for a location at $\phi = 45°$ is 14.0 MJ/m² in March. For that month, which has very little snow, estimate $\bar{H}_T$ for a south facing surface with $\beta = 60°$.

2.16 For Madison in October, for a south-facing surface sloped at 58°, estimate $\bar{R}$ and $\bar{H}_T$.

2.17 Estimate the monthly average radiation in February on a south facing vertical collector-storage wall at $\phi = 43.3°$N. The average radiation on a horizontal surface in February is 9.2 MJ/m².

2.18 Estimate the standard clear sky (i.e., 23 km visibility) beam and diffuse radiation on a horizontal surface for December 23 for Minneapolis ($\phi = 46.5°$, elevation 432 m) for each of the hours from sunrise to noon.

2.19 If the radiation on the surface of Problem 2.18 is 0.66 MJ/m² for the hour 10 to 11 a.m., estimate the diffuse on this horizontal surface.

2.20 Estimate the hourly beam radiation on the aperture of a collector which rotates continuously about a horizontal north–south axis so as to track the sun. The data base is measurements of normal incidence radiation measured with a pyrheliometer and integrated over the hour, and is indicated on the table below. The latitude is 38° and the date is January 7.

Hour	8–9	9–10	10–11	11–12	12–1	1–2	2–3	3–4
I_{bn}, MJ/m²	0.35	0.70	2.66	3.05	3.30	3.19	1.80	1.42

3.1 Verify the values of the blackbody spectral emissive power as given in Figure 3.4.1 for **a** $T = 1000$ K and $\lambda = 10$ μm; **b** $T = 400$ K and $\lambda = 5$ μm; and **c** $T = 6000$ K and $\lambda = 1$ μm.

3.2 What is the percentage of the blackbody radiation from a source at 300 K in the wavelength region from 8 to 14 μm? (This is the so-called "window" in the earth's atmosphere.)

3.3 Write a computer subroutine to calculate the fraction of the energy from a blackbody source at T in the wavelength interval a to b.

3.4 Calculate the energy transfer per unit area by radiation between two large parallel flat plates. The temperature and emittance of one plate are 500 K and 0.45 and for the other plate are 300 K and 0.2. What is the radiation heat transfer coefficient?

3.5 Calculate three different overall heat transfer coefficients for a plate facing up at 50 C when exposed to an ambient temperature of 10 C. Base your three results on three different approximations for the effective sky temperature. Plate emittance is 0.88, the relative humidity is 70 percent, and the wind heat transfer coeffieient is 25 W/m² C.

3.6 Consider two large flat plates spaced L mm apart. One plate is at 100 C and the other is at 50 C. Determine the convective heat transfer between the plates for the following conditions:

a Horizontal, heat flow up, $L = 20$ mm.
b Horizontal, heat flow up, $L = 50$ mm.
c Inclined at 45°, heat flow up, $L = 20$ mm.

3.7 Compute the equilibrium temperature of a thin polished copper plate 1 m × 1 m × 1 mm, under the following conditions:

a In earth orbit, with solar radiation normal to a side of the plate. Neglect the influence of the earth. See Table 4.5.1.
b Just above the earth's surface, with solar radiation normal to the plate and the sun directly overhead. See Table 4.5.1.

Assume **a** the sky is clear the transmits 0.80 of the solar radiation; **b** the "equivalent blackbody sky temperature" is 10 C less than the ambient temperature; **c** the ambient air temperature is 25 C; **d** the wind heat transfer coefficient is 23 W/m² C; and **e** the earth's surface is effectively a blackbody at 15 C.

3.8 Consider two thin circular disks, thermally isolated from each other, and suspended horizontally, side by side, in the same plane, inside a glass sphere on low conductance mounts. The sphere is filled with an inert gas, such as dry nitrogen, to prevent deterioration of the surfaces. Dimensions of the disks are identical. One disk is painted with black paint ($\alpha_b = 0.95$, $\varepsilon_b = 0.95$) and the other with white paint ($\alpha_w = 0.35$, $\varepsilon_w = 0.95$). The glass has a transmittance for solar radiation (τ_g) of 0.90 and an emittance for long-wave radiation of 0.88. The convection coefficient, h, between each of the disks and the glass cover is 16 W/m² C. (Note that the disks have two sides and that the edges can be neglected.) When exposed to an unknown solar radiation on a horizontal surface, G, the temperature of the white disks (T_w) is 5 C and the temperature of the black disc (T_b) is 15 C. $T_{ambient}$ is 0 C.

a Write the energy balances for the black and white disk assuming the glass cover is at a uniform temperature (T_c).
b Derive an expression for the combined convection and radiation heat transfer coefficient.
c Using the result of b, derive an expression giving the incident solar radiation as a function of the difference in temperature between the black and white disks.
d What is the incident solar radiation, G, for the conditions stated above?

3.9 Determine the convection heat transfer between two large flat plates (covers in a collector) separated by a distance of 20 mm and inclined at an angle of 60°. The lower plate is at 150 C and the upper plate is at 60 C.

3.10 What is the convective heat transfer for the conditions of Problem 3.9 when a "slat" type honeycomb is inserted in the space between the plates? The distance between the slats is 10 mm.

3.11 Determine the heat transfer coefficient for air flowing by forced convection in a 1 m wide by 2 m long by 15 mm deep channel. The flow rate is 0.012 kg/s. What is the heat transfer coefficient if the plate spacing is halved? What is the heat transfer coefficient if the mass flow rate is doubled? $T_a = 25$ C.

3.12 What is the convective heat transfer coefficient due to a 7 m/s wind over a free standing 2 m by 3 m collector array? Assume the air temperature is 20 C.

3.13 Estimate the heat transfer coefficient due to a 10 m/s wind flowing over a collector array. The collectors are mounted on a building with dimensions 10 m × 12 m × 4 m.

3.14 Estimate the pressure drop in a pebble bed with a 3 m by 4 m flow area and with a 2 m flow length. The flow rate through the bed is 1.1 kg/s. The pebbles are 0.02 m diameter river washed gravel with a void fraction of 0.45. Use an average air temperature of 40 C.

3.15 Estimate the rock diameter required for a pressure drop of 55 Pa for the conditions of Problem 3.14. Assume the pebble void fraction remains at 0.45 for all pebble sizes.

4.1 Consider a surface that has been prepared for use in outer space and has the following spectral characteristics:

$$\rho = 0.10 \quad \text{for} \quad 0 < \lambda < \lambda_c$$
$$\rho = 0.90 \quad \text{for} \quad \lambda_c < \lambda < \infty$$

For $\lambda_c = 1, 2,$ and 3 μm, calculate the equilibrium temperature of the plate. Assume the sun can be approximated by a blackbody at 6000 K and that the solar flux on the plate is 1353 W/m². Also assume the back side of the plate is perfectly insulated.

4.2 A selective surface for solar collector absorber plates has the characteristic

$$\alpha = \varepsilon = 0.95 \quad \text{for} \quad 0 < \lambda < 1.8 \ \mu m$$
$$\alpha = \varepsilon = 0.05 \quad \text{for} \quad 1.8 < \lambda < \infty \ \mu m$$

Assume the sun to be a blackbody emitter at a temperature of 6000 K. Calculate the absorptance of this surface. If the surface is at 150 C, calculate its emittance.

4.3 Determine the solar absorptance of the black chrome (after the humidity test) surface given in Figure 4.6.3. Use the extraterrestrial solar spectrum for this calculation.

4.4 What is the absorptance of the surface of Problem 4.3 for the terrestrial solar radiation distribution of Table 2.6.1?

4.5 What is the emittance of the surface of Problem 4.3 at a temperature of 350 C?

4.6 For curve C of Figure 4.6.2, calculate the emittance at temperatures of **a** 300 K; **b** 500 K; and **c** 1000 K.

5.1 Calculate the reflectance of one glass surface for angles of incidence of **a** 10°, **b** 30°, **c** 50°, and **d** 70°. (Index of refraction = 1.526.)

5.2 Calculate the transmission of three nonabsorbing glass covers at angles of 10 and 70° and compare your results to Figure 5.1.3.

5.3 For glass with $K = 20 \text{ m}^{-1}$ and 2.0 mm thick, calculate the transmission of two covers at **a** normal incidence and **b** at 50°.

5.4 Calculate the $(\tau\alpha)$ product for a two-glass-cover collector ($KL = 0.0370$ per plate) with a flat black collector plate surface for radiation incident on the collector at an angle of **a** 25° **b** 60°. $\alpha_n = 0.96$.

5.5 Estimate the transmittance of a single cover with $KL = 0.0370$ for diffuse radiation from the sky and for diffuse radiation from the ground. The slope of the cover is **a** 45° and **b** 90°.

5.6 What is the transmittance for solar radiation of a collector cover with index of refraction 1.60 at an angle of incidence of 58°? The cover is 2 mm thick and the extinction coefficient is 10 m^{-1}. If the refractive index is 1.40 what will be the transmittance?

5.7 Estimate the radiation absorbed by a collector under the following conditions: $I = 3.0 \text{MJ/m}^2$; $I_{b.T} = 4.1 \text{ MJ/m}^2$, $I_d = 0.4 \text{ MJ/m}^2$, $\theta_{bT} = 25°$, $\beta = 45°$, τ is given on Figure 5.3.1 with $KL = 0.037$ and one cover, $\alpha = 0.93$ and independent of angle, ground reflectance equal to 0.2.

5.8 Estimate the absorbed solar radiation and the monthly average transmittance-absorptance product for December for a vertical south-facing collector-storage wall at Albuquerque if the angular absorptance characteristic of the black surface is that shown in Figure 4.7.1 with $\alpha_n = 0.95$. The two covers have $KL = 0.0125$ per plate. The ground reflectance is 0.4.

6.1 Compare the value of U_L calculated with Equation 6.4.7 to the graphs of Fig. 6.4.4 at
  a $h_w = 5 \text{ W/m}^2 \text{ C}$, $\varepsilon_p = 0.95$, $T_p = 60 \text{ C}$, $T_a = 10 \text{ C}$, for 2 covers.
  b $h_w = 20 \text{ W/m}^2 \text{ C}$, $\varepsilon_p = 0.1$, $T_p = 100 \text{ C}$, $T_a = 40 \text{ C}$, for 1 cover.

6.2 Calculate the overall loss coefficient for a flat-plate solar collector located in Madison, Wisconsin, and tilted toward the equator with a slope equal to the latitude. Assume a single-glass cover 25 mm above the absorber plate (glass hemispherical emittance = 0.88), a wind speed of 6.5 m/s, an absorber plate long-wavelength emittance of 0.11, and 70 mm of rockwool insulation at the rear having a conductivity of 0.034 W/m C. The mean absorber plate temperature is 100 C and the ambient temperature is 25 C. Neglect edge effects and absorption of solar radiation by glass. The collectors are mounted on a house with a volume of 300 m³.

6.3 For a wind of 5 m/s, an ambient air temperature of 10 C, and an average plate temperature of 80 C, calculate the top loss coefficient for a collector having a single plastic cover with a transmittance for infrared radiation of 0.30 and an emittance of 0.63. The slope is 45°, the plate-cover spacing is 25 mm and the plate emittance is **a** 0.95, and **b** 0.10. The collector is 1 m by 2 m and is mounted on a rack on a flat roof.

6.4 A flat plate solar collector has 2 glass covers, a black absorber with $\varepsilon_p = 0.95$, mean plate temperature of 110 C at an ambient temperature of 10 C and a wind loss coefficient of 10 W/m² C. Estimate its top loss coefficient. If the back of the collector is insulated with 50 mm of mineral wool insulation of k of 0.035 W/m C, what is its overall loss coefficient? (Neglect edge effects.) The slope is 45°.

6.5 Calculate U_L for the collector in Figure 6.4.3a if it is insulated on the back with 50 mm of rockwool ($k = 0.045$ W/m C). Neglect edge losses.

6.6 A collector has one glass cover and the plate emittance is 0.10. Wind speed is 5 m/s, average plate temperature is 75 C and ambient temperature is 10 C. The back loss coefficient is 0.5 W/m² C. The collectors are mounted on a house 12 m long, 8 m wide, and 4 m high. Slope = 45°. Plate-cover spacing = 25 mm. What is U_L for this collector?

6.7 A tube and sheet collector is made entirely of copper. The $\frac{3}{8}$ in. diameter tubes are on 9 in. centers and soldered to the collector plate. The thickness of the copper plate is 0.027 in. The collector overall energy loss coefficient, U_L, is 1.4 BTU/hr ft² F. Calculate the fin efficiency factor, F.

6.8 Consider a flat-plate collector with a fin and tube type absorber plate. Assume $U_L = 8.0$ W/m² C, the plate is 0.5 mm thick, the tube center-to-center distance is 100 mm, and the heat transfer coefficient inside the 20 mm diameter tubes is 300 W/m² C. Assume bond conductance is high. Calculate the collector efficiency factor, F', for **a** copper fins, **b** aluminum fins, and **c** steel fins.

6.9 An aluminum collector absorber plate has a thickness of 1 mm and a conductivity of 211 W/m C. For a tube spacing of 200 mm, a tube diameter of 10 mm, $h_{fi} = 300$ W/m² C and a U_L of 4 W/m² C, F' is 0.88. If the plate were made of steel rather than aluminum (i.e., k is 48 W/m C), calculate F'.

6.10 A flat-plate water heating collector absorber plate is copper and is 1.0 mm thick. Tubes 10 mm in diameter are spaced 160 mm apart. The overall collector loss coefficient is 3.0 W/m² C and the inside heat transfer coefficient is 300 W/m² C. The solder bond between the plate and tubes is 5 mm wide and averages 2 mm thick. The solder has a conductivity of 20 W/m C. What is F' for the collector?

6.11 Estimate the useful output of a solar collector when $F_R U_L = 6.3$ W/m² C, $F_R(\tau\alpha) = 0.83$, $T_i = 56$ C, $T_a = 14$ C, $I_T = 3.4$ MJ/m².

6.12 The solar energy absorbed by a solar collector, S, and the ambient temperatures are given in the table below. The collector has $U_L = 5.2$ W/m² C and $F_R = 0.92$. Determine the useful output of the collector for the day in question with a constant inlet temperature of 35 C.

Hour	S, MJ/m^2	T_a, C	Hour	S, MJ/m^2	T_a, C
7–8	0.01	−3	12–1	3.42	9
8–9	0.40	0	1–2	3.21	11
9–10	1.90	4	2–3	1.54	5
10–11	2.85	5	3–4	1.07	1
11–12	3.02	7	4–5	0.52	−4

6.13 A rule of thumb for solar air heaters is that the air flow rate (at 20 C) should be 10 liters per second per square meter of collector area. For an air heater with this flow rate for which $F' = 0.75$ and $U_L = 6.0$ W/m^2 C, calculate F_R.

6.14 A solar water heater, when operating, has $F_R U_L = 5.5$ W/m^2 C. In an experiment, the pump is turned off, and the plate temperature is measured to be 118 C. The pump is then started, with the inlet water at 30 C. What is the useful gain from the collector if the solar radiation does not change?

6.15 Determine the mean plate and fluid temperatures for a water heating collector operating under the following conditions: $U_L = 4$ W/m^2 C, $F' = 0.9$, $\dot{m}/A_c = 0.015$ kg/s m^2, tube diameter $= 10$ mm, inside heat transfer coefficient $= 300$ W/m^2 C, inlet temperature $= 55$ C, ambient temperature $= 15$ C, incident radiation $= 1000$ W/m^2, $(\tau\alpha)_e = 0.85$, shading coefficient $= 3$ percent and dust coefficient $= 2$ percent.

6.16 Calculate $(\tau\alpha)_e$ for diffuse radiation incident on a single cover selective surface collector. The glass has a KL product of 0.0524 and the absorptance of the surface at normal incidence is 0.9. The ratio of α/α_n for the surface is the same as given in Figure 4.7.1. The plate emittance is 0.10.

6.17 For November 2 in Madison the total solar radiation on horizontal surface is 8 MJ/m^2 and the air temperature is 5 C. Estimate the steady-state efficiency and exit temperature of a type e (Figure 6.12.1) *air heater* at 11:30:

  Absorber plate $\alpha = 0.95$, $\varepsilon = 0.2$.
  Single glass cover, $\varepsilon_g = 0.88$, $KL = 0.0125$.
  Wind speed $= 5$ m/s, building volume 400 m^3.
  Air mass flow rate $= 0.016$ kg/s m^2.
  Air entering temperature $= 38°$C.
  Plate to cover spacing $= 20$ mm.
  Air passage depth $= 10$ mm.
  Collector width $= 1.3$ m.
  Collector length $= 3$ m.
  Polyurethane foam back insulation thickness $= 60$ mm.
  Collector tilt $= 53°$ (south facing).

6.18 Estimate the energy collection and efficiency of a module of a two-cover water heating flat-plate collector operating at a latitude of 30° in a space heating process. The date is March 11 and the hour is 1500 to 1600.

Assume $T_{sky} = T_{air}$, and $(\tau\alpha)_e = 1.02\tau\alpha$. All incident solar is beam radiation.

Ignore dust and shading and ignore edge losses. The design parameters of the module are as follows:

2 glass covers, $KL = 0.0125$ (per cover).

Cover spacing, 25 mm.

0.75 m wide, 5 m long.

Slope from horizontal $= 50°$.

Metal plate 1.0 mm thick with $k = 50$ W/m C.

Plate emittance, $\varepsilon_p = 0.95$.

Tubes are 10 mm diameter, on 100 mm centers.

$\alpha = 0.95$, independent of incident angle.

Back insulation is 80 mm thick; $k = 0.048$ W/m C.

Water flow rate is 0.03 kg/s (per module).

Heat transfer coefficient inside riser is 300 W/m^2 C.

For this hour, the operating conditions are as follows:

Solar radiation for the hour on a horizontal surface $= 1.4$ MJ/m^2.

Ambient temperature $= 2$ C.

Wind heat transfer coefficient $= 10$ W/m^2 C.

Temp of water entering collector $= 35$ C.

6.19 Estimate the hour-by-hour useful gain and day's efficiency for a flat-plate solar collector located in Boulder, CO and tilted toward the equator with a slope equal to the latitude. Use the hourly radiation data of January 10, from Table 5.2. The collector is the same as that in Problem 6.18, and the inlet water temperature is 25°C.

6.20 a Verify the calculations in Figure 6.4.3a, using the convection equations of Sections 3.11.

    b What will be the cover temperature, total loss rate and individual convection and radiation transfers if a set of convection supression slats with aspect ratio 0.3 is used between plate and cover?

6.21 In an industrial application for ventilating a paint spraying operation, ambient air is heated with flat-plate solar air heating collectors. The test results of the air heaters show that $F_R U_L = 4.0$ W/m^2 C and $F_R(\tau\alpha) = 0.61$. The ambient temperature is 12 C and the solar radiation incident on the collector surface is 665 W/m^2. Calculate the rate of useful energy collection and the collector efficiency.

6.22 Based on Equation 6.7.6, would it be better to have a surface in a one cover collector: **a** with $\alpha = 0.95$ and $\varepsilon = 0.20$, or **b** one with $\alpha = 0.90$ and $\varepsilon = 0.10$? Explain the basis of the choice.

6.23 Derive the F' expression for the air heater shown in Figure 6.12.1e. Assume U_b is negligible.

6.24 Calculate F', F'', and F_R for an air heater of the type shown in Figure 6.12.1e having the following characteristics:

$A = 2m^2$

$U_L = 5.0$ W/m^2 C

$\varepsilon_1 = \varepsilon_2 = 0.95$
$T_1 = 100\ C$
$T_2 = 40\ C$
Cross-sectional area of flow channel $= 0.02\ m^2$
Length of flow channel $= 2\ m$
Air flow rate $= 0.028\ kg/s$
Plate spacing $= 0.02\ m$
Evaluate all properties at 70 C

6.25 Calculate the useful energy output of an air heating collector constructed with flow under the absorber plate, with the following characteristics: Collector $(\tau\alpha)_e = 0.85$, air flow rate $= 0.03\ kg/s$, width 0.8 m,
length $= 2.1\ m$, air flow gap $= 10\ mm$, inside channel surfaces painted
with $\varepsilon = 0.95$, collector plate emittance $= 0.1$, outside wind coefficient $=$
10 W/m^2 C, inlet fluid temperature $= 30\ C$, and ambient temperature $=$
10 C. The incident solar radiation is 800 W/m^2. U_{be} is 0.60 W/m^2 C. Slope
is 45° and plate-cover spacing is 25 mm.

6.26 Water at 60 C is circulated through a flat-plate solar collector having
$(\tau\alpha)_e = 0.75$ and $U_L = 8.0$ W/m^2 C. If the ambient temperature is 0 C,
what is the minimum solar radiation required (disregarding any pumping
energy requirement) to collect useful heat?

6.27 A flat plate water heating collector has the following characteristics:
Plate: Copper, 0.026 cm thick, 15 mm diameter tubes on 80 mm centers,
$\alpha_n = 0.93$, $\varepsilon = 0.11$, $\dot{m}/A = 0.0080$ kg/m^2 s.
Covers: 1 glass, KL for the cover $= 0.0125$, spaced 25 mm from the
plate.
Insulation on back and edge: $U_{be} = 0.82$ W/m^2 C.
Collector orientation: 60° slope.
Conditions of operation: $T_a = 10\ C$, wind heat transfer coefficient $=$
20 W/m^2 C, $T_{in} = 35\ C$.
For an hour in which $I_T = 2.60$ MJ/m^2 and the radiation is nearly normal
to the plane of the collector, estimate Q_u/A. Assume $h_{fi} = 300$ W/m^{2}C.

7.1 The experimental data shown below are from a standard NBS/ASHRAE
collector test. For this collector, what are $F_R(\tau\alpha)_n$ and $F_R U_L$?

η_i	$(T_i - T_a)/G_T$
0.60	0.025 m^{2}C/W
0.28	0.10

7.2 Hour-long collector test gives the results shown in the table. For this
collector, what are $F_R(\tau\alpha)_n$ and $F_R U_L$? The collector area is 3.0 m^2.

Q_u, MJ	I_T, MJ/m^2	T_{in}, C	T_a, C
6.80	3.10	16.0	12.0
1.49	3.22	86.1	12.0

7.3   A collector has $F_R(\tau\alpha)_n = 0.73$ and $F_R U_L = 6.08 \text{ W/m}^2 \text{ C}$. When $T_i = T_a$, what is the instantaneous efficiency? If the instantaneous efficiency is 0, the ambient temperature is 30 C, and the inlet temperature is 150 C, what is the radiation on the collector?

7.4   An indoor experimental evaluation of a solar water heating collector produced the data in the table below. The collector has a selective coating with $\alpha = 0.87$–0.92 and $\varepsilon = 0.10$–0.20. The single glass cover has $\tau = 0.92$ at normal incidence. The collector dimensions are 0.914×2.133 m overall for a gross area of 1.95 m^2. Glass (aperture) area is 1.76 m^2 and the effective absorber area is 1.72 m^2. For all of the data in the table, the wind speed was 3.4 m/s. The average water flow rate is 0.0359 kg/s. For the conditions of these tests, what are $F_R(\tau\alpha)_n$ and $F_R U_L$? Base on gross collector area.

T_a, C	23.9	24.4	25.0	26.7	26.7	24.4	25.6	26.1
T_i, C	23.9	24.7	46.1	46.7	52.6	53.0	78.5	78.9
T_o, C	30.7	32.8	51.4	53.3	57.8	58.9	80.9	82.4
G_T, W/m^2	789	947	789	947	789	947	789	947
$\dot{m}$, kg/s	0.0363	0.0385	0.0357	0.0362	0.0357	0.0364	0.0348	0.0359

·7.5   The water heating collector of Problem 7.4 is to be operated at a flow rate of 0.020 kg/s. Estimate new values of $F_R(\tau\alpha)_n$ and $F_R U_L$ for this flow rate. Base calculations on the gross collector area.

7.6   From Example 5.8.1, determine the incidence angle modifier constant, b_o.

7.7   The figure below is from a report on water heater collector testing. What are $F_R(\tau\alpha)_n$ and $F_R U_L$ for this collector (in SI units)?

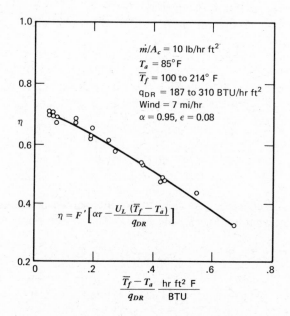

8.1 A concentrating collector is to have a tubular receiver with a plug in it so that the liquid being heated flows through an annulus. The inside and outside diameters of the steel outer tube are 0.054 m and 0.059 m respectively, and the outer diameter of the plug is 0.045 m. The assembly is surrounded by a glass tube cover. A collector module is 0.30 m in aperture width and 3.10 m long with the receiver length equal to the reflector length. The water flow rate through the tube is 0.0168 kg/s. If U_L is 7.5 W/m^{2}C(based on the absorbing surface area) and the average fluid temperature is 100 C, what are F' and F_R for this collector?

8.2 A CPC collector array has an acceptance half-angle θ_c of 15°. It is oriented along an east–west axis in a fixed position so that the slope of the array is 60°. The application is for heating at a location with latitude 43°. For the hour 10 to 11 on 26 January, the radiation on a horizontal plane is 0.36 MJ/m^2 diffuse and 1.20 MJ/m^2 beam. The CPC is not truncated. It is covered by a single glass cover with $KL = 0.0370$.

   a Estimate the absorbed radiation if the specular reflectance of the concentrator is 0.85. At normal incidence, α is 0.95 and its angular dependence is as shown in Figure 4.7.1.

   b Estimate the output of the collector per unit of aperture area for this hour if $U_L = 9.0$ W/m^{2}Cof absorbing area, F_R is 0.92, the inlet fluid temperature is 55 C, and T_a is -5 C.

8.3 A linear parabolic concentrator of width 1.80 m, length 10 m, and focal length 0.92 m is fitted with a flat receiver located in the focal plane. It is arranged to be continuously adjusted about a horizontal north-south axis. The latitude is 32° and the date is 16 March. The image in the focal plane, when the beam radiation is normal to the aperture, can be approximated as shown on the figure. The receiver is centered on the centerline

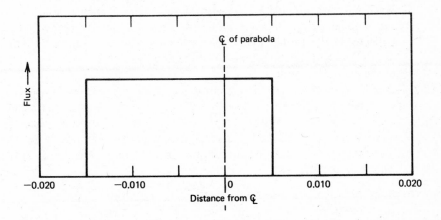

of the parabola, is 0.022 m wide, and 10.5 m long. The loss coefficient will be very near that of the Suntec collector (Figure 8.7.6). For this receiver, $\tau\alpha$ is 0.78. ρ of the concentrator is 0.87.

a   Estimate the absorbed energy for the hours 9 to 10 and 11 to 12 for this date, if I_{bn} for the hours is 2.06 and 3.14 MJ/m^2, respectively.

b   If the collector is used to boil water at 120 C, what will be the energy collected during these hours? The ambient temperatures are 2 and 9 C, respectively.

8.4   It has been suggested that a solar-thermal power plant should be based on a concentrator of the type shown in Figure 8.9.1 (or 8.1.1d) using very large fields of reflectors concentrating radiation onto a receiver on a central tower, thus transferring solar energy from a large area into a central location by optical means rather than piping hot fluids. Sketch such an arrangement, showing schematically the following dimensions: the distance L from a flat reflecting element to a spherical receiver of diameter D; and the width of the flat reflector W. If reflector size is limited by wind loading to $W = 10$ m, and if (in a large system) $L = 1$ km, what (in terms of D) must be the pointing accuracy of the reflector?

9.1   Rework Example 9.3.1, but with a tank containing 500 kg of water, and with $(UA)_s$ of 5.56 W/C. Comment on your results.

9.2   Develop a set of equations analogous to those in Section 9.3 for a two node (partially stratified) tank in which water to meet a load is replaced by water at a constant temperature from the mains.

9.3   Estimate the pressure drop for Example 9.4.1. Does the pressure drop satisfy the minimum recommended pressure drop of Table 13.2.1?

9.4   Estimate the equivalent thermal capacity of a phase change energy storage unit made with paraffin wax (see Table 13.10.1 for properties). The total mass of PCES material is 1300 kg.

10.1   A water heating system has a fully mixed tank of capacity 400 liters. The tank is located in a building having a temperature of 19 C. Collector area is 5.0 m^2. The tank area–loss coefficient product, $(UA)_T$, is 0.81 W/C, $F_R(\tau\alpha)$ is 0.76 and $F_R U_L$ is 4.80 W/m^2 C. In a particular hour when the ambient temperature is 12 C, the radiation on the plane of the collector is 18 MJ. The load on the system is 5.20 MJ. If the tank temperature is 43 C at the beginning of the hour, estimate the temperature at the end of the hour.

10.2   A fully mixed storage tank containing 1500 kg of water has a loss coefficient-area product of 11.1 W/C. The tank starts a particular 24-hr period at a temperature of 45 C. Q_u is added from a collector, and L is removed to a load. The loads L are indicated in the table below. The tank is located inside a building where the temperature is 20 C. The collector has

$$A_c = 30 \text{ m}^2$$
$$F_R(\tau\alpha) = 0.78$$
$$F_R U_L = 7.62 \text{ W/m}^2 \text{ C}$$

For this purpose, assume that $(\tau\alpha)$ is independent of angle of incidence. The incident radiation and outdoor ambient temperature are also given in the table.

	Hour Ending											
	1 am	2	3	4	5	6	7	8	9	10	11	12
Load, MJ	12	12	11	11	13	14	18	21	20	20	18	16
I_T, MJ/m²									1.09	1.75	2.69	3.78
T_a, C	−4	−4	−5	−3	−1	0	0	1	0	2	4	10

	Hour Ending											
	1 pm	2	3	4	5	6	7	8	9	10	11	12
Load	14	14	13	18	22	24	18	20	15	11	10	9
I_T	3.87	3.41	2.77	1.82	1.53							
T_a	10	8	8	6	4	4	3	5	6	6	7	6

a   Estimate the useful gain from the collector for the day.
b   Check the 24-hr energy balance on the tank.

10.3   Calculate the value of the collector heat exchanger correction factor, F'_R/F_R, for the following conditions.
Heat exchanger effectiveness = 0.70
Collector fluid flow rate = 0.70 kg/s
Collector fluid heat capacity = 3.350 kJ/kg C
Tank-heat exchanger flow rate = 0.70 kg/s
$F_R U_L = 3.75$ W/m² C
Collector area = 50 m²

10.4   Derive Equation 10.2.4.

10.5   A 5.2 m² water heating collector is connected by 20 m of 19 mm diameter piping on each of the inlet and outlet sides to the storage tank. The pipe is insulated with 18 mm of foam rubber insulation with conductivity 0.050 W/m C. The heat transfer coefficient on the outside of the insulated pipe is 25 W/m² C. F_R is 0.92, and $(\dot{m}C_p)_c$ is 295 W/C. The collector has $F_R(\tau\alpha) = 0.72$ and $F_R U_L = 4.65$ W/m² C. If the insulated piping is exposed to ambient temperature, what will be the equation for useful output of the collector?

10.6   An industrial process air heating system is shown in the sketch. The air being heated by the combination of solar plus auxiliary enters the heating section at 35 C, is partially heated by solar energy from the storage tank via the load heat exchanger, and then is further heated by the auxiliary exchanger to 55 C. The collector has an area of 80 m². $F'_R(\tau\alpha) = 0.78$, and can be considered as constant for this problem. $F'_R U_L = 4.45$ W/m² C. The fully mixed tank has a capacity of 5000 kg. It is well insulated, and $(UA)_s$ is 140 kJ/hr C. The tank is outdoors. The load heat exchanger has an effectiveness of 1, that is, it is very large, so the return temperature to the tank is always the same as the inlet air temperature. The air to be heated flows at constant rate for

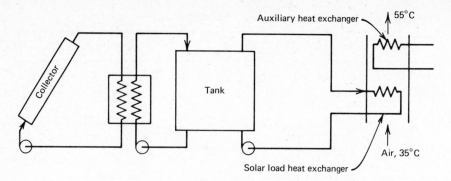

24 hr a day, at a rate of 0.83 kg/s and has a heat capacity of 1013 J/kg C. The flow rate of the water through the heat exchanger is 0.056 kg/s.

The radiation on the collector and ambient temperature data for a 12-hr period are shown on the table. The initial tank temperature is 47 C. For this period, how much energy is supplied to the load from the tank? How much auxiliary energy is needed?

Time	T_a, C	I_T, MJ/m^2
6–7	7	0
7–8	8	0
8–9	9	1.09
9–10	9	1.75
10–11	9	2.69
11–12	12	3.78
12–1	15	3.87
1–2	14	3.41
2–3	12	2.77
3–4	12	1.82
4–5	12	1.53
5–6	11	0

10.7 A liquid heating system with a one-cover selective collector with $\varepsilon = 0.2$ is set up to preheat water from the mains and supply it at T_s, to a boiler, as shown in the sketch. The characteristics of the system are as follows:

$F_R(\tau\alpha)_n = 0.73$

$F_R U_L = 5.10$ W/m^2 C

Heat exchanger effectiveness = 0.60

Capacitance rate on collector side = 1020 W/C

Capacitance rate on tank side = 1120 W/C

Collector area = 20 m^2

Tank capacity = 1000 liters

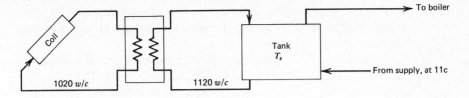

Tank $UA = 6.0$ W/C
Supply water temperature $= 11$ C
Initial tank temperature $= 35$ C

The meteorological data and load flow rate for a four-hour period are shown in the table below.

Make the following assumptions.

a   The controller turns the pump on whenever energy can be collected.
b   The $(\tau\alpha)$ for the hours in question are the same as $(\tau\alpha)_n$.
c   All quantities in the table are constant at their averages over each hour.
d   Tank is fully mixed.
e   The tank is in a room at a constant temperature of 20 C.

Hours	I_T, MJ/m^2	T_a, C	Load Flow Rate, kg
10–11	0.09	14	150
11–12	1.75	17	150
12–1	3.45	18	0
1–2	2.75	20	150

Calculate

a   The tank temperature at 2 p.m.
b   Integrated energy balance (over the 4 hours) on tank.

10.8   For problem 10.7 any one of the following might occur (each independently of the others). For each, indicate in what direction T_s at 2 p.m. would change and explain why.

a   The effectiveness of the collector heat exchanger goes to 0.90.
b   The energy absorbing surface deteriorates resulting in an increase in emittance to 0.5.
c   The pump runs at half speed.
d   The supply water temperature increases to 14 C.

11.1   Calculate the present worth of a cost that is expected to be $5,700 10 years hence if the market discount rate is **a** 8 percent/yr and **b** 12 percent/yr.

11.2   What is the annual loan payment on a loan of $20,000 which is borrowed at 8.5 percent for 15 years, if all payments are uniform?

11.3   What is the present worth factor for a series of payments over 18 years if the inflation rate is 7.8 percent per year and the market discount rate is 9.5 percent?

11.4   A collector is to be installed which has plastic glazing. It is expected that the glazing will have to be replaced every 3 years. The cost of its replacement now is $15/m^2; this is expected to inflate at 7 percent per year.

   What is the present worth of the cost of maintaining the glazing on a 75 m^2 collector array over a period of 20 years, if the market discount rate is 10 percent?

11.5   Calculate the present worth of a 10 period series of costs that starts at $1000 at the end of the first period, inflates at 10 percent/period, if the market discount rate is **a** 8 percent/period, and **b** 12 percent/period.

11.6   Show that Equation 11.5.1 is correct.

11.7   A solar energy system is to be paid for with a loan in the amount of $2700. The interest rate on the loan is 10.5 percent/yr, and the period is 8 years. The market discount rate is 9.5 percent/yr.

   a   Calculate the annual payment to the lender.

   b   Calculate the monthly payment to the lender (assuming $i = 10.5/12$ per month, compounded monthly).

   c   What is the present worth of the interest payments on the loan of part a?

11.8   In Examples 11.6.3 and 11.6.4, a fixed effective income tax rate of 45 percent was assumed. What would be the effect on the answer to these examples if the income tax rate was 45 percent during the first 10 years, and 70 percent in the last 10 years?

   (Note: This could be done quantitatively. Indicate from qualitative considerations what change you expect.)

11.9   A solar energy system costs $10,000 and supplies 80 percent of an annual energy used of 100 GJ. It is bought with a 20 percent downpayment with 80 percent borrowed from a bank at 9.5 percent/yr interest over 5 years. The purchaser's market discount rate is 11 percent/yr. Fuel costs are $11/GJ in the first year and are expected to go up at 15 percent/yr. Assume that real estate taxes, maintenance, and insurance are negligible, and that the mean effective income tax rate is 0.55. Assume all payments to be made at the end of the year.

   a   If the resale value is nil, what are the life cycle solar savings for a 10-year period? Use the method of Example 11.6.3.

   b   If the resale value of the equipment is equal to the initial purchase price, what would the solar savings be?

11.10   Repeat Problem 11.9, using the P_1, P_2 method.

11.11   For the conditions of Problem 11.9,

   a   What is payback time B of Section 11.7 without discounting fuel costs?

   b   What is payback time B if fuel costs are discounted?

11.12 A solar process is to deliver 70 percent of an annual energy requirement of 300 GJ. The solar process equipment cost (installed) is $18,000. The equipment is to be paid for by a 20 percent downpayment and 80 percent loan at 9 percent over 10 years. The market discount rate is 10 percent. The resale value is expected to be small. This is a nonincome-producing facility. Other costs are as follows:—Insurance, maintenance and operating power, $250 in first year, expected to inflate at 6 percent per year. The owner's federal income tax rate is 40 percent, and the installation is in a state where there is no state income tax. Auxiliary energy supplied is 90 GJ/yr, its initial cost is $8.80/GJ, and it is expected to rise at 10 percent/year.

What are the solar savings over a 15 year analysis period?

11.13 A life-cycle cost economic analysis is to be made for a solar process. The first costs is $11,000, and it supplies 110 GJ/yr of solar energy to the process (which is not an income producer). The equipment is durable, and it is expected that it can be resold at the end of 15 years for 80 percent of its installed first cost. Twenty percent of the first cost is paid as cash and the balance is financed by a loan at 9 percent interest over 10 years. Insurance, maintenance, parasitic power, and real estate taxes are negligible.

Real estate taxes are 2 percent of investment, expected to inflate at 6 percent per year.

Effective income tax bracket is 60 percent.

First year fuel cost (for auxiliary or for fuel-only operation) is $10/GJ and is expected to inflate at 14 percent per year.

Discount rate is 15 percent.

a What are the life cycle savings of this system over 15 years, compared to a fuel-only system?

b What are the life cycle savings of this system over 20 years, compared to a fuel-only system? (Note: the change is in the period of the analysis, not the period of the mortgage. Assume the same resale.)

c What is the return on investment? (This is a long calculation.)

11.14 A state's legislation provides tax credits for solar energy systems. The administrative rules state that the life cycle fuel savings must exceed the total present worth of the costs of the system within a 25-year period. A solar water heater installed on a house cost a total of $1550, and is expected to meet 60 percent of an annual load of 19 GJ. The back-up energy source is electricity, at 4.5c/kwh average. The rules specify that a market discount rate of 7 percent/year and an electric energy cost inflation rate of 12 percent/year shall be used in the calculations. Will this water heater meet the state's criterion?

11.15 If $\mathscr{F}$ versus A_c relationship for Problem 11.9 is as indicated on the sketch, estimate the optimum collector area. The area related costs are $250/m^2 and the fixed cost is $2,500. The down payment remains at 20 percent of the investment.

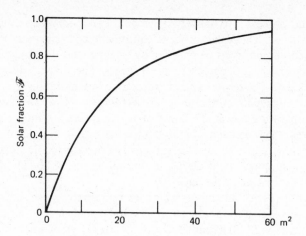

11.16 If the uncertainty in the fuel inflation rate is ± 5 percent (i.e., 10 to 20 percent) and the uncertainty in the predicted fraction by solar is 5 percent (i.e., at $A = 30 \text{ m}^2$, $= 0.80 \pm 0.05$), estimate the maximum and most probable uncertainty in the life cycle savings at the optimum collector area of Problem 11.15.

11.17 A solar heating system has cost and economic factors given below. It is paid for in cash (i.e., there is no mortgage). The total yearly heating load is 110 GJ/yr. The $\mathscr{F}$ versus area curve is given below.

Collector area dependent costs	$250/\text{m}^2$
Fixed cost	$500
1st year property taxes as percent of investment	1.5 percent
1st year insurance, maintenance, and operating expense (as percent of investment)	1.5 percent
Inflation rate for the above	11 percent
Income tax bracket	50 percent
Resale value (as percent of investment)	100 percent
1st year fuel costs	$15/\text{GJ}
Fuel inflation rate	14 percent

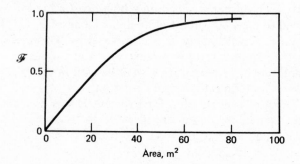

Term of analysis	10 years
Discount rate	12 percent

a   Calculate the economic parameters P_1 and P_2.

b   What is the optimum collector size under these assumptions?

c   What are the life cycle savings?

12.1   What would be the effect on the amount of solar energy delivered by a collector and tank if the same water heating load were concentrated between the hours of 1700 and 1800, rather than spread out uniformly through the 24 hours of the day?

12.2   Determine the collector area to supply 75 percent of the hot water requirements of a residence of a family of four in Boulder, CO, based on the meteorological data for the week of January 8 through 14 as given in Table 2.5.2. Each person requires 45 kg of water per day at a temperature of 60 C or above. Assume the load is uniformly distributed over the day from 7:00 to 21:00 hr. Whenever the storage temperature is less than 60 C, the auxiliary energy source supplies sufficient energy to heat the water coming from the storage tank to 60 C. Use the following collector and storage parameters in your analysis:

Storage mass to collector area	60 kg/m^2
Collector tilt	40° to the south
Collector loss coefficient,	4.0 W/m^2 C
Cover transmittance-absorptance product $(\tau\alpha)_n$	0.77
Collector efficiency factor F'	0.95
Flow rate through the collector	50 kg/m^2 hr
Temperature of supply water to bottom of storage tank	15 C
Height to diameter ratio of cylinder storage tank	3
Ambient temperature around tank	21 C

The system schematic is the same as Example 10.9.1. Note: This problem is best done by use of a computer program such as *TRNSYS*.

12.3   Consider the water heater and storage tank of Problem 12.2. Discuss qualitatively what will happen to the system performance if the following operation or design changes are made. Consider each independently. (Note: Whenever possible, use equations to justify *qualitative* conclusions.)

a   The cover glass is removed (e.g., by breakage).

b   The area of the collector is doubled.

c   The same total load is applied for each day but it is all required between 6 p.m. and midnight.

d   The storage tank design is changed so that water in the tank is thermally stratified rather than mixed.

14.1   A well-insulated residence in Madison, WI has an area-loss coefficient of 145 W/C. It has on it a collector of $A = 22$ m^2 facing south at a slope of 70°. The collector-heat exchanger parameters from the standard collector tests are $F'_R(\tau\alpha)_n = 0.79$ and $F'_R U_L = 3.80$ W/m^2 C. The collector has one cover. Storage tank capacity is 1650 liters. For the load heat exchanger $\varepsilon_L C_{min}/(UA)_h$ is 2.00. $(\overline{\tau\alpha})/(\tau\alpha)_n$ is 0.96. Table 2.16.1 shows estimates of average radiation on various surfaces for Madison, and also provides information on average temperature, and degree days (in C). The radiation data are all calculated from $\overline{H}$ data by the methods of Chapter 2, with $\rho = 0.2$, and so are conservative in the winter months. The water heating load for the building is 2.06 GJ/month. For the month of January:

a   What are the total loads?
b   What are X and Y?
c   What is the fraction carried by solar?
d   If storage volume is halved, what is f?
e   If $\varepsilon_L C_{min}/(UA)_h$ is reduced to 0.5 (in addition to halving the storage volume), what is f?

14.2   A space heating system in Madison, WI uses liquid heating collectors. The house, and system parameters and weather for a particular month are given below.

Days in month	30
Radiation on the plane of the collector, $\overline{H}_T$	15 MJ/m^2
Average ambient temperature	-2 C
Degree-days	553 C-days
House area-loss coefficient (UA)	350 W/C
$F'_R(\tau\alpha)_n$	0.78
$F'_R U_L$	7.0 W/m^2 C
$(\overline{\tau\alpha})/(\tau\alpha)_n$	0.95
Storage volume per unit collector area	180 liters/m^2
Load heat exchanger	standard size

a   For a collector area of 100 m^2, calculate the corrected value of X for use in f chart.
b   For a collector area of 100 m^2, calculate the corrected value of Y.
c   For this collector area, what is f?
d   If the collector area is 50 m^2, what is f?

14.3   A residential building in Madison, WI is to be fitted with a solar heating system of the standard configuration for air systems shown in Figure 14.2.2. The UA of the building is 400 W/C. The storage capacity is the standard storage capacity in the f-chart correlation. Air flow rate is also the standard value. The collector faces south with a slope of 50°. The collector to be used has the test results shown on the plot; these were measured at the same flow rate as is to be used in the application. For January, water heating loads are expected to be 2.15 GJ, and space heating loads correspond to 830 degree days. Average ambient temperature is -7.0 C. $\overline{H}_T$ on the collector is 10.5 MJ/m^2 per day. Assume $(\overline{\tau\alpha})/(\tau\alpha)_n$ is 0.96.

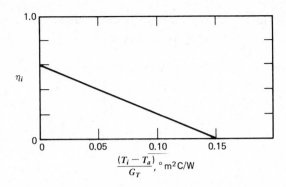

a What collector area is required to meet 0.5 of the January loads?

b Based on the data in Table 2.16.1, what is the annual solar fraction for the collector area of part a?

14.4 An air system in Madison, WI has $A_c = 40 \text{ m}^2$, $F_R(\tau\alpha)_n = 0.76$, $F_R U_L = 4.85 \text{ W/m}^2 \text{ C}$, 2 covers, $\beta = 50°$, $\gamma = 30°$, air flow rate $= 12 \text{ liters/m}^2 \text{ s}$, storage capacity $= 0.15 \text{ m}^3/\text{m}^2$ of rock of bulk (apparent) density $= 1700 \text{ kg/m}^3$. The building has $(UA)_h = 300 \text{ W/C}$. Calculate $\mathscr{F}$ for this system.

14.5 A residential building in Madison has an area-loss coefficient for thermal losses of 300 W/C. A liquid solar heating system is to be used on this building, with its collector mounted at a 60° slope and facing 30° west of south. A constant monthly hot water load of 1.9 GJ is expected. The collector area is to be 50 m², and the collector-heat exchanger parameters are $F'_R(\tau\alpha)_n = 0.72$ and $F'_R U_L = 3.5 \text{ W/m}^2 \text{ C}$. The collector has 2 covers. The storage capacity is to be 5000 liters. The load heat exchanger is to be sized so $\varepsilon_L C_{min}/(UA)_h = 2.0$. Estimate the annual solar fraction, $\mathscr{F}$.

14.6 For the heat pump characteristics shown in Figure 14.8.2 and the bin data for the month of January given in the table, determine the monthly work and auxiliary energy requirements for a house with a UA of 300 W/C. (The bins cover 2 C and are identified by their central temperatures.)

Bin	-26	-24	-22	-20	-18	-16	-14	-12	-10
hours	5.5	4	11.5	12	33.5	34	40	44	43
Bin	-8	-6	-4	-2	0	2	4	6	8
hours	77	76	92	110	103.5	46.5	8.5	3	0

14.7 If a solar system is added in parallel with the system of Problem 14.6, and supplies 30 percent of the heating load in January, what is the month's heat pump work and auxiliary energy requirements.

15.1 Using the shading plane idea, calculate the beam radiation incident on a window 2 m high × 3 m long, located 0.5 m below an infinitely long horizontal overhang with a projection of 0.7 m. The latitude is 38°, the window faces south, and the hour is one to two p.m. The normal beam radiation is 800 W/m². Do the calculation for **a** January 17, and **b** April 15.

15.2   For a vertical collector-storage wall located in Albuquerque and having dimensions and overhang as described in Problem 15.1, estimate the monthly average total radiation for the month of April.

15.3   A direct gain system has the same geometry as that of Figure 15.4.5, but is located in Minneapolis (latitude 44.9°). The ground reflectance is 0.6 in December and January, 0.4 in November and February, and 0.2 for other months.

    a   Estimate the beam, diffuse, ground-reflected, total shaded and total unshaded radiation on the receiver for the month of March

    b   Repeat for each month of the year and prepare a plot like Figure 15.4.5.

15.4   Redo Example 15.5.1, with these changes: The location is Madison, WI ($\phi = 43°$), the gap is 0.25 m, and the month is February ($\bar{H} = 9.12$ MJ/m², $\bar{T}_a = -6$ C). What is the net gain (or loss) from the window?

15.5   What would be the effect of adding night insulation to the collector-storage wall of Example 15.7.1? The resistance of the insulation is 1.6 m² C/W, and it is to be placed over the windows from 6 p.m. to 6 a.m. every day during the heating season.

18.1   For the conditions of problem 14.4 (but with $\gamma = 0$), use the generalized ϕ-charts to estimate the fraction by solar for the month of January. Assume that the collector receives air at a temperature of 20 C for the whole month. To simplify some of the calculations, the following information may be useful.

    For January; average day is 17, $\delta = -20.9$, $\omega_s = 69.12°$

    $\bar{R}_b = 2.67$, $\bar{K}_T = 0.44$, $\bar{H}_o = 13.23$ MJ/m², $\bar{H} = 5.85$ MJ/m²,

    $\bar{T}_a = -7$ C, $\bar{T}_{day} = -3$ C, $\bar{H}_d/\bar{H} = 0.39$, $\rho = 0.2$

Hours from Noon	r_t	r_d	θ_z	θ_T	$(\tau\alpha)/(\tau\alpha)_n$
0	0.179	0.167	63.9	13.9	0.96
$\frac{1}{2}$	0.176	0.165	64.3	15.7	0.96
$1\frac{1}{2}$	0.152	0.147	67.2	25.8	0.95
$2\frac{1}{2}$	0.108	0.113	72.6	38.8	0.94
$3\frac{1}{2}$	0.055	0.065	80.1	52.6	0.92

18.2   Repeat problem 18.1 using daily utilizability. Assume $(\overline{\tau\alpha})/(\tau\alpha)_n$ is 0.94.

18.3   In Example 15.5.1, the direct gain window was left uncovered during the night, which resulted in a large heat loss and a reduced net energy gain. Estimate the average net gain for the month of March that could be obtained with a perfect control system that would cover the window with a perfect insulator ($U = 0$) whenever the rate of energy gain is negative.

18.4 Using the generalized ϕ method, calculate the total energy collection for January in a location at 40°N latitude under the following conditions.

$$\overline{K}_T = 0.5$$
$$F_R U_L = 4.2 \text{ W/m}^2 \text{ C}$$
$$F_R(\tau\alpha) = 0.75 \text{ (constant)}$$
Collector tilt $= 55°$ (facing due south)
$$\overline{H}_o = 15.06 \text{ MJ/m}^2$$
$$\overline{T}_a = 0 \text{ C}$$
$$T_i = 50 \text{ C}$$

18.5 Repeat Problem 18.4 using the $\overline{\phi}$ method.

18.6 For Miami, estimate the contribution in July by solar for an absorption air conditioning system that has a COP of 0.7 when the temperature to the generator is greater than 70 C. The solar system characteristics are as given in Example 18.4.1. The collector tilt is 25° to the south. The collector area is 25 m². The cooling load in July is 5.1 GJ. Neglect storage losses.

18.7 For Problem 18.6, the tank has an area-loss coefficient product of 7 W/C. Estimate the reduction in performance due to tank losses, **a** if the tank is outside the building and **b** if the tank is inside the building.

APPENDIX B

Nomenclature

TABLE B-1

This Table of Nomenclature is a partial listing of symbols. Those that are used infrequently or in limited parts of the book are defined locally and do not appear on this list. In some cases section references are provided where there might be confusion about the significance of a symbol. A special table of radiation nomenclature is provided (Table B-2).

A	area, auxiliary, altitude
A_c	collector area
A_a	aperture area
A_r	receiver area
a, b	coefficients in empirical relationships
C	cost, concentration ratio (8.2), capacitance rate
C_A	cost per unit of collector area (11.1)
C_E	cost of equipment (11.1)
C_F	cost of energy from fuel (11.1)
C_p	specific heat
C_1, C_2	Planck's first and second radiation constants
C_b	bond conductance
D	diameter (defined locally), downpayment fraction
d	dust factor, market discount rate
E	energy, equation of time
e	emissive power, base of natural logarithm
F	fin efficiency factor (6.5), control function (defined locally)
F'	collector efficiency factor (6.5)
F''	collector flow factor (6.7)
F_R	collector heat removal factor (6.7)
F'_R	collector-heat exchanger factor (10.2)
F_{i-j}	diffuse energy leaving surface i that is incident on surface j without reflection/diffuse energy leaving surface i (view factor)
$\mathscr{F}$	annual fraction by solar energy (10.7)

678

f	monthly fraction by solar energy, focal length, fraction of days with various K_T (2.9)
f_i	shading factor (15.4)
$\bar{f}_i$	monthly average shading factor
g	gravitational constant
G_{sc}	solar constant, W/m^2
G	irradiance (see Table B-2)
H	daily irradiation (see Table B-2)
h	heat transfer coefficient, Planck's constant
I	hourly irradiation (see Table B-2), radiation intensity (3.7)
i	inflation rate
K	extinction coefficient
$K_{\tau\alpha}$	incidence angle modifier
k	thermal conductivity, Boltzmann's constant
k_T	hourly clearness index (2.9)
K_T	daily clearness index (2.9)
$\bar{K}_T$	monthly average daily clearness index (2.9)
L	longitude, length, distance, loss, load
l	length
m	mass, air mass, mean, mortgage interest rate
$\dot{m}$	flow rate
n	day of the year (1.6), index of refraction, hours of bright sunshine in a day
N	number of covers, day length, term of mortgage or economic analysis
N_p	payback time
P	power
p	pressure, vapor pressure
P_1	ratio, life cycle fuel savings/first year fuel energy cost (11.8)
P_2	ratio, owning cost/initial cost (11.8)
PWF	present worth factor (11.5)
Q	energy per unit time, energy
q	energy per unit time per unit area or length
r	radius, ratio
r_t	ratio, total radiation in hour/total in day (2.13)
r_d	ratio, diffuse radiation in hour/diffuse in day (2.13)
R	ratio of total radiation on tilted plane to that in plane of measurement (usually horizontal) (2.15), heat transfer resistance
$\bar{R}$	monthly average R (2.16)
R_b	ratio of beam radiation on tilted plane to that on plane of measurement (usually horizontal) (1.7)
$\bar{R}_b$	monthly average R_b (2.16)
R_d	ratio of diffuse radiation on tilted plane to that on plane of measurement (usually horizontal) (2.15)
ROI	return on investment
S	absorbed solar energy per unit area (5.8, 6.2)

s	shade factor
T	temperature
U	overall heat transfer coefficient
v	specific volume
W	distance between tubes
X	dimensionless collector loss ratio (14.2)
X'	modified dimensionless loss ratio (18.4)
$\overline{X}_c$	monthly average critical radiation ratio (18.3)
Y	dimensionless absorbed energy ratio (14.2)

Greek

α	absorptance, thermal diffusivity
β	slope
γ	surface azimuth angle, bond thickness, intercept factor
γ_s	solar azimuth angle (8.5)
δ	declination, thickness (defined locally), dispersion
δ_{ij}	delta-function: $\delta_{ij} = 1$ when $i = j$ and $\delta_{ij} = 0$ when $i \neq j$
ε	emittance, effectiveness
η	efficiency (defined locally)
θ	angle (defined locally), angle between surface normal and incident radiation
θ_c	acceptance half-angle of CPC
λ	wavelength
λ_c	cutoff wavelength of selective surface
μ	absolute viscosity, cosine of polar angle
v	kinematic viscosity, frequency
ρ	reflectance, ground reflectance, density
σ	Stefan–Boltzmann constant
τ	transmittance, time
ϕ	latitude, angle (defined locally), utilizability (18.1)
ψ	angle (defined locally), ground coverage of collector array
ω	hour angle, solid angle
ω_s	sunset (or sunrise) hour angle

Subscripts

a	air, ambient, absorbed, aperture, annual
b	blackbody, beam, back, bond, bed
c	collector, critical, cover, corrected
d	diffuse, day
e	effective, equivalent, edge
f	fin, fluid, fuel
g	glass, ground, glazing
i	incident, inlet
l	loss
m	moving, mean, maximum

n	normal, noon
o	overall, out, extraterrestrial
p	plate
r	radiation, reflected, receiver
s	storage, sunset, specular, scattered
T	tilted
t	top
u	useful
w	wind
z	zenith
λ	wavelength

Table B-2 RADIATION NOMENCLATURE (SEE SECTION 1.5)

G	irradiance (W/m^2)
H	irradiation for a day (J/m^2)
I	irradiation for an hour (J/m^2); can also be considered as an hourly average rate

Subscripts

b	beam
d	diffuse
n	normal
T	on tilted plane
o	extraterrestrial

The absence of subscript b or d means total radiation. The absence of subscript n or T means radiation on a horizontal surface. A "bar" above H or I means monthly average.

Examples

G_o	is extraterrestrial irradiance on horizontal plane.
I_d	is an hour's diffuse radiation on a horizontal plane.
$\bar{H}_b$	is a monthly average daily beam radiation on a horizontal surface.
$\bar{H}_T$	is monthly average daily radiation (beam plus diffuse) on a tilted plane.

APPENDIX   C

International System of Units*

SI UNITS

Basic units (name, symbol, quantity)

meter, m   length
kilogram, kg   mass
second, s   time
kelvin, K   thermodynamic temperature

Derived units

All other units are derived from basic and supplementary units. Some derived units have special names.

Decimal multiples of units

The following prefixes are recommended for use with SI units:
tera T 10^{12}, giga G 10^{9}
mega M 10^{6}, kilo k 10^{3}
milli m 10^{-3}, micro μ 10^{-6}
nano n 10^{-9}, pico p 10^{-12}
femto f 10^{-15}, atto a 10^{-18}
The use of the following prefixes should be limited:
hecto h 10^{2}, deca da 10
deci d 10^{-1} centi c 10^{-2}

SOME CONVERSIONS OF UNITS

Exact conversion factors are indicated by an asterisk(*)

---

* This table of conversion factors and constants is an edited version of a table supplied through the courtesy of the Division of Mechanical Engineering of CSIRO, Australia.

Length m, m/s

1 ft = 0.304 8* m
1 in = 25.4* mm
1 mile = 1.609 km
1 ft/min = 0.005 08* m/s
1 mile/hr = 0.447 0 m/s
1 km/hr = 0.277 78 m/s

Area m^2

$1\ ft^2 = 0.092\ 903\ 04^*\ m^2$
$1\ in.^2 = 0.000\ 645\ 16^*\ m^2$
$1\ mile^2 = 2.590\ km^2$

Volume m^3, m^3/kg, m^3/s

(Note: 1 liter = $10^{-3}\ m^3$)
$1\ ft^3 = 28.32$ liters
1 U.K. gal = 4.546 liters
1 U.S. gal = 3.785 liters
$1\ ft^3/lb = 0.062\ 43\ m^3/kg$
1 cfm = 0.471 9 liters/s
1 U.K. gpm = 0.075 77 liters/s
1 U.S. gpm = 0.063 01 liters/s
$1\ cfm/ft^2 = 5.080$ liters/s m^2

Force newton $N \equiv kg\ m/s^2$, N/m

pascal $Pa \equiv N/m^2$
1 lbf = 4.448 N
1 lbf/ft = 14.59 N/m
1 dyne/cm = 1 (mN)/m
1 mm H_2O = 9.806 65* Pa
1 bar = 10^5 Pa
1 psi = 6.894 kPa
1 in H_2O = 249.1 Pa
1 mm Hg = 133.3 Pa
1 at = kgf/cm^2 = 98.066 5* kPa
1 atm = 101.325* kPa

Energy joule $J \equiv Nm = Ws$, J/kg, J/kg C

1 kWh = 3.6* MJ
1 Btu = 1.055 kJ
1 Therm = 105.5 MJ
1 kcal = 4.186 8* kJ

1 Btu/lb = 2.326* kJ/kg
1 Btu/lb F = 4.186 8* kJ/kg C
1 Btu/ft^2 = 0.01136 MJ/m^2
1 cal/cm^2 = 0.04187 MJ/m^2

Power watt $\equiv$ J/s = N m/s, W/m^2, W/m^2 C, W/m C

1 Btu/h = 0.293 1 W
1 kcal/h = 1.163* W
1 hp = 0.745 7 kW
1 Ton refr. = 3.517 kW
1 W/ft^2 = 10.76 W/m^2
1 Btu/h ft^2 F = 5.678 W/m^2 C
1 Btu/h ft F = 1.731 W/m C
1 Btu/h ft^2 (F/in) = 0.1443 W/m C
1 Btu/ft^{2}h = 3.155 W/m^2

Viscosity Pa s = N s/m^2 = kg/m s

1 cP (centipoise) = 10^{-3} Pa s
1 lbf h/ft^2 = 0.172 4 MPa s

Mass kg, kg/m^3, kg/s, kg/s m^2

1 lb = 0.453 592 37* kg
1 oz = 28.35 g
1 lb/ft^3 = 16.02 kg/m^3
1 g/cm^3 = 10^3 kg/m^3
1 lb/h = 0.000 125 6 kg/s
1 lb/h ft^2 = 0.001 356 kg/s m^2

Plane angle rad

2π rad = 360* degrees

Diffusivity m^2/s

1 cST (centistoke) = 10^{-6} m^2/s
1 ft^2/h = 25.81 $\times$ 10^{-6} m^2/s

APPENDIX D

Tables of Monthly $\bar{R}_b$ as Function of $(\phi - \beta)$, ϕ, and γ

These tables are calculated by the methods outlined in Section 2.16. These $\bar{R}_b$ values are the basis of Figures 2.16.1. In the Northern Hemisphere use months across the top. In Southern Hemisphere use months across the bottom.

Table D-1 Values of $\bar{R}_b$ for (Latitude-Tilt) = 15.0

AZIMUTH ANGLE = 0

LATITUDE	JAN	FEB	MAR	APR	MAY	JUN	JUL	AUG	SEP	OCT	NOV	DEC
0	.82	.87	.95	1.03	1.10	1.14	1.12	1.06	.98	.90	.83	.80
10	.92	.95	.97	1.00	1.03	1.04	1.03	1.01	.99	.96	.93	.92
20	1.10	1.07	1.03	1.00	.98	.97	.98	.99	1.02	1.06	1.11	1.11
30	1.38	1.26	1.14	1.03	.96	.92	.95	1.02	1.09	1.22	1.34	1.42
35	1.59	1.40	1.21	1.06	.97	.92	.94	1.07	1.15	1.48	1.53	1.66
40	1.89	1.58	1.31	1.15	.98	.92	.96	1.07	1.22	1.48	1.80	1.99
45	2.32	1.83	1.43	1.21	1.00	.92	.96	1.11	1.31	1.96	2.17	2.50
50	3.00	2.18	1.60	1.29	1.03	.93	.97	1.16	1.42	1.96	2.74	3.33
55	4.19	2.72	1.82	1.39	1.05	.94	.99	1.23	1.57	2.36	3.69	4.91
60	6.79	3.59	2.13						1.77	2.98	5.56	8.86

AZIMUTH ANGLE = 15.0

LATITUDE	JAN	FEB	MAR	APR	MAY	JUN	JUL	AUG	SEP	OCT	NOV	DEC
0	.81	.87	.95	1.03	1.10	1.13	1.12	1.06	.98	.90	.83	.80
10	.93	.95	.98	1.00	1.03	1.04	1.03	1.01	.99	.96	.93	1.00
20	1.07	1.06	1.03	1.00	.98	.97	.98	.99	1.02	1.05	1.09	1.11
30	1.36	1.25	1.13	1.03	.97	.93	.95	1.02	1.08	1.21	1.33	1.43
35	1.57	1.38	1.21	1.06	.97	.93	.95	1.04	1.15	1.32	1.51	1.63
40	1.85	1.56	1.30	1.15	.99	.93	.95	1.07	1.21	1.46	1.77	1.96
45	2.27	1.84	1.42	1.21	1.01	.93	.96	1.11	1.31	1.65	2.12	2.45
50	2.92	2.14	1.58	1.29	1.03	.93	.97	1.17	1.41	1.92	2.67	3.25
55	4.08	2.65	1.79	1.39	1.06	.94	.97	1.17	1.56	2.30	3.59	4.78
60	6.58	3.50	2.09				.99	1.23	1.76	2.90	5.40	8.58

AZIMUTH ANGLE = 30.0

LATITUDE	JAN	FEB	MAR	APR	MAY	JUN	JUL	AUG	SEP	OCT	NOV	DEC
0	.81	.87	.94	1.01	1.08	1.11	1.10	1.04	.97	.89	.83	.80
10	.93	.95	.98	1.00	1.02	1.03	1.03	1.01	1.02	.95	.94	.92
20	1.32	1.06	1.03	1.03	.98	.97	.98	1.00	1.08	1.08	1.09	1.36
30	1.50	1.22	1.12	1.06	.97	.94	.95	1.04	1.13	1.28	1.33	1.56
35	1.75	1.34	1.19	1.09	.98	.94	.96	1.04	1.19	1.41	1.46	1.85
40	2.12	1.49	1.27	1.14	1.00	.94	.96	1.07	1.28	1.50	1.68	2.28
45	2.70	1.71	1.38	1.20	1.00	.94	.96	1.12	1.28	1.82	2.09	3.00
50	3.73	2.01	1.52	1.28	1.04	.95	.97	1.17	1.52	2.16	2.30	4.35
55	5.97	2.46	1.72	1.38	1.07	.95	1.00	1.23	1.72	2.69	4.91	7.77
60		3.21	2.00									

LATITUDE	JUL	AUG	SEP	OCT	NOV	DEC	JAN	FEB	MAR	APR	MAY	JUN

Table D-2   Values of $\bar{R}_b$ for (Latitude-Tilt) = 10.0

AZIMUTH ANGLE = 0

LATITUDE	JAN	FEB	MAR	APR	MAY	JUN	JUL	AUG	SEP	OCT	NOV	DEC
0	.88	.92	.97	1.03	1.08	1.10	1.09	1.05	1.00	.94	.89	.87
10	1.00	1.00	1.00	1.00	1.00	1.00	1.00	1.00	1.00	1.00	1.00	1.00
20	1.18	1.13	1.06	1.00	.94	.90	.92	.99	1.04	1.12	1.17	1.17
30	1.48	1.33	1.17	1.03	.94	.89	.91	.99	1.17	1.27	1.44	1.54
35	1.71	1.47	1.25	1.06	.95	.89	.92	1.03	1.24	1.35	1.64	1.79
40	2.02	1.66	1.34	1.14	.96	.89	.93	1.06	1.33	1.56	1.92	2.15
45	2.48	1.92	1.47	1.14	.98	.89	.93	1.10	1.44	1.76	2.31	2.70
50	3.20	2.29	1.64	1.21	1.00	.90	.95	1.15	1.59	2.04	2.91	3.58
55	4.48	2.85	1.87	1.28	1.03	.90	.95	1.21	1.80	2.46	3.92	5.26
60	7.23	3.77	2.19	1.38						3.10	5.90	9.47

AZIMUTH ANGLE = 15.0

LATITUDE	JAN	FEB	MAR	APR	MAY	JUN	JUL	AUG	SEP	OCT	NOV	DEC
0	.88	.92	.97	1.03	1.07	1.10	1.09	1.05	.99	.94	.89	.87
10	1.00	1.00	1.00	1.00	1.00	1.00	1.00	1.00	1.00	1.00	1.00	1.00
20	1.17	1.12	1.06	1.00	.94	.90	.92	.99	1.04	1.11	1.16	1.17
30	1.47	1.31	1.16	1.03	.94	.90	.92	1.01	1.16	1.26	1.42	1.52
35	1.69	1.45	1.24	1.06	.95	.89	.92	1.03	1.23	1.37	1.62	1.76
40	2.00	1.63	1.33	1.14	.96	.89	.92	1.06	1.32	1.52	1.88	2.11
45	2.43	1.89	1.45	1.14	.98	.90	.93	1.10	1.43	1.72	2.26	2.63
50	3.12	2.24	1.62	1.21	1.00	.90	.95	1.15	1.58	2.00	2.84	3.49
55	4.35	2.78	1.84	1.28	1.03	.91	.96	1.21	1.79	2.40	3.81	5.11
60	7.01	3.66	2.15	1.38						3.02	5.72	9.17

AZIMUTH ANGLE = 30.0

LATITUDE	JAN	FEB	MAR	APR	MAY	JUN	JUL	AUG	SEP	OCT	NOV	DEC
0	.89	.92	.97	1.02	1.06	1.08	1.07	1.04	.99	.94	.90	.88
10	1.00	1.00	1.00	1.00	1.00	1.00	1.00	1.00	1.00	1.00	1.00	1.00
20	1.16	1.11	1.05	1.00	.94	.91	.93	.99	1.05	1.10	1.14	1.17
30	1.41	1.28	1.15	1.03	.95	.91	.93	1.01	1.15	1.23	1.37	1.46
35	1.60	1.40	1.21	1.06	.96	.90	.94	1.03	1.21	1.33	1.54	1.67
40	1.87	1.56	1.30	1.14	.98	.91	.94	1.07	1.30	1.47	1.78	1.98
45	2.26	1.78	1.41	1.21	.99	.91	.95	1.11	1.41	1.63	2.12	2.44
50	2.88	2.10	1.56	1.28	1.02	.92	.95	1.16	1.56	1.89	2.63	3.21
55	3.97	2.57	1.76	1.38	1.04	.92	.97	1.22	1.73	2.24	3.49	4.65
60	6.35	3.35	2.04							2.79	5.20	8.29

LATITUDE	JUL	AUG	SEP	OCT	NOV	DEC	JAN	FEB	MAR	APR	MAY	JUN

Table D-3   Values of $\bar{R}_b$ for (Latitude-Tilt) = 5.0

AZIMUTH ANGLE= .0

AZIMUTH ANGLE= 15.0

AZIMUTH ANGLE= 30.0

Table D-4 Values of $\overline{R}_b$ for (Latitude-Tilt) = 0.0

AZIMUTH ANGLE = 0.0

LATITUDE	JAN	FEB	MAR	APR	MAY	JUN	JUL	AUG	SEP	OCT	NOV	DEC
0	1.00	1.00	1.00	1.00	1.00	1.00	1.00	1.00	1.00	1.00	1.00	1.00
10	1.13	1.08	1.03	.92	.93	.91	.92	.95	1.04	1.07	1.20	1.17
20	1.33	1.22	1.09	.97	.87	.85	.86	.94	1.07	1.17	1.30	1.37
30	1.60	1.43	1.20	1.03	.87	.81	.84	.96	1.12	1.35	1.60	1.74
35	1.91	1.59	1.28	1.03	.88	.81	.83	.96	1.17	1.48	1.82	2.02
40	2.26	1.79	1.39	1.06	.89	.80	.83	1.01	1.24	1.64	2.15	2.42
45	2.76	2.07	1.51	1.17	.90	.81	.84	1.04	1.33	1.86	2.55	3.02
50	3.55	2.46	1.68	1.17	.93	.81	.85	1.09	1.45	2.16	3.20	4.00
55	4.94	3.06	1.92	1.25	.95	.81	.86	1.09	1.60	2.60	4.30	5.85
60	7.95	4.03	2.25	1.34	.95	.82	.87	1.15	1.81	3.28	6.44	10.48

AZIMUTH ANGLE = 15.0

LATITUDE	JAN	FEB	MAR	APR	MAY	JUN	JUL	AUG	SEP	OCT	NOV	DEC
0	1.00	1.00	1.00	1.00	1.00	1.00	1.00	1.00	1.00	1.00	1.00	1.00
10	1.13	1.08	1.03	.92	.93	.91	.92	.95	1.04	1.07	1.19	1.14
20	1.32	1.21	1.09	.97	.89	.85	.87	.94	1.07	1.33	1.57	1.35
30	1.63	1.41	1.19	1.03	.88	.82	.85	.96	1.17	1.45	1.78	1.71
35	1.87	1.56	1.27	1.07	.88	.82	.84	.97	1.17	1.61	1.86	1.98
40	2.21	1.76	1.37	1.07	.90	.81	.84	1.01	1.24	1.82	2.49	2.36
45	2.62	2.02	1.49	1.17	.91	.82	.85	1.05	1.33	2.11	2.49	2.94
50	3.45	2.40	1.66	1.25	.93	.82	.86	1.05	1.44	2.53	3.12	3.89
55	4.79	2.97	1.88	1.25	.93	.83	.87	1.10	1.59	2.60	4.17	5.67
60	7.69	3.91	2.20	1.34	.96	.83	.88	1.16	1.80	3.18	6.24	10.15

AZIMUTH ANGLE = 30.0

LATITUDE	JAN	FEB	MAR	APR	MAY	JUN	JUL	AUG	SEP	OCT	NOV	DEC
0	1.00	1.00	1.00	1.00	1.00	1.00	1.00	1.00	1.00	1.00	1.00	1.00
10	1.12	1.07	1.02	.98	.94	.92	.93	.96	1.04	1.05	1.25	1.12
20	1.28	1.18	1.07	.98	.90	.87	.89	.95	1.05	1.15	1.50	1.31
30	1.56	1.36	1.17	1.01	.89	.85	.87	.97	1.15	1.40	1.69	1.62
35	1.77	1.49	1.24	1.03	.90	.84	.87	.97	1.22	1.54	1.94	1.82
40	2.08	1.66	1.33	1.07	.92	.84	.87	1.03	1.30	1.73	2.30	2.20
45	2.48	1.90	1.44	1.17	.93	.84	.88	1.06	1.41	1.98	2.86	2.71
50	3.16	2.23	1.60	1.25	.95	.84	.89	1.11	1.55	2.36	3.80	3.55
55	4.36	2.73	1.80	1.25	.98	.84	.89	1.17	1.74	2.93	5.15	5.15
60	6.95	3.56	2.09	1.35	.98	.85	.90	1.17	1.74	2.93	5.65	9.15

| LATITUDE | JUL | AUG | SEP | OCT | NOV | DEC | JAN | FEB | MAR | APR | MAY | JUN |

Table D-5 Values of $\bar{R}_b$ for (Latitude-Tilt) = − 5.0

AZIMUTH ANGLE= .0

LATITUDE	JAN	FEB	MAR	APR	MAY	JUN	JUL	AUG	SEP	OCT	NOV	DEC
0	1.05	1.03	1.00	.97	.95	.94	.94	.96	.99	1.02	1.04	1.05
10	1.18	1.15	1.09	.94	.88	.85	.86	.90	1.00	1.08	1.16	1.20
20	1.39	1.25	1.09	.94	.83	.76	.79	.91	1.03	1.20	1.36	1.43
25												
30	1.73	1.47	1.20	.97	.83	.75	.79	.92	1.14	1.37	1.68	1.82
35	1.99	1.62	1.28	1.00	.83	.75	.79	.94	1.23	1.67	2.04	2.53
40	2.35	1.83	1.38	1.03	.84	.76	.80	.97	1.32	1.89	2.64	3.17
45	2.87	2.12	1.51	1.08	.86	.76	.81	1.01	1.44	2.05	3.31	4.17
50	3.68	2.52	1.69	1.14	.88	.77	.81	1.05	1.58	2.65	3.44	6.08
55	5.12	3.13	1.92	1.21	.88	.77	.82	1.11	1.79			
60	8.22	4.12	2.25	1.30	.90	.77	.82	1.11	1.79	3.33	6.64	10.87

AZIMUTH ANGLE= 15.0

LATITUDE	JAN	FEB	MAR	APR	MAY	JUN	JUL	AUG	SEP	OCT	NOV	DEC
0	1.05	1.03	1.00	.97	.95	.94	.95	.97	.99	1.02	1.04	1.05
10	1.18	1.14	1.09	.95	.89	.86	.87	.90	1.00	1.09	1.16	1.20
20	1.37	1.24	1.09	.95	.85	.80	.82	.91	1.03	1.18	1.35	1.78
25												
30	1.70	1.40	1.27	1.01	.84	.77	.80	.93	1.16	1.38	1.85	3.07
35	1.93	1.60	1.37	1.04	.85	.77	.80	.95	1.32	1.48	2.47	2.47
40	2.29	1.80	1.66	1.15	.87	.77	.81	.98	1.43	1.85	2.52	3.04
45	2.58	2.04	1.89	1.22	.89	.77	.82	1.04	1.58	1.57	2.20	4.04
50	4.06	2.45	2.20	1.31	.91	.78	.83	1.06	1.78	2.57	4.30	5.89
55	4.96	3.04										
60	7.95	4.00									3.23	10.52

AZIMUTH ANGLE= 30.0

LATITUDE	JUL	AUG	SEP	OCT	NOV	DEC	JAN	FEB	MAR	APR	MAY	JUN
0	1.04	1.02	1.00	.98	.96	.95	.95	.97	.99	1.02	1.04	1.05
10	1.15	1.09	1.02	.95	.91	.87	.88	.92	1.00	1.07	1.14	1.17
20	1.33	1.21	1.08	.99	.87	.83	.85	.93	1.03	1.16	1.29	1.36
25												
30	1.61	1.52	1.17	1.01	.86	.80	.83	.94	1.14	1.31	1.55	1.68
35	1.83	1.69	1.24	1.05	.87	.79	.83	.97	1.29	1.42	1.74	1.93
40	2.13	1.93	1.33	1.09	.88	.79	.83	1.00	1.40	1.56	2.07	2.81
45	2.57	2.27	1.44	1.15	.90	.80	.84	1.03	1.54	1.01	2.37	3.68
50	3.27	2.78	1.60	1.23	.92	.80	.85	1.08	1.73	2.39	3.91	5.33
55	4.50											
60	7.17	3.63	2.09	1.32	.94	.80	.86	1.14	1.73	2.97	5.81	9.47

| LATITUDE | JUL | AUG | SEP | OCT | NOV | DEC | JAN | FEB | MAR | APR | MAY | JUN |

Table D-6 Values of $\bar{R}_b$ for (Latitude-Tilt) = −10.0

LATITUDE	JAN	FEB	MAR	APR	MAY	JUN	JUL	AUG	SEP	OCT	NOV	DEC
						AZIMUTH ANGLE= .0						
0	1.09	1.05	1.00	.94	.89	.87	.88	.92	.97	1.03	1.08	1.10
10	1.23	1.13	1.02	.91	.83	.79	.81	.88	.98	1.10	1.20	1.25
20	1.44	1.27	1.09	.91	.79	.74	.76	.86	1.01	1.21	1.39	1.49
30	1.79	1.49	1.20	.94	.78	.71	.74	.87	1.09	1.39	1.71	1.89
35	2.05	1.65	1.27	.97	.78	.70	.74	.88	1.14	1.52	1.94	2.19
40	2.42	1.86	1.38	1.00	.78	.70	.74	.90	1.21	1.69	2.25	2.62
45	2.95	2.15	1.51	1.04	.79	.70	.74	.93	1.30	1.91	2.71	3.25
50	3.79	2.56	1.68	1.10	.81	.70	.75	.96	1.41	2.22	3.40	4.30
55	5.25	3.17	1.91	1.17	.83	.71	.76	1.01	1.56	2.67	4.54	6.26
60	8.42	4.18	2.24	1.26	.85	.71	.77	1.06	1.76	3.36	6.79	11.18
						AZIMUTH ANGLE= 15.0						
0	1.09	1.05	1.00	.94	.90	.88	.89	.92	.98	1.03	1.08	1.10
10	1.22	1.13	1.02	.92	.84	.80	.81	.88	.98	1.09	1.19	1.24
20	1.42	1.26	1.08	.92	.80	.75	.77	.87	1.01	1.20	1.38	1.47
30	1.75	1.47	1.19	.95	.79	.72	.75	.88	1.09	1.37	1.67	1.85
35	2.01	1.62	1.26	.97	.79	.72	.75	.89	1.14	1.49	1.90	2.14
40	2.36	1.82	1.36	1.01	.80	.71	.75	.91	1.21	1.65	2.20	2.55
45	2.87	2.10	1.48	1.05	.81	.71	.75	.94	1.29	1.87	2.64	3.16
50	3.67	2.49	1.65	1.11	.82	.71	.76	.97	1.41	2.16	3.30	4.17
55	5.09	3.08	1.87	1.18	.84	.72	.77	1.02	1.55	2.59	4.40	6.07
60	8.15	4.05	2.19	1.27	.86	.72	.78	1.07	1.75	3.26	6.57	10.81
						AZIMUTH ANGLE= 30.0						
0	1.08	1.04	1.00	.95	.91	.89	.90	.93	.98	1.03	1.07	1.09
10	1.19	1.11	1.02	.93	.85	.82	.84	.90	.98	1.08	1.17	1.21
20	1.36	1.22	1.07	.93	.83	.78	.80	.88	1.01	1.17	1.33	1.41
30	1.65	1.40	1.16	.96	.82	.76	.78	.90	1.08	1.32	1.58	1.74
35	1.87	1.54	1.23	.98	.82	.75	.78	.91	1.13	1.43	1.78	1.99
40	2.18	1.71	1.32	1.02	.83	.75	.78	.93	1.19	1.57	2.04	2.35
45	2.63	1.95	1.43	1.06	.84	.75	.79	.96	1.27	1.76	2.42	2.90
50	3.35	2.30	1.58	1.12	.85	.75	.79	1.00	1.38	2.02	3.01	3.79
55	4.61	2.81	1.79	1.19	.87	.75	.80	1.04	1.52	2.40	3.99	5.48
60	7.34	3.67	2.08	1.28	.89	.75	.81	1.10	1.70	2.98	5.93	9.73
LATITUDE	JUL	AUG	SEP	OCT	NOV	DEC	JAN	FEB	MAR	APR	MAY	JUN

Table D-7 Values of $\bar{R}_b$ for (Latitude-Tilt) $= -15.0$

AZIMUTH ANGLE = .0

LATITUDE	JAN	FEB	MAR	APR	MAY	JUN	JUL	AUG	SEP	OCT	NOV	DEC
0	1.12	1.06	.98	.90	.83	.80	.81	.87	.95	1.03	1.11	1.14
10	1.26	1.14	1.07	.87	.74	.68	.70	.81	.99	1.10	1.23	1.29
20	1.48	1.28	1.18	.87	.74	.65	.70	.81	.99	1.21	1.42	1.54
30	1.83	1.50	1.26	.92	.72	.65	.68	.82	1.06	1.39	1.74	1.94
35	2.07	1.67	1.36	.96	.73	.64	.68	.83	1.11	1.52	1.98	2.25
40	2.47	1.87	1.46	1.00	.73	.64	.68	.85	1.16	1.69	2.30	2.68
45	3.06	2.17	1.66	1.05	.74	.64	.69	.88	1.27	1.91	2.76	3.33
50	3.86	2.58	1.88	1.12	.77	.65	.70	.91	1.38	2.22	3.45	4.40
55	5.35	3.19	2.21	1.21	.79	.65	.71	.95	1.52	2.67	4.61	6.00
60	8.57	4.20		1.21	.79		.71	1.00	1.72	3.36	6.89	11.39

AZIMUTH ANGLE = 15.0

LATITUDE	JAN	FEB	MAR	APR	MAY	JUN	JUL	AUG	SEP	OCT	NOV	DEC
0	1.12	1.06	.98	.90	.84	.81	.82	.88	.95	1.03	1.10	1.13
10	1.25	1.14	1.01	.88	.78	.74	.76	.84	.96	1.09	1.20	1.26
20	1.45	1.27	1.17	.88	.75	.69	.72	.82	.99	1.20	1.40	1.51
30	1.79	1.48	1.24	.91	.74	.67	.70	.83	1.06	1.37	1.71	1.90
35	2.05	1.63	1.34	.93	.74	.66	.69	.84	1.10	1.45	1.93	2.19
40	2.41	1.83	1.46	.97	.74	.66	.70	.86	1.18	1.65	2.20	2.61
45	2.93	2.11	1.63	1.01	.75	.66	.70	.89	1.26	1.87	2.68	3.24
50	3.74	2.50	1.85	1.04	.77	.66	.71	.92	1.37	2.16	3.35	4.26
55	5.18	3.00	2.16	1.13	.78	.66	.72	.96	1.52	2.59	4.47	6.19
60	8.28	4.07		1.22	.80		.72	1.01	1.71	3.25	6.66	11.02

AZIMUTH ANGLE = 30.0

LATITUDE	JAN	FEB	MAR	APR	MAY	JUN	JUL	AUG	SEP	OCT	NOV	DEC
0	1.10	1.05	.98	.91	.86	.83	.84	.89	.96	1.03	1.09	1.12
10	1.21	1.12	1.01	.89	.81	.77	.78	.85	.96	1.09	1.19	1.24
20	1.39	1.24	1.15	.92	.78	.73	.74	.86	.99	1.17	1.35	1.44
30	1.68	1.41	1.22	.95	.78	.71	.73	.87	1.05	1.32	1.61	1.77
35	1.91	1.54	1.30	.98	.77	.70	.73	.87	1.10	1.43	1.81	2.03
40	2.22	1.72	1.41	1.02	.78	.70	.74	.89	1.16	1.57	2.07	2.40
45	2.68	1.96	1.56	1.05	.79	.70	.74	.92	1.24	1.75	2.46	2.95
50	3.40	2.30	1.76	1.08	.80	.70	.75	.95	1.35	2.01	3.05	3.86
55	4.68	2.82	1.99	1.15	.82	.70	.76	.98	1.45	2.39	4.04	5.59
60	7.45	3.68	2.05	1.24	.84		.76	1.05	1.67	2.97	6.00	9.90

| LATITUDE | JUL | AUG | SEP | OCT | NOV | DEC | JAN | FEB | MAR | APR | MAY | JUN |

Table D-8 Values of $\bar{R}_b$ for (Latitude-Tilt) $= -20.0$

AZIMUTH ANGLE = 0

LATITUDE	JAN	FEB	MAR	APR	MAY	JUN	JUL	AUG	SEP	OCT	NOV	DEC
0	1.14	1.06	.96	.85	.76	.72	.74	.81	.92	1.03	1.12	1.17
10	1.28	1.15	.99	.83	.71	.65	.68	.78	.92	1.10	1.25	1.32
20	1.50	1.29	1.05	.83	.66	.61	.64	.76	.96	1.21	1.44	1.57
30	1.86	1.51	1.15	.85	.66	.59	.62	.77	1.02	1.38	1.76	1.98
35	2.13	1.67	1.23	.88	.67	.58	.62	.78	1.08	1.51	2.00	2.29
40	2.51	1.88	1.33	.91	.67	.58	.62	.80	1.14	1.60	2.38	2.73
45	3.05	2.17	1.45	.95	.68	.58	.63	.82	1.22	1.90	2.78	3.39
50	3.91	2.58	1.62	1.00	.69	.58	.63	.85	1.33	2.11	3.48	4.46
55	5.41	3.19	1.84	1.06	.71	.59	.64	.89	1.47	2.65	4.65	6.48
60	8.64	4.19	2.16	1.14	.73	.59	.65	.94	1.66	3.33	6.93	11.53

AZIMUTH ANGLE = 15.0

LATITUDE	JAN	FEB	MAR	APR	MAY	JUN	JUL	AUG	SEP	OCT	NOV	DEC
0	1.14	1.06	.96	.86	.77	.73	.75	.82	.92	1.03	1.12	1.16
10	1.27	1.14	.99	.84	.72	.67	.69	.79	.93	1.09	1.24	1.31
20	1.48	1.27	1.04	.84	.69	.63	.66	.77	.96	1.19	1.42	1.54
30	1.82	1.48	1.15	.86	.68	.60	.64	.78	1.03	1.36	1.72	1.93
35	2.08	1.63	1.22	.89	.69	.60	.64	.79	1.08	1.48	1.95	2.23
40	2.44	1.83	1.31	.92	.69	.60	.64	.81	1.14	1.65	2.26	2.65
45	2.96	2.11	1.43	.96	.71	.60	.64	.84	1.23	1.85	2.70	3.09
50	3.79	2.50	1.59	1.01	.72	.60	.65	.87	1.33	2.17	3.38	3.92
55	5.23	3.09	1.81	1.08	.73	.60	.65	.90	1.47	2.57	4.50	6.27
60	8.36	4.06	2.11	1.16	.74	.60	.66	.95	1.66	3.23	6.70	11.14

AZIMUTH ANGLE = 30.0

LATITUDE	JAN	FEB	MAR	APR	MAY	JUN	JUL	AUG	SEP	OCT	NOV	DEC
0	1.12	1.05	.97	.88	.80	.76	.78	.84	.93	1.02	1.10	1.14
10	1.23	1.12	.99	.86	.76	.71	.73	.81	.93	1.07	1.20	1.26
20	1.41	1.23	1.04	.86	.73	.67	.70	.80	.96	1.16	1.36	1.46
30	1.70	1.41	1.13	.89	.72	.65	.69	.81	1.02	1.31	1.62	1.80
35	1.93	1.54	1.27	.94	.73	.65	.68	.83	1.07	1.41	1.81	2.06
40	2.24	1.71	1.38	.98	.73	.64	.68	.85	1.13	1.55	2.08	2.43
45	2.70	1.95	1.53	1.03	.74	.64	.69	.87	1.21	1.74	2.47	2.99
50	3.43	2.29	1.73	1.10	.75	.64	.69	.90	1.31	1.97	3.06	3.91
55	4.72	2.81	2.00	1.18	.76	.64	.70	.94	1.44	2.37	4.06	5.85
60	7.51	3.66	—	1.18	.78	—	.70	.99	1.62	2.94	6.03	10.01

LATITUDE	JUL	AUG	SEP	OCT	NOV	DEC	JAN	FEB	MAR	APR	MAY	JUN

Table D-9 Values of $\bar{R}_b$ for (Latitude-Tilt) = −25.0

AZIMUTH ANGLE = .0

LATITUDE	JAN	FEB	MAR	APR	MAY	JUN	JUL	AUG	SEP	OCT	NOV	DEC
0	1.16	1.06	.93	.80	.69	.64	.66	.75	.88	1.02	1.13	1.19
10	1.30	1.14	.96	.78	.64	.58	.61	.72	.88	1.08	1.26	1.35
20	1.52	1.28	1.02	.78	.61	.54	.57	.70	.92	1.19	1.47	1.59
30	1.88	1.50	1.12	.80	.60	.52	.56	.71	.97	1.37	1.77	2.00
35	2.15	1.66	1.19	.82	.61	.52	.55	.71	1.03	1.49	2.01	2.31
40	2.52	1.87	1.29	.85	.61	.51	.56	.74	1.09	1.66	2.33	2.76
45	3.07	2.15	1.41	.89	.63	.51	.56	.76	1.17	1.88	2.79	3.42
50	3.92	2.55	1.57	.93	.64	.52	.57	.79	1.28	2.18	3.49	4.20
55	5.42	3.16	1.79	.99	.66	.52	.57	.82	1.41	2.61	4.65	6.52
60	8.65	4.15	2.10	1.07	.66	.52	.58	.87	1.59	3.28	6.93	11.57

AZIMUTH ANGLE = 15.0

LATITUDE	JAN	FEB	MAR	APR	MAY	JUN	JUL	AUG	SEP	OCT	NOV	DEC
0	1.15	1.05	.94	.81	.71	.66	.68	.76	.89	1.02	1.13	1.18
10	1.28	1.13	.96	.79	.65	.60	.62	.73	.89	1.07	1.24	1.33
20	1.49	1.26	1.01	.79	.63	.56	.59	.73	.92	1.17	1.43	1.56
30	1.83	1.47	1.11	.81	.62	.54	.58	.73	.99	1.34	1.73	1.95
35	2.09	1.62	1.18	.84	.63	.54	.57	.75	1.03	1.46	1.96	2.25
40	2.45	1.82	1.27	.87	.63	.53	.58	.75	1.07	1.62	2.27	2.68
45	2.98	2.09	1.39	.90	.64	.53	.58	.78	1.17	1.82	2.71	3.31
50	3.85	2.48	1.54	.95	.66	.53	.59	.80	1.28	2.11	3.38	4.35
55	5.06	3.06	1.75	1.01	.66	.54	.59	.84	1.41	2.53	4.50	6.31
60	8.36	4.01	2.05	1.09	.67	.54	.59	.88	1.59	3.18	6.69	11.18

AZIMUTH ANGLE = 30.0

LATITUDE	JAN	FEB	MAR	APR	MAY	JUN	JUL	AUG	SEP	OCT	NOV	DEC
0	1.13	1.04	.94	.83	.74	.70	.72	.79	.90	1.04	1.11	1.15
10	1.24	1.11	.96	.81	.70	.65	.67	.76	.93	1.06	1.20	1.28
20	1.41	1.22	1.01	.82	.68	.62	.64	.77	.99	1.14	1.36	1.48
30	1.71	1.39	1.09	.84	.67	.60	.63	.78	.99	1.29	1.62	1.81
35	1.93	1.52	1.16	.86	.67	.59	.63	.78	1.03	1.39	1.81	2.07
40	2.24	1.69	1.24	.89	.68	.59	.63	.82	1.16	1.53	2.06	2.44
45	2.70	1.93	1.34	.93	.69	.58	.63	.85	1.16	1.71	2.46	3.01
50	3.43	2.27	1.48	.98	.70	.58	.63	.88	1.26	1.96	3.06	3.93
55	4.72	2.77	1.67	1.04	.70	.58	.64	.88	1.38	2.32	4.05	5.67
60	7.51	3.61	1.94	1.12	.71	.58	.64	.93	1.55	2.89	6.03	10.03

LATITUDE	JUL	AUG	SEP	OCT	NOV	DEC	JAN	FEB	MAR	APR	MAY	JUN

Table D-10 Values of $\bar{R}_b$ for (Latitude-Tilt) = −30.0

AZIMUTH ANGLE = .0

LATITUDE	JAN	FEB	MAR	APR	MAY	JUN	JUL	AUG	SEP	OCT	NOV	DEC
0	1.17	1.05	.90	.74	.62	.56	.58	.69	.84	1.00	1.14	1.20
10	1.30	1.13	.92	.72	.57	.51	.53	.65	.87	1.06	1.26	1.36
20	1.52	1.26	.98	.74	.54	.47	.50	.64	.93	1.17	1.45	1.60
30	1.88	1.48	1.08	.76	.54	.45	.49	.65	.98	1.34	1.77	2.01
35	2.15	1.63	1.15	.79	.54	.45	.49	.66	1.04	1.62	2.00	2.32
40	2.52	1.84	1.24	.82	.55	.45	.49	.67	1.11	1.84	2.32	2.76
45	3.06	2.12	1.36	.87	.56	.45	.49	.69	1.21	2.13	2.78	3.42
50	3.91	2.51	1.51	.92	.57	.47	.50	.72	1.34	2.56	3.47	4.49
55	5.40	3.10	1.72	1.01	.60		.52	.75	1.53	3.23	4.62	6.51
60	8.62	4.10	2.03					.81			6.88	11.54

AZIMUTH ANGLE = 15.0

LATITUDE	JAN	FEB	MAR	APR	MAY	JUN	JUL	AUG	SEP	OCT	NOV	DEC
0	1.16	1.04	.90	.75	.63	.58	.60	.70	.84	.99	1.13	1.19
10	1.29	1.12	.92	.73	.59	.53	.56	.67	.85	1.05	1.24	1.34
20	1.49	1.24	.98	.74	.57	.50	.53	.66	.88	1.15	1.42	1.57
30	1.83	1.44	1.07	.76	.56	.48	.51	.67	.94	1.31	1.72	1.96
35	2.09	1.59	1.14	.78	.56	.47	.51	.68	.98	1.43	1.95	2.26
40	2.45	1.79	1.22	.81	.56	.47	.51	.69	1.04	1.58	2.25	2.68
45	2.97	2.06	1.34	.84	.58	.47	.51	.71	1.11	1.79	2.69	3.31
50	3.78	2.44	1.49	.89	.59	.47	.52	.74	1.21	2.06	3.35	4.35
55	5.22	3.01	1.69	.94	.62	.48	.54	.77	1.34	2.47	4.46	6.29
60	8.32	3.96	1.99	1.03				.82	1.53	3.12	6.65	11.15

AZIMUTH ANGLE = 30.0

LATITUDE	JAN	FEB	MAR	APR	MAY	JUN	JUL	AUG	SEP	OCT	NOV	DEC
0	1.13	1.03	.91	.78	.68	.63	.65	.74	.86	.99	1.10	1.16
10	1.24	1.09	.93	.77	.64	.59	.61	.71	.86	1.04	1.20	1.28
20	1.41	1.20	.97	.77	.62	.56	.59	.70	.89	1.12	1.35	1.48
30	1.70	1.37	1.06	.79	.62	.54	.57	.71	.94	1.26	1.60	1.81
35	1.92	1.49	1.12	.81	.62	.53	.57	.72	.99	1.36	1.80	2.07
40	2.23	1.66	1.19	.84	.63	.53	.57	.74	1.04	1.49	2.06	2.44
45	2.69	1.89	1.29	.87	.63	.52	.57	.76	1.11	1.67	2.44	3.00
50	3.41	2.22	1.43	.92	.64	.53	.58	.79	1.20	1.91	3.03	3.91
55	4.69	2.72	1.61	.97	.66			.82	1.32	2.26	4.01	5.65
60	7.46	3.56	1.88	1.06				.87	1.50	2.83	5.96	10.00

| LATITUDE | JUL | AUG | SEP | OCT | NOV | DEC | JAN | FEB | MAR | APR | MAY | JUN |

695

Table D-11 Values of $\bar{R}_b$ for Vertical Surface

AZIMUTH ANGLE = 0.0

LATITUDE	JAN	FEB	MAR	APR	MAY	JUN	JUL	AUG	SEP	OCT	NOV	DEC
0	.62	.38	.08	.00	.00	.00	.00	.0	.00	.28	.56	.69
10	.85	.58	.26	.01	.00	.00	.00	.00	.14	.48	.78	.94
20	1.16	.83	.46	.15	.01	.00	.00	.07	.32	.70	1.07	1.27
30	1.61	1.15	.69	.31	.12	.06	.09	.22	.52	.98	1.48	1.76
35	1.92	1.35	.82	.40	.19	.12	.15	.30	.63	1.16	1.76	2.11
40	2.39	1.61	.97	.50	.27	.19	.22	.39	.76	1.38	2.12	2.59
45	3.21	1.94	1.16	.61	.34	.25	.29	.48	.90	1.60	2.62	3.29
50	3.81	2.39	1.38	.73	.43	.32	.37	.58	1.07	2.00	3.36	4.40
55	5.35	3.05	1.66	.86	.51	.40	.44	.69	1.27	2.49	4.56	6.46
60	8.62	4.10	2.03	1.01	.60	.47	.52	.81	1.53	3.23	6.88	11.54

AZIMUTH ANGLE = 15.0

LATITUDE	JAN	FEB	MAR	APR	MAY	JUN	JUL	AUG	SEP	OCT	NOV	DEC
0	.60	.37	.17	.00	.00	.00	.00	.00	.11	.30	.54	.66
10	.83	.56	.30	.10	.06	.01	.03	.04	.21	.47	.76	.90
20	1.13	.80	.48	.21	.17	.10	.13	.14	.36	.68	1.04	1.23
30	1.55	1.11	.69	.36	.23	.16	.19	.27	.55	.95	1.43	1.70
35	1.86	1.31	.82	.44	.30	.22	.25	.34	.65	1.13	1.70	2.04
40	2.02	1.56	.96	.54	.37	.28	.32	.42	.78	1.39	2.05	2.50
45	2.82	1.88	1.14	.64	.45	.35	.39	.51	.92	1.59	2.53	3.18
50	3.68	2.31	1.35	.75	.53	.42	.46	.61	1.08	1.94	3.25	4.25
55	5.17	2.95	1.62	.88	.62	.48	.54	.71	1.28	2.41	4.41	6.24
60	8.32	3.96	1.99	1.03	.62	.48	.54	.82	1.53	3.12	6.65	11.15

AZIMUTH ANGLE = 30.0

LATITUDE	JAN	FEB	MAR	APR	MAY	JUN	JUL	AUG	SEP	OCT	NOV	DEC
0	.58	.44	.29	.12	.08	.00	.03	.04	.23	.39	.54	.63
10	.77	.58	.39	.22	.18	.12	.15	.17	.44	.52	.72	.83
20	1.03	.77	.53	.32	.27	.20	.23	.26	.59	.82	.96	1.11
30	1.40	1.04	.71	.44	.33	.25	.28	.36	.69	1.07	1.30	1.53
35	1.67	1.21	.82	.51	.38	.30	.34	.43	.80	1.25	1.54	1.83
40	2.03	1.43	.95	.60	.45	.35	.39	.50	.92	1.48	1.85	2.25
45	2.53	1.71	1.11	.69	.52	.41	.45	.58	1.07	1.79	2.28	2.85
50	3.30	2.10	1.30	.80	.59	.47	.52	.67	1.26	2.03	2.91	3.81
55	4.63	2.66	1.55	.89	.66	.53	.58	.76	1.55	2.83	5.96	10.00
60	7.46	3.56	1.88	1.06				.87	1.88			
LATITUDE	JUL	AUG	SEP	OCT	NOV	DEC	JAN	FEB	MAR	APR	MAY	JUN

696

APPENDIX   E

Properties of Materials

SOME PROPERTIES IN SI UNITS

Density kg/m³

Copper	8795
Steel	7850
Aluminium	2675
Glass, standard	2515
Concrete, typical, building	2400
Asbestos cement, sheet, 30 C	1900
Water, 4 C	1000
Ice, -1 C	918
Gypsum plaster, dry, 23 C	881
Oak, 14% wet	770
Pine, 15% wet	570
Pine fiberboard, 24 C	256
Cork board, dry, 18 C	144
Ebonite, expanded, 10 C	64
Mineral wool, batts, -2 C	32
Polyurethane, foam, rigid	24
Polystyrene, expanded, 10 C	16
Air, p_0, 20 C	1.204

Thermal conductivity, W/m C

Copper	385
Aluminium	211
Steel	47.6
Ice, -1 C	2.26
Concrete, typical, building	1.73
Glass, standard	1.05
Water, 20 C	0.596
Asbestos cement, sheet, 30 C	0.319
Gypsum plaster, dry, 23 C	0.170
Oak, 14% wet	0.160

Pine, 15 % wet   0.138
Pine fiberboard, 24 C   0.051 9
Cork board, dry, 18 C   0.041 8
Mineral wool, batts, -2 C   0.034 6
Polystyrene, expanded, 10 C   0.034 6
Ebonite, expanded, 10 C   0.030 3
Air, p_0, 20 C   0.026
Polyurethane, foam, rigid   0.024 5

Specific heat, kJ/kg C

Water, 20 C, p_0 4.19
Ice, -21 C to -1 C   2.10
Steam (c_p), 100 C, p_0 1.95
Air (c_p), 20 C, p_0 1.012
Concrete, 18 C   0.837

Heat of vaporization, kJ/kg

Water, 20 C   2454.0
Water, 100 C   2257.0
R12, 0 C, sat.   151.5
R22, 0 C, sat.   205.4
R11, 0 C, sat.   188.9
R500, 0 C, sat.   183.0
R717, 0 C, sat.   1263.3

Momentum diffusivity = kinematic viscosity, m^2/s

Air, 20 C, p_0 14.95×10^{-6}
Water, 20 C, p_0 1.01×10^{-6}

Heat (thermal) diffusivity, m^2/s

Air, 20 C, p_0 21.2×10^{-6}
Water, 20 C, p_0 0.142×10^{-6}

Mass diffusivity, m^2/s

Water vapor in air at 20 C, p_0 26.1×10^{-6}

Surface tension (N/m)

Water/air, 20 C, p_0 0.0728

MISCELLANEOUS INFORMATION

NTP (Normal temperature T_0 and pressure p_0):
  $T_0 = 273.15$ K $= 0$ C

$p_0 = 101.325$ kPa
Standard gravity g_0:
$g_0 = 9.806\ 65$ m/s^2
Velocity of sound in air:
344 m/s at p_0, 20 C, 50% R.H.
Gas constants:
$R_u = 8314.4$ J/kmol K (universal)
$R_a = 287.045$ J/kg K (air)
$R_v = 461.52$ J/kg K (water vapor)
Stefan–Boltzmann constant:
$\sigma = 5.669\ 7 \times 10^{-8}$ W/m^2 K^4

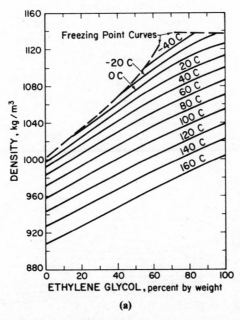

(a)

Figure E-1(a–f) Properties of aqueous ethylene glycol solutions. Adapted, with permission, from information supplied by Union Carbide. The manufacturer recommends only the use of inhibited ethylene glycol ("UCAR Thermofluid 17").

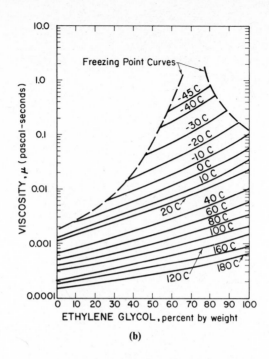

(b)

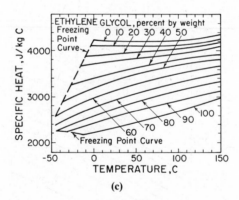

(c)

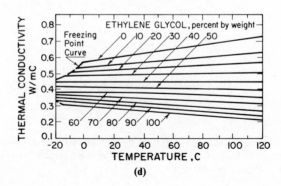

(d)

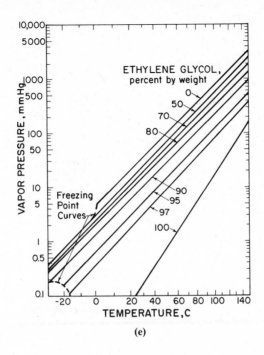

(e)

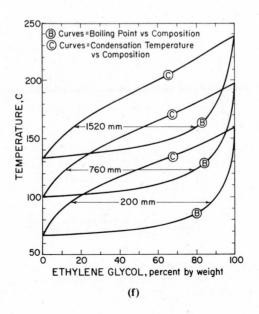

(f)

701

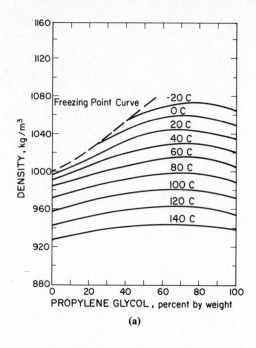

(a)

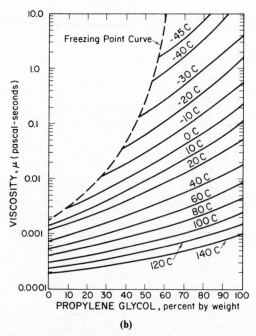

(b)

Figue E-2(a–f) Properties of aqueous propylene glycol solutions. Adapted, with permission, from information supplied by Union Carbide. The manufacturer recommends only the use of inhibited propylene glycol ("UCAR Foodfreeze 35").

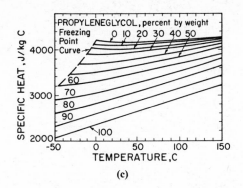

(c)

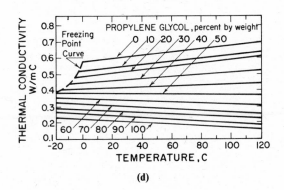

(d)

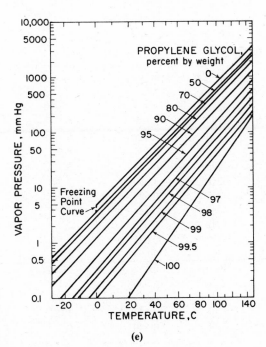

(e)

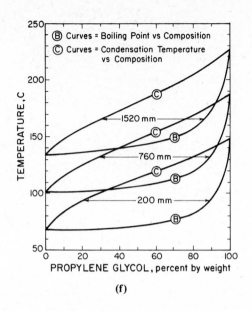

(f)

Properties of Saturated Water

T, C	ρ, kg/m^3	C_p, J/kg K	k, W/m K	μ, Pa s	α, m^2/s	Pr
0	1002	4218	0.552	17.9×10^{-4}	1.31×10^{-7}	13.06
20	1001	4182	0.597	10.1	1.43	7.02
40	995	4178	0.628	6.55	1.51	4.34
60	985	4184	0.651	4.71	1.55	3.02
80	974	4196	0.668	3.55	1.64	2.22
100	960	4216	0.680	2.82	1.68	1.74
120	945	4250	0.685	2.33	1.71	1.45
140	928	4283	0.684	1.99	1.72	1.24
160	910	4342	0.680	1.73	1.73	1.10
180	889	4417	0.675	1.54	1.72	1.00
200	867	4505	0.665	1.39	1.71	0.94
220	842	4610	0.652	1.26	1.68	0.89
240	816	4756	0.635	1.17	1.64	0.88
260	786	4949	0.611	1.08	1.58	0.87
280	753	5208	0.580	1.02	1.48	0.91
300	714	5728	0.540	0.96	1.32	1.02

Properties of Air at Atmospheric Pressure

T, C	ρ, kg/m^3	C_p, J/kg K	k, W/m K	μ, Pa s	α, m^2/s	Pr
0	1.292	1006	0.0242	1.72×10^{-5}	1.86×10^{-5}	0.72
20	1.204	1006	0.0257	1.81	2.12	0.71
40	1.127	1007	0.0272	1.90	2.40	0.70
60	1.059	1008	0.0287	1.99	2.69	0.70
80	0.999	1010	0.0302	2.09	3.00	0.70
100	0.946	1012	0.0318	2.18	3.32	0.69
120	0.898	1014	0.0333	2.27	3.66	0.69
140	0.854	1016	0.0345	2.34	3.98	0.69
160	0.815	1019	0.0359	2.42	4.32	0.69
180	0.779	1022	0.0372	2.50	4.67	0.69
200	0.746	1025	0.0386	2.57	5.05	0.68
220	0.715	1028	0.0399	2.64	5.43	0.68
240	0.688	1032	0.0412	2.72	5.80	0.68
260	0.662	1036	0.0425	2.79	6.20	0.68
280	0.638	1040	0.0437	2.86	6.59	0.68
300	0.616	1045	0.0450	2.93	6.99	0.68

APPENDIX F

Present Worth Factors (See Section 11.5)

Table F-1 Present Worth Factors for $N = 5$

d, Market Discount Rate, percent	i, Annual Inflation Rate, percent												
	0	1	2	3	4	5	6	7	8	9	10	11	12
0	5.000	5.101	5.204	5.309	5.416	5.526	5.637	5.751	5.867	5.985	6.105	6.228	6.353
1	4.853	4.950	5.049	5.150	5.253	5.358	5.466	5.575	5.686	5.799	5.915	6.033	6.153
2	4.713	4.807	4.902	4.999	5.098	5.199	5.302	5.407	5.514	5.623	5.734	5.847	5.962
3	4.580	4.669	4.761	4.854	4.950	5.047	5.146	5.246	5.349	5.454	5.561	5.669	5.780
4	4.452	4.538	4.626	4.716	4.808	4.901	4.996	5.093	5.192	5.293	5.395	5.500	5.606
5	4.329	4.413	4.497	4.584	4.672	4.762	4.853	4.947	5.042	5.139	5.238	5.338	5.441
6	4.212	4.292	4.374	4.457	4.542	4.629	4.717	4.807	4.898	4.992	5.087	5.183	5.282
7	4.100	4.177	4.256	4.336	4.418	4.501	4.586	4.673	4.761	4.851	4.942	5.036	5.131
8	3.993	4.067	4.143	4.220	4.299	4.379	4.461	4.545	4.630	4.716	4.804	4.894	4.986
9	3.890	3.961	4.035	4.109	4.185	4.263	4.342	4.422	4.504	4.587	4.672	4.759	4.847
10	3.791	3.860	3.931	4.003	4.076	4.151	4.227	4.304	4.383	4.464	4.545	4.629	4.714
11	3.696	3.763	3.831	3.900	3.971	4.043	4.117	4.191	4.268	4.345	4.424	4.505	4.586
12	3.605	3.669	3.735	3.802	3.870	3.940	4.011	4.083	4.157	4.231	4.308	4.385	4.464
13	3.517	3.580	3.643	3.708	3.774	3.841	3.909	3.979	4.050	4.122	4.196	4.271	4.347
14	3.433	3.493	3.555	3.617	3.681	3.746	3.812	3.879	3.948	4.018	4.089	4.161	4.235
15	3.352	3.410	3.470	3.530	3.592	3.655	3.719	3.784	3.850	3.917	3.986	4.056	4.127
16	3.274	3.331	3.388	3.447	3.506	3.567	3.629	3.691	3.755	3.821	3.887	3.954	4.023
17	3.199	3.254	3.309	3.366	3.424	3.482	3.542	3.603	3.665	3.728	3.792	3.857	3.924
18	3.127	3.180	3.234	3.288	3.344	3.401	3.459	3.518	3.577	3.638	3.700	3.764	3.828
19	3.058	3.109	3.161	3.214	3.268	3.323	3.379	3.436	3.493	3.552	3.612	3.673	3.736
20	2.991	3.040	3.091	3.142	3.194	3.247	3.301	3.357	3.413	3.470	3.528	3.587	3.647

Table F-2 Present Worth Factors for $N = 10$

d, Market Discount Rate, percent	i, Annual Inflation Rate, percent												
	0	1	2	3	4	5	6	7	8	9	10	11	12
0	10.000	10.462	10.950	11.464	12.006	12.578	13.181	13.816	14.487	15.193	15.937	16.722	17.549
1	9.471	9.901	10.354	10.831	11.335	11.865	12.425	13.014	13.635	14.289	14.979	15.705	16.470
2	8.983	9.383	9.804	10.248	10.716	11.209	11.728	12.275	12.851	13.458	14.097	14.770	15.479
3	8.530	8.903	9.295	9.709	10.144	10.603	11.085	11.594	12.129	12.692	13.286	13.910	14.567
4	8.111	8.459	8.825	9.210	9.615	10.042	10.492	10.965	11.462	11.986	12.537	13.117	13.727
5	7.722	8.046	8.388	8.748	9.126	9.524	9.942	10.383	10.846	11.334	11.847	12.386	12.953
6	7.360	7.664	7.983	8.319	8.672	9.043	9.434	9.845	10.277	10.731	11.208	11.710	12.238
7	7.024	7.308	7.607	7.921	8.251	8.598	8.962	9.346	9.749	10.172	10.618	11.085	11.577
8	6.710	6.976	7.256	7.550	7.859	8.184	8.525	8.883	9.259	9.655	10.070	10.507	10.965
9	6.418	6.667	6.930	7.205	7.495	7.798	8.118	8.453	8.805	9.174	9.562	9.970	10.398
10	6.145	6.379	6.625	6.884	7.155	7.440	7.739	8.053	8.382	8.728	9.091	9.472	9.872
11	5.889	6.110	6.341	6.584	6.838	7.105	7.386	7.680	7.989	8.313	8.652	9.009	9.383
12	5.650	5.858	6.075	6.303	6.543	6.793	7.057	7.333	7.622	7.926	8.244	8.578	8.929
13	5.426	5.622	5.826	6.041	6.266	6.502	6.749	7.008	7.280	7.565	7.864	8.177	8.505
14	5.216	5.400	5.593	5.795	6.007	6.229	6.462	6.705	6.961	7.228	7.509	7.803	8.111
15	5.019	5.193	5.374	5.565	5.765	5.974	6.193	6.422	6.662	6.914	7.177	7.453	7.743
16	4.833	4.997	5.169	5.349	5.537	5.734	5.940	6.156	6.383	6.619	6.867	7.127	7.399
17	4.659	4.814	4.976	5.146	5.323	5.509	5.704	5.908	6.121	6.344	6.577	6.822	7.077
18	4.494	4.641	4.794	4.955	5.123	5.298	5.482	5.674	5.875	6.085	6.305	6.536	6.776
19	4.339	4.478	4.623	4.775	4.934	5.100	5.273	5.455	5.644	5.843	6.050	6.267	6.494
20	4.192	4.324	4.462	4.606	4.756	4.913	5.077	5.248	5.428	5.615	5.811	6.016	6.230

Table F-3 Present Worth Factors for $N = 15$

i, Annual Inflation Rate, percent

d, Market Discount percent	0	1	2	3	4	5	6	7	8	9	10	11	12
0	15.000	16.097	17.293	18.599	20.024	21.579	23.276	25.129	27.152	29.361	31.772	34.405	37.280
1	13.865	14.851	15.926	17.098	18.375	19.767	21.285	22.942	24.748	26.718	28.867	31.212	33.770
2	12.849	13.738	14.706	15.759	16.906	18.156	19.517	21.000	22.616	24.377	26.297	28.389	30.669
3	11.938	12.741	13.614	14.563	15.596	16.719	17.942	19.273	20.722	22.300	24.017	25.888	27.925
4	11.118	11.845	12.634	13.492	14.423	15.435	16.536	17.733	19.035	20.451	21.991	23.667	25.491
5	10.380	11.039	11.754	12.530	13.372	14.286	15.279	16.357	17.529	18.802	20.187	21.691	23.327
6	9.712	10.311	10.960	11.664	12.426	13.254	14.151	15.125	16.182	17.329	18.575	19.929	21.399
7	9.108	9.654	10.244	10.883	11.575	12.325	13.138	14.019	14.974	16.010	17.134	18.354	19.677
8	8.559	9.057	9.595	10.177	10.807	11.488	12.225	13.024	13.889	14.826	15.842	16.943	18.137
9	8.061	8.516	9.007	9.538	10.111	10.731	11.402	12.127	12.912	13.761	14.681	15.678	16.757
10	7.606	8.023	8.473	8.958	9.481	10.046	10.657	11.317	12.030	12.802	13.636	14.539	15.516
11	7.191	7.574	7.986	8.430	8.909	9.425	9.982	10.584	11.233	11.935	12.694	13.514	14.400
12	6.811	7.163	7.541	7.949	8.387	8.860	9.369	9.919	10.511	11.151	11.842	12.587	13.393
13	6.462	6.786	7.135	7.509	7.912	8.345	8.812	9.314	9.856	10.440	11.070	11.749	12.483
14	6.142	6.441	6.762	7.107	7.477	7.875	8.303	8.764	9.260	9.794	10.370	10.990	11.659
15	5.847	6.124	6.420	6.738	7.079	7.445	7.839	8.262	8.717	9.206	9.733	10.300	10.911
16	5.575	5.831	6.105	6.399	6.714	7.051	7.413	7.803	8.220	8.670	9.153	9.672	10.231
17	5.324	5.561	5.815	6.087	6.378	6.689	7.024	7.382	7.767	8.180	8.623	9.100	9.612
18	5.092	5.312	5.547	5.799	6.069	6.357	6.665	6.996	7.351	7.731	8.139	8.577	9.048
19	4.876	5.081	5.300	5.533	5.783	6.050	6.336	6.641	6.969	7.320	7.696	8.099	8.532
20	4.675	4.867	5.070	5.288	5.519	5.767	6.032	6.315	6.618	6.942	7.289	7.661	8.059

Table F-4 Present Worth Factors for $N = 20$

Market Discount Rate, percent	i, Annual Inflation Rate, percent												
	0	1	2	3	4	5	6	7	8	9	10	11	12
0	20.000	22.019	24.297	26.870	29.778	33.066	36.786	40.995	45.762	51.160	57.275	64.203	72.052
1	18.046	19.802	21.780	24.009	26.524	29.362	32.568	36.190	40.284	44.913	50.150	56.074	62.778
2	16.351	17.885	19.608	21.546	23.728	26.186	28.958	32.084	35.612	39.594	44.093	49.174	54.917
3	14.877	16.221	17.727	19.417	21.317	23.453	25.857	28.564	31.613	35.050	38.926	43.299	48.232
4	13.590	14.771	16.092	17.571	19.231	21.093	23.185	25.536	28.180	31.156	34.506	38.279	42.531
5	12.462	13.503	14.665	15.965	17.419	19.048	20.874	22.922	25.222	27.806	30.710	33.977	37.651
6	11.470	12.391	13.417	14.562	15.840	17.269	18.868	20.659	22.665	24.916	27.442	30.277	33.463
7	10.594	11.411	12.320	13.332	14.459	15.717	17.122	18.692	20.448	22.414	24.617	27.086	29.856
8	9.818	10.546	11.353	12.250	13.247	14.358	15.596	16.977	18.519	20.242	22.169	24.325	26.740
9	9.129	9.779	10.498	11.296	12.181	13.164	14.258	15.476	16.834	18.349	20.039	21.928	24.040
10	8.514	9.096	9.739	10.450	11.238	12.112	13.082	14.160	15.359	16.694	18.182	19.841	21.693
11	7.963	8.487	9.063	9.700	10.403	11.182	12.044	13.001	14.063	15.243	16.556	18.018	19.647
12	7.469	7.941	8.460	9.031	9.661	10.356	11.125	11.977	12.920	13.967	15.129	16.421	17.857
13	7.025	7.451	7.919	8.433	8.998	9.622	10.310	11.070	11.910	12.841	13.872	15.017	16.287
14	6.623	7.009	7.432	7.896	8.406	8.966	9.583	10.263	11.014	11.844	12.762	13.779	14.906
15	6.259	6.610	6.994	7.414	7.874	8.379	8.934	9.545	10.217	10.959	11.779	12.685	13.687
16	5.929	6.249	6.597	6.978	7.395	7.851	8.352	8.902	9.506	10.172	10.905	11.714	12.608
17	5.628	5.920	6.238	6.585	6.963	7.376	7.829	8.325	8.870	9.468	10.126	10.851	11.650
18	5.353	5.620	5.911	6.227	6.572	6.947	7.358	7.807	8.298	8.838	9.430	10.081	10.798
19	5.101	5.347	5.613	5.902	6.216	6.558	6.932	7.339	7.784	8.272	8.806	9.392	10.036
20	4.870	5.096	5.340	5.605	5.893	6.205	6.545	6.916	7.320	7.762	8.245	8.774	9.355

Table F-5 Present Worth Factors for $N = 25$

d, Market Discount Rate, percent	i, Annual Inflation Rate, percent												
	0	1	2	3	4	5	6	7	8	9	10	11	12
0	25.000	28.243	32.030	36.459	41.646	47.727	54.864	63.249	73.106	84.701	98.347	114.413	133.334
1	22.023	24.752	27.929	31.633	35.958	41.014	46.933	53.869	62.003	71.550	82.762	95.935	111.419
2	19.523	21.832	24.510	27.622	31.245	35.470	40.401	46.164	52.906	60.800	70.051	80.897	93.621
3	17.413	19.375	21.644	24.272	27.322	30.867	34.994	39.804	45.417	51.974	59.639	68.606	79.104
4	15.622	17.298	19.229	21.459	24.038	27.028	30.498	34.531	39.224	44.693	51.071	58.516	67.213
5	14.094	15.532	17.184	19.085	21.277	23.810	26.740	30.137	34.079	38.660	43.990	50.197	57.431
6	12.783	14.024	15.444	17.072	18.943	21.098	23.585	26.458	29.784	33.639	38.112		49.350
7	11.654	12.729	13.954	15.356	16.961	18.803	20.923	23.364	26.183	29.440	33.210	37.578	42.645
8	10.675	11.611	12.674	13.885	15.269	16.851	18.666	20.750	23.148	25.912	29.103	32.791	37.058
9	9.823	10.641	11.568	12.620	13.817	15.182	16.743	18.530	20.580	22.936	25.648	28.774	32.382
10	9.077	9.796	10.607	11.525	12.566	13.749	15.097	16.636	18.396	20.412	22.727	25.388	28.452
11	8.422	9.056	9.769	10.574	11.482	12.512	13.682	15.012	16.530	18.264	20.248	22.523	25.134
12	7.843	8.405	9.035	9.743	10.540	11.440	12.459	13.615	14.929	16.425	18.133	20.086	22.321
13	7.330	7.830	8.388	9.014	9.716	10.506	11.398	12.406	13.548	14.846	16.322	18.005	19.926
14	6.873	7.320	7.817	8.372	8.993	9.689	10.473	11.356	12.353	13.483	14.764	16.220	17.878
15	6.464	6.865	7.309	7.803	8.355	8.971	9.662	10.439	11.314	12.301	13.417	14.683	16.119
16	6.097	6.457	6.856	7.298	7.790	8.338	8.950	9.636	10.406	11.272	12.249	13.353	14.602
17	5.766	6.092	6.451	6.848	7.288	7.776	8.321	8.929	9.609	10.372	11.230	12.197	13.288
18	5.467	5.762	6.086	6.444	6.839	7.277	7.763	8.304	8.907	9.582	10.339	11.189	12.146
19	5.195	5.463	5.758	6.081	6.437	6.830	7.266	7.749	8.286	8.886	9.556	10.306	11.148
20	4.948	5.192	5.460	5.753	6.075	6.430	6.822	7.255	7.735	8.269	8.864	9.529	10.272

Table F-6 Present Worth Factors for $N = 30$

d, Market Discount Rate, percent	i, Annual Inflation Rate, percent												
	0	1	2	3	4	5	6	7	8	9	10	11	12
0	30.000	34.785	40.568	47.575	56.085	66.439	79.058	94.461	113.283	136.307	164.494	199.021	241.333
1	25.808	29.703	34.389	40.042	46.878	55.164	65.225	77.462	92.367	110.545	132.735	159.843	192.981
2	22.396	25.589	29.412	34.002	39.529	46.201	54.270	64.050	75.922	90.353	107.916	129.313	155.400
3	19.600	22.235	25.374	29.126	33.624	39.029	45.541	53.404	62.914	74.435	88.413	105.392	126.034
4	17.292	19.481	22.076	25.163	28.846	33.254	38.541	44.900	52.563	61.813	73.000	86.545	102.965
5	15.372	17.203	19.363	21.919	24.955	28.571	32.891	38.065	44.276	51.746	60.748	71.613	84.744
6	13.765	15.307	17.116	19.246	21.765	24.751	28.302	32.537	37.601	43.668	50.953	59.716	70.272
7	12.409	13.716	15.241	17.028	19.131	21.612	24.549	28.037	32.190	37.147	43.076	50.182	58.715
8	11.258	12.372	13.667	15.176	16.942	19.017	21.461	24.351	27.778	31.851	36.704	42.499	49.433
9	10.274	11.230	12.335	13.618	15.111	16.856	18.904	21.313	24.157	27.523	31.518	36.271	41.937
10	9.427	10.253	11.202	12.299	13.569	15.046	16.771	18.792	21.166	23.965	27.273	31.192	35.848
11	8.694	9.411	10.232	11.175	12.262	13.520	14.982	16.687	18.681	21.022	23.776	27.027	30.873
12	8.055	8.682	9.395	10.211	11.147	12.225	13.472	14.918	16.603	18.572	20.879	23.590	26.786
13	7.496	8.046	8.670	9.379	10.190	11.119	12.188	13.423	14.855	16.520	18.464	20.738	23.407
14	7.003	7.489	8.037	8.658	9.363	10.169	11.091	12.151	13.375	14.792	16.438	18.356	20.599
15	6.566	6.997	7.482	8.028	8.646	9.347	10.147	11.063	12.115	13.327	14.729	16.356	18.250
16	6.177	6.562	6.992	7.475	8.019	8.633	9.331	10.126	11.035	12.078	13.279	14.667	16.275
17	5.829	6.174	6.558	6.987	7.468	8.009	8.621	9.315	10.104	11.007	12.041	13.231	14.605
18	5.517	5.827	6.171	6.554	6.981	7.460	7.999	8.608	9.298	10.083	10.979	12.005	13.184
19	5.235	5.515	5.825	6.168	6.550	6.976	7.453	7.990	8.596	9.282	10.061	10.951	11.968
20	4.979	5.233	5.513	5.822	6.165	6.545	6.970	7.446	7.980	8.583	9.265	10.040	10.922

APPENDIX G

Meteorological Data

The meteorological data for North American stations on the following pages are useful in designing systems by the methods of Chapters 14, 15 and 18. The data include: monthly average daily radiation on a horizontal surface, $\bar{H}$, in KJ/m^2; the monthly average clearness index, $\bar{K}_T$; the 24-hr average temperature, $\bar{T}_a$ in C; and the average number of degree days in the month, in degree-C days.

Most of the data in the table for U.S. stations are from Cinquemani et al. The Canadian data are from other sources. The table is arranged alphabetically by station; a list by states and provinces is included for convenience in locating data in particular geographic areas.

These data are subject to change as additional measurements are included, as further refinements are made in the processing of old data, and as the inevitable errors in compilations of this type are uncovered and corrected.

REFERENCE

Cinquemani, V., J. R. Owenby, and R. G. Baldwin, Report prepared for U.S. Department of Energy by the National Oceanic and Atmospheric Administration (1978). "Input Data for Solar Systems."

METEOROLOGICAL DATA FOR STATIONS BY STATES AND PROVINCES

Alabama

Birmingham
Mobile
Montgomery

Alaska

Annette
Barrow
Bethel
Fairbanks
Matanuska

Arizona

Page
Phoenix
Prescott
Tucson
Winslow
Yuma

Arkansas

Fort Smith
Little Rock

California

Arcata
Bakersfield
Daggett
Davis
El Toro
Fresno
Inyokern
Long Beach
Los Angeles
Mt. Shasta
Needles
Oakland
Pasadena
Point Mugu
Riverside
Sacramento
San Diego
San Francisco
San Jose
Santa Maria
Sunnyvale

Colorado

Boulder
Colorado Springs
Denver
Eagle
Grand Junction
Grand Lake
Pueblo

Connecticut

Hartford

Delaware

Wilmington

District of Columbia

Washington

Florida

Apalachicola
Daytona Beach
Gainesville
Jacksonville
Key West
Miami
Orlando
Pensacola
Tallahassee
Tampa

Georgia

Atlanta
Augusta
Griffin
Macon
Savannah

Hawaii

Hilo
Honolulu

Idaho

Boise
Pocatello
Twin Falls

Illinois

Chicago
Lemont
Moline
Peoria
Springfield

Indiana

Evansville
Fort Wayne
Indianapolis
South Bend

Iowa

Ames
Des Moines
Mason City
Sioux City

Kansas

Dodge City
Manhattan
Topeka
Wichita

Kentucky

Covington
Lexington
Louisville

Louisiana

Baton Rouge
Lake Charles
New Orleans
Shreveport

Maine

Caribou
Portland

Maryland

Annapolis
Baltimore
Patuxent River
Silver Hill

Massachusetts

Amherst
Blue Hill
Boston
Lynn
Natick

Michigan

Detroit
East Lansing
Flint
Grand Rapids
Lansing
Sault St. Marie
Traverse City

Minnesota

Duluth
International Falls
Minneapolis-St. Paul
Rochester
St. Cloud

Mississippi

Jackson
Meridian

Missouri

Columbia
Kansas City
St. Louis
Springfield

Montana

Billings
Dillon
Glasgow
Great Falls
Helena
Lewistown
Miles City
Missoula
Summit

Nebraska

Grand Island
Lincoln
North Omaha
North Platte
Scotts Bluff

Nevada

Elko
Ely
Las Vegas
Lovelock
Reno
Tonopah
Winnemucca

New Hampshire

Concord

New Jersey

Atlantic City
Lakehurst
Newark
Trenton

New Mexico

Albuquerque
Farmington
Roswell
Zuni

New York

Albany
Binghampton
Buffalo
Ithaca
Massena
New York
Rochester
Schenectady
Syracuse

North Carolina

Asheville
Cape Hatteras
Charlotte
Cherry Point
Greensboro
Raleigh
Raleigh-Durham

North Dakota

Bismarck
Fargo
Minot

Ohio

Akron
Cleveland
Columbus
Dayton
Put-In-Bay
Toledo
Youngstown

Oklahoma

Oklahoma City
Stillwater
Tulsa

Oregon

Astoria
Burns
Corvallis
Medford
North Bend
Pendleton
Portland
Redmond
Salem

Pennsylvania

Allentown
Avoca
Erie
Harrisburg
Philadelphia
Pittsburgh
State College

Rhode Island

Newport
Providence

South Carolina

Charleston
Columbia
Greenville-Spartanburg

South Dakota

Huron
Pierre
Rapid City
Sioux Falls

Tennessee

Chattanooga
Memphis
Nashville
Oak Ridge

Texas

Abilene
Amarillo
Big Spring
Brownsville
Corpus Christi
Dallas
El Paso
Forth Worth
Houston
Kingsville
Lubbock
Lufkin
Midland
Port Arthur
San Angelo
San Antonio
Waco
Wichita Falls

Utah

Cedar City
Salt Lake City

Vermont

Burlington

Virginia

Mt. Weather
Norfolk
Richmond
Roanoke

Washington

Olympia
Prosser
Pullman
Richland
Seattle
Spokane
Tacoma
Whidbey Island

West Virginia

Charleston
Parkersburg

Wisconsin

Eau Claire
Green Bay
La Crosse
Madison
Milwaukee

Wyoming

Casper
Cheyenne
Lander
Laramie

Rock Springs
Sheridan

Alberta

Edmonton
Lethbridge

British Columbia

Vancouver

Manitoba

Churchill
Winnipeg

New Brunswick

Moncton

Newfoundland

St. Johns

Northwest Territories

Aklavik
Kapuskasing

Ontario

Ottawa
Toronto

Quebec

Montreal

	JAN	FEB	MAR	APR	MAY	JUN	JUL	AUG	SEP	OCT	NOV	DEC
ABILENE TX LATITUDE 32.42												
HORIZ. RAD.	10484	13421	17887	20921	23122	25066	24277	22200	18131	14930	11439	9798
KT	.53	.55	.59	.58	.58	.61	.60	.59	.56	.57	.55	.54
AVE. TEMP.	6	8	13	18	22	27	29	29	24	19	12	8
DEGREE-DAYS	367	266	197	58	6	0	0	0	0	49	187	321
AKRON OH LATITUDE 40.92												
HORIZ. RAD.	4860	7371	10942	15401	18928	20872	20279	18110	14434	10300	5728	4008
KT	.34	.37	.41	.45	.48	.50	.50	.50	.49	.47	.36	.31
AVE. TEMP.	-3	-2	3	9	15	20	22	21	18	12	5	-1
DEGREE-DAYS	667	580	496	275	128	18	5	9	56	205	405	613
ALBANY NY LATITUDE 42.67												
HORIZ. RAD.	5701	8703	12539	16948	19991	22145	22217	19287	14941	10425	5960	4528
KT	.42	.46	.48	.51	.51	.54	.55	.54	.52	.50	.41	.38
AVE. TEMP.	-5	-4	1	8	14	19	22	21	17	10	4	-2
DEGREE-DAYS	749	646	544	302	141	22	25	12	75	234	423	673
ALBUQUERQUE NM LATITUDE 35.05												
HORIZ. RAD.	11536	15230	20060	25290	28805	30403	28243	25990	22377	17553	12866	10528
KT	.64	.66	.69	.71	.73	.74	.70	.70	.71	.67	.63	.63
AVE. TEMP.	1	4	8	12	17	22	25	23	20	13	6	1
DEGREE-DAYS	513	389	331	157	32	0	0	0	4	121	342	496
ALLENTOWN PA LATITUDE 40.65												
HORIZ. RAD.	5987	8665	12238	15998	18576	20166	20030	17546	14051	10509	6449	4885
KT	.41	.43	.46	.47	.47	.49	.50	.49	.48	.48	.41	.37
AVE. TEMP.	-2	-1	3	10	16	21	23	22	18	12	6	-1
DEGREE-DAYS	641	554	463	252	106	12	0	3	47	191	378	591
AMARILLO TX LATITUDE 35.23												
HORIZ. RAD.	10897	14112	18508	22915	25100	27159	25881	23865	19980	15928	11722	9892
KT	.62	.61	.63	.65	.63	.66	.64	.65	.64	.64	.61	.60
AVE. TEMP.	1	4	8	13	19	24	26	25	21	15	8	4
DEGREE-DAYS	499	393	334	153	45	6	0	0	11	114	312	457
AMES IA LATITUDE 42.03												
HORIZ. RAD.	7285	10592	13691	16872	20096	22650	22441	19259	15366	11471	7829	5987
KT	.53	.55	.52	.50	.51	.55	.56	.54	.53	.54	.52	.48
AVE. TEMP.	-7	-4	2	11	18	23	25	22	17	11	2	-4
DEGREE-DAYS	794	639	539	260	106	18	8	8	58	206	463	699
AMHERST MA LATITUDE 42.25												
HORIZ. RAD.	4857	7410	12560	14528	18045	21520	21604	18421	13816	10467	6405	5192
KT	.36	.39	.48	.43	.46	.52	.54	.52	.50	.50	.43	.42
AVE. TEMP.	-4	-3	2	8	14	19	21	20	16	11	4	-2
DEGREE-DAYS	713	611	515	300	139	23	4	12	68	221	403	644

717

Station (Latitude)	Parameter	JAN	FEB	MAR	APR	MAY	JUN	JUL	AUG	SEP	OCT	NOV	DEC
ANNAPOLIS MD (38.98)	HORIZ. RAD.	7326	10173	14235	17542	20431	23320	22692	19636	16035	12309	7913	6489
	KT	.47	.49	.52	.51	.52	.56	.54	.54	.53	.54	.47	.46
	AVE. TEMP.	1	2	6	12	17	22	24	23	20	14	8	4
	DEGREE-DAYS	526	454	376	183	58	22	0	0	16	137	293	484
ANNETTE AK (55.03)	HORIZ. RAD.	2638	4815	9881	15240	18296	18338	18338	14277	10802	5108	2470	1717
	KT	.44	.42	.51	.52	.49	.45	.47	.44	.47	.37	.34	.36
	AVE. TEMP.	0	2	3	4	7	11	15	13	11	8	4	4
	DEGREE-DAYS	527	440	460	370	269	177	128	117	183	312	418	501
APALACHICOLA FL (29.75)	HORIZ. RAD.	9679	12778	16728	21324	23726	22678	20582	19151	17425	15564	11804	9279
	KT	.46	.49	.53	.58	.60	.56	.51	.57	.55	.57	.53	.47
	AVE. TEMP.	12	13	15	19	23	26	27	27	26	21	16	13
	DEGREE-DAYS	204	161	97	17	0	0	0	0	0	12	88	177
ARCATA CA (40.98)	HORIZ. RAD.	6001	8998	12856	18006	20912	22263	20516	17723	15233	10618	6727	5329
	KT	.42	.46	.48	.53	.53	.54	.51	.50	.52	.49	.43	.41
	AVE. TEMP.	7	8	8	10	12	14	15	15	15	13	10	7
	DEGREE-DAYS	365	288	307	262	208	143	110	98	105	181	263	344
ASHEVILLE NC (35.43)	HORIZ. RAD.	8190	11024	14822	18925	20478	21047	20157	18461	15444	13022	9633	7463
	KT	.46	.48	.51	.54	.52	.51	.50	.50	.49	.53	.51	.45
	AVE. TEMP.	3	4	8	13	18	21	23	23	19	14	8	4
	DEGREE-DAYS	467	398	329	155	56	8	0	0	28	149	312	453
ASTORIA OR (46.20)	HORIZ. RAD.	3572	6189	9827	14223	18252	18450	19820	17010	13427	8093	4395	2958
	KT	.32	.37	.41	.44	.47	.45	.50	.49	.49	.43	.35	.30
	AVE. TEMP.	5	6	6	8	11	13	14	15	14	11	8	5
	DEGREE-DAYS	420	333	355	287	219	142	91	84	112	210	308	382
ATLANTA GA (33.65)	HORIZ. RAD.	8144	10996	14795	19136	21039	21720	20566	19390	16138	13618	10020	7652
	KT	.43	.46	.50	.50	.53	.53	.52	.52	.50	.53	.50	.44
	AVE. TEMP.	6	7	11	16	20	24	25	25	22	17	11	7
	DEGREE-DAYS	389	311	246	80	15	0	0	0	4	76	227	371
ATLANTIC CITY NJ (39.45)	HORIZ. RAD.	7410	10676	16119	18087	20640	23990	23781	19971	16412	12644	8834	6657
	KT	.48	.52	.59	.53	.52	.58	.59	.55	.55	.56	.53	.48
	AVE. TEMP.	0	1	5	11	16	21	23	23	20	15	11	3
	DEGREE-DAYS	520	461	407	242	100	8	0	0	13	111	277	470

	JAN	FEB	MAR	APR	MAY	JUN	JUL	AUG	SEP	OCT	NOV	DEC
AUGUSTA GA LATITUDE 33.37												
HORIZ. RAD.	8523	11522	15188	19616	21167	21605	20468	18920	15998	13841	10401	8182
KT	.45	.48	.51	.55	.53	.53	.51	.51	.50	.54	.51	.46
AVE. TEMP.	8	9	13	18	22	26	27	26	23	18	12	8
DEGREE-DAYS	334	264	192	50	6	0	0	0	0	58	191	321
AVOCA PA LATITUDE 41.33												
HORIZ. RAD.	5165	7815	11250	15197	18052	19974	19811	17174	13606	10175	5558	4175
KT	.36	.40	.42	.45	.46	.48	.49	.48	.47	.47	.36	.33
AVE. TEMP.	-3	-3	3	9	15	20	22	21	17	11	5	-2
DEGREE-DAYS	672	587	499	275	122	16	4	10	64	217	403	618
BAKERSFIELD CA LATITUDE 35.42												
HORIZ. RAD.	8698	12505	18099	23773	28476	31202	30455	27472	22605	16550	10694	7688
KT	.49	.55	.62	.67	.72	.76	.76	.74	.72	.67	.56	.47
AVE. TEMP.	9	11	14	17	21	25	29	28	25	19	13	9
DEGREE-DAYS	302	196	148	78	12	0	0	0	0	31	153	294
BALTIMORE MD LATITUDE 39.18												
HORIZ. RAD.	6661	9533	13190	16886	19451	21326	20691	18153	15097	11322	7494	5667
KT	.43	.46	.48	.49	.49	.52	.51	.50	.50	.50	.45	.40
AVE. TEMP.	1	2	6	12	18	22	25	24	20	14	8	2
DEGREE-DAYS	544	470	382	189	61	0	0	0	15	139	315	512
BATON ROUGE LA LATITUDE 30.53												
HORIZ. RAD.	8910	11963	15655	19080	21236	21861	19812	19030	16619	14766	10445	8362
KT	.43	.47	.50	.53	.54	.54	.49	.51	.50	.55	.48	.43
AVE. TEMP.	11	12	15	20	24	27	28	28	25	20	15	12
DEGREE-DAYS	251	186	116	18	0	0	0	0	0	30	116	212
BETHEL AK LATITUDE 60.78												
HORIZ. RAD.	1549	4689	11807	18589	19217	18799	15491	10634	8332	4899	1884	963
KT	.52	.59	.73	.69	.53	.46	.40	.35	.42	.47	.46	.50
AVE. TEMP.	-14	-13	-11	-1	5	10	12	11	7	-1	-8	-15
DEGREE-DAYS	1032	883	923	675	429	223	177	219	333	599	797	1044
BIG SPRING TX LATITUDE 32.25												
HORIZ. RAD.	11220	14403	19510	24408	23947	24827	23110	19593	21896	16161	12183	10885
KT	.57	.59	.64	.68	.60	.61	.58	.52	.68	.62	.58	.59
AVE. TEMP.	9	11	12	18	20	26	28	28	24	18	12	7
DEGREE-DAYS	362	260	192	53	20	0	0	0	0	40	192	321
BILLINGS MT LATITUDE 45.80												
HORIZ. RAD.	5516	8661	13499	17322	21708	24669	27053	22952	16683	11199	6371	4780
KT	.48	.51	.55	.53	.56	.60	.68	.61	.15	.58	.50	.47
AVE. TEMP.	-6	-3	-3	6	12	17	22	21	15	10	2	-3
DEGREE-DAYS	742	585	558	340	185	73	6	8	123	271	488	658

BINGHAMTON NY LATITUDE 42.22

	JAN	FEB	MAR	APR	MAY	JUN	JUL	AUG	SEP	OCT	NOV	DEC
HORIZ. RAD.	4378	6535	9774	14091	16977	19082	18830	16168	12837	8843	4696	3373
KT	.32	.34	.37	.42	.43	.46	.47	.45	.44	.42	.31	.27
AVE. TEMP.	-6	-5	0	7	13	18	21	20	16	10	3	-4
DEGREE-DAYS	741	657	581	338	178	42	12	22	96	253	447	682

BIRMINGHAM AL LATITUDE 33.57

	JAN	FEB	MAR	APR	MAY	JUN	JUL	AUG	SEP	OCT	NOV	DEC
HORIZ. RAD.	8019	10975	14709	18993	21074	21773	20539	19563	16508	13741	9736	7506
KT	.42	.46	.49	.53	.53	.53	.51	.53	.52	.54	.48	.43
AVE. TEMP.	7	8	12	17	21	25	27	26	23	17	11	7
DEGREE-DAYS	363	287	216	64	11	0	0	0	23	76	217	341

BISMARCK ND LATITUDE 46.78

	JAN	FEB	MAR	APR	MAY	JUN	JUL	AUG	SEP	OCT	NOV	DEC
HORIZ. RAD.	5298	8803	13257	16562	20974	23376	24782	21298	15372	10302	5757	4232
KT	.49	.54	.56	.51	.54	.57	.62	.61	.55	.55	.47	.44
AVE. TEMP.	-13	-11	-3	6	12	17	21	20	14	7	-1	-9
DEGREE-DAYS	978	801	687	367	188	68	10	19	140	313	602	851

BLUE HILL MA LATITUDE 42.22

	JAN	FEB	MAR	APR	MAY	JUN	JUL	AUG	SEP	OCT	NOV	DEC
HORIZ. RAD.	6531	9001	12728	15867	19720	21645	20933	18170	14737	10425	6615	5401
KT	.48	.47	.49	.47	.50	.52	.52	.51	.51	.49	.44	.44
AVE. TEMP.	-3	-3	1	6	13	18	21	20	16	11	5	-1
DEGREE-DAYS	668	583	513	312	151	30	3	8	62	203	378	608

BOISE ID LATITUDE 43.57

	JAN	FEB	MAR	APR	MAY	JUN	JUL	AUG	SEP	OCT	NOV	DEC
HORIZ. RAD.	5508	9530	14800	20733	25838	27955	29651	24928	19715	12913	7130	4962
KT	.43	.52	.58	.62	.66	.68	.74	.70	.70	.63	.50	.43
AVE. TEMP.	-1	4	8	9	14	18	23	22	17	11	4	3
DEGREE-DAYS	620	487	412	267	140	54	0	27	71	226	420	567

BOSTON MA LATITUDE 42.37

	JAN	FEB	MAR	APR	MAY	JUN	JUL	AUG	SEP	OCT	NOV	DEC
HORIZ. RAD.	5396	8053	11535	15046	18391	20622	19852	16870	14298	10096	5707	4574
KT	.40	.42	.44	.45	.47	.50	.49	.47	.50	.48	.38	.38
AVE. TEMP.	-1	0	3	9	15	20	23	22	18	13	8	1
DEGREE-DAYS	617	538	463	273	121	15	0	5	42	167	330	551

BOULDER CO LATITUDE 40.00

	JAN	FEB	MAR	APR	MAY	JUN	JUL	AUG	SEP	OCT	NOV	DEC
HORIZ. RAD.	8415	11221	16789	19259	19259	21981	21771	18380	17250	12980	9295	7620
KT	.56	.55	.62	.56	.49	.53	.54	.51	.58	.58	.57	.56
AVE. TEMP.	0	1	2	9	14	19	23	22	17	12	5	2
DEGREE-DAYS	551	459	449	268	131	49	23	20	77	204	383	503

BROWNSVILLE TX LATITUDE 25.92

	JAN	FEB	MAR	APR	MAY	JUN	JUL	AUG	SEP	OCT	NOV	DEC
HORIZ. RAD.	10359	12886	16545	19715	21870	24006	25109	23008	19224	16330	11968	9787
KT	.44	.51	.51	.53	.55	.60	.63	.61	.56	.56	.49	.44
AVE. TEMP.	15	17	20	23	26	28	28	28	27	24	19	16
DEGREE-DAYS	125	84	49	0	0	0	0	0	0	3	19	81

BUFFALO NY LATITUDE 42.93	JAN	FEB	MAR	APR	MAY	JUN	JUL	AUG	SEP	OCT	NOV	DEC
HORIZ. RAD.	3960	6201	10083	14923	18119	20470	20160	17173	13072	8902	4578	3215
KT	.30	.33	.39	.45	.46	.50	.50	.48	.46	.43	.31	.27
AVE. TEMP.	-5	-4	-4	13	13	19	21	20	16	11	4	-9
DEGREE-DAYS	711	632	567	335	178	32	7	18	77	233	420	639

BURLINGTON VT LATITUDE 44.47	JAN	FEB	MAR	APR	MAY	JUN	JUL	AUG	SEP	OCT	NOV	DEC
HORIZ. RAD.	4829	7703	11976	16452	20049	22131	22138	18995	14352	9474	4919	3635
KT	.39	.43	.48	.56	.52	.54	.55	.54	.51	.48	.36	.33
AVE. TEMP.	-8	-7	-2	8	13	18	21	20	15	9	3	-5
DEGREE-DAYS	830	722	618	367	184	35	11	27	106	279	467	730

BURNS OR LATITUDE 43.58	JAN	FEB	MAR	APR	MAY	JUN	JUL	AUG	SEP	OCT	NOV	DEC
HORIZ. RAD.	5561	8988	13472	18710	23293	25872	27918	23636	18386	11832	6736	4886
KT	.43	.49	.53	.56	.60	.63	.70	.67	.65	.58	.48	.43
AVE. TEMP.	-4	-1	2	7	11	15	20	19	15	8	4	-2
DEGREE-DAYS	686	529	498	347	223	114	17	38	126	305	487	639

CAPE HATTERAS NC LATITUDE 35.27	JAN	FEB	MAR	APR	MAY	JUN	JUL	AUG	SEP	OCT	NOV	DEC
HORIZ. RAD.	7781	10806	15053	20132	22264	23105	21797	19355	16688	12899	9906	7475
KT	.43	.47	.52	.57	.56	.56	.54	.55	.53	.52	.52	.45
AVE. TEMP.	8	8	10	14	19	23	25	25	23	18	13	8
DEGREE-DAYS	339	299	254	104	26	0	0	0	0	42	154	298

CARIBOU ME LATITUDE 46.87	JAN	FEB	MAR	APR	MAY	JUN	JUL	AUG	SEP	OCT	NOV	DEC
HORIZ. RAD.	4827	8242	12852	15984	17774	19863	19858	16882	12412	7736	4175	3613
KT	.44	.50	.54	.50	.46	.48	.50	.49	.46	.42	.34	.38
AVE. TEMP.	-11	-10	-4	3	10	15	18	17	12	7	0	-8
DEGREE-DAYS	935	811	713	472	263	94	47	68	182	365	560	842

CASPER WY LATITUDE 42.92	JAN	FEB	MAR	APR	MAY	JUN	JUL	AUG	SEP	OCT	NOV	DEC
HORIZ. RAD.	7754	11502	16355	20959	25008	28387	28765	25256	19855	13831	8684	6744
KT	.58	.63	.63	.66	.64	.69	.71	.71	.69	.67	.60	.57
AVE. TEMP.	-5	-3	1	6	11	17	22	21	15	9	1	-3
DEGREE-DAYS	720	594	586	372	216	82	9	9	127	298	518	668

CEDER CITY UT LATITUDE 37.70	JAN	FEB	MAR	APR	MAY	JUN	JUL	AUG	SEP	OCT	NOV	DEC
HORIZ. RAD.	10014	13389	18562	23746	28001	30708	28410	25436	22339	16566	11263	8915
KT	.61	.62	.66	.68	.71	.74	.71	.70	.73	.70	.64	.59
AVE. TEMP.	-2	1	4	8	13	18	23	23	17	11	4	-1
DEGREE-DAYS	625	496	458	298	156	48	0	3	63	236	437	589

CHARLESTON SC LATITUDE 32.90	JAN	FEB	MAR	APR	MAY	JUN	JUL	AUG	SEP	OCT	NOV	DEC
HORIZ. RAD.	8446	11296	15192	19660	21111	20926	20416	17991	15822	13536	10601	8179
KT	.44	.47	.50	.55	.53	.51	.51	.48	.49	.52	.52	.46
AVE. TEMP.	10	10	14	18	22	25	27	26	24	19	14	10
DEGREE-DAYS	267	218	157	29	0	0	0	0	0	28	117	242

	JAN	FEB	MAR	APR	MAY	JUN	JUL	AUG	SEP	OCT	NOV	DEC
CHARLESTON WV LATITUDE 38.37												
HORIZ. RAD.	5656	8018	11457	15386	18606	20155	19095	17186	14436	11034	6958	4995
KT	.35	.38	.41	.44	.47	.49	.47	.47	.48	.48	.40	.34
AVE. TEMP.				.13	.18	.22	.24	.23	.20	.14		.-2
DEGREE-DAYS	526	443	357	159	63	6	0	0	26	148	327	496
CHARLOTTE NC LATITUDE 35.22												
HORIZ. RAD.	8160	11020	14952	19236	21059	21802	20779	19236	16065	13317	9823	9631
KT	.45	.48	.51	.54	.53	.53	.52	.52	.51	.54	.51	.58
AVE. TEMP.				.16	.20	.24	.26	.25	.22	.16	.11	.-3
DEGREE-DAYS	394	327	256	81	19	0	0	0	6	84	233	388
CHATTANOOGA TN LATITUDE 35.03												
HORIZ. RAD.	7156	9744	13349	17588	19655	20784	19693	18499	15156	12577	8772	6587
KT	.40	.42	.46	.50	.50	.51	.49	.49	.48	.51	.46	.40
AVE. TEMP.				.16	.20	.24	.26	.26	.25	.16	.09	.05
DEGREE-DAYS	427	347	268	92	28	0	0	0	5	101	268	410
CHERRY POINT NC LATITUDE 34.90												
HORIZ. RAD.	8588	11635	15735	20358	21843	22002	20765	18547	16198	13274	10290	8148
KT	.47	.50	.54	.57	.55	.54	.52	.50	.51	.53	.53	.49
AVE. TEMP.				.15	.19	.23	.26	.25	.23	.18	.13	.-2
DEGREE-DAYS	339	299	254	104	26	0	0	0	23	42	154	298
CHEYENNE WY LATITUDE 41.15												
HORIZ. RAD.	8691	12118	16264	20093	22637	25626	25307	22307	18923	14093	9338	7615
KT	.61	.62	.61	.59	.58	.62	.63	.62	.65	.65	.60	.59
AVE. TEMP.	.-3	.-2	.0	.6	.11	.16	.21	.20	.15	.13	.2	.-2
DEGREE-DAYS	661	560	575	372	219	87	12	17	125	294	492	617
CHICAGO IL LATITUDE 41.98												
HORIZ. RAD.	5754	8620	12562	16558	20302	22777	22060	19513	15365	10996	6419	4557
KT	.42	.45	.48	.49	.52	.55	.55	.55	.53	.52	.42	.37
AVE. TEMP.	.-3	.-3	.3	.10	.16	.21	.24	.23	.19	.13	.5	.-1
DEGREE-DAYS	701	585	486	252	116	14	0	4	32	176	410	629
CLEVELAND OH LATITUDE 41.40												
HORIZ. RAD.	4408	6822	10467	15315	19079	20919	20745	17961	14067	9839	5290	3608
KT	.31	.35	.40	.45	.49	.51	.52	.50	.48	.46	.34	.28
AVE. TEMP.	.-1	.-1	.4	.9	.15	.20	.22	.21	.18	.12	.5	.0
DEGREE-DAYS	656	577	498	278	136	22	2	9	53	197	390	598
COLORADO SPRNGS CO LATITUDE 38.82												
HORIZ. RAD.	10108	13371	17591	21917	24158	26884	25102	22986	19964	15419	10716	8874
KT	.64	.64	.64	.63	.61	.65	.62	.63	.66	.67	.63	.62
AVE. TEMP.	.-2	.0	.8	.13	.18	.21	.21	.16	.10	.3	.-1	
DEGREE-DAYS	627	524	512	313	167	57	25	7	86	253	458	586

	JAN	FEB	MAR	APR	MAY	JUN	JUL	AUG	SEP	OCT	NOV	DEC
COLUMBIA MO LATITUDE 38.97												
HORIZ. RAD.	6948	9928	13378	17317	21334	23714	24015	21312	16461	12493	7975	5930
KT	.44	.48	.48	.50	.54	.57	.60	.59	.55	.55	.47	.42
AVE. TEMP.	-1	0	6	12	18	23	25	24	20	14	6	0
DEGREE-DAYS	615	488	406	174	61	6	0	0	23	137	352	554
COLUMBIA SC LATITUDE 33.95												
HORIZ. RAD.	8645	11582	15378	19824	21505	22095	20902	19324	16334	13748	10455	8195
KT	.46	.49	.52	.56	.55	.54	.52	.52	.51	.54	.53	.47
AVE. TEMP.	7	9	12	18	22	26	27	27	24	18	12	8
DEGREE-DAYS	338	274	200	46	22	0	0	0	0	62	189	327
COLUMBUS OH LATITUDE 40.00												
HORIZ. RAD.	5212	7681	11117	15354	18691	20572	19916	18619	14545	10726	6103	4394
KT	.35	.38	.41	.45	.48	.50	.49	.51	.49	.48	.37	.32
AVE. TEMP.	-1	0	4	11	16	21	23	22	18	12	5	0
DEGREE-DAYS	631	540	444	232	98	7	0	4	42	190	388	591
CONCORD NH LATITUDE 43.20												
HORIZ. RAD.	5215	7786	11049	14948	17956	19345	19005	16516	12940	9273	5251	4110
KT	.40	.42	.43	.45	.46	.47	.47	.46	.45	.45	.37	.35
AVE. TEMP.	-8	-5	0	6	13	18	21	20	15	10	3	-4
DEGREE-DAYS	764	659	563	347	175	32	9	25	101	271	450	692
CORPUS CHRISTI TX LATITUDE 27.77												
HORIZ. RAD.	10192	13022	16228	18640	21182	23762	24810	22593	19146	16073	11833	9587
KT	.46	.51	.51	.51	.54	.59	.62	.60	.57	.57	.51	.46
AVE. TEMP.	13	15	18	22	25	27	29	29	27	23	18	15
DEGREE-DAYS	169	111	67	20	0	0	0	0	0	4	45	122
CORVALLIS OR LATITUDE 44.55												
HORIZ. RAD.	5680	8380	11659	15872	18980	20849	20098	18549	14885	11233	6679	4908
KT	.46	.47	.47	.48	.49	.50	.50	.53	.53	.57	.49	.45
AVE. TEMP.	6	6	8	10	13	16	18	18	16	11	9	6
DEGREE-DAYS	451	341	336	248	163	80	34	31	67	203	328	413
COVINGTON KY LATITUDE 39.07												
HORIZ. RAD.	4229	5820	11764	16914	21268	24325	28051	22901	16705	9838	5862	3391
KT	.27	.28	.43	.49	.54	.59	.70	.63	.56	.43	.35	.24
AVE. TEMP.	-1	1	5	12	17	22	24	24	20	14	7	1
DEGREE-DAYS	584	493	401	189	77	5	0	0	24	151	353	539
DAGGETT CA LATITUDE 34.87												
HORIZ. RAD.	10875	14535	20114	25809	29407	31394	29546	27040	22788	17201	12315	9942
KT	.60	.63	.69	.73	.74	.76	.73	.73	.72	.69	.64	.59
AVE. TEMP.	8	11	14	18	22	27	31	30	26	20	13	9
DEGREE-DAYS	305	206	151	66	28	8	0	0	0	32	164	293

	JAN	FEB	MAR	APR	MAY	JUN	JUL	AUG	SEP	OCT	NOV	DEC
DALLAS TX LATITUDE 32.85												
HORIZ. RAD.	9323	12156	16136	18463	21433	24229	24083	22133	18012	14482	10627	8853
KT	.48	.50	.54	.51	.54	.59	.60	.59	.56	.56	.52	.49
AVE. TEMP.	8	10	13	19	23	28	30	30	26	20	13	9
DEGREE-DAYS	338	243	174	39	0	0	0	0	0	31	158	289
DAVIS CA LATITUDE 38.55												
HORIZ. RAD.	6615	10718	16830	22105	26628	29391	28888	25580	20850	14570	9043	6196
KT	.42	.51	.61	.64	.68	.71	.72	.70	.69	.63	.53	.43
AVE. TEMP.	7	9	11	14	17	21	23	22	21	17	11	4
DEGREE-DAYS	344	239	210	128	51	11	0	0	5	54	197	327
DAYTON OH LATITUDE 39.90												
HORIZ. RAD.	5552	8232	11637	15924	19286	21264	20537	18675	14958	10996	6402	4625
KT	.37	.40	.43	.46	.49	.51	.51	.52	.50	.49	.39	.34
AVE. TEMP.	-2	-1	4	11	16	22	24	23	19	13	5	-1
DEGREE-DAYS	636	538	448	229	92	7	0	4	35	171	387	587
DAYTONA BEACH FL LATITUDE 29.18												
HORIZ. RAD.	10877	13767	17567	21379	22326	20723	20247	19088	16769	14202	11752	9878
KT	.51	.53	.56	.58	.56	.51	.51	.51	.50	.51	.52	.49
AVE. TEMP.	15	15	18	21	24	26	27	27	26	23	18	15
DEGREE-DAYS	134	117	67	9	0	0	0	0	0	0	54	118
DENVER CO LATITUDE 39.75												
HORIZ. RAD.	10164	13714	18707	22975	26092	28628	27712	24870	21050	15808	10684	8822
KT	.67	.67	.69	.67	.66	.69	.69	.69	.71	.70	.65	.64
AVE. TEMP.	-1	1	3	9	14	19	23	22	17	11	4	0
DEGREE-DAYS	604	501	482	292	141	44	0	0	67	227	427	558
DES MOINES IA LATITUDE 41.53												
HORIZ. RAD.	6590	9768	13397	17666	21194	24114	23796	20745	16273	12218	7471	5526
KT	.47	.50	.51	.52	.54	.58	.59	.58	.56	.56	.49	.44
AVE. TEMP.	-7	-4	0	10	16	21	24	23	18	12	3	-4
DEGREE-DAYS	786	634	536	258	103	14	0	7	52	194	453	689
DETROIT MI LATITUDE 42.23												
HORIZ. RAD.	4737	7722	11351	15877	19474	21178	20830	17880	14223	9943	5422	3898
KT	.35	.41	.43	.47	.50	.51	.52	.50	.49	.47	.36	.32
AVE. TEMP.	-4	-3	-2	9	14	20	23	22	18	12	4	-2
DEGREE-DAYS	696	597	512	288	136	20	0	9	53	209	415	629
DILLON MT LATITUDE 45.25												
HORIZ. RAD.	5975	9604	14518	18600	22575	24326	27147	22962	17263	11613	6832	5108
KT	.50	.56	.59	.57	.58	.59	.68	.65	.63	.60	.52	.49
AVE. TEMP.	-7	-4	-1	1	10	14	19	18	13	7	3	-5
DEGREE-DAYS	772	614	609	398	252	132	30	47	181	344	553	708

	JAN	FEB	MAR	APR	MAY	JUN	JUL	AUG	SEP	OCT	NOV	DEC
DODGE CITY, KA			LATITUDE 37.77									
HORIZ. RAD.	9381	12734	16756	21402	23716	26763	26051	23326	19142	14761	10142	8306
KT	.57	.59	.60	.61	.60	.65	.65	.64	.63	.63	.58	.55
AVE. TEMP.	-1	1	5	12	17	23	26	25	20	13	5	0
DEGREE-DAYS	589	463	410	191	64	12	0	0	23	137	370	544
DULUTH, MN			LATITUDE 46.83									
HORIZ. RAD.	4410	7636	11740	15580	18642	20056	21044	17556	12427	8226	4321	3311
KT	.40	.47	.49	.48	.48	.49	.53	.51	.46	.44	.35	.35
AVE. TEMP.	-13	-11	-5	6	10	15	19	18	12	7	-2	-10
DEGREE-DAYS	973	823	715	440	269	108	37	58	177	339	610	872
EAGLE, CO			LATITUDE 39.65									
HORIZ. RAD.	8559	12234	17043	21933	25595	28473	27062	23649	20051	14835	9858	7841
KT	.56	.60	.62	.64	.65	.69	.67	.65	.67	.66	.60	.57
AVE. TEMP.	-8	-5	-1	5	11	15	19	18	13	7	-1	-7
DEGREE-DAYS	809	649	584	385	236	106	24	44	158	348	568	770
EAST LANSING, MI			LATITUDE 42.73									
HORIZ. RAD.	5066	8792	12937	15031	20222	22902	22609	19510	15617	10676	5694	4522
KT	.38	.47	.50	.45	.52	.55	.56	.55	.55	.51	.39	.38
AVE. TEMP.	-5	-4	0	6	13	19	21	20	16	10	3	-2
DEGREE-DAYS	730	638	553	308	156	27	5	15	74	234	443	653
EAU CLAIRE, WI			LATITUDE 44.87									
HORIZ. RAD.	5126	8471	12373	16182	19075	21244	21408	18393	13575	9375	5113	3868
KT	.42	.48	.50	.49	.49	.51	.53	.52	.48	.48	.38	.36
AVE. TEMP.	-11	-9	-3	7	13	19	21	20	15	9	0	-8
DEGREE-DAYS	918	772	649	342	163	36	8	21	112	281	550	809
ELKO, NV			LATITUDE 40.85									
HORIZ. RAD.	7818	11739	16604	21560	26140	28755	29767	26283	21479	15009	9216	7002
KT	.54	.59	.62	.63	.67	.70	.74	.73	.73	.69	.58	.53
AVE. TEMP.	-5	-2	1	6	11	15	19	18	14	8	3	-3
DEGREE-DAYS	720	557	517	358	226	106	15	33	138	312	503	673
EL PASO, TX			LATITUDE 31.80									
HORIZ. RAD.	12769	16798	21668	26823	29514	30443	27806	25927	22552	18601	14115	11697
KT	.64	.68	.71	.74	.75	.74	.69	.69	.73	.70	.67	.63
AVE. TEMP.	6	9	13	17	22	27	27	26	23	18	11	7
DEGREE-DAYS	368	258	182	49	20	0	0	0	23	51	223	355
EL TORO, CA			LATITUDE 33.67									
HORIZ. RAD.	10747	14027	18274	21886	23495	24901	26822	24457	19716	15399	11648	9865
KT	.57	.59	.61	.61	.59	.61	.67	.66	.62	.60	.58	.56
AVE. TEMP.	12	13	13	15	17	19	22	22	21	18	15	12
DEGREE-DAYS	207	166	155	98	52	21	0	0	5	36	108	189

ELY, NV — LATITUDE 39.28	JAN	FEB	MAR	APR	MAY	JUN	JUL	AUG	SEP	OCT	NOV	DEC
HORIZ. RAD.	9300	12949	18827	22797	26224	28519	27772	25312	21963	15975	10514	8201
KT	.60	.63	.66	.66	.67	.69	.69	.70	.73	.70	.63	.58
AVE. TEMP.	-4	-2	0	5	10	14	19	18	13	7	1	-3
DEGREE-DAYS	713	577	554	395	261	134	13	34	147	327	520	557

ERIE, PA — LATITUDE 42.08	JAN	FEB	MAR	APR	MAY	JUN	JUL	AUG	SEP	OCT	NOV	DEC
HORIZ. RAD.	3922	6546	10445	15423	18684	20957	20800	16512	13633	9387	4723	3151
KT	.28	.34	.40	.46	.48	.51	.52	.46	.47	.44	.31	.26
AVE. TEMP.	-4	-4	0	8	13	18	20	20	16	11	4	-2
DEGREE-DAYS	687	619	553	337	187	44	13	24	78	231	415	618

EVANSVILLE, IN — LATITUDE 38.05	JAN	FEB	MAR	APR	MAY	JUN	JUL	AUG	SEP	OCT	NOV	DEC
HORIZ. RAD.	6515	9343	13063	17032	20233	22502	21793	19691	15926	12336	7746	5660
KT	.40	.44	.47	.49	.51	.55	.54	.54	.52	.53	.44	.38
AVE. TEMP.	0	2	7	14	19	24	25	25	21	15	7	2
DEGREE-DAYS	558	453	363	146	53	0	0	0	19	131	335	512

FAIRBANKS, AK — LATITUDE 64.82	JAN	FEB	MAR	APR	MAY	JUN	JUL	AUG	SEP	OCT	NOV	DEC
HORIZ. RAD.	796	3182	9755	16118	19971	22064	18589	15197	7704	3600	1130	251
KT	.68	.57	.71	.64	.57	.54	.49	.52	.43	.45	.54	.60
AVE. TEMP.	-24	-19	-13	-1	8	14	15	12	6	-3	-16	-22
DEGREE-DAYS	1324	1050	956	602	305	117	82	169	343	686	1037	1298

FARGO, ND — LATITUDE 46.90	JAN	FEB	MAR	APR	MAY	JUN	JUL	AUG	SEP	OCT	NOV	DEC
HORIZ. RAD.	4709	8009	12460	16748	20822	22629	24059	20715	14796	9917	5190	3828
KT	.43	.49	.52	.52	.54	.55	.60	.60	.55	.53	.43	.41
AVE. TEMP.	-14	-12	-4	8	13	18	21	21	14	6	-2	-11
DEGREE-DAYS	1018	844	703	378	186	54	7	18	130	310	607	896

FARMINGTON, NM — LATITUDE 36.75	JAN	FEB	MAR	APR	MAY	JUN	JUL	AUG	SEP	OCT	NOV	DEC
HORIZ. RAD.	10719	14537	19218	24206	27823	30251	28125	25559	21952	16782	11885	9500
KT	.63	.66	.67	.69	.70	.73	.70	.70	.71	.70	.65	.61
AVE. TEMP.	-2	2	6	10	15	20	24	23	18	12	4	-1
DEGREE-DAYS	627	467	420	258	102	20	0	3	37	208	430	601

FLINT, MI — LATITUDE 42.97	JAN	FEB	MAR	APR	MAY	JUN	JUL	AUG	SEP	OCT	NOV	DEC
HORIZ. RAD.	4348	7223	10859	15196	18818	20579	20392	17647	13568	9406	4872	3506
KT	.33	.39	.42	.45	.48	.50	.51	.50	.48	.45	.34	.30
AVE. TEMP.	-5	-5	0	8	13	19	21	20	16	11	3	-3
DEGREE-DAYS	736	641	558	318	170	36	8	20	82	241	445	658

FORT SMITH, AR — LATITUDE 35.33	JAN	FEB	MAR	APR	MAY	JUN	JUL	AUG	SEP	OCT	NOV	DEC
HORIZ. RAD.	8440	11337	14886	18339	21700	23713	23439	21307	17040	13627	9663	7735
KT	.47	.50	.51	.52	.55	.58	.58	.58	.54	.55	.51	.47
AVE. TEMP.	4	6	10	17	21	26	28	27	23	17	10	5
DEGREE-DAYS	448	338	262	73	9	0	0	0	20	75	243	405

	JAN	FEB	MAR	APR	MAY	JUN	JUL	AUG	SEP	OCT	NOV	DEC
FORT WAYNE, IN LATITUDE 41.00												
HORIZ. RAD.	5166	7917	11145	15442	18974	20901	20280	18094	14454	10489	5860	4193
KT	.36	.40	.42	.45	.48	.51	.50	.50	.49	.48	.37	.32
AVE. TEMP.	-4	-2	4	10	15	21	23	22	18	12	5	-2
DEGREE-DAYS	684	582	491	262	120	13	0	7	50	202	413	627
FORT WORTH, TX LATITUDE 32.83												
HORIZ. RAD.	9089	12075	15908	18276	21373	24364	24369	22387	18276	14494	10525	8603
KT	.47	.50	.53	.51	.54	.59	.61	.60	.57	.56	.51	.48
AVE. TEMP.	7	9	13	18	22	27	29	29	25	19	13	8
DEGREE-DAYS	348	253	186	49	20	0	0	0	0	33	159	294
FRESNO, CA LATITUDE 36.77												
HORIZ. RAD.	7324	11348	17314	22324	26212	28723	28662	25817	21254	15386	9552	6240
KT	.43	.51	.61	.64	.66	.70	.71	.70	.69	.64	.52	.40
AVE. TEMP.	7	10	12	16	19	23	27	26	23	18	12	7
DEGREE-DAYS	339	235	191	101	28	5	0	0	0	50	192	331
GAINESVILLE, FL LATITUDE 29.65												
HORIZ. RAD.	11639	15366	18630	22566	24534	22776	21771	21268	18589	15407	13313	10634
KT	.55	.59	.59	.62	.62	.56	.54	.57	.56	.56	.60	.53
AVE. TEMP.	13	14	17	21	24	26	27	27	26	22	17	14
DEGREE-DAYS	164	133	73	11	0	0	0	0	0	27	69	143
GLASGOW, MT LATITUDE 48.22												
HORIZ. RAD.	4403	7618	12540	16888	20743	23232	24888	21142	15212	9957	5433	3794
KT	.44	.49	.54	.53	.54	.56	.63	.62	.58	.56	.48	.44
AVE. TEMP.	-12	-8	-3	6	12	17	21	21	14	8	-1	-7
DEGREE-DAYS	961	774	686	370	191	84	8	17	146	321	600	825
GRAND ISLAND, NE LATITUDE 40.97												
HORIZ. RAD.	7505	10407	14359	19207	22377	25448	25145	22011	17130	12911	8381	6462
KT	.52	.53	.54	.57	.57	.62	.62	.61	.58	.59	.53	.50
AVE. TEMP.	-5	-3	2	10	16	22	25	24	18	12	4	-3
DEGREE-DAYS	704	582	479	232	100	20	0	0	48	198	447	625
GRAND JUNCTION, CO LATITUDE 39.12												
HORIZ. RAD.	8980	12699	17630	22543	27008	29490	27977	24763	20818	15264	10419	8300
KT	.58	.61	.64	.65	.69	.71	.69	.68	.69	.62	.62	.59
AVE. TEMP.	-3	1	6	11	16	22	25	24	19	12	4	-1
DEGREE-DAYS	661	488	410	224	74	12	0	0	33	180	420	612
GRAND LAKE, CO LATITUDE 40.27												
HORIZ. RAD.	8876	13105	17710	21436	23111	26461	25121	21143	19929	15114	9797	7704
KT	.60	.65	.66	.63	.59	.64	.62	.59	.67	.68	.61	.57
AVE. TEMP.	-9	-7	-4	0	5	10	13	12	8	3	-3	-8
DEGREE-DAYS	864	734	720	525	381	250	153	174	280	446	653	820

	JAN	FEB	MAR	APR	MAY	JUN	JUL	AUG	SEP	OCT	NOV	DEC
GRAND RAPIDS MI	LATITUDE 42.90											
HORIZ. RAD.	4194	7357	11512	16024	19920	22204	21726	19024	14324	9735	5058	3526
KT	.32	.39	.45	.48	.51	.54	.54	.53	.50	.47	.35	.30
AVE. TEMP.	-5	-4	1	8	14	20	22	21	17	11	4	-3
DEGREE-DAYS	720	630	549	308	150	24	4	15	63	227	438	648
GREAT FALLS MT	LATITUDE 47.48											
HORIZ. RAD.	4772	8173	13283	16895	20968	23849	26432	21937	15644	10493	5647	3815
KT	.45	.51	.56	.53	.55	.58	.66	.64	.59	.58	.48	.42
AVE. TEMP.	-5	-1	0	6	12	16	21	20	14	9	1	-2
DEGREE-DAYS	767	597	594	360	204	90	10	23	144	291	507	663
GREEN BAY WI	LATITUDE 44.48											
HORIZ. RAD.	5121	8227	12532	16327	19512	21652	21432	18406	13823	9313	5278	3969
KT	.42	.46	.50	.50	.50	.52	.53	.52	.50	.47	.39	.36
AVE. TEMP.	-9	-8	-2	7	14	18	21	20	15	10	1	-6
DEGREE-DAYS	854	731	627	353	188	51	12	30	106	272	515	759
GREENSBORO NC	LATITUDE 36.08											
HORIZ. RAD.	8118	11009	14903	19103	21200	22166	21150	19255	16088	12953	9524	7475
KT	.47	.49	.52	.54	.54	.54	.55	.54	.52	.53	.51	.47
AVE. TEMP.	4	5	8	13	19	23	25	24	21	14	8	8
DEGREE-DAYS	453	379	302	113	33	0	0	0	13	116	278	437
GRNVLE-SPTNBRG SC	LATITUDE 34.90											
HORIZ. RAD.	9378	11304	16286	21394	23153	23320	23194	21478	17752	14946	10760	8582
KT	.52	.49	.56	.60	.59	.57	.58	.58	.56	.60	.56	.51
AVE. TEMP.	6	7	10	16	21	24	26	25	22	16	11	6
DEGREE-DAYS	391	321	250	80	16	0	0	0	5	81	233	381
GRIFFIN GA	LATITUDE 33.25											
HORIZ. RAD.	9964	12643	16245	21729	24158	24283	23404	21896	18296	15575	12058	8792
KT	.52	.53	.54	.61	.61	.59	.58	.59	.57	.60	.59	.50
AVE. TEMP.	6	8	11	16	21	24	25	25	22	17	11	7
DEGREE-DAYS	356	289	228	61	11	0	0	0	11	61	200	339
HARRISBURG PA	LATITUDE 40.22											
HORIZ. RAD.	6078	8750	12291	16008	18752	20480	20015	17597	14374	10602	6567	5076
KT	.41	.43	.45	.47	.48	.50	.50	.49	.48	.48	.41	.38
AVE. TEMP.	-1	0	5	12	17	22	24	23	19	13	7	0
DEGREE-DAYS	601	509	413	206	71	0	0	0	28	163	353	558
HARTFORD CT	LATITUDE 41.93											
HORIZ. RAD.	5419	8111	11105	14924	17801	19130	18714	16135	13102	9679	5644	4370
KT	.39	.42	.42	.44	.46	.46	.47	.45	.45	.45	.37	.35
AVE. TEMP.	-3	-3	2	9	15	20	23	21	17	11	5	-2
DEGREE-DAYS	692	594	506	288	126	13	0	7	59	213	395	634

HELENA, MT — LATITUDE 46.58

	JAN	FEB	MAR	APR	MAY	JUN	JUL	AUG	SEP	OCT	NOV	DEC
HORIZ. RAD.	4760	8044	13000	16874	21111	23149	24486	21906	16030	10512	5915	4135
KT	.43	.49	.54	.52	.55	.56	.66	.63	.59	.56	.48	.43
AVE. TEMP.	-8	-4	-1	6	11	15	20	19	13	7	0	-5
DEGREE-DAYS	808	616	592	372	223	108	18	32	169	339	555	718

HILO, HI — LATITUDE 19.73

	JAN	FEB	MAR	APR	MAY	JUN	JUL	AUG	SEP	OCT	NOV	DEC
HORIZ. RAD.	11709	15473	18735	18024	18735	24004	22331	20491	18902	14177	12629	10998
KT	.44	.51	.54	.48	.48	.61	.58	.54	.53	.45	.46	.43
AVE. TEMP.	21	21	21	22	22	23	23	24	24	23	22	21
DEGREE-DAYS	0	0	0	0	0	0	0	0	0	0	0	0

HONOLULU, HI — LATITUDE 21.30

	JAN	FEB	MAR	APR	MAY	JUN	JUL	AUG	SEP	OCT	NOV	DEC
HORIZ. RAD.	15198	17668	21604	23404	25833	25749	25749	25623	23990	21227	17836	15533
KT	.59	.59	.63	.63	.66	.65	.66	.68	.68	.69	.66	.63
AVE. TEMP.	22	22	22	23	24	25	26	26	26	25	24	23
DEGREE-DAYS	0	0	0	0	0	0	0	0	0	0	0	0

HOUSTON, TX — LATITUDE 29.97

	JAN	FEB	MAR	APR	MAY	JUN	JUL	AUG	SEP	OCT	NOV	DEC
HORIZ. RAD.	8766	11737	14724	17277	20143	21541	20747	19137	16694	14477	10486	8280
KT	.42	.46	.47	.47	.51	.53	.52	.51	.50	.53	.47	.42
AVE. TEMP.	11	13	16	21	24	27	28	29	26	22	16	13
DEGREE-DAYS	231	163	105	13	0	0	0	0	0	13	86	185

HURON, SD — LATITUDE 44.38

	JAN	FEB	MAR	APR	MAY	JUN	JUL	AUG	SEP	OCT	NOV	DEC
HORIZ. RAD.	5540	8451	12639	17364	21240	23845	24772	21476	16091	11216	6550	4600
KT	.45	.48	.50	.53	.55	.58	.62	.61	.58	.56	.48	.42
AVE. TEMP.	-8	-2	2	14	14	19	23	22	16	10	0	-7
DEGREE-DAYS	904	733	620	320	152	40	5	27	94	268	543	789

INDIANAPOLIS, IN — LATITUDE 39.73

	JAN	FEB	MAR	APR	MAY	JUN	JUL	AUG	SEP	OCT	NOV	DEC
HORIZ. RAD.	5624	8477	11773	15870	19157	21201	20500	18652	15026	11088	6572	4728
KT	.37	.41	.43	.46	.49	.51	.51	.51	.50	.49	.40	.34
AVE. TEMP.	-1	3	6	11	17	22	24	23	19	13	5	0
DEGREE-DAYS	639	533	436	215	88	6	0	3	35	168	388	587

INTRNTNL FALLS, MN — LATITUDE 48.57

	JAN	FEB	MAR	APR	MAY	JUN	JUL	AUG	SEP	OCT	NOV	DEC
HORIZ. RAD.	4037	7519	11870	16385	19477	21033	21801	18366	12725	7988	3921	3084
KT	.41	.49	.52	.52	.51	.51	.55	.54	.49	.45	.35	.37
AVE. TEMP.	-17	-14	-6	3	10	16	19	17	12	6	-4	-13
DEGREE-DAYS	1087	902	764	447	257	93	37	62	202	371	668	969

INYOKERN, CA — LATITUDE 35.65

	JAN	FEB	MAR	APR	MAY	JUN	JUL	AUG	SEP	OCT	NOV	DEC
HORIZ. RAD.	13063	17542	24199	29349	33033	35001	32824	30897	27130	20263	15324	12351
KT	.74	.77	.83	.83	.84	.85	.81	.84	.87	.83	.81	.76
AVE. TEMP.	7	11	14	18	23	27	32	31	27	20	13	8
DEGREE-DAYS	341	218	148	71	6	0	0	0	0	24	176	334

		JAN	FEB	MAR	APR	MAY	JUN	JUL	AUG	SEP	OCT	NOV	DEC
ITHACA	NY LATITUDE 42.45												
HORIZ. RAD.		5108	8499	11807	14570	19636	22566	22399	19259	14905	10383	5233	4145
KT		.38	.45	.45	.43	.50	.55	.56	.54	.52	.49	.35	.34
AVE. TEMP.		-5	-4	.0	.7	18	18	20	19	15	10	4	-2
DEGREE-DAYS		723	646	562	332	176	39	11	22	87	243	423	654
JACKSON	MS LATITUDE 32.32												
HORIZ. RAD.		8551	11648	15538	19388	22026	22973	21665	20207	17128	14429	10232	8044
KT		.43	.48	.51	.54	.56	.56	.54	.54	.53	.55	.49	.44
AVE. TEMP.		8	10	13	19	23	26	28	27	24	19	13	9
DEGREE-DAYS		316	246	174	41	3	0	0	0	0	51	167	280
JACKSONVILLE	FL LATITUDE 30.42												
HORIZ. RAD.		10213	13214	17270	21060	22202	21392	20451	19227	16369	13881	11303	9279
KT		.49	.52	.56	.58	.56	.52	.51	.51	.50	.51	.52	.48
AVE. TEMP.		12	13	16	20	23	26	27	27	25	21	16	12
DEGREE-DAYS		193	157	98	13	0	0	0	0	0	11	89	176
KANSAS CITY	MO LATITUDE 39.28												
HORIZ. RAD.		7353	10154	13652	17875	21252	23601	23857	21136	16483	12396	8367	6372
KT		.47	.49	.50	.52	.54	.57	.59	.58	.55	.55	.50	.45
AVE. TEMP.		-2	1	5	13	18	23	26	25	20	15	6	.0
DEGREE-DAYS		641	496	414	174	62	7	0	0	23	131	357	563
KEY WEST	FL LATITUDE 24.55												
HORIZ. RAD.		13690	17165	20515	23948	24241	22734	22357	20975	18631	16495	13900	12225
KT		.57	.61	.62	.64	.62	.57	.57	.55	.54	.56	.55	.53
AVE. TEMP.		21	21	23	25	27	28	29	29	28	26	23	21
DEGREE-DAYS		9	14	23	25	0	0	0	0	0	0	20	10
KINGSVILLE	TX LATITUDE 27.52												
HORIZ. RAD.		10354	13178	16282	18871	21156	23105	23963	21807	18438	15774	11738	9639
KT		.46	.51	.51	.51	.54	.57	.60	.58	.55	.56	.50	.46
AVE. TEMP.		13	15	18	23	26	29	30	30	27	23	18	15
DEGREE-DAYS		177	111	67	24	0	0	0	0	0	6	56	140
LA CROSSE	WI LATITUDE 43.87												
HORIZ. RAD.		5463	8679	12493	16186	19439	21625	21569	18911	14095	9800	5605	4193
KT		.43	.48	.49	.49	.50	.52	.54	.53	.50	.48	.40	.37
AVE. TEMP.		-9	-7	-1	8	15	20	23	22	17	11		-6
DEGREE-DAYS		842	700	584	290	124	22	6	0	72	234	493	744
LAKE CHARLES	LA LATITUDE 30.22												
HORIZ. RAD.		8267	11460	14906	17822	20989	22361	20289	18810	16855	15674	10402	8008
KT		.40	.45	.48	.49	.53	.55	.51	.50	.51	.58	.47	.41
AVE. TEMP.		11	12	15	20	23	27	28	27	25	21	15	12
DEGREE-DAYS		231	170	111	14	0	0	0	0	0	20	98	188

		JAN	FEB	MAR	APR	MAY	JUN	JUL	AUG	SEP	OCT	NOV	DEC
LAKEHURST	NJ LATITUDE 40.03												
HORIZ. RAD.		6351	9044	12585	16523	18976	20140	19328	17395	14306	10846	7052	5389
KT		.42	.45	.46	.48	.48	.49	.48	.48	.48	.49	.43	.40
AVE. TEMP.		-1	0		11		21	24	23	19	13		1
DEGREE-DAYS		584	504	424	228	87	5	0	0	29	158	327	539
LANDER	WY LATITUDE 42.80												
HORIZ. RAD.		9629	13439	18882	23153	24492	28260	27130	24283	19426	14905	9922	8247
KT		.72	.72	.70	.69	.63	.68	.68	.68	.68	.78	.68	.69
AVE. TEMP.		-6	-3		6	12	16	21	21	15	8	0	-4
DEGREE-DAYS		782	614	579	368	212	83	5	8	125	313	558	723
LANSING	MI LATITUDE 42.78												
HORIZ. RAD.		5652	8917	12895	14988	20892	23153	22776	20138	15784	10801	5861	4689
KT		.42	.48	.50	.45	.54	.56	.57	.57	.55	.52	.40	.39
AVE. TEMP.		-5	-4		8	14	19	22	21	16	11	4	-3
DEGREE-DAYS		730	638	553	308	156	27	5	15	74	234	443	653
LARAMIE	WY LATITUDE 41.30												
HORIZ. RAD.		9378	12476	17751	20850	22943	26251	24827	22022	17584	13355	9504	7662
KT		.66	.64	.67	.62	.59	.63	.68	.67	.60	.62	.61	.60
AVE. TEMP.		-4	-4		2	9	14	18	17	12	6	-1	-4
DEGREE-DAYS		763	650	653	453	296	143	39	56	192	374	577	715
LAS VEGAS	NV LATITUDE 36.08												
HORIZ. RAD.		11099	15202	20695	26318	30033	31525	29376	26725	23121	17475	12319	9993
KT		.64	.68	.72	.75	.76	.77	.73	.73	.74	.72	.66	.62
AVE. TEMP.		6	9	12	17	23	28	31	30	26	19	11	
DEGREE-DAYS		358	257	180	70	6	0	0	0	0	41	198	341
LEMONT	IL LATITUDE 41.67												
HORIZ. RAD.		7159	9713	13649	16328	20808	23153	22064	20347	16076	11095	6574	5484
KT		.51	.50	.52	.48	.53	.56	.55	.57	.55	.52	.43	.44
AVE. TEMP.		-3	-2		10	16	21	24	23	19	13	4	-1
DEGREE-DAYS		701	585	486	252	116	14	0	4	32	176	410	629
LEWISTOWN	MT LATITUDE 47.05												
HORIZ. RAD.		4766	7856	12806	16391	20508	23368	25964	21580	15574	10271	5701	4120
KT		.44	.48	.54	.51	.53	.57	.65	.62	.58	.56	.47	.44
AVE. TEMP.		-7	-5	-3	4	10	14	19	18	12	8	0	-4
DEGREE-DAYS		791	641	646	415	265	147	39	52	193	336	547	698
LEXINGTON	KY LATITUDE 38.03												
HORIZ. RAD.		6195	8847	12478	16787	19827	21530	21000	19126	15458	11850	7460	5510
KT		.38	.41	.45	.48	.50	.53	.52	.54	.51	.51	.43	.37
AVE. TEMP.		0	2	6	13	18	23	25	24	20	14	7	2
DEGREE-DAYS		553	462	374	168	59	4	0	0	22	137	340	508

	JAN	FEB	MAR	APR	MAY	JUN	JUL	AUG	SEP	OCT	NOV	DEC
LINCOLN NE LATITUDE 40.85												
HORIZ. RAD.	7955	10676	14528	17751	20766	22817	22483	21268	17249	13607	8667	7201
KT	.55	.54	.54	.52	.53	.55	.56	.59	.59	.62	.55	.55
AVE. TEMP.	-14	2	-4	11	16	22	25	24	19	16	4	-1
DEGREE-DAYS	687	564	463	223	95	17	0	3	42	167	405	592
LITTLE ROCK AR LATITUDE 34.73												
HORIZ. RAD.	8299	11381	14898	18280	21895	23907	23064	21115	17228	13940	9615	7646
KT	.46	.49	.51	.58	.55	.58	.57	.57	.53	.56	.50	.45
AVE. TEMP.	4	-2	10	16	21	26	27	27	23	17	10	5
DEGREE-DAYS	439	344	264	77	12	0	0	0	3	79	245	403
LONGBEACH CA LATITUDE 33.82												
HORIZ. RAD.	10528	13789	18271	21991	23430	24286	26101	23830	19304	15053	11389	9610
KT	.56	.58	.61	.62	.59	.59	.65	.64	.61	.59	.57	.55
AVE. TEMP.	12	13	14	16	18	20	22	23	22	19	16	13
DEGREE-DAYS	188	152	137	82	39	13	0	0	4	27	86	164
LOS ANGELES CA LATITUDE 33.93												
HORIZ. RAD.	10510	13778	18370	22141	23374	24049	26188	23600	19082	14946	11393	9630
KT	.56	.58	.62	.62	.59	.59	.65	.64	.60	.59	.57	.56
AVE. TEMP.	12	13	13	15	16	18	20	20	20	18	15	13
DEGREE-DAYS	184	150	148	108	63	39	11	8	13	43	88	155
LOUISVILLE KY LATITUDE 38.18												
HORIZ. RAD.	6191	8958	12506	16646	19518	21603	20854	19069	15448	11828	7409	5537
KT	.38	.42	.45	.48	.50	.52	.52	.52	.51	.51	.43	.38
AVE. TEMP.	0	2	7	13	18	22	24	24	20	14	7	1
DEGREE-DAYS	546	454	367	159	58	20	0	0	19	134	333	506
LOVELOCK NV LATITUDE 40.07												
HORIZ. RAD.	9124	13326	18799	24574	28993	31204	31594	28191	23007	16468	10546	8108
KT	.61	.65	.69	.72	.74	.76	.78	.78	.77	.74	.65	.60
AVE. TEMP.	-2	4	4	9	14	19	23	22	17	11	4	-1
DEGREE-DAYS	582	426	393	250	124	38	0	6	55	202	398	433
LUBBOCK TX LATITUDE 33.65												
HORIZ. RAD.	11700	15113	19997	24602	27191	28876	27371	25063	20656	16662	12666	10606
KT	.62	.64	.67	.69	.69	.70	.68	.67	.65	.65	.63	.61
AVE. TEMP.	4	9	16	20	25	27	26	24	18	16	9	8
DEGREE-DAYS	446	347	282	106	16	0	0	0	24	90	270	408
LUFKIN TX LATITUDE 31.23												
HORIZ. RAD.	9010	12134	15617	18429	21185	23326	22770	21155	17372	15306	10932	8712
KT	.44	.48	.51	.51	.53	.57	.57	.56	.53	.57	.51	.46
AVE. TEMP.	9	11	14	20	23	27	28	28	25	20	14	10
DEGREE-DAYS	283	206	142	31	0	0	0	0	0	29	142	244

LYNN, MA — LATITUDE 42.47

	JAN	FEB	MAR	APR	MAY	JUN	JUL	AUG	SEP	OCT	NOV	DEC
HORIZ. RAD.	4941	8750	12560	16496	19008	22525	22566	17668	14234	9671	5568	4187
KT	.36	.46	.48	.49	.49	.54	.56	.50	.50	.46	.38	.35
AVE. TEMP.	-1	.4	3	9	15	20	23	22	18	13	7	1
DEGREE-DAYS	604	540	470	285	116	20	0	5	33	176	335	546

MACON, GA — LATITUDE 32.70

	JAN	FEB	MAR	APR	MAY	JUN	JUL	AUG	SEP	OCT	NOV	DEC
HORIZ. RAD.	8726	11572	15472	19704	21394	21783	20261	19495	16329	14153	10664	8273
KT	.45	.48	.51	.55	.54	.53	.50	.52	.51	.54	.52	.46
AVE. TEMP.	9	10	14	19	23	26	27	27	24	19	13	9
DEGREE-DAYS	302	235	166	37	3	0	0	0	0	46	169	288

MADISON, WI — LATITUDE 43.13

	JAN	FEB	MAR	APR	MAY	JUN	JUL	AUG	SEP	OCT	NOV	DEC
HORIZ. RAD.	5847	9125	12892	15870	19784	22107	21953	19385	14747	10338	5722	4414
KT	.45	.49	.50	.48	.51	.53	.55	.55	.52	.50	.40	.38
AVE. TEMP.	-7	-6	0	7	13	19	21	20	15	10	1	-5
DEGREE-DAYS	830	696	599	328	165	40	8	22	96	263	505	742

MANHATTAN, KA — LATITUDE 39.20

	JAN	FEB	MAR	APR	MAY	JUN	JUL	AUG	SEP	OCT	NOV	DEC
HORIZ. RAD.	8039	11053	14444	18129	22064	23069	22232	22023	17166	12225	9504	6531
KT	.52	.53	.53	.53	.56	.56	.55	.61	.57	.54	.57	.46
AVE. TEMP.	-1	1	5	13	18	23	26	25	20	14	6	0
DEGREE-DAYS	625	482	401	169	61	7	0	0	26	134	358	562

MASON CITY, IA — LATITUDE 43.15

	JAN	FEB	MAR	APR	MAY	JUN	JUL	AUG	SEP	OCT	NOV	DEC
HORIZ. RAD.	6284	9490	13255	17234	21510	23989	23653	20800	15950	11468	6808	5030
KT	.48	.51	.52	.52	.55	.58	.59	.59	.56	.56	.47	.43
AVE. TEMP.	-10	-8	-2	8	14	20	22	21	16	10	1	-7
DEGREE-DAYS	875	723	620	322	147	36	0	17	92	254	523	773

MASSENA, NY — LATITUDE 44.93

	JAN	FEB	MAR	APR	MAY	JUN	JUL	AUG	SEP	OCT	NOV	DEC
HORIZ. RAD.	4896	7855	12245	17044	20543	22782	22568	19163	14391	9427	5089	3755
KT	.41	.45	.50	.52	.53	.55	.56	.55	.52	.48	.38	.35
AVE. TEMP.	-10	-9	-2	6	12	18	21	19	15	9	2	-7
DEGREE-DAYS	870	751	644	380	194	43	12	32	107	284	485	773

MATANUSKA, AK — LATITUDE 61.57

	JAN	FEB	MAR	APR	MAY	JUN	JUL	AUG	SEP	OCT	NOV	DEC
HORIZ. RAD.	1340	3852	10132	14905	18254	19343	17124	13147	8290	4187	1591	628
KT	.52	.52	.65	.56	.51	.48	.45	.44	.42	.41	.43	.39
AVE. TEMP.	-11	-7	-3	2	7	12	14	12	8	1	-6	-10
DEGREE-DAYS	914	714	689	477	310	168	129	169	288	526	738	904

MEDFORD, OR — LATITUDE 42.38

	JAN	FEB	MAR	APR	MAY	JUN	JUL	AUG	SEP	OCT	NOV	DEC
HORIZ. RAD.	4618	8369	12854	18597	23080	25851	28094	24067	18033	11144	5723	3821
KT	.34	.44	.49	.55	.59	.63	.70	.67	.63	.53	.38	.31
AVE. TEMP.	3	5	7	10	14	18	22	21	18	12	6	3
DEGREE-DAYS	489	369	348	247	139	52	6	12	49	200	358	470

		JAN	FEB	MAR	APR	MAY	JUN	JUL	AUG	SEP	OCT	NOV	DEC
MEMPHIS TN (LATITUDE 35.05)	HORIZ. RAD.	7748	10722	14505	18598	21392	23204	22380	20700	16693	13670	9269	7134
	KT	.43	.47	.50	.52	.54	.56	.56	.56	.53	.55	.48	.43
	AVE. TEMP.	5	7	11	17	22	26	28	27	23	17	10	6
	DEGREE-DAYS	422	330	254	73	12			0	4	79	235	384
MERIDIAN MS (LATITUDE 32.35)	HORIZ. RAD.	8445	11488	15069	18859	21105	22275	20695	19739	16502	14273	10176	7938
	KT	.43	.47	.50	.52	.53	.54	.51	.53	.54	.54	.49	.43
	AVE. TEMP.	8	10	13	19	22	26	27	27	24	18	12	8
	DEGREE-DAYS	319	246	173	44	4	0	0	0	0	62	184	294
MIAMI FL (LATITUDE 25.78)	HORIZ. RAD.	12000	14912	18196	21098	20923	19383	20013	18497	16527	14784	12695	11566
	KT	.51	.54	.56	.57	.53	.48	.50	.49	.48	.51	.52	.52
	AVE. TEMP.	19	19	21	23	25	27	27	28	27	25	22	20
	DEGREE-DAYS	29	37	9	0	0	0	0	0	0	0	0	31
MIDLAND TX (LATITUDE 31.93)	HORIZ. RAD.	12271	15691	20868	24880	27579	29081	27116	25082	20926	17269	13348	11346
	KT	.62	.64	.68	.69	.70	.71	.68	.67	.64	.65	.63	.61
	AVE. TEMP.	6	8	12	17	22	26	27	27	23	18	11	7
	DEGREE-DAYS	368	268	194	54	20			0	0	45	198	329
MILES CITY MT (LATITUDE 46.43)	HORIZ. RAD.	5187	8458	13449	17500	21515	24353	26072	22437	16386	10903	6254	4532
	KT	.47	.51	.56	.54	.56	.59	.65	.64	.60	.58	.50	.47
	AVE. TEMP.	-9	-6	-1	7	13	18	24	23	15	9		-6
	DEGREE-DAYS	854	675	599	328	160	65	5	9	121	282	543	741
MILWAUKEE WI (LATITUDE 42.95)	HORIZ. RAD.	5441	8358	12357	16373	20070	22438	22264	19509	14870	10304	5954	4295
	KT	.41	.45	.48	.49	.51	.54	.55	.55	.52	.50	.41	.36
	AVE. TEMP.	-7	-5	-1	7	13	18	21	20	16	11		-4
	DEGREE-DAYS	786	661	579	338	193	50	28	20	78	244	475	703
MINN-ST. PAUL MN (LATITUDE 44.88)	HORIZ. RAD.	5266	8669	12524	16364	19716	21875	22357	19146	14239	9756	5452	4010
	KT	.44	.50	.48	.50	.51	.53	.56	.54	.51	.50	.50	.37
	AVE. TEMP.	-11	-9	-2	7	14	19	22	21	16	10	0	-7
	DEGREE-DAYS	916	759	637	340	159	42	8	14	108	276	552	806
MINOT ND (LATITUDE 48.22)	HORIZ. RAD.	4355	7444	11852	16576	20955	22412	23805	20433	14493	9642	4976	3518
	KT	.43	.48	.51	.52	.55	.54	.60	.60	.55	.54	.44	.41
	AVE. TEMP.	-13	-11	-5	5	12	17			13	8	-2	-10
	DEGREE-DAYS	983	812	713	398	213	83	15	39	159	326	618	866

734

	JAN	FEB	MAR	APR	MAY	JUN	JUL	AUG	SEP	OCT	NOV	DEC
MISSOULA	MT LATITUDE 46.92											
HORIZ. RAD.	3539	6516	11139	15686	20230	21937	26415	21346	15410	9221	4655	3033
KT	.33	.40	.47	.49	.53	.53	.66	.62	.57	.50	.38	.32
AVE. TEMP.	-6	-3	1	7	11	15	19	18	13	7	0	-4
DEGREE-DAYS	761	588	546	352	221	112	22	39	167	360	545	694
MOBILE	AL LATITUDE 30.68											
HORIZ. RAD.	9399	12479	15974	19540	21246	21205	19467	18629	16449	14739	10839	8616
KT	.46	.49	.52	.54	.54	.52	.49	.50	.50	.55	.50	.45
AVE. TEMP.	11	12	15	20	24	27	28	28	25	20	15	12
DEGREE-DAYS	251	187	123	22	0	0	0	0	0	22	117	214
MOLINE	IL LATITUDE 41.45											
HORIZ. RAD.	6073	9215	12695	16564	19905	22351	22001	19459	15401	11302	6748	4913
KT	.43	.47	.48	.49	.51	.54	.55	.54	.53	.52	.44	.39
AVE. TEMP.	-6	-4	2	10	16	22	24	23	18	12	4	-3
DEGREE-DAYS	749	611	504	242	102	11	0	6	44	191	430	661
MONTGOMERY	AL LATITUDE 32.30											
HORIZ. RAD.	8531	11496	15214	19621	21534	22384	20893	19812	16657	14318	10389	8164
KT	.43	.47	.50	.54	.54	.55	.52	.53	.51	.55	.50	.45
AVE. TEMP.	10	14	18	22	24	26	27	27	24	19	13	9
DEGREE-DAYS	309	233	166	42	4	0	0	0	0	52	170	284
MT. SHASTA	CA LATITUDE 41.30											
HORIZ. RAD.	6364	9731	14186	19929	24809	27647	29252	25115	19694	13109	7482	5730
KT	.45	.50	.53	.59	.63	.67	.73	.70	.67	.61	.48	.45
AVE. TEMP.	3	4	6	12	16	20	20	19	16	11	5	2
DEGREE-DAYS	541	423	424	312	206	99	21	36	81	234	388	508
MT. WEATHER	VA LATITUDE 39.07											
HORIZ. RAD.	7201	11472	14151	17333	21269	21981	21353	18003	15700	11765	8457	7034
KT	.46	.55	.51	.50	.54	.53	.53	.50	.52	.52	.50	.50
AVE. TEMP.	-1	0	3	10	15	20	20	21	18	12	5	0
DEGREE-DAYS	615	535	453	251	102	13	0	3	44	189	370	573
NASHVILLE	TN LATITUDE 36.12											
HORIZ. RAD.	6578	9349	12822	17518	20709	22278	21462	19712	15865	12641	8072	5908
KT	.38	.45	.44	.50	.52	.54	.53	.54	.54	.52	.43	.37
AVE. TEMP.	3	4	9	15	20	24	26	25	22	16	9	5
DEGREE-DAYS	460	373	291	98	25	0	0	0	6	100	277	424
NATICK	MA LATITUDE 42.28											
HORIZ. RAD.	6363	9713	13649	16370	20933	17124	21478	19133	15030	10928	6154	5610
KT	.47	.51	.52	.49	.54	.41	.53	.54	.52	.52	.41	.46
AVE. TEMP.	-2	-1	3	9	15	20	23	22	17	12	6	0
DEGREE-DAYS	672	585	487	282	123	17	0	6	56	203	375	608

	JAN	FEB	MAR	APR	MAY	JUN	JUL	AUG	SEP	OCT	NOV	DEC
NEEDLES CA LATITUDE 34.85												
HORIZ. RAD.	11177	15358	20713	26295	30095	31677	28839	25850	22863	17449	12754	10365
KT	.62	.66	.71	.74	.76	.77	.72	.70	.72	.70	.66	.62
AVE. TEMP.	11	14	16	21	26	31	35	34	30	23	16	12
DEGREE-DAYS	234	145	83	23	2	6	0	0	0	6	91	212
NEWARK NJ LATITUDE 40.70												
HORIZ. RAD.	6261	9000	12582	16440	19147	20375	19973	17759	14446	10792	6766	5157
KT	.43	.45	.47	.48	.49	.50	.50	.49	.49	.49	.43	.39
AVE. TEMP.	0	0	5	11	17	22	25	24	20	14	8	1
DEGREE-DAYS	579	504	420	222	79	0	0	0	19	135	313	526
NEW ORLEANS LA LATITUDE 29.98												
HORIZ. RAD.	9472	12619	16056	20204	22331	22741	20581	19482	17178	15151	11038	8845
KT	.45	.49	.51	.55	.56	.56	.51	.52	.52	.55	.50	.45
AVE. TEMP.	11	13	15	20	23	26	27	27	25	20	15	12
DEGREE-DAYS	224	166	104	16	0	0	0	0	0	22	179	327
NEWPORT RI LATITUDE 41.48												
HORIZ. RAD.	6489	9671	13816	16538	20472	22525	21645	18799	15909	11430	7326	5903
KT	.46	.50	.52	.49	.52	.54	.54	.52	.55	.53	.48	.46
AVE. TEMP.	0	0	3	8	12	17	21	21	18	13	8	2
DEGREE-DAYS	567	531	487	340	191	55	0	9	43	171	330	501
NEW YORK NY LATITUDE 40.77												
HORIZ. RAD.	5679	8183	11770	15479	18569	19410	19155	16834	13774	10161	6048	4585
KT	.39	.41	.44	.45	.47	.47	.48	.47	.47	.46	.38	.35
AVE. TEMP.	1	1	5	11	17	22	25	24	20	15	9	2
DEGREE-DAYS	579	510	443	252	104	5	0	0	23	137	308	518
NORFOLK VA LATITUDE 36.90												
HORIZ. RAD.	7698	10576	14537	19029	21421	22701	21032	19069	15839	12291	9207	7080
KT	.45	.48	.51	.54	.54	.55	.52	.52	.51	.51	.51	.46
AVE. TEMP.	4	5	8	14	19	23	25	24	22	16	10	5
DEGREE-DAYS	422	367	296	126	29	0	0	0	5	78	223	391
NORTH BEND OR LATITUDE 43.42												
HORIZ. RAD.	4977	7995	12006	17135	21076	22628	23920	20267	15632	10130	5954	4322
KT	.38	.44	.47	.51	.54	.55	.60	.57	.55	.49	.42	.37
AVE. TEMP.	8	8	8	10	12	14	15	15	15	13	10	8
DEGREE-DAYS	351	286	312	265	205	135	104	93	112	174	248	319
NORTH OMAHA NE LATITUDE 41.37												
HORIZ. RAD.	7195	10124	13874	17686	21252	24088	23906	21092	15584	11914	7310	5802
KT	.51	.52	.52	.52	.54	.58	.59	.59	.53	.55	.47	.45
AVE. TEMP.	-5	-3	2	10	17	22	25	23	19	12	4	-2
DEGREE-DAYS	730	576	481	217	82	11	0	3	39	167	417	637

	JAN	FEB	MAR	APR	MAY	JUN	JUL	AUG	SEP	OCT	NOV	DEC
NORTH PLATTE NE LATITUDE 41.13												
HORIZ. RAD.	7858	10876	15128	19565	22558	25722	25843	22579	17764	13359	8617	6869
KT	.55	.55	.57	.59	.58	.62	.64	.63	.61	.61	.55	.53
AVE. TEMP.	-2	-2	2	9	15	20	24	23	18	11	3	-3
DEGREE-DAYS	717	574	529	290	132	36	4	4	78	244	258	658
OAKLAND CA LATITUDE 37.73												
HORIZ. RAD.	8033	11547	16528	21814	25096	26670	26358	23295	19306	13755	9330	7343
KT	.49	.54	.59	.63	.64	.65	.65	.64	.63	.58	.53	.49
AVE. TEMP.	9	11	12	13	15	17	17	18	18	16	13	10
DEGREE-DAYS	282	204	194	150	107	63	44	41	33	75	162	260
OAK RIDGE TN LATITUDE 36.02												
HORIZ. RAD.	6950	9964	13649	18714	21562	22817	21813	19929	17459	13355	8834	6741
KT	.40	.44	.47	.53	.55	.55	.54	.54	.56	.55	.47	.42
AVE. TEMP.	3	5	9	15	19	23	25	24	21	15	8	4
DEGREE-DAYS	463	383	306	122	43	0	0	0	11	120	298	444
OKLAHOMA CITY OK LATITUDE 35.40												
HORIZ. RAD.	9089	11973	15890	19581	21768	24331	24155	22134	17638	13989	10225	8233
KT	.51	.52	.55	.55	.55	.59	.60	.60	.56	.57	.54	.50
AVE. TEMP.	4	6	9	15	20	25	27	27	23	16	9	5
DEGREE-DAYS	486	369	296	100	20	0	0	0	7	82	263	431
OLYMPIA WA LATITUDE 46.97												
HORIZ. RAD.	3051	5707	9590	14245	18520	19217	21707	17577	13130	7221	3851	2514
KT	.28	.35	.40	.44	.48	.47	.54	.51	.49	.39	.32	.27
AVE. TEMP.	3	6	8	9	12	15	18	17	15	10	6	4
DEGREE-DAYS	479	373	376	280	189	109	49	57	110	248	362	439
ORLANDO FL LATITUDE 28.55												
HORIZ. RAD.	11341	14112	17957	21541	22573	20783	20443	18989	16985	14803	12439	10510
KT	.52	.53	.57	.59	.57	.51	.51	.50	.51	.53	.54	.51
AVE. TEMP.	16	17	19	22	25	27	28	28	27	24	19	17
DEGREE-DAYS	108	99	49	5	25	0	0	0	0	0	38	92
PAGE AZ LATITUDE 36.63												
HORIZ. RAD.	12560	15994	22023	25874	29098	29601	28470	24953	21604	16831	12979	10174
KT	.73	.72	.77	.74	.74	.72	.71	.68	.70	.70	.71	.65
AVE. TEMP.	-1	2	5	9	15	20	24	22	18	12	5	0
DEGREE-DAYS	591	447	396	240	107	21	6	0	41	189	390	562
PARKERSBURG WV LATITUDE 39.27												
HORIZ. RAD.	5987	8457	12644	15868	20389	22650	22022	20264	16496	11890	6992	5526
KT	.39	.41	.46	.46	.52	.55	.55	.56	.55	.52	.42	.39
AVE. TEMP.	1	1	6	13	18	22	24	23	20	14	7	2
DEGREE-DAYS	553	471	381	178	67	4	0	0	26	149	333	515

	JAN	FEB	MAR	APR	MAY	JUN	JUL	AUG	SEP	OCT	NOV	DEC
PASADENA CA LATITUDE 34.15												
HORIZ. RAD.	10508	13942	18380	21310	23822	24283	26544	25078	20180	15323	11346	9880
KT	.57	.59	.62	.60	.60	.59	.66	.68	.63	.61	.57	.57
AVE. TEMP.	.12	.13	.13	.15	.17	.19	.23	.23	.22	.19	.15	.13
DEGREE-DAYS	191	151	141	94	47	26	23	23	6	29	91	166
PATUXENT RIVER MD LATITUDE 39.17												
HORIZ. RAD.	6902	9784	13405	17457	20006	21483	20618	18466	15402	11587	8021	6093
KT	.44	.47	.49	.51	.51	.52	.51	.51	.51	.51	.48	.43
AVE. TEMP.	1	2	6	.12	.18	.23	.25	.24	.20	.15	9	3
DEGREE-DAYS	526	454	376	183	58	23	0	0	16	137	293	484
PENDLETON OR LATITUDE 45.68												
HORIZ. RAD.	3951	6964	11844	17054	21852	24335	27189	22630	17048	10306	4975	3325
KT	.34	.41	.48	.52	.56	.59	.68	.65	.62	.54	.39	.33
AVE. TEMP.	1	4	7	.10	.15	.19	.23	.22	.18	.11	5	2
DEGREE-DAYS	568	406	365	235	122	39	23	27	54	213	393	504
PENSACOLA FL LATITUDE 30.47												
HORIZ. RAD.	10467	13439	16956	21310	23529	23781	22483	21310	18003	16495	11639	9378
KT	.50	.53	.55	.59	.59	.58	.56	.57	.55	.61	.53	.48
AVE. TEMP.	.11	.12	.15	.20	.23	.26	.27	.27	.25	.21	.15	.12
DEGREE-DAYS	237	179	117	21	23	20	27	20	25	18	105	199
PEORIA IL LATITUDE 40.67												
HORIZ. RAD.	6824	9545	13481	17668	21310	23990	23571	21017	17040	12518	7745	5819
KT	.47	.48	.50	.52	.54	.58	.59	.58	.58	.57	.49	.44
AVE. TEMP.	5	2	3	.11	.16	.22	.24	.23	.19	.13	4	2
DEGREE-DAYS	709	580	477	231	100	26	24	23	39	182	418	637
PHOENIX AZ LATITUDE 33.43												
HORIZ. RAD.	11591	15595	20588	26725	30375	31087	28219	26019	22873	17892	13057	10577
KT	.61	.65	.69	.75	.77	.76	.70	.70	.71	.70	.65	.60
AVE. TEMP.	.10	.13	.15	.19	.24	.29	.32	.31	.28	.22	.15	.11
DEGREE-DAYS	238	162	103	33	24	3	32	31	20	9	101	216
PHILADELPHIA PA LATITUDE 39.88												
HORIZ. RAD.	6302	9017	12577	16273	18838	20555	19953	17869	14542	10878	7028	5339
KT	.42	.44	.46	.47	.48	.50	.49	.49	.50	.49	.43	.39
AVE. TEMP.	1	1	5	.11	.17	.22	.24	.23	.20	.14	8	3
DEGREE-DAYS	563	484	398	204	68	22	20	20	21	138	313	513
PIERRE SD LATITUDE 44.38												
HORIZ. RAD.	6015	9024	13692	18316	22316	24909	25854	22613	16979	11936	7067	5018
KT	.49	.51	.55	.56	.57	.60	.65	.64	.61	.60	.52	.46
AVE. TEMP.	9	6	1	.8	.14	.20	.24	.23	.17	.10	1	6
DEGREE-DAYS	851	694	606	312	148	41	3	6	84	251	520	749

	JAN	FEB	MAR	APR	MAY	JUN	JUL	AUG	SEP	OCT	NOV	DEC
PITTSBURGH PA LATITUDE 40.50												
HORIZ. RAD.	6615	8917	13481	16747	20389	23404	22901	20180	16998	12309	7703	5945
KT	.45	.45	.50	.49	.52	.57	.57	.56	.58	.56	.48	.45
AVE. TEMP.	0	0	4	11	17	22	24	23	19	13	7	1
DEGREE-DAYS	592	513	424	212	89	6	0	3	32	166	348	546
POCATELLO ID LATITUDE 42.92												
HORIZ. RAD.	6119	10010	15564	20659	25878	28143	29505	25415	20080	13655	7816	5415
KT	.46	.54	.60	.62	.66	.68	.73	.71	.70	.66	.54	.46
AVE. TEMP.	-4	0	2	7	11	17	20	21	15	9	2	-2
DEGREE-DAYS	720	554	510	328	187	77	0	11	107	286	488	656
POINT MUGU CA LATITUDE 34.12												
HORIZ. RAD.	10523	13845	18564	22142	22902	23317	24041	21959	18246	14709	11421	9717
KT	.57	.59	.63	.62	.58	.57	.60	.59	.57	.57	.58	.56
AVE. TEMP.	11	11	11	11	12	13	14	14	14	14	13	12
DEGREE-DAYS	222	199	224	215	198	167	148	143	124	138	153	202
PORT ARTHUR TX LATITUDE 29.95												
HORIZ. RAD.	9076	12150	15356	18266	21232	22824	20952	19705	17330	14997	10814	8562
KT	.43	.47	.49	.50	.54	.56	.52	.52	.52	.55	.49	.43
AVE. TEMP.	11	13	16	20	24	27	28	28	26	21	16	12
DEGREE-DAYS	233	168	112	18	0	0	0	0	0	19	102	190
PORTLAND ME LATITUDE 43.65												
HORIZ. RAD.	5110	7739	11004	14798	17788	19425	18829	16580	13140	9333	5212	4118
KT	.40	.42	.43	.45	.46	.47	.47	.47	.46	.46	.37	.36
AVE. TEMP.	-5	-4	0	6	11	17	20	19	15	9	4	-3
DEGREE-DAYS	777	655	572	594	212	59	15	31	111	274	440	677
PORTLAND OR LATITUDE 45.60												
HORIZ. RAD.	3518	6289	10157	14841	18876	20116	23121	18995	13808	8212	4398	2949
KT	.30	.37	.42	.46	.49	.49	.58	.54	.50	.43	.34	.29
AVE. TEMP.	3	6	8	10	14	17	19	19	17	12	7	5
DEGREE-DAYS	463	346	332	240	147	71	27	31	66	193	328	418
PRESCOTT AZ LATITUDE 34.65												
HORIZ. RAD.	11533	15149	20168	25820	29837	31345	26209	23739	22182	17512	12938	10518
KT	.63	.65	.69	.73	.75	.76	.65	.64	.70	.70	.66	.62
AVE. TEMP.	5	5	5	10	16	21	24	23	20	14	8	6
DEGREE-DAYS	481	381	357	219	92	18	0	0	13	141	320	454
PROSSER WA LATITUDE 46.25												
HORIZ. RAD.	4899	9295	14696	21813	25791	28470	29601	25288	19176	11472	5694	4187
KT	.44	.56	.61	.67	.67	.69	.74	.73	.71	.61	.45	.43
AVE. TEMP.	-1	4	6	10	14	18	21	20	16	10	4	0
DEGREE-DAYS	603	428	376	240	127	47	7	16	66	231	413	543

		JAN	FEB	MAR	APR	MAY	JUN	JUL	AUG	SEP	OCT	NOV	DEC
PROVIDENCE	RI LATITUDE 41.73												
	HORIZ. RAD.	5745	8381	11710	15592	18784	20150	19241	17007	13719	10290	6100	4750
	KT	.41	.43	.44	.46	.48	.49	.48	.47	.47	.48	.40	.38
	AVE. TEMP.	-2	-1	3	8	14	19	21	21	17	12	6	0
	DEGREE-DAYS	631	554	484	295	144	20	6	26	52	194	362	577
PUEBLO	CO LATITUDE 38.28												
	HORIZ. RAD.	10149	13296	17747	22198	24542	27626	26234	23855	20195	15445	10825	8877
	KT	.63	.62	.64	.64	.62	.67	.65	.64	.67	.66	.62	.60
	AVE. TEMP.	-1	1	4	11	16	21	25	24	19	12	5	1
	DEGREE-DAYS	601	471	431	225	82	16	25	0	31	186	403	551
PULLMAN	WA LATITUDE 46.73												
	HORIZ. RAD.	5150	7620	12434	19091	22692	28721	29558	23110	17920	10718	6154	4020
	KT	.47	.46	.52	.59	.59	.70	.74	.67	.66	.57	.50	.42
	AVE. TEMP.	-2	3	3	9	11	15	19	18	15	7	3	0
	DEGREE-DAYS	637	481	460	317	204	105	26	42	114	271	452	572
PUT-IN-BAY	OH LATITUDE 41.65												
	HORIZ. RAD.	5024	8332	12225	15449	20683	22692	23738	21604	16705	12351	6574	4647
	KT	.36	.43	.46	.46	.53	.55	.59	.60	.57	.58	.43	.37
	AVE. TEMP.	-2	-1	2	9	15	21	24	23	19	14	6	0
	DEGREE-DAYS	661	576	501	283	122	13	0	0	22	154	367	589
RALEIGH	NC LATITUDE 35.78												
	HORIZ. RAD.	7875	10703	14478	18661	20522	21156	20151	18287	15629	12545	9217	7213
	KT	.45	.47	.50	.53	.52	.51	.50	.50	.50	.51	.49	.45
	AVE. TEMP.	4	5	10	15	20	24	25	25	22	16	10	5
	DEGREE-DAYS	422	354	279	100	27	0	0	0	7	103	250	410
RALEIGH-DURHAM	NC LATITUDE 35.87												
	HORIZ. RAD.	9001	11848	15658	20222	21645	22441	23153	20138	16621	13230	10048	8164
	KT	.51	.52	.54	.57	.55	.54	.57	.55	.53	.54	.54	.51
	AVE. TEMP.	4	5	10	15	20	24	25	25	21	16	10	5
	DEGREE-DAYS	422	354	279	100	27	0	0	0	7	103	250	410
RAPID CITY	SD LATITUDE 44.15												
	HORIZ. RAD.	6159	9380	13946	18035	21415	24187	25229	22275	17227	12071	7339	5407
	KT	.49	.52	.55	.55	.55	.59	.63	.63	.64	.60	.53	.49
	AVE. TEMP.	-5	-2	0	7	13	18	23	22	16	10	2	-2
	DEGREE-DAYS	742	610	582	340	177	74	7	6	106	263	493	663
REDMOND	OR LATITUDE 44.27												
	HORIZ. RAD.	5574	8794	13507	19098	23600	25959	27762	23478	17977	11342	6492	4818
	KT	.45	.49	.54	.58	.61	.63	.69	.67	.64	.57	.47	.44
	AVE. TEMP.	-1	2	4	7	11	15	19	18	14	9	4	1
	DEGREE-DAYS	599	454	454	343	236	122	31	57	129	286	433	544

	JAN	FEB	MAR	APR	MAY	JUN	JUL	AUG	SEP	OCT	NOV	DEC
RENO NV LATITUDE 39.50												
HORIZ. RAD.	9084	13050	18719	24506	28635	30658	30552	27302	22672	16240	10354	8007
KT	.59	.63	.68	.71	.73	.74	.76	.75	.75	.72	.62	.57
AVE. TEMP.			4		12	16	20	19	15	10	4	
DEGREE-DAYS	570	434	426	303	182	81	9	28	93	253	415	551
RICHLAND WA LATITUDE 46.28												
HORIZ. RAD.	3600	8416	13942	19551	21604	27088	24116	25162	16245	9588	5192	4145
KT	.32	.50	.58	.60	.56	.66	.60	.72	.60	.51	.41	.42
AVE. TEMP.	0	4	8	12	17	20	24	22	18	12	6	2
DEGREE-DAYS	571	395	327	183	72	17	7	16	43	202	383	503
RICHMOND VA LATITUDE 37.83												
HORIZ. RAD.	7171	9954	13737	17772	19997	21250	20138	18165	15297	11720	8319	6432
KT	.44	.46	.49	.51	.51	.51	.50	.50	.50	.50	.47	.43
AVE. TEMP.	3	4	8	14	19	23	25	25	21	15	9	4
DEGREE-DAYS	474	398	316	126	36	3	0	0	12	113	267	448
RIVERSIDE CA LATITUDE 33.95												
HORIZ. RAD.	11514	15366	20013	22651	26084	28470	28177	25874	22400	17040	13356	11304
KT	.62	.65	.67	.63	.66	.69	.70	.70	.70	.67	.67	.65
AVE. TEMP.	11	12	13	15	18	20	24	24	22	18	14	11
DEGREE-DAYS	226	173	157	93	41	12	0	0	3	34	118	208
ROANOKE VA LATITUDE 37.32												
HORIZ. RAD.	7496	10207	14028	17948	20018	21357	20385	18387	15414	12259	8679	6705
KT	.45	.47	.50	.51	.51	.52	.51	.50	.50	.52	.49	.44
AVE. TEMP.	3	4	8	13	18	20	24	23	20	14	8	3
DEGREE-DAYS	493	418	339	157	56	20	0	0	18	131	305	470
ROCHESTER MN LATITUDE 43.92												
HORIZ. RAD.	5413	8543	12278	16000	19243	21584	21664	18864	14186	9870	5609	4200
KT	.43	.47	.49	.48	.49	.52	.54	.53	.50	.49	.40	.37
AVE. TEMP.	-11	-8	-2	7	13	19	21	20	15	10	0	-7
DEGREE-DAYS	897	748	641	342	162	43	12	19	103	269	540	794
ROCHESTER NY LATITUDE 43.12												
HORIZ. RAD.	4134	6350	10253	15198	18231	20619	20210	17239	13161	8874	4584	3188
KT	.31	.34	.40	.46	.47	.50	.50	.48	.46	.43	.32	.27
AVE. TEMP.	-4	-4	8	14	19	22	25	21	17	11	5	-2
DEGREE-DAYS	706	626	551	315	158	26	5	14	70	221	408	632
ROCK SPRINGS WY LATITUDE 41.60												
HORIZ. RAD.	8341	12363	17365	22061	26605	29217	28908	25424	20797	14821	9378	7384
KT	.59	.64	.66	.65	.68	.71	.72	.71	.71	.69	.61	.58
AVE. TEMP.	-7	-5	-2	4	10	15	20	19	14	7	-1	-5
DEGREE-DAYS	789	647	622	415	252	110	10	27	149	349	572	730

ROSWELL NM LATITUDE 33.40, SACRAMENTO CA LATITUDE 38.52, ST. CLOUD MN LATITUDE 45.57, ST. LOUIS MO LATITUDE 38.75, SALEM OR LATITUDE 44.92, SALT LAKE CITY UT LATITUDE 40.77, SAN ANGELO TX LATITUDE 31.37, SAN ANTONIO TX LATITUDE 29.53

	JAN	FEB	MAR	APR	MAY	JUN	JUL	AUG	SEP	OCT	NOV	DEC
ROSWELL NM — LATITUDE 33.40												
HORIZ. RAD.	11877	15579	20512	25167	27909	29624	27698	25442	21710	17331	12840	10803
KT	.62	.65	.69	.70	.70	.72	.69	.68	.68	.67	.64	.61
AVE. TEMP.	6	10	15	20	25	-0	26	25	21	15	8	4
DEGREE-DAYS	463	344	271	103	11	-0	0	0	9	108	302	443
SACRAMENTO CA — LATITUDE 38.52												
HORIZ. RAD.	6774	10661	16551	22739	27632	30458	30506	26878	21639	14923	8874	6110
KT	.42	.50	.60	.66	.70	.78	.76	.74	.71	.65	.52	.42
AVE. TEMP.	10	10	12	15	18	21	24	23	23	17	12	8
DEGREE-DAYS	343	237	207	126	67	11	0	0	3	56	200	331
ST. CLOUD MN — LATITUDE 45.57												
HORIZ. RAD.	7117	10509	15324	17709	20892	22650	23237	20557	15072	10090	6113	5150
KT	.61	.61	.63	.54	.54	.55	.58	.59	.55	.52	.47	.50
AVE. TEMP.	-12	-9	-2	6	13	18	21	20	14	4	0	-8
DEGREE-DAYS	966	804	673	368	180	47	10	21	127	299	583	847
ST. LOUIS MO — LATITUDE 38.75												
HORIZ. RAD.	7120	10051	13672	17752	21237	23748	23260	20615	16560	12481	8152	6022
KT	.45	.48	.49	.51	.54	.58	.58	.57	.55	.54	.48	.42
AVE. TEMP.	0	2	6	14	19	24	26	25	21	15	7	4
DEGREE-DAYS	581	465	379	151	57	6	0	0	19	124	333	523
SALEM OR — LATITUDE 44.92												
HORIZ. RAD.	3769	6674	10748	15552	19722	20980	24314	20141	15075	8732	4657	3148
KT	.31	.38	.43	.47	.51	.51	.61	.57	.54	.44	.35	.30
AVE. TEMP.	4	6	6	11	13	16	19	19	17	12	7	5
DEGREE-DAYS	451	344	341	253	164	74	24	29	67	203	330	415
SALT LAKE CITY UT — LATITUDE 40.77												
HORIZ. RAD.	7253	11221	16505	21498	26811	29063	29395	25576	20920	14678	8942	6467
KT	.50	.56	.62	.63	.68	.70	.73	.71	.71	.67	.56	.49
AVE. TEMP.	-1	1	4	10	15	19	25	24	18	11	4	0
DEGREE-DAYS	637	492	437	263	132	49	0	3	58	223	432	598
SAN ANGELO TX — LATITUDE 31.37												
HORIZ. RAD.	10915	13714	18227	21004	23045	24810	24089	22310	18240	15170	11846	10154
KT	.54	.55	.59	.58	.58	.61	.60	.60	.56	.57	.55	.54
AVE. TEMP.	8	10	14	20	24	28	29	29	25	20	13	9
DEGREE-DAYS	321	229	159	41	0	0	0	0	0	41	166	288
SAN ANTONIO TX — LATITUDE 29.53												
HORIZ. RAD.	10162	13097	16456	18298	21501	23481	24072	22098	18589	15322	11450	9614
KT	.48	.51	.52	.50	.54	.58	.60	.59	.56	.56	.51	.48
AVE. TEMP.	11	12	16	20	24	27	28	28	26	18	15	12
DEGREE-DAYS	257	172	108	17	0	0	0	0	0	18	99	207

742

SAN DIEGO, CA — LATITUDE 32.73

	JAN	FEB	MAR	APR	MAY	JUN	JUL	AUG	SEP	OCT	NOV	DEC
HORIZ. RAD.	11073	14371	18517	21980	22730	23404	24814	23348	19491	15586	12060	10257
KT	.57	.59	.61	.61	.57	.57	.62	.63	.60	.60	.59	.57
AVE. TEMP.	12	13	14	15	17	18	20	21	21	18	15	13
DEGREE-DAYS	174	132	122	80	44	29	3	0	9	24	78	143

SAN FRANCISCO, CA — LATITUDE 37.78

	JAN	FEB	MAR	APR	MAY	JUN	JUL	AUG	SEP	OCT	NOV	DEC
HORIZ. RAD.	8031	11454	16514	21790	25258	26975	27142	24020	19770	13915	9322	7290
KT	.49	.53	.59	.63	.64	.65	.67	.66	.65	.59	.53	.49
AVE. TEMP.	10	12	12	13	14	15	15	15	17	16	14	11
DEGREE-DAYS	288	214	207	162	117	67	52	47	37	76	162	263

SAN JOSE, CA — LATITUDE 37.33

	JAN	FEB	MAR	APR	MAY	JUN	JUL	AUG	SEP	OCT	NOV	DEC
HORIZ. RAD.	7531	11129	14770	20292	24016	25062	25522	22468	19205	13054	8577	7238
KT	.45	.51	.52	.58	.61	.61	.63	.61	.63	.55	.48	.47
AVE. TEMP.	11	13	15	15	15	17	17	18	18	16	13	10
DEGREE-DAYS	273	201	177	125	67	23	6	4	12	57	89	261

SANTA MARIA, CA — LATITUDE 34.90

	JAN	FEB	MAR	APR	MAY	JUN	JUL	AUG	SEP	OCT	NOV	DEC
HORIZ. RAD.	9690	12948	17953	21801	24294	26654	26569	23898	19637	15360	11049	9123
KT	.53	.56	.61	.61	.61	.65	.66	.65	.62	.62	.57	.54
AVE. TEMP.	10	11	11	12	13	15	16	16	17	15	13	10
DEGREE-DAYS	250	202	210	168	136	92	62	57	53	88	150	227

SAVANNAH, GA — LATITUDE 32.13

	JAN	FEB	MAR	APR	MAY	JUN	JUL	AUG	SEP	OCT	NOV	DEC
HORIZ. RAD.	10299	13146	16872	21645	23571	23194	22441	20975	16872	14570	11095	8959
KT	.52	.53	.55	.60	.60	.57	.56	.56	.52	.55	.53	.49
AVE. TEMP.	10	11	14	19	23	26	27	27	25	20	14	10
DEGREE-DAYS	268	211	142	35	23	0	0	0	0	33	141	254

SAULT ST. MARIE, MI — LATITUDE 46.47

	JAN	FEB	MAR	APR	MAY	JUN	JUL	AUG	SEP	OCT	NOV	DEC
HORIZ. RAD.	3686	6847	11673	15699	19158	20551	20826	17281	11906	7638	3765	2870
KT	.39	.41	.49	.49	.50	.50	.52	.50	.41	.47	.41	.30
AVE. TEMP.	-9	-10	-4	3	9	14	17	17	13	7	0	-6
DEGREE-DAYS	875	774	706	447	276	112	53	69	162	324	537	773

SCHENECTADY, NY — LATITUDE 42.83

	JAN	FEB	MAR	APR	MAY	JUN	JUL	AUG	SEP	OCT	NOV	DEC
HORIZ. RAD.	5442	8416	11471	14234	17333	18799	18589	16705	12560	9169	5401	4354
KT	.41	.45	.44	.43	.44	.45	.46	.47	.44	.44	.37	.37
AVE. TEMP.	-5	-4	0	8	15	20	23	21	17	11	4	-2
DEGREE-DAYS	744	641	543	302	136	20	11	11	76	234	420	656

SCOTTS BLUFF, NE — LATITUDE 41.87

	JAN	FEB	MAR	APR	MAY	JUN	JUL	AUG	SEP	OCT	NOV	DEC
HORIZ. RAD.	7669	10787	14838	18930	21940	25383	25918	22692	18146	12995	8208	6527
KT	.55	.56	.56	.58	.56	.61	.64	.64	.63	.61	.54	.52
AVE. TEMP.	-4	-2	2	8	14	19	23	22	16	10	3	-2
DEGREE-DAYS	691	552	529	313	156	51	0	4	89	255	480	644

LATITUDE values and monthly climate data (HORIZ. RAD., KT, AVE. TEMP., DEGREE-DAYS)

	JAN	FEB	MAR	APR	MAY	JUN	JUL	AUG	SEP	OCT	NOV	DEC
SEATTLE WA LATITUDE 47.45												
HORIZ. RAD.	3266	5694	11053	16579	20975	21813	23738	19134	13733	7913	4438	2679
KT	.31	.36	.47	.52	.55	.53	.60	.55	.52	.43	.38	.29
AVE. TEMP.	5	7	7	10	13	16	18	18	16	12	8	6
DEGREE-DAYS	410	319	329	238	143	69	31	32	68	184	297	372
SHERIDAN WY LATITUDE 44.85												
HORIZ. RAD.	5873	8945	13673	17446	21367	24470	26432	22766	17044	11409	6705	5009
KT	.49	.51	.55	.53	.55	.59	.66	.65	.61	.58	.50	.47
AVE. TEMP.	-6	-3	-1	6	12	16	21	21	14	9	1	-4
DEGREE-DAYS	758	608	586	357	208	93	16	17	136	296	527	681
SHREVEPORT LA LATITUDE 32.42												
HORIZ. RAD.	9462	11681	15826	19551	22943	22776	23529	21771	17375	14905	10174	8290
KT	.48	.48	.52	.58	.58	.56	.59	.58	.54	.57	.49	.45
AVE. TEMP.	8	10	14	18	22	26	28	28	25	19	13	9
DEGREE-DAYS	307	231	162	36	22	0	0	0	0	39	194	272
SILVER HILL MD LATITUDE 38.83												
HORIZ. RAD.	7620	10216	14234	18338	21478	23237	21604	19217	16621	12351	8457	6824
KT	.48	.49	.51	.53	.55	.56	.54	.53	.55	.54	.50	.48
AVE. TEMP.	1	2	7	13	18	23	25	24	21	15	9	3
DEGREE-DAYS	506	431	343	147	40	0	0	0	8	106	283	476
SIOUX CITY IA LATITUDE 42.40												
HORIZ. RAD.	6453	9551	13283	17906	21576	24101	24084	20941	16131	11781	7293	5326
KT	.48	.50	.51	.53	.55	.58	.60	.59	.56	.56	.49	.44
AVE. TEMP.	-8	-5	1	10	16	21	24	23	17	10	2	-5
DEGREE-DAYS	809	647	548	263	105	18	0	6	63	210	478	715
SIOUX FALLS SD LATITUDE 43.57												
HORIZ. RAD.	6044	9103	13076	17510	21492	23832	24396	20933	16000	11409	6895	5006
KT	.47	.50	.51	.53	.55	.58	.61	.59	.57	.56	.49	.44
AVE. TEMP.	-10	-7	-1	8	14	20	23	22	16	10	1	-7
DEGREE-DAYS	875	709	603	315	144	36	6	10	92	258	532	775
SOUTH BEND IN LATITUDE 41.70												
HORIZ. RAD.	4718	7486	11264	15745	19549	21812	21023	18911	14655	10318	5642	3862
KT	.34	.39	.43	.47	.50	.53	.52	.53	.50	.48	.37	.31
AVE. TEMP.	-3	-3	3	9	15	20	23	22	18	12	4	-2
DEGREE-DAYS	706	602	512	282	136	19	23	13	54	204	423	634
SPOKANE WA LATITUDE 47.67												
HORIZ. RAD.	3575	6876	11810	16966	21767	23638	26754	22040	16289	9543	4514	2896
KT	.34	.43	.50	.53	.57	.57	.67	.64	.61	.53	.39	.32
AVE. TEMP.	-3	0	3	8	13	16	21	20	15	9	4	-1
DEGREE-DAYS	682	510	474	315	182	80	12	26	109	296	492	620

	JAN	FEB	MAR	APR	MAY	JUN	JUL	AUG	SEP	OCT	NOV	DEC
SPRINGFIELD IL — LATITUDE 39.83												
HORIZ. RAD.	6636	9770	12972	17194	21171	23795	23358	20494	16500	12124	7679	5562
KT	.44	.48	.48	.50	.54	.58	.58	.57	.55	.54	.47	.41
AVE. TEMP.	-3	-1	4	12	17	23	24	24	20	14	5	-1
DEGREE-DAYS	659	538	441	202	73	7	0	4	27	157	385	594
SPRINGFIELD MO — LATITUDE 37.23												
HORIZ. RAD.	7758	10506	14016	18209	21356	23552	23414	21262	16804	12981	8799	6840
KT	.46	.48	.49	.52	.54	.57	.58	.58	.55	.55	.49	.45
AVE. TEMP.	3	4	7	14	18	23	25	25	21	15	7	-1
DEGREE-DAYS	553	436	367	153	52	6	0	3	19	126	325	499
STATE COLLEGE PA — LATITUDE 40.80												
HORIZ. RAD.	5820	8457	12434	15616	19551	22776	22105	19908	15114	11513	6489	5024
KT	.40	.43	.46	.46	.50	.55	.55	.53	.51	.53	.41	.38
AVE. TEMP.	-2	-2	2	9	15	20	22	21	17	11	4	-1
DEGREE-DAYS	654	572	489	267	115	13	3	8	61	214	400	609
STILLWATER OK — LATITUDE 36.15												
HORIZ. RAD.	8667	11974	16245	19091	21017	24827	24827	22692	19050	14737	10759	8541
KT	.50	.53	.56	.54	.53	.60	.62	.62	.61	.53	.58	.53
AVE. TEMP.	2	4	8	16	20	25	27	27	22	17	9	4
DEGREE-DAYS	481	358	287	97	21	0	0	0	6	81	258	429
SUMMIT MT — LATITUDE 48.32												
HORIZ. RAD.	5108	6783	11221	17333	19343	20641	23446	21353	14821	9043	4271	3182
KT	.51	.44	.49	.55	.50	.50	.59	.62	.56	.51	.38	.37
AVE. TEMP.	-8	-5	-4	1	6	11	14	13	8	4	-2	-6
DEGREE-DAYS	854	691	715	518	377	253	143	171	302	457	647	777
SUNNYVALE CA — LATITUDE 37.42												
HORIZ. RAD.	8371	11774	16857	22060	25839	27837	27706	24594	19969	14168	9570	7494
KT	.50	.54	.60	.63	.65	.67	.69	.67	.65	.60	.54	.49
AVE. TEMP.	10	11	13	14	16	19	20	20	20	17	13	10
DEGREE-DAYS	267	194	179	127	68	28	7	8	7	50	153	253
SYRACUSE NY — LATITUDE 43.12												
HORIZ. RAD.	4370	6484	10105	15025	17908	20177	19948	17064	13225	8822	4525	3238
KT	.33	.35	.39	.45	.46	.49	.50	.48	.46	.43	.31	.28
AVE. TEMP.	-5	-4	1	8	14	19	22	21	17	11	5	-2
DEGREE-DAYS	713	628	548	308	151	26	6	10	67	218	400	636
TACOMA WA — LATITUDE 47.25												
HORIZ. RAD.	2970	5618	9640	14680	19451	20449	25515	18343	13025	7447	3827	2396
KT	.28	.35	.41	.46	.51	.50	.64	.53	.49	.41	.32	.26
AVE. TEMP.	4	6	6	10	13	16	18	18	16	12	8	5
DEGREE-DAYS	432	333	341	250	159	83	36	41	88	207	318	398

	JAN	FEB	MAR	APR	MAY	JUN	JUL	AUG	SEP	OCT	NOV	DEC
TALLAHASSEE FL LATITUDE 30.43												
HORIZ. RAD.	9948	12911	16791	20689	21970	21367	19841	19014	16944	14955	11445	9225
KT	.48	.51	.54	.57	.55	.52	.49	.51	.51	.55	.52	.47
AVE. TEMP.	11	12	16	19	23	26	27	27	27	20	15	12
DEGREE-DAYS	227	179	104	19	0	0	0	0	0	17	113	209
TAMPA FL LATITUDE 27.97												
HORIZ. RAD.	11470	14293	18087	21660	22677	20966	19891	18761	16933	15280	12572	10616
KT	.52	.54	.57	.59	.57	.52	.50	.50	.50	.54	.54	.51
AVE. TEMP.	16	18	19	21	24	26	27	27	26	23	19	16
DEGREE-DAYS	113	98	50	5	0	0	0	0	0	0	39	94
TOLEDO OH LATITUDE 41.57												
HORIZ. RAD.	4935	7722	11311	15707	19485	21316	20986	18340	14477	10340	5648	4033
KT	.35	.40	.43	.46	.50	.52	.52	.51	.50	.48	.37	.32
AVE. TEMP.	-4	-3	2	9	15	20	22	22	18	12	4	-3
DEGREE-DAYS	692	589	503	277	127	18	3	10	55	211	423	637
TONOPAH NV LATITUDE 38.07												
HORIZ. RAD.	10417	14460	20166	25545	29251	31641	30674	27668	23183	17256	11697	9383
KT	.64	.68	.72	.74	.74	.77	.76	.76	.76	.74	.67	.63
AVE. TEMP.	-1	2	5	9	14	18	23	22	18	11	4	-3
DEGREE-DAYS	599	473	437	284	149	51	20	22	80	226	420	570
TOPEKA KA LATITUDE 39.07												
HORIZ. RAD.	7728	10679	14264	18630	21738	24132	24149	21676	17210	13013	8757	6622
KT	.49	.51	.52	.54	.55	.58	.60	.60	.57	.57	.52	.47
AVE. TEMP.	-2	1	5	13	18	23	26	25	20	13	5	0
DEGREE-DAYS	637	492	414	183	66	27	26	0	31	144	368	572
TRAVERSE CITY MI LATITUDE 44.73												
HORIZ. RAD.	3527	6441	11360	15948	19623	21704	21674	18264	13225	8556	4276	2914
KT	.29	.37	.46	.49	.51	.53	.54	.52	.48	.43	.32	.27
AVE. TEMP.	-6	-6	-2	6	12	18	20	20	15	10	2	-3
DEGREE-DAYS	761	689	625	372	215	58	18	37	99	262	468	673
TRENTON NJ LATITUDE 40.22												
HORIZ. RAD.	7243	10215	14360	17752	20557	22859	22608	19636	16286	12309	8164	6489
KT	.49	.51	.53	.52	.52	.55	.56	.54	.55	.55	.51	.48
AVE. TEMP.	0	1	5	11	16	21	24	23	19	13	7	1
DEGREE-DAYS	567	492	410	213	75	21	0	0	22	140	312	518
TUCSON AZ LATITUDE 32.12												
HORIZ. RAD.	12473	16252	21158	26818	30318	30978	26569	24774	22457	18180	13714	11301
KT	.63	.66	.69	.74	.77	.76	.66	.66	.69	.69	.66	.61
AVE. TEMP.	10	11	14	18	22	27	30	28	26	20	14	10
DEGREE-DAYS	242	182	131	42	0	0	0	0	0	14	116	221

	JAN	FEB	MAR	APR	MAY	JUN	JUL	AUG	SEP	OCT	NOV	DEC
TULSA OK LATITUDE 36.20												
HORIZ. RAD.	8304	11101	14816	18189	20680	22932	23044	21170	16714	13208	9390	7482
KT	.48	.50	.51	.52	.52	.56	.57	.57	.54	.54	.51	.47
AVE. TEMP.	3	5	9	16	20	25	28	27	23	17	10	4
DEGREE-DAYS	489	370	293	98	16	1	0	0	6	79	260	434
TWIN FALLS ID LATITUDE 40.58												
HORIZ. RAD.	6824	10048	14863	19343	23111	24786	25205	22609	18087	11974	7369	5485
KT	.46	.50	.55	.57	.59	.61	.63	.63	.56	.54	.46	.41
AVE. TEMP.	-1	1	4	9	13	17	22	21	16	10	4	0
DEGREE-DAYS	644	490	454	290	161	73	0	12	99	260	442	589
WACO TX LATITUDE 31.62												
HORIZ. RAD.	9449	12442	16201	18296	20128	23974	24177	22222	18169	14768	10858	9111
KT	.47	.50	.53	.51	.51	.59	.60	.59	.56	.56	.51	.49
AVE. TEMP.	8	10	14	20	24	28	30	30	26	21	14	10
DEGREE-DAYS	310	223	156	31	0	0	0	0	0	28	134	262
WASHINGTON DC LATITUDE 38.85												
HORIZ. RAD.	6492	9253	12768	16557	19499	21573	20627	18356	15208	11392	7387	5460
KT	.41	.44	.46	.48	.50	.52	.51	.51	.50	.50	.43	.38
AVE. TEMP.	1	2	6	13	18	23	25	24	21	15	9	3
DEGREE-DAYS	484	423	348	160	41	0	0	0	18	121	288	463
WHIDBEY ISLAND WA LATITUDE 48.35												
HORIZ. RAD.	3211	6037	10416	15265	19980	20656	22483	18074	13316	7434	4048	2644
KT	.32	.39	.45	.48	.52	.50	.57	.53	.51	.42	.36	.31
AVE. TEMP.	4	5	6	9	12	14	16	16	14	10	7	5
DEGREE-DAYS	462	367	371	280	200	124	77	74	132	252	353	424
WICHITA KA LATITUDE 37.65												
HORIZ. RAD.	8896	12010	15951	20229	23104	25697	25405	23057	18341	14184	9883	7830
KT	.54	.56	.57	.58	.59	.62	.63	.63	.60	.60	.56	.52
AVE. TEMP.	1	3	6	14	19	24	27	26	21	15	7	1
DEGREE-DAYS	581	447	373	153	50	4	0	0	18	117	337	526
WICHITA FALLS TX LATITUDE 33.97												
HORIZ. RAD.	9783	12744	16704	20006	22894	25210	24587	22348	18179	14656	10864	9066
KT	.52	.54	.56	.56	.58	.61	.61	.60	.57	.58	.55	.52
AVE. TEMP.	5	7	11	18	22	27	30	30	25	19	12	5
DEGREE-DAYS	405	297	227	62	7	0	0	0	0	51	205	358
WILMINGTON DE LATITUDE 39.67												
HORIZ. RAD.	6485	9386	13042	16798	19409	21365	20687	18324	14955	11166	7315	5545
KT	.43	.46	.48	.49	.49	.52	.51	.51	.50	.50	.44	.40
AVE. TEMP.	0	1	5	11	17	22	24	23	20	14	8	1
DEGREE-DAYS	568	488	403	212	71	0	0	0	18	141	322	522

Solar radiation, temperature, and degree-day data by location and month.

	JAN	FEB	MAR	APR	MAY	JUN	JUL	AUG	SEP	OCT	NOV	DEC
WINNEMUCCA, NV — LATITUDE 40.90												
HORIZ. RAD.	7837	11662	16707	22329	26802	29159	30389	26647	21643	15004	9187	7017
KT	.54	.59	.63	.66	.68	.71	.76	.74	.75	.69	.58	.54
AVE. TEMP.	-2	0	3	7	12	17	22	20	15	9	3	-1
DEGREE-DAYS	634	481	472	332	199	83	23	23	111	288	462	596
WINSLOW, AZ — LATITUDE 35.02												
HORIZ. RAD.	11174	15060	20201	25913	29446	30774	26635	24293	21879	17169	12703	10150
KT	.62	.65	.69	.73	.74	.75	.66	.66	.69	.69	.66	.61
AVE. TEMP.	2	4	7	12	17	22	26	24	21	14	6	1
DEGREE-DAYS	558	403	348	193	69	8	0	0	11	140	363	537
YOUNGSTOWN, OH — LATITUDE 41.27												
HORIZ. RAD.	4370	6656	10102	14507	17999	19964	19677	17095	13549	9662	5181	3578
KT	.31	.34	.38	.43	.46	.48	.49	.48	.46	.45	.33	.28
AVE. TEMP.	-4	-3	3	9	14	19	22	21	17	11	5	-2
DEGREE-DAYS	677	596	512	288	143	23	5	12	66	213	412	623
YUMA, AZ — LATITUDE 32.67												
HORIZ. RAD.	12440	16379	21781	27383	30963	31935	27843	26435	23277	18417	13786	11350
KT	.64	.67	.72	.76	.78	.78	.69	.71	.72	.71	.67	.63
AVE. TEMP.	12	15	17	21	25	29	34	33	30	24	17	13
DEGREE-DAYS	171	107	54	13	0	0	0	0	0	0	60	153
ZUNI, NM — LATITUDE 35.10												
HORIZ. RAD.	11193	14722	19151	24594	28066	29531	25699	23586	21505	16982	12350	10130
KT	.62	.64	.65	.69	.71	.72	.64	.64	.68	.68	.64	.61
AVE. TEMP.	-1	1	4	9	14	22	22	21	17	11	11	0
DEGREE-DAYS	598	473	437	282	147	38	0	7	51	216	415	568
CHURCHILL, MA — LATITUDE 58.75												
HORIZ. RAD.	2721	6280	12770	18630	21352	22189	21143	15909	9420	4815	2512	1466
KT	.68	.69	.74	.67	.59	.55	.55	.51	.45	.42	.48	.51
AVE. TEMP.	-27	-26	-19	-10	-2	5	11	11	5	-1	-11	-21
DEGREE-DAYS	1421	1265	1183	872	641	375	200	208	378	601	900	1249
EDMONTON, AT — LATITUDE 53.57												
HORIZ. RAD.	3726	7368	13063	17291	21310	21478	22022	17124	12476	7871	4647	2763
KT	.54	.60	.64	.58	.57	.52	.56	.52	.53	.54	.57	.50
AVE. TEMP.	-14	-11	-5	1	11	14	16	15	10	4	-4	-10
DEGREE-DAYS	1006	844	739	425	222	123	41	100	228	410	675	891
KAPUSKASING, OT — LATITUDE 49.42												
HORIZ. RAD.	4605	7955	12979	15491	17166	20096	20096	16747	11304	6699	3350	3350
KT	.49	.54	.58	.40	.45	.49	.51	.49	.44	.39	.32	.42
AVE. TEMP.	-18	-16	-9	0	11	14	16	15	10	4	-4	-14
DEGREE-DAYS	1132	964	868	543	322	123	41	95	225	420	692	1004

	JAN	FEB	MAR	APR	MAY	JUN	JUL	AUG	SEP	OCT	NOV	DEC
LETHBRIDGE AT LATITUDE 49.63												
HORIZ. RAD.	5024	8792	14234	17584	21771	24283	25539	21771	15491	10049	5862	3768
KT	.55	.60	.63	.56	.57	.59	.64	.64	.60	.59	.56	.48
AVE. TEMP.	-8	-7	-2	5	11	14	17	16	12	7	0	-4
DEGREE-DAYS	832	717	644	387	224	118	31	62	177	339	562	709
MONCTON NB LATITUDE 46.12												
HORIZ. RAD.	4187	7536	12142	15909	18421	18841	19678	17166	12979	8792	4605	3768
KT	.37	.45	.50	.49	.48	.46	.49	.49	.48	.46	.37	.38
AVE. TEMP.	-8	-8	-3	6	10	15	17	16	13	7	-1	-5
DEGREE-DAYS	823	742	663	438	260	95	34	58	153	339	495	746
MONTREAL QU LATITUDE 45.50												
HORIZ. RAD.	4605	8374	13397	16747	19678	20515	21352	18421	12979	8374	4187	3350
KT	.39	.49	.55	.51	.51	.50	.53	.53	.47	.43	.32	.33
AVE. TEMP.	-9	-7	-2	5	12	17	18	17	15	8	1	-6
DEGREE-DAYS	870	767	653	380	176	38	5	24	92	289	490	773
OTTAWA OT LATITUDE 45.45												
HORIZ. RAD.	6029	9546	14025	16872	20808	23362	22901	19636	14863	9253	5150	4563
KT	.51	.56	.57	.52	.54	.57	.57	.56	.54	.48	.40	.44
AVE. TEMP.	-11	-10	-3	5	12	16	17	16	14	8	0	-8
DEGREE-DAYS	902	801	684	393	189	50	14	45	123	315	520	816
ST.JOHNS NF LATITUDE 47.52												
HORIZ. RAD.	3350	6280	10049	13397	16747	18003	18421	14234	11722	7117	3350	2930
KT	.32	.39	.42	.42	.44	.44	.45	.41	.44	.39	.28	.32
AVE. TEMP.	-4	-4	-2	1	5	10	15	15	11	6	2	-1
DEGREE-DAYS	701	650	659	515	394	240	103	100	190	362	462	618
TORONTO OT LATITUDE 43.67												
HORIZ. RAD.	5066	7746	12225	15575	20012	21771	22064	18170	14403	9378	5150	3978
KT	.40	.43	.48	.47	.51	.53	.55	.51	.51	.46	.37	.35
AVE. TEMP.	-4	-3	0	6	12	17	18	17	15	10	2	-1
DEGREE-DAYS	685	622	563	342	166	34	4	10	84	244	422	617
VANCOUVER BC LATITUDE 48.98												
HORIZ. RAD.	3182	4354	7829	14360	19593	20180	22817	16663	10550	6699	3978	2345
KT	.33	.29	.34	.46	.51	.49	.57	.46	.41	.39	.37	.29
AVE. TEMP.	3	3	6	9	12	15	16	16	14	10	6	4
DEGREE-DAYS	479	402	376	278	172	87	45	48	122	253	365	437
WINNIPEG MA LATITUDE 49.90												
HORIZ. RAD.	5484	9420	15197	18380	21352	21980	23780	19762	13355	8583	4982	3851
KT	.61	.65	.68	.59	.56	.53	.60	.58	.52	.51	.48	.50
AVE. TEMP.	-17	-15	-7	3	11	15	17	17	12	6	-4	-13
DEGREE-DAYS	1116	955	814	452	225	82	21	39	179	379	695	976

Author Index

Subject Index